# Planning for Success

## Bittinger Student Organizer

### Success can be planned!

This text is designed to help you succeed, so use it to your advantage! Learn to work smarter (not necessarily harder!) in your math course by studying more efficiently and by making the most of the helpful learning resources available to you with this text and through your course (see the Preface and Study Tips for more information).

### Schedule the time you need to succeed!

At the start of the course, use the weekly planner on the reverse side to schedule time to study. Decide that success in this math course is a priority and give yourself 2 to 3 hours of study time (including homework) for each hour of class instruction time that you have each week.

### Study the Study Tips!

Even the best students can learn to study more efficiently. Read ahead, check off the Study Tips on this list that work best for you, and review them often as you progress through the course. One way to use the Study Tips is by category. For example, if you feel that you can make better use of your time, cover all the suggestions on time management. Before you take a test, revisit all the tips on test taking.

## Record Important Contacts

on this page, including your instructor, tutor, and campus math lab. Talk with your classmates and exchange contact information with at least two people so that you stay in touch about class assignments and help each other with study questions, etc.

| CONTACT | NAME | EMAIL | PHONE | FAX | OFFICE HOURS | LOCATION |
|---|---|---|---|---|---|---|
| Instructor | | | | | | |
| Campus Tutor | | | | | | |
| Campus Math Lab | | | | | | |
| Classmate | | | | | | |
| Classmate | | | | | | |

**Supplements recommended by the instructor:**

**Online resources** (Web address, access code, password, etc.):

## Study Tips

### Learning Resources
Using This Textbook (pp. 9, 14)
Exercises (p. 29)
Textbook Supplements (p. 42)
Help Sessions (p. 101)
Reading a Math Text (p. 120)
Using the Answer Section (p. 138)
Form a Study Group (p. 157)
Skill Maintenance Exercises (p. 266)
Learning Resources on Campus (p. 279)
Use Your E-mail (p. 393)
Being a Tutor (p. 585)
Studying the Examples (p. 659)

### Time Management
Time Management (pp. 19, 21, 187)
Sleep Well (p. 109)
Organization (p. 113)
Pace Yourself (p. 162)
Manage Your Schedule (p. 290)
Registering for Future Courses (p. 433)
Keep Your Focus (p. 528)
Finishing a Chapter (p. 555)

### Test-Taking Tips
Preparing for and Taking Tests (p. 76)
Quiz–Test Follow-Up (p. 307)
Don't Get Stuck on a Question (p. 324)
Learn from Your Mistakes (p. 475)
Working with a Classmate (p. 564)
Beginning to Study for the Final Exam (p. 651)
Summing It All Up (p. 685)
Get Some Rest (p. 691)
Ask to See Your Final (p. 700)
Preparing for a Final Exam (p. 710)

### Other Helpful Tips
Highlighting (p. 28)
Solving Applied Problems (p. 62)
Learn Definitions (p. 90)
Understand Your Mistakes (p. 130)
Take a Power Nap (p. 169)
Use Your Text for Notes (p. 179)
Asking Questions (p. 192)
Use a Pencil (p. 203)
Use a Second Text (p. 207)
Fill in Your Blanks (p. 226)
Show Your Work (p. 233)
Plan for Your Absences (p. 243)
Check Your Answers (p. 250)
Visualize (p. 259)
Neatness Counts (p. 317)
Know Your Calculator (p. 336)
Homework Tips (p. 345)
Put Math to Use (p. 356)
Double-Check the Numbers (p. 361)
Five Steps for Problem Solving (pp. 370, 484)
Keeping Math Relevant (p. 402)
Master New Skills (p. 412)
Does More Than One Solution Exist? (p. 420)
If You Miss a Class (p. 439)
Include Correct Units (p. 460)
Aim for Mastery (p. 467)
Don't Be Overwhelmed (p. 497)
Memorizing (p. 518)
Using a Different Approach (p. 532)
Professors' Mistakes (p. 541)
Create Your Own Glossary (p. 548)
Working Mentally (p. 597)
Sketches and Drawings (p. 601)
Do Extra Problems (p. 617)
Key Terms (p. 628)
Stay Involved (p. 642)
Don't Give Up Now! (p. 679)

# Scheduling Success

## Plan to succeed!

On this page, plan a typical week. Consider issues such as class time, study time, work time, travel time, family time, and relaxation time.

## Important Dates

Mid-Term Exam

_____

Final Exam

_____

Holidays

_____

_____

_____

_____

Other
(Assignments, Quizzes, etc.)

_____

_____

_____

_____

_____

_____

_____

_____

_____

### Weekly Planner

| TIME | Sun. | Mon. | Tues. | Wed. | Thurs. | Fri. | Sat. |
|------|------|------|-------|------|--------|------|------|
| 6:00 AM | | | | | | | |
| 6:30 | | | | | | | |
| 7:00 | | | | | | | |
| 7:30 | | | | | | | |
| 8:00 | | | | | | | |
| 8:30 | | | | | | | |
| 9:00 | | | | | | | |
| 9:30 | | | | | | | |
| 10:00 | | | | | | | |
| 10:30 | | | | | | | |
| 11:00 | | | | | | | |
| 11:30 | | | | | | | |
| 12:00 PM | | | | | | | |
| 12:30 | | | | | | | |
| 1:00 | | | | | | | |
| 1:30 | | | | | | | |
| 2:00 | | | | | | | |
| 2:30 | | | | | | | |
| 3:00 | | | | | | | |
| 3:30 | | | | | | | |
| 4:00 | | | | | | | |
| 4:30 | | | | | | | |
| 5:00 | | | | | | | |
| 5:30 | | | | | | | |
| 6:00 | | | | | | | |
| 6:30 | | | | | | | |
| 7:00 | | | | | | | |
| 7:30 | | | | | | | |
| 8:00 | | | | | | | |
| 8:30 | | | | | | | |
| 9:00 | | | | | | | |
| 9:30 | | | | | | | |
| 10:00 | | | | | | | |
| 10:30 | | | | | | | |
| 11:00 | | | | | | | |
| 11:30 | | | | | | | |
| 12:00 AM | | | | | | | |

# Prealgebra

## SIXTH EDITION

**Marvin L. Bittinger**
*Indiana University Purdue University Indianapolis*

**David J. Ellenbogen**
*Community College of Vermont*

**Barbara L. Johnson**
*Indiana University Purdue University Indianapolis*

**Addison-Wesley**

Boston   Columbus   Indianapolis   New York   San Francisco   Upper Saddle River   Amsterdam   Cape Town
Dubai   London   Madrid   Milan   Munich   Paris   Montréal   Toronto   Delhi   Mexico City
São Paulo   Sydney   Hong Kong   Seoul   Singapore   Taipei   Tokyo

| | |
|---|---|
| *Editorial Director* | Christine Hoag |
| *Editor in Chief* | Maureen O'Connor |
| *Executive Editor* | Cathy Cantin |
| *Executive Content Editor* | Kari Heen |
| *Associate Content Editor* | Christine Whitlock |
| *Assistant Editor* | Jonathan Wooding |
| *Production Manager* | Ron Hampton |
| *Composition* | PremediaGlobal |
| *Production Services* | Sally Lifland/Lifland et al., Bookmakers |
| *Art Editor and Photo Researcher* | Geri Davis/The Davis Group, Inc. |
| *Associate Media Producer* | Nathaniel Koven |
| *Content Development Manager* | Rebecca Williams (MathXL) |
| *Senior Content Developer* | Mary Durnwald (TestGen) |
| *Executive Marketing Manager* | Michelle Renda |
| *Marketing Coordinator* | Alicia Frankel |
| *Manufacturing Manager* | Evelyn Beaton |
| *Senior Manufacturing Buyer* | Carol Melville |
| *Senior Media Buyer* | Ginny Michaud |
| *Text Designer* | Geri Davis/The Davis Group, Inc. |
| *Senior Designer/Cover Design* | Beth Paquin |
| *Cover Photograph* | Corbis/All rights reserved. |

**Photo Credits**

Photo credits appear on page P-1.

**Library of Congress Cataloging-in-Publication Data**

Bittinger, Marvin L.
  Prealgebra / Marvin L. Bittinger, David J. Ellenbogen, Barbara L.
Johnson. — 6th ed.
     p. cm.
Includes index.
  1.  Mathematics.   I. Ellenbogen, David.  II.  Johnson, Barbara L.
III.  Title.
   QA39.2.B585 2012
   513'.14—dc22                                                2010023187

1 2 3 4 5 6 7 8 9 10—CRK—15 14 13 12 11

© 2012, 2008, 2004, 2000. Pearson Education, Inc.

**Addison-Wesley**
is an imprint of

www.pearsonhighered.com

ISBN-13: 978-0-321-73154-8
ISBN-10: 0-321-73154-9

# Contents

# Index of Applications

# Authors' Note to Students

Welcome to *Prealgebra*. Having a solid grasp of the mathematical skills taught in this book will enrich your life in many ways, both personally and professionally, including increasing your earning power and enabling you to make wise decisions about your personal finances.

As we wrote this text, we were guided by the desire to do everything possible to help you learn its concepts and skills. The material in this book has been developed and refined with feedback from users of the five previous editions so that you can benefit from their class-tested strategies for success. Regardless of your past experiences in mathematics courses, we encourage you to consider this course as a fresh start and to approach it with a positive attitude.

One of the most important things you can do to ensure your success in this course is to allow enough time for it. This includes time spent in class and time spent out of class studying and doing homework. To help you derive the greatest benefit from this textbook, from your study time, and from the many other learning resources available to you, we have included an organizer card at the front of the book. This card serves as a handy reference for contact information for your instructor, fellow students, and campus learning resources, as well as a weekly planner. It also includes a list of the Study Tips that appear throughout the text. You might find it helpful to read all of these tips as you begin your course work.

Knowing that your time is both valuable and limited, we have designed this objective-based text to help you learn quickly and efficiently. You are led through the development of each concept, then presented with one or more examples of the corresponding skills, and finally given the opportunity to use these skills by doing the interactive margin exercises that appear on the page beside the examples. For quick assessment of your understanding, you can check your answers with the answers placed at the bottom of the page. This innovative feature, along with illustrations designed to help you visualize mathematical concepts and the extensive exercise sets keyed to section objectives, gives you the support and reinforcement you need to be successful in your math course.

To help apply and retain your knowledge, take advantage of the new Skill to Review exercises when they appear at the beginning of a section and the comprehensive mid-chapter reviews, summary and reviews, and cumulative reviews. Read through the list of supplementary material available to students that appears in the preface to make sure you get the most out of your learning experience, and investigate other learning resources that may be available to you.

Give yourself the best opportunity to succeed by spending the time required to learn. We hope you enjoy learning this material and that you will find it of benefit.

Best wishes for success!
Marv Bittinger
David Ellenbogen
Barbara Johnson

**Related Bittinger Paperback Titles**

- Bittinger: *Fundamental College Mathematics,* 5th Edition
- Bittinger: *Basic College Mathematics,* 11th Edition
- Bittinger/Penna: *Basic College Mathematics with Early Integers,* 2nd Edition
- Bittinger/Ellenbogen/Beecher/Johnson: *Prealgebra and Introductory Algebra,* 3rd Edition
- Bittinger: *Introductory Algebra,* 11th Edition
- Bittinger: *Intermediate Algebra,* 11th Edition
- Bittinger/Beecher: *Introductory and Intermediate Algebra,* 4th Edition

**Accuracy**

Students rely on accurate textbooks, and our users value the Bittinger reputation for accuracy. All Bittinger titles go through an exhaustive checking process to ensure accuracy in the problem sets, mathematical art, and accompanying supplements.

# Preface

## New in This Edition

To maximize retention of the concepts and skills presented, five highly effective review features are included in the 6th edition. Student success is increased when review is integrated throughout each chapter.

### Five Types of Integrated Review

Skill to Review exercises, found at the beginning of most sections, link to a section objective. These exercises offer a just-in-time review of a previously presented skill that relates to new material in the section. For convenient studying, section and objective references are followed by two practice exercises for immediate review and reinforcement. Exercise answers are given at the bottom of the page for immediate feedback.

Skill Maintenance Exercises, found in each exercise set, review concepts from other sections in the text to prepare students for their final examination. Section and objective references appear next to each Skill Maintenance exercise. All Skill Maintenance answers are included in the text.

A Mid-Chapter Review reinforces understanding of the mathematical concepts and skills just covered before students move on to new material. Section and objective references are included. Exercise types include Concept Reinforcement, Guided Solutions, Mixed Review, and Understanding Through Discussion and Writing. Answers to all exercises in the Mid-Chapter Review are given at the back of the book.

The Chapter Summary and Review at the end of each chapter is expanded to provide more comprehensive in-text practice and review.

- **Key Terms, Properties, and Formulas** are highlighted, with page references for convenient review.
- **Concept Reinforcement** offers true/false questions to enhance students' understanding of mathematical concepts.
- Important Concepts are listed by section objectives, followed by *worked-out examples* for reference and review and *similar practice exercises* for students to solve.
- **Review Exercises**, including Synthesis exercises and two new multiple-choice exercises, are organized by objective and cover the whole chapter.
- **Understanding Through Discussion and Writing** exercises strengthen understanding by giving students a chance to express their thoughts in spoken or written form.

Section and objective references for all exercises are included. Answers to all exercises in the Summary and Review are given at the back of the book.

**Chapter Tests**, including Synthesis questions and a new multiple-choice question, allow students to review and test their comprehension of chapter skills prior to taking an instructor's exam. Answers to all questions in the Chapter Tests are given at the back of the book. Section and objective references for each question are included with the answers.

A Cumulative Review after every chapter starting with Chapter 2 revisits skills and concepts from all preceding chapters to help students recall previously learned material and prepare for exams. Answers to all Cumulative Review exercises are coded by section and objective at the back of the book to help students identify areas where additional practice is needed.

## Other New Elements

A new design enhances the Bittinger guided-learning approach. Margin exercises are now located next to examples for easier navigation, and answers for those exercises are given at the bottom of the page for immediate feedback.

Content changes include the following:

- The section on Rounding and Estimating; Order has been moved to follow the sections on Multiplication and Division to improve the flow of Chapter 1.
- Section 3.4 now includes applications of multiplying fractions.
- Sections 4.6 and 4.7 now include applications involving adding, subtracting, multiplying, and dividing mixed numerals.
- Section 4.8 is a new section on Order of Operations and Complex Fractions.
- Section 5.1 now covers order and rounding as well as decimal notation.
- Section 5.5 is now Using Fraction Notation with Decimal Notation.
- Section 8.2 is now Solving Percent Problems Using Percent Equations.
- Section 8.5 is a new section on Percent of Increase or Decrease.
- Chapter 10 has been expanded and now covers both exponents and polynomials. Topics have been rearranged so that it begins with exponents and scientific notation, followed by addition, subtraction, multiplication, and factoring of polynomials.
- Over 125 new examples and over 2300 new exercises have been added.

# Hallmark Features

**Revised!**   The **Bittinger Student Organizer** card at the front of the text helps students keep track of important contacts and dates and provides a weekly planner to help schedule time for classes, studying, and homework. A helpful list of study tips found in each chapter is also included.

**New!**   **Chapter Openers** feature motivating real-world applications that are revisited later in the chapters. This feature engages students and prepares them for the upcoming chapter material. (See pages 87, 153, and 223.)

**New!**   **Real-Data Applications** encourage students to see and interpret the mathematics that appears every day in the world around them. (See pages 170, 278, and 369.) Over 600 applications are new to this edition, and many are drawn from the fields of business and economics, life and physical sciences, social sciences, medicine, and areas of general interest such as sports and daily life.

**Study Tips** appear throughout the text to give students pointers on how to develop good study habits as they progress through the course, encouraging them to get involved in the learning process. (See pages 157, 233, and 307.) For easy reference, a list of all Study Tips, organized by category and page number, is included in the Bittinger Student Organizer.

**Caution Boxes** are found at relevant points throughout the text. The heading "*Caution!*" alerts students to coverage of a common misconception or an error often made in performing a particular mathematics operation or skill. (See pages 164, 180, and 275.)

**Revised!**   Optional **Calculator Corners** are located where appropriate throughout the text. These streamlined Calculator Corners are written to be accessible to students and to represent current calculators. A calculator icon indicates exercises suitable for calculator use. (See pages 244, 280, and 418.)

# Immediate Practice and Assessment in Each Section

OBJECTIVES ➡ SKILL TO REVIEW ➡ EXPOSITION ➡ EXAMPLES WITH DETAILED
ANNOTATIONS AND VISUAL ART PIECES ➡ MARGIN EXERCISES ➡ EXERCISE SETS

**Objective Boxes** begin each section. A boxed list of objectives is keyed by letter not only to section subheadings, but also to the section exercise sets and the Mid-Chapter Review and the Summary and Review exercises, as well as to the answers to the questions in the Chapter Tests and Cumulative Reviews. This correlation enables students to easily find appropriate review material if they need help with a particular exercise or skill at the objective level. (See pages 177, 232, and 275.)

**New!** **Skill to Review** exercises, found at the beginning of most sections, link to a section objective and offer students a just-in-time review of a previously presented skill that relates to new material in the section. For convenient studying, objective references are followed by two practice exercises for immediate review and reinforcement. Answers to these exercises are given at the bottom of the page for immediate feedback. (See pages 201, 224, and 333.)

**Revised!** **Annotated Examples** provide annotations and color highlighting to lead students through the structured steps of the examples. The level of detail in these annotations is a significant reason for students' success with this book. This edition contains over 125 new examples. (See pages 185, 250, and 334.)

**Revised!** The **art and photo program** is designed to help students visualize mathematical concepts and real-data applications. Many applications include source lines and feature graphs and drawings similar to those students see in the media. Color is used consistently in geometric shapes and fractional parts to convey meaning. For example, a red border emphasizes perimeter, a black outline with a yellow fill emphasizes area, three-dimensional figures illustrating volume are blue, and fractional parts are shown in purple. (See pages 208, 272, and 286.)

**Revised!** **Margin Exercises**, now located next to examples for easier navigation, accompany examples throughout the text and give students the opportunity to work similar problems for immediate practice and reinforcement of the concept just learned. Answers are now available at the bottom of the page. (See pages 185, 234, and 257.)

## Exercise Sets

To give students ample opportunity to practice what they have learned, each section is followed by an extensive exercise set *keyed by letter to the section objectives* for easy review and remediation. In addition, students also have the opportunity to synthesize the objectives from the current section with those from preceding sections. **For Extra Help** icons, shown at the beginning of each exercise set, indicate supplementary learning resources that students may need.

- **Skill Maintenance Exercises**, found in each exercise set, review concepts from other sections in the text to prepare students for their final examination. Section and objective codes appear next to each Skill Maintenance exercise for easy reference. All Skill Maintenance answers are included in the text. (See pages 189, 230, and 409.)
- **Vocabulary Reinforcement Exercises** provide an integrated review of key terms that students must know to communicate effectively in the language of mathematics. These appear once per chapter in the Skill Maintenance portion of an exercise set. (See pages 143, 205, and 287.)
- **Synthesis Exercises** help build critical-thinking skills by requiring students to use what they know to synthesize, or combine, learning objectives from the current section with those from previous sections. These are available in most exercise sets. (See pages 176, 240, and 501.)

## Mid-Chapter Review

**New!**   A **Mid-Chapter Review** gives students the opportunity to reinforce their understanding of the mathematical skills and concepts just covered before they move on to new material. Section and objective references are included for convenient studying, and answers to all the Mid-Chapter Review exercises are included in the text. The types of exercises are as follows:

- **Concept Reinforcement** are true/false questions that enhance students' understanding of mathematical concepts. These are also available in the Summary and Review at the end of the chapter. (See pages 118, 190, and 262.)
- **Guided Solutions** present worked-out problems with blanks for students to fill in the correct expressions to complete the solution. (See pages 118, 190, and 262.)
- **Mixed Review** provides free-response exercises, similar to those in the preceding sections in the chapter, reinforcing mastery of skills and concepts. (See pages 118, 190, and 263.)
- **Understanding Through Discussion and Writing** lets students demonstrate their understanding of mathematical concepts by expressing their thoughts in spoken and written form. This type of exercise is also found in each Chapter Summary and Review. (See pages 119, 191, and 263.)

## Matching Feature

**Translating for Success** problem sets give extra practice with the important "Translate" step of the process for solving word problems. After translating each of ten problems into its appropriate equation or inequality, students are asked to choose from fifteen possible translations, encouraging them to comprehend the problem before matching. (See pages 131, 210, and 281.)

## End-of-Chapter Material

**Revised!**   The **Chapter Summary and Review** at the end of each chapter is expanded to provide more comprehensive in-text practice and review. Section and objective references and answers to all the Chapter Summary and Review exercises are included in the text. (See pages 144, 214, and 295.)

- **Key Terms, Properties, and Formulas** are highlighted, with page references for convenient review. (See pages 144, 214, and 295.)
- **Concept Reinforcement** offers true/false questions to enhance student understanding of mathematical concepts. (See pages 144, 214, and 295.)
- **New! Important Concepts** are listed by section objectives, followed by *a worked-out example* for reference and review and *a similar practice exercise* for students to solve. (See pages 144–146, 214–216, and 295–297.)
- **Review Exercises,** including Synthesis exercises and two new multiple-choice exercises, covering the whole chapter are organized by objective. (See pages 146, 216, and 298.)
- **Understanding Through Discussion and Writing** exercises strengthen understanding by giving students a chance to express their thoughts in spoken or written form. (See pages 148, 218, and 300.)

**Chapter Tests**, including Synthesis questions and a new multiple-choice question, allow students to review and test their comprehension of chapter skills prior to taking an instructor's exam. Answers to all questions in the Chapter Test are given at the back of the book. Section and objective references for each question are included with the answers. (See pages 149, 219, and 301.)

**New!**   A **Cumulative Review** now follows every chapter starting with Chapter 2; this review revisits skills and concepts from all preceding chapters to help students recall previously learned material and prepare for exams. Answers to all Cumulative Review exercises are coded by section and objective at the back of the book to help students identify areas where additional practice is needed. (See pages 221, 303, and 389.)

# For Extra Help

## Student Supplements

**New! MyWorkBook** (ISBN: 978-0-321-73102-9)

MyWorkBook can be packaged with the textbook or with the MyMathLab access kit and includes the following resources for each section of the text:

- Key vocabulary terms and vocabulary practice problems
- Guided examples with stepped-out solutions and similar practice exercises, keyed to the text by learning objective
- References to textbook examples and section lecture videos for additional help
- Additional exercises with ample space for students to show their work, keyed to the text by learning objective

**Student's Solutions Manual** (ISBN: 978-0-321-73095-4)
By Judith Penna

Contains completely worked-out annotated solutions for all the odd-numbered exercises in the text. Also includes fully worked-out annotated solutions for all the exercises (odd- and even-numbered) in the Mid-Chapter Reviews, the Summary and Reviews, the Chapter Tests, and the Cumulative Reviews.

**Chapter Test Prep Videos**

Chapter Tests can serve as practice tests to help you study. Watch instructors work through step-by-step solutions to all the Chapter Test exercises from the textbook. These videos are available on YouTube (search using BittingerPrealgebra) and in MyMathLab. They are also included on the Video Resources on DVD described below and available for purchase at www.MyPearsonStore.com.

**Video Resources on DVD Featuring Chapter Test Prep Videos**
(ISBN: 978-0-321-73096-1)

- Complete set of lectures covering every objective of every section in the textbook
- Complete set of Chapter Test Prep videos (see above)
- All videos include optional English and Spanish subtitles
- Ideal for distance learning or supplemental instruction
- DVD-ROM format for student use at home or on campus

**InterAct Math Tutorial Website** (www.interactmath.com)

Get practice and tutorial help online! This interactive tutorial website provides algorithmically generated practice exercises that correlate directly to the exercises in the textbook. Students can retry an exercise as many times as they like with new values each time for unlimited practice and mastery. Every exercise is accompanied by an interactive guided solution that provides helpful feedback for incorrect answers, and students can also view a worked-out sample problem that steps them through an exercise similar to the one they're working on.

## Instructor Supplements

**Annotated Instructor's Edition** (ISBN: 978-0-321-73091-6)

Includes answers to all exercises printed in blue on the same page as the exercises. Also includes the student answer section, for easy reference.

**Instructor's Solutions Manual** (ISBN: 978-0-321-73092-3)
By Judith Penna

Contains brief solutions to the even-numbered exercises in the exercise sets. Also includes fully worked-out annotated solutions for all the exercises (odd- and even-numbered) in the Mid-Chapter Reviews, the Summary and Reviews, the Chapter Tests, and the Cumulative Reviews.

**Instructor Resource Manual with Printed Test Forms**
(ISBN: 978-0-321-73098-5) By Laurie Hurley

- Features resources and teaching tips designed to help both new and adjunct faculty with course preparation and classroom management.
- **New!** Includes a mini-lecture for each section of the text with objectives, key examples, and teaching tips.
- Additional resources include general first-time advice, sample syllabi, teaching tips, collaborative learning activities, correlation guide, video index, and transparency masters.
- Contains one diagnostic test, plus two cumulative tests per chapter, beginning with Chapter 2.
- Provides eight test forms for every chapter and six test forms for the final exam.
- For the chapter tests, four free-response tests are modeled after the chapter tests in the main text, two test forms are designed for a 50-minute class period, and two are multiple-choice.
- For the final exams, two test forms are organized by chapter, two test forms have questions scrambled, and two are multiple-choice.
- Also includes extra practice exercises for select sections.

# Additional Media Supplements

**MyMathLab** | **MyMathLab® Online Course (access code required)**

MyMathLab® is a text-specific, easily customizable online course that integrates interactive multimedia instruction with textbook content. MyMathLab gives you the tools you need to deliver all or a portion of your course online, whether your students are in a lab setting or working from home.

- **Interactive homework exercises,** correlated to your textbook at the objective level, are algorithmically generated for unlimited practice and mastery. Most exercises are free-response and provide guided solutions, sample problems, and tutorial learning aids for extra help.

- **Personalized homework** assignments that you can design to meet the needs of your class. MyMathLab tailors the assignment for each student based on test or quiz scores. Each student receives a homework assignment that contains only the problems he or she still needs to master.

- **Personalized Study Plan,** generated when students complete a test or quiz or homework, indicates which topics have been mastered and links to tutorial exercises for topics students have not mastered. You can customize the Study Plan so that the topics available match your course content.

- **Multimedia learning aids,** such as video lectures and podcasts, animations, interactive games, and a complete multimedia textbook, help students independently improve their understanding and performance. You can assign these multimedia learning aids as homework to help your students grasp the concepts.

- **Homework and Test Manager** lets you assign homework, quizzes, and tests that are automatically graded. Select just the right mix of questions from the MyMathLab exercise bank, instructor-created custom exercises, and/or TestGen® test items.

- **Gradebook,** designed specifically for mathematics and statistics, automatically tracks students' results, lets you stay on top of student performance, and gives you control over how to calculate final grades. You can also add offline (paper-and-pencil) grades to the gradebook.

- **MathXL Exercise Builder** allows you to create static and algorithmic exercises for your online assignments. You can use the library of sample exercises as an easy starting point, or you can edit any course-related exercise.

- **Pearson Tutor Center** (www.pearsontutorservices.com) access is automatically included with MyMathLab. The Tutor Center is staffed by qualified math instructors who provide textbook-specific tutoring for students via toll-free phone, fax, email, and interactive Web sessions.

Students do their assignments in the Flash®-based MathXL Player, which is compatible with almost any browser (Firefox®, Safari™, or Internet Explorer®) on almost any platform (Macintosh® or Windows®). MyMathLab is powered by CourseCompass™, Pearson Education's online teaching and learning environment, and by MathXL®, our online homework, tutorial, and assessment system. MyMathLab is available to qualified adopters. For more information, visit www.mymathlab.com or contact your Pearson representative.

**Math XL** | **MathXL® Online Course (access code required)**

MathXL® is a powerful online homework, tutorial, and assessment system that accompanies Pearson Education's textbooks in mathematics or statistics.

With MathXL, instructors can

- create, edit, and assign online homework and tests using algorithmically generated exercises correlated at the objective level to the textbook.
- create and assign their own online exercises and import TestGen tests for added flexibility.
- maintain records of all student work tracked in MathXL's online gradebook.

With MathXL, students can

- take chapter tests in MathXL and receive personalized study plans and/or personalized homework assignments based on their test results.
- use the study plan and/or the homework to link directly to tutorial exercises for the objectives they need to study.
- access supplemental animations and video clips directly from selected exercises.

MathXL is available to qualified adopters. For more information, visit our website at www.mathxl .com, or contact your Pearson representative.

**TestGen®** (www.pearsoned.com/testgen) enables instructors to build, edit, and print tests using a computerized bank of questions developed to cover all the objectives of the text. TestGen is algorithmically based, allowing instructors to create multiple but equivalent versions of the same question or test with the click of a button. Instructors can also modify test bank questions or add new questions. The software and test bank are available for download from Pearson Education's online catalog.

**PowerPoint® Lecture Slides** present key concepts and definitions from the text. Slides are available to download from within MyMathLab and from Pearson Education's online catalog.

**Pearson Math Adjunct Support Center** (www.pearsontutorservices.com/math-adjunct.html) is staffed by qualified instructors with more than 100 years of combined experience at both the community college and university levels. Assistance is provided for faculty in the following areas: suggested syllabus consultation, tips on using materials packed with your book, book-specific content assistance, and teaching suggestions, including advice on classroom strategies.

# Acknowledgments

Our deepest appreciation to all of you who helped to shape this text by reviewing and spending time with us on your campuses. In particular, we would like to thank the following reviewers of the fifth and sixth editions:

Rosanne Benn, *Prince George's Community College*
Wayne Browne, *Oklahoma State University–Oklahoma City*
Gail Burkett, *Palm Beach Community College*
John Close, *Salt Lake Community College*
Lucio Della Vecchia, *Daytona Beach Community College*
Babette Dickelman, *Jackson Community College*
Vidya Nahar, *Athens Technical College*
Linda Spears, *Rock Valley College*
James Vogel, *Sanford-Brown College*
David Whittlesey, *Valencia Community College*

The endless hours of hard work by Sally Lifland, Jane Hoover, and Geri Davis have led to products of which we are immensely proud. We also want to thank Judy Penna for writing the *Student's* and *Instructor's Solutions Manuals* and for her strong leadership in the preparation of the printed supplements. Other strong support has come from Laurie Hurley for the *Instructor's Resource Manual with Printed Test Forms*, accuracy checkers Holly Martinez, Gary Williams, Jeremy Pletcher, Mindy Pergel, David Kedrowsky, and David Johnson, and proofreader Patty LaGree. Michelle Lanosga assisted with applications research. We also wish to recognize Tom Atwater, who wrote video scripts and presented videos along with Patty Schwarzkopf, Margaret Donlan, Clem Vance, and authors Judy Penna, Barbara Johnson, and David Ellenbogen.

In addition, a number of people at Pearson have contributed in special ways to the development and production of this textbook including the Developmental Math team: Vice President, Executive Director of Development Carol Trueheart, Senior Development Editor Dawn Nuttall, Production Manager Ron Hampton, Senior Designer Beth Paquin, Associate Editor Christine Whitlock, Assistant Editor Jonathan Wooding, and Associate Media Producer Nathaniel Koven and Ann Broomhead from the Pearson Tutor Center. Executive Editor Cathy Cantin and Executive Marketing Manager Michelle Renda encouraged our vision and provided marketing insight. Kari Heen, Executive Content Editor, deserves special recognition for overseeing every phase of the project and keeping it moving.

# Whole Numbers

# Real-World Application

Kingda Ka, in Six Flags Great Adventure, New Jersey, and Top Thrill Dragster, in Cedar Point, Ohio, are the two fastest roller coasters in the world. Kingda Ka is 3118 ft long, and Top Thrill Dragster is 2800 ft long. How much longer is Kingda Ka than Top Thrill Dragster?

*Source:* ultimaterollercoaster.com

***This problem appears as Exercise 1 in Section 1.8.***

# 1.1

## Standard Notation

### OBJECTIVES

**a** Give the meaning of digits in standard notation.

**b** Convert from standard notation to expanded notation.

**c** Convert between standard notation and word names.

---

**TO THE STUDENT**

At the front of the text, you will find the Bittinger Student Organizer card. This pullout card will help you keep track of important dates and useful contact information. You can also use it to plan time for class, study, work, and relaxation. By managing your time wisely, you will provide yourself the best possible opportunity to be successful in this course.

---

We study mathematics in order to be able to solve problems. In this section, we study how numbers are named. We begin with the concept of place value.

### a Place Value

Consider the numbers in the following table.

**Three Most Populous Countries in the World**

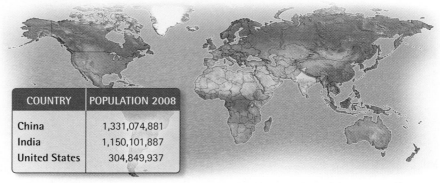

| COUNTRY | POPULATION 2008 |
|---------|-----------------|
| China | 1,331,074,881 |
| India | 1,150,101,887 |
| United States | 304,849,937 |

SOURCE: msnu.edu

A **digit** is a number 0, 1, 2, 3, 4, 5, 6, 7, 8, or 9 that names a place-value location. For large numbers, digits are separated by commas into groups of three, called **periods**. Each period has a name: *ones, thousands, millions, billions, trillions,* and so on. To understand the population of India in the table above, we can use a **place-value chart**, as shown below.

| PLACE-VALUE CHART | | | | | | | | | | | | | | |
|---|---|---|---|---|---|---|---|---|---|---|---|---|---|---|
| Periods → | Trillions | | | Billions | | | Millions | | | Thousands | | | Ones | | |
| | | | | | | 1 | 1 | 5 | 0 | 1 | 0 | 1 | 8 | 8 | 7 |
| | Hundreds | Tens | Ones | Hundreds | Tens | Ones | Hundreds | Tens | Ones | Hundreds | Tens | Ones | Hundreds | Tens | Ones |

1 billion  150 millions  101 thousands  887 ones

**EXAMPLES** In each of the following numbers, what does the digit 8 mean?

1. 278,342      8 thousands
2. 872,342      8 hundred thousands
3. 28,343,399,223      8 billions
4. 1,023,850      8 hundreds
5. 98,413,099      8 millions
6. 6328      8 ones

Do Margin Exercises 1–6 (in the margin at left).

---

What does the digit 2 mean in each number?

**1.** 526,555

**2.** 265,789

**3.** 42,789,654

**4.** 24,789,654

**5.** 8924

**6.** 5,643,201

*Answers*

1. 2 ten thousands    2. 2 hundred thousands
3. 2 millions    4. 2 ten millions    5. 2 tens
6. 2 hundreds

---

**EXAMPLE 7** *Hurricane Relief.* Private donations for relief for Hurricanes Katrina and Rita, which struck the Gulf Coast of the United States in 2005, totaled $3,378,185,879. What does each digit name?

**Source:** The Center on Philanthropy at Indiana University

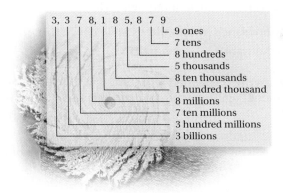

3, 3 7 8, 1 8 5, 8 7 9
- 9 ones
- 7 tens
- 8 hundreds
- 5 thousands
- 8 ten thousands
- 1 hundred thousand
- 8 millions
- 7 ten millions
- 3 hundred millions
- 3 billions

Do Exercise 7.

## b Converting from Standard Notation to Expanded Notation

To answer questions such as "How many?", "How much?", and "How tall?", we often use whole numbers. The set, or collection, of **whole numbers** is

0, 1, 2, 3, 4, 5, 6, 7, 8, 9, 10, 11, 12, . . . .

The set goes on indefinitely. There is no largest whole number, and the smallest whole number is 0. Each whole number can be named using various notations. The set 1, 2, 3, 4, 5, . . . , without 0, is called the set of **natural numbers**.

Consider the data in the table below showing the Advanced Placement exams taken most frequently by the class of 2009.

| EXAM | NUMBER TAKEN |
|------|--------------|
| U.S. History | 292,004 |
| English Literature and Composition | 273,691 |
| English Language and Composition | 265,395 |
| Calculus AB | 182,423 |

SOURCE: The College Board

The number of Calculus exams taken was 182,423. This number is expressed in **standard notation**. We write **expanded notation** for 182,423 as follows:

$$182,423 = 1 \text{ hundred thousand} + 8 \text{ ten thousands}$$
$$+ 2 \text{ thousands} + 4 \text{ hundreds}$$
$$+ 2 \text{ tens} + 3 \text{ ones}.$$

**7. Federal Payroll.** In December 2009, the payroll for civilian employees of the federal government was $15,471,672,417. What does each digit name?

**Source:** U.S. Census Bureau

*Answer*

7. 1 ten billion; 5 billions; 4 hundred millions; 7 ten millions; 1 million; 6 hundred thousands; 7 ten thousands; 2 thousands; 4 hundreds; 1 ten; 7 ones

**EXAMPLE 8**   Write expanded notation for 1815 ft, the height of the CN Tower in Toronto, Canada.

1815 = 1 thousand + 8 hundreds + 1 ten + 5 ones

**EXAMPLE 9**   Write expanded notation for 3400.

3400 = 3 thousands + 4 hundreds + 0 tens + 0 ones,   or

3 thousands + 4 hundreds

**EXAMPLE 10**   Write expanded notation for 273,691, the number of Advanced Placement English Literature and Composition exams taken by the class of 2009.

273,691 = 2 hundred thousands + 7 ten thousands
       + 3 thousands + 6 hundreds + 9 tens + 1 one

Do Exercises 8–12.

## c Converting Between Standard Notation and Word Names

We often use **word names** for numbers. When we pronounce a number, we are speaking its word name. Russia won 72 medals in the 2008 Summer Olympics in Beijing, China. A word name for 72 is "seventy-two." Word names for some two-digit numbers like 36, 51, and 72 use hyphens. Others like 17 use only one word, "seventeen."

**2008 Summer Olympics Medal Count**

| COUNTRY | GOLD | SILVER | BRONZE | TOTAL |
|---|---|---|---|---|
| United States of America | 36 | 38 | 36 | 110 |
| People's Republic of China | 51 | 21 | 28 | 100 |
| Russia | 23 | 21 | 28 | 72 |
| Great Britain | 19 | 13 | 15 | 47 |
| Australia | 14 | 15 | 17 | 46 |

SOURCE: beijing2008.cn

**EXAMPLES** Write a word name.

**11.** 36, the number of gold medals won by the United States

Thirty-six

**12.** 15, the number of silver medals won by Australia

Fifteen

**13.** 100, the total number of medals won by the People's Republic of China

One hundred

Do Exercises 13–15.

Do Exercises 13–15.

For word names for larger numbers, we begin at the left with the largest period. The number named in the period is followed by the name of the period; then a comma is written and the next number and period are named. Note that the name of the ones period is not included in the word name for a whole number.

**EXAMPLE 14** Write a word name for 46,605,314,732.

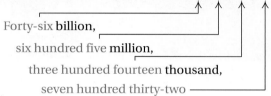

Forty-six **billion,**

six hundred five **million,**

three hundred fourteen **thousand,**

seven hundred thirty-two

The word "and" *should not* appear in word names for whole numbers. Although we commonly hear such expressions as "two hundred *and* one," the use of "and" is not, strictly speaking, correct in word names for whole numbers. For decimal notation, it is appropriate to use "and" for the decimal point. For example, 317.4 is read as "three hundred seventeen *and* four tenths."

Do Exercises 16–19.

**EXAMPLE 15** Write standard notation.

Five hundred six **million,**

three hundred forty-five **thousand,**

two hundred twelve

Standard notation is 506,345,212.

Do Exercise 20.

Write a word name. (Refer to the table on the previous page.)

**13.** 46, the total number of medals won by Australia

**14.** 13, the number of silver medals won by Great Britain

**15.** 28, the number of bronze medals won by Russia

Write a word name.

**16.** 204

**17.** $42,446, the median starting salary offered to accounting majors who graduated from college in 2009
**Source:** CollegeRecruiter.com

**18.** 1,879,204

**19.** 6,830,586,905, the world population in 2010
**Source:** U.S. Census Bureau

**20.** Write standard notation.

Two hundred thirteen million, one hundred five thousand, three hundred twenty-nine

*Answers*

**13.** Forty-six  **14.** Thirteen  **15.** Twenty-eight  **16.** Two hundred four  **17.** Forty-two thousand, four hundred forty-six
**18.** One million, eight hundred seventy-nine thousand, two hundred four  **19.** Six billion, eight hundred thirty million, five hundred eighty-six thousand, nine hundred five
**20.** 213,105,329

**a)** What does the digit 5 mean in each number?

**1.** 235,888        **2.** 253,777        **3.** 1,488,526        **4.** 500,736

*Movie Receipts.* Box-office receipts on the opening weekend of *Shrek the Third* were $121,629,270.
**Source:** Box Office Mojo

What digit names the number of:

**5.** thousands?        **6.** ten millions?

**7.** tens?        **8.** hundred thousands?

**b)** Write expanded notation.

**9.** 5702        **10.** 3097        **11.** 93,986        **12.** 38,453

*Stair-Climbing Races.* The figure below shows the number of stairs in four buildings in which stair-climbing races are held. In Exercises 13–16, write expanded notation for the number of stairs in each race.

**Stair-Climbing Races**

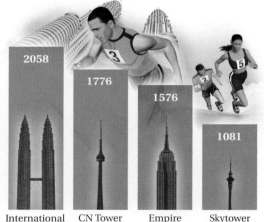

2058   1776   1576   1081

International Towerthon, Kuala Lumpur, Malaysia    CN Tower Stair Climb, Toronto, Ontario, Canada    Empire State Building Run-Up, New York    Skytower Vertical Challenge, Auckland, New Zealand

SOURCE: towerrunning.com

**13.** 2058 steps in the International Towerthon, Kuala Lumpur, Malaysia

**14.** 1776 steps in the CN Tower Stair Climb, Toronto, Ontario, Canada

**15.** 1576 steps in the Empire State Building Run-Up, New York City, New York

**16.** 1081 steps in the Skytower Vertical Challenge, Auckland, New Zealand

*Population Projections.* The table below shows the projected population in 2050 for six countries. In Exercises 17–22, write expanded notation for the population of the given country.

**PROJECTED POPULATION IN 2050**

| COUNTRY | POPULATION |
|---------|-----------|
| China | 1,424,161,948 |
| Hong Kong | 6,172,725 |
| Japan | 99,886,568 |
| Monaco | 32,964 |
| Surinam | 617,249 |
| United States | 420,080,587 |

SOURCES: International Programs Center, U.S. Census Bureau, U.S. Department of Commerce

**17.** 1,424,161,948 in China

**18.** 6,172,725 in Hong Kong

**19.** 99,886,568 in Japan

**20.** 32,964 in Monaco

**21.** 617,249 in Surinam

**22.** 420,080,587 in the United States

 Write a word name.

**23.** 85

**24.** 48

**25.** 88,000

**26.** 45,987

**27.** 123,765

**28.** 111,013

**29.** 7,754,211,577

**30.** 43,550,651,808

Write a word name for the number in each sentence.

**31.** *Wilderness Areas.* In 2009, in the most sweeping land-protection law passed in fifteen years, all development was banned on huge swaths of land in nine states. The largest was in California, where 700,634 acres were newly designated as wilderness.
**Source:** The Wilderness Society

**32.** *Busiest Airport.* In 2008, the world's busiest airport, Hartsfield in Atlanta, had 90,039,280 passengers.
**Source:** Airports Council International World Headquarters, Geneva, Switzerland

**33.** *Auto Racing.* Helio Castroneves, winner of the 2009 Indianapolis 500 auto race, won a record prize of $3,048,005.
Source: ESPN

**34.** *College Football.* The 2010 Fiesta Bowl game was attended by 73,227 fans.
Source: www.mmbolding.com/bowls/Fiesta_2010.htm

Write standard notation.

**35.** Two million, two hundred thirty-three thousand, eight hundred twelve

**36.** Three hundred fifty-four thousand, seven hundred two

**37.** Eight billion

**38.** Seven hundred million

**39.** Fifty thousand, three hundred twenty-four

**40.** Twenty-six billion

**41.** Six hundred thirty-two thousand, eight hundred ninety-six

**42.** Seventeen thousand, one hundred twelve

Write standard notation for the number in each sentence.

**43.** *Ice Cream Purchases.* Americans buy one billion, six hundred million gallons of ice cream and frozen desserts each year.
Source: International Dairy Foods Association

**44.** *Learning a Language.* There are two hundred million Chinese children studying English.
Source: U.S. Department of Education

**45.** *Pacific Ocean.* The area of the Pacific Ocean is sixty-four million, one hundred eighty-six thousand square miles.

**46.** The average distance from the sun to Neptune is two billion, seven hundred ninety-three million miles.

## Synthesis

*To the student and the instructor:* The Synthesis exercises found at the end of every exercise set challenge students to combine concepts or skills studied in that section or in preceding parts of the text. Exercises marked with a ▦ symbol are meant to be solved using a calculator.

**47.** How many whole numbers between 100 and 400 contain the digit 2 in their standard notation?

**48.** ▦ What is the largest number that you can name on your calculator? How many digits does that number have? How many periods?

# 1.2 Addition

## a Addition of Whole Numbers

Addition of whole numbers corresponds to combining or putting things together.

We combine two sets.      This is the resulting set.

3 PDAs      4 PDAs      7 PDAs

We say that the **sum** of 3 and 4 is 7. The numbers added are called **addends**. The addition that corresponds to the figure above is

$$3 \quad + \quad 4 \quad = \quad 7.$$

Addend   Addend   Sum

To add whole numbers, we add the ones digits first, then the tens, then the hundreds, then the thousands, and so on.

**EXAMPLE 1** Add: $6878 + 4995$.

Place values are lined up in columns.

$$\begin{array}{r} \overset{\phantom{6}\phantom{8}\phantom{7}1}{6\ 8\ 7\ 8} \\ +\ 4\ 9\ 9\ 5 \\ \hline 3 \end{array}$$

Add ones. We get 13 ones, or 1 ten + 3 ones. Write 3 in the ones column and 1 above the tens. This is called *carrying*, or *regrouping*.

$$\begin{array}{r} \overset{\phantom{6}1\ 1}{6\ 8\ 7\ 8} \\ +\ 4\ 9\ 9\ 5 \\ \hline 7\ 3 \end{array}$$

Add tens. We get 17 tens, so we have 10 tens + 7 tens. This is also 1 hundred + 7 tens. Write 7 in the tens column and 1 above the hundreds.

$$\begin{array}{r} \overset{1\ 1\ 1}{6\ 8\ 7\ 8} \\ +\ 4\ 9\ 9\ 5 \\ \hline 8\ 7\ 3 \end{array}$$

Add hundreds. We get 18 hundreds, or 10 hundreds + 8 hundreds. This is also 1 thousand + 8 hundreds. Write 8 in the hundreds column and 1 above the thousands.

$$\begin{array}{r} \overset{1\ 1\ 1}{6\ 8\ 7\ 8} \\ +\ 4\ 9\ 9\ 5 \\ \hline 1\ 1\ 8\ 7\ 3 \end{array}$$

Add thousands. We get 11 thousands.

We show you these steps for explanation. You need write only this.

$$\begin{array}{r} \overset{1\ 1\ 1}{6\ 8\ 7\ 8} \\ +\ 4\ 9\ 9\ 5 \\ \hline 1\ 1\ 8\ 7\ 3 \end{array}$$ — Addends — Sum

**EXAMPLE 2** Add: 391 + 276 + 789 + 498.

$$
\begin{array}{r}
\overset{2}{\phantom{0}} \\
3\ 9\ 1 \\
2\ 7\ 6 \\
7\ 8\ 9 \\
+\ 4\ 9\ 8 \\
\hline
4
\end{array}
$$

Add ones. We get 24, so we have 2 tens + 4 ones. Write 4 in the ones column and 2 above the tens.

$$
\begin{array}{r}
\overset{3}{\phantom{0}}\overset{2}{\phantom{0}} \\
3\ 9\ 1 \\
2\ 7\ 6 \\
7\ 8\ 9 \\
+\ 4\ 9\ 8 \\
\hline
5\ 4
\end{array}
$$

Add tens. We get 35 tens, so we have 30 tens + 5 tens. This is also 3 hundreds + 5 tens. Write 5 in the tens column and 3 above the hundreds.

$$
\begin{array}{r}
\overset{3}{\phantom{0}}\overset{2}{\phantom{0}} \\
3\ 9\ 1 \\
2\ 7\ 6 \\
7\ 8\ 9 \\
+\ 4\ 9\ 8 \\
\hline
1\ 9\ 5\ 4
\end{array}
$$

Add hundreds. We get 19 hundreds.

Do Exercises 1–4.

Add.

**1.** 6203 + 3542

**2.**
$$
\begin{array}{r}
7\ 9\ 6\ 8 \\
+\ 5\ 4\ 9\ 7 \\
\hline
\end{array}
$$

**3.**
$$
\begin{array}{r}
9\ 8\ 0\ 4 \\
+\ 6\ 3\ 7\ 8 \\
\hline
\end{array}
$$

**4.**
$$
\begin{array}{r}
1\ 9\ 3\ 2 \\
6\ 7\ 2\ 3 \\
9\ 8\ 7\ 8 \\
+\ 8\ 9\ 4\ 1 \\
\hline
\end{array}
$$

How do we perform an addition of three numbers, like 2 + 3 + 6? We could do it by adding 3 and 6, and then 2. We can show this with parentheses:

$2 + (3 + 6) = 2 + 9 = 11.$     Parentheses tell what to do first.

We could also add 2 and 3, and then 6:

$(2 + 3) + 6 = 5 + 6 = 11.$

Either way the result is 11. It does not matter how we group the numbers. This illustrates the **associative law of addition**, $a + (b + c) = (a + b) + c$. We can also add whole numbers in any order. That is, $2 + 3 = 3 + 2$. This illustrates the **commutative law of addition**, $a + b = b + a$. Together, the commutative and associative laws tell us that to add more than two numbers, we can use any order and grouping we wish. Adding 0 to a number does not change the number: $a + 0 = 0 + a = a$. That is, $6 + 0 = 0 + 6 = 6$, or $198 + 0 = 0 + 198 = 198$. We say that 0 is the **additive identity**.

*Answers*

**1.** 9745   **2.** 13,465   **3.** 16,182
**4.** 27,474

---

### Calculator Corner

**Adding Whole Numbers**   This is the first of a series of *optional* discussions on using a calculator. A calculator is *not* a requirement for this textbook. Check with your instructor about whether you are allowed to use a calculator in the course.

There are many kinds of calculators and different instructions for their usage. We have included instructions here for a low-cost calculator. Be sure to consult your user's manual as well.

To add whole numbers on a calculator, we use the ⊞ and ⊟ keys. For example, to find 314 + 259 + 478, we press 3 1 4 + 2 5 9 + 4 7 8 =. The display reads ⎡ 1051 ⎤, so 314 + 259 + 478 = 1051.

**Exercises:**   Use a calculator to find each sum.

**1.** 73 + 48

**2.** 925 + 677

**3.** 826 + 415 + 691

**4.** 253 + 490 + 121

## b Finding Perimeter

Addition can be used when finding perimeter.

> **PERIMETER**
>
> The distance around an object is its **perimeter**.

▌ **EXAMPLE 3**   Find the perimeter of the figure.

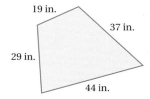

We add the lengths of the sides:

    Perimeter = 29 in. + 19 in. + 37 in. + 44 in.

We carry out the addition as follows.

$$
\begin{array}{r}
\overset{2}{2}\ 9 \\
1\ 9 \\
3\ 7 \\
+\ 4\ 4 \\
\hline
1\ 2\ 9
\end{array}
$$

The perimeter of the figure is 129 in.

*Do Exercises 5 and 6.*

▌ **EXAMPLE 4**   Find the perimeter of the octagonal (eight-sided) resort swimming pool.

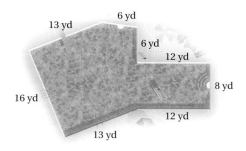

    Perimeter = 13 yd + 6 yd + 6 yd + 12 yd + 8 yd
                  + 12 yd + 13 yd + 16 yd

The perimeter of the pool is 86 yd.

*Do Exercise 7.*

Find the perimeter of each figure.

**5.**

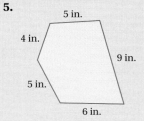

**6.**

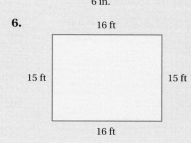

Solve.

**7. Index Cards.**   Two standard sizes for index cards are 3 in. (inches) by 5 in. and 5 in. by 8 in. Find the perimeter of each type of card.

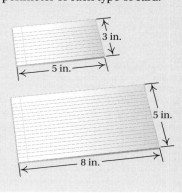

*Answers*

**5.** 29 in.　**6.** 62 ft　**7.** 16 in.; 26 in.

**a** Add.

**1.** 
$$
\begin{array}{r}
3\ 6\ 4 \\
+\ \ 2\ 3 \\
\hline
\end{array}
$$

**2.** 
$$
\begin{array}{r}
1\ 5\ 2\ 1 \\
+\ \ \ 3\ 4\ 8 \\
\hline
\end{array}
$$

**3.** 
$$
\begin{array}{r}
8\ 6 \\
+\ 7\ 8 \\
\hline
\end{array}
$$

**4.** 
$$
\begin{array}{r}
7\ 3 \\
+\ 6\ 9 \\
\hline
\end{array}
$$

**5.** 
$$
\begin{array}{r}
1\ 7\ 1\ 6 \\
+\ 3\ 4\ 8\ 2 \\
\hline
\end{array}
$$

**6.** 
$$
\begin{array}{r}
7\ 5\ 0\ 3 \\
+\ 2\ 6\ 8\ 3 \\
\hline
\end{array}
$$

**7.** 
$$
\begin{array}{r}
9\ 9 \\
+\ \ 1 \\
\hline
\end{array}
$$

**8.** 
$$
\begin{array}{r}
9\ 9\ 9 \\
+\ \ 1\ 1 \\
\hline
\end{array}
$$

**9.** $8113 + 390$

**10.** $271 + 3338$

**11.** $356 + 4910$

**12.** $280 + 34{,}902$

**13.** $3870 + 92 + 7 + 497$

**14.** $10{,}120 + 12{,}989 + 5738$

**15.** 
$$
\begin{array}{r}
4\ 8\ 2\ 5 \\
+\ 1\ 7\ 8\ 3 \\
\hline
\end{array}
$$

**16.** 
$$
\begin{array}{r}
3\ 6\ 5\ 4 \\
+\ 2\ 7\ 0\ 0 \\
\hline
\end{array}
$$

**17.** 
$$
\begin{array}{r}
2\ 3{,}4\ 4\ 3 \\
+\ 1\ 0{,}9\ 8\ 9 \\
\hline
\end{array}
$$

**18.** 
$$
\begin{array}{r}
4\ 5{,}8\ 7\ 9 \\
+\ 2\ 1{,}7\ 8\ 6 \\
\hline
\end{array}
$$

**19.** 
$$
\begin{array}{r}
7\ 7{,}5\ 4\ 3 \\
+\ 2\ 3{,}7\ 6\ 7 \\
\hline
\end{array}
$$

**20.** 
$$
\begin{array}{r}
9\ 9{,}9\ 9\ 9 \\
+\ \ \ \ 1\ 1\ 2 \\
\hline
\end{array}
$$

**21.** 
$$
\begin{array}{r}
4\ 5 \\
2\ 5 \\
3\ 6 \\
4\ 4 \\
+\ 8\ 0 \\
\hline
\end{array}
$$

**22.** 
$$
\begin{array}{r}
3\ 8 \\
2\ 7 \\
3\ 2 \\
1\ 4 \\
+\ 7\ 6 \\
\hline
\end{array}
$$

**23.** 
$$
\begin{array}{r}
1\ 2{,}0\ 7\ 0 \\
2{,}9\ 5\ 4 \\
+\ \ 3{,}4\ 0\ 0 \\
\hline
\end{array}
$$

**24.** 
$$
\begin{array}{r}
4\ 2{,}4\ 8\ 7 \\
8\ 3{,}1\ 4\ 1 \\
+\ 3\ 6{,}7\ 1\ 2 \\
\hline
\end{array}
$$

**25.** 
$$
\begin{array}{r}
4\ 8\ 3\ 5 \\
7\ 2\ 9 \\
9\ 2\ 0\ 4 \\
8\ 9\ 8\ 6 \\
+\ 7\ 9\ 3\ 1 \\
\hline
\end{array}
$$

**26.** 
$$
\begin{array}{r}
9\ 8\ 9 \\
5\ 6\ 6 \\
8\ 3\ 4 \\
9\ 2\ 0 \\
+\ 7\ 0\ 3 \\
\hline
\end{array}
$$

**b** Find the perimeter of each figure.

**27.**

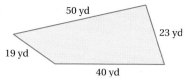
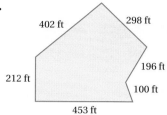

50 yd

23 yd

19 yd

40 yd

**28.**

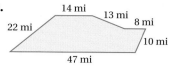
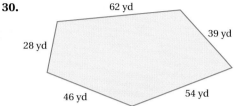

14 mi

13 mi

8 mi

22 mi

10 mi

47 mi

**29.**

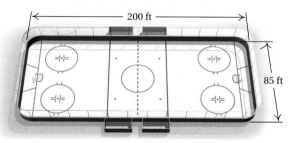

402 ft

298 ft

196 ft

212 ft

100 ft

453 ft

**30.**

62 yd

39 yd

28 yd

46 yd

54 yd

**31.** Find the perimeter of a standard hockey rink.

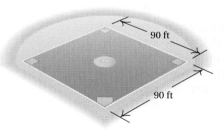

200 ft

85 ft

**32.** In Major League Baseball, how far does a batter travel in circling the bases when a home run has been hit?

90 ft

90 ft

## Skill Maintenance

The exercises that follow begin an important feature called *Skill Maintenance exercises*. These exercises provide an ongoing review of topics previously covered in the book. You will see them in virtually every exercise set. It has been found that this kind of continuing review can significantly improve your performance on a final examination.

**33.** What does the digit 8 mean in 486,205?   [1.1a]

**34.** Write a word name for the number in the following sentence:   [1.1c]

The population of the world is projected to be 9,346,399,468 in 2050.

**Source:** U.S. Census Bureau

## Synthesis

**35.** A fast way to add all the numbers from 1 to 10 inclusive is to pair 1 with 9, 2 with 8, and so on. Use a similar approach to add all numbers from 1 to 100 inclusive.

# 1.3 Subtraction

## OBJECTIVE

**a** Subtract whole numbers.

## **a** Subtraction of Whole Numbers

Subtraction is finding the difference of two numbers. Suppose you pick 6 pints of blueberries and give your neighbor 2 pints.

The subtraction that represents this is

$$6 \quad - \quad 2 \quad = \quad 4.$$

Minuend   Subtrahend   Difference

The **minuend** is the number from which another number is being subtracted. The **subtrahend** is the number being subtracted. The **difference** is the result of subtracting the subtrahend from the minuend.

In the subtraction above, note that the difference, 4, is the number we add to 2 to get 6. This illustrates the relationship between addition and subtraction and leads us to the following definition of subtraction.

> **SUBTRACTION**
>
> The difference $a - b$ is that unique whole number $c$ for which $a = c + b$.

We see that $6 - 2 = 4$ because $4 + 2 = 6$.

To subtract numbers, we subtract the ones digits first, then the tens digits, then the hundreds, then the thousands, and so on.

**EXAMPLE 1**   Subtract: $9768 - 4320$.

$$\begin{array}{r} 9\ 7\ 6\ 8 \\ -\ 4\ 3\ 2\ 0 \\ \hline 8 \end{array}$$   Subtract ones.

$$\begin{array}{r} 9\ 7\ 6\ 8 \\ -\ 4\ 3\ 2\ 0 \\ \hline 4\ 8 \end{array}$$   Subtract tens.

$$\begin{array}{r} 9\ 7\ 6\ 8 \\ -\ 4\ 3\ 2\ 0 \\ \hline 4\ 4\ 8 \end{array}$$   Subtract hundreds.

$$\begin{array}{r} 9\ 7\ 6\ 8 \\ -\ 4\ 3\ 2\ 0 \\ \hline 5\ 4\ 4\ 8 \end{array}$$   Subtract thousands.

We show these steps for explanation. You need write only this.

$$\begin{array}{r} 9\ 7\ 6\ 8 \\ -\ 4\ 3\ 2\ 0 \\ \hline 5\ 4\ 4\ 8 \end{array}$$

Because subtraction is defined in terms of addition, we can use addition to *check* subtraction.

*Subtraction:*                          *Check by Addition:*

```
    9  7  6  8                              5  4  4  8
  − 4  3  2  0          ?                 + 4  3  2  0
    5  4  4  8                              9  7  6  8
```

Do Exercise 1.

**EXAMPLE 2**   Subtract: $348 - 165$.

We have

$$
\begin{array}{rl}
3 \text{ hundreds } + 4 \text{ tens } + 8 \text{ ones } = & 2 \text{ hundreds } + 14 \text{ tens } + 8 \text{ ones} \\
- 1 \text{ hundred } - 6 \text{ tens } - 5 \text{ ones } = & -1 \text{ hundred } - 6 \text{ tens } - 5 \text{ ones} \\
= & 1 \text{ hundred } + 8 \text{ tens } + 3 \text{ ones} \\
= & 183.
\end{array}
$$

First, we subtract the ones.

```
    3  4  8       Subtract ones.
  − 1  6  5
          3
```

We cannot subtract the tens because there is no whole number that when added to 6 gives 4. To complete the subtraction, we must *borrow* 1 hundred from 3 hundreds and regroup it with the 4 tens. Then we can do the subtraction 14 tens − 6 tens = 8 tens.

```
    2  14
    3̶  4̶  8       Borrow one hundred. That is, 1 hundred = 10 tens, and
  − 1  6  5       10 tens + 4 tens = 14 tens. Write 2 above the hundreds
          3       column and 14 above the tens.
```

```
    2  14
    3̶  4̶  8       Subtract tens; subtract hundreds.
  − 1  6  5
    1  8  3
```

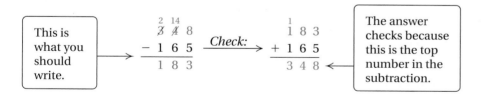

This is what you should write.

```
    2  14
    3̶  4̶  8     Check:
  − 1  6  5
    1  8  3
```

```
    1
    1  8  3
  + 1  6  5
    3  4  8
```

The answer checks because this is the top number in the subtraction.

---

**1.** Subtract. Check by adding.

```
    7  8  9  3
  − 4  0  9  2
```

---

**Calculator Corner**

**Subtracting Whole Numbers**   To subtract whole numbers on a calculator, we use the − and = keys. For example, to find $63 - 47$, we press 6 3 − 4 7 = . The calculator displays 16, so $63 - 47 = 16$. We can check this result by adding the subtrahend, 47, and the difference, 16. To do this, we press 1 6 + 4 7 = . The sum is the minuend, 63, so the subtraction is correct.

**Exercises:**   Use a calculator to perform each subtraction. Check by adding.

**1.** $57 - 29$

**2.** $81 - 34$

**3.** $145 - 78$

**4.** $612 - 493$

**5.**
```
    4  9  7  6
  − 2  8  4  8
```

**6.**
```
  1  2 , 4  0  6
  −    9  8  1  3
```

---

***Answer***

**1.** 3801; *check:* 3801 + 4092 = 7893

**EXAMPLE 3**  Subtract: 6246 − 1879.

$$
\begin{array}{r}
\phantom{-}3\ 16\\
6\ 2\ \cancel{4}\ \cancel{6}\\
-\ 1\ 8\ 7\ 9\\
\hline
7
\end{array}
$$

We cannot subtract 9 ones from 6 ones, but we can subtract 9 ones from 16 ones. We borrow 1 ten to get 16 ones.

$$
\begin{array}{r}
13\\
1\ \cancel{3}\ 16\\
6\ \cancel{2}\ \cancel{4}\ \cancel{6}\\
-\ 1\ 8\ 7\ 9\\
\hline
6\ 7
\end{array}
$$

We cannot subtract 7 tens from 3 tens, but we can subtract 7 tens from 13 tens. We borrow 1 hundred to get 13 tens.

$$
\begin{array}{r}
11\ 13\\
5\ \cancel{1}\ \cancel{3}\ 16\\
\cancel{6}\ \cancel{2}\ \cancel{4}\ \cancel{6}\\
-\ 1\ 8\ 7\ 9\\
\hline
4\ 3\ 6\ 7
\end{array}
$$

We cannot subtract 8 hundreds from 1 hundred, but we can subtract 8 hundreds from 11 hundreds. We borrow 1 thousand to get 11 hundreds. Finally, we subtract the thousands.

This is what you should write.

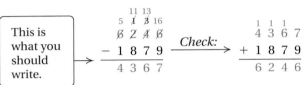

$$
\begin{array}{r}
11\ 13\\
5\ \cancel{1}\ \cancel{3}\ 16\\
\cancel{6}\ \cancel{2}\ \cancel{4}\ \cancel{6}\\
-\ 1\ 8\ 7\ 9\\
\hline
4\ 3\ 6\ 7
\end{array}
$$

*Check:*

$$
\begin{array}{r}
1\ 1\ 1\\
4\ 3\ 6\ 7\\
+\ 1\ 8\ 7\ 9\\
\hline
6\ 2\ 4\ 6
\end{array}
$$

The answer checks because this is the top number in the subtraction.

Do Exercises 2 and 3.

Subtract. Check by adding.

**2.**
$$
\begin{array}{r}
8\ 6\ 8\ 6\\
-\ 2\ 3\ 5\ 8\\
\hline
\end{array}
$$

**3.**
$$
\begin{array}{r}
7\ 1\ 4\ 5\\
-\ 2\ 3\ 9\ 8\\
\hline
\end{array}
$$

**EXAMPLE 4**  Subtract: 902 − 477.

$$
\begin{array}{r}
8\ 9\ 12\\
\cancel{9}\ \cancel{0}\ \cancel{2}\\
-\ 4\ 7\ 7\\
\hline
4\ 2\ 5
\end{array}
$$

We cannot subtract 7 ones from 2 ones. We have 9 hundreds, or 90 tens. We borrow 1 ten to get 12 ones. We then have 89 tens.

Do Exercises 4 and 5.

Subtract.

**4.**
$$
\begin{array}{r}
7\ 0\\
-\ 1\ 4\\
\hline
\end{array}
$$

**5.**
$$
\begin{array}{r}
5\ 0\ 3\\
-\ 2\ 9\ 8\\
\hline
\end{array}
$$

**EXAMPLE 5**  Subtract: 8003 − 3667.

$$
\begin{array}{r}
7\ 9\ 9\ 13\\
\cancel{8}\ \cancel{0}\ \cancel{0}\ \cancel{3}\\
-\ 3\ 6\ 6\ 7\\
\hline
4\ 3\ 3\ 6
\end{array}
$$

We have 8 thousands, or 800 tens. We borrow 1 ten to get 13 ones. We then have 799 tens.

**EXAMPLES**

**6.** Subtract: 6000 − 3762.

$$
\begin{array}{r}
5\ 9\ 9\ 10\\
\cancel{6}\ \cancel{0}\ \cancel{0}\ \cancel{0}\\
-\ 3\ 7\ 6\ 2\\
\hline
2\ 2\ 3\ 8
\end{array}
$$

**7.** Subtract: 6024 − 2968.

$$
\begin{array}{r}
11\\
5\ 9\ \cancel{1}\ 14\\
\cancel{6}\ 0\ \cancel{2}\ \cancel{4}\\
-\ 2\ 9\ 6\ 8\\
\hline
3\ 0\ 5\ 6
\end{array}
$$

Do Exercises 6–9.

Subtract.

**6.**
$$
\begin{array}{r}
7\ 0\ 0\ 7\\
-\ 6\ 3\ 4\ 9\\
\hline
\end{array}
$$

**7.**
$$
\begin{array}{r}
6\ 0\ 0\ 0\\
-\ 3\ 1\ 4\ 9\\
\hline
\end{array}
$$

**8.**
$$
\begin{array}{r}
9\ 0\ 3\ 5\\
-\ 7\ 4\ 8\ 9\\
\hline
\end{array}
$$

**9.**
$$
\begin{array}{r}
2\ 0\ 0\ 1\\
-\ \ \ 1\ 2\ 4\\
\hline
\end{array}
$$

*Answers*

**2.** 6328; *check:* 6328 + 2358 = 8686
**3.** 4747; *check:* 4747 + 2398 = 7145
**4.** 56  **5.** 205  **6.** 658  **7.** 2851
**8.** 1546  **9.** 1877

**a** Subtract. Check by adding.

1.  
$$\begin{array}{r} 6\,5 \\ -\,2\,1 \end{array}$$

2.  
$$\begin{array}{r} 8\,7 \\ -\,3\,4 \end{array}$$

3.  
$$\begin{array}{r} 8\,6\,6 \\ -\,3\,3\,3 \end{array}$$

4.  
$$\begin{array}{r} 5\,2\,6 \\ -\,3\,2\,3 \end{array}$$

5. $86 - 47$

6. $73 - 28$

7. $51 - 37$

8. $64 - 19$

9.  
$$\begin{array}{r} 5\,6\,3 \\ -\,1\,9\,4 \end{array}$$

10.  
$$\begin{array}{r} 7\,9\,5 \\ -\,3\,9\,8 \end{array}$$

11.  
$$\begin{array}{r} 3\,9\,1 \\ -\,3\,6\,5 \end{array}$$

12.  
$$\begin{array}{r} 3\,1\,6 \\ -\,2\,4\,7 \end{array}$$

13. $981 - 747$

14. $887 - 698$

15. $683 - 266$

16. $342 - 217$

17.  
$$\begin{array}{r} 7\,7\,6\,9 \\ -\,2\,3\,8\,7 \end{array}$$

18.  
$$\begin{array}{r} 6\,4\,3\,1 \\ -\,2\,8\,9\,6 \end{array}$$

19.  
$$\begin{array}{r} 4\,5\,1\,2 \\ -\,1\,7\,3\,4 \end{array}$$

20.  
$$\begin{array}{r} 8\,3\,6\,4 \\ -\,5\,3\,7\,5 \end{array}$$

21. $5318 - 2249$

22. $9241 - 5643$

23. $3947 - 2858$

24. $7583 - 3641$

25.  
$$\begin{array}{r} 1\,2{,}6\,4\,7 \\ -\;\;\;4{,}8\,9\,9 \end{array}$$

26.  
$$\begin{array}{r} 1\,6{,}2\,2\,2 \\ -\;\;\;5{,}8\,8\,8 \end{array}$$

27.  
$$\begin{array}{r} 5\,1{,}3\,4\,2 \\ -\,4\,7{,}1\,9\,8 \end{array}$$

28.  
$$\begin{array}{r} 3\,2{,}1\,9\,4 \\ -\,2\,9{,}2\,3\,6 \end{array}$$

29.  
$$\begin{array}{r} 8\,0 \\ -\,2\,4 \end{array}$$

30.  
$$\begin{array}{r} 9\,0 \\ -\,7\,8 \end{array}$$

31.  
$$\begin{array}{r} 6\,9\,0 \\ -\,2\,3\,6 \end{array}$$

32.  
$$\begin{array}{r} 8\,0\,3 \\ -\,4\,1\,8 \end{array}$$

**33.**
```
   7 6 4 0
 − 3 8 0 9
```

**34.**
```
   5 2 8 0
 − 3 0 9 1
```

**35.**
```
   6 8 0 8
 − 3 0 5 9
```

**36.**
```
   9 4 0 5
 −   2 5 8
```

**37.**
```
   2 3 0 0
 −   1 0 9
```

**38.**
```
   7 5 0 0
 − 3 6 0 4
```

**39.**
```
   6 0 0 7
 − 1 5 8 9
```

**40.**
```
   8 0 0 3
 −   5 9 9
```

**41.** $90{,}237 − 47{,}209$

**42.** $84{,}703 − 298$

**43.** $101{,}734 − 5760$

**44.** $15{,}017 − 7809$

**45.**
```
   7 0 0 0
 − 2 7 9 4
```

**46.**
```
   8 0 0 1
 − 6 5 4 3
```

**47.**
```
   3 9,0 0 0
 − 3 7,6 9 5
```

**48.**
```
   1 7,0 0 0
 − 1 1,5 9 8
```

**49.** $10{,}008 − 19$

**50.** $40{,}006 − 147$

**51.** $50{,}001 − 1984$

**52.** $30{,}004 − 6749$

## Skill Maintenance

Add.  [1.2a]

**53.**
```
   9 4 6
 +   7 8
```

**54.**
```
   9 0 7 8
 + 3 6 5 4
```

**55.**
```
   5 7,8 7 7
 + 3 2,4 0 6
```

**56.**
```
   8 0 0 4
   6 7 8 9
   7 7 2 0
 + 6 8 5 1
```

**57.** $567 + 778$

**58.** $901 + 23$

**59.** $12{,}885 + 9807$

**60.** $9909 + 1011$

**61.** Write a word name for 6,375,602.  [1.1c]

**62.** What does the digit 7 mean in 6,375,602?  [1.1a]

## Synthesis

**63.** Fill in the missing digits to make the subtraction true:
$9{,}\boxed{\phantom{0}}48{,}621 − 2{,}097{,}\boxed{\phantom{0}}81 = 7{,}251{,}140.$

**64.** 🖩 Subtract: $3{,}928{,}124 − 1{,}098{,}947.$

# 1.4 Multiplication

## (a) Multiplication of Whole Numbers

### Repeated Addition

The multiplication $3 \times 5$ corresponds to this repeated addition.

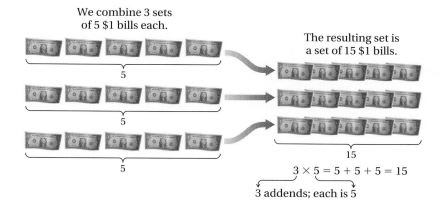

We combine 3 sets of 5 $1 bills each.

The resulting set is a set of 15 $1 bills.

$$3 \times 5 = 5 + 5 + 5 = 15$$

3 addends; each is 5

The numbers that we multiply are called **factors**. The result of the multiplication is called a **product**.

$$
\begin{array}{ccc}
3 & \times \quad 5 & = \quad 15 \\
\downarrow & \downarrow & \downarrow \\
\text{Factor} & \text{Factor} & \text{Product}
\end{array}
$$

### Rectangular Arrays

Multiplications can also be thought of as rectangular arrays. Each of the following corresponds to the multiplication $3 \times 5$.

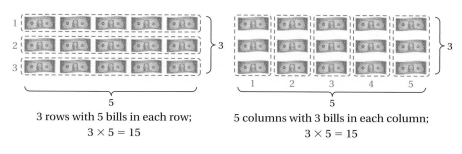

3 rows with 5 bills in each row;
$3 \times 5 = 15$

5 columns with 3 bills in each column;
$3 \times 5 = 15$

When you write a multiplication sentence corresponding to a real-world situation, you should think of either a rectangular array or repeated addition. In some cases, it may help to think both ways.

We have used an "$\times$" to denote multiplication. A dot "$\cdot$" is also commonly used. (Use of the dot is attributed to the German mathematician Gottfried Wilhelm von Leibniz in 1698.) Parentheses are also used to denote multiplication. For example,

$$3 \times 5 = 3 \cdot 5 = (3)(5) = 3(5) = 15.$$

The product of 0 and any whole number is 0: $0 \cdot a = a \cdot 0 = 0$. For example, $0 \cdot 3 = 3 \cdot 0 = 0$. Multiplying a number by 1 does not change the number: $1 \cdot a = a \cdot 1 = a$. For example, $1 \cdot 3 = 3 \cdot 1 = 3$. We say that 1 is the **multiplicative identity**.

**EXAMPLE 1** Multiply: $5 \times 734$.

We have

```
      7 3 4
  ×       5
  ─────────
      2 0   ← Multiply the 4 ones by 5: 5 × 4 = 20.
    1 5 0   ← Multiply the 3 tens by 5: 5 × 30 = 150.
  3 5 0 0   ← Multiply the 7 hundreds by 5: 5 × 700 = 3500.
  ─────────
  3 6 7 0   ← Add.
```

Instead of writing each product on a separate line, we can use a shorter form.

```
        2
    7   3   4
  ×           5
  ───────────
              0
```
Multiply the ones by 5: $5 \cdot (4 \text{ ones}) = 20 \text{ ones} = 2 \text{ tens} + 0 \text{ ones}$. Write 0 in the ones column and 2 above the tens.

```
    1   2
    7   3   4
  ×           5
  ───────────
          7   0
```
Multiply the 3 tens by 5 and add 2 tens: $5 \cdot (3 \text{ tens}) = 15 \text{ tens}$, $15 \text{ tens} + 2 \text{ tens} = 17 \text{ tens} = 1 \text{ hundred} + 7 \text{ tens}$. Write 7 in the tens column and 1 above the hundreds.

```
    1   2
    7   3   4
  ×           5
  ───────────
  3   6   7   0
```
Multiply the 7 hundreds by 5 and add 1 hundred: $5 \cdot (7 \text{ hundreds}) = 35 \text{ hundreds}$, $35 \text{ hundreds} + 1 \text{ hundred} = 36 \text{ hundreds}$.

```
    1   2
    7   3   4 ⎫
  ×           5 ⎬
  ───────────  ⎭
  3   6   7   0
```
You should write only this.

Do Exercises 1–4.

Multiplication of whole numbers is based on a property called the **distributive law**. It says that to multiply a number by a sum, $a \cdot (b + c)$, we can multiply each addend by $a$ and then add like this: $(a \cdot b) + (a \cdot c)$. Thus, $a \cdot (b + c) = (a \cdot b) + (a \cdot c)$. For example, consider the following.

$$4 \cdot (2 + 3) = 4 \cdot 5 = 20 \qquad \text{Adding first; then multiplying}$$

$$4 \cdot (2 + 3) = (4 \cdot 2) + (4 \cdot 3) = 8 + 12 = 20 \qquad \text{Multiplying first; then adding}$$

The results are the same, so $4 \cdot (2 + 3) = (4 \cdot 2) + (4 \cdot 3)$.

Multiply.

**1.**
```
    5 8
  ×   2
```

**2.**
```
    3 7
  ×   4
```

**3.**
```
    8 2 3
  ×     6
```

**4.**
```
    1 3 4 8
  ×       5
```

*Answers*

**1.** 116    **2.** 148    **3.** 4938    **4.** 6740

Let's find the product $51 \times 32$. Since $32 = 2 + 30$, we can think of this product as

$$51 \times 32 = 51 \times (2 + 30) = (51 \times 2) + (51 \times 30).$$

That is, we multiply 51 by 2, then we multiply 51 by 30, and finally we add. We can write our work this way.

```
        5 1
      × 3 2
      ─────
      1 0 2      Multiplying by 2
    1 5 3 0      Multiplying by 30. (We write a 0 and then multiply 51 by 3.)
```

You may have learned that such a 0 need not be written. You may omit it if you wish. If you do omit it, remember, when multiplying by tens, to start writing the answer in the tens place.

We add to obtain the product.

```
        5 1
      × 3 2
      ─────
      1 0 2
    1 5 3 0
    ───────
    1 6 3 2      Adding to obtain the product
```

**EXAMPLE 2** Multiply: $457 \times 683$.

```
      5   2
      6  8  3
    × 4  5  7
    ─────────
    4  7  8  1      Multiplying 683 by 7
```

```
      4    1
      5    2
      6    8    3
    × 4    5    7
    ───────────────
      4    7    8    1
    3 4    1    5    0      Multiplying 683 by 50
```

```
      3     1
      4     1
      5     2
      6     8     3
    × 4     5     7
    ─────────────────
      4     7     8     1
    3 4     1     5     0      ⌐ Multiplying 683 by 400. (We write
  2 7 3     2     0     0  ←  ⌐ 00 and then multiply 683 by 4.)
  ─────────────────────
  3 1 2 ,   1     3     1      Adding
```

Do Exercises 5–8.

**TIME MANAGEMENT**

- **A rule of thumb on study time.** Budget about 2–3 hours for homework and study for every hour you spend in class each week.

- **Scheduling your time.** Make an hour-by-hour schedule of your typical week. Include work, school, home, sleep, study, and leisure times. Try to schedule time for study when you are most alert. Choose a setting that will enable you to focus and concentrate. Plan for success and it will happen!

Multiply.

**5.**
```
    4 5
  × 2 3
  ─────
```

**6.** $48 \times 63$

**7.**
```
    7 4 6
  ×   6 2
  ───────
```

**8.** $245 \times 837$

*Answers*

**5.** 1035  **6.** 3024  **7.** 46,252  **8.** 205,065

1.4 Multiplication  **21**

Multiply.

**9.** 4 7 2
× 3 0 6

**10.** 408 × 704

**11.** 2 3 4 4
× 6 0 0 5

**12.** 1 0 0 6
× 7 0 3

**EXAMPLE 3** Multiply: 306 × 274.

Note that 306 = 3 hundreds + 6 ones.

```
      2 7 4
    × 3 0 6  ┌ Multiplying by 6
    1 6 4 4  ←┌ Multiplying by 3 hundreds. (We write 00
    8 2 2 0 0 ←┘ and then multiply 274 by 3.)
    8 3,8 4 4   Adding
```

Do Exercises 9–12.

**EXAMPLE 4** Multiply: 360 × 274.

Note that 360 = 3 hundreds + 6 tens.

```
      2 7 4  ┌ Multiplying by 6 tens. (We write 0 and
    × 3 6 0  │ then multiply 274 by 6.)
    1 6 4 4 0 ←┌ Multiplying by 3 hundreds. (We write 00
    8 2 2 0 0 ←┘ and then multiply 274 by 3.)
    9 8,6 4 0   Adding
```

Multiply.

**13.** 4 7 2
× 8 3 0

**14.** 2 3 4 4
× 7 4 0 0

**15.** 100 × 562

**16.** 1000 × 562

Do Exercises 13–16.

When we multiply two numbers, we can change the order of the numbers without changing their product. For example, $3 \cdot 6 = 18$ and $6 \cdot 3 = 18$. This illustrates the **commutative law of multiplication:** $a \cdot b = b \cdot a$.

Do Exercise 17.

**17.** **a)** Find $23 \cdot 47$.

**b)** Find $47 \cdot 23$.

**c)** Compare your answers to parts (a) and (b).

To multiply three or more numbers, we generally first group them so that we multiply two at a time. Consider $2 \cdot 3 \cdot 4$. We can group these numbers as $2 \cdot (3 \cdot 4)$ or as $(2 \cdot 3) \cdot 4$. The parentheses tell what to do first:

$$2 \cdot (3 \cdot 4) = 2 \cdot (12) = 24.$$ We multiply 3 and 4 and then that product by 2.

We can also multiply 2 and 3 and then that product by 4:

$$(2 \cdot 3) \cdot 4 = (6) \cdot 4 = 24.$$

Either way we get 24. It does not matter how we group the numbers. This illustrates the **associative law of multiplication:** $a \cdot (b \cdot c) = (a \cdot b) \cdot c$.

Multiply.

**18.** $5 \cdot 2 \cdot 4$

**19.** $4 \cdot 2 \cdot 6$

Do Exercises 18 and 19.

*Answers*

**9.** 144,432  **10.** 287,232
**11.** 14,075,720  **12.** 707,218
**13.** 391,760  **14.** 17,345,600
**15.** 56,200  **16.** 562,000
**17.** (a) 1081; (b) 1081; (c) same
**18.** 40  **19.** 48

## b Finding Area

The area of a rectangular region can be considered to be the number of square units needed to fill it. Here is a rectangle 4 cm (centimeters) long and 3 cm wide. It takes 12 square centimeters (sq cm) to fill it.

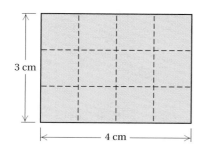

This is a square centimeter (a square unit).

In this case, we have a rectangular array of 3 rows, each of which contains 4 squares. The number of square units is given by $3 \cdot 4$, or 12. That is, $A = l \cdot w = 3 \, \text{cm} \cdot 4 \, \text{cm} = 12 \, \text{sq cm}$.

**EXAMPLE 5**  *Table Tennis.*  Find the area of a standard table tennis table that has dimensions of 9 ft by 5 ft.

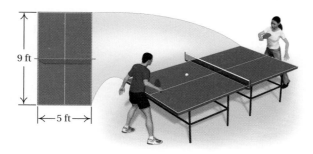

If we think of filling the rectangle with square feet, we have a rectangular array. The length $l = 9$ ft and the width $w = 5$ ft. Thus the area $A$ is given by the formula

$$A = l \cdot w = 9 \, \text{ft} \cdot 5 \, \text{ft} = 45 \, \text{sq ft}.$$

Do Exercise 20.

**20. Professional Pool Table.**  The playing area of a pool table used in professional tournaments is 50 in. by 100 in. (There are 6-in. wide rails on the outside that are not included in the playing area.) Determine the playing area.

*Answer*

**20.**  5000 sq in.

**1.4** **Exercise Set**

For Extra Help

*MyMathLab*

Math XL
PRACTICE

WATCH

DOWNLOAD

READ

REVIEW

**a**    Multiply.

1.    $\begin{array}{r} 6\,5 \\ \times\quad 8 \\ \hline \end{array}$

2.    $\begin{array}{r} 8\,7 \\ \times\quad 4 \\ \hline \end{array}$

3.    $\begin{array}{r} 9\,4 \\ \times\quad 6 \\ \hline \end{array}$

4.    $\begin{array}{r} 7\,6 \\ \times\quad 9 \\ \hline \end{array}$

5. $3 \cdot 509$

6. $7 \cdot 806$

7. $7(9229)$

8. $4(7867)$

9. $90(53)$

10. $60(78)$

11. $(47)(85)$

12. $(34)(87)$

13.    $\begin{array}{r} 8\,7 \\ \times\,1\,0 \\ \hline \end{array}$

14.    $\begin{array}{r} 2\,3\,4\,0 \\ \times\,1\,0\,0\,0 \\ \hline \end{array}$

15.    $\begin{array}{r} 9\,6 \\ \times\,2\,0 \\ \hline \end{array}$

16.    $\begin{array}{r} 8\,0\,0 \\ \times\,7\,0\,0 \\ \hline \end{array}$

17.    $\begin{array}{r} 6\,4\,3 \\ \times\quad 7\,2 \\ \hline \end{array}$

18.    $\begin{array}{r} 7\,7\,7 \\ \times\quad 7\,7 \\ \hline \end{array}$

19.    $\begin{array}{r} 4\,4\,4 \\ \times\quad 3\,3 \\ \hline \end{array}$

20.    $\begin{array}{r} 5\,4\,9 \\ \times\quad 8\,8 \\ \hline \end{array}$

21.    $\begin{array}{r} 5\,6\,4 \\ \times\,4\,5\,8 \\ \hline \end{array}$

22.    $\begin{array}{r} 4\,3\,2 \\ \times\,3\,7\,5 \\ \hline \end{array}$

23.    $\begin{array}{r} 8\,5\,3 \\ \times\,9\,3\,6 \\ \hline \end{array}$

24.    $\begin{array}{r} 3\,4\,6 \\ \times\,6\,5\,9 \\ \hline \end{array}$

25.    $\begin{array}{r} 6\,4\,2\,8 \\ \times\,3\,2\,2\,4 \\ \hline \end{array}$

26.    $\begin{array}{r} 8\,9\,2\,8 \\ \times\,3\,1\,7\,2 \\ \hline \end{array}$

27.    $\begin{array}{r} 3\,4\,8\,2 \\ \times\quad 1\,0\,4 \\ \hline \end{array}$

28.    $\begin{array}{r} 6\,4\,0\,8 \\ \times\,6\,0\,6\,4 \\ \hline \end{array}$

29.    $\begin{array}{r} 8\,7\,6 \\ \times\,3\,4\,5 \\ \hline \end{array}$

30.    $\begin{array}{r} 3\,5\,5 \\ \times\,2\,9\,9 \\ \hline \end{array}$

31.    $\begin{array}{r} 7\,8\,8\,9 \\ \times\,6\,2\,2\,4 \\ \hline \end{array}$

32.    $\begin{array}{r} 6\,5\,2\,1 \\ \times\,3\,4\,4\,9 \\ \hline \end{array}$

**33.**  5 6 0 8
     × 4 5 0 0

**34.**  4 5 0 6
     × 7 8 0 0

**35.**  5 0 0 6
     × 4 0 0 8

**36.**  6 0 0 9
     × 2 0 0 3

**b** Find the area of each region.

**37.**

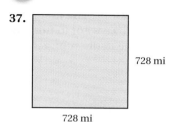

728 mi

728 mi

**38.**  129 yd

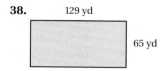

65 yd

**39.** Find the area of the region formed by the base lines on a Major League Baseball diamond.

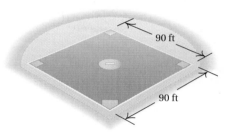

90 ft

90 ft

**40.** Find the area of a standard-sized hockey rink.

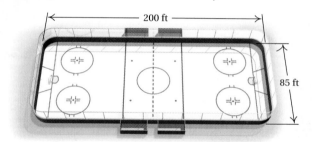

200 ft

85 ft

## Skill Maintenance

Add.  [1.2a]

**41.**  4 9 0 8
      5 6 6 7
    + 2 1 1 0

**42.**  9 8 7 6
       8 7 6
        7 6
    +      6

**43.**  3 4 0 , 7 9 8
     +   8 6 , 6 7 9

**44.**  8 8 , 7 7 7
     + 2 2 , 3 3 3

Subtract.  [1.3a]

**45.**  4 9 0 8
     − 3 6 6 7

**46.**  9 8 7 6
     −   9 8 7

**47.**  3 4 0 , 7 9 8
     −   8 6 , 6 7 9

**48.**  8 8 , 7 7 7
     − 2 2 , 3 3 3

## Synthesis

**49.** 🖩 An 18-story office building is box-shaped. Each floor measures 172 ft by 84 ft with a 20-ft by 35-ft rectangular area lost to an elevator and a stairwell. How much area is available as office space?

# 1.5

## Division

### OBJECTIVE

a   Divide whole numbers.

**SKILL TO REVIEW**
Objective 1.3a:
Subtract whole numbers.

Subtract.

**1.**   5 6 4
        − 3 9 7

**2.**   7 0 3 5
        − 2 9 4 4

### a  Division of Whole Numbers

#### Repeated Subtraction

Division of whole numbers applies to two kinds of situations. The first is repeated subtraction. Suppose we have 20 doughnuts, and we want to find out how many sets of 5 there are. One way to do this is to repeatedly subtract sets of 5 as follows.

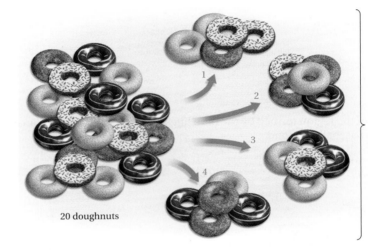

How many sets of 5 doughnuts each?

20 doughnuts

Since there are 4 sets of 5 doughnuts each, we have

$$20 \div 5 = 4.$$

Dividend   Divisor   Quotient

The division $20 \div 5$ is read "20 divided by 5." The **dividend** is 20, the **divisor** is 5, and the **quotient** is 4. We divide the *dividend* by the *divisor* to get the *quotient*.

We can also express the division $20 \div 5 = 4$ as

$$\frac{20}{5} = 4 \quad \text{or} \quad 5\overline{)20}^{\,4}.$$

#### Rectangular Arrays

We can also think of division in terms of rectangular arrays. Consider again the 20 doughnuts and division by 5. We can arrange the doughnuts in a rectangular array with 5 rows and ask, "How many are in each row?"

We can also consider a rectangular array with 5 doughnuts in each column and ask, "How many columns are there?" The answer is still 4.

In each case, we are asking, "What do we multiply 5 by in order to get 20?"

Missing factor               Quotient

$$5 \cdot \square = 20 \qquad 20 \div 5 = \square$$

This leads us to the following definition of division.

*Answers*

*Skill to Review:*
**1.** 167    **2.** 4091

## DIVISION

The quotient $a \div b$, where $b \neq 0$, is that unique number $c$ for which $a = b \cdot c$.

This definition shows the relation between division and multiplication. We see, for instance, that

$$20 \div 5 = 4 \quad \text{because} \quad 20 = 5 \cdot 4.$$

This relation allows us to use multiplication to check division.

**EXAMPLE 1** Divide. Check by multiplying.

**a)** $16 \div 8$ **b)** $\dfrac{36}{4}$ **c)** $7\overline{)56}$

We do so as follows.

**a)** $16 \div 8 = 2$     *Check*: $8 \cdot 2 = 16$.

**b)** $\dfrac{36}{4} = 9$     *Check*: $4 \cdot 9 = 36$.

**c)** $7\overline{)56}$ with $8$ above     *Check*: $7 \cdot 8 = 56$.

Do Exercises 1–3.

Divide. Check by multiplying.
**1.** $9\overline{)45}$

**2.** $27 \div 3$

**3.** $\dfrac{48}{6}$

Let's consider some basic properties of division.

## DIVIDING BY 1

Any number divided by 1 is that same number: $a \div 1 = \dfrac{a}{1} = a$.

For example, $6 \div 1 = 6$ and $\dfrac{15}{1} = 15$.

## DIVIDING A NUMBER BY ITSELF

Any nonzero number divided by itself is 1: $a \div a = \dfrac{a}{a} = 1, \quad a \neq 0$.

For example, $7 \div 7 = 1$ and $\dfrac{22}{22} = 1$.

## DIVIDENDS OF 0

Zero divided by any nonzero number is 0: $0 \div a = \dfrac{0}{a} = 0, \quad a \neq 0$.

For example, $0 \div 14 = 0$ and $\dfrac{0}{3} = 0$.

***Answers***
**1.** 5; *check:* $9 \cdot 5 = 45$    **2.** 9; *check:* $3 \cdot 9 = 27$
**3.** 8; *check:* $6 \cdot 8 = 48$

Why can't we divide by 0? Suppose the number 4 could be divided by 0. Then if ▢ were the answer, we would have

$$4 \div 0 = \boxed{\phantom{x}},$$

and since 0 times any number is 0, we would have

$$4 = \boxed{\phantom{x}} \cdot 0 = 0. \qquad \text{False!}$$

Thus, the only possible number that could be divided by 0 would be 0 itself. But such a division would give us any number we wish. For instance,

$$
\left.
\begin{aligned}
0 \div 0 = 8 \quad &\text{because} \quad 0 = 8 \cdot 0; \\
0 \div 0 = 3 \quad &\text{because} \quad 0 = 3 \cdot 0; \\
0 \div 0 = 7 \quad &\text{because} \quad 0 = 7 \cdot 0.
\end{aligned}
\right\} \quad \text{All true!}
$$

We avoid the preceding difficulties by agreeing to exclude division by 0.

Do Exercises 4–7.

### Division with a Remainder

Suppose we have 22 cans of soda and want to pack them in cartons of 6 cans each. We could fill 3 cartons and have 4 cans left over.

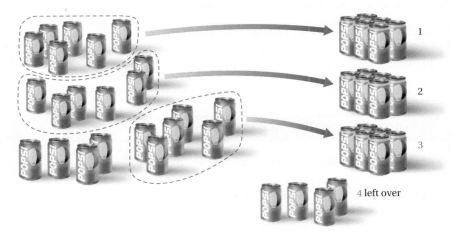

1

2

3

4 left over

We can think of this as the following division. The leftover cans are the **remainder**.

$$
\begin{array}{r}
3 \;\leftarrow \text{Quotient} \\
6\overline{)2\;2} \\
\underline{1\;8} \\
4 \;\leftarrow \text{Remainder}
\end{array}
$$

---

**Sidebar (left column):**

Divide, if possible. If not possible, write "not defined."

**4.** $\dfrac{9}{9}$

**5.** $5 \div 0$

**6.** $0 \div 20$

**7.** $\dfrac{8}{1}$

---

*Answers*

**4.** 1  **5.** Not defined
**6.** 0  **7.** 8

We express the result as

$$22 \div 6 = 3 \text{ R } 4.$$

Dividend    Divisor    Quotient    Remainder

Note that

Quotient $\cdot$ Divisor + Remainder = Dividend.

Thus we have

$$3 \cdot 6 = 18 \qquad \text{Quotient} \cdot \text{Divisor}$$

and    $18 + 4 = 22.$    Adding the remainder. The result is the dividend.

We now show a procedure for dividing whole numbers.

**EXAMPLE 2**   Divide and check: $4\overline{)3457}$.

First, we try to divide the first digit of the dividend, 3, by the divisor, 4. Since $3 \div 4$ is not a whole number, we consider the first *two* digits of the dividend.

$$
\begin{array}{r}
8 \phantom{457} \\
4\overline{)3\ 4\ 5\ 7} \\
3\ 2 \phantom{57} \\
\hline
2 \phantom{57}
\end{array}
$$

Since $4 \cdot 8 = 32$ and 32 is smaller than 34, we write an 8 in the quotient above the 4. We also write 32 below 34 and subtract.

What if we had chosen a number other than 8 for the first digit of the quotient? Suppose we had used 7 instead of 8 and subtracted $4 \cdot 7$, or 28, from 34. The result would have been $34 - 28$, or 6. Because 6 is larger than the divisor, 4, we know that there is at least one more factor of 4 in 34, and thus 7 is too small. If we had used 9 instead of 8, then we would have tried to subtract $4 \cdot 9$, or 36, from 34. That difference is not a whole number, so we know 9 is too large. When we subtract, the difference must be smaller than the divisor.

Let's continue dividing.

$$
\begin{array}{r}
8\ 6 \phantom{57} \\
4\overline{)3\ 4\ 5\ 7} \\
3\ 2\ \downarrow \phantom{7} \\
\hline
2\ 5 \phantom{7} \\
2\ 4 \phantom{7} \\
\hline
1 \phantom{7}
\end{array}
$$

Now bring down the 5 in the dividend and consider $25 \div 4$. Since $4 \cdot 6 = 24$ and 24 is smaller than 25, we write 6 in the quotient above the 5. We also write 24 below 25 and subtract. The difference, 1, is smaller than the divisor, so we know that 6 is the correct choice.

$$
\begin{array}{r}
8\ 6\ 4 \\
4\overline{)3\ 4\ 5\ 7} \\
3\ 2 \phantom{\ 5\ 7} \\
\hline
2\ 5 \phantom{\ 7} \\
2\ 4\ \downarrow \\
\hline
1\ 7 \\
1\ 6 \\
\hline
1 \leftarrow
\end{array}
$$

Bring down the 7 and consider $17 \div 4$. Since $4 \cdot 4 = 16$ and 16 is smaller than 17, we write 4 in the quotient above the 7. We also write 16 below 17 and subtract.

The remainder is 1.

Check:   $864 \cdot 4 = 3456$ and $3456 + 1 = 3457$.

The answer is 864 R 1.

Do Exercises 8–10.

Divide and check.

**8.** $3\overline{)239}$

**9.** $5\overline{)5864}$

**10.** $6\overline{)3855}$

***Answers***

**8.** 79 R 2; *check:* $3 \cdot 79 + 2 = 239$
**9.** 1172 R 4; *check:* $5 \cdot 1172 + 4 = 5864$
**10.** 642 R 3; *check:* $6 \cdot 642 + 3 = 3855$

**EXAMPLE 3**  Divide: 8904 ÷ 42.

Because 42 is close to 40, we think of the divisor as 40 when we make our choices of digits in the quotient.

$$
\begin{array}{r}
2\phantom{000} \\
42\overline{)8904} \\
84\phantom{00}\downarrow \\
\hline
50\phantom{0}
\end{array}
$$
← *Think:* 89 ÷ 40. We try 2. Multiply 42 · 2 and subtract. Then bring down the 0.

$$
\begin{array}{r}
2\ 1\phantom{00} \\
42\overline{)8904} \\
84\phantom{000} \\
\hline
50\phantom{0} \\
42\downarrow \\
\hline
84
\end{array}
$$
← *Think:* 50 ÷ 40. We try 1. Multiply 42 · 1 and subtract. Then bring down the 4.

$$
\begin{array}{r}
2\ 1\ 2 \\
42\overline{)8904} \\
84\phantom{000} \\
\hline
50\phantom{0} \\
42\phantom{0} \\
\hline
84 \\
84 \\
\hline
0
\end{array}
$$
← *Think:* 84 ÷ 40. We try 2. Multiply 2 · 42 and subtract.

The remainder is 0, so the answer is 212.

Do Exercises 11 and 12.

Divide.

**11.** $45\overline{)6030}$

**12.** $52\overline{)3288}$

-------------------- *Caution!* --------------------

Be careful to keep the digits lined up correctly when you divide.

--------------------------------------------------------

### Calculator Corner

**Dividing Whole Numbers**   To divide whole numbers on a calculator, we use the ÷ and = keys. For example, to divide 711 by 9, we press [7] [1] [1] [÷] [9] [=]. The display reads [ 79 ], so 711 ÷ 9 = 79.

When we enter 453 ÷ 15, the display reads [ 30.2 ]. Note that the result is not a whole number. This tells us that there is a remainder. The number 30.2 is expressed in decimal notation. The symbol "." is called a decimal point. Decimal notation will be studied in Chapter 5. Although it is possible to use the number to the right of the decimal point to find the remainder, we will not do so here.

**Exercises:**   Use a calculator to perform each division.

**1.** $19\overline{)532}$

**2.** $7\overline{)861}$

**3.** 9367 ÷ 29

**4.** 12,276 ÷ 341

## Zeros in Quotients

**EXAMPLE 4**  Divide: $6341 \div 7$.

```
      9
 7 ) 6 3 4 1   ← Think: 63 ÷ 7 = 9. The first digit in the quotient
   6 3 ↓         is 9. We do not write the 0 when we find 63 − 63.
       4         Bring down the 4.
```

```
      9 0
 7 ) 6 3 4 1
   6 3   ↓
       4 1   ← Think: 4 ÷ 7. If we subtract a group of 7's, such
                as 7, 14, 21, etc., from 4, we do not get a whole
                number, so the next digit in the quotient is 0.
                Bring down the 1.
```

```
      9 0 5
 7 ) 6 3 4 1
   6 3
       4 1   ← Think: 41 ÷ 7. We try 5. Multiply 7 · 5 and
       3 5      subtract.
         6   ← The remainder is 6.
```

The answer is 905 R 6.

Do Exercises 13 and 14.

**EXAMPLE 5**  Divide: $8169 \div 34$.

Because 34 is close to 30, we think of the divisor as 30 when we make our choices of digits in the quotient.

```
         2
 3 4 ) 8 1 6 9   ← Think: 81 ÷ 30. We try 2. Multiply 34 · 2 and subtract.
     6 8 ↓            Then bring down the 6.
     1 3 6
```

```
         2 4
 3 4 ) 8 1 6 9
     6 8 |
     1 3 6   ← Think: 136 ÷ 30. We try 4. Multiply 34 · 4 and
     1 3 6 ↓     subtract. The difference is 0, so we do not write it.
           9     Bring down the 9.
```

```
         2 4 0
 3 4 ) 8 1 6 9
     6 8
     1 3 6
     1 3 6
           9 ⌐ Think: 9 ÷ 34. If we subtract a group of 34's, such as
           0 ⌐    34 or 68, from 9, we do not get a whole number, so the
           9 ←    last digit in the quotient is 0.
                The remainder is 9.
```

The answer is 240 R 9.

Do Exercises 15 and 16.

**TO THE INSTRUCTOR AND THE STUDENT**

This section presents a review of division of whole numbers. Students who are successful should go on to Section 1.6. Those who have trouble should study developmental unit D near the back of this text and then repeat Section 1.5.

Divide.

**13.** $6 ) \overline{4\ 8\ 4\ 6}$

**14.** $7 ) \overline{7\ 6\ 1\ 6}$

Divide.

**15.** $2\ 7 ) \overline{9\ 7\ 2\ 4}$

**16.** $5\ 6 ) \overline{4\ 4,8\ 4\ 7}$

**Answers**

**13.** 807 R 4    **14.** 1088
**15.** 360 R 4    **16.** 800 R 47

**1.5** **Exercise Set**

For Extra Help

**MyMathLab**

*Math XL*
PRACTICE

WATCH

DOWNLOAD

READ

REVIEW

**a** Divide, if possible. If not possible, write "not defined."

**1.** $72 \div 6$

**2.** $54 \div 9$

**3.** $\dfrac{23}{23}$

**4.** $\dfrac{37}{37}$

**5.** $22 \div 1$

**6.** $\dfrac{56}{1}$

**7.** $\dfrac{0}{7}$

**8.** $\dfrac{0}{32}$

**9.** $\dfrac{16}{0}$

**10.** $74 \div 0$

**11.** $\dfrac{48}{8}$

**12.** $\dfrac{20}{4}$

Divide.

**13.** $277 \div 5$

**14.** $699 \div 3$

**15.** $864 \div 8$

**16.** $869 \div 8$

**17.** $4 \overline{)1\ 2\ 2\ 8}$

**18.** $3 \overline{)2\ 1\ 2\ 4}$

**19.** $6 \overline{)4\ 5\ 2\ 1}$

**20.** $9 \overline{)9\ 1\ 1\ 0}$

**21.** $297 \div 4$

**22.** $389 \div 2$

**23.** $738 \div 8$

**24.** $881 \div 6$

**25.** $5 \overline{)8\ 5\ 1\ 5}$

**26.** $3 \overline{)6\ 0\ 2\ 7}$

**27.** $9 \overline{)8\ 8\ 8\ 8}$

**28.** $8 \overline{)4\ 1\ 3\ 9}$

**29.** $127{,}000 \div 10$

**30.** $127{,}000 \div 100$

**31.** $127{,}000 \div 1000$

**32.** $4260 \div 10$

**33.** $70 \overline{)\ 3\ 6\ 9\ 2}$

**34.** $20 \overline{)\ 5\ 7\ 9\ 8}$

**35.** $30 \overline{)\ 8\ 7\ 5}$

**36.** $40 \overline{)\ 9\ 8\ 7}$

**37.** $852 \div 21$

**38.** $942 \div 23$

**39.** $85 \overline{)\ 7\ 6\ 7\ 2}$

**40.** $54 \overline{)\ 2\ 7\ 2\ 9}$

**41.** $111 \overline{)\ 3\ 2\ 1\ 9}$

**42.** $102 \overline{)\ 5\ 6\ 1\ 2}$

**43.** $8 \overline{)\ 8\ 4\ 3}$

**44.** $7 \overline{)\ 7\ 4\ 9}$

**45.** $5 \overline{)\ 8\ 0\ 4\ 7}$

**46.** $9 \overline{)\ 7\ 2\ 7\ 3}$

**47.** $5 \overline{)\ 5\ 0\ 3\ 6}$

**48.** $7 \overline{)\ 7\ 0\ 7\ 4}$

**49.** $1058 \div 46$

**50.** $7242 \div 24$

**51.** $3425 \div 32$

**52.** $48 \overline{)\ 4\ 8\ 9\ 9}$

**53.** $24 \overline{)\ 8\ 8\ 8\ 0}$

**54.** $36 \overline{)\ 7\ 5\ 6\ 3}$

**55.** $28 \overline{)\ 1\ 7,0\ 6\ 7}$

**56.** $36 \overline{)\ 2\ 8,9\ 2\ 9}$

**57.** $80 \overline{)\ 2\ 4,3\ 2\ 0}$

**58.** $90 \overline{)\ 8\ 8,5\ 6\ 0}$

**59.** $285 \overline{)\ 9\ 9\ 9,9\ 9\ 9}$

**60.** $3\ 0\ 6\ \overline{)\ 8\ 8\ 8,8\ 8\ 8}$

**61.** $4\ 5\ 6\ \overline{)\ 3,6\ 7\ 9,9\ 2\ 0}$

**62.** $8\ 0\ 3\ \overline{)\ 5,6\ 2\ 2,6\ 0\ 6}$

## Skill Maintenance

In each of Exercises 63–70, fill in the blank with the correct term from the given list. Some of the choices may not be used and some may be used more than once.

**63.** The distance around an object is its _____. [1.2b]

**64.** The _____ is the number from which another number is being subtracted. [1.3a]

**65.** For large numbers, _____ are separated by commas into groups of three, called _____. [1.1a]

**66.** In the sentence $28 \div 7 = 4$, the _____ is 28. [1.5a]

**67.** In the sentence $10 \times 1000 = 10,000$, 10 and 1000 are called _____ and 10,000 is called the _____. [1.4a]

**68.** The number 0 is called the _____ identity. [1.2a]

**69.** The sentence $3 \times (6 \times 2) = (3 \times 6) \times 2$ illustrates the _____ law of multiplication. [1.4a]

**70.** We can use the following statement to check division: quotient · _____ + _____ = _____. [1.5a]

associative
commutative
addends
factors
area
perimeter
minuend
subtrahend
product
digits
periods
additive
multiplicative
dividend
quotient
remainder
divisor

## Synthesis

**71.** Complete the following table.

| $a$ | $b$ | $a \cdot b$ | $a + b$ |
|-----|-----|-------------|---------|
|     | 68  | 3672        |         |
| 84  |     |             | 117     |
|     |     | 32          | 12      |

**72.** Find a pair of factors whose product is 36 and:
**a)** whose sum is 13.
**b)** whose difference is 0.
**c)** whose sum is 20.
**d)** whose difference is 9.

**73.** A group of 1231 college students is going to take buses for a field trip. Each bus can hold 42 students. How many buses are needed?

**74.** ▦ Fill in the missing digits to make the equation true:

$$34,584,132 \div 76\square = 4\square,386.$$

# Mid-Chapter Review

## Concept Reinforcement

Determine whether each statement is true or false.

_____ **1.** If $a - b = c$, then $b = a + c$. [1.3a]

_____ **2.** We can think of the multiplication $4 \times 3$ as a rectangular array containing 4 rows with 3 items in each row. [1.4a]

_____ **3.** We can think of the multiplication $4 \times 3$ as a rectangular array containing 3 columns with 4 items in each column. [1.4a]

_____ **4.** The product of two whole numbers is always greater than either of the factors. [1.4a]

_____ **5.** Zero divided by any nonzero number is 0. [1.5a]

_____ **6.** Any number divided by 1 is the number 1. [1.5a]

## Guided Solutions

Fill in each blank with the number that creates a correct statement or solution.

**7.** Write a word name for 95,406,237. [1.1c]

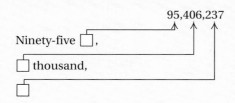

**8.** Subtract: $604 - 497$. [1.3a]

$$
\begin{array}{r}
\square\ \square\ \square \\
6\ \cancel{0}\ \cancel{4} \\
-\ 4\ 9\ 7 \\
\hline
\square\ \square\ \square
\end{array}
$$

## Mixed Review

In each of the following numbers what does the digit 6 mean? [1.1a]

**9.** 2698      **10.** 61,204      **11.** 146,237      **12.** 586

Consider the number 306,458,129. What digit names the number of: [1.1a]

**13.** tens      **14.** millions      **15.** ten thousands      **16.** hundreds

Write expanded notation. [1.1b]

**17.** 5602      **18.** 69,345

Write a word name. [1.1c]

**19.** 136      **20.** 64,325

Write standard notation. [1.1c]

**21.** Three hundred eight thousand, seven hundred sixteen

**22.** Four million, five hundred sixty-seven thousand, two hundred sixteen

Add.  [1.2a]

**23.**  3 1 6
       + 4 8 2

**24.**  5 9 3
       + 4 3 7

**25.**  2 6 3 8
       + 5 2 8 4

**26.**  4 6 1 7
         2 4 3 6
       +   4 8 1

Subtract.  [1.3a]

**27.**  7 8 6
       − 3 2 1

**28.**  6 2 4
       − 2 8 5

**29.**  3 6 0 2
       − 1 7 4 8

**30.**  5 0 0 4
       −   6 7 6

Multiply.  [1.4a]

**31.**  3 6
       ×   6

**32.**  5 6 7
       ×   2 8

**33.**  4 0 7
       × 3 2 5

**34.**  9 4 3 5
       ×   6 0 2

Divide.  [1.5a]

**35.** 4 ) 1 0 1 2

**36.** 3 8 ) 4 2 6 1

**37.** 6 0 ) 1 3 9 9

**38.** 5 6 ) 8 0 9 5

**39.** Find the perimeter of the figure.  [1.2b]

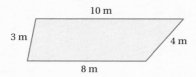

10 m

3 m

4 m

8 m

**40.** Find the area of the region.  [1.4b]

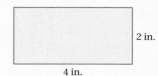

2 in.

4 in.

# Understanding Through Discussion and Writing

***To the student and the instructor:*** The Discussion and Writing exercises are meant to be answered with one or more sentences. They can be discussed and answered collaboratively by the entire class or by small groups.

**41.** Explain in your own words what the associative law of addition means.  [1.2a]

**42.** Is subtraction commutative? That is, is there a commutative law of subtraction? Why or why not?  [1.3a]

**43.** Describe a situation that corresponds to each multiplication:  4 · $150;  $4 · 150.  [1.4a]

**44.** Suppose a student asserts that "0 ÷ 0 = 0 because nothing divided by nothing is nothing." Devise an explanation to persuade the student that the assertion is false.  [1.5a]

# 1.6 Rounding and Estimating; Order

## a Rounding

We round numbers in various situations when we do not need an exact answer. For example, we might round to see if we are being charged the correct amount in a store. We might also round to check if an answer to a problem is reasonable or to check a calculation done by hand or on a calculator.

To understand how to round, we first look at some examples using the number line. The number line displays numbers at equally spaced intervals.

**EXAMPLE 1** Round 47 to the nearest ten.

47 is between 40 and 50. Since 47 is closer to 50, we round up to 50.

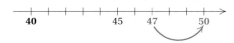

**EXAMPLE 2** Round 42 to the nearest ten.

42 is between 40 and 50. Since 42 is closer to 40, we round down to 40.

**EXAMPLE 3** Round 45 to the nearest ten.

45 is halfway between 40 and 50. We could round 45 down to 40 or up to 50. We agree to round up to 50.

> When a number is halfway between rounding numbers, round up.

Do Margin Exercises 1–7.

We round whole numbers according to the following rule.

### ROUNDING WHOLE NUMBERS

To round to a certain place:

a) Locate the digit in that place.

b) Consider the next digit to the right.

c) If the digit to the right is 5 or higher, round up. If the digit to the right is 4 or lower, round down.

d) Change all digits to the right of the rounding location to zeros.

## OBJECTIVES

**a** Round to the nearest ten, hundred, or thousand.

**b** Estimate sums, differences, products, and quotients by rounding.

**c** Use < or > for ☐ to write a true sentence in a situation like 6 ☐ 10.

**SKILL TO REVIEW**
Objective 1.1a: Give the meaning of digits in standard notation.

In the number 145,627 what digit names the number of:
1. Tens?
2. Thousands?

Round to the nearest ten.
1. 37

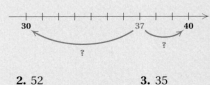

2. 52          3. 35

4. 73          5. 75

6. 88          7. 64

***Answers***
*Skill to Review:*
1. 2    2. 5

*Margin Exercises:*
1. 40    2. 50    3. 40    4. 70    5. 80
6. 90    7. 60

**EXAMPLE 4**   Round 6485 to the nearest ten.

a)  Locate the digit in the tens place, 8.

$$6\ 4\ \underset{\uparrow}{8}\ 5$$

b)  Consider the next digit to the right, 5.

$$6\ 4\ 8\ \underset{\uparrow}{5}$$

c)  Since that digit, 5, is 5 or higher, round 8 tens up to 9 tens.

d)  Change all digits to the right of the tens digit to zeros.

$$6\ 4\ 9\ 0\ \leftarrow \text{This is the answer.}$$

Do Exercises 8–11.

Round to the nearest ten.
   **8.** 137              **9.** 473

**10.** 235              **11.** 285

**EXAMPLE 5**   Round 6485 to the nearest hundred.

a)  Locate the digit in the hundreds place, 4.

$$6\ \underset{\uparrow}{4}\ 8\ 5$$

b)  Consider the next digit to the right, 8.

$$6\ 4\ \underset{\uparrow}{8}\ 5$$

c)  Since that digit, 8, is 5 or higher, round 4 hundreds up to 5 hundreds.

d)  Change all digits to the right of hundreds to zeros.

$$6\ 5\ 0\ 0\ \leftarrow \text{This is the answer.}$$

Do Exercises 12–15.

Round to the nearest hundred.
**12.** 641              **13.** 759

**14.** 1871             **15.** 9325

**EXAMPLE 6**   Round 6485 to the nearest thousand.

a)  Locate the digit in the thousands place, 6.

$$\underset{\uparrow}{6}\ 4\ 8\ 5$$

b)  Consider the next digit to the right, 4.

$$6\ \underset{\uparrow}{4}\ 8\ 5$$

c)  Since that digit, 4, is 4 or lower, round down, meaning that 6 thousands stays as 6 thousands.

d)  Change all digits to the right of thousands to zeros.

$$6\ 0\ 0\ 0\ \leftarrow \text{This is the answer.}$$

Do Exercises 16–19.

Round to the nearest thousand.
**16.** 7896             **17.** 8459

**18.** 19,343           **19.** 68,500

----

*Caution!*

----

7000 is not a correct answer to Example 6. It is incorrect to round from the ones digit over, as follows:

$$6485 \nrightarrow 6490 \nrightarrow 6500 \nrightarrow 7000.$$

Note that 6485 is closer to 6000 than it is to 7000.

----

*Answers*

**8.** 140      **9.** 470      **10.** 240      **11.** 290
**12.** 600      **13.** 800      **14.** 1900      **15.** 9300
**16.** 8000      **17.** 8000      **18.** 19,000
**19.** 69,000

Sometimes rounding involves changing more than one digit in a number.

**EXAMPLE 7** Round 78,595 to the nearest ten.

a) Locate the digit in the tens place, 9.

7 8,5 9 5
       ↑

b) Consider the next digit to the right, 5.

7 8,5 9 5
         ↑

c) Since that digit, 5, is 5 or higher, round 9 tens to 10 tens. To carry this out, we think of 10 tens as 1 hundred + 0 tens and increase the hundreds digit by 1, to get 6 hundreds + 0 tens. We then write 6 in the hundreds place and 0 in the tens place.

d) Change the digit to the right of the tens digit to zero.

7 8,6 0 0 ← This is the answer.

Note that if we round this number to the nearest hundred, we get the same answer.

Do Exercises 20 and 21.

## b Estimating

Estimating can be done in many ways. In general, an estimate made by rounding to the nearest ten is more accurate than one rounded to the nearest hundred, and an estimate rounded to the nearest hundred is more accurate than one rounded to the nearest thousand, and so on.

**EXAMPLE 8** Estimate this sum by first rounding to the nearest ten:

78 + 49 + 31 + 85.

We round each number to the nearest ten. Then we add.

```
  7 8        8 0
  4 9        5 0
  3 1        3 0
+ 8 5      + 9 0
          ───────
           2 5 0  ← Estimated answer
```

Do Exercises 22 and 23.

**EXAMPLE 9** Estimate the difference by first rounding to the nearest thousand: 9324 − 2849.

We have

```
  9 3 2 4      9 0 0 0
− 2 8 4 9    − 3 0 0 0
           ───────────
              6 0 0 0  ← Estimated answer
```

Do Exercises 24 and 25.

20. Round 48,968 to the nearest ten, hundred, and thousand.

21. Round 269,582 to the nearest ten, hundred, and thousand.

22. Estimate the sum by first rounding to the nearest ten. Show your work.

```
  7 4
  2 3
  3 5
+ 6 6
```

23. Estimate the sum by first rounding to the nearest hundred. Show your work.

```
  6 5 0
  6 8 5
  2 3 8
+ 1 6 8
```

24. Estimate the difference by first rounding to the nearest hundred. Show your work.

```
  9 2 8 5
− 6 7 3 9
```

25. Estimate the difference by first rounding to the nearest thousand. Show your work.

```
  2 3,2 7 8
− 1 1,6 9 8
```

*Answers*

20. 48,970; 49,000; 49,000
21. 269,580; 269,600; 270,000
22. 70 + 20 + 40 + 70 = 200
23. 700 + 700 + 200 + 200 = 1800
24. 9300 − 6700 = 2600
25. 23,000 − 12,000 = 11,000

**EXAMPLE 10** Estimate the following product by first rounding to the nearest ten and then to the nearest hundred: $683 \times 457$.

*Nearest ten*

$$
\begin{array}{r}
6\ 8\ 0 \\
\times\quad 4\ 6\ 0 \\
\hline
4\ 0\ 8\ 0\ 0 \\
2\ 7\ 2\ 0\ 0\ 0 \\
\hline
3\ 1\ 2,8\ 0\ 0
\end{array}
$$
$683 \approx 680$
$457 \approx 460$

*Nearest hundred*

$$
\begin{array}{r}
7\ 0\ 0 \\
\times\quad 5\ 0\ 0 \\
\hline
3\ 5\ 0,0\ 0\ 0
\end{array}
$$
$683 \approx 700$
$457 \approx 500$

*Exact*

$$
\begin{array}{r}
6\ 8\ 3 \\
\times\quad 4\ 5\ 7 \\
\hline
4\ 7\ 8\ 1 \\
3\ 4\ 1\ 5\ 0 \\
2\ 7\ 3\ 2\ 0\ 0 \\
\hline
3\ 1\ 2,1\ 3\ 1
\end{array}
$$

We see that rounding to the nearest ten gives a better estimate than rounding to the nearest hundred.

Do Exercise 26.

**26.** Estimate the product by first rounding to the nearest ten and then to the nearest hundred. Show your work.

$$
\begin{array}{r}
8\ 3\ 7 \\
\times\ 2\ 4\ 5
\end{array}
$$

**EXAMPLE 11** Estimate the following quotient by first rounding to the nearest ten and then to the nearest hundred: $12,238 \div 175$.

*Nearest ten*

$$
\begin{array}{r}
6\ 8 \\
1\ 8\ 0\ )\overline{1\ 2,2\ 4\ 0} \\
\underline{1\ 0\ 8\ 0} \\
1\ 4\ 4\ 0 \\
\underline{1\ 4\ 4\ 0} \\
0
\end{array}
$$

*Nearest hundred*

$$
\begin{array}{r}
6\ 1 \\
2\ 0\ 0\ )\overline{1\ 2,2\ 0\ 0} \\
\underline{1\ 2\ 0\ 0} \\
2\ 0\ 0 \\
\underline{2\ 0\ 0} \\
0
\end{array}
$$

Do Exercise 27.

**27.** Estimate the quotient by first rounding to the nearest hundred. Show your work.

$64,534 \div 349$

In the sentence $7 - 5 = 2$, the equals sign indicates that $7 - 5$ is the *same* as 2. When we round to make an estimate, the outcome is rarely the same as the exact result. Thus we cannot use an equals sign when we round. Instead, we use the symbol $\approx$. This symbol means "**is approximately equal to.**" In Example 9, for instance, we could write

$$9324 - 2849 \approx 6000.$$

The next two examples show how estimating can be used in making a purchase.

**EXAMPLE 12** *Microwave Ovens.* Ellen manages a small apartment building and is planning to purchase a new over-the-range microwave oven for each of the 12 units in the building. One model that she is considering costs $248. Estimate, by rounding to the nearest ten, the total cost of the purchase.

We have

$$
\begin{array}{r}
2\ 5\ 0 \\
\times \quad 1\ 0 \\
\hline
2\ 5\ 0\ 0.
\end{array}
$$

The microwave ovens will cost about $2500.

Do Exercise 28.

**EXAMPLE 13** *Purchasing a New Car.* Jon and Joanna are shopping for a new car. They are considering buying an Astra 5-Door XE. The base price of the car is $16,495. A 4-speed automatic transmission package can be added to this, as well as several other options, as shown in the chart below. Jon and Joanna want to stay within a budget of $20,000.

Estimate, by rounding to the nearest hundred, the cost of the Astra with the automatic transmission package and all other options and determine whether this will fit within their budget.

| ASTRA 5-DOOR XE | PRICE |
|---|---|
| Base price | $16,495 |
| 4-speed automatic transmission | $1,325 |
| 16-in. twin-spoke machined alloy wheels | $350 |
| Air conditioning | $960 |
| Dual-panel power sunroof (Requires purchase of air conditioning) | $1,200 |
| Heated cloth front seats | $250 |
| StabiliTrak Stability Control | $495 |

SOURCE: General Motors

First, we list the base price of the car and then the cost of each of the options. We then round each number to the nearest hundred and add.

$$
\begin{array}{r}
1\ 6{,}4\ 9\ 5 \\
1\ 3\ 2\ 5 \\
3\ 5\ 0 \\
9\ 6\ 0 \\
1\ 2\ 0\ 0 \\
2\ 5\ 0 \\
+ \quad 4\ 9\ 5 \\
\end{array}
\qquad
\begin{array}{r}
1\ 6{,}5\ 0\ 0 \\
1\ 3\ 0\ 0 \\
4\ 0\ 0 \\
1\ 0\ 0\ 0 \\
1\ 2\ 0\ 0 \\
3\ 0\ 0 \\
+ \quad 5\ 0\ 0 \\
\hline
2\ 1{,}2\ 0\ 0 \leftarrow \text{Estimated cost}
\end{array}
$$

The estimated cost is $21,200. This exceeds Jon and Joanna's budget of $20,000, so they will have to forgo at least one option.

Do Exercises 29 and 30.

**28. Microwave Ovens.** Suppose Ellen chooses a smaller model of microwave oven that costs $198. Estimate, by rounding to the nearest ten, the total cost of 12 ovens.

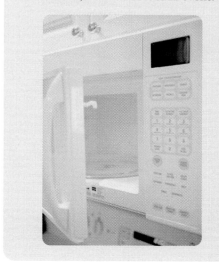

Refer to the chart at left to do Margin Exercises 29 and 30.

**29.** By eliminating at least one option, determine how Jon and Joanna can buy an Astra and stay within their budget. Keep in mind that purchasing the sunroof also requires the purchase of air conditioning.

**30.** Elizabeth and C.J. are also considering buying an Astra 5-Door XE. Estimate, by rounding to the nearest hundred, the cost of this car with automatic transmission, air conditioning, and StabiliTrak Stability Control.

*Answers*

**28.** $2000   **29.** Eliminate either the automatic transmission or the sunroof. There are other correct answers as well.   **30.** $19,300

## (c) Order

We know that 2 is not the same as 5. We express this by the sentence $2 \neq 5$. We also know that 2 is less than 5. We symbolize this by the expression $2 < 5$. We can see this order on the number line: 2 is to the left of 5. The number 0 is the smallest whole number.

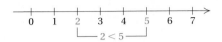

---

### ORDER OF WHOLE NUMBERS

For any whole numbers $a$ and $b$:

1. $a < b$ (read "$a$ is less than $b$") is true when $a$ is to the left of $b$ on the number line.
2. $a > b$ (read "$a$ is greater than $b$") is true when $a$ is to the right of $b$ on the number line.

We call $<$ and $>$ **inequality symbols**.

---

**EXAMPLE 14**   Use $<$ or $>$ for ☐ to write a true sentence:  7 ☐ 11.

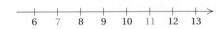

Since 7 is to the left of 11 on the number line,  $7 < 11$.

**EXAMPLE 15**   Use $<$ or $>$ for ☐ to write a true sentence:  92 ☐ 87.

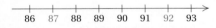

Since 92 is to the right of 87 on the number line,  $92 > 87$.

A sentence like $8 + 5 = 13$ is called an **equation**. It is a *true* equation. The equation $4 + 8 = 11$ is a *false* equation. A sentence like $7 < 11$ is called an **inequality**. The sentence $7 < 11$ is a *true* inequality. The sentence $23 > 69$ is a *false* inequality.

Do Exercises 31–36.

---

### STUDY TIPS

#### TEXTBOOK SUPPLEMENTS

Are you aware of all the supplements that exist for this textbook? See the preface for a description of each supplement.

---

Use $<$ or $>$ for ☐ to write a true sentence. Draw the number line if necessary.

**31.** 8 ☐ 12

**32.** 12 ☐ 8

**33.** 76 ☐ 64

**34.** 64 ☐ 76

**35.** 217 ☐ 345

**36.** 345 ☐ 217

---

*Answers*

**31.** $<$    **32.** $>$    **33.** $>$    **34.** $<$
**35.** $<$    **36.** $>$

**1.6** **Exercise Set**

For Extra Help

*MyMathLab*

Math XL
PRACTICE

WATCH

DOWNLOAD

READ

REVIEW

**a** Round to the nearest ten.

**1.** 48

**2.** 532

**3.** 463

**4.** 8945

**5.** 731

**6.** 54

**7.** 895

**8.** 798

Round to the nearest hundred.

**9.** 146

**10.** 874

**11.** 957

**12.** 650

**13.** 9079

**14.** 4645

**15.** 32,839

**16.** 198,402

Round to the nearest thousand.

**17.** 5876

**18.** 4500

**19.** 7500

**20.** 2001

**21.** 45,340

**22.** 735,562

**23.** 373,405

**24.** 6,713,255

**b** Estimate each sum or difference by first rounding to the nearest ten. Show your work.

**25.**
$$\begin{array}{r} 7\ 8 \\ +\ 9\ 2 \\ \hline \end{array}$$

**26.**
$$\begin{array}{r} 6\ 2 \\ 9\ 7 \\ 4\ 6 \\ +\ 8\ 1 \\ \hline \end{array}$$

**27.**
$$\begin{array}{r} 8\ 0\ 7\ 4 \\ -\ 2\ 3\ 4\ 7 \\ \hline \end{array}$$

**28.**
$$\begin{array}{r} 6\ 7\ 3 \\ -\ \ \ \ 2\ 8 \\ \hline \end{array}$$

Estimate each sum by first rounding to the nearest ten. State if the given sum seems to be incorrect when compared to the estimate.

**29.**
$$\begin{array}{r} 4\ 5 \\ 7\ 7 \\ 2\ 5 \\ +\ 5\ 6 \\ \hline 3\ 4\ 3 \end{array}$$

**30.**
$$\begin{array}{r} 4\ 1 \\ 2\ 1 \\ 5\ 5 \\ +\ 6\ 0 \\ \hline 1\ 7\ 7 \end{array}$$

**31.**
$$\begin{array}{r} 6\ 2\ 2 \\ 7\ 8 \\ 8\ 1 \\ +\ 1\ 1\ 1 \\ \hline 9\ 3\ 2 \end{array}$$

**32.**
$$\begin{array}{r} 8\ 3\ 6 \\ 3\ 7\ 4 \\ 7\ 9\ 4 \\ +\ 9\ 3\ 8 \\ \hline 3\ 9\ 4\ 7 \end{array}$$

Estimate each sum or difference by first rounding to the nearest hundred. Show your work.

**33.**
$$\begin{array}{r} 7\ 3\ 4\ 8 \\ +\ 9\ 2\ 4\ 7 \\ \hline \end{array}$$

**34.**
$$\begin{array}{r} 5\ 6\ 8 \\ 4\ 7\ 2 \\ 9\ 3\ 8 \\ +\ 4\ 0\ 2 \\ \hline \end{array}$$

**35.**
$$\begin{array}{r} 6\ 8\ 5\ 2 \\ -\ 1\ 7\ 4\ 8 \\ \hline \end{array}$$

**36.**
$$\begin{array}{r} 9\ 4\ 3\ 8 \\ -\ 2\ 7\ 8\ 7 \\ \hline \end{array}$$

Estimate each sum by first rounding to the nearest hundred. State if the given sum seems to be incorrect when compared to the estimate.

**37.**
```
    2 1 6
      8 4
    7 4 5
  + 5 9 5
  ---------
    1 6 4 0
```

**38.**
```
    4 8 1
    7 0 2
    6 2 3
  + 1 0 4 3
  -----------
    1 8 4 9
```

**39.**
```
    7 5 0
    4 2 8
      6 3
  + 2 0 5
  ---------
    1 4 4 6
```

**40.**
```
    3 2 6
    2 7 5
    7 5 8
  + 9 4 3
  ---------
    2 3 0 2
```

Estimate each sum or difference by first rounding to the nearest thousand. Show your work.

**41.**
```
    9 6 4 3
    4 8 2 1
    8 9 4 3
  + 7 0 0 4
```

**42.**
```
    7 6 4 8
    9 3 4 8
    7 8 4 2
  + 2 2 2 2
```

**43.**
```
    9 2,1 4 9
  - 2 2,5 5 5
```

**44.**
```
    8 4,8 9 0
  - 1 1,1 1 0
```

Estimate each product by first rounding to the nearest ten. Show your work.

**45.**
```
    4 5
  × 6 7
```

**46.**
```
    5 1
  × 7 8
```

**47.**
```
    3 4
  × 2 9
```

**48.**
```
    6 3
  × 5 4
```

Estimate each product by first rounding to the nearest hundred. Show your work.

**49.**
```
    8 7 6
  × 3 4 5
```

**50.**
```
    3 5 5
  × 2 9 9
```

**51.**
```
    4 3 2
  × 1 9 9
```

**52.**
```
    7 8 9
  × 4 3 4
```

Estimate each quotient by first rounding to the nearest ten. Show your work.

**53.** $347 \div 73$

**54.** $454 \div 87$

**55.** $8452 \div 46$

**56.** $1263 \div 29$

Estimate each quotient by first rounding to the nearest hundred. Show your work.

**57.** $1165 \div 236$

**58.** $3641 \div 571$

**59.** $8358 \div 295$

**60.** $32,854 \div 748$

*Planning a Kitchen.* Perfect Kitchens offers custom kitchen packages with three choices for each of four items: cabinets, countertops, appliances, and flooring. The chart below lists the price for each choice. Customers design their kitchens by making one selection from each group of items.

| CABINETS | TYPE | PRICE |
|---|---|---|
| (a) | Oak | $7450 |
| (b) | Cherry | 8820 |
| (c) | Painted | 9630 |
| COUNTERTOPS | TYPE | PRICE |
| (d) | Laminate | $1595 |
| (e) | Solid surface | 2870 |
| (f) | Granite | 3528 |
| APPLIANCES | PRICE RANGE | PRICE |
| (g) | Low | $1540 |
| (h) | Medium | 3575 |
| (i) | High | 6245 |
| FLOORING | TYPE | PRICE |
| (j) | Vinyl | $625 |
| (k) | Travertine | 985 |
| (l) | Hardwood | 1160 |

**61.** Estimate the cost of remodeling a kitchen with choices (a), (d), (g), and (j) by rounding to the nearest hundred dollars.

**62.** Estimate the cost of a kitchen with choices (c), (f), (i), and (l) by rounding to the nearest hundred dollars.

**63.** Sara and Ben are planning to remodel their kitchen and have a budget of $17,700. Estimate by rounding to the nearest hundred dollars the cost of their kitchen remodeling project if they choose options (b), (e), (i), and (k). Can they afford their choices?

**64.** The Davidsons must make a final decision on the kitchen choices for their new home. The allotted kitchen budget is $16,000. Estimate by rounding to the nearest hundred dollars the kitchen cost if they choose options (a), (f), (h), and (l). Does their budget allotment cover the cost?

**65.** Suppose you are planning a new kitchen and must stay within a budget of $14,500. Decide on the options you would like and estimate the cost by rounding to the nearest hundred dollars. Does your budget support your choices?

**66.** Suppose you are planning a new kitchen and must stay within a budget of $18,500. Decide on the options you would like and estimate the cost by rounding to the nearest hundred dollars. Does your budget support your choices?

**67.** *Company Cars.* A publishing company buys a Honda Accord LX for each of its 112 sales representatives. Each car costs $21,160 with an additional $670 per car in destination charges.

**a)** Estimate the total cost of the purchase by rounding the cost of each car, the destination charge, and the number of sales representatives to the nearest hundred.

**b)** Estimate the total cost of the purchase by rounding the cost of each car to the nearest thousand and the destination charge and the number of sales representatives to the nearest hundred.

**Source:** Honda

**68.** *Airline Tickets.* A travel club of 176 people decides to fly from Chicago to Seattle. The cost of a round-trip ticket is $535.

**a)** Estimate the total cost of the trip by rounding the cost of the airfare and the number of travelers to the nearest ten.

**b)** Estimate the total cost of the trip by rounding the cost of the airfare and the number of travelers to the nearest hundred.

**69.** *Banquet Attendance.* Tickets to the annual awards banquet for the Riviera Swim Club cost $28 each. Ticket sales for the banquet totaled $2716. Estimate the number of people who attended the banquet by rounding the cost of a ticket to the nearest ten and the total sales to the nearest hundred.

**70.** *School Fundraiser.* For a school fundraiser, Charlotte sells trash bags at a price of $11 per roll. If her sales total $2211, estimate the number of rolls sold by rounding the price per roll to the nearest ten and the total sales to the nearest hundred.

---

**C**  Use < or > for ☐ to write a true sentence. Draw the number line if necessary.

**71.** 0 ☐ 17

**72.** 32 ☐ 0

**73.** 34 ☐ 12

**74.** 28 ☐ 18

**75.** 1000 ☐ 1001

**76.** 77 ☐ 117

**77.** 133 ☐ 132

**78.** 999 ☐ 997

**79.** 460 ☐ 17

**80.** 345 ☐ 456

**81.** 37 ☐ 11

**82.** 12 ☐ 32

*New Book Titles.* The number of new book titles published in the United States in each of three recent years is shown in the table below. Use this table to do Exercises 83 and 84.

| YEAR | NEW BOOK TITLES |
|------|-----------------|
| 2005 | 172,000 |
| 2007 | 284,370 |
| 2009 | 275,232 |

SOURCE: R. R. Bowker

**83.** Write an inequality to compare the number of new titles published in 2005 and in 2007.

**84.** Write an inequality to compare the number of new titles published in 2007 and in 2009.

**85.** *Wind-Power Capacity.* Wind-power capacity in the United States has increased from 1663 megawatts installed in 2003 to 9453 megawatts installed in 2009. Write an inequality to compare these numbers of megawatts of wind power installed.

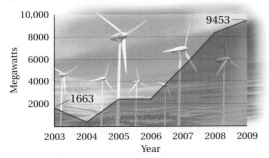

**U.S. Wind-Power Boom**

SOURCE: American Wind Energy Association

**86.** *Life Expectancy.* The life expectancy of a female in the United States in 2015 is predicted to be about 82 yr and that of a male about 76 yr. Write an inequality to compare these life expectancies.

**Life Expectancy in the United States**

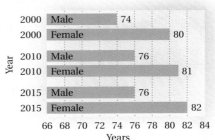

SOURCE: U.S. Census Bureau

## Skill Maintenance

Add. [1.2a]

**87.**
$$\begin{array}{r} 6\,7,7\,8\,9 \\ +\ 1\,8,9\,6\,5 \\ \hline \end{array}$$

**88.**
$$\begin{array}{r} 9\,0\,0\,2 \\ +\ 4\,5\,8\,7 \\ \hline \end{array}$$

Subtract. [1.3a]

**89.**
$$\begin{array}{r} 6\,7,7\,8\,9 \\ -\ 1\,8,9\,6\,5 \\ \hline \end{array}$$

**90.**
$$\begin{array}{r} 9\,0\,0\,2 \\ -\ 4\,5\,8\,7 \\ \hline \end{array}$$

Multiply. [1.4a]

**91.**
$$\begin{array}{r} 4\,6 \\ \times\ 3\,7 \\ \hline \end{array}$$

**92.**
$$\begin{array}{r} 3\,0\,6 \\ \times\ 5\,8 \\ \hline \end{array}$$

Divide. [1.5a]

**93.** $328 \div 6$

**94.** $4784 \div 23$

## Synthesis

**95.–98.** ▦ Use a calculator to find the sums and the differences in each of Exercises 41–44. Then compare your answers with those found using estimation. Even when using a calculator it is possible to make an error if you press the wrong buttons, so it is a good idea to check by estimating.

# 1.7 Solving Equations

## OBJECTIVES

**a** Solve simple equations by trial.

**b** Solve equations like $t + 28 = 54$, $28 \cdot x = 168$, and $98 \cdot 2 = y$.

Find a number that makes each sentence true.

**1.** $8 = 1 + \square$

**2.** $\square + 2 = 7$

**3.** Determine whether 7 is a solution of $\square + 5 = 9$.

**4.** Determine whether 4 is a solution of $\square + 5 = 9$.

Solve by trial.

**5.** $n + 3 = 8$

**6.** $x - 2 = 8$

**7.** $45 \div 9 = y$

**8.** $10 + t = 32$

**Answers**

**1.** 7    **2.** 5    **3.** No    **4.** Yes    **5.** 5
**6.** 10    **7.** 5    **8.** 22

## a Solutions by Trial

Let's find a number that we can put in the blank to make this sentence true:

$$9 = 3 + \square.$$

We are asking "9 is 3 plus what number?" The answer is 6.

$$9 = 3 + 6$$

Do Exercises 1 and 2.

A sentence with = is called an **equation**. A **solution** of an equation is a number that makes the sentence true. Thus, 6 is a solution of

$$9 = 3 + \square \quad \text{because} \quad 9 = 3 + 6 \text{ is true.}$$

However, 7 is not a solution of

$$9 = 3 + \square \quad \text{because} \quad 9 = 3 + 7 \text{ is false.}$$

Do Exercises 3 and 4.

We can use a letter in an equation instead of a blank:

$$9 = 3 + n.$$

We call $n$ a **variable** because it can represent any number. If a replacement for a variable makes an equation true, it is a **solution** of the equation.

> ### SOLUTIONS OF AN EQUATION
>
> A **solution of an equation** is a replacement for the variable that makes the equation true. When asked to **solve** an equation, we find all its solutions.

**EXAMPLE 1** Solve $y + 12 = 27$ by trial.

We replace $y$ with several numbers.

If we replace $y$ with 13, we get a false equation: $13 + 12 = 27$.
If we replace $y$ with 14, we get a false equation: $14 + 12 = 27$.
If we replace $y$ with 15, we get a true equation: $15 + 12 = 27$.

No other replacement makes the equation true, so the solution is 15.

**EXAMPLES** Solve.

**2.** $7 + n = 22$
(7 plus what number is 22?)
The solution is 15.

**3.** $63 = 3 \cdot x$
(63 is 3 times what number?)
The solution is 21.

Do Exercises 5–8.

## b Solving Equations

We now begin to develop more efficient ways to solve certain equations. When an equation has a variable alone on one side and a calculation on the other side, we can find the solution by carrying out the calculation.

**EXAMPLE 4** Solve: $x = 245 \times 34$.

To solve the equation, we carry out the calculation.

$$
\begin{array}{r}
2\ 4\ 5 \\
\times\ \ 3\ 4 \\
\hline
9\ 8\ 0 \\
7\ 3\ 5\ 0 \\
\hline
8\ 3\ 3\ 0
\end{array}
$$

$x = 245 \times 34$

$x = 8330$

The solution is 8330.

Do Exercises 9–12.

Look at the equation

$$x + 12 = 27.$$

We can get $x$ alone by subtracting 12 *on both sides*. Thus,

$$x + 12 - 12 = 27 - 12 \qquad \text{Subtracting 12 on both sides}$$
$$x + 0 = 15 \qquad\qquad \text{Carrying out the subtraction}$$
$$x = 15.$$

---

**SOLVING** $x + a = b$

To solve $x + a = b$, subtract $a$ on both sides.

---

If we can get an equation in a form with the variable alone on one side, we can "see" the solution.

**EXAMPLE 5** Solve: $t + 28 = 54$.

We have

$$t + 28 = 54$$
$$t + 28 - 28 = 54 - 28 \qquad \text{Subtracting 28 on both sides}$$
$$t + 0 = 26$$
$$t = 26.$$

To check the answer, we substitute 26 for $t$ in the original equation.

Check:
$$
\begin{array}{r}
t + 28 = 54 \\
\hline
26 + 28\ \overset{?}{\vphantom{|}}\ 54 \\
54\ \Big| \qquad \text{TRUE} \qquad \text{Since } 54 = 54 \text{ is true, 26 checks.}
\end{array}
$$

The solution is 26.

Do Exercises 13 and 14.

**EXAMPLE 6** Solve: $182 = 65 + n$.

We have

$$182 = 65 + n$$
$$182 - 65 = 65 + n - 65 \quad \text{Subtracting 65 on both sides}$$
$$117 = 0 + n \quad \text{65 plus } n \text{ minus 65 is } 0 + n.$$
$$117 = n.$$

Check: 
$$\frac{182 = 65 + n}{182 \ ? \ 65 + 117}$$
$$| \ 182 \qquad \text{TRUE}$$

The solution is 117.

Do Exercise 15.

**15.** Solve: $155 = t + 78$. Be sure to check.

**EXAMPLE 7** Solve: $7381 + x = 8067$.

We have

$$7381 + x = 8067$$
$$7381 + x - 7381 = 8067 - 7381 \quad \text{Subtracting 7381 on both sides}$$
$$x = 686.$$

Check: 
$$\frac{7381 + x = 8067}{7381 + 686 \ ? \ 8067}$$
$$8067 \ | \qquad \text{TRUE}$$

The solution is 686.

Do Exercises 16 and 17.

Solve. Be sure to check.
**16.** $4566 + x = 7877$

**17.** $8172 = h + 2058$

We now learn to solve equations like $8 \cdot n = 96$. Look at

$$8 \cdot n = 96.$$

We can get $n$ alone by dividing by 8 *on both sides*. Thus,

$$\frac{8 \cdot n}{8} = \frac{96}{8} \quad \text{Dividing by 8 on both sides}$$
$$n = 12. \quad \text{8 times } n \text{ divided by 8 is } n.$$

To check the answer, we substitute 12 for $n$ in the original equation.

Check: 
$$\frac{8 \cdot n = 96}{8 \cdot 12 \ ? \ 96}$$
$$96 \ | \qquad \text{TRUE}$$

Since $96 = 96$ is a true equation, 12 is the solution of the equation.

---

**SOLVING** $a \cdot x = b$

To solve $a \cdot x = b$, divide by $a$ on both sides.

---

**EXAMPLE 8**  Solve: $10 \cdot x = 240$.

We have

$$10 \cdot x = 240$$

$$\frac{10 \cdot x}{10} = \frac{240}{10} \qquad \text{Dividing by 10 on both sides}$$

$$x = 24.$$

Check:
$$\frac{10 \cdot x = 240}{10 \cdot 24 \ ? \ 240}$$
$$240 \ \bigm| \quad \text{TRUE}$$

The solution is 24.

**EXAMPLE 9**  Solve: $5202 = 9 \cdot t$.

We have

$$5202 = 9 \cdot t$$

$$\frac{5202}{9} = \frac{9 \cdot t}{9} \qquad \text{Dividing by 9 on both sides}$$

$$578 = t.$$

Check:
$$\frac{5202 = 9 \cdot t}{5202 \ ? \ 9 \cdot 578}$$
$$\bigm| \ 5202 \quad \text{TRUE}$$

The solution is 578.

Do Exercises 18–20.

**EXAMPLE 10**  Solve: $14 \cdot y = 1092$.

We have

$$14 \cdot y = 1092$$

$$\frac{14 \cdot y}{14} = \frac{1092}{14} \qquad \text{Dividing by 14 on both sides}$$

$$y = 78.$$

The check is left to the student. The solution is 78.

**EXAMPLE 11**  Solve: $n \cdot 56 = 4648$.

We have

$$n \cdot 56 = 4648$$

$$\frac{n \cdot 56}{56} = \frac{4648}{56} \qquad \text{Dividing by 56 on both sides}$$

$$n = 83.$$

The check is left to the student. The solution is 83.

Do Exercises 21 and 22.

Solve. Be sure to check.

**18.** $8 \cdot x = 64$

**19.** $144 = 9 \cdot n$

**20.** $5152 = 8 \cdot t$

Solve. Be sure to check.

**21.** $18 \cdot y = 1728$

**22.** $n \cdot 48 = 4512$

*Answers*

**18.** 8   **19.** 16   **20.** 644   **21.** 96   **22.** 94

**a**  Solve by trial.

**1.** $x + 0 = 14$

**2.** $x - 7 = 18$

**3.** $y \cdot 17 = 0$

**4.** $56 \div m = 7$

**b** Solve. Be sure to check.

**5.** $x = 12{,}345 + 78{,}555$

**6.** $t = 5678 + 9034$

**7.** $908 - 458 = p$

**8.** $9007 - 5667 = m$

**9.** $16 \cdot 22 = y$

**10.** $34 \cdot 15 = z$

**11.** $t = 125 \div 5$

**12.** $w = 256 \div 16$

**13.** $13 + x = 42$

**14.** $15 + t = 22$

**15.** $12 = 12 + m$

**16.** $16 = t + 16$

**17.** $10 + x = 89$

**18.** $20 + x = 57$

**19.** $61 = 16 + y$

**20.** $53 = 17 + w$

**21.** $3 \cdot x = 24$

**22.** $6 \cdot x = 42$

**23.** $112 = n \cdot 8$

**24.** $162 = 9 \cdot m$

**25.** $3 \cdot m = 96$

**26.** $4 \cdot y = 96$

**27.** $715 = 5 \cdot z$

**28.** $741 = 3 \cdot t$

**29.** $8322 + 9281 = x$

**30.** $9281 - 8322 = y$

**31.** $47 + n = 84$

**32.** $56 + p = 92$

**33.** $45 \cdot 23 = x$

**34.** $23 \cdot 78 = y$

**35.** $x + 78 = 144$

**36.** $z + 67 = 133$

**37.** $6 \cdot p = 1944$  **38.** $4 \cdot w = 3404$  **39.** $5 \cdot x = 3715$  **40.** $9 \cdot x = 1269$

**41.** $x + 214 = 389$  **42.** $x + 221 = 333$  **43.** $567 + x = 902$  **44.** $438 + x = 807$

**45.** $234 \cdot 78 = y$  **46.** $10{,}534 \div 458 = q$  **47.** $18 \cdot x = 1872$  **48.** $19 \cdot x = 6080$

**49.** $40 \cdot x = 1800$  **50.** $20 \cdot x = 1500$  **51.** $2344 + y = 6400$  **52.** $9281 = 8322 + t$

**53.** $m = 7006 - 4159$  **54.** $n = 3004 - 1745$  **55.** $165 = 11 \cdot n$  **56.** $660 = 12 \cdot n$

**57.** $58 \cdot m = 11{,}890$  **58.** $233 \cdot x = 22{,}135$  **59.** $491 - 34 = y$  **60.** $512 - 63 = z$

## Skill Maintenance

Divide.  [1.5a]

**61.** $1283 \div 9$  **62.** $1278 \div 9$  **63.** $1\,7\,\overline{)\,5\,6\,7\,8}$  **64.** $1\,7\,\overline{)\,5\,6\,8\,9}$

Use $>$ or $<$ for ☐ to write a true sentence.  [1.6c]

**65.** $123$ ☐ $789$  **66.** $342$ ☐ $339$  **67.** $688$ ☐ $0$  **68.** $0$ ☐ $11$

**69.** Round 6,375,602 to the nearest thousand.  [1.6a]  **70.** Round 6,375,602 to the nearest ten.  [1.6a]

## Synthesis

Solve.

**71.** 🖩 $23{,}465 \cdot x = 8{,}142{,}355$  **72.** 🖩 $48{,}916 \cdot x = 14{,}332{,}388$

# 1.8

## Applications and Problem Solving

### OBJECTIVE

**a** Solve applied problems involving addition, subtraction, multiplication, or division of whole numbers.

### (a) A Problem-Solving Strategy

One of the most important ways in which we use mathematics is as a tool in solving problems. To solve a problem, we use the following five-step strategy.

---

**FIVE STEPS FOR PROBLEM SOLVING**

1. **Familiarize** yourself with the problem situation. If the problem is presented in words, this means to read and reread it carefully until you understand what you are being asked to find. Some or all of the following can also be helpful.
   a) Make a drawing, if it makes sense to do so.
   b) Make a written list of the known facts and a list of what you wish to find out.
   c) Assign a letter, or *variable*, to the unknown.
   d) Organize the information in a chart or a table.
   e) Find further information. Look up a formula, consult a reference book or an expert in the field, or do research on the Internet.
   f) Guess or estimate the answer and check your guess or estimate.
2. **Translate** the problem to an equation using the variable.
3. **Solve** the equation.
4. **Check** to see whether your possible solution actually fits the problem situation and is thus really a solution of the problem. Although you may have solved an equation, the solution of the equation might not be a solution of the original problem.
5. **State** the answer clearly using a complete sentence and appropriate units.

---

The first of these five steps, becoming familiar with the problem, is probably the most important. It provides a solid foundation for translating the problem to an equation that represents the situation accurately.

**EXAMPLE 1** *Community Colleges and New Jobs.* Community colleges are playing an increasingly large role in educating America's work force. The numbers of new jobs requiring a community college degree that will be created between 2004 and 2014 are shown in the table on the next page. Find the total number of new jobs for registered nurses, nursing aides and orderlies, and medical assistants.

**New Jobs Created, 2004–2014**

| JOB | NUMBER |
|---|---|
| Registered nurse | 703,000 |
| Customer service | 471,000 |
| Nursing aide; orderly | 325,000 |
| Heavy-truck driver | 223,000 |
| Maintenance; repair | 202,000 |
| Medical assistant | 202,000 |
| Executive secretary/assistant | 192,000 |
| Sales representative | 187,000 |
| Carpenter | 186,000 |

SOURCES: U.S. Bureau of Labor Statistics;
College Board Center for Innovative Thought

1. **Familiarize.**  First, we assign a letter, or variable, to the number we wish to find. We let $n$ = the total number of new jobs created for registered nurses, nursing aides and orderlies, and medical assistants. Key words such as "total," "in all," and "all together" usually tell us to combine quantities. Since we are combining numbers, we will add.

2. **Translate.**  We translate to an equation:

| Registered nurses' jobs | plus | Nursing aides' and orderlies' jobs | plus | Medical assistants' jobs | is | Total number of jobs |
|---|---|---|---|---|---|---|
| ↓ | ↓ | ↓ | ↓ | ↓ | ↓ | ↓ |
| 703,000 | + | 325,000 | + | 202,000 | = | $n$. |

3. **Solve.**  We solve the equation by carrying out the addition.

$$\begin{array}{r} 7\,0\,3,0\,0\,0 \\ 3\,2\,5,0\,0\,0 \\ +\ 2\,0\,2,0\,0\,0 \\ \hline 1,2\,3\,0,0\,0\,0 \end{array}$$

$$703,000 + 325,000 + 202,000 = n$$
$$1,230,000 = n$$

4. **Check.**  We check 1,230,000 in the original problem. There are many ways in which this can be done. For example, we can repeat the calculation. (We leave this to the student.) Another way is to check whether the answer is reasonable. In this case, we would expect the total to be greater than the number of each individual type of new job, and it is. We can also estimate the expected result by rounding. Here we round to the nearest hundred thousand:

$$703,000 + 325,000 + 202,000 \approx 700,000 + 300,000 + 200,000$$
$$\approx 1,200,000.$$

Since $1,200,000 \approx 1,230,000$, our answer seems reasonable. If the estimate had differed greatly from the possible solution found in step (3), we would suspect that the possible solution is incorrect.

5. **State.**  The total number of new jobs created for registered nurses, nursing aides and orderlies, and medical assistants between 2004 and 2014 is 1,230,000.

Do Exercises 1–3.

Refer to the table above to do Margin Exercises 1–3.

1. Find the total number of new jobs that will be created for customer-service representatives, executive secretaries/assistants, and sales representatives.

2. Find the total number of new jobs that will be created for heavy-truck drivers, maintenance and repair workers, and carpenters.

3. Find the total number of new jobs listed in the table.

*Answers*

1. 850,000 new jobs   2. 611,000 new jobs
3. 2,691,000 new jobs

$99

**EXAMPLE 2** *Checking Account Balance.* The balance in Francisco's checking account is $573. He uses his debit card to buy a juicer that costs $99. Find the new balance in his checking account.

1. **Familiarize.** We first make a drawing or at least visualize the situation. We let $B$ = the new balance in Francisco's checking account. We start with $573 and take away $99.

Take away
$99

$573                                New balance

2. **Translate.** We translate to an equation:

| Money in the account | minus | Money spent | is | New balance |
|:---:|:---:|:---:|:---:|:---:|
| 573 | − | 99 | = | B. |

3. **Solve.** This equation tells us what to do. We subtract.

$$
\begin{array}{r}
{\scriptstyle 16} \\
{\scriptstyle 4\ \ 6\ \ 13} \\
5\ 7\ 3 \\
-\ \ \ 9\ 9 \\
\hline
4\ 7\ 4
\end{array}
$$

$573 - 99 = B$

$474 = B$

4. **Check.** To check our answer of $474, we can repeat the calculation. We can also note that the answer should be less than the original amount, $573, and it is. Another way to check is to add the money spent, $99, to the new balance, $474: $99 + $474 = $573. We get the original balance, so the answer checks. We can also estimate:

$573 − $99 ≈ $570 − $100 = $470 ≈ $474.

This tells us that the answer is reasonable.

5. **State.** The new balance in Francisco's checking account is $474.

Do Exercise 4.

In the real world, problems may not be stated in written words. You must still become familiar with the situation before you can solve the problem.

**EXAMPLE 3** *Travel Distance.* Abigail is driving from Indianapolis to Salt Lake City to attend a family reunion. The distance from Indianapolis to Salt Lake City is 1634 mi. In the first two days, she travels 1154 mi to Denver. How much farther must she travel?

1. **Familiarize.** We first make a drawing or at least visualize the situation. We let $d$ = the remaining distance to Salt Lake City.

**4. Checking Account Balance.** The balance in Laura's checking account is $457. She uses her debit card to buy a digital-picture frame that costs $49. Find the new balance in her checking account.

*Answer*

4. $408

**2. Translate.** We want to determine how many more miles Abigail must travel. We translate to an equation:

$$\underbrace{\text{Distance already traveled}}_{1154} \underbrace{\text{plus}}_{+} \underbrace{\text{Distance to go}}_{d} \underbrace{\text{is}}_{=} \underbrace{\text{Total distance of trip}}_{1634.}$$

**3. Solve.** To solve the equation, we subtract 1154 on both sides.

$$1154 + d = 1634$$
$$1154 + d - 1154 = 1634 - 1154$$
$$d = 480$$

$$\begin{array}{r} {\scriptstyle 5\ 13} \\ 1\ \cancel{6}\ \cancel{3}\ 4 \\ -\ 1\ 1\ 5\ 4 \\ \hline 4\ 8\ 0 \end{array}$$

**4. Check.** We check our answer of 480 mi in the original problem. This number should be less than the total distance, 1634 mi, and it is. We can add the distance traveled, 1154, and the distance left to go, 480: $1154 + 480 = 1634$. We can also estimate:

$$1634 - 1154 \approx 1600 - 1200$$
$$= 400 \approx 480.$$

The answer, 480 mi, checks.

**5. State.** Abigail must travel 480 mi farther to Salt Lake City.

> Do Exercise 5.

**5. Reading Assignment.** William has been assigned 234 pages of reading for his history class. He has read 86 pages. How many more pages does he have to read?

**EXAMPLE 4** *Total Cost of Chairs.* What is the total cost of 6 Adirondack chairs if each one costs $169?

**1. Familiarize.** We first make a drawing or at least visualize the situation. We let $C =$ the cost of 6 chairs.

$^{\$}169 \qquad ^{\$}169 \qquad ^{\$}169 \qquad ^{\$}169 \qquad ^{\$}169 \qquad ^{\$}169$

***Answer***

**5.** 148 pages

**2. Translate.** We translate to an equation:

| Number of chairs | times | Cost of each chair | is | Total cost |
|:---:|:---:|:---:|:---:|:---:|
| 6 | × | $169 | = | C. |

**3. Solve.** This sentence tells us what to do. We multiply.

$$
\begin{array}{r}
1\ 6\ 9 \\
\times\quad 6 \\
\hline
1\ 0\ 1\ 4
\end{array}
$$

$$6 \times 169 = C$$
$$1014 = C$$

**4. Check.** We have an answer, 1014, that is much greater than the cost of any individual chair, which is reasonable. We can repeat our calculation. We can also check by estimating:

$$6 \times 169 \approx 6 \times 170 = 1020 \approx 1014.$$

The answer checks.

**5. State.** The total cost of 6 chairs is $1014.

> Do Exercise 6.

**6. Total Cost of Gas Grills.** What is the total cost of 14 gas grills, each with 520 sq in. of total cooking surface, if each one costs $398?

**EXAMPLE 5** *Area of an Oriental Rug.* The dimensions of the oriental rug in the Fosters' front hallway are 42 in. by 66 in. What is the area of the rug?

**1. Familiarize.** We first make a drawing to visualize the situation. We let $A$ = the area of the rug and use the formula for the area of a rectangle, $A$ = length · width = $l \cdot w$. Since we usually consider length to be larger than width, we will let $l$ = 66 in. and $w$ = 42 in.

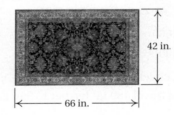

42 in.

66 in.

**2. Translate.** We substitute in the formula:

$$A = l \cdot w = 66 \cdot 42.$$

**3. Solve.** We carry out the multiplication.

$$
\begin{array}{r}
6\ 6 \\
\times\ 4\ 2 \\
\hline
1\ 3\ 2 \\
2\ 6\ 4\ 0 \\
\hline
2\ 7\ 7\ 2
\end{array}
$$

$$A = 66 \cdot 42$$
$$A = 2772$$

**4. Check.** We can repeat the calculation. We can also round and estimate:

$$66 \times 42 \approx 70 \times 40 = 2800 \approx 2772.$$

The answer checks.

**5. State.** The area of the rug is 2772 sq in.

> Do Exercise 7.

**7. Bed Sheets.** The dimensions of a flat sheet for a queen-size bed are 90 in. by 102 in. What is the area of the sheet?

*Answers*

**6.** $5572   **7.** 9180 sq in.

**EXAMPLE 6** *Packages of Paper Towels.* A paper-products company produces 3304 rolls of paper towels. How many 12-roll packages can be filled? How many rolls will be left over?

1. **Familiarize.** We first make a drawing. We let $n$ = the number of 12-roll packages that can be filled. The problem can be considered as repeated subtraction, taking successive sets of 12 rolls and putting them into $n$ packages.

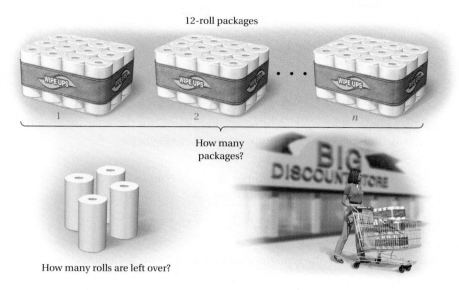

12-roll packages

1    2    n

How many packages?

How many rolls are left over?

2. **Translate.** We translate to an equation:

Number of rolls   divided by   Number in each package   is   Number of packages

$$3304 \div 12 = n.$$

3. **Solve.** We solve the equation by carrying out the division.

$$
\begin{array}{r}
2\ 7\ 5 \\
1\ 2\ \overline{)\ 3\ 3\ 0\ 4} \\
2\ 4 \\
\hline
9\ 0 \\
8\ 4 \\
\hline
6\ 4 \\
6\ 0 \\
\hline
4
\end{array}
$$

$$3304 \div 12 = n$$
$$275\,\text{R}\,4 = n$$

4. **Check.** We can check by multiplying the number of packages by 12 and adding the remainder, 4:

$$12 \cdot 275 = 3300,$$
$$3300 + 4 = 3304.$$

5. **State.** Thus, 275 twelve-roll packages of paper towels can be filled. There will be 4 rolls left over.

Do Exercise 8.

8. **Packages of Paper Towels.** The paper-products company in Example 6 also produces 6-roll packages. How many 6-roll packages can be filled with 2269 rolls of paper towels? How many rolls will be left over?

*Answer*

8. 378 packages with 1 roll left over

**EXAMPLE 7** *Automobile Mileage.* The 2009 Toyota Matrix gets 21 miles to the gallon (mpg) in city driving. How many gallons will it use in 3843 mi of city driving?

Source: Toyota

1. **Familiarize.** We first make a drawing. We let $g$ = the number of gallons of gasoline used in 3843 mi of city driving.

3843 mi to drive

2. **Translate.** Repeated addition applies here. Thus the following multiplication applies to the situation.

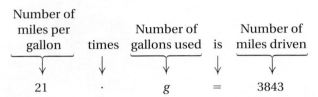

$$21 \cdot g = 3843$$

3. **Solve.** To solve the equation, we divide by 21 on both sides.

$$21 \cdot g = 3843$$
$$\frac{21 \cdot g}{21} = \frac{3843}{21}$$
$$g = 183$$

```
        1 8 3
2 1 ) 3 8 4 3
      2 1
      1 7 4
      1 6 8
          6 3
          6 3
            0
```

4. **Check.** To check, we multiply 183 by 21.

```
    1 8 3
×     2 1
    1 8 3
  3 6 6 0
  3 8 4 3
```

The answer checks.

5. **State.** The Toyota Matrix will use 183 gal of gasoline.

Do Exercise 9.

## Multistep Problems

Sometimes we must use more than one operation to solve a problem, as in the following example.

**EXAMPLE 8** *Weight Loss.* To lose one pound, you must burn about 3500 calories in excess of the calories you consume. The chart on the next page shows how long a person must engage in several types of exercise in order to burn 100 calories. For how long would a person have to run at a brisk pace in order to lose one pound?

---

9. **Automobile Mileage.** The 2009 Toyota Matrix gets 29 miles to the gallon (mpg) in highway driving. How many gallons will it use in 2291 mi of highway driving?

Source: Toyota

*Answer*

9. 79 gal

1. **Familiarize.** This is a multistep problem. We begin by visualizing the situation.

| ONE POUND 3500 CALORIES | | | |
|---|---|---|---|
| 100 cal 8 min | 100 cal 8 min | ... | 100 cal 8 min |

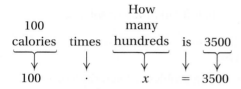

To burn 100 calories, you must:

• Run for 8 minutes at a brisk pace, or
• Swim for 2 minutes at a brisk pace, or
• Bicycle for 15 minutes at 9 mph, or
• Do aerobic exercises for 15 minutes, or
• Golf, walking, for 20 minutes, or
• Play tennis, singles, for 11 minutes

We will first find how many hundreds are in 3500. This will tell us how many times a person must run for 8 min in order to lose one pound. Then we will multiply to find the total number of minutes required for the weight loss.

We let $x =$ the number of hundreds in 3500.

2. **Translate.** We translate to an equation. Repeated addition applies here, so we will multiply.

$$
\underbrace{100}_{100} \text{ calories} \quad \underbrace{\text{times}}_{\cdot} \quad \underbrace{\text{How many hundreds}}_{x} \quad \underbrace{\text{is}}_{=} \quad \underbrace{3500}_{3500}
$$

3. **Solve.** We divide by 100 on both sides of the equation.

$$100 \cdot x = 3500$$
$$\frac{100 \cdot x}{100} = \frac{3500}{100}$$
$$x = 35$$

$$
\begin{array}{r}
3\,5 \\
100\,)\overline{3\,5\,0\,0} \\
3\,0\,0 \\
\hline
5\,0\,0 \\
5\,0\,0 \\
\hline
0
\end{array}
$$

We know that running for 8 min will burn 100 calories. This must be done 35 times in order to lose one pound. We let $t =$ the time it takes to lose one pound. Thus we have the following.

$$t = 35 \times 8$$
$$t = 280$$

$$
\begin{array}{r}
3\,5 \\
\times \quad 8 \\
\hline
2\,8\,0
\end{array}
$$

4. **Check.** $280 \div 8 = 35$, so there are 35 8's in 280 min, and $35 \cdot 100 = 3500$, the number of calories that must be burned in order to lose one pound. The answer checks.

5. **State.** You must run for 280 min, or 4 hr 40 min, at a brisk pace in order to lose one pound.

Do Exercise 10.

10. **Weight Loss.** Use the information in Example 8 to determine how long an individual must swim at a brisk pace in order to lose one pound.

*Answer*

**10.** 70 min, or 1 hr 10 min

You will find it helpful to look for the words, phrases, and concepts in the table below as you familiarize yourself with applied problems. They will be useful when you translate the problems to equations.

**KEYWORDS, PHRASES, AND CONCEPTS**

| ADDITION (+) | SUBTRACTION (−) | MULTIPLICATION (·) | DIVISION (÷) |
|---|---|---|---|
| add | subtract | multiply | divide |
| added to | subtracted from | multiplied by | divided by |
| sum | difference | product | quotient |
| total | minus | times | repeated subtraction |
| plus | less than | of | missing factor |
| more than | decreased by | repeated addition | finding equal quantities |
| increased by | take away | rectangular arrays | |
| | how much more | | |

The following tips will also be helpful in problem solving.

**PROBLEM-SOLVING TIPS**

1. Look for patterns when solving problems. Each time you study an example in the text, you may observe a pattern for problems found later in the exercise sets or in other practical situations.

2. When translating in mathematics, consider the dimensions of the variables and constants in the equation. The variables that represent length should all be in the same unit, those that represent money should all be in dollars or all in cents, and so on.

3. Make sure that units appear in the answer whenever appropriate and that you completely answer the original problem.

**STUDY TIPS**

**SOLVING APPLIED PROBLEMS**

Don't be discouraged if, at first, you find the exercises in this section to be more challenging than those in earlier sections or if you have had difficulty working applied problems in the past. Your skill will improve with each problem you solve. After you have done your homework for this section, you might want to work extra problems from the text. As you gain experience solving applied problems, you will find yourself becoming comfortable with them.

# Translating for Success

1. *Brick-Mason Expense.* A commercial contractor is building 30 two-unit condominiums in a retirement community. The brick-mason expense for each building is $10,860. What is the total cost of bricking the buildings?

2. *Heights.* Dean's sons are on the high school basketball team. Their heights are 73 in., 69 in., and 76 in. How much taller is the tallest son than the shortest son?

3. *Account Balance.* James has $423 in his checking account. Then he deposits $73 and uses his debit card for purchases of $76 and $69. How much is left in the account?

4. *Purchasing Camcorder.* A camcorder is on sale for $423. Jenny has only $69. How much more does she need to buy the camcorder?

5. *Purchasing Coffee Makers.* Sara purchases 8 coffee makers for the newly remodeled bed-and-breakfast hotel that she manages. If she pays $52 for each coffee maker, what is the total cost of her purchase?

The goal of these matching questions is to practice step (2), *Translate*, of the five-step problem-solving process. Translate each word problem to an equation and select a correct translation from equations A–O.

A. $8 \cdot 52 = n$

B. $69 \cdot n = 76$

C. $73 - 76 - 69 = n$

D. $423 + 73 - 76 - 69 = n$

E. $30 \cdot 10{,}860 = n$

F. $15 \cdot n = 195$

G. $69 + n = 423$

H. $n = 10{,}860 - 300$

I. $n = 423 \div 69$

J. $30 \cdot n = 10{,}860$

K. $15 \cdot 195 = n$

L. $n = 52 - 8$

M. $69 + n = 76$

N. $15 \div 195 = n$

O. $52 + n = 60$

*Answers on page A-2*

6. *Hourly Rate.* Miller Auto Repair charges $52 per hour for labor. Jackson Auto Care charges $60 per hour. How much more does Jackson charge than Miller?

7. *College Band.* A college band with 195 members marches in a 15-row formation in the homecoming halftime performance. How many members are in each row?

8. *Shoe Purchase.* A professional football team purchases 15 pairs of shoes at $195 a pair. What is the total cost of this purchase?

9. *Loan Payment.* Kendra's uncle loans her $10,860, interest free, to buy a car. The loan is to be paid off in 30 payments. How much is each payment?

10. *College Enrollment.* At the beginning of the fall term, the total enrollment in Lakeview Community College was 10,860. By the end of the first two weeks, 300 students had withdrawn. How many students were then enrolled?

**a** Solve.

*Roller Coasters.* The table below shows the lengths of the fastest roller coasters in the world. Use this information to do Exercises 1–4.

**Fastest Roller Coasters**

| ROLLER COASTER | LENGTH (in feet) |
|---|---|
| Kingda Ka (128 mph) Six Flags Great Adventure, New Jersey | 3118 |
| Top Thrill Dragster (120 mph) Cedar Point, Ohio | 2800 |
| Dodonpa (107 mph) Fuji-Q Highland, Japan | 3901 |
| Steel Dragon 2000 (95 mph) Nagashima Spa Land, Japan | 8133 |

SOURCE: ultimaterollercoaster.com

**1.** How much longer is Kingda Ka than Top Thrill Dragster?

**2.** How much longer is Steel Dragon 2000 than Kingda Ka?

**3.** The Steel Dragon 2000 is the longest roller coaster in the world. It is 683 feet longer than the second-longest roller coaster, The Ultimate, in Lightwater Valley, UK. How long is The Ultimate?

**4.** The longest roller coaster in the United States is Millennium Force, in Cedar Point, Ohio. It is 3795 feet longer than Top Thrill Dragster, in the same amusement park. How long is Millennium Force?

**5.** *Caffeine Content.* An 8-oz serving of Red Bull energy drink contains 76 milligrams of caffeine. An 8-oz serving of brewed coffee contains 19 more milligrams of caffeine than the energy drink. How many milligrams of caffeine does the 8-oz serving of coffee contain?
**Source:** The Mayo Clinic

**6.** *Caffeine Content.* Hershey's 6-oz milk chocolate almond bar contains 25 milligrams of caffeine. A 20-oz bottle of Coca-Cola has 32 more milligrams of caffeine than the Hershey bar. How many milligrams of caffeine does the 20-oz bottle of Coca-Cola have?
**Source:** *National Geographic*, "Caffeine," by T. R. Reid, January 2005

**7.** A carpenter drills 216 holes in a rectangular array in a pegboard. There are 12 holes in each row. How many rows are there?

**8.** Lou arranges 504 entries on a spreadsheet in a rectangular array that has 36 rows. How many entries are in each row?

**9.** *TV Watching.* On average, people age 12 and older spent 1704 hr watching TV in 2008. This is 202 hr more than in 2000. Determine the number of hours people in this age group spent watching TV in 2000.

Source: U.S. Census Bureau

**10.** *Drilling Activity.* In 1988, there were 554 rotary rigs drilling for crude oil in the United States. This was 150 more rigs than were active in 2009. Find the number of active rotary oil rigs in 2009.

Source: Energy Information Administration

**11.** *Boundaries Between Countries.* The boundary between mainland United States and Canada including the Great Lakes is 3987 mi long. The length of the boundary between the United States and Mexico is 1933 mi. How much longer is the Canadian border?

Source: U.S. Geological Survey

**12.** *Longest Rivers.* The longest river in the world is the Amazon in South America at about 4225 mi. The longest river in the United States is the Missouri–Mississippi at about 3860 mi. How much longer is the Amazon?

**13.** *Pixels.* A high-definition television (HDTV) screen consists of small rectangular dots called *pixels*. How many pixels are there on a screen that has 1080 rows with 1920 pixels in each row?

Pixel

**14.** *Crossword.* The *USA Today* crossword puzzle is a rectangle containing 15 rows with 15 squares in each row. How many squares does the puzzle have altogether?

**15.** *Associate's Degrees.* About 728,000 associate's degrees were earned in the United States in 2007. Of these, 453,000 were earned by women. How many men earned associate's degrees in 2007?

Source: U.S. Department of Education

**16.** *Bachelor's Degrees.* About 649,000 bachelor's degrees were earned by men in the United States in 2007. This is 226,000 fewer than the number of bachelor's degrees earned by women the same year. How many bachelor's degrees were earned by women in 2007?

Source: U.S. Department of Education

**17.** There are 24 hr in a day and 7 days in a week. How many hours are there in a week?

**18.** There are 60 min in an hour and 24 hr in a day. How many minutes are there in a day?

*Housing Costs.*    The graph below shows the average monthly rent for a one-bedroom apartment in several cities in June 2008. Use this graph to do Exercises 19–24.

**Average Rent for a One-Bedroom Apartment**

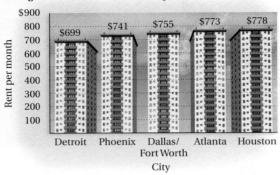

SOURCE: Apartments.com

**19.** How much higher is the average monthly rent in Houston than in Dallas/Fort Worth?

**20.** How much lower is the average monthly rent in Phoenix than in Atlanta?

**21.** Phil, Scott, and Julio plan to rent a one-bedroom apartment in Detroit immediately after graduation, sharing the rent equally. What average monthly rent can each of them expect to pay?

**22.** Maria and her sister Theresa share a one-bedroom apartment in Houston, dividing the average monthly rent equally between them. How much does each pay?

**23.** On average, how much rent would a tenant pay for a one-bedroom apartment in Atlanta during a 12-month period?

**24.** On average, how much rent would a tenant pay for a one-bedroom apartment in Dallas/Fort Worth during a 6-month period?

**25.** *Colonial Population.*    Before the establishment of the U.S. Census in 1790, it was estimated that the Colonial population in 1780 was 2,780,400. This was an increase of 2,628,900 from the population in 1680. What was the Colonial population in 1680?

**Source:** *Time Almanac,* 2005

**26.** *Interstate Speed Limits.*    The speed limit for passenger cars on interstate highways in rural areas in Montana is 75 mph. This is 10 mph faster than the speed limit for trucks on the same roads. What is the speed limit for trucks?

**27.** *Motorcycle Sales.* Motorcycle sales totaled 521,000 in the United States in 2008. Sales in 2006 exceeded this total by 669,000. How many motorcycles were sold in 2006?

Source: Motorcycle Industry Council

**28.** *Summer Olympics.* There were 43 events in the first modern Olympic games in Athens, Greece, in 1896. There were 259 more events in the 2008 Summer Olympics in Beijing, China. How many events were there in 2008?

Sources: *USA Today* research; beijing2008.cn

**29.** *Yard-Sale Profit.* Ruth made $312 at her yard sale and divided the money equally among her four grandchildren. How much did each child receive?

**30.** *Paper Measures.* A quire of paper consists of 25 sheets, and a ream of paper consists of 500 sheets. How many quires are in a ream?

**31.** *Parking Rates.* The most expensive parking in the United States occurs in midtown New York City, where the average rate is $585 per month. This is $545 per month more than in the city with the least expensive rate, Bakersfield, California. What is the average monthly parking rate in Bakersfield?

Source: Colliers International

**32.** *Trade Balance.* In 2008, foreign visitors spent $142,713,000,000 on travel and tourism-related activities in the United States, while Americans spent $112,340,000,000 traveling abroad. How much more was spent by visitors to the United States than by Americans traveling abroad?

Source: U.S. Department of Commerce

**33.** *Refrigerator Purchase.* Gourmet Deli has a chain of 24 restaurants. It buys a commercial refrigerator for each store at a cost of $1019 each. Determine the total cost of the purchase.

**34.** *Microwave Purchase.* Bridgeway College is constructing new dorms, in which each room has a small kitchen. It buys 96 microwave ovens at $88 each. Determine the total cost of the purchase.

**35.** *"Seinfeld."*  A local television station plans to air the 177 episodes of the long-running comedy series "Seinfeld." If the station airs 5 episodes per week, how many full weeks will pass before it must begin re-airing previously shown episodes? How many unaired episodes will be shown the following week before the previously aired episodes are rerun?

**36.** *"Everybody Loves Raymond."*  The popular television comedy series "Everybody Loves Raymond" had 208 scripted episodes and 2 additional episodes consisting of clips from previous shows. A local television station plans to air the 208 scripted episodes, showing 5 episodes per week. How many full weeks will pass before it must begin re-airing episodes? How many unaired episodes will be shown the following week before the previously aired episodes are rerun?

**37.** *Automobile Mileage.*  The 2010 Hyundai Tucson gets 31 miles to the gallon (mpg) in highway driving. How many gallons will it use in 7750 mi of highway driving?
Source: Hyundai

**38.** *Automobile Mileage.*  The 2010 Volkswagen Jetta (5 cylinder) gets 23 miles to the gallon (mpg) in city driving. How many gallons will it use in 3795 mi of city driving?
Source: Volkswagen of America, Inc.

**39.** *Crossword.*  The *Los Angeles Times* crossword puzzle is a rectangle containing 441 squares arranged in 21 rows. How many columns does the puzzle have?

**40.** *Mailing Labels.*  A box of mailing labels contains 750 labels on 25 sheets. How many labels are on each sheet?

**41.** *High School Court.*  The standard basketball court used by high school players has dimensions of 50 ft by 84 ft.
   **a)** What is its area?
   **b)** What is its perimeter?

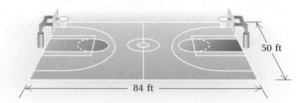

**42.** *College Court.*  The standard basketball court used by college players has dimensions of 50 ft by 94 ft.
   **a)** What is its area?
   **b)** What is its perimeter?
   **c)** How much greater is the area of a college court than a high school court? (See Exercise 41.)

**43.** *Loan Payments.*   Dana borrows $5928 for a used car. The loan is to be paid off in 24 equal monthly payments. How much is each payment (excluding interest)?

**44.** *Home Improvement Loan.*   The Van Reken family borrows $7824 to build a sunroom on the back of their home. The loan is to be paid off in equal monthly payments of $163 (excluding interest). How many months will it take to pay off the loan?

**45.** *Map Drawing.*   A map has a scale of 215 mi to the inch. How far apart *in reality* are two cities that are 3 in. apart on the map? How far apart *on the map* are two cities that, in reality, are 1075 mi apart?

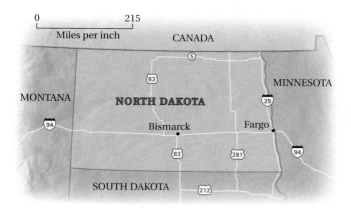

**46.** *Map Drawing.*   A map has a scale of 288 mi to the inch. How far apart *on the map* are two cities that, in reality, are 2016 mi apart? How far apart *in reality* are two cities that are 8 in. apart on the map?

**47.** Copies of this book are usually shipped from the warehouse in cartons containing 24 books each. How many cartons are needed to ship 1344 books?

**48.** The H. J. Heinz Company ships 16-oz bottles of ketchup in cartons containing 12 bottles each. How many cartons are needed to ship 528 bottles of ketchup?

**49.** Elena buys 5 video games at $64 each and pays for them with $10 bills. How many $10 bills does it take?

**50.** Pedro buys 5 video games at $64 each and pays for them with $20 bills. How many $20 bills does it take?

**51.** The balance in Meg's bank account is $568. She uses her debit card for purchases of $46, $87, and $129. Then she deposits $94 in the account after returning a book. How much is left in her account?

**52.** The balance in Dylan's bank account is $749. He uses his debit card for purchases of $34 and $65. Then he makes a deposit of $123 from his paycheck. What is the new balance?

Refer to the information in Example 8 to do Exercises 53–56.

**53.** For how long must you do aerobic exercises in order to lose one pound?

**54.** For how long must you bicycle at 9 mph in order to lose one pound?

**55.** For how long must you play golf in order to lose one pound?

**56.** For how long must you play tennis singles in order to lose one pound?

**57.** *Seating Configuration.* The seats in the Boeing 737-500 airplanes in United Airlines' North American fleet are configured with 2 rows of 4 seats across in first class and 16 rows of 6 seats across in economy class. Determine the total seating capacity of the plane.
**Source:** United Airlines

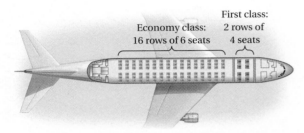

Economy class:
16 rows of 6 seats

First class:
2 rows of
4 seats

**58.** *Seating Configuration.* The seats in the Airbus 320 airplanes in United Airlines' North American fleet are configured with 3 rows of 4 seats across in first class and 21 rows of 6 seats across in economy class. Determine the total seating capacity of the plane.
**Source:** United Airlines

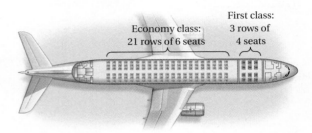

Economy class:
21 rows of 6 seats

First class:
3 rows of
4 seats

**59.** *Bones in the Hands and Feet.* There are 27 bones in each human hand and 26 bones in each human foot. How many bones are there in all in the hands and feet?

**60.** An office for adjunct instructors at a community college has 6 bookshelves, each of which is 3 ft wide. The office is moved to a new location that has dimensions of 16 ft by 21 ft. Is it possible for the bookshelves to be put side by side on the 16-ft wall?

## Skill Maintenance

Round 234,562 to the nearest:   [1.6a]

**61.** Hundred.

**62.** Ten.

**63.** Thousand.

Estimate each sum or difference by rounding to the nearest thousand.   [1.6b]

**64.** 2783 + 4602 + 5797 + 8111

**65.** 28,430 − 11,977

**66.** 5800 − 2100

**67.** 2100 + 5800

Estimate each product by rounding to the nearest hundred.   [1.6b]

**68.** 787 · 363

**69.** 887 · 799

**70.** 10,362 · 4531

## Synthesis

**71.** 📱 *Speed of Light.* Light travels about 186,000 miles per second (mi/sec) in a vacuum such as in outer space. In ice it travels about 142,000 mi/sec, and in glass it travels about 109,000 mi/sec. In 18 sec, how many more miles will light travel in a vacuum than in ice? than in glass?

**72.** Carney Community College has 1200 students. Each instructor teaches 4 classes and each student takes 5 classes. There are 30 students and 1 instructor in each classroom. How many instructors are there at Carney Community College?

# Exponential Notation and Order of Operations

## a Writing Exponential Notation

Consider the product $3 \cdot 3 \cdot 3 \cdot 3$. Such products occur often enough that mathematicians have found it convenient to create a shorter notation, called **exponential notation**, for them. For example,

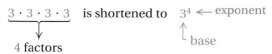

We read exponential notation as follows.

| NOTATION | WORD DESCRIPTION |
|----------|------------------|
| $3^4$ | "three to the fourth power," or "the fourth power of three" |
| $5^3$ | "five-cubed," or "the cube of five," or "five to the third power," or "the third power of five" |
| $7^2$ | "seven squared," or "the square of seven," or "seven to the second power," or "the second power of seven" |

The wording "seven squared" for $7^2$ is derived from the fact that a square with side $s$ has area $A$ given by $A = s^2$.

An expression like $3 \cdot 5^2$ is read "three times five squared," or "three times the square of five."

**EXAMPLE 1** Write exponential notation for $10 \cdot 10 \cdot 10 \cdot 10 \cdot 10$.

Exponential notation is $10^5$.    5 is the *exponent*.
10 is the *base*.

**EXAMPLE 2** Write exponential notation for $2 \cdot 2 \cdot 2$.

Exponential notation is $2^3$.

Do Exercises 1–4.

## OBJECTIVES

a Write exponential notation for products such as $4 \cdot 4 \cdot 4$.

b Evaluate exponential notation.

c Simplify expressions using the rules for order of operations.

d Remove parentheses within parentheses.

Write exponential notation.

1. $5 \cdot 5 \cdot 5 \cdot 5$

2. $5 \cdot 5 \cdot 5 \cdot 5 \cdot 5 \cdot 5$

3. $10 \cdot 10$

4. $10 \cdot 10 \cdot 10 \cdot 10$

*Answers*

1. $5^4$    2. $5^6$    3. $10^2$    4. $10^4$

## b Evaluating Exponential Notation

We evaluate exponential notation by rewriting it as a product and then computing the product.

**EXAMPLE 3** Evaluate: $10^3$.

$$10^3 = 10 \cdot 10 \cdot 10 = 1000$$

---------- *Caution!* ----------

$10^3$ does not mean $10 \cdot 3$.

-------------------------------

**EXAMPLE 4** Evaluate: $5^4$.

$$5^4 = 5 \cdot 5 \cdot 5 \cdot 5 = 625$$

Do Exercises 5–8.

Do Exercises 5–8.

When working with exponential notation, you may be asked to do several things.

- **Write exponential notation.**    The exponential notation for $2 \cdot 2 \cdot 2$ is $2^3$.

- **Write the meaning of an exponential expression.**    The meaning of $5^2$ is $5 \cdot 5$.

- **Evaluate an exponential expression.**    The value of $10^2$ is 100, since $10^2$ means $10 \cdot 10$.

## c Simplifying Expressions

Suppose we have a calculation like the following:

$$3 + 4 \cdot 8.$$

How do we find the answer? Do we add 3 to 4 and then multiply by 8, or do we multiply 4 by 8 and then add 3? In the first case, the answer is 56. In the second, the answer is 35. We agree to compute as in the second case:

$$3 + 4 \cdot 8 = 3 + 32 = 35.$$

Now consider the calculation $7 \cdot 14 - (12 + 18)$. What do the parentheses mean? To deal with these questions, we must make some agreement regarding the order in which we perform operations. The rules are as follows.

---

**RULES FOR ORDER OF OPERATIONS**

1. Do all calculations within parentheses ( ), brackets [ ], or braces { } before operations outside.
2. Evaluate all exponential expressions.
3. Do all multiplications and divisions in order from left to right.
4. Do all additions and subtractions in order from left to right.

---

It is worth noting that these are the rules that computers and most scientific calculators use to do computations.

**EXAMPLE 5** Simplify: $16 \div 8 \cdot 2$.

There are no parentheses or exponents, so we begin with the third step.

$$16 \div 8 \cdot 2 = 2 \cdot 2$$
$$= 4$$

Doing all multiplications and divisions in order from left to right

---

**Evaluate.**

**5.** $10^4$

**6.** $10^2$

**7.** $8^3$

**8.** $2^5$

---

### Calculator Corner

**Exponential Notation**

Many calculators have a $\boxed{y^x}$ or $\boxed{\wedge}$ key for raising a base to a power. To find $16^3$, for example, we press $\boxed{1}\,\boxed{6}\,\boxed{y^x}\,\boxed{3}\,\boxed{=}$ or $\boxed{1}\,\boxed{6}\,\boxed{\wedge}\,\boxed{3}\,\boxed{=}$. The result is 4096.

**Exercises:** Use a calculator to find each of the following.

**1.** $3^5$

**2.** $5^6$

**3.** $12^4$

**4.** $2^{11}$

---

*Answers*

**5.** 10,000    **6.** 100    **7.** 512    **8.** 32

**EXAMPLE 6** Simplify: $7 \cdot 14 - (12 + 18)$.

$$7 \cdot 14 - (12 + 18) = 7 \cdot 14 - 30 \qquad \text{Carrying out operations inside parentheses}$$
$$= 98 - 30 \qquad \text{Doing all multiplications and divisions}$$
$$= 68 \qquad \text{Doing all additions and subtractions}$$

Do Exercises 9–12.

Do Exercises 9–12.

Simplify.

**9.** $93 - 14 \cdot 3$

**10.** $104 \div 4 + 4$

**11.** $25 \cdot 26 - (56 + 10)$

**12.** $75 \div 5 + (83 - 14)$

**EXAMPLE 7** Simplify and compare: $23 - (10 - 9)$ and $(23 - 10) - 9$.

We have

$$23 - (10 - 9) = 23 - 1 = 22;$$
$$(23 - 10) - 9 = 13 - 9 = 4.$$

We can see that $23 - (10 - 9)$ and $(23 - 10) - 9$ represent different numbers. Thus subtraction is not associative.

Do Exercises 13 and 14.

Simplify and compare.

**13.** $64 \div (32 \div 2)$ and $(64 \div 32) \div 2$

**14.** $(28 + 13) + 11$ and $28 + (13 + 11)$

**EXAMPLE 8** Simplify: $7 \cdot 2 - (12 + 0) \div 3 - (5 - 2)$.

$$7 \cdot 2 - (12 + 0) \div 3 - (5 - 2)$$
$$= 7 \cdot 2 - 12 \div 3 - 3 \qquad \text{Carrying out operations inside parentheses}$$
$$= 14 - 4 - 3 \qquad \text{Doing all multiplications and divisions in order from left to right}$$
$$\left.\begin{array}{l} = 10 - 3 \\ = 7 \end{array}\right\} \qquad \text{Doing all additions and subtractions in order from left to right}$$

Do Exercise 15.

**15.** Simplify:

$9 \times 4 - (20 + 4) \div 8 - (6 - 2)$.

**EXAMPLE 9** Simplify: $15 \div 3 \cdot 2 \div (10 - 8)$.

$$15 \div 3 \cdot 2 \div (10 - 8)$$
$$= 15 \div 3 \cdot 2 \div 2 \qquad \text{Carrying out operations inside parentheses}$$
$$\left.\begin{array}{l} = 5 \cdot 2 \div 2 \\ = 10 \div 2 \\ = 5 \end{array}\right\} \qquad \text{Doing all multiplications and divisions in order from left to right}$$

Do Exercises 16–18.

Simplify.

**16.** $5 \cdot 5 \cdot 5 + 26 \cdot 71 - (16 + 25 \cdot 3)$

**17.** $30 \div 5 \cdot 2 + 10 \cdot 20 + 8 \cdot 8 - 23$

**18.** $95 - 2 \cdot 2 \cdot 2 \cdot 5 \div (24 - 4)$

*Answers*

**9.** 51   **10.** 30   **11.** 584   **12.** 84   **13.** 4; 1
**14.** 52; 52   **15.** 29   **16.** 1880   **17.** 253
**18.** 93

**EXAMPLE 10**  Simplify: $4^2 \div (10 - 9 + 1)^3 \cdot 3 - 5$.

$$4^2 \div (10 - 9 + 1)^3 \cdot 3 - 5$$
$$= 4^2 \div (1 + 1)^3 \cdot 3 - 5 \qquad \text{Subtracting inside parentheses}$$
$$= 4^2 \div 2^3 \cdot 3 - 5 \qquad \text{Adding inside parentheses}$$
$$= 16 \div 8 \cdot 3 - 5 \qquad \text{Evaluating exponential expressions}$$
$$= 2 \cdot 3 - 5 \left.\vphantom{\begin{matrix}a\\b\end{matrix}}\right\} \quad \text{Doing all multiplications and}$$
$$= 6 - 5 \qquad\qquad \text{divisions in order from left to right}$$
$$= 1 \qquad\qquad\qquad \text{Subtracting}$$

Do Exercises 19–21.

**EXAMPLE 11**  Simplify: $2^9 \div 2^6 \cdot 2^3$.

$$2^9 \div 2^6 \cdot 2^3 = 512 \div 64 \cdot 8 \qquad \text{There are no parentheses. Evaluating exponential expressions}$$
$$= 8 \cdot 8 \left.\vphantom{\begin{matrix}a\\b\end{matrix}}\right\} \quad \text{Doing all multiplications and}$$
$$= 64 \qquad\qquad \text{divisions in order from left to right}$$

Do Exercise 22.

**Simplify.**

**19.** $5^3 + 26 \cdot 71 - (16 + 25 \cdot 3)$

**20.** $(1 + 3)^3 + 10 \cdot 20 + 8^2 - 23$

**21.** $81 - 3^2 \cdot 2 \div (12 - 9)$

**22.** Simplify: $2^3 \cdot 2^8 \div 2^9$.

---

**Calculator Corner**

**Order of Operations**  To determine whether a calculator is programmed to follow the rules for order of operations, we can enter a simple calculation that requires using those rules. For example, we enter $\boxed{3}\ \boxed{+}\ \boxed{4}\ \boxed{\times}\ \boxed{2}\ \boxed{=}$. If the result is 11, we know that the rules for order of operations have been followed. That is, the multiplication $4 \times 2 = 8$ was performed first and then 3 was added to produce a result of 11. If the result is 14, we know that the calculator performs operations as they are entered rather than following the rules for order of operations. That means, in this case, that 3 and 4 were added first to get 7 and then that sum was multiplied by 2 to produce the result of 14. For such calculators, we would have to enter the operations in the order in which we want them performed. In this case, we would press $\boxed{4}\ \boxed{\times}\ \boxed{2}\ \boxed{+}\ \boxed{3}\ \boxed{=}$.

Many calculators have parenthesis keys that can be used to enter an expression containing parentheses. To enter $5(4 + 3)$, for example, we press $\boxed{5}\ \boxed{(}\ \boxed{4}\ \boxed{+}\ \boxed{3}\ \boxed{)}\ \boxed{=}$. The result is 35.

**Exercises:**  Simplify.

**1.** $84 - 5 \cdot 7$
**2.** $80 + 50 \div 10$
**3.** $3^2 + 9^2 \div 3$
**4.** $4^4 \div 64 - 4$
**5.** $15 \cdot 7 - (23 + 9)$
**6.** $(4 + 3)^2$

---

### Averages

In order to find the average of a set of numbers, we use addition and then division. For example, the average of 2, 3, 6, and 9 is found as follows.

The number of addends is 4.

$$\text{Average} = \frac{2 + 3 + 6 + 9}{4} = \frac{20}{4} = 5$$

Divide by 4.

The fraction bar acts as a pair of grouping symbols so

$$\frac{2 + 3 + 6 + 9}{4} \quad \text{is equivalent to} \quad (2 + 3 + 6 + 9) \div 4.$$

Thus we are using order of operations when we compute an average.

---

**AVERAGE**

The **average** of a set of numbers is the sum of the numbers divided by the number of addends.

---

**EXAMPLE 12** *Average Height of Waterfalls.* The heights of the four highest waterfalls in the world are shown in the figure below. Determine the average height of the four falls.

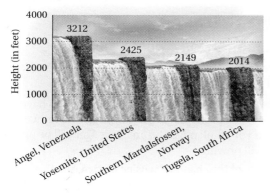

**World's Highest Waterfalls**

Height (in feet)

- Angel, Venezuela: 3212
- Yosemite, United States: 2425
- Southern Mardalsfossen, Norway: 2149
- Tugela, South Africa: 2014

SOURCE: World Almanac

The average is given by

$$\frac{3212 + 2425 + 2149 + 2014}{4} = \frac{9800}{4} = 2450.$$

The average height of the world's four highest waterfalls is 2450 ft.

Do Exercise 23.

### (d) Removing Parentheses within Parentheses

When parentheses occur within parentheses, we can make them different shapes, such as [ ] (also called "brackets") and { } (also called "braces"). All of these have the same meaning. When parentheses occur within parentheses, computations in the innermost ones are to be done first.

**23. Average Number of Career Hits.** The numbers of career hits of five Hall of Fame baseball players are given in the bar graph below. Find the average number of career hits of all five.

**Career Hits**

- Mel Ott: 2876
- Jake Beckley: 2930
- Dave Winfield: 3110
- Eddie Murray: 3255
- Carl Yastrzemski: 3419

SOURCES: Associated Press; Major League Baseball

*Answer*

**23.** 3118 hits

**EXAMPLE 13** Simplify: $[25 - (4 + 3) \cdot 3] \div (11 - 7)$.

$$[25 - (4 + 3) \cdot 3] \div (11 - 7)$$

$= [25 - 7 \cdot 3] \div (11 - 7)$     Doing the calculations in the innermost parentheses first

$= [25 - 21] \div (11 - 7)$     Doing the multiplication in the brackets

$= 4 \div 4$     Subtracting

$= 1$     Dividing

**EXAMPLE 14** Simplify: $16 \div 2 + \{40 - [13 - (4 + 2)]\}$.

$$16 \div 2 + \{40 - [13 - (4 + 2)]\}$$

$= 16 \div 2 + \{40 - [13 - 6]\}$     Doing the calculations in the innermost parentheses first

$= 16 \div 2 + \{40 - 7\}$     Again, doing the calculations in the innermost brackets

$= 16 \div 2 + 33$     Subtracting inside the braces

$= 8 + 33$     Doing all multiplications and divisions in order from left to right

$= 41$     Adding

Do Exercises 24 and 25.

Simplify.

**24.** $9 \times 5 + \{6 \div [14 - (5 + 3)]\}$

**25.** $[18 - (2 + 7) \div 3] - (31 - 10 \times 2)$

---

*Answers*

**24.** 46    **25.** 4

Write exponential notation.

**1.** $3 \cdot 3 \cdot 3 \cdot 3$

**2.** $2 \cdot 2 \cdot 2 \cdot 2 \cdot 2$

**3.** $5 \cdot 5$

**4.** $13 \cdot 13 \cdot 13$

**5.** $7 \cdot 7 \cdot 7 \cdot 7 \cdot 7$

**6.** $9 \cdot 9$

**7.** $10 \cdot 10 \cdot 10$

**8.** $1 \cdot 1 \cdot 1 \cdot 1$

 Evaluate.

**9.** $7^2$

**10.** $5^3$

**11.** $9^3$

**12.** $8^2$

**13.** $12^4$

**14.** $10^5$

**15.** $3^5$

**16.** $2^6$

**c** Simplify.

**17.** $12 + (6 + 4)$

**18.** $(12 + 6) + 18$

**19.** $52 - (40 - 8)$

**20.** $(52 - 40) - 8$

**21.** $1000 \div (100 \div 10)$

**22.** $(1000 \div 100) \div 10$

**23.** $(256 \div 64) \div 4$

**24.** $256 \div (64 \div 4)$

**25.** $(2 + 5)^2$

**26.** $2^2 + 5^2$

**27.** $16 \cdot 24 + 50$

**28.** $23 + 18 \cdot 20$

**29.** $83 - 7 \cdot 6$

**30.** $10 \cdot 7 - 4$

**31.** $10 \cdot 10 - 3 \cdot 4$

**32.** $90 - 5 \cdot 5 \cdot 2$

**33.** $4^3 \div 8 - 4$

**34.** $8^2 - 8 \cdot 2$

**35.** $17 \cdot 20 - (17 + 20)$

**36.** $1000 \div 25 - (15 + 5)$

**37.** $5(6 - 3) - (18 - 4)$

**38.** $11(9 - 7) - (13 - 10)$

**39.** $15(7 - 3) \div 12$

**40.** $8(12 - 7) \div 20$

**41.** $(11 - 8)^2 - (18 - 16)^2$

**42.** $(32 - 27)^3 + (19 + 1)^3$

**43.** $4^2 + 8^2 \div 2^2$

**44.** $6^2 - 3^4 \div 3^3$

**45.** $10^3 - 10 \cdot 6 - (4 + 5 \cdot 6)$

**46.** $7^2 + 20 \cdot 4 - (28 + 9 \cdot 2)$

**47.** $6 \cdot 11 - (7 + 3) \div 5 - (6 - 4)$

**48.** $8 \times 9 - (12 - 8) \div 4 - (10 - 7)$

**49.** $120 - 3^3 \cdot 4 \div (5 \cdot 6 - 6 \cdot 4)$

**50.** $80 - 2^4 \cdot 15 \div (7 \cdot 5 - 45 \div 3)$

**51.** $3 \cdot (2 + 8)^2 - 5(4 - 3)^2$

**52.** $7 \cdot (10 - 3)^2 - 2(3 + 1)^2$

**53.** $2^3 \cdot 2^8 \div 2^6$

**54.** $2^7 \div 2^5 \cdot 2^4 \div 2^2$

**55.** $3(5^2) - 6(9 - 2) + 4(7 - 1)$

**56.** $8(3^2) - 5(8 - 6) - 3(9 - 6)$

**57.** $24 \div (7 - 5)^2 \cdot 3^2$

**58.** $2^4 \cdot (8 - 5)^3 \div 6$

**59.** Find the average of $64, $97, and $121.

**60.** Find the average of four test grades of 86, 92, 80, and 78.

**61.** Find the average of 320, 128, 276, and 880.

**62.** Find the average of $1025, $775, $2062, $942, and $3721.

**d)**   Simplify.

**63.** $2[3 + 5(9 + 1)]$

**64.** $6[2(8 - 4) - 7]$

**65.** $6 + 3[15 - 2(7 - 3) + 1]$

**66.** $18 - 2[10 + 3(5 + 6) - 40]$

**67.** $25 \div 5 + 10[3(9 - 2) - 2(10 - 7)]$

**68.** $72 \div 2 - 3[6(9 - 4) - 5(2 + 3)]$

**69.** $30 - \{50 - [4 + 5(3 + 4)]\}$

**70.** $6 + \{7 + [2 \cdot 9 - 3(2 + 3)]\}$

**71.** $15 - 2 \cdot (9 - 7) - 3[5 - 2(3 - 1)]$

**72.** $25 - 3(8 - 5) - 3[18 - 2(3 + 5)]$

**73.** $6(7 - 2) \div 5[8 - (7 - 1)]$

**74.** $10(12 - 5) \div 5[3 + 4(8 - 7)]$

**75.** $8 \times 13 + \{42 \div [18 - (6 + 5)]\}$

**76.** $72 \div 6 - \{2 \times [9 - (4 \times 2)]\}$

**77.** $[14 - (3 + 5) \div 2] - [18 \div (8 - 2)]$

**78.** $[92 \times (6 - 4) \div 8] + [7 \times (8 - 3)]$

**79.** $(82 - 14) \times [(10 + 45 \div 5) - (6 \cdot 6 - 5 \cdot 5)]$

**80.** $(18 \div 2) \cdot \{[(9 \cdot 9 - 1) \div 2] - [5 \cdot 20 - (7 \cdot 9 - 2)]\}$

**81.** $2 + (4 + 3^2) - 2\{18 - 3[7 - (8 - 6)] + 1\}$

**82.** $(2 + 6^2) - 5^2 + 4\{3(9 - 2) - 5[6 - 2(21 - 19)]\}$

**83.** $1 + \{7[4(5 - 2)^3 - 8(2^4 - 3 \cdot 5)] - 3 - 2\}$

**84.** $5^3 - 3\{5 + (6 - 2)^2 + 2[8 - (11 - 7) + 6^2 \div 3^2]\}$

## Skill Maintenance

Solve. [1.7b]

**85.** $x + 341 = 793$

**86.** $4197 + x = 5032$

**87.** $7 \cdot x = 91$

**88.** $1554 = 42 \cdot y$

**89.** $3240 = y + 898$

**90.** $6000 = 1102 + t$

**91.** $25 \cdot t = 625$

**92.** $10,000 = 100 \cdot t$

## Synthesis

Each of the answers in Exercises 93–95 is incorrect. First find the correct answer. Then place as many parentheses as needed in the expression in order to make the incorrect answer correct.

**93.** $1 + 5 \cdot 4 + 3 = 36$

**94.** $12 \div 4 + 2 \cdot 3 - 2 = 2$

**95.** $12 \div 4 + 2 \cdot 3 - 2 = 4$

**96.** Use one occurrence each of 1, 2, 3, 4, 5, 6, 7, 8, and 9 and any of the symbols $+$, $-$, $\cdot$, $\div$, and ( ) to represent 100.

# Summary and Review

## Key Terms and Properties

digit, p. 2
periods, p. 2
place-value chart, p. 2
whole numbers, p. 3
natural numbers, p. 3
sum, p. 9
addend, p. 9
additive identity, p. 10
perimeter, p. 11

minuend, p. 14
subtrahend, p. 14
difference, p. 14
factor, p. 19
product, p. 19
multiplicative identity, p. 20
dividend, p. 26
divisor, p. 26
quotient, p. 26

remainder, p. 28
rounding, p. 37
equation, p. 42
inequality, p. 42
solution of an equation, p. 48
exponential notation, p. 71
base, p. 71
exponent, p. 71

| | |
|---|---|
| *Associative Law of Addition:* | $a + (b + c) = (a + b) + c$ |
| *Commutative Law of Addition:* | $a + b = b + a$ |
| *Associative Law of Multiplication:* | $a \cdot (b \cdot c) = (a \cdot b) \cdot c$ |
| *Commutative Law of Multiplication:* | $a \cdot b = b \cdot a$ |

## Concept Reinforcement

Determine whether each statement is true or false.

_____ **1.** $a > b$ is true when $a$ is to the right of $b$ on the number line.   [1.6c]

_____ **2.** Any nonzero number divided by itself is 1.   [1.5a]

_____ **3.** For any whole number $a$, $a \div 0 = 0$.   [1.5a]

_____ **4.** Every equation is true.   [1.7a]

_____ **5.** The rules for order of operations tell us to multiply and divide before adding and subtracting.   [1.9c]

_____ **6.** The average of three numbers is the middle number.   [1.9c]

## Important Concepts

**Objective 1.1a**   Give the meaning of digits in standard notation.

**Example**   What does the digit 7 mean in 2,379,465?

2,3 7 9,465

7 means 7 ten thousands.

**Practice Exercise**

1. What does the digit 2 mean in 432,079?

**Objective 1.2a**   Add whole numbers.

**Example**   Add: 7368 + 3547.

$$
\begin{array}{r}
{\scriptstyle 1\ 1} \\
7\ 3\ 6\ 8 \\
+\ 3\ 5\ 4\ 7 \\
\hline
1\ 0,9\ 1\ 5
\end{array}
$$

**Practice Exercise**

2. Add: 36,047 + 29,255.

## Objective 1.3a   Subtract whole numbers.

**Example**   Subtract: $8045 - 2897$.

$$
\begin{array}{r}
{\scriptstyle 7\ \ 9\ \ \overset{13}{\cancel{3}}\ \ 15} \\
\cancel{8}\,\cancel{0}\,\cancel{4}\,\cancel{5} \\
-\ 2\,8\,9\,7 \\
\hline
5\,1\,4\,8
\end{array}
$$

**Practice Exercise**

3. Subtract: $4805 - 1568$.

## Objective 1.4a   Multiply whole numbers.

**Example**   Multiply: $57 \times 315$.

$$
\begin{array}{r}
{\scriptstyle \ \ \ \ \ \ \ 2} \\
{\scriptstyle \ \ \ 1\ \ 3} \\
3\,1\,5 \\
\times\ \ \ 5\,7 \\
\hline
2\,2\,0\,5 \leftarrow 315 \times 7 \\
1\,5\,7\,5\,0 \leftarrow 315 \times 50 \\
\hline
1\,7{,}9\,5\,5
\end{array}
$$

**Practice Exercise**

4. Multiply: $329 \times 684$.

## Objective 1.5a   Divide whole numbers.

**Example**   Divide: $6463 \div 26$.

$$
\begin{array}{r}
2\,4\,8 \\
26\overline{)6\,4\,6\,3} \\
5\,2\phantom{\,4\,3} \\
\hline
1\,2\,6\phantom{\,3} \\
1\,0\,4\phantom{\,3} \\
\hline
2\,2\,3 \\
2\,0\,8 \\
\hline
1\,5
\end{array}
$$

The answer is 248 R 15.

**Practice Exercise**

5. Divide: $8519 \div 27$.

## Objective 1.6a   Round to the nearest ten, hundred, or thousand.

**Example**   Round 6471 to the nearest hundred.

6 4 **7** 1

   The digit 4 is in the hundreds place. We consider the next digit to the right. Since the digit, 7, is 5 or higher, we round 4 hundreds up to 5 hundreds. Then we change all digits to the right of the hundreds digit to zeros. The answer is 6500.

**Example**   Round to the nearest thousand.

6 **4** 7 1

   The digit 6 is in the thousands place. We consider the next digit to the right. Since the digit, 4, is 4 or lower, we round 6 thousands down, meaning that 6 thousands stays as 6 thousands. Change all digits to the right of the thousands digit to zeros. The answer is 6000.

**Practice Exercises**

6. Round 36,468 to the nearest hundred.

7. Round 36,468 to the nearest thousand.

**Objective 1.6c**  Use < or > for ☐ to write a true sentence in a situation like 6 ☐ 10.

**Example**  Use < or > for ☐ to write a true sentence:
34 ☐ 29.
Since 34 is to the right of 29 on the number line,

$$34 > 29.$$

**Practice Exercise**

**8.** Use < or > for ☐ to write a true sentence:
78 ☐ 81.

**Objective 1.7b**  Solve equations like $t + 28 = 54$, $28 \cdot x = 168$, and $98 \cdot 2 = y$.

**Example**  Solve: $y + 12 = 27$.

$$y + 12 = 27$$
$$y + 12 - 12 = 27 - 12$$
$$y + 0 = 15$$
$$y = 15$$

The solution is 15.

**Practice Exercise**

**9.** Solve: $24 \cdot x = 864$.

**Objective 1.9b**  Evaluate exponential notation.

**Example**  Evaluate: $5^4$.

$$5^4 = 5 \cdot 5 \cdot 5 \cdot 5 = 625$$

**Practice Exercise**

**10.** Evaluate: $6^3$.

## Review Exercises

The review exercises that follow are for practice. Answers are given at the back of the book. If you miss an exercise, restudy the objective indicated in red next to the exercise or on the direction line that precedes it.

**1.** What does the digit 8 mean in 4,678,952?  [1.1a]

**2.** In 13,768,940, what digit tells the number of millions?  [1.1a]

Write expanded notation.  [1.1b]

**3.** 2793

**4.** 56,078

**5.** 4,007,101

Write a word name.  [1.1c]

**6.** 67,819

**7.** 2,781,427

**8.** 4,817,941, the number of vehicles sold globally by Toyota Motor Corporation in the first six months of 2008
**Source:** *Indianapolis Star,* July 24, 2008

Write standard notation.  [1.1c]

**9.** Four hundred seventy-six thousand, five hundred eighty-eight

**10.** *Subway Ridership.*  Ridership on the New York City Subway system totaled one billion, six hundred twenty million in 2008.
**Source:** Metropolitan Transit Authority

Add.  [1.2a]

**11.** $7304 + 6968$

**12.** $27{,}609 + 38{,}415$

**13.** $2703 + 4125 + 6004 + 8956$

**14.**
$$\begin{array}{r} 9\,1{,}4\,2\,6 \\ +\quad 7{,}4\,9\,5 \\ \hline \end{array}$$

Subtract.  [1.3a]

**15.** $8045 - 2897$

**16.** $9001 - 7312$

**17.** $6003 - 3729$

**18.**
$$\begin{array}{r} 3\,7{,}4\,0\,5 \\ -\ 1\,9{,}6\,4\,8 \\ \hline \end{array}$$

Multiply.  [1.4a]

**19.** $17{,}000 \cdot 300$

**20.** $7846 \cdot 800$

**21.** $726 \cdot 698$

**22.** $587 \cdot 47$

**23.**
$$\begin{array}{r} 8\,3\,0\,5 \\ \times\quad 6\,4\,2 \\ \hline \end{array}$$

Divide.  [1.5a]

**24.** $63 \div 5$

**25.** $80 \div 16$

**26.** $7\,)\overline{6\,3\,9\,4}$

**27.** $3073 \div 8$

**28.** $6\,0\,)\overline{2\,8\,6}$

**29.** $4266 \div 79$

**30.** $3\,8\,)\overline{1\,7{,}1\,7\,6}$

**31.** $1\,4\,)\overline{7\,0{,}1\,1\,2}$

**32.** $52{,}668 \div 12$

**33.** $32{,}000 \div 100$

Round 345,759 to the nearest:  [1.6a]

**34.** Hundred.

**35.** Ten.

**36.** Thousand.

Use $<$ or $>$ for $\square$ to write a true sentence.  [1.6c]

**37.** $67\ \square\ 56$

**38.** $1\ \square\ 23$

Estimate each sum, difference, or product by first rounding to the nearest hundred. Show your work.  [1.6b]

**39.** $41{,}348 + 19{,}749$

**40.** $38{,}652 - 24{,}549$

**41.** $396 \cdot 748$

Solve.  [1.7b]

**42.** $46 \cdot n = 368$

**43.** $47 + x = 92$

**44.** $1 \cdot y = 58$

**45.** $24 = x + 24$

**46.** Write exponential notation: $4 \cdot 4 \cdot 4$.  [1.9a]

Evaluate.  [1.9b]

**47.** $10^4$

**48.** $6^2$

Simplify.  [1.9c, d]

**49.** $8 \cdot 6 + 17$

**50.** $10 \cdot 24 - (18 + 2) \div 4 - (9 - 7)$

**51.** $(80 \div 16) \times [(20 - 56 \div 8) + (8 \cdot 8 - 5 \cdot 5)]$

**52.** Find the average of 157, 170, and 168.  [1.9c]

Solve.  [1.8a]

**53.** *Computer Workstation.*  Natasha has $196 and wants to buy a computer workstation for $698. How much more does she need?

**54.** Toni has $406 in her checking account. She is paid $78 for a part-time job and deposits that in her checking account. How much is then in her account?

**55.** *Lincoln-Head Pennies.*  In 1909, the first Lincoln-head pennies were minted. Seventy-three years later, these pennies were first minted with a decreased copper content. In what year was the copper content reduced?

**56.** A beverage company packed 228 cans of soda into 12-can cartons. How many cartons did they fill?

**57.** An apartment builder bought 13 gas stoves at $425 each and 13 refrigerators at $620 each. What was the total cost?

**58.** An apple farmer keeps bees in her orchard to help pollinate the apple blossoms. The bees from an average beehive can pollinate 30 surrounding trees during one growing season. A farmer has 420 trees. How many beehives does she need to pollinate all of them?
**Source:** Jordan Orchards, Westminster, PA

**59.** *Olympic Trampoline.* Shown below is an Olympic trampoline. Determine the area and the perimeter of the trampoline. [1.2b], [1.4b]
**Source:** International Trampoline Industry Association, Inc.

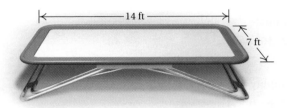

**60.** A chemist has 2753 mL of alcohol. How many 20-mL beakers can be filled? How much will be left over?

**61.** A family budgeted $7825 for food and clothing and $2860 for entertainment. The yearly income of the family was $38,283. How much of this income remained after these two allotments?

**62.** Simplify: $7 + (4 + 3)^2$. [1.9c]
  **A.** 32              **B.** 56
  **C.** 151            **D.** 196

**63.** Simplify: $7 + 4^2 + 3^2$. [1.9c]
  **A.** 32               **B.** 56
  **C.** 130            **D.** 196

**64.** $[46 - (4 - 2) \cdot 5] \div 2 + 4$ [1.9d]
  **A.** 6                **B.** 20
  **C.** 114            **D.** 22

## Synthesis

**65.** ▦ Determine the missing digit $d$. [1.4a]

$$\begin{array}{r} 9\ d \\ \times\ \ d\ 2 \\ \hline 8\ 0\ 3\ 6 \end{array}$$

**66.** ▦ Determine the missing digits $a$ and $b$. [1.5a]

$$\begin{array}{r} 9\ a\ 1 \\ 2\ b\ 1\ \overline{)\ 2\ 3\ 6{,}4\ 2\ 1} \end{array}$$

**67.** A mining company estimates that a crew must tunnel 2000 ft into a mountain to reach a deposit of copper ore. Each day, the crew tunnels about 500 ft. Each night, about 200 ft of loose rocks roll back into the tunnel. How many days will it take the mining company to reach the copper deposit? [1.8a]

# Understanding Through Discussion and Writing

**1.** Is subtraction associative? Why or why not? [1.3a]

**2.** Explain how estimating and rounding can be useful when shopping for groceries. [1.6b]

**3.** Write a problem for a classmate to solve. Design the problem so that the solution is "The driver still has 329 mi to travel." [1.8a]

**4.** Consider the expressions $9 - (4 \cdot 2)$ and $(3 \cdot 4)^2$. Are the parentheses necessary in each case? Explain. [1.9c]

**Test**  For Extra Help

**1.** In the number 546,789, which digit tells the number of hundred thousands?

**2.** Write expanded notation: 8843.

**3.** Write a word name: 38,403,277.

Add.

**4.**
$$\begin{array}{r} 6\ 8\ 1\ 1 \\ +\ 3\ 1\ 7\ 8 \\ \hline \end{array}$$

**5.**
$$\begin{array}{r} 4\ 5,8\ 8\ 9 \\ +\ 1\ 7,9\ 0\ 2 \\ \hline \end{array}$$

**6.**
$$\begin{array}{r} 1\ 2\ 3\ 9 \\ 8\ 4\ 3 \\ 3\ 0\ 1 \\ +\ \ \ 7\ 8\ 2 \\ \hline \end{array}$$

**7.**
$$\begin{array}{r} 6\ 2\ 0\ 3 \\ +\ 4\ 3\ 1\ 2 \\ \hline \end{array}$$

Subtract.

**8.**
$$\begin{array}{r} 7\ 9\ 8\ 3 \\ -\ 4\ 3\ 5\ 3 \\ \hline \end{array}$$

**9.**
$$\begin{array}{r} 2\ 9\ 7\ 4 \\ -\ 1\ 9\ 3\ 5 \\ \hline \end{array}$$

**10.**
$$\begin{array}{r} 8\ 9\ 0\ 7 \\ -\ 2\ 0\ 5\ 9 \\ \hline \end{array}$$

**11.**
$$\begin{array}{r} 2\ 3,0\ 6\ 7 \\ -\ 1\ 7,8\ 9\ 2 \\ \hline \end{array}$$

Multiply.

**12.**
$$\begin{array}{r} 4\ 5\ 6\ 8 \\ \times\ \ \ \ \ \ \ 9 \\ \hline \end{array}$$

**13.**
$$\begin{array}{r} 8\ 8\ 7\ 6 \\ \times\ \ \ 6\ 0\ 0 \\ \hline \end{array}$$

**14.**
$$\begin{array}{r} 6\ 5 \\ \times\ 3\ 7 \\ \hline \end{array}$$

**15.**
$$\begin{array}{r} 6\ 7\ 8 \\ \times\ 7\ 8\ 8 \\ \hline \end{array}$$

Divide.

**16.** $15 \div 4$

**17.** $420 \div 6$

**18.** $89\,\overline{)\,8\ 6\ 3\ 3}$

**19.** $4\ 4\,\overline{)\,3\ 5,4\ 2\ 8}$

Round 34,528 to the nearest:

**20.** Thousand.

**21.** Ten.

**22.** Hundred.

Estimate each sum, difference, or product by first rounding to the nearest hundred. Show your work.

**23.**
$$\begin{array}{r} 2\ 3,6\ 4\ 9 \\ +\ 5\ 4,7\ 4\ 6 \\ \hline \end{array}$$

**24.**
$$\begin{array}{r} 5\ 4,7\ 5\ 1 \\ -\ 2\ 3,6\ 4\ 9 \\ \hline \end{array}$$

**25.**
$$\begin{array}{r} 8\ 2\ 4 \\ \times\ 4\ 8\ 9 \\ \hline \end{array}$$

Use < or > for ☐ to write a true sentence.

**26.** 34 ☐ 17

**27.** 117 ☐ 157

Solve.

**28.** $28 + x = 74$

**29.** $169 \div 13 = n$

**30.** $38 \cdot y = 532$

**31.** $381 = 0 + a$

Solve.

**32.** *Calorie Content.* An 8-oz serving of whole milk contains 146 calories. This is 63 calories more than the number of calories in an 8-oz serving of skim milk. How many calories are in an 8-oz serving of skim milk?

**Source:** *American Journal of Clinical Nutrition*

**33.** A box contains 5000 staples. How many staplers can be filled from the box if each stapler holds 250 staples?

**34.** *Largest States.* The following table lists the five largest states in terms of their land area. Find the total land area of these states.

| STATE | AREA (in square miles) |
|---|---|
| Alaska | 571,951 |
| Texas | 261,797 |
| California | 155,959 |
| Montana | 145,552 |
| New Mexico | 121,356 |

SOURCES: U.S. Department of Commerce; U.S. Census Bureau

**35.** *Pool Tables.* The Bradford™ pool table made by Brunswick Billiards comes in three sizes of playing area, 50 in. by 100 in., 44 in. by 88 in., and 38 in. by 76 in.

**Source:** Brunswick Billiards

a) Determine the perimeter and the playing area of each table.
b) By how much does the area of the largest table exceed the area of the smallest table?

**36.** *Hostess Ding Dongs®.* Hostess packages its Ding Dong snack cakes in 12-packs. How many 12-packs can it fill with 22,231 cakes? How many will be left over?

**37.** *Office Supplies.* Morgan manages the office of a small graphics firm. He buys 3 black inkjet cartridges at $15 each and 2 photo inkjet cartridges at $25 each. How much does the purchase cost?

**38.** Write exponential notation: $12 \cdot 12 \cdot 12 \cdot 12$.

Evaluate.

**39.** $7^3$

**40.** $10^5$

Simplify.

**41.** $35 - 1 \cdot 28 \div 4 + 3$

**42.** $10^2 - 2^2 \div 2$

**43.** $(25 - 15) \div 5$

**44.** $2^4 + 24 \div 12$

**45.** $8 \times \{(20 - 11) \cdot [(12 + 48) \div 6 - (9 - 2)]\}$

**46.** Find the average of 97, 99, 87, and 89.

**A.** 93  **B.** 124  **C.** 186  **D.** 372

## Synthesis

**47.** An open cardboard container is 8 in. wide, 12 in. long, and 6 in. high. How many square inches of cardboard are used?

**48.** Use trials to find the single-digit number $a$ for which
$$359 - 46 + a \div 3 \times 25 - 7^2 = 339.$$

**49.** Cara spends $229 a month to repay her student loan. If she has already paid $9160 on the 10-yr loan, how many payments remain?

# Introduction to Integers and Algebraic Expressions

## Real-World Application

The tallest mountain in the world, when measured from base to peak, is Mauna Kea (White Mountain) in Hawaii. From its base 19,684 ft below sea level in the Hawaiian Trough, it rises 33,480 ft. What is the elevation of the peak?

*Source: The Guinness Book of Records*

***This problem appears as Exercise 76 in Exercise Set 2.3.***

# 2.1

# Integers and the Number Line

## OBJECTIVES

**a** State the integer that corresponds to a real-world situation.

**b** Form a true sentence using < or >.

**c** Find the absolute value of any integer.

**d** Find the opposite of any integer.

---

### SKILL TO REVIEW

Objective 1.6c: Use < or > for □ to write a true sentence in a situation like 6 □ 10.

Use < or > for □ to write a true sentence.

**1.** 0 □ 10          **2.** 51 □ 15

In this section, we extend the set of whole numbers to form the set of *integers*. You have probably already used negative numbers. For example, the outside temperature could drop to *negative five* degrees and a credit card statement could indicate activity of *negative forty-eight* dollars.

To describe integers, we start with the whole numbers, 0, 1, 2, 3, and so on. For each number 1, 2, 3, and so on, we obtain a new number the same number of units to the left of zero on a number line.

For the number 1, there is the *opposite* number $-1$ (negative 1).

For the number 2, there is the *opposite* number $-2$ (negative 2).

For the number 3, there is the *opposite* number $-3$ (negative 3), and so on.

The **integers** consist of the whole numbers and these new numbers. We picture them on the number line as follows.

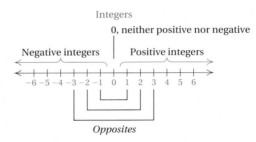

The integers to the left of zero on the number line are called **negative integers** and those to the right of zero are called **positive integers**. Zero is neither positive nor negative. We call $-1$ and 1 **opposites** of each other. Similarly, $-2$ and 2 are opposites, $-3$ and 3 are opposites, $-100$ and 100 are opposites, and 0 is its own opposite.

---

**INTEGERS**

The **integers**: $\ldots, -5, -4, -3, -2, -1, 0, 1, 2, 3, 4, 5, \ldots$

---

## **a** Integers and the Real World

Integers correspond to many real-world problems and situations. The following examples will help you get ready to translate problem situations to mathematical language.

Stoneware bottles, leather shoes, and glass window panes carpet the wreck site of the SS *Republic* 1700 feet deep. The remains of a wooden shipping crate are visible in the background.

**EXAMPLE 1** *Shipwreck Exploration.* Tell which integer corresponds to this situation: Odyssey Marine Exploration discovered the wreck of the SS *Republic* 1700 ft below sea level.

**Source:** Odyssey Marine Exploration

1700 ft below sea level corresponds to the integer $-1700$.

*Answers*

*Skill to Review:*

1. <     2. >

**EXAMPLE 2** *Stock Price Change.* Tell which integers correspond to this situation: Hal owns a stock whose price decreased from $27 per share to $11 per share over a recent time period. He owns another stock whose price increased from $20 per share to $22 per share over the same time period.

The integer −16 corresponds to the decrease in the value of the first stock. The integer 2 represents the increase in the value of the second stock.

Do Exercises 1–5.

## b   Order on the Number Line

Numbers are written in order on the number line, increasing as we move to the right. For any two numbers on the line, the one to the left is *less than* the one to the right.

Since the symbol < means "is less than," the sentence −5 < 9 is read "−5 is less than 9." The symbol > means "is greater than," so the sentence −4 > −8 is read "−4 is greater than −8."

**EXAMPLES**   Use either < or > for ☐ to form a true sentence.

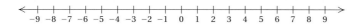

**3.** −9 ☐ 2          Since −9 is to the left of 2, we have −9 < 2.

**4.** 7 ☐ −13         Since 7 is to the right of −13, we have 7 > −13.

**5.** −19 ☐ −6        Since −19 is to the left of −6, we have −19 < −6.

Do Exercises 6–9.

## c   Absolute Value

From the number line, we see that some integers, like 4 and −4, are the same distance from zero. We call the distance of a number from zero the **absolute value** of the number. Since distance is always a nonnegative number, absolute value is always nonnegative.

The distance of −4 from 0 is 4.          The distance of 4 from 0 is 4.
The absolute value of −4 is 4.          The absolute value of 4 is 4.

```
        −4        0        4
      |←4 units→|←4 units→|
```

---

**ABSOLUTE VALUE**

The **absolute value** of a number is its distance from zero on the number line. We use the symbol $|x|$ to represent the absolute value of a number $x$.

---

Tell which integers correspond to each situation.

1. **Temperature High and Low.** The highest recorded temperature in Nevada is 125°F on June 29, 1994, in Laughlin. The lowest recorded temperature in Nevada is 50°F below zero on January 8, 1937, in San Jacinto.
   **Source:** National Climatic Data Center, Asheville, NC, and Storm Phillips, STORMFAX, INC.

2. **Stock Decrease.** The price of a stock decreased from $41 per share to $38 per share over a recent period.

3. At 10 sec before liftoff, ignition occurs. At 148 sec after liftoff, the first stage is detached from the rocket.

4. The halfback gained 8 yd on first down. The quarterback was sacked for a 5-yd loss on second down.

5. A submarine dove 120 ft, rose 50 ft, and then dove 80 ft.

Use either < or > for ☐ to write a true sentence.

6. −3 ☐ 7

7. −8 ☐ −5

8. 7 ☐ −10

9. −4 ☐ −20

To find the absolute value of a number:

a) If a number is negative, its absolute value is its opposite.

b) If a number is positive or zero, its absolute value is the same as the number.

**EXAMPLES** Find the absolute value of each number.

**Find the absolute value.**

**10.** $|18|$          **11.** $|-9|$

**12.** $|-29|$         **13.** $|52|$

**6.** $|-3|$    The distance of $-3$ from 0 is 3, so $|-3| = 3$.

**7.** $|25|$    The distance of 25 from 0 is 25, so $|25| = 25$.

**8.** $|0|$    The distance of 0 from 0 is 0, so $|0| = 0$.

Do Exercises 10–13.

## d) Opposites

Given a number on one side of 0 on the number line, we can get a number on the other side by *reflecting* the number across zero. For example, the *reflection* of 2 is $-2$. We can read $-2$ as "negative 2" or "the opposite of 2."

**NOTATION FOR OPPOSITES**

The **opposite** of a number $x$ is written $-x$ (read "the opposite of $x$").

**EXAMPLE 9** If $x$ is $-3$, find $-x$.

To find the opposite of $x$ when $x$ is $-3$, we reflect $-3$ to the other side of 0.

When $x = -3$, $-x = -(-3)$. We have $-(-3) = 3$. The opposite of $-3$ is 3.

**EXAMPLE 10** Find $-x$ when $x$ is 0.

When we try to reflect 0 "to the other side of 0," we go nowhere:

$$-x = 0 \quad \text{when} \quad x \text{ is } 0. \qquad \text{The opposite of 0 is 0; } -0 = 0.$$

In Examples 9 and 10, the variable was replaced with a number. When this occurs, we say that we are **evaluating** the expression.

**In each case, draw a number line, if necessary.**

**14.** Find $-x$ when $x$ is 1.

**15.** Find $-x$ when $x$ is $-2$.

**16.** Evaluate $-x$ when $x$ is 0.

**EXAMPLE 11** Evaluate $-x$ when $x$ is 4.

To find the opposite of $x$ when $x$ is 4, we reflect 4 to the other side of 0.

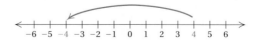

We have $-(4) = -4$. The opposite of 4 is $-4$.

**Answers**

**10.** 18   **11.** 9   **12.** 29   **13.** 52   **14.** $-1$
**15.** 2   **16.** 0

Do Exercises 14–16.

A negative number is sometimes said to have a negative *sign*. A positive number is said to have a positive sign, even though it rarely is written in.

**EXAMPLES** Determine the sign of each number.

**12.** $-7$     Negative          **13.** $23$     Positive

Replacing a number with its opposite, or *additive inverse*, is sometimes called *changing the sign*.

**EXAMPLES** Change the sign (find the opposite, or additive inverse) of each number.

**14.** $-6$     $-(-6) = 6$          **15.** $-10$     $-(-10) = 10$

**16.** $0$     $-(0) = 0$          **17.** $14$     $-(14) = -14$

Do Exercises 17–20.

Change the sign. (Find the opposite, or additive inverse.)

**17.** $-4$          **18.** $-13$

**19.** $39$          **20.** $0$

**EXAMPLE 18** If $x$ is 2, find $-(-x)$.

We replace $x$ with 2:

$-(-x)$          Read "the opposite of the opposite of $x$"

$= -(-2)$.          We copy the expression, replacing $x$ with 2

The opposite of the opposite of 2 is 2, or $-(-2) = 2$.

**EXAMPLE 19** Evaluate $-(-x)$ for $x = -4$.

We replace $x$ with $-4$:

$-(-x)$

$= -(-(-4))$          Using an extra set of parentheses to avoid notation like $--4$

$= -(\ \ 4\ \ )$          Changing the sign of $-4$

$= -4$.          Changing the sign of 4

Thus, $-(-(-4)) = -4$.

**When we change a number's sign twice, we return to the original number.**

Do Exercises 21–24.

**21.** If $x$ is 7, find $-(-x)$.

**22.** If $x$ is 1, find $-(-x)$.

**23.** Evaluate $-(-x)$ for $x = -6$.

**24.** Evaluate $-(-x)$ for $x = -2$.

It is important not to confuse parentheses with absolute-value symbols.

**EXAMPLE 20** Evaluate $-|-x|$ for $x = 2$.

We replace $x$ with 2:

$-|-x|$

$= -|-2|$          Replacing $x$ with 2

$= -2$.          The absolute value of $-2$ is 2.

Thus, $-|-2| = -2$.

Note that $-(-2) = 2$, whereas $-|-2| = -2$.

Do Exercises 25 and 26.

**25.** Find $-|-7|$.

**26.** Find $-|-39|$.

**a** Tell which integers correspond to each situation.

**1.** *Death Valley.* With an elevation of 282 ft below sea level, Badwater Basin in Death Valley in California has the lowest elevation in the United Sates.
Source: Desert USA

**2.** *Pollution Fine.* The Massey Energy Company, the nation's fourth largest coal producer, was fined $20 million for water pollution in 2008.
Source: Environmental Protection Agency

**3.** At tax time, Janine received an $820 refund while David owed $541.

**4.** *Oceanography.* At a depth of 2438 meters, researchers found the first hydrothermal vent ever seen by humans. This depth is approximately 8000 ft below sea level.
Source: Office of Naval Research

**5.** *Highest and Lowest Temperatures.* The highest temperature ever created on Earth was 950,000,000°F. The lowest temperature ever created was approximately 460°F below zero.
Source: The Guinness Book of Records

**6.** *Extreme Climate.* Verkhoyansk, a river port in northeast Siberia, has the most extreme climate on the planet. Its average monthly winter temperature is 59°F below zero, and its average monthly summer temperature is 57°F.
Source: The Guinness Book of Records

**7.** The recycling program for Colchester once received $40 for a ton of office paper. More recently, they've had to pay $15 to get rid of a ton of office paper.

**8.** During a video game, Sara intercepted a missile worth 20 points, lost a starship worth 150 points, and captured a base worth 300 points.

**b** Use either $<$ or $>$ for ☐ to form a true sentence.

**9.** $-8$ ☐ $0$    **10.** $7$ ☐ $0$    **11.** $9$ ☐ $0$    **12.** $-7$ ☐ $0$    **13.** $8$ ☐ $-8$

**14.** $6$ ☐ $-6$    **15.** $-6$ ☐ $-4$    **16.** $-1$ ☐ $-7$    **17.** $-8$ ☐ $-5$    **18.** $-5$ ☐ $-3$

**19.** $-13$ ☐ $-9$    **20.** $-5$ ☐ $-11$    **21.** $-3$ ☐ $-4$    **22.** $-6$ ☐ $-5$

Find the absolute value.

**23.** $|57|$      **24.** $|11|$      **25.** $|0|$      **26.** $|-4|$      **27.** $|-24|$

**28.** $|-36|$      **29.** $|53|$      **30.** $|54|$      **31.** $|-8|$      **32.** $|-79|$

 Find $-x$ when $x$ is each of the following.

**33.** $-7$      **34.** $-6$      **35.** $7$      **36.** $6$      **37.** $0$      **38.** $-1$

Change the sign. (Find the opposite, or additive inverse.)

**39.** $-21$      **40.** $-67$      **41.** $53$      **42.** $0$      **43.** $-1$      **44.** $16$

Evaluate $-(-x)$ when $x$ is each of the following.

**45.** $7$      **46.** $-8$      **47.** $-9$      **48.** $3$      **49.** $-17$      **50.** $-19$

**51.** $23$      **52.** $0$      **53.** $-1$      **54.** $73$      **55.** $85$      **56.** $-37$

Evaluate $-|-x|$ when $x$ is each of the following.

**57.** $345$      **58.** $729$      **59.** $0$      **60.** $1$      **61.** $-8$      **62.** $-3$

## Skill Maintenance

**63.** Add: $327 + 498$.   [1.2a]

**64.** Evaluate: $5^3$.   [1.9b]

**65.** Multiply: $209 \cdot 34$.   [1.4a]

**66.** Solve: $300 \cdot x = 1200$.   [1.7b]

**67.** Simplify: $7(9 - 3)$.   [1.9c]

**68.** Simplify: $9^2 - 3[2 + (10 - 8)]$.   [1.9d]

## Synthesis

Use either $<$, $>$, or $=$ for $\square$ to write a true sentence.

**69.** $|-5| \square |-2|$      **70.** $|4| \square |-7|$      **71.** $|-8| \square |8|$

Solve. Consider only integer replacements.

**72.** $|x| = 7$      **73.** $|x| < 2$

**74.** Simplify $-(-x)$, $-(-(-x))$, and $-(-(-(-x)))$.

**75.** List these integers in order from least to greatest.

$$2^{10}, \ -5, \ |-6|, \ 4, \ |3|, \ -100, \ 0, \ 2^7, \ 7^2, \ 10^2$$

# 2.2

## Addition of Integers

### OBJECTIVE

**a** Add integers without using the number line.

### **a** Addition

To explain addition of integers, we can use the number line. Once our understanding is developed, we will streamline our approach.

#### Addition on the Number Line

To find $a + b$, we start at $a$ and then move according to $b$.

**a)** If $b$ is positive, we move from $a$ to the right.

**b)** If $b$ is negative, we move from $a$ to the left.

**c)** If $b$ is 0, we stay at $a$.

**EXAMPLE 1** Add: $2 + (-5)$.

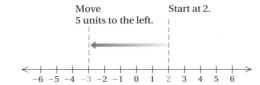

$2 + (-5) = -3$

**EXAMPLE 2** Add: $-1 + (-3)$.

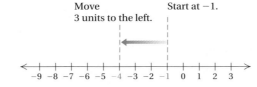

$-1 + (-3) = -4$

**EXAMPLE 3** Add: $-4 + 9$.

$-4 + 9 = 5$

Do Exercises 1–7.

You may have noticed a pattern in Example 2 and Margin Exercises 2 and 6. When two negative integers are added, the result is negative.

> **ADDING NEGATIVE INTEGERS**
>
> To add two negative integers, add their absolute values and change the sign (making the answer negative).

Add using the number line.

**1.** $1 + (-4)$

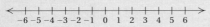

**2.** $-3 + (-2)$

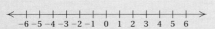

**3.** $-3 + 7$

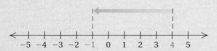

**4.** $-5 + 5$

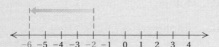

For each illustration, write a corresponding addition sentence.

**5.**

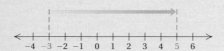

**6.**

**7.**

**Answers**

**1.** $-3$ **2.** $-5$ **3.** 4 **4.** 0
**5.** $4 + (-5) = -1$ **6.** $-2 + (-4) = -6$
**7.** $-3 + 8 = 5$

**EXAMPLES** Add.

**4.** $-5 + (-7) = -12$     *Think*: Add the absolute values: $5 + 7 = 12$. Make the answer negative, $-12$.

**5.** $-8 + (-2) = -10$     We can visualize the number line without actually drawing it.

> Do Exercises 8–11.

When the number 0 is added to any number, that number remains unchanged. For this reason, the number 0 is referred to as the **additive identity**.

**EXAMPLES** Add.

**6.** $-4 + 0 = -4$     **7.** $0 + (-9) = -9$     **8.** $17 + 0 = 17$

> Do Exercises 12–14.

When we add a positive integer and a negative integer, as in Examples 1 and 3, the sign of the number with the greater absolute value is the sign of the answer.

---

**ADDING POSITIVE AND NEGATIVE INTEGERS**

To add a positive integer and a negative integer, find the difference of their absolute values.

**a)** If the negative integer has the greater absolute value, the answer is negative.

**b)** If the positive integer has the greater absolute value, the answer is positive.

---

**EXAMPLES** Add.

**9.** $3 + (-5) = -2$     *Think*: The absolute values are 3 and 5. The difference is 2. Since the negative number has the larger absolute value, the answer is *negative*, $-2$.

**10.** $11 + (-8) = 3$     *Think*: The absolute values are 11 and 8. The difference is 3. The positive number has the larger absolute value, so the answer is *positive*, 3.

**11.** $-7 + 4 = -3$

**12.** $-6 + 10 = 4$

> Do Exercises 15–18.

Sometimes $-a$ is referred to as the **additive inverse** of $a$ because adding any number to its additive inverse always results in the additive identity, 0.

---

**ADDING OPPOSITES**

For any integer $a$,

$$a + (-a) = -a + a = 0.$$

(The sum of any number and its additive inverse, or opposite, is 0.)

---

Add. Do not use the number line except as a check.

**8.** $-5 + (-6)$

**9.** $-9 + (-3)$

**10.** $-20 + (-14)$

**11.** $-11 + (-11)$

Add.

**12.** $0 + (-17)$

**13.** $49 + 0$

**14.** $-56 + 0$

Add, using the number line only as a check.

**15.** $-4 + 6$

**16.** $-7 + 3$

**17.** $5 + (-7)$

**18.** $10 + (-7)$

*Answers*

**8.** $-11$   **9.** $-12$   **10.** $-34$   **11.** $-22$
**12.** $-17$   **13.** $49$   **14.** $-56$   **15.** $2$
**16.** $-4$   **17.** $-2$   **18.** $3$

Add, using the number line only as a check.

**19.** $5 + (-5)$

**20.** $-6 + 6$

**21.** $-10 + 10$

**22.** $89 + (-89)$

Add.
**23.** $-5 + (-10)$

**24.** $18 + (-11)$

**25.** $-13 + 13$

**26.** $-20 + 7$

Add.
**27.** $(-15) + (-37) + 25 + 42 + (-59) + (-14)$

**28.** $42 + (-81) + (-28) + 24 + 18 + (-31)$

**29.** $-8 + 17 + 14 + (-27) + 31 + (-12)$

**EXAMPLES**   Add.

**13.** $-8 + 8 = 0$

**14.** $14 + (-14) = 0$

Do Exercises 19–22.

In summary, to add integers, look first at the signs of the numbers you are adding. This tells you whether you should add or subtract to find the sum.

> **RULES FOR ADDITION OF INTEGERS**
>
> 1. *Positive numbers*:  Add the same way as arithmetic numbers. The answer is positive.
> 2. *Negative numbers*:  Add absolute values. The answer is negative.
> 3. *A positive and a negative number*:  Subtract absolute values.
>    a) If the positive number has the greater absolute value, the answer is positive.
>    b) If the negative number has the greater absolute value, the answer is negative.
>    c) If the numbers have the same absolute value, they are additive inverses and the answer is 0.
> 4. *One number is zero*:  The sum is the other number.

**EXAMPLES**   Add.

**15.** $-12 + (-7) = -19$    Two negative numbers; add absolute values.
**16.** $-20 + 36 = 16$    A positive and a negative number; subtract absolute values. The answer is positive.

Do Exercises 23–26.

Suppose we wish to add several numbers, positive and negative:

$$15 + (-2) + 7 + 14 + (-5) + (-12).$$

Because of the commutative and associative laws for addition, we can group the positive numbers together and the negative numbers together and add them separately. Then we add the two results.

**EXAMPLE 17**   Add: $15 + (-2) + 7 + 14 + (-5) + (-12)$.

First add the positive numbers:  $15 + 7 + 14 = 36$.

Then add the negative numbers:  $-2 + (-5) + (-12) = -19$.

Finally, add the results:  $36 + (-19) = 17$.

We can also add in any other order we wish, say, from left to right:

$$
\begin{aligned}
15 + (-2) + 7 + 14 + (-5) + (-12) &= 13 + 7 + 14 + (-5) + (-12) \\
&= 20 + 14 + (-5) + (-12) \\
&= 34 + (-5) + (-12) \\
&= 29 + (-12) \\
&= 17.
\end{aligned}
$$

Do Exercises 27–29.

**2.2** Exercise Set

For Extra Help

**MyMathLab**

Math XL
PRACTICE

WATCH

DOWNLOAD

READ

REVIEW

**a**    Add, using the number line.

**1.** $-7 + 2$      **2.** $1 + (-5)$      **3.** $-9 + 5$      **4.** $8 + (-3)$      **5.** $-3 + 9$

**6.** $9 + (-9)$      **7.** $-7 + 7$      **8.** $-8 + (-5)$      **9.** $-3 + (-1)$      **10.** $-2 + (-9)$

Add. Use the number line only as a check.

**11.** $-3 + (-9)$      **12.** $-3 + (-7)$      **13.** $-6 + (-5)$      **14.** $-10 + (-14)$

**15.** $-15 + 0$      **16.** $0 + (-11)$      **17.** $0 + 42$      **18.** $27 + 0$

**19.** $9 + (-4)$      **20.** $-7 + 8$      **21.** $-10 + 6$      **22.** $6 + (-13)$

**23.** $5 + (-5)$      **24.** $10 + (-10)$      **25.** $-2 + 2$      **26.** $-3 + 3$

**27.** $-4 + (-5)$      **28.** $10 + (-12)$      **29.** $13 + (-6)$      **30.** $-3 + 14$

**31.** $-25 + 25$      **32.** $40 + (-40)$      **33.** $63 + (-18)$      **34.** $85 + (-65)$

**35.** $-11 + 8$      **36.** $0 + (-34)$      **37.** $-19 + 19$      **38.** $-10 + 3$

**39.** $-16 + 6$      **40.** $-15 + 5$      **41.** $-17 + (-7)$      **42.** $-15 + (-5)$

**43.** $11 + (-16)$      **44.** $-8 + 14$      **45.** $-15 + (-6)$      **46.** $-8 + 8$

**47.** $-15 + (-15)$      **48.** $-25 + (-25)$      **49.** $-11 + 17$      **50.** $19 + (-19)$

**51.** $-15 + (-7) + 1$

**52.** $23 + (-5) + 4$

**53.** $30 + (-10) + 5$

**54.** $40 + (-8) + 5$

**55.** $-23 + (-9) + 15$

**56.** $-25 + 25 + (-9)$

**57.** $40 + (-40) + 6$

**58.** $63 + (-18) + 12$

**59.** $12 + (-65) + (-12)$

**60.** $-35 + (-63) + 35$

**61.** $-24 + (-37) + (-19) + (-45) + (-35)$

**62.** $75 + (-14) + (-17) + (-5)$

**63.** $28 + (-44) + 17 + 31 + (-94)$

**64.** $27 + (-54) + (-32) + 65 + 46$

**65.** $-19 + 73 + (-23) + 19 + (-73)$

**66.** $35 + (-51) + 29 + 51 + (-35)$

## Skill Maintenance

**67.** Add: $587 + 6094$.  [1.2a]

**68.** Subtract: $3046 - 2973$.  [1.3a]

**69.** Write in expanded notation: 39,417.  [1.1b]

**70.** Multiply: $42 \cdot 56$.  [1.4a]

**71.** Divide: $288 \div 9$.  [1.5a]

**72.** Round to the nearest ten: 3496.  [1.6a]

## Synthesis

Add.

**73.** $-|27| + (-|-13|)$

**74.** $|-32| + (-|15|)$

**75.** ▦ $-3496 + (-2987)$

**76.** ▦ $497 + (-3028)$

**77.** For what numbers $x$ is $-x$ positive?

**78.** For what numbers $x$ is $-x$ negative?

Tell whether each sum is positive, negative, or zero.

**79.** If $n$ is positive and $m$ is negative, then $-n + m$ is _____ .

**80.** If $n = m$ and $n$ is negative, then $-n + (-m)$ is _____ .

**81.** If $n$ is negative and $m$ is less than $n$, then $n + m$ is _____ .

**82.** If $n$ is positive and $m$ is greater than $n$, then $n + m$ is _____ .

# 2.3 Subtraction of Integers

## (a) Subtraction

We now consider subtraction of integers. To find the difference $a - b$, we look for a number to add to $b$ that gives us $a$.

> **THE DIFFERENCE**
>
> The difference $a - b$ is the number that when added to $b$ gives $a$.

For example, $45 - 17 = 28$ because $28 + 17 = 45$. Let's consider an example in which the answer is a negative number.

**EXAMPLE 1** Subtract: $5 - 8$.

*Think:* $5 - 8$ is the number that when added to 8 gives 5. What number can we add to 8 to get 5? The number must be negative. The number is $-3$:

$$5 - 8 = -3.$$

That is, $5 - 8 = -3$ because $8 + (-3) = 5$.

Do Exercises 1-3.

The definition of $a - b$ above does not always provide the most efficient way to subtract. To understand a faster way to subtract, consider finding $5 - 8$ using the number line. We start at 5. Then we move 8 units to the *left* to do the subtracting. Note that this is the same as adding the opposite of 8, or $-8$, to 5.

$$5 - 8 = -3$$

Look for a pattern in the following table.

| SUBTRACTIONS | ADDING AN OPPOSITE |
|---|---|
| $5 - 8 = -3$ | $5 + (-8) = -3$ |
| $-6 - 4 = -10$ | $-6 + (-4) = -10$ |
| $-7 - (-10) = 3$ | $-7 + 10 = 3$ |
| $-7 - (-2) = -5$ | $-7 + 2 = -5$ |

Do Exercises 4-7.

Perhaps you have noticed that we can subtract by adding the opposite of the number being subtracted. This can always be done.

## OBJECTIVES

**(a)** Subtract integers and simplify combinations of additions and subtractions.

**(b)** Solve applied problems involving addition and subtraction of integers.

Subtract.

**1.** $-6 - 4$
*Think:* What number can be added to 4 to get $-6$:
$\boxed{\phantom{x}} + 4 = -6?$

**2.** $-7 - (-10)$
*Think:* What number can be added to $-10$ to get $-7$:
$\boxed{\phantom{x}} + (-10) = -7?$

**3.** $-7 - (-2)$
*Think:* What number can be added to $-2$ to get $-7$:
$\boxed{\phantom{x}} + (-2) = -7?$

Complete the addition and compare with the subtraction.

**4.** $4 - 6 = -2$;
$4 + (-6) = $ _____

**5.** $-3 - 8 = -11$;
$-3 + (-8) = $ _____

**6.** $-5 - (-9) = 4$;
$-5 + 9 = $ _____

**7.** $-5 - (-3) = -2$;
$-5 + 3 = $ _____

*Answers*

**1.** $-10$  **2.** 3  **3.** $-5$  **4.** $-2$  **5.** $-11$
**6.** 4  **7.** $-2$

## SUBTRACTING BY ADDING THE OPPOSITE

To subtract, add the opposite, or additive inverse, of the number being subtracted:

$$a - b = a + (-b).$$

This is the method generally used for quick subtraction of integers.

**EXAMPLES** Equate each subtraction with a corresponding addition. Then write the equation in words.

**2.** $-12 - 30$;

$-12 - 30 = -12 + (-30)$     Adding the opposite of 30

Negative twelve minus thirty is negative twelve plus negative thirty.

**3.** $-20 - (-17)$;

$-20 - (-17) = -20 + 17$     Adding the opposite of $-17$

Negative twenty minus negative seventeen is negative twenty plus seventeen.

Do Exercises 8-10.

Once the subtraction has been rewritten as addition, we add as in Section 2.2.

**EXAMPLES** Subtract.

**4.** $2 - 6 = 2 + (-6)$     The opposite of 6 is $-6$. We change the subtraction to addition and add the opposite. Instead of subtracting 6, we add $-6$.

$\phantom{2 - 6} = -4$

**5.** $4 - (-9) = 4 + 9$     The opposite of $-9$ is 9. We change the subtraction to addition and add the opposite. Instead of subtracting $-9$, we add 9.

$\phantom{4 - (-9)} = 13$

**6.** $-4 - 8 = -4 + (-8)$     We change the subtraction to addition and add the opposite. Instead of subtracting 8, we add $-8$.

$\phantom{-4 - 8} = -12$

**7.** $10 - 7 = 10 + (-7)$     We change the subtraction to addition and add the opposite. Instead of subtracting 7, we add $-7$.

$\phantom{10 - 7} = 3$

**8.** $-4 - (-9) = -4 + 9$     Instead of subtracting $-9$, we add 9.

$\phantom{-4 - (-9)} = 5$     To check, note that $5 + (-9) = -4$.

**9.** $-7 - (-3) = -7 + 3$     Instead of subtracting $-3$, we add 3.

$\phantom{-7 - (-3)} = -4$     *Check*: $-4 + (-3) = -7$.

Do Exercises 11-16.

---

Equate each subtraction with a corresponding addition. Then write the equation in words.

**8.** $3 - 10$

**9.** $-12 - (-9)$

**10.** $-12 - 10$

Subtract.

**11.** $7 - 11$

**12.** $-6 - 10$

**13.** $13 - 8$

**14.** $-7 - (-9)$

**15.** $-8 - (-2)$

**16.** $5 - (-8)$

*Answers*

**8.** $3 - 10 = 3 + (-10)$; three minus ten is three plus negative ten.
**9.** $-12 - (-9) = -12 + 9$; negative twelve minus negative nine is negative twelve plus nine.   **10.** $-12 - 10 = -12 + (-10)$; negative twelve minus ten is negative twelve plus negative ten.   **11.** $-4$   **12.** $-16$
**13.** 5   **14.** 2   **15.** $-6$   **16.** 13

When several additions and subtractions occur together, we can make them all additions. The commutative law for addition can then be used.

**EXAMPLE 10** Simplify: $-3 - (-5) - 9 + 4 - (-6)$.

$$-3 - (-5) - 9 + 4 - (-6) = -3 + 5 + (-9) + 4 + 6 \quad \text{Adding opposites}$$
$$= -3 + (-9) + 5 + 4 + 6 \quad \text{Using a commutative law}$$
$$= -12 + 15$$
$$= 3$$

Do Exercises 17 and 18.

Simplify.

**17.** $-6 - (-2) - (-4) - 12 + 3$

**18.** $9 - (-6) + 7 - 9 - 8 - (-20)$

## b Applications and Problem Solving

We need addition and subtraction of integers to solve a variety of applied problems.

**EXAMPLE 11** *Temperature Changes.* Denver, Colorado, experienced its greatest temperature change in one day on January 25, 1872. From the low of $-20°F$ (degrees Fahrenheit), the temperature increased $66°$. If it later decreased $10°$, what was the final temperature?

**Source:** Based on information from www.examiner.com

We let $T$ represent the final temperature.

*Rewriting:* Final temperature is starting temperature plus increase minus decrease

*Translating:* $\quad T \quad = \quad -20 \quad + \quad 66 \quad - \quad 10$

Carrying out the addition and subtraction, we have

$$T = -20 + 66 - 10 = 46 - 10 = 36.$$

The final temperature was $36°F$.

Do Exercise 19.

**19. Temperature Extremes.** In Churchill, Manitoba, Canada, the average daily low temperature in January is $-31°C$ (degrees Celsius). The average daily low temperature in Key West, Florida, is $50°$ warmer. What is the average daily low temperature in Key West, Florida?

---

### STUDY TIPS

#### HELP SESSIONS

*Make the most of tutoring sessions by doing what you can ahead of time and knowing the topics with which you need help.*

- **Work on the topics *before* you go to the help or tutoring session.** Going to a session unprepared is a waste of time and, in many cases, money. Attend class, study the textbook, work exercises, and mark trouble spots. *Then* use the help and tutoring sessions to work on the trouble spots.

- **Ask questions.** The more you talk to your tutor, the more the tutor can help you.

- **Try being a "tutor" yourself.** Explaining a topic to someone else is often the best way to master it.

*Answers*

**17.** $-9$ **18.** 25 **19.** 19°C

**a** Subtract.

**1.** $3 - 7$

**2.** $5 - 10$

**3.** $0 - 7$

**4.** $0 - 8$

**5.** $-8 - (-2)$

**6.** $-6 - (-8)$

**7.** $-10 - (-10)$

**8.** $-8 - (-8)$

**9.** $12 - 16$

**10.** $14 - 19$

**11.** $20 - 27$

**12.** $26 - 7$

**13.** $-9 - (-3)$

**14.** $-6 - (-9)$

**15.** $-11 - (-11)$

**16.** $-14 - (-14)$

**17.** $7 - 7$

**18.** $9 - 9$

**19.** $7 - (-7)$

**20.** $4 - (-4)$

**21.** $8 - (-3)$

**22.** $-7 - 4$

**23.** $-6 - 8$

**24.** $6 - (-10)$

**25.** $-4 - (-9)$

**26.** $-14 - 2$

**27.** $2 - 9$

**28.** $2 - 8$

**29.** $-6 - (-5)$

**30.** $-4 - (-3)$

**31.** $8 - (-10)$

**32.** $5 - (-6)$

**33.** $0 - 10$

**34.** $0 - 23$

**35.** $-5 - (-2)$

**36.** $-3 - (-1)$

**37.** $-7 - 14$

**38.** $-9 - 16$

**39.** $0 - (-5)$

**40.** $0 - (-1)$

**41.** $-8 - 0$

**42.** $-9 - 0$

**43.** $7 - (-5)$

**44.** $7 - (-4)$

**45.** $6 - 25$

**46.** $18 - 63$

**47.** $-42 - 26$

**48.** $-18 - 63$

**49.** $-72 - 9$

**50.** $-49 - 3$

**51.** $24 - (-92)$

**52.** $48 - (-73)$

**53.** $-50 - (-50)$

**54.** $-70 - (-70)$

**55.** $-30 - (-85)$

**56.** $-25 - (-15)$

Simplify.

**57.** $7 - (-5) + 4 - (-3)$

**58.** $-5 - (-8) + 3 - (-7)$

**59.** $-31 + (-28) - (-14) - 17$

**60.** $-43 - (-19) - (-21) + 25$

**61.** $-34 - 28 + (-33) - 44$

**62.** $39 + (-88) - 29 - (-83)$

**63.** $-93 - (-84) - 41 - (-56)$

**64.** $84 + (-99) + 44 - (-18) - 43$

**65.** $-5 - (-30) + 30 + 40 - (-12)$

**66.** $14 - (-50) + 20 - (-32)$

**67.** $132 - (-21) + 45 - (-21)$

**68.** $81 - (-20) - 14 - (-50) + 53$

 Solve.

**69.** *Reading.*   Before falling asleep, Alicia read from the top of page 37 to the top of page 62 of her book. How many pages did she read?

**70.** *Writing.*   During a weekend retreat, James wrote from the bottom of page 29 to the bottom of page 37 of his memoirs. How many pages did he write?

**71.** Through exercise, Rod went from 8 lb above his "ideal" body weight to 9 lb below it. How many pounds did Rod lose?

**72.** Laura has a charge of $477 on her credit card, but she then returns a sweater that cost $129. How much does she now owe on her credit card?

**73.** *Temperature Changes.*   One day the temperature in Lawrence, Kansas, is 32° at 6:00 A.M. It rises 15° by noon, but falls 50° by midnight after a cold front moves in. What is the final temperature?

**74.** *Stock Price Changes.*   On a recent day, the price of a stock opened at a value of $61. It rose $5, dropped $7, and rose $4. Find the value of the stock at the end of the day.

**75.** *Profit.*   Treasure Tea lost $5000 in 2009. In 2010, the store made a profit of $8000. How much more did the store make in 2010 than in 2009?

**76.** *Tallest Mountain.*   The tallest mountain in the world, when measured from base to peak, is Mauna Kea (White Mountain) in Hawaii. From its base 19,684 ft below sea level in the Hawaiian Trough, it rises 33,480 ft. What is the elevation of the peak?

**Source:** *The Guinness Book of Records*

**77.** *Temperature Records.* The greatest recorded temperature change in one 24-hr period occured between January 23 and January 24, 1916, in Browning, Montana, where the temperature fell from 44°F to −56°F. Find the difference between these temperatures.

Source: *The Guinness Book of Records*

**78.** *Surface Temperature on Mercury.* Surface temperatures on Mercury vary from 840°F on the equator when the planet is closest to the sun to −290°F at night. Find the difference between these temperatures.

Source: Ian Ridpath, *Stors and Planets*, Princeton University Press, 1998

**79.** *Difference in Elevation.* At its highest point, the elevation of Denver, Colorado, is 5672 ft above sea level. At its lowest point, the elevation of New Orleans, Louisiana, is 4 ft below sea level. Find the difference between these elevations.

Source: *Information Please Almonoc*

**80.** *Golf.* As a result of coaching, Cedric's average golf score improved from 3 over par to 2 under. By how many strokes did his score change?

**81.** *Account Balance.* Leah has $460 in her checking account. She writes a check for $530, makes a deposit of $75, and then writes a check for $90. What is the balance in the account?

**82.** *Credit Card Bills.* On August 1, Lyle's credit card bill shows that he owes $470. During August, he sends a check to the credit card company for $45, charges another $160 in merchandise, and then pays off another $500 of his bill. What is the new balance of Lyle's account at the end of August (excluding interest for August)?

**83.** *Offshore Oil.* In 1998, the elevation of the world's deepwater drilling record was −7718 ft. In 2009, the deepwater drilling record was 2293 ft deeper. What was the elevation of the deepwater drilling record in 2009?

Source: www.deepwater.com.FactsandFirsts.cfm

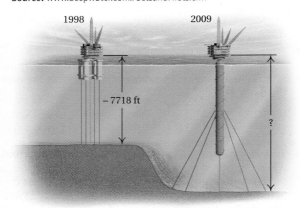

**84.** *Oceanography.* The deepest point in the Pacific Ocean is the Marianas Trench, with a depth of 11,033 m. The deepest point in the Atlantic Ocean is the Puerto Rico Trench, with a depth of 8648 m. What is the difference in the elevation of the two trenches?

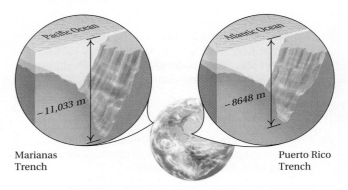

Marianas Trench

Puerto Rico Trench

**85.** *Toll Roads.* The E-Z Pass program allows drivers in the Northeast to travel certain toll roads without having to stop to pay. Instead, a transponder attached to the vehicle is scanned as the vehicle rolls through a toll booth. Recently the Ramones began a trip to New York City with a balance of $12 in their E-Z Pass account. They accumulated $15 in tolls on their trip, and because they overspent their balance, the Ramones had to pay $80 in fines and administrative fees. By how much were the Ramones in debt as a result of their travel on toll roads?

Source: State of New Jersey

**86.** *Toll Roads.* The Murrays began a trip with $13 in their E-Z Pass account (see Exercise 85). They accumulated $20 in tolls and had to pay $80 in fines and administrative fees. By how much were the Murrays in debt as a result of their travel on toll roads?

## Skill Maintenance

Evaluate.

**87.** $4^3$   [1.9b]

**88.** $68 \cdot 72$   [1.4a]

**89.** $1^7$   [1.9b]

**90.** $143 \cdot 29$   [1.4a]

**91.** How many 12-oz cans of soda can be filled with 96 oz of soda?   [1.8a]

**92.** A case of soda contains 24 bottles. If each bottle contains 12 oz, how many ounces of soda are in the case?   [1.8a]

Simplify.   [1.9c]

**93.** $5 + 4^2 + 2 \cdot 7$

**94.** $45 \div (2^2 + 11)$

**95.** $(9 + 7)(9 - 7)$

**96.** $(13 - 2)(13 + 2)$

## Synthesis

Subtract.

**97.** ▦ $123{,}907 - 433{,}789$

**98.** ▦ $23{,}011 - (-60{,}432)$

For Exercises 99–104, tell whether each statement is true or false for all integers $a$ and $b$. If false, give an example to show why.

**99.** $a - 0 = 0 - a$

**100.** $0 - a = a$

**101.** If $a \neq b$, then $a - b \neq 0$.

**102.** If $a = -b$, then $a + b = 0$.

**103.** If $a + b = 0$, then $a$ and $b$ are opposites.

**104.** If $a - b = 0$, then $a = -b$.

**105.** If $a - 54$ is $-37$, find the value of $a$.

**106.** If $x - 48$ is $-15$, find the value of $x$.

**107.** Maureen kept track of the weekly changes in the stock market over a period of 5 weeks. By how many points (pts) had the market risen or fallen over this time?

| WEEK 1 | WEEK 2 | WEEK 3 | WEEK 4 | WEEK 5 |
|--------|--------|--------|--------|--------|
| Down 13 pts | Down 16 pts | Up 36 pts | Down 11 pts | Up 19 pts |

**108.** *Blackjack Counting System.*   Players of the casino game of blackjack make use of many card-counting systems in which a negative count means that players have an advantage. One such system is called *High–Low*, first developed by Harvey Dubner in 1963. Each card counts as $-1, 0$, or $1$ as follows:

2, 3, 4, 5, 6     count as +1;

7, 8, 9     count as 0;

10, J, Q, K, A     count as $-1$.

**Source:** Patterson, Jerry L. *Casino Gambling.* New York: Perigee, 1982

**a)** Find the total count on the sequence of cards

K, A, 2, 4, 5, 10, J, 8, Q, K, 5.

**b)** Does the player have a winning edge?

# 2.4

## Multiplication of Integers

### OBJECTIVES

**a** Multiply integers.

**b** Find products of three or more integers and simplify powers of integers.

### a Multiplication

Multiplication of integers is like multiplication of whole numbers. The difference is that we must determine whether the answer is positive or negative.

#### Multiplication of a Positive Integer and a Negative Integer

To see how to multiply a positive integer and a negative integer, consider the following pattern.

This number decreases by 1 each time.

$$4 \cdot 5 = 20$$
$$3 \cdot 5 = 15$$
$$2 \cdot 5 = 10$$
$$1 \cdot 5 = 5$$
$$0 \cdot 5 = 0$$
$$-1 \cdot 5 = -5$$
$$-2 \cdot 5 = -10$$
$$-3 \cdot 5 = -15$$

This number decreases by 5 each time.

Do Exercise 1.

According to this pattern, it looks as though the product of a negative integer and a positive integer is negative. To confirm this, use repeated addition:

$$-1 \cdot 5 = 5 \cdot (-1) = -1 + (-1) + (-1) + (-1) + (-1) = -5$$
$$-2 \cdot 5 = 5 \cdot (-2) = -2 + (-2) + (-2) + (-2) + (-2) = -10$$
$$-3 \cdot 5 = 5 \cdot (-3) = -3 + (-3) + (-3) + (-3) + (-3) = -15$$

> **MULTIPLYING A POSITIVE INTEGER AND A NEGATIVE INTEGER**
>
> To multiply a positive integer and a negative integer, multiply their absolute values. The answer is negative.

**EXAMPLES**  Multiply.

**1.** $8(-5) = -40$   **2.** $50(-1) = -50$   **3.** $-7 \cdot 6 = -42$

Do Exercises 2–4.

#### Multiplication of Two Negative Integers

How do we multiply two negative integers? Again we look for a pattern.

This number decreases by 1 each time.

$$4 \cdot (-5) = -20$$
$$3 \cdot (-5) = -15$$
$$2 \cdot (-5) = -10$$
$$1 \cdot (-5) = -5$$
$$0 \cdot (-5) = 0$$
$$-1 \cdot (-5) = 5$$
$$-2 \cdot (-5) = 10$$
$$-3 \cdot (-5) = 15$$

This number increases by 5 each time.

Do Exercise 5.

**1.** Complete, as in the example.

$$4 \cdot 10 = 40$$
$$3 \cdot 10 = 30$$
$$2 \cdot 10 =$$
$$1 \cdot 10 =$$
$$0 \cdot 10 =$$
$$-1 \cdot 10 =$$
$$-2 \cdot 10 =$$
$$-3 \cdot 10 =$$

Multiply.
**2.** $-3 \cdot 6$    **3.** $20 \cdot (-5)$

**4.** $12(-1)$

**5.** Complete, as in the example.

$$3 \cdot (-10) = -30$$
$$2 \cdot (-10) = -20$$
$$1 \cdot (-10) =$$
$$0 \cdot (-10) =$$
$$-1 \cdot (-10) =$$
$$-2 \cdot (-10) =$$
$$-3 \cdot (-10) =$$

*Answers*

**1.** 20; 10; 0; −10; −20; −30   **2.** −18
**3.** −100   **4.** −12   **5.** −10; 0; 10; 20; 30

According to the pattern, the product of two negative integers is positive. This leads to the second part of the rule for multiplying integers.

> ### MULTIPLYING TWO NEGATIVE INTEGERS
>
> To multiply two negative integers, multiply their absolute values. The answer is positive.

**EXAMPLES** Multiply.

**4.** $(-2)(-4) = 8$

**5.** $(-10)(-7) = 70$

Do Exercises 6–8.

The following is another way to state the rules for multiplication.

> To multiply two integers:
> a) Multiply the absolute values.
> b) If the signs are the same, the answer is positive.
> c) If the signs are different, the answer is negative.

**EXAMPLES** Multiply.

**6.** $(-3)(-5) = 15$

**7.** $8(-10) = -80$

Do Exercises 9–12.

## Multiplication by Zero

No matter how many times 0 is added to itself, the answer is 0.

> The product of 0 and any integer is 0:
> $a \cdot 0 = 0.$

**EXAMPLES** Multiply.

**8.** $-19 \cdot 0 = 0$

**9.** $0(-7) = 0$

Do Exercises 13 and 14.

## b) Multiplication of More Than Two Integers

When multiplying more than two real numbers, we can choose order and grouping as we please, using the commutative and associative laws.

**EXAMPLES** Multiply.

**10. a)** $-8 \cdot 2(-3) = -16(-3)$    Multiplying the first two numbers

$= 48$    Multiplying the results

**b)** $-8 \cdot 2(-3) = 24 \cdot 2$    Multiplying the negative numbers

$= 48$    The result is the same as above.

Multiply.

**6.** $(-3)(-4)$

**7.** $-9(-5)$

**8.** $(-1)(-6)$

Multiply.

**9.** $3(-6)$

**10.** $(-6)(-5)$

**11.** $(-1)(-50)$

**12.** $(-8) \cdot 11$

Multiply.

**13.** $0(-5)$

**14.** $-23 \cdot 0$

*Answers*

**6.** 12   **7.** 45   **8.** 6   **9.** −18   **10.** 30
**11.** 50   **12.** −88   **13.** 0   **14.** 0

**11.** $7(-1)(-4)(-2) = (-7)8$     Multiplying the first two numbers and the last two numbers

$$= -56$$

**12.** $-5 \cdot (-2) \cdot (-3) \cdot (-6) = 10 \cdot 18$    Each pair of negative numbers gives a positive product.

$$= 180$$

**13.** $(-3)(-5)(-2)(-3)(-1) = 15 \cdot 6 \cdot (-1)$

$$= 90(-1) = -90$$

We can see the following pattern in the results of Examples 11–13.

> The product of an even number of negative integers is positive.
> The product of an odd number of negative integers is negative.

**Multiply.**

**15.** $5 \cdot (-3) \cdot 2$

**16.** $-2 \cdot (-5) \cdot (-4) \cdot (-3)$

**17.** $(-4)(-5)(-2)(-3)(-1)$

**18.** $(-1)(-1)(-2)(-3)(-1)(-1)$

Do Exercises 15–18.

### Powers of Integers

A positive number raised to any power is positive. When a negative number is raised to a power, the sign of the result depends upon whether the exponent is even or odd.

**EXAMPLES** Simplify.

**14.** $(-7)^2 = (-7)(-7) = 49$    The result is positive.

**15.** $(-4)^3 = (-4)(-4)(-4)$

$$= 16(-4)$$

$$= -64$$    The result is negative.

**16.** $(-3)^4 = (-3)(-3)(-3)(-3)$

$$= 9 \cdot 9$$

$$= 81$$    The result is positive.

**17.** $(-2)^5 = (-2)(-2)(-2)(-2)(-2)$

$$= 4 \cdot 4 \cdot (-2)$$

$$= 16(-2)$$

$$= -32$$    The result is negative.

Perhaps you noted the following.

> When a negative number is raised to an even exponent, the result is positive.
> When a negative number is raised to an odd exponent, the result is negative.

**Simplify.**

**19.** $(-2)^3$      **20.** $(-9)^2$

**21.** $(-1)^9$      **22.** $2^5$

Do Exercises 19–22.

When an integer is multiplied by $-1$, the result is the opposite of that integer.

> For any integer $a$,
> $$-1 \cdot a = -a.$$

*Answers*

**15.** $-30$    **16.** $120$    **17.** $-120$    **18.** $6$
**19.** $-8$    **20.** $81$    **21.** $-1$    **22.** $32$

**EXAMPLE 18** Simplify: $-7^2$.

In the expression $-7^2$, the base is 7, not $-7$. We can regard $-7^2$ as $-1 \cdot 7^2$:

$$-7^2 = -1 \cdot 7^2$$
$$= -1 \cdot 7 \cdot 7 \quad \text{The rules for order of operations tell us to square first.}$$
$$= -1 \cdot 49$$
$$= -49.$$

Compare Examples 14 and 18 and note that $(-7)^2 \neq -7^2$. In fact, the expressions $(-7)^2$ and $-7^2$ are not even read the same way: $(-7)^2$ is read "negative seven squared," whereas $-7^2$ is read "the opposite of seven squared."

Do Exercises 23–25.

**23.** Simplify: $-5^2$.

**24.** Simplify: $(-5)^2$.

**25.** Write $(-8)^2$ and $-8^2$ in words.

## Calculator Corner

**Exponential Notation**   When using a calculator to calculate expressions like $(-39)^4$ or $-39^4$, it is important to use the correct sequence of keystrokes.

**Calculators with** $\boxed{+/-}$ **key:**   On some calculators, a $\boxed{+/-}$ key must be pressed after a number is entered to make the number negative. For these calculators, appropriate keystrokes for $(-39)^4$ are

$$\boxed{3}\ \boxed{9}\ \boxed{+/-}\ \boxed{x^y}\ \boxed{4}\ \boxed{=}.$$

To calculate $-39^4$, we must first raise 39 to the power 4. Then the sign of the result must be changed. This can be done with the keystrokes

$$\boxed{3}\ \boxed{9}\ \boxed{x^y}\ \boxed{4}\ \boxed{=}\ \boxed{+/-}$$

or by multiplying $39^4$ by $-1$:

$$\boxed{3}\ \boxed{9}\ \boxed{x^y}\ \boxed{4}\ \boxed{=}\ \boxed{\times}\ \boxed{1}\ \boxed{+/-}\ \boxed{=}.$$

**Calculators with** $\boxed{(-)}$ **key:**   On some calculators, the $\boxed{(-)}$ key is pressed before a number to indicate that the number is negative. This is similar to the way the expression is written on paper. For these calculators, $(-39)^4$ is found by pressing

$$\boxed{(}\ \boxed{(-)}\ \boxed{3}\ \boxed{9}\ \boxed{)}\ \boxed{\wedge}\ \boxed{4}\ \boxed{ENTER =}$$

and $-39^4$ is found by pressing

$$\boxed{(-)}\ \boxed{3}\ \boxed{9}\ \boxed{\wedge}\ \boxed{4}\ \boxed{ENTER =}.$$

You can either experiment or consult a user's manual if you are unsure of the proper keystrokes for your calculator.

**Exercises:**   Use a calculator to determine each of the following.

**1.** $(-23)^6$

**2.** $(-17)^5$

**3.** $(-104)^3$

**4.** $(-4)^{10}$

**5.** $-9^6$

**6.** $-7^6$

**7.** $-6^5$

**8.** $-3^9$

*Answers*

**23.** $-25$   **24.** 25   **25.** Negative eight squared; the opposite of eight squared

**a** Multiply.

**1.** $-2 \cdot 8$

**2.** $-7 \cdot 3$

**3.** $10 \cdot (-6)$

**4.** $12 \cdot (-2)$

**5.** $8 \cdot (-6)$

**6.** $8 \cdot (-3)$

**7.** $-10 \cdot 3$

**8.** $-9 \cdot 8$

**9.** $-3(-5)$

**10.** $-8 \cdot (-2)$

**11.** $-9 \cdot (-2)$

**12.** $(-8)(-9)$

**13.** $(-6)(-7)$

**14.** $-8 \cdot (-3)$

**15.** $-10(-2)$

**16.** $-9(-8)$

**17.** $12(-10)$

**18.** $15(-8)$

**19.** $-23 \cdot 0$

**20.** $-38 \cdot 0$

**21.** $(-72)(-1)$

**22.** $41(-3)$

**23.** $(-20)17$

**24.** $(-1)(-43)$

**25.** $-8(-50)$

**26.** $(-25) \cdot 8$

**27.** $0(-14)$

**28.** $0(-38)$

**b** Multiply.

**29.** $3 \cdot (-8) \cdot (-1)$

**30.** $(-7) \cdot (-4) \cdot (-1)$

**31.** $7(-4)(-3)5$

**32.** $9(-2)(-6)7$

**33.** $-2(-5)(-7)$

**34.** $(-2)(-5)(-3)(-5)$

**35.** $(-5)(-2)(-3)(-1)$

**36.** $-6(-5)(-9)$

**37.** $(-15)(-29) \cdot 0 \cdot 8$

**38.** $19(-7)(-8) \cdot 0 \cdot 6$

**39.** $(-7)(-1)(7)(-6)$

**40.** $(-5)6(-4)5$

Simplify.

**41.** $(-6)^2$      **42.** $(-8)^2$      **43.** $(-5)^3$      **44.** $(-2)^4$

**45.** $(-10)^4$      **46.** $(-1)^5$      **47.** $-2^4$      **48.** $(-2)^6$

**49.** $(-3)^5$      **50.** $-10^4$      **51.** $(-1)^{12}$      **52.** $(-1)^{13}$

**53.** $-11^2$      **54.** $-2^6$      **55.** $-4^3$      **56.** $-2^5$

Write each of the following expressions in words.

**57.** $-8^4$      **58.** $(-6)^8$      **59.** $(-9)^{10}$      **60.** $-5^4$

## Skill Maintenance

**61.** Round 532,451 to the nearest hundred.   [1.6a]

**62.** Write standard notation for sixty million.   [1.1c]

**63.** Divide: $2880 \div 36$.   [1.5a]

**64.** Multiply: $75 \times 34$.   [1.4a]

**65.** Simplify: $10 - 2^3 + 6 \div 2$.   [1.9c]

**66.** Simplify: $2 \cdot 5^2 - 3 \cdot 2^3 \div (3 + 3^2)$.   [1.9c]

**67.** A rectangular rug measures 5 ft by 8 ft. What is the area of the rug?   [1.4b], [1.8a]

**68.** How many 12-egg cartons can be filled with 2880 eggs? [1.8a]

**69.** A ferry can accommodate 12 cars and 53 cars are waiting. How many trips will be required to ferry them all?   [1.8a]

**70.** An elevator can hold 16 people and 50 people are waiting to go up. How many trips will be required to transport all of them?   [1.8a]

## Synthesis

Simplify.

**71.** $(-3)^5(-1)^{379}$      **72.** $(-2)^3 \cdot [(-1)^{29}]^{46}$      **73.** $-9^4 + (-9)^4$      **74.** $-5^2(-1)^{29}$

**75.** $|(-2)^5 + 3^2| - (3 - 7)^2$      **76.** $|-12(-3)^2 - 5^3 - 6^2 - (-5)^2|$

**77.** ▦ $-47^2$      **78.** ▦ $-53^2$      **79.** ▦ $(-19)^4$      **80.** ▦ $(-23)^4$

**81.** ▦ $(73 - 86)^3$      **82.** ▦ $(-49 + 34)^3$      **83.** ▦ $-935(238 - 243)^3$      **84.** ▦ $(-17)^4(129 - 133)^5$

**85.** Jo wrote seven checks for $13 each. If she had a balance of $68 in her account, what was her balance after writing the checks?

**86.** After diving 95 m below the surface, a diver rises at a rate of 7 meters per minute for 9 min. What is the diver's new elevation?

**87.** What must be true of $m$ and $n$ if $[(-5)^m]^n$ is to be **(a)** negative? **(b)** positive?

**88.** What must be true of $m$ and $n$ if $-mn$ is to be **(a)** positive? **(b)** zero? **(c)** negative?

# 2.5

## Division of Integers and Order of Operations

### OBJECTIVES

**a** Divide integers.

**b** Use the rules for order of operations with integers.

**SKILL TO REVIEW**
Objective 1.9c: Simplify expressions using the rules for order of operations.

Simplify.

**1.** $5^2 - 2(10 - 3)$

**2.** $2[21 - (11 - 3)]$

We now consider division of integers. Because of the way in which division is defined, its rules are similar to those for multiplication.

### a Division of Integers

> **THE QUOTIENT**
>
> The quotient $\dfrac{a}{b}$ (or $a \div b$, or $a/b$) is the number, if there is one, that when multiplied by $b$ gives $a$.

Let's use the definition to divide integers.

**EXAMPLES** Divide, if possible. Check each answer.

**1.** $14 \div (-7) = -2$     *Think:* What number multiplied by $-7$ gives 14? The number is $-2$. *Check:* $(-2)(-7) = 14$.

**2.** $\dfrac{-32}{-4} = 8$     *Think:* What number multiplied by $-4$ gives $-32$? The number is 8. *Check:* $8(-4) = -32$.

**3.** $-21 \div 7 = -3$     *Think:* What number multiplied by 7 gives $-21$? The number is $-3$. *Check:* $(-3) \cdot 7 = -21$.

**4.** $\dfrac{0}{-5} = 0$     *Think:* What number multiplied by $-5$ gives 0? The number is 0. *Check:* $0(-5) = 0$.

**5.** $\dfrac{-5}{0}$ is **not defined**.     *Think:* What number multiplied by 0 gives $-5$? There is no such number because the product of 0 and *any* number is 0.

The rules for determining the sign of a quotient are the same as those for determining the sign of a product. We state them together.

> To multiply or divide two integers:
>
> **a)** Multiply or divide the absolute values.
>
> **b)** If the signs are the same, the answer is positive.
>
> **c)** If the signs are different, the answer is negative.

Divide.

**1.** $6 \div (-2)$
*Think:* What number multiplied by $-2$ gives 6?

**2.** $\dfrac{-15}{-3}$
*Think:* What number multiplied by $-3$ gives $-15$?

**3.** $-24 \div 8$
*Think:* What number multiplied by 8 gives $-24$?

**4.** $\dfrac{0}{-4}$       **5.** $\dfrac{30}{-5}$

**6.** $\dfrac{-45}{9}$

Do Margin Exercises 1–6.

### Dividing by 0

Recall that, in general, $a \div b$ and $b \div a$ are different numbers. In Example 4, we divided *into* 0. In Example 5, we attempted to divide *by* 0. Since any number times 0 gives 0, not $-5$, we say that $-5 \div 0$ is **not defined** or is **undefined**. Also, since *any* number times 0 gives 0, $0 \div 0$ is also not defined.

*Answers*

*Skill to Review:*
**1.** 11    **2.** 26

*Margin Exercises:*
**1.** $-3$    **2.** 5    **3.** $-3$    **4.** 0
**5.** $-6$    **6.** $-5$

## EXCLUDING DIVISION BY 0

Division by 0 is not defined:

$a \div 0$, or $\dfrac{a}{0}$, is undefined for all real numbers $a$.

### Dividing 0 by Other Numbers

Note that $0 \div 8 = 0$ because $0 = 0 \cdot 8$.

## DIVIDENDS OF 0

Zero divided by any nonzero real number is 0:

$\dfrac{0}{a} = 0, \quad a \neq 0.$

**EXAMPLES** Divide.

**6.** $0 \div (-6) = 0$      **7.** $\dfrac{0}{12} = 0$      **8.** $\dfrac{-3}{0}$ is undefined.

Do Exercises 7–9.

Divide, if possible.

**7.** $34 \div 0$

**8.** $0 \div (-4)$

**9.** $-52 \div 0$

## (b) Rules for Order of Operations

When more than one operation appears in a calculation or problem, we apply the rules that were first used in Section 1.9. We repeat them here for review, now including absolute-value symbols.

### RULES FOR ORDER OF OPERATIONS

1. Do all calculations within parentheses, brackets, braces, absolute-value symbols, numerators, or denominators.
2. Evaluate all exponential expressions.
3. Do all multiplications and divisions in order from left to right.
4. Do all additions and subtractions in order from left to right.

**EXAMPLE 9** Simplify: $17 - 10 \div 2 \cdot 4$.

With no grouping symbols or exponents, we begin with the third rule.

$$
\begin{aligned}
17 - 10 \div 2 \cdot 4 &= 17 - 5 \cdot 4 \\
&= 17 - 20 \\
&= -3
\end{aligned}
$$

Carrying out all multiplications and divisions in order from left to right

*Answers*

**7.** Undefined    **8.** 0    **9.** Undefined

**EXAMPLES** Simplify.

**10.** $|(-2)^3 \div 4| - 5(-2)$

We first simplify within the absolute-value symbols.

$$|(-2)^3 \div 4| - 5(-2) = |-8 \div 4| - 5(-2) \qquad (-2)(-2)(-2) = -8$$
$$= |-2| - 5(-2) \qquad \text{Dividing}$$
$$= 2 - 5(-2) \qquad |-2| = 2$$
$$= 2 - (-10) \qquad \text{Multiplying}$$
$$= 12 \qquad \text{Subtracting;}$$
$$2 - (-10) = 2 + 10$$

**11.** $2^4 + 51 \cdot 4 - (37 + 23 \cdot 2)$

$2^4 + 51 \cdot 4 - (37 + 23 \cdot 2)$

$= 2^4 + 51 \cdot 4 - (37 + 46)$     Carrying out all operations inside parentheses first, following the rules for order of operations within the parentheses

$= 2^4 + 51 \cdot 4 - 83$     Adding inside parentheses

$= 16 + 51 \cdot 4 - 83$     Evaluating exponential expressions

$= 16 + 204 - 83$     Doing all multiplications

$= 220 - 83$     Doing all additions and subtractions in order from left to right

$= 137$

Always regard a fraction bar as a grouping symbol. It separates any calculations in the numerator from those in the denominator.

Simplify.

**10.** $5 - (-7)(-3)^2$

**11.** $(-2) \cdot |3 - 2^2| + 5$

**12.** $52 \cdot 5 + 5^3 - (4^2 - 48 \div 4)$

**13.** $\dfrac{(-5)(-9)}{1 - 2 \cdot 2}$

**EXAMPLE 12** Simplify: $\dfrac{5 - (-3)^2}{-2}$.

$$\left. \begin{array}{l} \dfrac{5 - (-3)^2}{-2} = \dfrac{5 - 9}{-2} \\[2mm] = \dfrac{-4}{-2} \end{array} \right\} \quad \begin{array}{l} \text{Calculating within the numerator:} \\ (-3)^2 = (-3)(-3) = 9 \text{ and } 5 - 9 = -4 \end{array}$$

$$= 2 \qquad \text{Dividing}$$

Do Exercises 10-13.

---

**Calculator Corner**

**Grouping Symbols**    On calculators, grouping symbols may appear as ⎡(⎤ and ⎡)⎤ or ⎡[(...⎤ and ⎡...)]⎤. We often need grouping symbols when simplifying expressions written in fraction form. For example, the fraction bar in the expression

$$\frac{38 + 142}{2 - 47}$$

acts as a grouping symbol. To simplify, we add parentheses around the numerator and around the denominator, pressing ⎡(⎤ ⎡3⎤ ⎡8⎤ ⎡+⎤ ⎡1⎤ ⎡4⎤ ⎡2⎤ ⎡)⎤ ⎡÷⎤ ⎡(⎤ ⎡2⎤ ⎡−⎤ ⎡4⎤ ⎡7⎤ ⎡)⎤ ⎡=⎤. The result is −4. Without the grouping symbols, we would be simplifying a different expression:

$$38 + \frac{142}{2} - 47.$$

**Exercises:**    Use a calculator with grouping symbols to simplify each of the following.

**1.** $\dfrac{38 - 178}{5 + 30}$        **2.** $\dfrac{311 - 17^2}{2 - 13}$        **3.** $785 - \dfrac{285 - 5^4}{17 + 3 \cdot 51}$

---

*Answers*

**10.** 68    **11.** 3    **12.** 381    **13.** −15

**a** Divide, if possible, and check each answer by multiplying. If an answer is undefined, state so.

**1.** $36 \div (-6)$

**2.** $\dfrac{42}{-7}$

**3.** $\dfrac{26}{-2}$

**4.** $24 \div (-12)$

**5.** $\dfrac{-16}{8}$

**6.** $-22 \div (-2)$

**7.** $\dfrac{-48}{-12}$

**8.** $-72 \div (-9)$

**9.** $\dfrac{-72}{8}$

**10.** $\dfrac{-50}{25}$

**11.** $-100 \div (-50)$

**12.** $\dfrac{-400}{8}$

**13.** $-108 \div 9$

**14.** $\dfrac{-128}{8}$

**15.** $\dfrac{200}{-25}$

**16.** $-651 \div (-31)$

**17.** $\dfrac{-56}{0}$

**18.** $\dfrac{0}{-5}$

**19.** $\dfrac{88}{-11}$

**20.** $\dfrac{-145}{-5}$

**21.** $-\dfrac{276}{12}$

**22.** $-\dfrac{217}{7}$

**23.** $\dfrac{0}{-2}$

**24.** $\dfrac{-13}{0}$

**25.** $\dfrac{19}{-1}$

**26.** $\dfrac{-17}{1}$

**27.** $-41 \div 1$

**28.** $23 \div (-1)$

**b** Simplify, if possible. If an answer is undefined, state so.

**29.** $8 - 2 \cdot 3 - 9$

**30.** $8 - (2 \cdot 3 - 9)$

**31.** $(8 - 2 \cdot 3) - 9$

**32.** $(8 - 2)(3 - 9)$

**33.** $16 \cdot (-24) + 50$

**34.** $10 \cdot 20 - 15 \cdot 24$

**35.** $40 - 3^2 - 2^3$

**36.** $2^4 + 2^2 - 10$

**37.** $4 \cdot (6 + 8)/(4 + 3)$

**38.** $4^3 + 10 \cdot 20 + 8^2 - 23$

**39.** $4 \cdot 5 - 2 \cdot 6 + 4$

**40.** $5^3 + 4 \cdot 9 - (8 + 9 \cdot 3)$

**41.** $1 - (-2)^2 \cdot 3 \div 6$

**42.** $-6 + (-3)^2 + 6 \div (-2)$

**43.** $18 - (-3)^3 - 3^2 \cdot 5$

**44.** $9 - (-2)^3 - 50 \div 2$

**45.** $\dfrac{9^2 - 1}{1 - 3^2}$

**46.** $\dfrac{100 - 6^2}{(-5)^2 - 3^2}$

**47.** $8(-7) + 6(-5)$

**48.** $10(-5) \div 1(-1)$

**49.** $20 \div 5(-3) + 3$

**50.** $14 \div 2(-6) + 7$

**51.** $18 - 0(3^2 - 5^2 \cdot 7 - 4)$

**52.** $9 \cdot 0 \div 5 \cdot 4$

**53.** $-4^2 + 6$

**54.** $-5^2 + 7$

**55.** $-8^2 - 3$

**56.** $-9^2 - 11$

**57.** $4 \cdot 5^2 \div 10$

**58.** $(2 - 5)^2 \div (-9)$

**59.** $(3 - 8)^2 \div (-1)$

**60.** $3 - 3^2$

**61.** $12 - 20^3$

**62.** $20 + 4^3 \div (-8)$

**63.** $2 \times 10^3 - 5000$

**64.** $-7(3^4) + 18$

**65.** $6[9 - (3 - 4)]$

**66.** $8[(6 - 13) - 11]$

**67.** $-1000 \div (-100) \div 10$

**68.** $256 + (-32) \div (-4)$

**69.** $-7 - 3[-80 \div (2 - 10)]$

**70.** $-1 - 5[3 - (7 - 4^2)]$

**71.** $-2[3 - (7 - 9)^3]$

**72.** $-10[(2 - 8)^2 - 6]$

**73.** $8 - |7 - 9| \cdot 3$

**74.** $|8 - 7 - 9| \cdot 2 + 1$

**75.** $9 - |7 - 3^2|$

**76.** $9 - |5 - 7|^3$

**77.** $\dfrac{6^3 - 7 \cdot 3^4 - 2^5 \cdot 9}{(1 - 2^3)^3 + 7^3}$

**78.** $\dfrac{6 \div 2 \cdot 4^2 - 7^2 + 1}{(7 - 4)^3 - 2 \cdot 5 - 4}$

**79.** $\dfrac{2 \cdot 3^2 \div (3^2 - (2 + 1))}{5^2 - 6^2 - 2^2(-3)}$

**80.** $\dfrac{5 \cdot 6^2 \div (2^2 \cdot 5) - 7^2}{3^2 - 4^2 - (-2)^3 - 2}$

**81.** $\dfrac{(-5)^3 + 17}{10(2 - 6) - 2(5 + 2)}$

**82.** $\dfrac{(3 - 5)^2 - (7 - 13)}{(2 - 5)3 + 2 \cdot 4}$

**83.** $\dfrac{2 \cdot 4^3 - 4 \cdot 32}{19^3 - 17^4}$

**84.** $\dfrac{-16 \cdot 28 \div 2^2}{5 \cdot 25 - 5^3}$

## Skill Maintenance

**85.** Fabrikant Fine Diamonds ran a 4-in. by 7-in. advertisement in the *New York Times*. Find the area of the ad.   [1.4b], [1.8a]

**86.** A hotel has 4 floors with 62 rooms on each floor. How many rooms are there in the hotel?   [1.8a]

**87.** Cindi's Ford Focus gets 32 mpg (miles per gallon). How many gallons will it take to travel 384 mi?   [1.8a]

**88.** Craig's Chevy Blazer gets 14 mpg. How many gallons will it take to travel 378 mi?   [1.8a]

**89.** A 7-oz bag of tortilla chips contains 1050 calories. How many calories are in a 1-oz serving?   [1.8a]

**90.** A 7-oz bag of tortilla chips contains 8 g (grams) of fat per ounce. How many grams of fat are in a carton containing 12 bags of chips?   [1.8a]

**91.** There are 18 sticks in a large pack of Trident gum. If 4 people share the pack equally, how many whole pieces will each person receive? How many extra pieces will remain?   [1.8a]

**92.** A bag of Ricola throat lozenges contains 24 cough drops. If 5 people share the bag equally, how many lozenges will each person receive? How many extra lozenges will remain?   [1.8a]

## Synthesis

Simplify, if possible.

**93.** $\dfrac{9 - 3^2}{2 \cdot 4^2 - 5^2 \cdot 9 + 8^2 \cdot 7}$

**94.** $\dfrac{7^3 \cdot 9 - 6^2 \cdot 8 + 4^3 \cdot 6}{5^2 - 25}$

**95.** $\dfrac{(25 - 4^2)^3}{17^2 - 16^2} \cdot ((-6)^2 - 6^2)$

**96.** $\dfrac{(7 - 8)^{37}}{7^2 - 8^2} \cdot (98 - 7^2 \cdot 2)$

**97.** $\dfrac{19 - 17^2}{13^2 - 34}$

**98.** $\dfrac{195 + (-15)^3}{195 - 7 \cdot 5^2}$

**99.** $28^2 - 36^2/4^2 + 17^2$

**100.** $9^3 - 36^3/12^2 + 9^2$

**101.** Write down the keystrokes needed to calculate $\dfrac{15^2 - 5^3}{3^2 + 4^2}$.

**102.** Write down the keystrokes needed to calculate $\dfrac{16^2 - 24 \cdot 23}{3 \cdot 4 + 5^2}$.

**103.** Evaluate the expression for which the keystrokes are as follows: $\boxed{4}\ \boxed{-}\ \boxed{1}\ \boxed{0}\ \boxed{\div}\ \boxed{2}\ \boxed{+}\ \boxed{6}$.

**104.** Evaluate the expression for which the keystrokes are $\boxed{4}\ \boxed{-}\ \boxed{1}\ \boxed{6}\ \boxed{\div}\ \boxed{(}\ \boxed{2}\ \boxed{+}\ \boxed{6}\ \boxed{)}$.

Determine the sign of each expression if $m$ is negative and $n$ is positive.

**105.** $\dfrac{-n}{m}$

**106.** $\dfrac{-n}{-m}$

**107.** $-\left(\dfrac{-n}{m}\right)$

**108.** $-\left(\dfrac{n}{-m}\right)$

**109.** $-\left(\dfrac{-n}{-m}\right)$

# Mid-Chapter Review

## Concept Reinforcement

Determine whether each statement is true or false.

_____ 1. Every integer is either positive or negative.   [2.1a]

_____ 2. If $a > b$, then $a$ lies to the left of $b$ on the number line.   [2.1b]

_____ 3. The absolute value of a number is always nonnegative.   [2.1c]

## Guided Solutions

Fill in each blank with the number that creates a correct statement or solution.

4. Evaluate $-x$ and $-(-x)$ when $x = -4$.   [2.1d]

$$-x = -(\square) = \square;$$
$$-(-x) = -(-(\square)) = -(\square) = \square$$

Subtract.   [2.3a]

5. $5 - 13 = 5 + (\square) = \square$

6. $-6 - (-7) = -6 + \square = \square$

## Mixed Review

7. State the integers that correspond to this situation.

Jerilyn deposited $450 in her checking account. Later that week she wrote a check for $79.   [2.1a]

8. Change the sign of 9.   [2.1d]

Use either $<$ or $>$ for $\square$ to write a true sentence.   [2.1b]

9. $-6 \; \square \; 6$

10. $-5 \; \square \; -3$

11. $-10 \; \square \; 0$

12. $-20 \; \square \; -30$

Find the absolute value.   [2.1c]

13. $|38|$

14. $|-18|$

15. $|0|$

16. $|-12|$

Find the opposite, or additive inverse, of the number.   [2.1d]

17. $-56$

18. $3$

19. $0$

20. $-49$

21. Find $-x$ when $x$ is $-19$.   [2.1d]

22. Evaluate $-(-x)$ when $x$ is 23.   [2.1d]

Compute and simplify. [2.2a], [2.3a], [2.4a], [2.5a], [2.5b]

**23.** $7 + (-9)$

**24.** $-6 + (-10)$

**25.** $36 + (-36)$

**26.** $-8 + (-9)$

**27.** $-9 + 10$

**28.** $19 + (-17)$

**29.** $2 - 28$

**30.** $-8 - (-4)$

**31.** $-3 - 10$

**32.** $5 - (-11)$

**33.** $0 - (-6)$

**34.** $12 - 24$

**35.** $-12 \cdot 3$

**36.** $6(-9)$

**37.** $(-13)(-2)$

**38.** $(-2)(-41)$

**39.** $(-9)^2$

**40.** $-9^2$

**41.** $-75 \div (-3)$

**42.** $-20 \div 4$

**43.** $17 - (-25) + 15 - (-18)$

**44.** $-9 + (-3) + 16 - (-10)$

**45.** $(-7)(-2)(-1)(-3)$

**46.** $3 - 6 \cdot 5 - 11$

**47.** $-5^2 + 6[1 - (3 - 4)]$

**48.** $\dfrac{6^2 - 3(5 - 9)}{7^2 - (-5)^2}$

Solve. [2.3b]

**49.** *Temperature Change.* In a chemistry lab, Ben works with a substance whose initial temperature is 25°C. During an experiment, the temperature falls to −8°C. Find the difference between the two temperatures.

**50.** *Stock Price Change.* The price of a stock opened at $56. During the day, it dropped $6, then rose $2, and dropped $8. Find the value of the stock at the end of the day.

# Understanding Through Discussion and Writing

**51.** A student states "−45 is bigger than −21." What mistake do you think the student is making? [2.1b]

**52.** Is subtraction of integers associative? Why or why not? [2.3a]

**53.** Explain in your own words why the sum of two negative numbers is always negative. [2.2a]

**54.** If a negative number is subtracted from a positive number, will the result always be positive? Why or why not? [2.3a]

# 2.6

# Introduction to Algebra and Expressions

## OBJECTIVES

**a** Evaluate an algebraic expression by substitution.

**b** Use the distributive law to find equivalent expressions.

## STUDY TIPS

### READING A MATH TEXT

A math text does not read like a magazine or novel. On one hand, most assigned readings in a math text consist of only a few pages. On the other hand, every sentence and word is important and should make sense. It is also common to read a math text with a pencil in hand and to work problems while reading.

In this section, we will write *equivalent expressions* by making use of the *distributive law*.

## a  Algebraic Expressions

In arithmetic, we work with expressions such as

$$37 + 86, \qquad 7 \cdot 8, \qquad 19 - 7, \quad \text{and} \quad \frac{3}{8}.$$

In algebra, we use both numbers and letters and work with *algebraic expressions* such as

$$x + 86, \qquad 7 \cdot t, \qquad 19 - y, \quad \text{and} \quad \frac{a}{b}.$$

When a letter can stand for various numbers, we call the letter a **variable**. A number or a letter that stands for just one number is called a **constant**. Let $c =$ the speed of light. Then $c$ is a constant. Let $a =$ the speed of a car. Then $a$ is a variable since the value of $a$ can vary.

An **algebraic expression** consists of variables, constants, numerals, and operation signs. When we replace a variable with a number, we say that we are **substituting** for the variable. Carrying out the **operations** of addition, subtraction, and so on, is called **evaluating the expression**.

**EXAMPLE 1** Evaluate $x + y$ for $x = 37$ and $y = 29$.

We substitute 37 for $x$ and 29 for $y$ and carry out the addition:

$$x + y = 37 + 29 = 66.$$

The number 66 is called the **value** of the expression.

Algebraic expressions involving multiplication, like "8 times $a$," can be written as $8 \times a$, $8 \cdot a$, $8(a)$, or simply $8a$. Two letters written together without an operation symbol, such as $ab$, also indicate multiplication.

**1.** Evaluate $a + b$ for $a = 38$ and $b = 26$.

**2.** Evaluate $x - y$ for $x = 57$ and $y = 29$.

**3.** Evaluate $5t$ for $t = -14$.

**EXAMPLE 2** Evaluate $3y$ for $y = -14$.

$$3y = 3(-14) = -42 \qquad \text{Parentheses are required here.}$$

Do Exercises 1–3.

Algebraic expressions involving division can also be written several ways. For example, "8 divided by $t$" can be written as $8 \div t$, $8/t$, or $\dfrac{8}{t}$.

**EXAMPLE 3** Evaluate $\dfrac{a}{b}$ and $\dfrac{-a}{-b}$ for $a = 35$ and $b = 7$.

We substitute 35 for $a$ and 7 for $b$:

$$\frac{a}{b} = \frac{35}{7} = 5; \qquad \frac{-a}{-b} = \frac{-35}{-7} = 5.$$

*Answers*

**1.** 64    **2.** 28    **3.** −70

**EXAMPLE 4** Evaluate $-\dfrac{a}{b}$, $\dfrac{-a}{b}$, and $\dfrac{a}{-b}$ for $a = 15$ and $b = 3$.

We substitute 15 for $a$ and 3 for $b$:

$$-\frac{a}{b} = -\frac{15}{3} = -5; \qquad \frac{-a}{b} = \frac{-15}{3} = -5; \qquad \frac{a}{-b} = \frac{15}{-3} = -5.$$

Examples 3 and 4 illustrate the following.

> $\dfrac{-a}{-b}$ and $\dfrac{a}{b}$ represent the same number.
>
> $-\dfrac{a}{b}$, $\dfrac{-a}{b}$, and $\dfrac{a}{-b}$ all represent the same number.

For each number, find two equivalent expressions with negative signs in different places.

**4.** $\dfrac{-6}{x}$      **5.** $-\dfrac{m}{n}$

**6.** $\dfrac{r}{-4}$

**7.** Evaluate $\dfrac{a}{-b}$, $\dfrac{-a}{b}$, and $-\dfrac{a}{b}$ for $a = 28$ and $b = 4$.

**EXAMPLE 5** Evaluate $\dfrac{9C}{5} + 32$ for $C = 20$.

This expression can be used to find the Fahrenheit temperature that corresponds to 20 degrees Celsius:

$$\frac{9C}{5} + 32 = \frac{9 \cdot 20}{5} + 32 = \frac{180}{5} + 32 = 36 + 32 = 68.$$

Do Exercise 8.

**8.** Find the Fahrenheit temperature that corresponds to 10 degrees Celsius (see Example 5).

**EXAMPLE 6** Evaluate $5x^2$ for $x = 3$ and $x = -3$.

The rules for order of operations specify that the replacement for $x$ be squared. That result is then multiplied by 5:

$$\left.\begin{array}{l} 5x^2 = 5(3)^2 = 5(9) = 45; \\ 5x^2 = 5(-3)^2 = 5(9) = 45. \end{array}\right\}$$

You can always use parentheses when substituting. They are usually necessary when substituting a negative number.

Example 6 illustrates that when opposites are raised to an even power, the results are the same.

Do Exercises 9 and 10.

**9.** Evaluate $3x^2$ for $x = 4$ and $x = -4$.

**10.** Evaluate $a^4$ for $a = 3$ and $a = -3$.

**EXAMPLE 7** Evaluate $(-x)^2$ and $-x^2$ for $x = 7$.

When we evaluate $(-x)^2$ for $x = 7$, we have

$$(-x)^2 = (-7)^2 = (-7)(-7) = 49. \qquad \text{Substitute 7 for } x. \text{ Then evaluate the power.}$$

To evaluate $-x^2$, we again substitute 7 for $x$. We must recall that taking the opposite of a number is the same as multiplying that number by $-1$.

$$-x^2 = -1 \cdot x^2 \qquad \text{The opposite of a number is the same as multiplying by } -1.$$

$$-7^2 = -1 \cdot 7^2 \qquad \text{Substituting 7 for } x$$

$$= -1 \cdot 49 = -49 \qquad \text{Using the rules for order of operations; calculating the power before multiplying}$$

Example 7 shows that $(-x)^2 \neq -x^2$.

Do Exercises 11–13.

**11.** Evaluate $(-x)^2$ and $-x^2$ for $x = 3$.

**12.** Evaluate $(-x)^2$ and $-x^2$ for $x = 2$.

**13.** Evaluate $x^5$ for $x = 2$ and $x = -2$.

***Answers***

**4.** $-\dfrac{6}{x}; \dfrac{6}{-x}$   **5.** $\dfrac{-m}{n}; \dfrac{m}{-n}$   **6.** $\dfrac{-r}{4}; -\dfrac{r}{4}$
**7.** $-7; -7; -7$   **8.** 50   **9.** 48; 48
**10.** 81; 81   **11.** 9; $-9$   **12.** 4; $-4$
**13.** 32; $-32$

2.6 Introduction to Algebra and Expressions    **121**

Complete each table by evaluating each expression for the given values.

**14.**

| | $3x + 2x$ | $5x$ |
|---|---|---|
| $x = 4$ | | |
| $x = -2$ | | |
| $x = 0$ | | |

**15.**

| | $4x - x$ | $3x$ |
|---|---|---|
| $x = 2$ | | |
| $x = -2$ | | |
| $x = 0$ | | |

---

## (b) Equivalent Expressions and the Distributive Law

Some pairs of algebraic expressions are *equivalent*.

**EXAMPLE 8**   Evaluate $x + x$ and $2x$ for $x = 3$ and $x = -5$.

We substitute 3 for $x$ in $x + x$ and again in $2x$:

$$x + x = 3 + 3 = 6; \qquad 2x = 2 \cdot 3 = 6.$$

Next we repeat the procedure, substituting $-5$ for $x$:

$$x + x = -5 + (-5) = -10; \qquad 2x = 2(-5) = -10.$$

The results can be shown in a table. It appears that $x + x$ and $2x$ represent the same number.

| | $x + x$ | $2x$ |
|---|---|---|
| $x = 3$ | 6 | 6 |
| $x = -5$ | $-10$ | $-10$ |

Do Exercises 14 and 15.

Example 8 suggests that $x + x$ and $2x$ represent the same number for any replacement of $x$. This is in fact true, so we can say that $x + x$ and $2x$ are **equivalent expressions**.

> **EQUIVALENT EXPRESSIONS**
>
> Two expressions that have the same value for all allowable replacements are called **equivalent**.

In Examples 3 and 7 we saw that the expressions $\dfrac{-a}{-b}$ and $\dfrac{a}{b}$ are equivalent but that the expressions $(-x)^2$ and $-x^2$ are *not* equivalent.

An important concept, known as the **distributive law**, is useful for finding equivalent algebraic expressions. The distributive law involves two operations: multiplication and either addition or subtraction.

To understand how the distributive law works, consider the following:

$$\begin{array}{r} 4\,5 \\ \times \quad 7 \\ \hline 3\,5 \\ 2\,8\,0 \\ \hline 3\,1\,5 \end{array}$$

$3\,5 \longleftarrow$ This is $7 \cdot 5$.
$2\,8\,0 \longleftarrow$ This is $7 \cdot 40$.
$3\,1\,5 \longleftarrow$ This is the sum $7 \cdot 40 + 7 \cdot 5$.

To carry out the multiplication, we actually added two products. That is,

$$7 \cdot 45 = 7(40 + 5) = 7 \cdot 40 + 7 \cdot 5.$$

The distributive law says that if we want to multiply a sum of several numbers by a number, we can either add within the grouping symbols and then multiply, or multiply each of the terms separately and then add.

> **THE DISTRIBUTIVE LAW**
>
> For any numbers $a$, $b$, and $c$,
>
> $$a(b + c) = ab + ac.$$

---

### Calculator Corner

**Evaluating Powers**

To evaluate an expression like $-x^3$ for $x = -14$ with a calculator, we must keep in mind the rules for order of operations. On some calculators, this expression is evaluated by pressing

$\boxed{1}\ \boxed{4}\ \boxed{+/-}\ \boxed{x^y}\ \boxed{3}\ \boxed{=}\ \boxed{+/-}$.

Other calculators use the keystrokes

$\boxed{(-)}\ \boxed{(}\ \boxed{(-)}\ \boxed{1}\ \boxed{4}\ \boxed{)}\ \boxed{\wedge}\ \boxed{3}$

$\boxed{\text{ENTER} =}$. The result should be 2744. Consult your owner's manual or an instructor, or simply experiment if your calculator behaves differently.

**Exercises:**   Evaluate.

**1.** $-a^5$ for $a = -3$

**2.** $-x^5$ for $x = -4$

**3.** $-x^5$ for $x = 2$

**4.** $-x^5$ for $x = 5$

---

*Answers*

**14.** $20, 20; -10, -10; 0, 0$
**15.** $6, 6; -6, -6; 0, 0$

**EXAMPLE 9** Evaluate $a(b + c)$ and $ab + ac$ for $a = 3$, $b = 4$, and $c = 2$.

We have

$$a(b + c) = 3(4 + 2) = 3 \cdot 6 = 18 \quad \text{and}$$
$$ab + ac = 3 \cdot 4 + 3 \cdot 2 = 12 + 6 = 18.$$

The parentheses in the statement of the distributive law tell us to multiply both $b$ and $c$ by $a$. Without the parentheses, we would have $ab + c$. To see that $a(b + c) \neq ab + c$, note that $3(4 + 2) = 18$, but $3 \cdot 4 + 2 = 14$.

**EXAMPLE 10** Use the distributive law to write an expression equivalent to $2(l + w)$.

$$2(l + w) = 2 \cdot l + 2 \cdot w \qquad \text{Note that the } + \text{ sign between } l \text{ and } w \text{ now appears between } 2 \cdot l \text{ and } 2 \cdot w.$$
$$= 2l + 2w. \qquad \text{Try to go directly to this step.}$$

Do Exercises 16 and 17.

Do Exercises 16 and 17.

Use the distributive law to write an equivalent expression.

**16.** $5(a + b)$

**17.** $6(x + y + z)$

Since subtraction can be regarded as addition of the opposite, it follows that the distributive law is true for subtraction as well as addition.

**EXAMPLE 11** Use the distributive law to write an expression equivalent to each of the following:

**a)** $7(a - b)$;      **b)** $9(x - 5)$;      **c)** $(a - 7)b$;

**d)** $-4(x - 2y + 3z)$;      **e)** $-(2x - 3y)$

**a)** $7(a - b) = 7 \cdot a - 7 \cdot b$
$$= 7a - 7b \qquad \text{Try to go directly to this step.}$$

**b)** $9(x - 5) = 9x - 9(5)$
$$= 9x - 45 \qquad \text{Again, try to go directly to this step.}$$

**c)** $(a - 7)b = b(a - 7) \qquad$ Using a commutative law
$$= b \cdot a - b \cdot 7 \qquad \text{Using the distributive law}$$
$$= ab - 7b \qquad \text{Using a commutative law to write } ba \text{ alphabetically and } b \cdot 7 \text{ with the constant first}$$

**d)** $-4(x - 2y + 3z) = -4 \cdot x - (-4)(2y) + (-4)(3z) \qquad$ Using the distributive law
$$= -4x - (-4 \cdot 2)y + (-4 \cdot 3)z \qquad \text{Using an associative law (twice)}$$
$$= -4x - (-8y) + (-12z)$$
$$= -4x + 8y - 12z$$

**e)** $-(2x - 3y) = -1(2x - 3y) \qquad$ The opposite of a number is the same as multiplying by $-1$.
$$= -1 \cdot 2x - (-1)(3y) \qquad \text{Using the distributive law}$$
$$= -2x - (-3y) \qquad \text{Using an associative law (twice)}$$
$$= -2x + 3y$$

Do Exercises 18–22.

Do Exercises 18–22.

Use the distributive law to write an equivalent expression.

**18.** $4(x - y)$

**19.** $3(a - b + c)$

**20.** $(m - 4)6$

**21.** $-8(2a - b + 3c)$

**22.** $-(c - 8d)$

*Answers*

**16.** $5a + 5b$    **17.** $6x + 6y + 6z$
**18.** $4x - 4y$    **19.** $3a - 3b + 3c$
**20.** $6m - 24$    **21.** $-16a + 8b - 24c$
**22.** $-c + 8d$

**a**    Evaluate.

**1.** $10n$, for $n = 2$
(The cost, in cents, of sending 2 text messages)

**2.** $99n$, for $n = 2$
(The cost, in cents, of downloading 2 songs)

**3.** $\dfrac{x}{y}$, for $x = 6$ and $y = -3$

**4.** $\dfrac{m}{n}$, for $m = 18$ and $n = 2$

**5.** $\dfrac{2d}{c}$, for $c = 6$ and $d = 3$

**6.** $\dfrac{5y}{z}$, for $y = 15$ and $z = -25$

**7.** $\dfrac{72}{r}$, for $r = 4$
(The approximate doubling time, in years, for an
investment earning 4% interest per year)

**8.** $\dfrac{72}{i}$, for $i = 2$
(The approximate doubling time, in years, for an
investment earning 2% interest per year)

**9.** $3 - 5 \cdot x$, for $x = 2$

**10.** $9 - 2 \cdot x$, for $x = 5$

**11.** $2l + 2w$, for $l = 3$ and $w = 4$
(The perimeter, in feet, of a 3-ft by 4-ft rectangle)

**12.** $3(a + b)$, for $a = 2$ and $b = 4$

**13.** $2(l + w)$, for $l = 3$ and $w = 4$
(The perimeter, in feet, of a 3-ft by 4-ft rectangle)

**14.** $3a + 3b$, for $a = 2$ and $b = 4$

**15.** $7a - 7b$, for $a = -1$ and $b = 2$

**16.** $4x - 4y$, for $x = -5$ and $y = 1$

**17.** $7(a - b)$, for $a = -1$ and $b = 2$

**18.** $4(x - y)$, for $x = -5$ and $y = 1$

**19.** $16t^2$, for $t = 5$
(The distance, in feet, that an object falls in 5 sec)

**20.** $\dfrac{49t^2}{10}$, for $t = 10$
(The distance, in meters, that an object falls in 10 sec)

**21.** $a + (b - a)^2$, for $a = 6$ and $b = 10$

**22.** $(x + y)^2$, for $x = 2$ and $y = 10$

**23.** $9a + 9b$, for $a = 13$ and $b = -13$

**24.** $8x + 8y$, for $x = 17$ and $y = -17$

**25.** $\dfrac{n^2 - n}{2}$, for $n = 9$
(For determining the number of handshakes possible among 9 people)

**26.** $\dfrac{5(F - 32)}{9}$, for $F = 50$
(For converting 50 degrees Fahrenheit to degrees Celsius)

**27.** $1 - x^2$, for $x = -2$

**28.** $4 - y^2$, for $y = -1$

**29.** $m^2 - n^2$, for $m = 6$ and $n = 5$

**30.** $a^2 + b^2$, for $a = 3$ and $b = 4$

**31.** $a^3 - a^2$, for $a = -10$

**32.** $x^2 - x^3$, for $x = -10$

For each expression, write two equivalent expressions with negative signs in different places.

**33.** $-\dfrac{5}{t}$

**34.** $\dfrac{7}{-x}$

**35.** $\dfrac{-n}{b}$

**36.** $-\dfrac{3}{r}$

**37.** $\dfrac{9}{-p}$

**38.** $\dfrac{-u}{5}$

**39.** $\dfrac{-14}{w}$

**40.** $\dfrac{-23}{m}$

Evaluate $\dfrac{-a}{b}$, $\dfrac{a}{-b}$, and $-\dfrac{a}{b}$ for the given values.

**41.** $a = 45, b = 9$

**42.** $a = 40, b = 2$

**43.** $a = 81, b = 3$

**44.** $a = 56, b = 7$

Evaluate.

**45.** $(-3x)^2$ and $-3x^2$, for $x = 2$

**46.** $(-2x)^2$ and $-2x^2$, for $x = 3$

**47.** $5x^2$, for $x = 3$ and $x = -3$

**48.** $2x^2$, for $x = 5$ and $x = -5$

**49.** $x^3$, for $x = 6$ and $x = -6$

**50.** $x^6$, for $x = 2$ and $x = -2$

**51.** $x^8$, for $x = 1$ and $x = -1$

**52.** $x^5$, for $x = 3$ and $x = -3$

**53.** $a^5$, for $a = 2$ and $a = -2$

**54.** $a^7$, for $a = 1$ and $a = -1$

**b**  Use the distributive law to write an equivalent expression.

**55.** $5(a + b)$

**56.** $7(x + y)$

**57.** $4(x + 1)$

**58.** $6(a + 1)$

**59.** $2(b + 5)$

**60.** $3(x - 6)$

**61.** $7(1 - t)$

**62.** $4(1 - y)$

**63.** $6(5x - 2)$

**64.** $9(6m - 7)$

**65.** $8(x + 7 + 6y)$

**66.** $4(5x + 8 + 3p)$

**67.** $-7(y - 2)$

**68.** $-9(y - 7)$

**69.** $(x + 2)3$

**70.** $(x + 4)2$

**71.** $-4(x - 3y - 2z)$

**72.** $8(2x - 5y - 8z)$

**73.** $8(a - 3b + c)$

**74.** $-6(a + 2b - c)$

**75.** $4(x - 3y - 7z)$

**76.** $5(9x - y + 8z)$

**77.** $(4a - 5b + c - 2d)5$

**78.** $(9a - 4b + 3c - d)7$

**79.** $-1(3m + 2n)$

**80.** $-1(6a + 7b)$

**81.** $-1(2a - 3b + 4)$

**82.** $-1(7x - 8y - 9)$

**83.** $-(x - y - z)$

**84.** $-(a - b - c)$

## Skill Maintenance

**85.** Write a word name for 23,043,921.   [1.1c]

**86.** Multiply: $17 \cdot 53$.   [1.4a]

**87.** Estimate by rounding to the nearest ten. Show your work.   [1.6b]

$$\begin{array}{r} 5\ 2\ 8\ 3 \\ -\ 2\ 4\ 7\ 5 \\ \hline \end{array}$$

**88.** Divide: $2982 \div 3$.   [1.5a]

**89.** On March 9, it snowed 12 in., but on March 10, the sun melted 7 in. How much snow remained?   [1.8a]

**90.** For Tania's graduation party, her husband ordered three buckets of chicken wings at $12 apiece and 3 trays of nachos at $9 a tray. How much did he pay for the wings and nachos?   [1.8a]

## Synthesis

**91.** A car's catalytic converter works most efficiently after it is heated to about 370°C. To what Fahrenheit temperature does this correspond? (*Hint*: See Example 5.)

**92.** Evaluate $\dfrac{9C}{5} + 32$ for $C = 10$ and for $C = 20$.

(See Example 5.) When the Celsius temperature is doubled, is the corresponding Fahrenheit temperature also doubled?

Evaluate.

**93.** $\blacksquare$ $a - b^3 + 17a$, for $a = 19$ and $b = -16$

**94.** $\blacksquare$ $x^2 - 23y + y^3$, for $x = 18$ and $y = -21$

**95.** $\blacksquare$ $r^3 + r^2t - rt^2$, for $r = -9$ and $t = 7$

**96.** $\blacksquare$ $a^3b - a^2b^2 + ab^3$, for $a = -8$ and $b = -6$

**97.** $a^{1996} - a^{1997}$, for $a = -1$

**98.** $x^{1492} - x^{1493}$, for $x = -1$

**99.** $(m^3 - mn)^m$, for $m = 4$ and $n = 6$

**100.** $5a^{3a-4}$, for $a = 2$

Replace the blanks with $\boxed{+}$, $\boxed{-}$, $\boxed{\times}$, or $\boxed{\div}$ to make each statement true.

**101.** $\blacksquare$ $-32 \ \square \ (88 \ \square \ 29) = -1888$

**102.** $\blacksquare$ $59 \ \square \ 17 \ \square \ 59 \ \square \ 8 = 1475$

Classify each statement as true or false. If false, write an example showing why.

**103.** For any choice of $x$, $x^2 = (-x)^2$.

**104.** For any choice of $x$, $x^3 = -x^3$.

**105.** For any choice of $x$, $x^6 + x^4 = (-x)^6 + (-x)^4$.

**106.** For any choice of $x$, $(-3x)^2 = 9x^2$.

## 2.7

# Like Terms and Perimeter

## OBJECTIVES

**a** Combine like terms.

**b** Determine the perimeter of a polygon.

One common way in which equivalent expressions are formed is by *combining like terms*.

## a Combining Like Terms

A **term** is a number, a variable, a product of numbers and/or variables, or a quotient of numbers and/or variables. Terms are separated by addition signs. If there are subtraction signs, we can find an equivalent expression that uses addition signs.

**EXAMPLE 1** What are the terms of $3xy - 4y + \dfrac{2}{z}$?

$$3xy - 4y + \frac{2}{z} = 3xy + (-4y) + \frac{2}{z} \qquad \text{Separating parts with } + \text{ signs}$$

The terms are $3xy$, $-4y$, and $\dfrac{2}{z}$.

What are the terms of each expression?

1. $5x - 4y + 3$

2. $-4y - 2x + \dfrac{x}{y}$

Do Exercises 1 and 2.

Terms in which the variable factors are exactly the same, such as $9x$ and $-4x$, are called **like**, or **similar**, **terms**. For example, $3y^2$ and $7y^2$ are like terms, whereas $5x$ and $6x^2$ are not. Constants, like 7 and 3, are also like terms.

**EXAMPLES** Identify the like terms.

**2.** $7x + 5x^2 + 2x + 8 + 5x^3 + 1$

$7x$ and $2x$ are like terms;  8 and 1 are like terms.

**3.** $5ab + a^3 - a^2b - 2ab + 7a^3$

$5ab$ and $-2ab$ are like terms;  $a^3$ and $7a^3$ are like terms.

Identify the like terms.

3. $9a^3 + 4ab + a^3 + 3ab + 7$

4. $3xy - 5x^2 + y^2 - 4xy + y$

Do Exercises 3 and 4.

When an algebraic expression contains like terms, an equivalent expression can be formed by **combining**, or **collecting**, **like terms**. To combine like terms, we use the distributive law.

**EXAMPLE 4** Combine like terms to form an equivalent expression.

**a)** $4x + 3x$            **b)** $6mn - 7mn$

**c)** $7y - 2 - 6y + 5$     **d)** $2a^5 + 9ab + 3 + a^5 - 7 - 4ab$

**a)** $4x + 3x = (4 + 3)x$     Using the distributive law (in "reverse")

           $= 7x$            We usually go directly to this step.

**b)** $6mn - 7mn = (6 - 7)mn$     Try to do this mentally.

           $= -1mn$, or simply $-mn$

**c)** $7y - 2 - 6y + 5 = 7y + (-2) + (-6y) + 5$     Rewriting as addition

                  $= 7y + (-6y) + (-2) + 5$     Using a commutative law

                  $= 1y + 3$, or simply $y + 3$

**Answers**

1. $5x$; $-4y$; 3     2. $-4y$; $-2x$; $\dfrac{x}{y}$

3. $9a^3$ and $a^3$; $4ab$ and $3ab$     4. $3xy$ and $-4xy$

**d)** $2a^5 + 9ab + 3 + a^5 - 7 - 4ab$

$= 2a^5 + 9ab + 3 + a^5 + (-7) + (-4ab)$

$= 2a^5 + a^5 + 9ab + (-4ab) + 3 + (-7)$  Rearranging terms

$= 3a^5 + 5ab + (-4)$  Think of $a^5$ as $1a^5$; $2a^5 + a^5 = 3a^5$

$= 3a^5 + 5ab - 4$

Do Exercises 5-7.

Combine like terms to form an equivalent expression.

**5.** $2a + 7a$

**6.** $5x^2 + 9 - 4x^2 + 3$

**7.** $4m - 2n^2 + 5 + n^2 + m - 9$

## b Perimeter

> ### PERIMETER OF A POLYGON
>
> A **polygon** is a closed geometric figure with three or more sides. The **perimeter** of a polygon is the distance around it, or the sum of the lengths of its sides.

**EXAMPLE 5**  Find the perimeter of this polygon.

A polygon with five sides is called a pentagon.

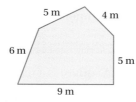

We add the lengths of all sides. Since all the units are the same, we are effectively combining like terms.

Perimeter $= 6\,m + 5\,m + 4\,m + 5\,m + 9\,m$

$= (6 + 5 + 4 + 5 + 9)\,m$  Using the distributive law

$= 29\,m$  Try to go directly to this step.

-------- *Caution!* --------

When units of measurement are given in the statement of a problem, as in Example 5, the solution should also contain units of measurement.

Do Exercises 8 and 9.

A **rectangle** is a polygon with four sides and four 90° angles. Opposite sides of a rectangle have the same measure. The symbol ⌐ or ⌐ indicates a 90° angle. A 90° angle is often referred to as a **right angle**.

**EXAMPLE 6**  Find the perimeter of a rectangle that is 3 cm by 4 cm.

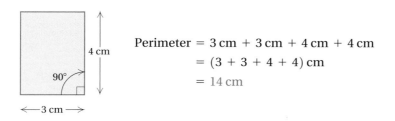

Perimeter $= 3\,cm + 3\,cm + 4\,cm + 4\,cm$

$= (3 + 3 + 4 + 4)\,cm$

$= 14\,cm$

Do Exercise 10.

Find the perimeter of each polygon.

**8.**

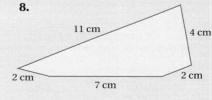

**9.**

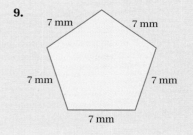

**10.** Find the perimeter of a rectangle that is 2 cm by 4 cm.

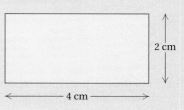

***Answers***

**5.** $9a$  **6.** $x^2 + 12$  **7.** $5m - n^2 - 4$

**8.** 26 cm  **9.** 35 mm  **10.** 12 cm

The perimeter of the rectangle in Example 6 is $2 \cdot 3 \text{ cm} + 2 \cdot 4 \text{ cm}$, or equivalently $2(3 \text{ cm} + 4 \text{ cm})$. This can be generalized, as follows.

## STUDY TIPS

### UNDERSTAND YOUR MISTAKES

When you receive a graded quiz, test, or assignment back from your instructor, it is important to review and understand what your mistakes were. View any mistake as another opportunity to learn, and purpose not to make the same error again.

---

**PERIMETER OF A RECTANGLE**

The **perimeter $P$ of a rectangle** of length $l$ and width $w$ is given by

$$P = 2l + 2w, \quad \text{or} \quad P = 2 \cdot (l + w).$$

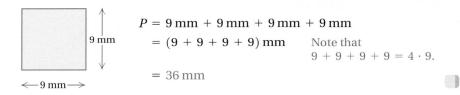

---

**EXAMPLE 7**   A common door size is 3 ft by 7 ft. Find the perimeter of such a door.

$$P = 2l + 2w \qquad \text{We could also use } P = 2(l + w).$$
$$= 2 \cdot 7 \text{ ft} + 2 \cdot 3 \text{ ft}$$
$$= (2 \cdot 7) \text{ ft} + (2 \cdot 3) \text{ ft} \qquad \text{Try to do this mentally.}$$
$$= 14 \text{ ft} + 6 \text{ ft}$$
$$= 20 \text{ ft} \qquad \text{Combining like terms}$$

The perimeter of the door is 20 ft.

**11.** Find the perimeter of a 4-ft by 8-ft sheet of plywood.

Do Exercise 11.

A **square** is a rectangle in which all sides have the same length.

**EXAMPLE 8**   Find the perimeter of a square with sides of length 9 mm.

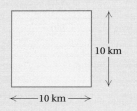

**12.** Find the perimeter of a square with sides of length 10 km.

$$P = 9 \text{ mm} + 9 \text{ mm} + 9 \text{ mm} + 9 \text{ mm}$$
$$= (9 + 9 + 9 + 9) \text{ mm} \qquad \text{Note that}$$
$$9 + 9 + 9 + 9 = 4 \cdot 9.$$
$$= 36 \text{ mm}$$

Do Exercise 12.

---

**PERIMETER OF A SQUARE**

The **perimeter $P$ of a square** is four times $s$, the length of a side:

$$P = s + s + s + s$$
$$= 4s.$$

---

**EXAMPLE 9**   Find the perimeter of a square garden with sides of length 12 ft.

$$P = 4s$$
$$= 4 \cdot 12 \text{ ft}$$
$$= 48 \text{ ft}$$

**13.** Find the perimeter of a square sandbox with sides of length 6 ft.

The perimeter of the garden is 48 ft.

*Answers*

**11.** 24 ft   **12.** 40 km   **13.** 24 ft

Do Exercise 13.

# Translating for Success

1. **Wood Costs.** It costs $8 for the wood for each birdhouse that Annette builds. If she used $120 worth of wood, how many birdhouses did she build?

2. **Elevation.** Genine started hiking a 10-mi trail at an elevation that was 150 ft below sea level. At the end of the trail, she was 75 ft above sea level. How many feet higher was she at the end of the trail than at the beginning?

3. **Community Service.** In order to fulfill the requirements for a sociology class, Glen must log 120 hr of community service. So far, he has spent 75 hr volunteering at a youth center. How many more hours must he serve?

4. **Disaster Relief.** Each package that is prepared for a disaster relief effort contains 15 meal bars. How many packages can be filled from a donation of 750 bars?

5. **Perimeter.** A rectangular building lot is 75 ft wide and 150 ft long. What is the perimeter of the lot?

The goal of these matching questions is to practice step (2), *Translate*, of the five-step problem-solving process. Translate each word problem to an equation and select a correct translation from equations A–O.

**A.** $75 + x = 120$

**B.** $15 \div 750 = x$

**C.** $-10 - 15 = x$

**D.** $8 \cdot 120 = x$

**E.** $150 - 75 = x$

**F.** $15 \cdot 750 = x$

**G.** $75 - (-150) = x$

**H.** $2 \cdot 150 + 2 \cdot 75 = x$

**I.** $-10 - (-15) = x$

**J.** $8 \cdot x = 120$

**K.** $750 \div 15 = x$

**L.** $15 - (-10) = x$

**M.** $75 + 120 = x$

**N.** $75 - 150 = x$

**O.** $75 = 120 + x$

*Answers on page A-4*

6. **Account Balance.** Lorenzo had $75 in his checking account. He then wrote a check for $150. What was the balance in his account?

7. **Laptop Computers.** Great Graphics purchased a laptop computer for each of its 15 employees. If each laptop cost $750, how much did the computers cost?

8. **Basketball.** A basketball team scored 75 points in one game. In the next game, the team scored a record 120 points. How many points did the team score in the two games?

9. **Pizza Sales.** A youth club sold 120 pizzas for a fundraiser. If each pizza sold for $8, how much money was taken in?

10. **Temperature.** The temperature in Fairbanks was $-10°$ at 6:00 P.M. and fell another 15° by midnight. What was the temperature at midnight?

**a**  List the terms of each expression.

**1.** $2a + 5b - 7c$

**2.** $4x - 6y + 7z$

**3.** $mn - 6n + 8$

**4.** $7rs - s - 5$

**5.** $3x^2y - 4y^2 - 2z^3$

**6.** $4a^3b + ab^2 - 9b^3$

Combine like terms to form an equivalent expression.

**7.** $5x + 9x$

**8.** $9a + 7a$

**9.** $10a - 15a$

**10.** $-17x + x$

**11.** $2x + 6y + x$

**12.** $3t - y + 7t$

**13.** $27a + 70 - 40a - 8$

**14.** $42x - 6 - x + 2$

**15.** $9 + 5t + 7y - t - y - 13$

**16.** $8 - 4a + 9b + 5a - 3b - 15$

**17.** $a + 3b + 5a - 2 + b$

**18.** $x + 7y + 5 - 2y + 3x$

**19.** $-8 + 11a - 5b - 10a - 7b + 7$

**20.** $8x - 5x + 6 + 3y - y - 4$

**21.** $8x^2 + 3y - x^2$

**22.** $8y^3 - 3z + 4y^3$

**23.** $11x^4 + 2y^3 - 4x^4 - y^3$

**24.** $13a^5 + 9b^4 - 2a^5 - 4b^4$

**25.** $9a^2 - 4a + a - 3a^2$

**26.** $3a^2 + 7a^3 - a^2 + 5 + a^3$

**27.** $x^3 - 5x^2 + 2x^3 - 3x^2 + 4$

**28.** $9xy + 4y^2 - 2xy + 2y^2 - 1$

**29.** $9x^3y + 4xy^3 - 5xy^3 + 3xy$

**30.** $8a^2b - 3ab^2 - 7a^2b + 2ab$

**31.** $3a^6 - b^4 + 2a^6b^4 - 7a^6 - 2b^4$

**32.** $3x^4 - 2y^4 + 8x^4y^4 - 7x^4 + y^4$

**b** Find the perimeter of each polygon.

**33.**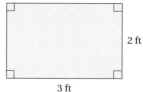

2 ft

3 ft

**34.**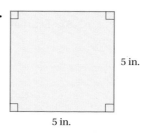

5 in.

5 in.

**35.**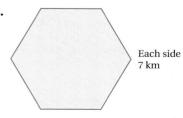

Each side 7 km

**36.**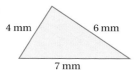

4 mm    6 mm

7 mm

**37.**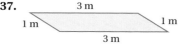

3 m

1 m    1 m

3 m

**38.**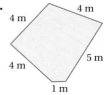

4 m    4 m

4 m    5 m

4 m

1 m

*Tennis Court.* A tennis court contains many rectangles. Use the diagram of a regulation tennis court to calculate the perimeters in Exercises 39–42.

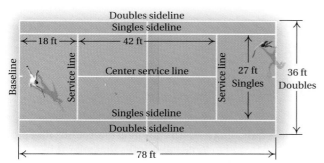

**39.** The perimeter of a singles court.

**40.** The perimeter of a doubles court.

**41.** The perimeter of the rectangle formed by the service lines and the singles sidelines.

**42.** The perimeter of the rectangle formed by a service line, a baseline, and the singles sidelines.

**43.** Find the perimeter of a rectangular 8-ft by 10-ft bedroom.

**44.** Find the perimeter of a rectangular 3-ft by 4-ft doghouse.

**45.** Find the perimeter of a checkerboard that is 14 in. on each side.

**46.** Find the perimeter of a square skylight that is 2 m on each side.

**47.** Find the perimeter of a square frame that is 65 cm on each side.

**48.** Find the perimeter of a square garden that is 12 yd on each side.

**49.** Find the perimeter of a 12-ft by 20-ft rectangular deck.

**50.** Find the perimeter of a 40-ft by 35-ft rectangular backyard.

## Skill Maintenance

**51.** A box of Shaw's Corn Flakes contains 510 grams (g) of corn flakes. A serving of corn flakes weighs 30 g. How many servings are in one box?   [1.8a]

**52.** Estimate the difference by rounding to the nearest ten. [1.6b]

$$\begin{array}{r} 7\ 0\ 4 \\ -\ 4\ 8\ 6 \\ \hline \end{array}$$

Simplify.   [1.9c]

**53.** $5 + 3 \cdot 2^3$

**54.** $(9 - 7)^4 - 3^2$

**55.** $12 \div 3 \cdot 2$

**56.** $27 \div 3(2 + 1)$

**57.** $15 - 3 \cdot 2 + 7$

**58.** $30 - 4^2 \div 8 \cdot 2$

Solve.   [1.7b]

**59.** $25 = t + 9$

**60.** $19 = x + 6$

**61.** $45 = 3x$

**62.** $50 = 2t$

## Synthesis

Simplify. (Multiply and then combine like terms.)

**63.** $5(x + 3) + 2(x - 7)$

**64.** $3(a - 7) + 7(a + 4)$

**65.** $2(3 - 4a) + 5(a - 7)$

**66.** $7(2 - 5x) + 3(x - 8)$

**67.** $-5(2 + 3x + 4y) + 7(2x - y)$

**68.** $3(4 - 2x) + 5(9x - 3y + 1)$

**69.** In order to save energy, Andrea plans to run a bead of caulk sealant around 3 exterior doors and 13 windows. Each window measures 3 ft by 4 ft, each door measures 3 ft by 7 ft, and there is no need to caulk the bottom of each door. If each cartridge of caulk seals 56 ft and costs $6, how much will it cost Andrea to seal the windows and doors?

**70.** Jacie is attaching lace trim to small tablecloths that are 5 ft by 5 ft, and to large tablecloths that are 7 ft by 7 ft. If the lace costs $2 per yard, how much will the trim cost for 6 small tablecloths and 6 large tablecloths?

**71.** ▦ A square wooden rack is used to store the 15 numbered pool balls as well as the cue ball. If a pool ball has a diameter of 57 mm, find the inside perimeter of the storage rack.

**72.** A rectangular box is used to store six Christmas ornaments. Find the perimeter of such a box if each ornament has a diameter of 72 mm.

# 2.8  Solving Equations

A *solution* of an equation is a replacement for the variable that makes the equation true. We can *solve* many equations, or find their solutions, using the addition principle and the division principle.

## a  The Addition Principle

The solution of the equation $x = 10$ is simply $10$. The solution of the equation $x + 5 = 15$ is also $10$, since $10 + 5 = 15$ is true. Because their solutions are identical, $x = 10$ and $x + 5 = 15$ are said to be **equivalent equations**.

> ### EQUIVALENT EQUATIONS
>
> Equations with the same solutions are called **equivalent equations**.

It is important to be able to distinguish between equivalent *expressions* and equivalent *equations*.

- $6a$ and $4a + 2a$ are equivalent *expressions* because, for any replacement of $a$, both expressions represent the same number.
- $3x = 15$ and $4x = 20$ are equivalent *equations* because any solution of one equation is also a solution of the other equation.

**EXAMPLE 1**  Classify each pair as either equivalent equations or equivalent expressions:

**a)** $5x + 1;\quad 2x - 4 + 3x + 5$      **b)** $x = -7;\quad x + 2 = -5.$

**a)** First note that these are expressions, not equations. To see if they are equivalent, we combine like terms in the second expression:

$$2x - 4 + 3x + 5 = (2 + 3)x + (-4 + 5) \qquad \text{Regrouping and using the distributive law}$$
$$= 5x + 1.$$

We see that $2x - 4 + 3x + 5$ and $5x + 1$ are *equivalent expressions*.

**b)** First note that both $x = -7$ and $x + 2 = -5$ are equations. The solution of $x = -7$ is $-7$. We substitute to see if $-7$ is also the solution of $x + 2 = -5$:

$$x + 2 = -5$$
$$-7 + 2 = -5 \quad \text{TRUE}$$

Since $x = -7$ and $x + 2 = -5$ have the same solution, they are *equivalent equations*.

> Do Margin Exercises 1 and 2.

There are principles that enable us to begin with one equation and create an equivalent equation similar to $x = 15$, for which the solution is obvious. One such principle, the *addition principle*, is stated on the next page. It tells us that we can add the same number to both sides of an equation without changing the solutions of the equation.

## OBJECTIVES

**a** Use the addition principle to solve equations.

**b** Use the division principle to solve equations.

**c** Decide which principle should be used to solve an equation.

**d** Solve equations that require use of both the addition principle and the division principle.

---

**SKILL TO REVIEW**
Objective 1.7b: Solve equations like $t + 28 = 54, 28 \cdot x = 168$, and $98 \cdot 2 = y$.

Solve.

**1.** $x + 13 = 87$

**2.** $17 \cdot y = 357$

Classify each pair as equivalent expressions or equivalent equations.

**1.** $a - 5 = -3;\quad a = 2$

**2.** $a - 9 + 6a;\quad 7a - 9$

***Answers***

*Skill to Review:*
1. 74   2. 21

*Margin Exercises:*
1. Equivalent equations
2. Equivalent expressions

> ## THE ADDITION PRINCIPLE
>
> For any numbers $a$, $b$, and $c$,
>
> $$a = b \quad \text{is equivalent to} \quad a + c = b + c.$$

**EXAMPLE 2** Solve: $x - 7 = -2$.

We have

$$x - 7 = -2$$
$$x - 7 + 7 = -2 + 7 \qquad \text{Using the addition principle:}$$
$$\text{adding 7 to both sides}$$
$$x + 0 = 5 \qquad \text{Adding 7 "undoes" the subtraction of 7.}$$
$$x = 5. \qquad \text{This equation has the same solution as}$$
$$x - 7 = -2.$$

The solution appears to be 5. To check, we use the original equation.

Check:
$$\begin{array}{c} x - 7 = -2 \\ \hline 5 - 7 \ ? \ -2 \\ -2 \ | \qquad \text{TRUE} \end{array}$$

The solution is 5.

Do Exercises 3 and 4.

We can subtract by adding the opposite of the number being subtracted. Because of this, the addition principle allows us to subtract the same number from both sides of an equation.

**EXAMPLE 3** Solve: $23 = t + 7$.

We have

$$23 = t + 7$$
$$23 - 7 = t + 7 - 7 \qquad \text{Using the addition principle to add } -7$$
$$\text{or to subtract 7 on both sides}$$
$$16 = t + 0 \qquad \text{Subtracting 7 "undoes" the addition of 7.}$$
$$16 = t. \qquad \text{The solution of } 23 = t + 7 \text{ is also 16.}$$

The solution is 16. The check is left to the student.

To visualize the addition principle, think of a jeweler's balance. When both sides of the balance hold equal amounts of weight, the balance is level. If weight is added or removed, equally, on both sides, the balance remains level.

Do Exercises 5 and 6.

**Solve.**

**3.** $x - 5 = 19$

**4.** $x - 9 = -12$

**Solve.**

**5.** $42 = x + 17$

**6.** $a + 8 = -6$

**Answers**

**3.** 24   **4.** −3   **5.** 25   **6.** −14

## b The Division Principle

In Section 1.7, we solved $8n = 96$ by dividing both sides by 8:

$$8 \cdot n = 96$$

$$\frac{8 \cdot n}{8} = \frac{96}{8} \qquad \text{Dividing both sides by 8}$$

$$n = 12. \qquad \text{8 times } n \text{, divided by 8, is } n. \; 96 \div 8 \text{ is 12.}$$

Both $8n = 96$ and $n = 12$ have the solution 12. We can divide both sides of an equation by any nonzero number in order to find an equivalent equation.

> **THE DIVISION PRINCIPLE**
>
> For any numbers $a$, $b$, and $c$ ($c \neq 0$),
>
> $$a = b \quad \text{is equivalent to} \quad \frac{a}{c} = \frac{b}{c}.$$

In Chapter 3, after we have discussed multiplication of fractions, we will use an equivalent form of this principle: the multiplication principle.

**EXAMPLE 4**   Solve: $9x = 63$.

We have

$$9x = 63$$

$$\frac{9x}{9} = \frac{63}{9} \qquad \begin{array}{l} \text{Using the division principle to} \\ \text{divide both sides by 9} \end{array}$$

$$x = 7.$$

Check:   $\dfrac{9x = 63}{9 \cdot 7 \; ? \; 63}$   Check in the original equation.

$63 \;|\;$   TRUE

The solution is 7.

Do Exercises 7 and 8.

**EXAMPLE 5**   Solve: $48 = -8n$.

It is important to distinguish between multiplication by a negative number, as we have in $-8n$, and subtraction, as we had in $x - 5 = 19$ (Margin Exercise 3). To undo multiplication by $-8$, we use the division principle:

$$48 = -8n$$

$$\frac{48}{-8} = \frac{-8n}{-8} \qquad \text{Dividing both sides by } -8$$

$$-6 = n.$$

Check:   $\dfrac{48 = -8n}{48 \; ? \; -8(-6)}$

$|\; 48 \qquad$ TRUE

The solution is $-6$.

Do Exercises 9 and 10.

Solve.

**7.** $7x = 42$

**8.** $-24 = 3t$

Solve.

**9.** $63 = -7n$

**10.** $-6x = 72$

*Answers*

**7.** 6   **8.** $-8$   **9.** $-9$   **10.** $-12$

Our goal in equation solving is to have the variable by itself on one side of the equation. In an equation like $-x = 7$, the variable is not by itself—there is an opposite sign in front of $x$. We can solve an equation like $-x = 7$ by dividing or by multiplying both sides of the equation by $-1$.

**EXAMPLE 6** Solve: $-x = 7$.

One way to solve this equation is to note that $-x = -1 \cdot x$. Then we can divide both sides by $-1$:

$$-x = 7$$
$$-1 \cdot x = 7$$
$$\frac{-1 \cdot x}{-1} = \frac{7}{-1} \qquad \text{Using the division principle}$$
$$x = -7.$$

Check:
$$\frac{-x = 7}{-(-7) \; ? \; 7}$$
$$7 \; | \qquad \text{TRUE}$$

Be sure to check in the original equation.

The solution is $-7$.

Another way to solve the equation in Example 6 is to remember that when an expression is multiplied or divided by $-1$, its sign is changed. Here we multiply on both sides by $-1$ to change the sign of $-x$:

$$-x = 7$$
$$(-1)(-x) = (-1) \cdot 7 \qquad \text{Multiplying both sides by } -1$$
$$x = -7. \qquad \text{Note that } (-1)(-x) \text{ is the same as } (-1)(-1)x.$$

Solve.

**11.** $-x = 23$

**12.** $-t = -3$

Do Exercises 11 and 12.

## (c) Selecting the Correct Approach

It is important for you to be able to determine which principle should be used to solve a particular equation.

**EXAMPLE 7** Solve: $39 = -3 + t$.

Note that $-3$ is added to $t$. To undo addition of $-3$, we subtract $-3$ or simply add 3 on both sides:

$$3 + 39 = 3 + (-3) + t \qquad \text{Using the addition principle}$$
$$42 = 0 + t$$
$$42 = t.$$

Check:
$$\frac{39 = -3 + t}{39 \; ? \; -3 + 42}$$
$$| \; 39 \qquad \text{TRUE}$$

The solution is 42.

*Answers*

**11.** $-23$    **12.** 3

**EXAMPLE 8** Solve: $39 = -3t$.

Here $t$ is multiplied by $-3$. To undo multiplication by $-3$, we divide by $-3$ on both sides:

$$39 = -3t$$

$$\frac{39}{-3} = \frac{-3t}{-3} \qquad \text{Using the division principle}$$

$$-13 = t.$$

Check: 
$$\begin{array}{c|c} \hline 39 = -3t \\ \hline 39 \; ? \; -3(-13) \\ \phantom{39} \;\big|\; 39 \qquad \text{TRUE} \end{array}$$

The solution is $-13$.

Do Exercises 13–15.

Solve.

**13.** $-2x = -52$

**14.** $-2 + x = -52$

**15.** $x \cdot 7 = -28$

## d  Using the Principles Together

Suppose we want to determine whether 7 is the solution of $5x - 8 = 27$. To check, we replace $x$ with 7 and simplify.

Check: 
$$\begin{array}{c|c} \hline 5x - 8 = 27 \\ \hline 5 \cdot 7 - 8 \; ? \; 27 \\ 35 - 8 \;\big|\; \\ 27 \;\big|\; \qquad \text{TRUE} \end{array}$$

This shows that 7 *is* the solution.

Do Exercises 16 and 17.

**16.** Determine whether $-9$ is the solution of $7x + 8 = -55$.

**17.** Determine whether $-6$ is the solution of $4x + 3 = -25$.

In the check above, note that the rules for order of operations require that we multiply before we subtract (or add). Thus, to evaluate $5x - 8$,

we *select* a value: $\qquad x$

then *multiply* by 5: $\qquad 5x$

and then *subtract* 8: $\qquad 5x - 8.$

In Example 9, which follows, these steps are reversed to solve for $x$:

We will *add* 8: $\qquad 5x - 8 + 8$

then *divide* by 5: $\qquad \dfrac{5x}{5}$

and then *isolate* $x$: $\qquad x.$

In general, the *last* step performed when calculating is the *first* step to be reversed when finding a solution.

When $x$ is by itself on one side of an equation, we say that it is *isolated*. Our goal in the remaining examples in this section will be to isolate the term containing the variable (using the addition principle) and then to isolate the variable itself (using the division principle).

*Answers*

**13.** 26  **14.** $-50$  **15.** $-4$
**16.** Yes  **17.** No

**EXAMPLE 9** Solve: $5x - 8 = 27$.

We first note that the term containing $x$ is $5x$. To isolate $5x$, we add 8 on both sides:

$$5x - 8 = 27$$
$$5x - 8 + 8 = 27 + 8 \qquad \text{Using the addition principle}$$
$$5x + 0 = 35 \qquad \text{Try to do this step mentally.}$$
$$5x = 35. \qquad \text{We have isolated } 5x.$$

Next, we isolate $x$ by dividing by 5 on both sides:

$$5x = 35$$
$$\frac{5x}{5} = \frac{35}{5} \qquad \text{Using the division principle}$$
$$1x = 7 \qquad \text{Try to do this step mentally.}$$
$$x = 7. \qquad \text{We have isolated } x.$$

The check was performed on the previous page. The solution is 7.

**18.** Solve: $2x - 9 = 43$.

Do Exercise 18.

**EXAMPLE 10** Solve: $38 = -9t + 2$.

We first isolate $-9t$ by subtracting 2 on both sides:

$$38 = -9t + 2$$
$$38 - 2 = -9t + 2 - 2 \qquad \text{Subtracting 2 (or adding } -2 \text{) from both sides}$$
$$36 = -9t + 0 \qquad \text{Try to do this step mentally.}$$
$$36 = -9t.$$

Now that we have isolated $-9t$ on one side of the equation, we can divide by $-9$ to isolate $t$:

$$36 = -9t$$
$$\frac{36}{-9} = \frac{-9t}{-9} \qquad \text{Dividing both sides by } -9$$
$$-4 = t. \qquad \text{Simplifying}$$

Check: 
$$\begin{array}{c|c} \hline 38 = -9t + 2 \\ \hline 38 \; ? \; -9 \cdot (-4) + 2 \\ \phantom{38 \; ?} \; 36 + 2 \\ \phantom{38 \; ?} \; 38 \qquad \text{TRUE} \end{array}$$

The solution is $-4$.

**19.** Solve: $-3n + 2 = 47$.

Do Exercise 19.

**a**  Classify each pair as either equivalent expressions or equivalent equations.

**1.** $x = -1$;  $x + 5 = 4$

**2.** $4x + 1$;  $6 + 4x - 5$

**3.** $7a - 3$;  $13 + 7a - 16$

**4.** $t = -2$;  $5t = -10$

**5.** $4r + 3$;  $9 - r + 5r - 6$

**6.** $2r - 7$;  $r - 10 + r + 3$

**7.** $x - 9 = 8$;  $x = 20 - 3$

**8.** $t + 4 = 19$;  $t = 9 + 6$

**9.** $3(t + 2)$;  $5 + 3t + 1$

**10.** $2x = -14$;  $x - 2 = -9$

**11.** $x + 4 = -8$;  $2x = -24$

**12.** $4(x - 7)$;  $3x - 28 + x$

Solve.

**13.** $x - 6 = -9$

**14.** $x - 5 = -7$

**15.** $x - 4 = -12$

**16.** $x - 7 = 5$

**17.** $a + 7 = 25$

**18.** $x + 9 = -3$

**19.** $-8 = n + 7$

**20.** $38 = a + 12$

**21.** $24 = t - 8$

**22.** $-9 = x + 3$

**23.** $-12 = x + 5$

**24.** $17 = n - 6$

**25.** $-5 + a = 12$

**26.** $3 = 17 + x$

**27.** $-8 = -8 + t$

**28.** $-7 + t = -7$

**b**  Solve.

**29.** $6x = 60$

**30.** $-8t = 40$

**31.** $-3t = 42$

**32.** $3x = 24$

**33.** $-7n = -35$

**34.** $64 = -2t$

**35.** $0 = 6x$

**36.** $-5n = -65$

**37.** $55 = -5t$

**38.** $-x = 83$

**39.** $-x = 56$

**40.** $-2x = 0$

**41.** $n(-4) = -48$

**42.** $-x = -475$

**43.** $-x = -390$

**44.** $n(-7) = 42$

Solve.

**45.** $t - 6 = -2$

**46.** $3t = -45$

**47.** $6x = -54$

**48.** $x + 9 = -15$

**49.** $15 = -x$

**50.** $-13 = x - 4$

**51.** $-21 = x + 5$

**52.** $-42 = -x$

**53.** $35 = -7t$

**54.** $7 + t = -18$

**55.** $-17x = 68$

**56.** $-34 = x - 10$

**57.** $12 + t = -160$

**58.** $-48 = t(-12)$

**59.** $-27 = x - 23$

**60.** $-135 = -9t$

**d** Solve.

**61.** $5x - 1 = 34$

**62.** $7x - 3 = 25$

**63.** $4t + 2 = 14$

**64.** $3t + 5 = 26$

**65.** $6a + 1 = -17$

**66.** $8a + 3 = -37$

**67.** $2x - 9 = -23$

**68.** $3x - 5 = -35$

**69.** $-2x + 1 = 17$

**70.** $-4t + 3 = -17$

**71.** $-8t - 3 = -67$

**72.** $-7x - 4 = -46$

**73.** $-x + 9 = -15$

**74.** $-x - 6 = 8$

**75.** $7 = 2x - 5$

**76.** $9 = 4x - 7$

**77.** $13 = 3 + 2x$

**78.** $33 = 5 - 4x$

**79.** $13 = 5 - x$

**80.** $12 = 7 - x$

## Skill Maintenance

In each of Exercises 81–88, fill in the blank with the correct term from the given list. Some of the choices may not be used and some may be used more than once.

81. A(n) _____ is a closed geometric figure.   [2.7b]

82. Terms are _____ if they have the same variable factor(s).   [2.7a]

83. Numbers we multiply together are called _____.   [1.4a]

84. Equations are _____ if they have the same solutions.   [2.8a]

85. The result of an addition is a(n) _____.   [1.2a]

86. A(n) _____ is a letter that can stand for various numbers.   [2.6a]

87. The _____ of a number is its distance from zero on the number line.   [2.1c]

88. We _____ for a variable when we replace it with a number.   [2.6a]

absolute value

opposite

constant

variable

factors

addends

evaluate

substitute

polygon

perimeter

equivalent

similar

sum

product

## Synthesis

Solve.

**89.** $2x - 7x = -40$

**90.** $9 + x - 5 = 23$

**91.** $2x - 7 + x = 5 - 12$

**92.** $3 - 6x + 5 = 2(4)$

**93.** $n + n = -2 - 3 \cdot 6 \div 2 + 1$

**94.** $10 \div 5 \cdot 2 + 3 = 5t - 4t$

**95.** $17 - 3^2 = 4 + t - 5^2$

**96.** $(-9)^2 = 2^3 t + (3 \cdot 6 + 1)t$

**97.** $(-7)^2 - 5 = t + 4^3$

**98.** ▦ $(-42)^3 = 14^2 t$

**99.** ▦ $x - (19)^3 = -18^3$

**100.** ▦ $23^2 = x + 22^2$

**101.** ▦ $35^3 = -125t$

**102.** ▦ $248 = 24 - 32x$

**103.** ▦ $529 - 143x = -1902$

# Summary and Review

## Key Terms and Properties

integers, p. 88
absolute value, p. 89
opposite, p. 90
additive identity, p. 95

additive inverse, p. 95
variable, p. 120
constant, p. 120
algebraic expression, p. 120

value, p. 120
equivalent expressions, p. 122
combining like terms, p. 128
equivalent equations, p. 135

*Rules for Addition of Real Numbers,* p. 96

*Multiplying Two Nonzero Real Numbers,* p. 107

*Rules for Order of Operations,* p. 113

*Distributive Law,* p. 122

*Perimeter of a Rectangle,* p. 130

*Perimeter of a Square,* p. 130

*The Addition Principle,* p. 136

*The Division Principle,* p. 137

## Concept Reinforcement

Determine whether each statement is true or false.

_____ 1. The opposite of the opposite of a number is the original number. [2.1d]

_____ 2. The product of an even number of negative numbers is positive. [2.4b]

_____ 3. The expression $2(x + 3)$ is equivalent to the expression $2 \cdot x + 3$. [2.6b]

## Important Concepts

**Objective 2.1c** Find the absolute value of any integer.

**Example** Find. **a)** $|-97|$ **b)** $|35|$

**a)** The number is negative, so we write the opposite.

$$|-97| = 97$$

**b)** The number is positive, so the absolute value is the same as the number.

$$|35| = 35$$

**Practice Exercise**

1. Find. **a)** $|-17|$ **b)** $|300|$

**Objective 2.2a** Add integers without using the number line.

**Example** Add without using the number line: $-15 + 9$.

To add a negative number and a positive number, subtract the absolute values: $15 - 9 = 6$. The negative number has the larger absolute value, so the answer is negative.

$$-15 + 9 = -6$$

**Example** Add without using the number line: $-8 + (-9)$.

To add two negative numbers, add the absolute values: $8 + 9 = 17$. The answer is negative.

$$-8 + (-9) = -17$$

**Practice Exercise**

2. Add without using the number line: $37 + (-16)$.

**Objective 2.3a** Subtract integers and simplify combinations of additions and subtractions.

**Example** Subtract: $8 - 12$.

We add the opposite of the number being subtracted.

$$8 - 12 = 8 + (-12)$$
$$= -4$$

**Practice Exercise**

**3.** Subtract: $6 - (-8)$.

---

**Objective 2.4a** Multiply integers.

**Example** Multiply: $-6(-4)$.

The signs are the same, so the answer is positive.

$$-6(-4) = 24$$

**Example** Multiply: $-4(3)$.

The signs are different, so the answer is negative.

$$-4(3) = -12$$

**Practice Exercise**

**4.** Multiply: $6(-15)$.

---

**Objective 2.5a** Divide integers.

**Example** Divide: $-36 \div (-4)$.

The signs are the same, so the answer is positive.

$$-36 \div (-4) = 9$$

**Example** Divide: $-30 \div 10$.

The signs are different, so the answer is negative.

$$-30 \div 10 = -3$$

**Practice Exercise**

**5.** Divide: $99 \div (-9)$.

---

**Objective 2.5b** Use the rules for order of operations with integers.

**Example** Simplify: $3^2 - 24 \div 2 - (4 + 2 \cdot 8)$.

$$3^2 - 24 \div 2 - (4 + 2 \cdot 8)$$
$$= 3^2 - 24 \div 2 - (4 + 16)$$
$$= 3^2 - 24 \div 2 - 20$$
$$= 9 - 24 \div 2 - 20$$
$$= 9 - 12 - 20$$
$$= -3 - 20$$
$$= -23$$

**Practice Exercise**

**6.** Simplify: $4 - 8^2 \div (10 - 6)$.

---

**Objective 2.6b** Use the distributive law to find equivalent expressions.

**Example** Use the distributive law to write an expression equivalent to $-2(3x - 4)$.

$$-2(3x - 4) = -2(3x) - (-2)(4)$$
$$= -6x - (-8)$$
$$= -6x + 8$$

**Practice Exercise**

**7.** Use the distributive law to write an expression equivalent to $5(6x - 8y - z)$.

**Objective 2.7a**   Combine like terms.

**Example**   Combine like terms to form an expression equivalent to $6 - 30x + 12x - 7$.

$$6 - 30x + 12x - 7 = 6 + (-30x) + 12x + (-7)$$
$$= -30x + 12x + 6 + (-7)$$
$$= -18x - 1$$

**Practice Exercise**

8. Combine like terms to form an expression equivalent to $8a - b + 9a - 6b$.

---

**Objective 2.8d**   Solve equations that require use of both the addition principle and the division principle.

**Example**   Solve: $-3x + 1 = 16$.

$$-3x + 1 = 16$$
$$-3x + 1 - 1 = 16 - 1$$
$$-3x = 15$$
$$\frac{-3x}{-3} = \frac{15}{-3}$$
$$x = -5$$

The solution is $-5$.

**Practice Exercise**

9. Solve: $-19 = 5x + 11$.

---

# Review Exercises

1. Tell which integers correspond to this situation:

   David has a debt of $45 and Joe has $72 in his savings account.   [2.1a]

Use either $<$ or $>$ for $\square$ to form a true statement.   [2.1b]

2. $0 \, \square \, -5$      3. $-7 \, \square \, 6$      4. $-4 \, \square \, -19$

Find the absolute value.   [2.1c]

5. $|-39|$         6. $|23|$           7. $|0|$

8. Find $-x$ when $x = -72$.   [2.1d]

9. Find $-(-x)$ when $x = 59$.   [2.1d]

Compute and simplify.

10. $-14 + 5$   [2.2a]          11. $-5 + (-6)$   [2.2a]

12. $14 + (-8)$   [2.2a]          13. $0 + (-24)$   [2.2a]

14. $17 - 29$   [2.3a]          15. $9 - (-14)$   [2.3a]

16. $-8 - (-7)$   [2.3a]          17. $-3 - (-3)$   [2.3a]

18. $-3 + 7 + (-8)$   [2.3a]

19. $8 - (-9) - 7 + 2$   [2.3a]

**20.** $-23 \cdot (-4)$   [2.4a]

**21.** $7(-12)$   [2.4a]

**22.** $2(-4)(-5)(-1)$   [2.4b]

**23.** $15 \div (-5)$   [2.5a]

**24.** $\dfrac{-55}{11}$   [2.5a]

**25.** $\dfrac{0}{7}$   [2.5a]

Simplify.   [2.5b]

**26.** $625 \div (-25) \div 5$

**27.** $-16 \div 4 - 30 \div (-5)$

**28.** $9[(7 - 14) - 13]$

**29.** $(-3)|4 - 3^2| - 5$

**30.** $[-12(-3) - 2^3] - (-9)(-10)$

**31.** Evaluate $3a + b$ for $a = 4$ and $b = -5$.   [2.6a]

**32.** Evaluate $\dfrac{-x}{y}$, $\dfrac{x}{-y}$, and $-\dfrac{x}{y}$ for $x = 30$ and $y = 5$.   [2.6a]

Use the distributive law to write an equivalent expression.
[2.6b]

**33.** $4(5x + 9)$          **34.** $3(2a - 4b + 5)$

**35.** $-10(2x + y)$

Combine like terms.   [2.7a]

**36.** $5a + 12a$                    **37.** $-7x + 13x$

**38.** $9m + 14 - 12m - 8$

**39.** Find the perimeter of a rectangular frame that is 8 in. by 10 in.   [2.7b]

**40.** Find the perimeter of a square pane of glass that is 25 cm on each side.   [2.7b]

Solve.   [2.8a, b, c, d]

**41.** $x - 9 = -17$                    **42.** $-4t = 36$

**43.** $13 = -x$                    **44.** $56 = 6x - 10$

**45.** $-x + 3 = -12$                    **46.** $18 = 4 - 2x$

Solve.

**47.** On the first, second, and third downs, a football team had these gains and losses: 5-yd gain, 12-yd loss, and 15-yd gain, respectively. Find the total gain (or loss). [2.3b]

**48.** Kaleb's total assets are $170. He borrows $300. What are his total assets now? [2.3b]

**49.** Evaluate $-|-x|$ when $x = -10$. [2.1d]
 **A.** $-10$        **B.** $-20$
 **C.** $100$        **D.** $10$

**50.** Simplify: $-3 \cdot 4 - 12 \div 4$. [2.5b]
 **A.** $-16$        **B.** $-15$
 **C.** $0$          **D.** $6$

## Synthesis

**51.** The sum of two numbers is 800. The difference is 6. Find the numbers. [2.3b]

**52.** The following are examples of consecutive integers: $4, 5, 6, 7, 8;\ -13, -12, -11, -10$. [2.3b], [2.4b]
 **a)** Express the number 8 as the sum of 16 consecutive integers.
 **b)** Find the product of the 16 consecutive integers in part (a).

Simplify. [2.5b]

**53.** 📱 $87 \div 3 \cdot 29^3 - (-6)^6 + 1957$

**54.** 📱 $1969 + (-8)^5 - 17 \cdot 15^3$

**55.** 📱 $\dfrac{113 - 17^3}{15 + 8^3 - 507}$

**56.** For what values of $x$ will $8 + x^3$ be negative? [2.6a]

**57.** For what values of $x$ is $|x| > x$? [2.1b, c]

# Understanding Through Discussion and Writing

**1.** What rule have we developed that would tell you the sign of $(-7)^8$ and $(-7)^{11}$ without doing the computations? Explain. [2.4b]

**2.** Does $-x$ always represent a negative number? Why or why not? [2.1d]

**3.** Jake enters $18/2 \cdot 3$ on his calculator and expects the result to be 3. What mistake is he making? [2.5b]

**4.** Does $-x^2$ always represent a negative number? Why or why not? [2.6a]

Test

For Extra Help

CHAPTER
Test Prep
VIDEOS

Step-by-step test solutions are found on the Chapter Test Prep Videos available via the Video Resources on DVD, in *MyMathLab* , and on YouTube (search "BittingerPrealgebra" and click on "Channels").

**1.** Tell which integers correspond to this situation: The Tee Shop sold 542 fewer muscle shirts than expected in January and 307 more than expected in February.

**2.** Use either $<$ or $>$ for $\square$ to form a true statement.
$$-14 \ \square \ -21$$

**3.** Find the absolute value: $|-739|$.

**4.** Find $-(-x)$ when $x = -19$.

Compute and simplify.

**5.** $6 + (-17)$

**6.** $-9 + (-12)$

**7.** $-8 + 17$

**8.** $0 - 12$

**9.** $7 - 22$

**10.** $-5 - 19$

**11.** $-8 - (-27)$

**12.** $31 - (-3) - 5 + 9$

**13.** $(-4)^3$

**14.** $27(-10)$

**15.** $-9 \cdot 0$

**16.** $-72 \div (-9)$

**17.** $\dfrac{-56}{7}$

**18.** $8 \div 2 \cdot 2 - 3^2$

**19.** $29 - (3 - 5)^2$

**20.** *Antarctica Highs and Lows.* The continent of Antarctica, which lies in the southern hemisphere, experiences winter in July. The average high temperature is $-67°F$ and the average low temperature is $-81°F$. How much higher is the average high than the average low?

Source: National Climatic Data Center

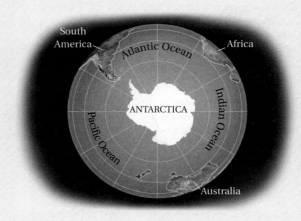

**21.** Evaluate $\dfrac{a - b}{6}$ for $a = -8$ and $b = 10$.

**22.** Use the distributive law to write an expression equivalent to $7(2x + 3y - 1)$.

**23.** Combine like terms.
$9x - 14 - 5x - 3$

**24.** Find the perimeter of a square garden that is 5 ft on each side.

Solve.

**25.** $-7x = -35$

**26.** $a + 9 = -3$

**27.** $95 = -x$

**28.** $3t - 7 = 5$

**29.** Use the distributive law to write an expression equivalent to $-2(n - 6m)$.

    **A.** $-2n - 6m$         **B.** $-2n - 12m$

    **C.** $-2n + 12m$         **D.** $2n + 12m$

# Synthesis

**30.** Monty plans to attach trim around the doorway and along the base of the walls in a 12-ft by 14-ft room. If the doorway is 3 ft by 7 ft, how many feet of trim are needed? (Only three sides of a doorway get trim.)

Simplify.

**31.** $9 - 5[x + 2(3 - 4x)] + 14$

**32.** $15x + 3(2x - 7) - 9(4 + 5x)$

**33.** ▦ $49 \cdot 14^3 \div 7^4 + 1926^2 \div 6^2$

**34.** ▦ $3487 - 16 \div 4 \cdot 4 \div 2^8 \cdot 14^4$

# Cumulative Review

1. Write standard notation for the number written in words in the following sentence: In 2008, Volunteers of America served two million, two hundred ninety-four thousand, three hundred ninety-two people.

   **Source:** Volunteers of America

2. Write a word name for 5,380,001,437.

Add.

3.   $\begin{array}{r} 1\,5,8\,9\,2 \\ +\ \ \ 2,9\,3\,5 \\ \hline \end{array}$

4.   $\begin{array}{r} 7\,9\,8\,9 \\ 7\,8\,9 \\ +\ \ \ \ \ 7\,9 \\ \hline \end{array}$

5. $-16 + 72$

6. $-30 + (-72)$

Subtract.

7.   $\begin{array}{r} 8\,2\,7\,6 \\ -\ \ \ 4\,3\,0 \\ \hline \end{array}$

8.   $\begin{array}{r} 3\,0\,0\,6 \\ -\ \ \ 5\,7\,8 \\ \hline \end{array}$

9. $7 - (-7)$

10. $-18 - 25$

Multiply.

11.   $\begin{array}{r} 6\,2\,1 \\ \times\ \ \ 2\,7 \\ \hline \end{array}$

12.   $\begin{array}{r} 2\,5\,0\,5 \\ \times\ 3\,3\,0\,0 \\ \hline \end{array}$

13. $43 \cdot (-8)$

14. $-12(-6)$

Divide.

15. $6\,3\,\overline{)\,6\,5\,5\,2}$

16. $6\,2\,\overline{)\,3\,8\,4\,4}$

17. $0 \div (-67)$

18. $60 \div (-12)$

19. Round 427,931 to the nearest thousand.

20. Round 5309 to the nearest hundred.

Estimate each sum or product by rounding to the nearest hundred. Show your work.

21.   $\begin{array}{r} 7\,4\,9,5\,5\,9 \\ +\ 3\,0\,1,3\,6\,2 \\ \hline \end{array}$

22.   $\begin{array}{r} 7\,4\,9 \\ \times\ 5\,3\,1 \\ \hline \end{array}$

23. Use $<$ or $>$ for $\square$ to form a true sentence:
   $-26\ \square\ 19.$

24. Find the absolute value: $|-279|$.

Simplify.

**25.** $35 - 25 \div 5 + 2 \times 3$

**26.** $\{17 - [8 - (5 - 2 \times 2)]\} \div (3 + 12 \div 6)$

**27.** $10 \div 1(-5) - 6^2$

**28.** $5^3$

**29.** Evaluate $\dfrac{x - y}{5}$ for $x = 11$ and $y = -4$.

**30.** Evaluate $7x^2$ for $x = -2$.

Use the distributive law to write an equivalent expression.

**31.** $-2(x + 5)$

**32.** $6(3x - 2y + 4)$

Simplify.

**33.** $12 + (-14)$    **34.** $(-2)^3$

**35.** $23 - 38$    **36.** $-64 \div (-2)$

**37.** $-12 - (-25)$    **38.** $(-2)(-3)(-5)$

**39.** $3 - (-8) + 2 - (-3)$    **40.** $16 \div 2(-8) + 7$

Solve.

**41.** $x + 8 = 35$    **42.** $-12t = 36$

**43.** $6 - x = -9$    **44.** $-39 = 4x - 7$

Solve.

**45.** In the movie *Little Big Man,* Dustin Hoffman plays a character who ages from 17 to 121. This represents the greatest age range depicted by one actor in one film. How many years did Hoffman's character age?
**Source:** Guinness Book of World Records

**46.** Nine of the ten largest hotels in the United States are in Las Vegas. Of these, the four largest are the MGM Grand, the Luxor, the Excalibur, and the Flamingo. These have 5005 rooms, 4400 rooms, 4032 rooms, and 3240 rooms, respectively. What is the total number of rooms in these four hotels?
**Source:** www.lasvegas-nv.com

**47.** Surface temperatures on Mars vary from $-128°C$ during polar night to $27°C$ at the equator during midday at the closest point in orbit to the sun. Find the difference between the highest value and the lowest value in this temperature range.
**Source:** Mars Institute

**48.** Eastside Appliance sells a refrigerator for $900 and $30 tax with no delivery charge. Westside Appliance sells the same model for $860 and $28 tax plus a $25 delivery charge. Which is the better buy?

**49.** Write an equivalent expression by combining like terms: $7x - 9 + 3x - 5$.

## Synthesis

**50.** A soft-drink distributor has 166 loose cans of cola. The distributor wishes to form as many 24-can cases as possible and then, with any remaining cans, as many six-packs as possible. How many cases will be filled? How many six-packs? How many loose cans will remain?

**51.** Simplify: $a - \{3a - [4a - (2a - 4a)]\}$.

**52.** ▦ Simplify: $37 \cdot 64 \div 4^2 \cdot 2 - (7^3 - (-4)^5)$.

**53.** Find two solutions of $5|x| - 2 = 13$.

# Fraction Notation: Multiplication and Division

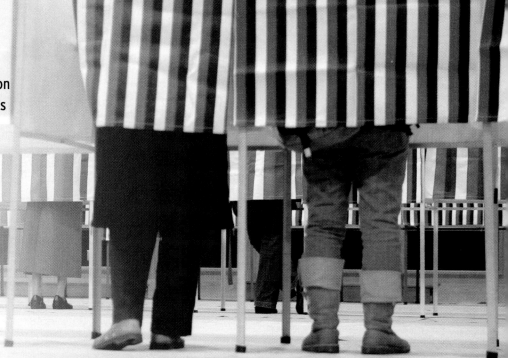

## Real-World Application

Of every 100 voters in the 2004 presidential election, 16 voters were in the 18–29 age group. In the 2008 presidential election, this number increased to 18.5 voters. What is the ratio of voters in the 18–29 age group to all voters in the 2004 presidential election? in the 2008 election?

*Sources*: The Center for Information & Research on Civic Learning & Engagement; www.civicyouth.org

***This problem appears as Exercise 41 in Section 3.3.***

# 3.1

## Multiples and Divisibility

### OBJECTIVES

**a** Find some multiples of a number and determine whether a number is divisible by another number.

**b** Test to see if a number is divisible by 2, 3, 5, 6, 9, or 10.

---

**SKILL TO REVIEW**
Objective 1.5a: Divide whole numbers.

Divide.

1. $329 \div 8$
2. $23 \overline{)1\,0\,8\,1}$

---

1. Show that each of the numbers 5, 45, and 100 is a multiple of 5.

2. Show that each of the numbers 10, 60, and 110 is a multiple of 10.

3. Multiply by 1, 2, 3, and so on, to find ten multiples of 5.

---

In this section, we discuss *multiples* and *divisibility* in order to be able to simplify fractions like $\frac{117}{225}$.

### **a** Multiples

A **multiple** of a number is a product of that number and an integer. For example, some multiples of 2 are

  2 (because $2 = 1 \cdot 2$);
  4 (because $4 = 2 \cdot 2$);
  6 (because $6 = 3 \cdot 2$);
  8 (because $8 = 4 \cdot 2$);
 10 (because $10 = 5 \cdot 2$).

We can also find multiples of 2 by counting by twos: 2, 4, 6, 8, and so on.

**EXAMPLE 1** Show that each of the numbers 3, 6, 9, and 15 is a multiple of 3.

Each of 3, 6, 9, and 15 can be expressed as a product of 3 and some integer:

$$3 = 1 \cdot 3; \quad 6 = 2 \cdot 3; \quad 9 = 3 \cdot 3; \quad 15 = 5 \cdot 3.$$

Do Margin Exercises 1 and 2.

**EXAMPLE 2** Multiply by 1, 2, 3, and so on, to find eight multiples of six.

| | |
|---|---|
| $1 \cdot 6 = 6$ | $5 \cdot 6 = 30$ |
| $2 \cdot 6 = 12$ | $6 \cdot 6 = 36$ |
| $3 \cdot 6 = 18$ | $7 \cdot 6 = 42$ |
| $4 \cdot 6 = 24$ | $8 \cdot 6 = 48$ |

Do Exercise 3.

---

> **DIVISIBILITY**
>
> A number $b$ is said to be **divisible** by another number $a$ if $b$ is a multiple of $a$.

Thus,

6 is divisible by 2 because 6 is a multiple of 2 ($6 = 3 \cdot 2$);

27 is divisible by 3 because 27 is a multiple of 3 ($27 = 9 \cdot 3$);

100 is divisible by 25 because 100 is a multiple of 25 ($100 = 4 \cdot 25$).

Saying that $b$ is divisible by $a$ means that if we divide $b$ by $a$, the remainder is 0. When this happens, we sometimes say that $a$ divides $b$ "evenly."

---

*Answers*

*Skill to Review:*
1. 41 R 1   2. 47

*Margin Exercises:*
1. $5 = 1 \cdot 5; 45 = 9 \cdot 5; 100 = 20 \cdot 5$
2. $10 = 1 \cdot 10; 60 = 6 \cdot 10; 110 = 11 \cdot 10$
3. 5, 10, 15, 20, 25, 30, 35, 40, 45, 50

---

**EXAMPLE 3** Determine **(a)** whether 45 is divisible by 9 and **(b)** whether 45 is divisible by 4.

a) We divide 45 by 9 to determine whether 9 is a factor of 45.

$$\begin{array}{r} 5 \\ 9\overline{)45} \\ \underline{45} \\ 0 \end{array} \leftarrow \text{Remainder is 0.}$$

Because the remainder is 0, 45 is divisible by 9.

b) We divide 45 by 4 to determine whether 4 is a factor of 45.

$$\begin{array}{r} 11 \\ 4\overline{)45} \\ \underline{4} \\ 5 \\ \underline{4} \\ 1 \end{array} \leftarrow \text{Not 0}$$

Since the remainder is not 0, 45 is not divisible by 4.

Do Exercises 4–6.

4. Determine whether 16 is divisible by 2.

5. Determine whether 125 is divisible by 5.

6. Determine whether 125 is divisible by 6.

## b  Tests for Divisibility

We now look at quick ways of checking for divisibility by 2, 3, 5, 6, 9, and 10 without actually performing long division. Tests do exist for divisibility by 4, 7, and 8, but they can be as difficult to perform as the actual long division.

To test for divisibility by 2, 5, and 10, we examine the ones digit.

### Divisibility by 2

All even numbers are divisible by 2.

> **BY 2**
>
> A number is **divisible by 2** (is *even*) if it has a ones digit of 0, 2, 4, 6, or 8 (that is, it has an even ones digit).

To see why this test works, start counting by 2's: 2, 4, 6, 8, 10, 12, 14, 16, 18, 20, 22, . . . . Note that the ones digit will always be 0, 2, 4, 6, or 8, no matter how high we count.

**EXAMPLES** Determine whether each number is divisible by 2.

4. 355   *is not* divisible by 2;  5 is not even.
5. 4786  *is* divisible by 2;  6 is even.
6. 8990  *is* divisible by 2;  0 is even.
7. 4261  *is not* divisible by 2;  1 is not even.

Do Exercises 7–10.

### Calculator Corner

**Divisibility**  We can use a calculator to determine whether one number is divisible by another number. For example, to determine whether 387 is divisible by 9, we press ⓪ ⑧ ⑦ ⌹ ⑨ ⌸. The display is 43 . Since 43 contains no digits to the right of the decimal point, we know that 387 is divisible by 9. On the other hand, since 387 ÷ 10 = 38.7, we know that 387 is *not* divisible by 10.

**Exercises:**  For each pair of numbers, determine whether the first number is divisible by the second number.

1. 731; 17
2. 1502; 79
3. 1053; 36
4. 4183; 47

Determine whether each number is divisible by 2.

| 7. 84 | 8. 59 |
|---|---|
| 9. 998 | 10. 2225 |

*Answers*

4. Yes   5. Yes   6. No   7. Yes
8. No   9. Yes   10. No

## Divisibility by 5

To determine the test for divisibility by 5, we start counting by 5's: 5, 10, 15, 20, 25, 30, 35, .... Note that the ones digit will always be 5 or 0, no matter how high we count.

> **BY 5**
>
> A number is **divisible by 5** if its ones digit is 0 or 5.

**EXAMPLES** Determine whether each number is divisible by 5.

**8.** 220 *is* divisible by 5 because the ones digit is 0.

**9.** 475 *is* divisible by 5 because the ones digit is 5.

**10.** 6514 *is not* divisible by 5 because the ones digit is neither 0 nor 5.

Do Exercises 11–14.

Determine whether each number is divisible by 5.

**11.** 5780          **12.** 3427

**13.** 34,678          **14.** 7775

## Divisibility by 10

> **BY 10**
>
> A number is **divisible by 10** if its ones digit is 0.

We know that this test works because the product of 10 and *any* number has a ones digit of 0.

**EXAMPLES** Determine whether each number is divisible by 10.

**11.** 3440 *is* divisible by 10 because the ones digit is 0.

**12.** 3447 *is not* divisible by 10 because the ones digit is not 0.

Do Exercises 15–18.

Determine whether each number is divisible by 10.

**15.** 305          **16.** 847

**17.** 300          **18.** 8760

## Divisibility by 3

To test for divisibility by 3 and 9, we examine the sum of a number's digits.

> **BY 3**
>
> A number is **divisible by 3** if the sum of its digits is divisible by 3.

An explanation of why this test works is outlined in Exercise 69 at the end of this section.

Determine whether each number is divisible by 3.

**19.** 111          **20.** 1111

**21.** 309          **22.** 17,216

**EXAMPLES** Determine whether each number is divisible by 3.

**13.** 18          $1 + 8 = 9$

**14.** 93          $9 + 3 = 12$          Each *is* divisible by 3 because the sum of its digits *is* divisible by 3.

**15.** 201          $2 + 0 + 1 = 3$

**16.** 256          $2 + 5 + 6 = 13$          The sum of the digits, 13, *is not* divisible by 3, so 256 *is not* divisible by 3.

Do Exercises 19–22.

**Answers**

11. Yes    12. No    13. No    14. Yes
15. No     16. No    17. Yes   18. Yes
19. Yes    20. No    21. Yes   22. No

## Divisibility by 9

The test for divisibility by 9 is similar to the test for divisibility by 3.

---

**BY 9**

A number is **divisible by 9** if the sum of its digits is divisible by 9.

---

**EXAMPLES**   Determine whether each number is divisible by 9.

**17.** 6984

Because $6 + 9 + 8 + 4 = 27$ and 27 is divisible by 9, 6984 *is* divisible by 9.

**18.** 322

Because $3 + 2 + 2 = 7$ and 7 is not divisible by 9, 322 *is not* divisible by 9.

> Do Exercises 23–26.

Determine whether each number is divisible by 9.

**23.** 16                    **24.** 117

**25.** 309                  **26.** 29,223

## Divisibility by 6

A number divisible by 6 is a multiple of 6. But $6 = 2 \cdot 3$, so the number is also a multiple of 2 and 3. Since 2 and 3 have no factors in common, a number is divisible by 6 if it is divisible by 2 *and* by 3.

---

**BY 6**

A number is **divisible by 6** if its ones digit is 0, 2, 4, 6, or 8 (is even) and the sum of its digits is divisible by 3.

---

**EXAMPLES**   Determine whether each number is divisible by 6.

**19.** 720

Because 720 is even, it is divisible by 2. Also, $7 + 2 + 0 = 9$, so 720 is divisible by 3. Thus, 720 *is* divisible by 6.

720     $7 + 2 + 0 = 9$
↑                    ↑
Even        Divisible by 3

**20.** 73

73 *is not* divisible by 6 because it is not even.

**21.** 256

Although 256 is even, it *is not* divisible by 6 because the sum of its digits, $2 + 5 + 6$, or 13, is not divisible by 3.

> Do Exercises 27–30.

Determine whether each number is divisible by 6.

**27.** 420                  **28.** 106

**29.** 321                  **30.** 444

*Answers*

**23.** No    **24.** Yes    **25.** No    **26.** Yes
**27.** Yes    **28.** No    **29.** No    **30.** Yes

**a**   Multiply by 1, 2, 3, and so on, to find ten multiples of each number.

**1.** 7          **2.** 4          **3.** 20          **4.** 50          **5.** 3          **6.** 8

**7.** 12         **8.** 15         **9.** 10          **10.** 11         **11.** 25         **12.** 100

**13.** Determine whether 83 is divisible by 3.          **14.** Determine whether 29 is divisible by 2.

**15.** Determine whether 525 is divisible by 7.          **16.** Determine whether 346 is divisible by 8.

**17.** Determine whether 8127 is divisible by 9.          **18.** Determine whether 4144 is divisible by 4.

**b**   For Exercises 19–30, answer Yes or No and give a reason based on the tests for divisibility.

**19.** Determine whether 84 is divisible by 3.          **20.** Determine whether 467 is divisible by 9.

**21.** Determine whether 5553 is divisible by 5.          **22.** Determine whether 2004 is divisible by 6.

**23.** Determine whether 671,500 is divisible by 10.          **24.** Determine whether 6120 is divisible by 5.

**25.** Determine whether 1773 is divisible by 9.          **26.** Determine whether 3286 is divisible by 3.

**27.** Determine whether 21,687 is divisible by 2.          **28.** Determine whether 64,091 is divisible by 10.

**29.** Determine whether 32,109 is divisible by 6.          **30.** Determine whether 9840 is divisible by 2.

For Exercises 31–38, test each number for divisibility by 2, 3, 5, 6, 9, and 10.

**31.** 6825          **32.** 12,600          **33.** 119,117          **34.** 2916

**35.** 127,575          **36.** 25,088          **37.** 9360          **38.** 143,507

To answer Exercises 39–44, consider the following numbers. Use the tests for divisibility.

| 46 | 300 | 85 | 256 |
| 224 | 36 | 711 | 8064 |
| 19 | 45,270 | 13,251 | 1867 |
| 555 | 4444 | 254,765 | 21,568 |

**39.** Which of the above are divisible by 3?

**40.** Which of the above are divisible by 2?

**41.** Which of the above are divisible by 10?

**42.** Which of the above are divisible by 5?

**43.** Which of the above are divisible by 6?

**44.** Which of the above are divisible by 9?

To answer Exercises 45–50, consider the following numbers.

| 56 | 200 | 75 | 35 |
| 324 | 42 | 812 | 402 |
| 784 | 501 | 2345 | 111,111 |
| 55,555 | 3009 | 2001 | 1005 |

**45.** Which of the above are divisible by 2?

**46.** Which of the above are divisible by 3?

**47.** Which of the above are divisible by 5?

**48.** Which of the above are divisible by 10?

**49.** Which of the above are divisible by 9?

**50.** Which of the above are divisible by 6?

## Skill Maintenance

Solve.

**51.** $16 \cdot t = 848$   [1.7b], [2.8b]

**52.** $m + 9 = 14$   [1.7b], [2.8a]

**53.** $23 + x = 15$   [1.7b], [2.8a]

**54.** $24 \cdot m = -576$   [1.7b], [2.8b]

Solve.   [1.8a]

**55.** Marty's automobile has a 5-speed transmission and gets 33 mpg in city driving. How many gallons of gas will it use to travel 1485 mi of city driving?

**56.** There are 60 min in 1 hr. How many minutes are there in 72 hr?

Evaluate.   [1.9b], [2.4b]

**57.** $5^3$

**58.** $(-2)^4$

Write in exponential notation.   [1.9a]

**59.** $9 \cdot 9 \cdot 9$

**60.** $3 \cdot 3 \cdot 3 \cdot 3 \cdot 3 \cdot 3$

## Synthesis

**61.** 🖩 Find the largest five-digit number that is divisible by 47.

**62.** 🖩 Find the largest six-digit number that is divisible by 53.

Find the smallest number that is simultaneously a multiple of the given numbers.

**63.** 2, 3, and 5

**64.** 3, 5, and 7

**65.** 6, 10, and 14

**66.** 🖩 17, 43, and 85

**67.** 30, 70, and 120

**68.** 25, 100, and 175

**69.** To help see why the tests for division by 3 and 9 work, note that any four-digit number *abcd* can be rewritten as $1000 \cdot a + 100 \cdot b + 10 \cdot c + d$, or $999a + 99b + 9c + a + b + c + d$.

  **a)** Explain why $999a + 99b + 9c$ is divisible by both 9 and 3 for all choices of *a, b, c,* and *d*.

  **b)** Explain why the four-digit number *abcd* is divisible by 9 if $a + b + c + d$ is divisible by 9 and is divisible by 3 if $a + b + c + d$ is divisible by 3.

**70.** A passenger in a taxicab asks for the cab number. The driver says abruptly, "Sure—it's the smallest multiple of 11 that, when divided by 2, 3, 4, 5, or 6, has a remainder of 1." What is the number?

Following are the rules for divisibility by 4, 8, 7, and 11. Use these for Exercises 71 and 72.

| NUMBER | RULE | TEST 23,904,328 FOR DIVISIBILITY |
| --- | --- | --- |
| 4 | A number is divisible by 4 if the number named by its last two digits is divisible by 4. | The number named by the last two digits is 28. Since 28 is divisible by 4, 23,904,328 is divisible by 4. |
| 8 | A number is divisible by 8 if the number named by its last three digits is divisible by 8. | The number named by the last three digits is 328. Since 328 is divisible by 8, 23,904,328 is divisible by 8. |
| 7 | Divide the number into groups of three digits, starting at the right. Start with the group at the right, subtract the next group to the left, add the next group to the left, and so on. If the resulting number is divisible by 7, the original number is divisible by 7. | The number is already divided into groups of three by commas. Calculate $328 - 904 + 23 = -553$. Since $-553$ is divisible by 7, 23,904,328 is divisible by 7. |
| 11 | Find the sum of the odd-numbered digits (the 1st digit plus the 3rd plus the 5th and so on). Find the sum of the even-numbered digits. Subtract these sums. If the difference is divisible by 11, then the original number is divisible by 11. | Add the odd-numbered digits: $8 + 3 + 0 + 3 = 14$. Add the even-numbered digits: $2 + 4 + 9 + 2 = 17$. Subtract: $17 - 14 = 3$. Since 3 is *not* divisible by 11, 23,904,328 is not divisible by 11. |

**71.** Test 332,986,412 for divisibility by 4, 8, 7, and 11.

**72.** Test 6,637,105,860 for divisibility by 4, 8, 7, and 11.

# 3.2

# Factorizations

In Section 3.1, we saw that both 28 and 35 are multiples of 7. This is the same as saying that 7 is a *factor* of both 28 and 35. When a number is expressed as a product of factors, we say that we have *factored* the original number. Thus, "factor" can be used as either a noun or a verb. The ability to factor is an important skill needed for a solid understanding of fractions.

## a Factors and Factorizations

From the equation $3 \cdot 4 = 12$, we can say that 3 and 4 are **factors** of 12. Since $12 = 12 \cdot 1$, we know that 12 and 1 are also factors of 12.

---

**FACTORS AND FACTORIZATIONS**

- In the product $a \cdot b$, $a$ and $b$ are called **factors**.
- A number $c$ is a **factor** of $a$ if $a$ is divisible by $c$.
- A **factorization** of $a$ expresses $a$ as a product of two or more numbers.

---

For example, each of the following gives a factorization of 12.

$12 = 4 \cdot 3$ ⟵ This factorization shows that 4 and 3 are factors of 12.

$12 = 12 \cdot 1$ ⟵ This factorization shows that 12 and 1 are factors of 12.

$12 = 6 \cdot 2$ ⟵ This factorization shows that 6 and 2 are factors of 12.

$12 = 2 \cdot 3 \cdot 2$ ⟵ This factorization shows that 2 and 3 are factors of 12.

Thus, 1, 2, 3, 4, 6, and 12 are all factors of 12. Note that since $n = n \cdot 1$, every number has a factorization, and every number has itself and 1 as factors.

**EXAMPLE 1** Find all the factors of 18.

Beginning at 1, we check all positive integers to see if they are factors of 18. If they are, we write the factorization. We stop when we have already included the next integer in a factorization.

1 is a factor of every number.     $1 \cdot 18$

2 is a factor of 18.               $2 \cdot 9$

3 is a factor of 18.               $3 \cdot 6$

4 is *not* a factor of 18.

5 is *not* a factor of 18.

6 is the next integer, but we have already listed 6 as a factor in the product $3 \cdot 6$.

We need check no additional numbers, because any integer greater than 6 must be paired with a factor less than 6.

We now write the factors of 18 beginning with 1, going down the list of factorizations writing the first factor, then up the list of factorizations writing the second factor:

1,   2,   3,   6,   9,   18.

Do Margin Exercises 1–4.

## OBJECTIVES

**a** Find the factors of a number.

**b** Given a number from 1 to 100, tell whether it is prime, composite, or neither.

**c** Find the prime factorization of a composite number.

---

**SKILL TO REVIEW**
Objective 1.9a: Write exponential notation for products such as $4 \cdot 4 \cdot 4$.

Write in exponential notation.
1. $2 \cdot 2 \cdot 2$       2. $5 \cdot 5 \cdot 5 \cdot 5$

Find all the factors of each number.
1. 10                    2. 45

3. 62                    4. 24

***Answers***

*Skill to Review:*
1. $2^3$    2. $5^4$

*Margin Exercises:*
1. 1, 2, 5, 10    2. 1, 3, 5, 9, 15, 45
3. 1, 2, 31, 62    4. 1, 2, 3, 4, 6, 8, 12, 24

##  b Prime and Composite Numbers

> **PRIME AND COMPOSITE NUMBERS**
>
> A natural number that has exactly two different factors, only itself and 1, is called a **prime number**.
>  • The number 1 is *not* prime.
>  • A natural number, other than 1, that is not prime is **composite**.

**EXAMPLE 2**  Determine which of the numbers 1, 2, 7, 8, 9, 11, 18, 27, 39, 43, 56, 59, and 77 are prime, which are composite, and which are neither.

The number 1 is not prime. It does not have *two* different factors.

The number 2 is prime. It has only the factors 2 and 1.

The numbers 7, 11, 43, and 59 are prime. Each has only two factors, itself and 1.

The number 8 is not prime. It has the factors 1, 2, 4, and 8 and is composite.

The numbers 9, 18, 27, 39, 56, and 77 are composite. Each has more than two factors.

Thus, we have

Prime:         2, 7, 11, 43, 59;
Composite:   8, 9, 18, 27, 39, 56, 77;
Neither:       1.

We can make several observations about prime numbers.

 • Because 0 is not a natural number, it is neither prime nor composite.
 • The number 1 is not prime because it does not have two different factors.
 • The number 2 is the smallest prime and the only even prime, since 2 is a factor of all even numbers.
 • To determine whether an odd number is prime, check divisibility by prime numbers beginning with 3 and 5. If you reach a point where the quotient is less than the divisor and none of the primes up to that point are factors, the number you are checking is prime.

Do Exercise 5.

Lists of numbers known to be prime are available in math handbooks. The following is a table of the prime numbers from 2 to 157. Although you need not memorize the entire list, remembering at least the first nine or ten is important.

> **A TABLE OF PRIMES FROM 2 TO 157**
>
> 2, 3, 5, 7, 11, 13, 17, 19, 23, 29, 31, 37, 41, 43, 47, 53, 59, 61, 67, 71, 73, 79, 83, 89, 97, 101, 103, 107, 109, 113, 127, 131, 137, 139, 149, 151, 157

There are infinitely many prime numbers. It takes extensive computer operations to determine whether very large numbers are prime. In 2009, the largest known prime was $2^{43112609} - 1$. This number has almost 13 million digits!

Source: http://primes.utm.edu//largest.html

**5.** Classify each number as prime, composite, or neither.

1, 2, 6, 12, 13, 19, 41, 65, 73, 99

*Answer*

**5.** 2, 13, 19, 41, 73 are prime; 6, 12, 65, 99 are composite; 1 is neither

## (c) Prime Factorizations

When we factor a composite number into a product of primes, we find the **prime factorization** of the number. We can do this by making a series of successive divisions or by using a *factor tree*.

To use division, we consider the prime numbers 2, 3, 5, 7, 11, 13, and so on, and determine whether a given number is divisible by the primes.

**EXAMPLE 3**  Find the prime factorization of 39.

We check for divisibility by the first prime, 2. Since 39 is not even, 2 is not a factor of 39. Since the sum of the digits in 39 is 12 and 12 is divisible by 3, we know that 39 is divisible by 3. We then perform the division.

$$\begin{array}{r} 13 \\ 3\overline{)39} \end{array} \quad R = 0 \qquad \text{\textit{A remainder of 0 confirms that 3 is a factor of 39.}}$$

Because 13 is prime, we can now write the prime factorization:

$$39 = 3 \cdot 13.$$

**EXAMPLE 4**  Find the prime factorization of 220.

We consider the first prime, 2. Since 220 is even, it must have 2 as a factor.

$$\begin{array}{r} 110 \\ 2\overline{)220} \end{array} \quad R = 0 \qquad 220 = 2 \cdot 110$$

Because 110 is also even, we divide again by 2.

$$\begin{array}{r} 55 \\ 2\overline{)110} \end{array} \quad R = 0 \qquad 220 = 2 \cdot 2 \cdot 55$$

Since 55 is odd, it is not divisible by 2. We move to the next prime, 3. The sum of the digits is 10 and 10 is not divisible by 3, so we move to the next prime, 5. Since the ones digit of 55 is 5, we divide by 5.

$$\begin{array}{r} 11 \\ 5\overline{)55} \end{array} \quad R = 0 \qquad 220 = 2 \cdot 2 \cdot 5 \cdot 11$$

Because 11 is prime, we are finished. The prime factorization is

$$220 = 2 \cdot 2 \cdot 5 \cdot 11.$$

We abbreviate our procedure as follows.

$$\begin{array}{r} 11 \\ 5\overline{)55} \\ 2\overline{)110} \\ 2\overline{)220} \end{array}$$

$$220 = 2 \cdot 2 \cdot 5 \cdot 11$$

A factorization like $2 \cdot 2 \cdot 5 \cdot 11$ could also be expressed as $5 \cdot 2 \cdot 2 \cdot 11$ or $2 \cdot 5 \cdot 11 \cdot 2$ or $2^2 \cdot 5 \cdot 11$ or $11 \cdot 2^2 \cdot 5$. The prime factors are the same in each case. For this reason, we agree that any of these may be considered "the" prime factorization of 220.

**TO THE STUDENT**

At the front of the text, you will find the Bittinger Student Organizer card. This pullout card will help you keep track of important dates and useful contact information. You can also use it to plan time for class, study, work, and relaxation. By managing your time wisely, you will provide yourself the best possible opportunity to be successful in this course.

> Every composite number has just one (unique) prime factorization.

This result is sometimes called the Fundamental Theorem of Arithmetic.

**EXAMPLE 5**   Find the prime factorization of 187.

We check for divisibility by 2, 3, and 5, and find that 187 is not divisible by any of these numbers. The next prime number, 7, does not divide 187 evenly. However, when we divide by 11, the remainder is 0, so 11 is a factor of 187.

$$\begin{array}{r} 17 \\ 11\overline{)187} \end{array}$$   We can write $187 = 11 \cdot 17$.

Because 17 is prime, we can factor no further. The complete factorization is

$187 = 11 \cdot 17$.    All factors are prime.

**EXAMPLE 6**   Find the prime factorization of 72.

$$\begin{array}{r} 3 \\ 3\overline{)9} \\ 2\overline{)18} \\ 2\overline{)36} \\ 2\overline{)72} \end{array}$$   ← 3 is prime, so we stop dividing.

← Begin here and work upward.

$72 = 2 \cdot 2 \cdot 2 \cdot 3 \cdot 3$

Another way to find the prime factorization of 72 is to use a **factor tree** as follows. Begin by determining any factorization you can, and then continue factoring. Each of the following trees gives the same prime factorization.

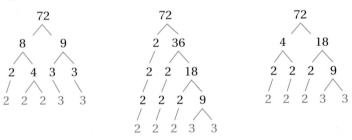

**EXAMPLE 7**   Find the prime factorization of 130 and list all factors of 130.

We can use a string of divisions or a factor tree.

$$\begin{array}{r} 13 \\ 5\overline{)\,65} \\ 2\overline{)130} \end{array}$$

$$\begin{array}{c} 130 \\ \diagdown \\ 10 \;\cdot\; 13 \end{array}$$

$130 = 2 \cdot 5 \cdot 13$

From the prime factorization we can easily identify 2, 5, and 13 as factors. We can form the products $2 \cdot 5$, $2 \cdot 13$, and $5 \cdot 13$ to find that 10, 26, and 65 are also factors. A complete list of factors is 1, 2, 5, 10, 13, 26, 65, and 130.

Do Exercises 6–13.

Finding a number's prime factorization can be quite challenging, especially when the prime factors themselves are large. This difficulty is used worldwide as a way of securing transactions over the Internet.

---

> **Caution!**
>
> Finding the prime factorization of a number is not the same as finding all of its factors. A prime factorization of a number is a product; the factors of a number are a list. Compare the following for the number 72:
>
> Prime factorization of 72:
>
> $2 \cdot 2 \cdot 2 \cdot 3 \cdot 3$;
>
> Factors of 72:
>
> 1, 2, 3, 4, 6, 8, 9, 12, 18, 24, 36, 72.

---

Find the prime factorization of each number.

**6.** 6      **7.** 12

**8.** 45      **9.** 98

**10.** 126      **11.** 144

**12.** 91      **13.** 1925

**Answers**

**6.** $2 \cdot 3$   **7.** $2 \cdot 2 \cdot 3$   **8.** $3 \cdot 3 \cdot 5$
**9.** $2 \cdot 7 \cdot 7$   **10.** $2 \cdot 3 \cdot 3 \cdot 7$
**11.** $2 \cdot 2 \cdot 2 \cdot 2 \cdot 3 \cdot 3$   **12.** $7 \cdot 13$
**13.** $5 \cdot 5 \cdot 7 \cdot 11$

**a**  List all the factors of each number.

**1.** 18                **2.** 16                **3.** 54                **4.** 48

**5.** 9                 **6.** 4                 **7.** 13                **8.** 11

**9.** 98                **10.** 100              **11.** 255              **12.** 120

**b**  Classify each number as prime, composite, or neither.

**13.** 17               **14.** 24               **15.** 22               **16.** 31

**17.** 48               **18.** 43               **19.** 53               **20.** 54

**21.** 1                **22.** 2                **23.** 81               **24.** 37

**25.** 47               **26.** 51               **27.** 29               **28.** 49

**c**  Find the prime factorization of each number.

**29.** 27       **30.** 16       **31.** 14       **32.** 15       **33.** 80

**34.** 32       **35.** 25       **36.** 40       **37.** 62       **38.** 169

**39.** 100      **40.** 110      **41.** 143      **42.** 50       **43.** 121

**44.** 170      **45.** 273      **46.** 675      **47.** 175      **48.** 196

**49.** 209      **50.** 217      **51.** 1200     **52.** 1800     **53.** 693

**54.** 675      **55.** 2884     **56.** 484      **57.** 1122     **58.** 6435

# Skill Maintenance

Multiply.

**59.** $-2 \cdot 13$   [2.4a]

**60.** $(-8)(-32)$   [2.4a]

Add.

**61.** $-17 + 25$   [2.2a]

**62.** $-9 + (-14)$   [2.2a]

Divide.

**63.** $53 \div 53$   [1.5a]

**64.** $-98 \div 1$   [2.5a]

**65.** $0 \div 22$   [1.5a]

**66.** $0 \div (-42)$   [2.5a]

Solve.   [1.8a]

**67.** *Morel Mushrooms.*   During the spring, Kate's Country Market sold 43 pounds of fresh morel mushrooms. The mushrooms sold for $22 a pound. Find the total amount Kate took in from the sale of the mushrooms.

**68.** Sandy can type 62 words per minute. How long will it take her to type 12,462 words?

# Synthesis

Find the prime factorization of each number.

**69.** ▦ 136,097

**70.** ▦ 102,971

**71.** ▦ 473,073,361

**72.** Describe an arrangement of 54 objects that corresponds to the factorization $54 = 6 \times 9$.

**73.** Describe an arrangement of 24 objects that corresponds to the factorization $24 = 2 \cdot 3 \cdot 4$.

**74.** Two numbers are **relatively prime** if there is no prime number that is a factor of both numbers. For example, 10 and 21 are relatively prime but 15 and 18 are not. List five pairs of composite numbers that are relatively prime.

**75.** *Factors and Sums.*   In the table below, the top number in each column has been factored in such a way that the sum of the factors is the bottom number in the column. For example, in the first column, 56 has been factored as $7 \cdot 8$, and $7 + 8 = 15$, the bottom number. Such thinking will be important in understanding the meaning of a factor and in algebra.

| PRODUCT | 56 | 63 | 36 | 72 | 140 | 96 | 48 | 168 | 110 | 90 | 432 | 63 |
|---------|----|----|----|----|-----|----|----|-----|-----|----|-----|----|
| FACTOR  | 7  |    |    |    |     |    |    |     |     |    |     |    |
| FACTOR  | 8  |    |    |    |     |    |    |     |     |    |     |    |
| SUM     | 15 | 16 | 20 | 38 | 24  | 20 | 14 | 29  | 21  | 19 | 42  | 24 |

Find the missing numbers in the table.

The study of arithmetic begins with the set of whole numbers

0, 1, 2, 3, 4, 5, 6, 7, 8, 9, 10, 11, and so on.

We also need to be able to use fractional parts of numbers such as halves, thirds, fourths, and so on. Here is an example.

The tread depth of an IRL Indy Car Series tire is $\frac{3}{32}$ in. Tires for a normal car have a tread depth of $\frac{10}{32}$ in. when new and are considered bald at $\frac{2}{32}$ in.

**Sources:** Indy500.com; *Consumer Reports*

## a Identifying Numerators and Denominators

Numbers like those above and the ones below are written in **fraction notation**. The top number is called the **numerator** and the bottom number is called the **denominator**.

$$\frac{1}{2}, \quad \frac{7}{8}, \quad \frac{-8}{5}, \quad \frac{x}{y}, \quad -\frac{4}{25}, \quad \frac{2a}{7b}$$

**EXAMPLE 1** Identify the numerator and the denominator.

$\dfrac{7}{8}$ ← Numerator
     ← Denominator

Do Exercises 1–4.

Let's look at various situations that involve fractions.

### *Fractions as a Partition of an Object Divided into Equal Parts*

Consider a candy bar divided into 5 equal sections. If you eat 2 sections, you have eaten $\frac{2}{5}$ of the candy bar.

The denominator 5 tells us the unit, $\frac{1}{5}$. The numerator 2 tells us the number of equal parts we are considering, 2.

For each fraction, identify the numerator and the denominator.

1. $\frac{3}{20}$ of people age 5 and older in Maryland speak a language other than English at home.
   **Source:** 2006 American Community Survey

2. $\frac{19}{100}$ of the U.S. population in 2050 is projected to be foreign-born.
   **Source:** Pew Research Center

3. $\dfrac{5a}{7m}$

4. $\dfrac{-22}{3}$

*Answers*

1. Numerator: 3; denominator: 20
2. Numerator: 19; denominator: 100
3. Numerator: 5a; denominator: 7m
4. Numerator: −22; denominator: 3

**EXAMPLE 2** What part is shaded?

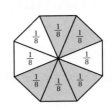

There are 8 equal parts. This tells us the unit, $\frac{1}{8}$. The *denominator* is 8. We have 5 of the units shaded. This tells us the *numerator*, 5. Thus,

$$\frac{5}{8} \quad \begin{array}{l} \leftarrow 5 \text{ units are shaded.} \\ \leftarrow \text{The unit is } \frac{1}{8}. \end{array}$$

is shaded.

**EXAMPLE 3** What part of a cup is filled?

The measuring cup is divided into 3 parts of the same size, and 2 of them are filled. This is $2 \cdot \frac{1}{3}$, or $\frac{2}{3}$. Thus, $\frac{2}{3}$ (read *two-thirds*) of a cup is filled.

Do Exercises 5–8.

**EXAMPLE 4** What part of an inch is indicated?

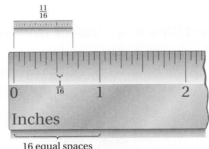

Each inch on the ruler shown above is divided into 16 equal parts. The highlighting extends to the 11th mark. Thus, $\frac{11}{16}$ of an inch is indicated.

Do Exercise 9.

---

What part is shaded?

**5.**

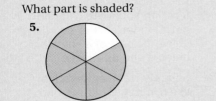

**6.** 1 mile

**7.**

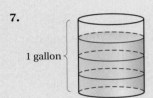

1 gallon

**8.**

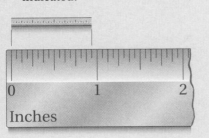

**9.** What part of an inch is indicated?

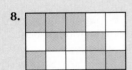

---

*Answers*

5. $\frac{5}{6}$   6. $\frac{1}{3}$   7. $\frac{3}{4}$   8. $\frac{8}{15}$   9. $\frac{15}{16}$

Fractions greater than or equal to 1, such as $\frac{24}{24}$, $\frac{11}{6}$, and $\frac{5}{4}$, correspond to situations like the following.

**EXAMPLE 5** What part is shaded?

**a)**

The rectangle is divided into 24 equal parts. Thus, the unit is $\frac{1}{24}$. The denominator is 24. All 24 equal parts are shaded. This tells us that the numerator is 24. Thus, $\frac{24}{24}$ is shaded.

**b)**

Each circle is divided into 6 parts. Thus, the unit is $\frac{1}{6}$. The denominator is 6. We see that 11 of the equal units are shaded. This tells us that the numerator is 11. Thus, $\frac{11}{6}$ is shaded.

**EXAMPLE 6** *Ice Cream Roll-up Cake.* What part of an ice cream roll-up cake is shaded?

3 ice cream roll-up cakes

Each cake is divided into 6 equal slices. The unit is $\frac{1}{6}$. The *denominator* is 6. We see that 13 of the slices are shaded. This tells us that the *numerator* is 13. Thus, $\frac{13}{6}$ are shaded.

> Do Exercises 10–12.

Fractions larger than or equal to 1, such as $\frac{13}{6}$ or $\frac{9}{9}$, are sometimes referred to as "improper" fractions. We will not use this terminology because notation such as $\frac{27}{8}$, $\frac{11}{3}$ and $\frac{4}{4}$ is quite "proper" and very common in algebra.

### Fractions as Ratios

A **ratio** is a quotient of two quantities. We can express a ratio with fraction notation. (We will consider ratios in more detail in Chapter 7.)

What part is shaded?

**10.**

**11.** 1 mile
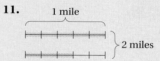
2 miles

**12.**
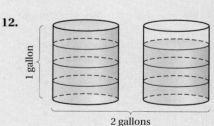
2 gallons

*Answers*

10. $\frac{15}{15}$    11. $\frac{8}{5}$    12. $\frac{7}{4}$

**EXAMPLE 7** *States East and West of the Mississippi River.* What part of this group of states is east of the Mississippi River? west of the Mississippi River?

| | |
|---|---|
| Arkansas | Wisconsin |
| Illinois | West Virginia |
| Alabama | South Dakota |
| Nebraska | |

There are 7 states in the set. We know that 4 of them, Illinois, Alabama, Wisconsin, and West Virginia, are east of the Mississippi River. Thus, 4 of 7, or $\frac{4}{7}$, are east of the Mississippi River. The 3 remaining states are west of the Mississippi River. Thus, $\frac{3}{7}$ are west of the Mississippi River.

**13.** What part of the set of states in Example 7 is north of Nashville, Tennessee?

Do Exercise 13.

**EXAMPLE 8** *Baseball Standings.* The following are the final standings in the National League West for 2009, when the division was won by the Los Angeles Dodgers. Find the ratio of Dodger wins to losses, wins to total games, and losses to total games.

| WEST | W | L | Pct. | GB | STRK | LAST 10 | HOME | ROAD |
|---|---|---|---|---|---|---|---|---|
| LA Dodgers* | 95 | 67 | .586 | — | W2 | 4-6 | 50-31 | 45-36 |
| Colorado | 92 | 70 | .568 | 3.0 | L2 | 6-4 | 51-30 | 41-40 |
| San Francisco | 88 | 74 | .543 | 7.0 | W1 | 6-4 | 52-29 | 36-45 |
| San Diego | 75 | 87 | .463 | 20.0 | L1 | 6-4 | 42-39 | 33-48 |
| Arizona | 70 | 92 | .432 | 25.0 | W1 | 4-6 | 36-45 | 34-47 |

*clinched division

SOURCE: Major League Baseball

The Dodgers won 95 games and lost 67 games. They played a total of 95 + 67, or 162, games. Thus, we have the following.

The ratio of wins to losses is $\frac{95}{67}$.

The ratio of wins to total games is $\frac{95}{162}$.

The ratio of losses to total games is $\frac{67}{162}$.

**14. Baseball Standings.** Refer to the table in Example 8. The Colorado Rockies finished second in the National League West in 2009. Find the ratio of Colorado wins to losses, wins to total games, and losses to total games.

**Source:** Major League Baseball

Do Exercise 14.

## **b** Some Fraction Notation for Integers

### *Fraction Notation for 1*

The number 1 corresponds to situations like those shown here. If we divide an object into $n$ parts and take $n$ of them, we get all of the object (1 whole object).

Since a negative number divided by a negative number is a positive number, the following is stated for all nonzero integers.

1

$\frac{2}{2}$

$\frac{4}{4}$

---

**THE NUMBER 1 IN FRACTION NOTATION**

$$\frac{n}{n} = 1, \quad \text{for any integer } n \text{ that is not 0.}$$

---

*Answers*

**13.** $\frac{5}{7}$   **14.** $\frac{92}{70}; \frac{92}{162}; \frac{70}{162}$

**EXAMPLES** Simplify. Assume any variables are nonzero.

**9.** $\dfrac{5}{5} = 1$  **10.** $\dfrac{-9}{-9} = 1$  **11.** $\dfrac{23x}{23x} = 1$

Do Exercises 15–20.

## Fraction Notation for 0

Consider the fraction $\frac{0}{4}$. This corresponds to dividing an object into 4 parts and taking none of them. We get 0.

> **THE NUMBER 0 IN FRACTION NOTATION**
>
> $\dfrac{0}{n} = 0$,  for any integer $n$ that is not 0.

**EXAMPLES** Simplify.

**12.** $\dfrac{0}{1} = 0$  **13.** $\dfrac{0}{9} = 0$

**14.** $\dfrac{0}{-23} = 0$  **15.** $\dfrac{0}{5a} = 0$,  assuming that $a \neq 0$

Fraction notation with a denominator of 0, such as $n/0$, does not represent a number because we cannot speak of an object being divided into *zero* parts. (If it is not divided at all, then we say it is undivided and remains in one part.)

> **A DENOMINATOR OF 0**
>
> $\dfrac{n}{0}$ is not defined. We say that $\dfrac{n}{0}$ is *undefined*.

Do Exercises 21–26.

## Other Integers

Consider the fraction $\frac{4}{1}$. This corresponds to taking 4 objects and dividing each into 1 part. (In other words, we do not divide them.) We have 4 objects.

> **ANY INTEGER IN FRACTION NOTATION**
>
> Any integer divided by 1 is the original integer. That is,
>
> $\dfrac{n}{1} = n$,  for any integer $n$.

**EXAMPLES** Simplify.

**16.** $\dfrac{2}{1} = 2$  **17.** $\dfrac{-9}{1} = -9$  **18.** $\dfrac{3x}{1} = 3x$

Do Exercises 27–30.

---

Simplify.

**15.** $\dfrac{1}{1}$  **16.** $\dfrac{4}{4}$

**17.** $\dfrac{a}{a}$  **18.** $\dfrac{-100}{-100}$

**19.** $\dfrac{2347}{2347}$  **20.** $\dfrac{54n}{54n}$

Simplify, if possible. Assume that $x \neq 0$.

**21.** $\dfrac{0}{1}$  **22.** $\dfrac{0}{-6}$

**23.** $\dfrac{0}{4x}$  **24.** $\dfrac{4-4}{567}$

**25.** $\dfrac{15}{0}$  **26.** $\dfrac{-4}{3-3}$

Simplify.

**27.** $\dfrac{8}{1}$  **28.** $\dfrac{-10}{1}$

**29.** $\dfrac{-346}{1}$  **30.** $\dfrac{24-1}{23-22}$

***Answers***

**15.** 1  **16.** 1  **17.** 1  **18.** 1  **19.** 1
**20.** 1  **21.** 0  **22.** 0  **23.** 0  **24.** 0
**25.** Undefined  **26.** Undefined  **27.** 8
**28.** −10  **29.** −346  **30.** 23

**a** Identify the numerator and the denominator of each fraction.

1. $\dfrac{3}{4}$

2. $\dfrac{-9}{10}$

3. $\dfrac{7}{-9}$

4. $\dfrac{15}{8}$

5. $\dfrac{2x}{3y}$

6. $\dfrac{9a}{2b}$

What part of each object or set of objects is shaded?

7.

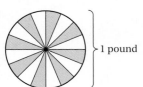

$1

8.

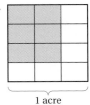

1 acre

9.

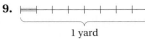

1 yard

10.

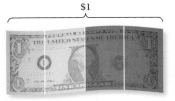

1 mile

11.

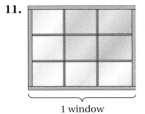

1 window

12.

1 square yard

13.

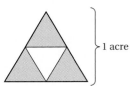

1 acre

14.

1 year

15.

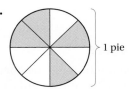

1 pie

16.

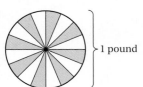

1 pound

17.

18.

**19.**

1 quart

**20.**

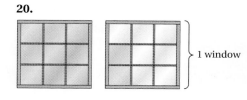

1 window

**21.**

1 spool

**22.**

1 quart

**23.**

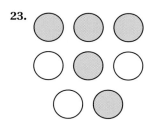

**24.**

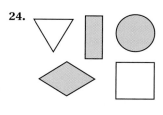

**25.**

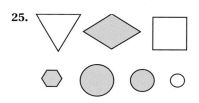

**26.**

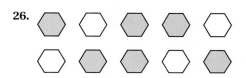

What part of an inch is indicated?

**27.**

**28.**

**29.**

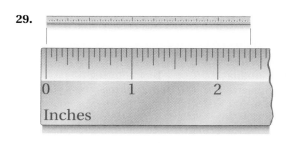

**30.**

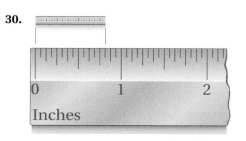

For each of Exercises 31–34, give fraction notation for the amount of gas **(a)** in the tank and **(b)** used from a full tank.

**31.**

**32.**

**33.**

**34.**

**35.** For the following set of people, what is the ratio of
   **a)** women to the total number of people?
   **b)** women to men?
   **c)** men to the total number of people?
   **d)** men to women?

**36.** For the following set of nuts and bolts, what is the ratio of
   **a)** nuts to bolts?
   **b)** bolts to nuts?
   **c)** nuts to the total number of elements?
   **d)** total number of elements to nuts?

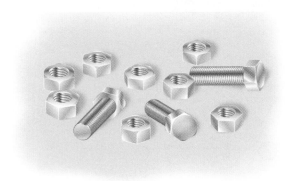

For Exercises 37 and 38, use the following bar graph, which lists a state and the number of registered nurses per 100,000 residents in that state.

**Registered Nurses per 100,000 Residents**

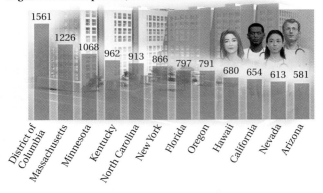

District of Columbia 1561, Massachusetts 1226, Minnesota 1068, Kentucky 962, North Carolina 913, New York 866, Florida 797, Oregon 791, Hawaii 680, California 654, Nevada 613, Arizona 581

SOURCES: www.statehealthfacts.org; Kaiser Family Foundation, 2008

**37.** What is the ratio of registered nurses to 100,000 residents in the given state?
   **a)** Minnesota    **b)** Hawaii
   **c)** Florida    **d)** Kentucky
   **e)** New York    **f)** District of Columbia

**38.** What is the ratio of registered nurses to 100,000 residents in the given state?
   **a)** North Carolina    **b)** California
   **c)** Massachusetts    **d)** Nevada
   **e)** Arizona    **f)** Oregon

**39.** Bryce delivers car parts to auto service centers. On Thursday he had 15 deliveries scheduled. By noon he had delivered only 4 orders. What is the ratio of

**a)** orders delivered to total number of orders?
**b)** orders delivered to orders not delivered?
**c)** orders not delivered to total number of orders?

**40.** *Gas Mileage.* A Mazda MX-5 will travel 308 mi on 11 gal of gasoline in highway driving. What is the ratio of

**a)** miles driven to gasoline used?
**b)** gasoline used to miles driven?

**Source:** *Road & Track*, August 2008, p. 52

**41.** *18–29 Voter Age Group.* Of every 100 voters in the 2004 presidential election, 16 voters were in the 18–29 age group. In the 2008 presidential election, this number increased to 18.5 voters. What is the ratio of voters in the 18–29 age group to all voters in the 2004 presidential election? in the 2008 election?

**Sources:** The Center for Information & Research on Civic Learning & Engagement; www.civicyouth.org

**42.** *Air-Conditioned Houses.* In 1971, of every 1000 new houses in the United States, 360 houses had air conditioning. In 2006, this ratio rose to 890 houses of every 1000 new houses. What is the ratio of new houses with air conditioning to all new houses in 1971? in 2006?

**Source:** U.S. Census Bureau

**b** Simplify. Assume that all variables are nonzero.

**43.** $\dfrac{0}{8}$

**44.** $\dfrac{8}{8}$

**45.** $\dfrac{8-1}{9-8}$

**46.** $\dfrac{16}{1}$

**47.** $\dfrac{20}{20}$

**48.** $\dfrac{-20}{1}$

**49.** $\dfrac{-45}{-45}$

**50.** $\dfrac{11-1}{10-9}$

**51.** $\dfrac{0}{-238}$

**52.** $\dfrac{19x}{19x}$

**53.** $\dfrac{19x}{1}$

**54.** $\dfrac{0}{-16}$

**55.** $\dfrac{13t}{13t}$

**56.** $\dfrac{-27m}{1}$

**57.** $\dfrac{-87}{1}$

**58.** $\dfrac{-98}{-98}$

**59.** $\dfrac{0}{2a}$

**60.** $\dfrac{0}{18}$

**61.** $\dfrac{52}{0}$

**62.** $\dfrac{8-8}{1247}$

**63.** $\dfrac{7n}{1}$

**64.** $\dfrac{1317}{0}$

**65.** $\dfrac{5}{6-6}$

**66.** $\dfrac{13}{10-10}$

## Skill Maintenance

Multiply.

**67.** $-7(30)$    [2.4a]

**68.** $23 \cdot (-14)$    [2.4a]

**69.** $(-71)(-12)0$    [2.4b]

**70.** $32(-29)0$    [2.4b]

Solve.    [1.8a]

**71.** *New York Road Runners.* The New York Road Runners, a group that runs the New York City Marathon, had only 22,190 members in 1983. The membership increased to 46,655 in 2008. How many more members were there in 2008 than in 1983?
Source: New York Road Runners

**72.** *Gas Mileage.* The 2009 Chevrolet Corvette Z51 gets 26 mpg in highway driving. How many gallons will it use in 1846 mi of highway driving?
Source: *Road & Track*, August 2008, p. 52

## Synthesis

What part of each object is shaded?

**73.**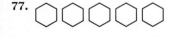

**74.**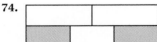

**75.**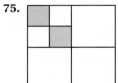

**76.**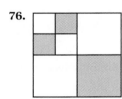

Shade or mark each figure to show $\frac{3}{5}$.

**77.**

**78.**

**79.**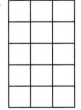

**80.**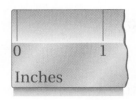

**81.** The year 2006 began on a Sunday. What fractional part of 2006 were Mondays?

**82.** The year 2007 began on a Monday. What fractional part of 2007 were Mondays?

**83.** The surface of Earth is 3 parts water and 1 part land. What fractional part of Earth is water? land?

**84.** A couple had 3 sons, each of whom had 3 daughters. If each daughter gave birth to 3 sons, what fractional part of the couple's descendants is female?

# 3.4 Multiplication and Applications

## a Multiplication by an Integer

We can find $3 \cdot \frac{1}{4}$ by thinking of repeated addition. We add three $\frac{1}{4}$'s.

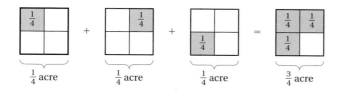

$\frac{1}{4}$ acre     $\frac{1}{4}$ acre     $\frac{1}{4}$ acre     $\frac{3}{4}$ acre

We see that $3 \cdot \frac{1}{4} = \frac{1}{4} + \frac{1}{4} + \frac{1}{4} = \frac{3}{4}$.

Do Margin Exercises 1 and 2.

To multiply a fraction by an integer,

a) multiply the top number (the numerator) by the integer and

$$6 \cdot \frac{4}{5} = \frac{6 \cdot 4}{5} = \frac{24}{5}$$

b) keep the same denominator.

**EXAMPLES** Multiply.

1. $5 \times \frac{3}{8} = \frac{5 \times 3}{8} = \frac{15}{8}$   We generally replace the $\times$ symbol with $\cdot$

Skip this step when you feel comfortable doing so.

2. $\frac{2}{5} \cdot 13 = \frac{2 \cdot 13}{5} = \frac{26}{5}$

3. $-10 \cdot \frac{1}{3} = \frac{-10}{3}$, or $-\frac{10}{3}$   Recall that $\frac{-a}{b} = -\frac{a}{b}$.

4. $a \cdot \frac{4}{7} = \frac{4a}{7}$   Recall that $a \cdot 4 = 4 \cdot a$.

Do Exercises 3–6.

**SKILL TO REVIEW**
Objective 1.4a: Multiply whole numbers.

Multiply.

1. $24 \cdot 17$     2. $5(13)$

1. Find $2 \cdot \frac{1}{3}$.

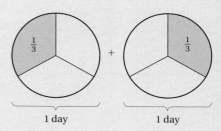

1 day     1 day

2. Find $5 \cdot \frac{1}{8}$.

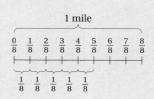

Multiply.

3. $7 \times \frac{2}{3}$     4. $(-11) \cdot \frac{3}{10}$

5. $34 \cdot \frac{2}{5}$     6. $x \cdot \frac{4}{9}$

**Answers**

*Skill to Review:*
1. 408   2. 65

*Margin Exercises:*
1. $\frac{2}{3}$   2. $\frac{5}{8}$   3. $\frac{14}{3}$   4. $-\frac{33}{10}$, or $\frac{-33}{10}$
5. $\frac{68}{5}$   6. $\frac{4x}{9}$

## b Multiplication Using Fraction Notation

**7.** Draw diagrams like the ones at right to illustrate $\frac{1}{4}$ and $\frac{1}{2} \cdot \frac{1}{4}$.

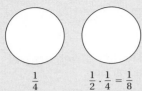

$\frac{1}{4}$          $\frac{1}{2} \cdot \frac{1}{4} = \frac{1}{8}$

To illustrate the meaning of an expression like $\frac{1}{2} \cdot \frac{1}{3}$, we first represent $\frac{1}{3}$ and then shade half of that region. Note that $\frac{1}{2} \cdot \frac{1}{3}$ is the same as $\frac{1}{2}$ of $\frac{1}{3}$.

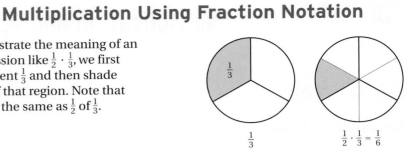

$\frac{1}{3}$          $\frac{1}{2} \cdot \frac{1}{3} = \frac{1}{6}$

Do Exercise 7.

To visualize $\frac{2}{5} \cdot \frac{3}{4}$, we first represent $\frac{3}{4}$. This is shown by the shading on the left below. Next, we divide the shaded area into 5 equal parts (using horizontal lines) and take 2 of them. That is shown by the darker shading on the right below.

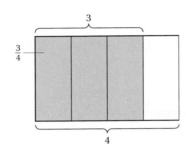 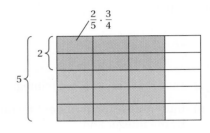

**8.** Draw diagrams like the ones at right to illustrate $\frac{1}{3}$ and $\frac{4}{5} \cdot \frac{1}{3}$.

$\frac{1}{3}$          $\frac{4}{5} \cdot \frac{1}{3} = \frac{4}{15}$

The entire object has now been divided into 20 parts, and we have shaded 6 of them twice. Thus,

$$\frac{2}{5} \cdot \frac{3}{4} = \frac{6}{20}. \quad \begin{array}{l} \leftarrow \text{This is the product of the numerators.} \\ \leftarrow \text{This is the product of the denominators.} \end{array}$$

Do Exercise 8.

Notice that the product of two fractions is the product of the numerators over the product of the denominators.

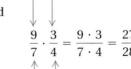

To multiply a fraction by a fraction,

**a)** multiply the numerators and

$$\frac{9}{7} \cdot \frac{3}{4} = \frac{9 \cdot 3}{7 \cdot 4} = \frac{27}{28}$$

**b)** multiply the denominators.

**EXAMPLES** Multiply.

**5.** $\dfrac{5}{6} \cdot \dfrac{7}{4} = \dfrac{5 \cdot 7}{6 \cdot 4} = \dfrac{35}{24}$

Skip this step when you feel comfortable doing so.

**6.** $\dfrac{3}{5} \cdot \dfrac{7}{8} = \dfrac{3 \cdot 7}{5 \cdot 8} = \dfrac{21}{40}$

*Answers*

**7.**

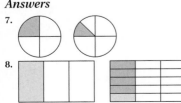

**8.**

**7.** $\left(-\dfrac{4}{x}\right)\left(-\dfrac{y}{7}\right) = \dfrac{4y}{7x}$    A negative times a negative is positive.

**8.** $(-6)\left(-\dfrac{4}{5}\right) = \dfrac{-6}{1} \cdot \dfrac{-4}{5} = \dfrac{24}{5}$    We can always write $n$ as $\dfrac{n}{1}$.

Do Exercises 9–12.

Multiply.

**9.** $\dfrac{3}{8} \cdot \dfrac{5}{7}$    **10.** $\dfrac{4}{3} \cdot \dfrac{8}{5}$

**11.** $\left(-\dfrac{3}{10}\right)\left(-\dfrac{1}{10}\right)$    **12.** $(-7)\dfrac{a}{b}$

## (c) Applications and Problem Solving

Many problems that can be solved by multiplying fractions can be thought of in terms of rectangular arrays.

**EXAMPLE 9**  A dude ranch owns a square mile of land. The owner gives $\frac{4}{5}$ of it to her daughter who, in turn, gives $\frac{2}{3}$ of her share to her son. How much land goes to the daughter's son?

1. **Familiarize.**  We first make a drawing to help solve the problem. The land may not be square, but we can think of it as a square.
   The daughter gets $\frac{4}{5}$ of the land. We shade $\frac{4}{5}$.

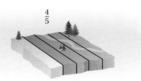

Her son gets $\frac{2}{3}$ of her part. We "raise" that.

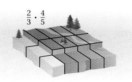

2. **Translate.**  We let $n =$ the part of the land that goes to the daughter's son. We are taking "two-thirds of four-fifths." The word "of" corresponds to multiplication. Thus, the following multiplication sentence corresponds to the situation:

$$\frac{2}{3} \cdot \frac{4}{5} = n.$$

3. **Solve.**  The number sentence tells us what to do. We multiply:

$$\frac{2}{3} \cdot \frac{4}{5} = n, \quad \text{or} \quad \frac{8}{15} = n.$$

4. **Check.**  We can check this in the figure above, where we see that 8 of 15 equally sized parts will go to the daughter's son.

5. **State.**  The daughter's son gets $\frac{8}{15}$ of a square mile of land.

Do Exercise 13.

### STUDY TIPS

**USE YOUR TEXT FOR NOTES**

If you own your text, consider using it as a notebook. Since many instructors closely parallel the book, it is often useful to make notes on the appropriate page as he or she is lecturing.

**13.** Camp Mohawk uses $\frac{3}{4}$ of its extra land for recreational purposes. Of that, $\frac{1}{2}$ is used for swimming pools. What part of the extra land is used for swimming pools?

*Answers*

**9.** $\dfrac{15}{56}$  **10.** $\dfrac{32}{15}$  **11.** $\dfrac{3}{100}$

**12.** $-\dfrac{7a}{b}$, or $\dfrac{-7a}{b}$  **13.** $\dfrac{3}{8}$

The area of a rectangular region is found by multiplying length by width. That is true whether length and width are whole numbers or not. Remember, the area of a rectangular region is given by the formula

$$A = l \cdot w \quad (Area = length \cdot width).$$

**EXAMPLE 10** The length of a computer's Delete key is $\frac{9}{16}$ in. The width is $\frac{3}{8}$ in. What is the area?

1. **Familiarize.** Recall that area is length times width. We make a drawing, letting $A =$ the area of the computer key.

2. **Translate.** Next, we translate.

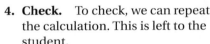

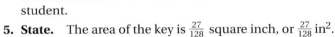

3. **Solve.** The sentence tells us what to do. We multiply:

$$\frac{9}{16} \cdot \frac{3}{8} = \frac{9 \cdot 3}{16 \cdot 8} = \frac{27}{128}.$$

4. **Check.** To check, we can repeat the calculation. This is left to the student.

5. **State.** The area of the key is $\frac{27}{128}$ square inch, or $\frac{27}{128}$ in². 

> Do Exercise 14.

**EXAMPLE 11** A recipe for oatmeal chocolate chip cookies calls for $\frac{3}{4}$ cup of rolled oats. Monica is making $\frac{1}{2}$ of the recipe. How much oats should she use?

1. **Familiarize.** We first make a drawing or at least visualize the situation. We let $n =$ the amount of oats that Monica should use.

$\frac{3}{4}$ cup in the recipe    $\frac{1}{2} \cdot \frac{3}{4}$ cup in $\frac{1}{2}$ the recipe

2. **Translate.** We are finding $\frac{1}{2}$ of $\frac{3}{4}$, so the multiplication sentence $\frac{1}{2} \cdot \frac{3}{4} = n$ corresponds to the situation.

3. **Solve.** We carry out the multiplication:

$$\frac{1}{2} \cdot \frac{3}{4} = \frac{1 \cdot 3}{2 \cdot 4} = \frac{3}{8}.$$

Thus, $\frac{3}{8} = n$.

4. **Check.** We check by repeating the calculation. This is left to the student.

5. **State.** Monica should use $\frac{3}{8}$ cup of oats.

> Do Exercise 15.

---

14. The length of a rectangular cornfield is $\frac{15}{16}$ mi. The width is $\frac{5}{8}$ mi. What is the area of the cornfield?

---

**Caution!**

When stating the solution of an applied problem, be sure to include appropriate units. Recall that units for area are square inches, square meters, and so on.

---

15. Of the students at Overton Junior College, $\frac{1}{8}$ participate in sports and $\frac{3}{5}$ of these play football. What fractional part of the students play football?

*Answers*

14. $\frac{75}{128}$ mi²   15. $\frac{3}{40}$

**a** Multiply.

**1.** $3 \cdot \dfrac{1}{8}$

**2.** $2 \cdot \dfrac{1}{5}$

**3.** $(-5) \times \dfrac{1}{6}$

**4.** $(-4) \times \dfrac{1}{7}$

**5.** $\dfrac{2}{3} \cdot 7$

**6.** $\dfrac{2}{5} \cdot 7$

**7.** $(-1)\dfrac{7}{9}$

**8.** $(-1)\dfrac{4}{11}$

**9.** $\dfrac{5}{6} \cdot x$

**10.** $\left(-\dfrac{7}{8}\right)y$

**11.** $\dfrac{2}{5}(-3)$

**12.** $\dfrac{3}{5}(-4)$

**13.** $a \cdot \dfrac{2}{7}$

**14.** $b \cdot \dfrac{3}{8}$

**15.** $-17 \times \dfrac{m}{6}$

**16.** $\dfrac{n}{7} \cdot 30$

**17.** $-3 \cdot \dfrac{-2}{5}$

**18.** $-4 \cdot \dfrac{-5}{7}$

**19.** $-\dfrac{2}{7}(-x)$

**20.** $-\dfrac{3}{4}(-a)$

**b** Multiply.

**21.** $\dfrac{2}{5} \cdot \dfrac{2}{3}$

**22.** $\dfrac{3}{4} \cdot \dfrac{3}{5}$

**23.** $\left(-\dfrac{1}{4}\right) \times \dfrac{1}{10}$

**24.** $\left(-\dfrac{1}{3}\right) \times \dfrac{1}{10}$

**25.** $\dfrac{2}{3} \times \dfrac{1}{5}$

**26.** $\dfrac{3}{5} \times \dfrac{1}{5}$

**27.** $\dfrac{2}{y} \cdot \dfrac{x}{9}$

**28.** $\left(-\dfrac{3}{4}\right)\left(-\dfrac{3}{5}\right)$

**29.** $\left(-\dfrac{3}{4}\right)\left(-\dfrac{3}{4}\right)$

**30.** $\dfrac{3}{b} \cdot \dfrac{a}{7}$

**31.** $\dfrac{2}{3} \cdot \dfrac{7}{13}$

**32.** $\dfrac{3}{11} \cdot \dfrac{4}{5}$

**33.** $\dfrac{1}{10}\left(\dfrac{-3}{5}\right)$

**34.** $\dfrac{3}{10}\left(\dfrac{-7}{5}\right)$

**35.** $\dfrac{7}{8} \cdot \dfrac{a}{8}$

**36.** $\dfrac{4}{5} \cdot \dfrac{7}{x}$

**37.** $\dfrac{1}{y} \cdot 100$

**38.** $\dfrac{b}{10} \cdot 13$

**39.** $\dfrac{-21}{4} \cdot \dfrac{7}{5}$

**40.** $\dfrac{-8}{3} \cdot \dfrac{20}{9}$

**c** Solve.

**41.** *Serving of Cheesecake.* At the Cheesecake Factory, a piece of cheesecake is $\frac{1}{12}$ of a cheesecake. How much of the cheesecake is $\frac{1}{2}$ piece?

**Source:** The Cheesecake Factory

**42.** *Tossed Salad.* The recipe for a tossed salad calls for $\frac{3}{4}$ cup of sliced almonds. How much is needed to make $\frac{1}{2}$ of the recipe?

**43.** *Basketball: High School to Pro.* One of 35 high school basketball players plays college basketball. One of 75 college players plays professional basketball. What fractional part of high school basketball players play professional basketball?

Source: National Basketball Association

**45.** *Hair Bows.* It takes $\frac{5}{3}$ yd of ribbon to make a hair bow. How much ribbon is needed to make 8 bows?

**47.** *Floor Tiling.* The floor of a room is being covered with tile. An area $\frac{3}{5}$ of the length and $\frac{3}{4}$ of the width is covered. What fraction of the floor has been tiled?

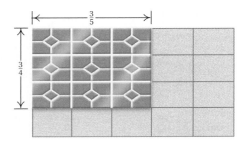

**44.** *Football: High School to Pro.* One of 42 high school football players plays college football. One of 85 college players plays professional football. What fractional part of high school football players play professional football?

Source: National Football League

**46.** A rectangular table top measures $\frac{4}{5}$ m long by $\frac{3}{5}$ m wide. What is its area?

**48.** A gasoline can holds $\frac{5}{2}$ gal. How much will the can hold when it is $\frac{1}{2}$ full?

## Skill Maintenance

Divide.   [1.5a], [2.5a]

**49.** $7140 \div 35$

**50.** $32{,}200 \div 46$

**51.** $-65 \div (-5)$

**52.** $540 \div (-6)$

Simplify.   [1.9c]

**53.** $8 \cdot 12 - (63 \div 9 + 13 \cdot 3)$

**54.** $(10 - 3)^4 + 10^3 \cdot 4 - 10 \div 5$

## Synthesis

Multiply. Write each answer using fraction notation.

**55.** 🖩 $\dfrac{341}{517} \cdot \dfrac{209}{349}$

**56.** 🖩 $\left(\dfrac{57}{61}\right)^3$

**57.** $\left(\dfrac{2}{5}\right)^3\left(-\dfrac{7}{9}\right)$

**58.** $\left(-\dfrac{1}{2}\right)^5\left(\dfrac{3}{5}\right)$

**59.** *Forestry.* A chain saw holds $\frac{1}{5}$ gal of fuel. Chain saw fuel is $\frac{1}{16}$ two-cycle oil and $\frac{15}{16}$ unleaded gasoline. How much two-cycle oil is in a freshly filled chain saw?

**60.** Evaluate $-\frac{2}{3}xy$ for $x = \frac{2}{5}$ and $y = -\frac{1}{7}$.

# 3.5 Simplifying

## a Multiplying by 1

OBJECTIVES

**a** Multiply by 1 to find an equivalent expression using a different denominator.

**b** Simplify fraction notation.

**c** Test to determine whether two fractions are equivalent.

Recall the following:

$$1 = \frac{1}{1} = \frac{2}{2} = \frac{3}{3} = \frac{4}{4} = \frac{-13}{-13} = \frac{45}{45} = \frac{100}{100} = \frac{n}{n}.$$

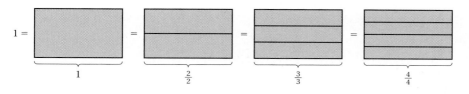

$$1 = \qquad = \qquad = \qquad = \qquad$$

Any nonzero number divided by itself is 1. (See Section 1.5.)

Now recall that for any whole number $a$, we have $1 \cdot a = a \cdot 1 = a$. This holds for fractions as well.

### MULTIPLICATIVE IDENTITY FOR FRACTIONS

When we multiply a number by 1, we get the same number:

$$\frac{3}{5} = \frac{3}{5} \cdot 1 = \frac{3}{5} \cdot \frac{4}{4} = \frac{12}{20}.$$

Since $\frac{3}{5} = \frac{12}{20}$, we know that $\frac{3}{5}$ and $\frac{12}{20}$ are two names for the same number. We also say that $\frac{3}{5}$ and $\frac{12}{20}$ are **equivalent fractions**.

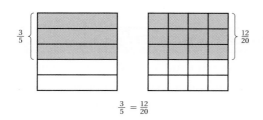

$$\frac{3}{5} = \frac{12}{20}$$

Do Margin Exercises 1–4.

Suppose we want to rename $\frac{2}{3}$, using a denominator of 15. We can multiply by 1 to find a number equivalent to $\frac{2}{3}$:

$$\frac{2}{3} = \frac{2}{3} \cdot \frac{5}{5} = \frac{2 \cdot 5}{3 \cdot 5} = \frac{10}{15}.$$

We chose $\frac{5}{5}$ for 1 because $15 \div 3$ is 5.

**EXAMPLE 1** Find a number equivalent to $\frac{1}{4}$ with a denominator of 24.

Since $24 \div 4 = 6$, we multiply by 1, using $\frac{6}{6}$:

$$\frac{1}{4} = \frac{1}{4} \cdot \frac{6}{6} = \frac{1 \cdot 6}{4 \cdot 6} = \frac{6}{24}.$$

**SKILL TO REVIEW**

Objective 3.2c: Find the prime factorization of a composite number.

Find the prime factorization.

**1.** 84 **2.** 2250

Multiply.

**1.** $\frac{1}{2} \cdot \frac{8}{8}$ **2.** $\frac{3}{7} \cdot \frac{a}{a}$

**3.** $-\frac{8}{25} \cdot \frac{4}{4}$ **4.** $\frac{8}{3}\left(\frac{-2}{-2}\right)$

*Answers*

*Skill to Review:*
**1.** $2 \cdot 2 \cdot 3 \cdot 7$ **2.** $2 \cdot 3 \cdot 3 \cdot 5 \cdot 5 \cdot 5$

*Margin Exercises:*
**1.** $\frac{8}{16}$ **2.** $\frac{3a}{7a}$ **3.** $-\frac{32}{100}$ **4.** $\frac{-16}{-6}$

**EXAMPLE 2** Find a number equivalent to $\frac{2}{5}$ with a denominator of $-35$.

Since $-35 \div 5 = -7$, we multiply by 1, using $\frac{-7}{-7}$:

$$\frac{2}{5} = \frac{2}{5}\left(\frac{-7}{-7}\right) = \frac{2(-7)}{5(-7)} = \frac{-14}{-35}.$$

**EXAMPLE 3** Find an expression equivalent to $\frac{9}{8}$ with a denominator of $8a$.

Since $8a \div 8 = a$, we multiply by 1, using $\frac{a}{a}$:

$$\frac{9}{8} \cdot \frac{a}{a} = \frac{9a}{8a}.$$

Do Exercises 5-10.

Do Exercises 5-10.

## (b) Simplifying Fraction Notation

All of the following are names for eight-ninths:

$$\frac{8}{9}, \frac{-8}{-9}, \frac{16}{18}, \frac{80}{90}, \frac{-24}{-27}.$$

We say that $\frac{8}{9}$ is **simplest** because it has the smallest positive denominator. Note that 8 and 9 have no factor in common other than 1.

To simplify fraction notation, we reverse the process of multiplying by 1:

$$\frac{12}{18} = \frac{2 \cdot 6}{3 \cdot 6} \quad \begin{array}{l} \longleftarrow \text{Factoring the numerator} \\ \longleftarrow \text{Factoring the denominator} \end{array}$$

$$= \frac{2}{3} \cdot \frac{6}{6} \qquad \text{Factoring the fraction}$$

$$= \frac{2}{3} \cdot 1 \qquad \frac{6}{6} = 1$$

$$= \frac{2}{3} \qquad \text{Removing the factor 1: } \frac{2}{3} \cdot 1 = \frac{2}{3}$$

---

**SIMPLIFYING FRACTION NOTATION**

1. Factor the numerator and factor the denominator.
2. Identify any common factors in the numerator and denominator.
3. Use the common factors to remove a factor equal to 1.

---

**EXAMPLES** Simplify.

**4.** $\dfrac{-8}{20} = \dfrac{-2 \cdot 4}{5 \cdot 4} = \dfrac{-2}{5} \cdot \dfrac{4}{4} = \dfrac{-2}{5}$    Removing a factor equal to 1: $\dfrac{4}{4} = 1$

**5.** $\dfrac{2}{6} = \dfrac{1 \cdot 2}{3 \cdot 2} = \dfrac{1}{3} \cdot \dfrac{2}{2} = \dfrac{1}{3}$

> Writing 1 allows for pairing of factors in the numerator and the denominator.

**6.** $\dfrac{30}{6} = \dfrac{5 \cdot 6}{1 \cdot 6} = \dfrac{5}{1} \cdot \dfrac{6}{6} = \dfrac{5}{1} = 5$

> We could also simplify $\frac{30}{6}$ by doing the division $30 \div 6$. That is, $\frac{30}{6} = 30 \div 6 = 5$.

**7.** $-\dfrac{15}{10} = -\dfrac{3 \cdot 5}{2 \cdot 5} = -\dfrac{3}{2} \cdot \dfrac{5}{5} = -\dfrac{3}{2}$    Removing a factor equal to 1: $\dfrac{5}{5} = 1$

---

*Answers*

**5.** $\dfrac{12}{9}$    **6.** $\dfrac{-18}{-24}$    **7.** $\dfrac{9x}{10x}$
**8.** $\dfrac{30}{2}$    **9.** $\dfrac{-56}{49}$    **10.** $\dfrac{3}{-6}$

**8.** $\dfrac{4x}{15x} = \dfrac{4 \cdot x}{15 \cdot x} = \dfrac{4}{15} \cdot \dfrac{x}{x} = \dfrac{4}{15}$    Removing a factor equal to 1: $\dfrac{x}{x} = 1$

Do Exercises 11–16.

The use of prime factorizations can be helpful for simplifying.

**EXAMPLE 9**   Simplify: $\dfrac{90}{84}$.

$$\dfrac{90}{84} = \dfrac{2 \cdot 3 \cdot 3 \cdot 5}{2 \cdot 2 \cdot 3 \cdot 7}$$   Factoring the numerator and the denominator into primes

$$= \dfrac{2 \cdot 3 \cdot 3 \cdot 5}{2 \cdot 3 \cdot 2 \cdot 7}$$   Changing the order so that like primes are above and below each other

$$= \dfrac{2}{2} \cdot \dfrac{3}{3} \cdot \dfrac{3 \cdot 5}{2 \cdot 7}$$   Factoring the fraction

$$= \dfrac{3 \cdot 5}{2 \cdot 7}$$   Removing factors of 1

$$= \dfrac{15}{14}$$

**EXAMPLE 10**   Simplify: $\dfrac{105}{135}$.

Since both 105 and 135 end in 5, we know that 5 is a factor of both the numerator and the denominator:

$$\dfrac{105}{135} = \dfrac{21 \cdot 5}{27 \cdot 5} = \dfrac{21}{27} \cdot \dfrac{5}{5} = \dfrac{21}{27}.$$   To find the 21, we divided 105 by 5. To find the 27, we divided 135 by 5.

A fraction is not "simplified" if common factors of the numerator and the denominator remain. Because 21 and 27 are both divisible by 3, we must simplify further:

$$\dfrac{105}{135} = \dfrac{21}{27} = \dfrac{7 \cdot 3}{9 \cdot 3} = \dfrac{7}{9} \cdot \dfrac{3}{3} = \dfrac{7}{9}.$$   To find the 7, we divided 21 by 3. To find the 9, we divided 27 by 3.

**EXAMPLE 11**   Simplify: $\dfrac{322}{434}$.

Since 322 and 434 are both even, we know that 2 is a common factor:

$$\dfrac{322}{434} = \dfrac{2 \cdot 161}{2 \cdot 217} = \dfrac{2}{2} \cdot \dfrac{161}{217} = \dfrac{161}{217}$$   Removing a factor equal to 1: $\dfrac{2}{2} = 1$

$$= \dfrac{7 \cdot 23}{7 \cdot 31} = \dfrac{7}{7} \cdot \dfrac{23}{31} = \dfrac{23}{31}.$$   7 is also a common factor; removing a factor equal to 1

We found the common factor 7 by focusing first on 161. After determining that 7 is a factor of 161, we checked to see if 7 is also a factor of 217.

Do Exercises 17–23.

Simplify.

**11.** $\dfrac{8}{14}$    **12.** $\dfrac{-10}{12}$

**13.** $\dfrac{40}{8}$    **14.** $\dfrac{4a}{3a}$

**15.** $-\dfrac{50}{30}$    **16.** $\dfrac{x}{3x}$

Simplify.

**17.** $\dfrac{-35}{40}$    **18.** $\dfrac{801}{702}$

**19.** $\dfrac{24}{21}$    **20.** $\dfrac{429}{561}$

**21.** $\dfrac{280}{960}$    **22.** $\dfrac{1332}{2880}$

**23.** Simplify each fraction in this circle graph.

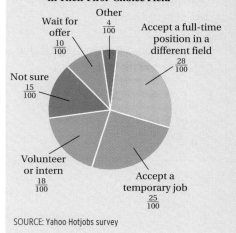

**What College Graduates Will Do If They Are Not Offered a Full-Time Job in Their First-Choice Field**

Other $\dfrac{4}{100}$

Wait for offer $\dfrac{10}{100}$

Accept a full-time position in a different field $\dfrac{28}{100}$

Not sure $\dfrac{15}{100}$

Volunteer or intern $\dfrac{18}{100}$

Accept a temporary job $\dfrac{25}{100}$

SOURCE: Yahoo Hotjobs survey

***Answers***

**11.** $\dfrac{4}{7}$   **12.** $\dfrac{-5}{6}$   **13.** 5   **14.** $\dfrac{4}{3}$   **15.** $-\dfrac{5}{3}$
**16.** $\dfrac{1}{3}$   **17.** $\dfrac{-7}{8}$   **18.** $\dfrac{89}{78}$   **19.** $\dfrac{8}{7}$   **20.** $\dfrac{13}{17}$
**21.** $\dfrac{7}{24}$   **22.** $\dfrac{37}{80}$   **23.** $\dfrac{28}{100} = \dfrac{7}{25}; \dfrac{25}{100} = \dfrac{1}{4};$
$\dfrac{18}{100} = \dfrac{9}{50}; \dfrac{15}{100} = \dfrac{3}{20}; \dfrac{10}{100} = \dfrac{1}{10}; \dfrac{4}{100} = \dfrac{1}{25}$

## Canceling

Canceling is a shortcut that you may have used for removing a factor that equals 1 when working with fraction notation. With concern, we mention it as a possibility for speeding up your work. Canceling may be done only when removing common factors in numerators and denominators. Each common factor allows us to remove a factor equal to 1 in a product.

Generally, slashes are used to indicate factors equal to 1 that have been removed. For instance, Example 11 might have been done faster as follows:

$$\frac{322}{434} = \frac{2 \cdot 161}{2 \cdot 217} \qquad \text{Factoring the numerator and the denominator}$$

$$= \frac{\cancel{2} \cdot 161}{\cancel{2} \cdot 217} \qquad \begin{array}{l}\text{When a factor equal to 1 is noted,} \\ \text{it is "canceled" as shown: } \frac{2}{2} = 1.\end{array}$$

$$= \frac{161}{217} = \frac{7 \cdot 23}{7 \cdot 31} = \frac{23}{31}.$$

------ *Caution!* ------

The difficulty with canceling is that it is often applied incorrectly in situations like the following:

$$\frac{2 + 3}{\cancel{2}} = 3; \qquad \frac{\cancel{4} + 1}{\cancel{4} + 2} = \frac{1}{2}; \qquad \frac{1\cancel{5}}{\cancel{5}4} = \frac{1}{4}.$$
$$\quad \text{Wrong!} \qquad\qquad \text{Wrong!} \qquad\qquad \text{Wrong!}$$

The correct answers are

$$\frac{2 + 3}{2} = \frac{5}{2}; \qquad \frac{4 + 1}{4 + 2} = \frac{5}{6}; \qquad \frac{15}{54} = \frac{3 \cdot 5}{3 \cdot 18} = \frac{5}{18}.$$

In each incorrect case, the numbers canceled did not form a factor equal to 1. Factors are parts of products. For example, in $2 \cdot 3$, the numbers 2 and 3 are factors, but in $2 + 3$, 2 and 3 are terms, not factors.

- **If you cannot factor, do not cancel! If in doubt, do not cancel!**
- **Only factors can be canceled, and factors are never separated by + or − signs.**

---

## (c) A Test for Equality

When denominators are the same, we say that fractions have a **common denominator**. One way to compare fractions like $\frac{2}{4}$ and $\frac{3}{6}$ is to find a common denominator and compare numerators. We can multiply each fraction by 1, using the other denominator to write 1.

The denominator is 6.

$$\frac{3}{6} = \frac{3}{6} \cdot \frac{4}{4} = \frac{3 \cdot 4}{6 \cdot 4} = \frac{12}{24}$$

$$\frac{2}{4} = \frac{2}{4} \cdot \frac{6}{6} = \frac{2 \cdot 6}{4 \cdot 6} = \frac{12}{24}$$    Both denominators are 24.

The denominator is 4.

Because $\frac{12}{24} = \frac{12}{24}$ is true, it follows that $\frac{3}{6} = \frac{2}{4}$ is also true.

The "key" to the above work is that $3 \cdot 4$ and $2 \cdot 6$ are equal. Had these products differed, we would have shown that $\frac{3}{6}$ and $\frac{2}{4}$ were *not* equal.

## A TEST FOR EQUALITY

We multiply these two numbers: $3 \cdot 4$.

We multiply these two numbers: $6 \cdot 2$.

$$\frac{3}{6} \;\square\; \frac{2}{4}$$

We call $3 \cdot 4$ and $6 \cdot 2$ **cross products**. Since the cross products are the same—that is, $3 \cdot 4 = 6 \cdot 2$—we know that

$$\frac{3}{6} = \frac{2}{4}.$$

In the sentence $a \neq b$, the symbol $\neq$ means "is not equal to."

**EXAMPLE 12**  Use $=$ or $\neq$ for $\square$ to write a true sentence:

$$\frac{6}{7} \;\square\; \frac{7}{8}.$$

We multiply these two numbers: $6 \cdot 8 = 48$.

We multiply these two numbers: $7 \cdot 7 = 49$.

$$\frac{6}{7} \;\square\; \frac{7}{8}$$

Because $48 \neq 49$, $\frac{6}{7}$ and $\frac{7}{8}$ do not name the same number. Thus, $\frac{6}{7} \neq \frac{7}{8}$.

**EXAMPLE 13**  Use $=$ or $\neq$ for $\square$ to write a true sentence:

$$\frac{6}{10} \;\square\; \frac{3}{5}.$$

We multiply these two numbers: $6 \cdot 5 = 30$.

We multiply these two numbers: $10 \cdot 3 = 30$.

$$\frac{6}{10} \;\square\; \frac{3}{5}$$

Because the cross products are the same, we have $\frac{6}{10} = \frac{3}{5}$.

Remembering that $\dfrac{-a}{b}$, $\dfrac{a}{-b}$, and $-\dfrac{a}{b}$ all represent the same number can be helpful when checking for equality.

**EXAMPLE 14**  Use $=$ or $\neq$ for $\square$ to write a true sentence:

$$\frac{-6}{8} \;\square\; -\frac{9}{12}.$$

We rewrite $-\frac{9}{12}$ as $\frac{-9}{12}$ and then check cross products:

$-6 \cdot 12 = -72$

$-9 \cdot 8 = -72$

$$\frac{-6}{8} \;\square\; \frac{-9}{12}$$

Because the cross products are the same, we have $\frac{-6}{8} = -\frac{9}{12}$.

> Do Exercises 24–26.

## STUDY TIPS

### TIME MANAGEMENT

*Keep on Schedule.* Your course syllabus provides a plan for the semester's schedule. Use a write-on calendar, daily planner, laptop computer, or personal digital assistant to outline your time for the semester. Be sure to note deadlines involving projects and exams so that you can begin a task early, breaking it down into small segments that can be accomplished easily.

Use $=$ or $\neq$ for $\square$ to write a true sentence.

**24.** $\dfrac{2}{6} \;\square\; \dfrac{3}{9}$    **25.** $\dfrac{2}{3} \;\square\; \dfrac{14}{20}$

**26.** $-\dfrac{10}{15} \;\square\; \dfrac{8}{-12}$

*Answers*

**24.** $=$    **25.** $\neq$    **26.** $=$

# 3.5

## Exercise Set

For Extra Help

**MyMathLab**

Math XL
PRACTICE    WATCH    DOWNLOAD    READ    REVIEW

**a** Find an equivalent expression for each number, using the denominator, indicated. Use multiplication by 1.

**1.** $\dfrac{1}{2} = \dfrac{?}{10}$

**2.** $\dfrac{1}{6} = \dfrac{?}{12}$

**3.** $\dfrac{3}{4} = \dfrac{?}{-48}$

**4.** $\dfrac{2}{9} = \dfrac{?}{-18}$

**5.** $\dfrac{7}{10} = \dfrac{?}{50}$

**6.** $\dfrac{3}{8} = \dfrac{?}{48}$

**7.** $\dfrac{11}{5} = \dfrac{?}{5t}$

**8.** $\dfrac{5}{3} = \dfrac{?}{3a}$

**9.** $\dfrac{5}{1} = \dfrac{?}{4}$

**10.** $\dfrac{7}{1} = \dfrac{?}{5}$

**11.** $-\dfrac{17}{18} = -\dfrac{?}{54}$

**12.** $-\dfrac{11}{16} = -\dfrac{?}{256}$

**13.** $\dfrac{3}{-8} = \dfrac{?}{-40}$

**14.** $\dfrac{7}{-8} = \dfrac{?}{-32}$

**15.** $\dfrac{-7}{22} = \dfrac{?}{132}$

**16.** $\dfrac{-10}{21} = \dfrac{?}{126}$

**17.** $\dfrac{1}{8} = \dfrac{?}{8x}$

**18.** $\dfrac{1}{3} = \dfrac{?}{3a}$

**19.** $\dfrac{-10}{7} = \dfrac{?}{7a}$

**20.** $\dfrac{-4}{3} = \dfrac{?}{3n}$

**21.** $\dfrac{4}{9} = \dfrac{?}{9ab}$

**22.** $\dfrac{8}{11} = \dfrac{?}{11xy}$

**23.** $\dfrac{4}{9} = \dfrac{?}{27b}$

**24.** $\dfrac{8}{11} = \dfrac{?}{55y}$

**b** Simplify.

**25.** $\dfrac{2}{4}$

**26.** $\dfrac{3}{6}$

**27.** $-\dfrac{6}{9}$

**28.** $\dfrac{-9}{12}$

**29.** $\dfrac{10}{25}$

**30.** $\dfrac{8}{10}$

**31.** $\dfrac{24}{8}$

**32.** $\dfrac{36}{9}$

**33.** $\dfrac{27}{36}$

**34.** $\dfrac{30}{40}$

**35.** $-\dfrac{24}{14}$

**36.** $-\dfrac{16}{10}$

**37.** $\dfrac{3n}{4n}$

**38.** $\dfrac{7x}{8x}$

**39.** $\dfrac{-17}{51}$

**40.** $-\dfrac{13}{26}$

**41.** $\dfrac{-100}{20}$

**42.** $\dfrac{-150}{25}$

**43.** $\dfrac{420}{480}$

**44.** $\dfrac{180}{240}$

**45.** $\dfrac{-540}{810}$

**46.** $\dfrac{-1000}{1080}$

**47.** $\dfrac{12x}{30x}$

**48.** $\dfrac{54n}{90n}$

**49.** $\dfrac{153}{136}$

**50.** $\dfrac{117}{91}$

**51.** $\dfrac{132}{143}$

**52.** $\dfrac{91}{259}$

**53.** $\dfrac{221}{247}$

**54.** $\dfrac{299}{403}$

**55.** $\dfrac{3ab}{8ab}$

**56.** $\dfrac{6xy}{7xy}$

**57.** $\dfrac{9xy}{6x}$

**58.** $\dfrac{10ab}{15a}$

**59.** $\dfrac{-18a}{20ab}$

**60.** $\dfrac{-19x}{38xy}$

**c** Use = or ≠ for ☐ to write a true sentence.

**61.** $\dfrac{3}{4}$ ☐ $\dfrac{9}{12}$

**62.** $\dfrac{4}{8}$ ☐ $\dfrac{3}{6}$

**63.** $\dfrac{1}{5}$ ☐ $\dfrac{2}{9}$

**64.** $\dfrac{1}{4}$ ☐ $\dfrac{2}{9}$

**65.** $\dfrac{3}{8}$ ☐ $\dfrac{6}{16}$

**66.** $\dfrac{2}{6}$ ☐ $\dfrac{6}{18}$

**67.** $\dfrac{2}{5}$ ☐ $\dfrac{2}{7}$

**68.** $\dfrac{3}{10}$ ☐ $\dfrac{3}{11}$

**69.** $\dfrac{-3}{10}$ ☐ $\dfrac{-4}{12}$

**70.** $\dfrac{-2}{9}$ ☐ $\dfrac{-8}{36}$

**71.** $-\dfrac{12}{9}$ ☐ $\dfrac{-8}{6}$

**72.** $\dfrac{-8}{7}$ ☐ $-\dfrac{16}{14}$

**73.** $\dfrac{5}{-2}$ ☐ $-\dfrac{17}{7}$

**74.** $-\dfrac{10}{3}$ ☐ $\dfrac{24}{-7}$

**75.** $\dfrac{305}{145}$ ☐ $\dfrac{122}{58}$

**76.** $\dfrac{425}{165}$ ☐ $\dfrac{130}{66}$

## Skill Maintenance

Solve.  [1.8a]

**77.** *Tiananmen Square.* Tiananmen Square in Beijing, China, is the largest public square in the world. The length is 963 yd and the width is 547 yd. What is its area? its perimeter?

**78.** A landscaper buys 13 small maple trees and 17 small oak trees for a project. A maple costs $23 and an oak costs $37. How much is spent altogether for the trees?

Multiply.  [2.4a]

**79.** $-12(-5)$

**80.** $-5(-13)$

**81.** $-9 \cdot 7$

**82.** $-8 \cdot 8$

Solve.  [1.7b]

**83.** $30 \cdot x = 150$

**84.** $10{,}947 = 123 \cdot y$

**85.** $5280 = 1760 + t$

**86.** $x + 2368 = 11{,}369$

## Synthesis

Simplify. Use the list of prime numbers on p. 162.

**87.** $\dfrac{391}{667}$

**88.** $\dfrac{209ab}{247ac}$

**89.** $-\dfrac{1073x}{555y}$

**90.** $-\dfrac{187a}{289b}$

**91.** ▦ $\dfrac{4247}{4619}$

**92.** ▦ $\dfrac{3473}{3197}$

**93.** Sociologists have found that 4 of 10 people are shy. Write fraction notation for **(a)** the part of the population that is shy; and **(b)** the part that is not shy. Simplify.

**94.** Sociologists estimate that 3 of 20 people are left-handed. In a crowd of 460 people, how many would you expect to be left-handed?

**95.** ▦ *Batting Averages.* For the 2009 season, Hanley Ramirez of the Florida Marlins won the National League batting title with 197 hits in 576 times at bat. Joe Mauer of the Minnesota Twins won the American League title with 191 hits in 523 times at bat. Did they have the same fraction of hits in times at bat (batting average)? Why or why not?

**Source:** Major League Baseball

**96.** ▦ On a test of 82 questions, Taylor got 63 correct. On another test of 100 questions, she got 77 correct. Did she get the same portion of each test correct? Why or why not?

# Mid-Chapter Review

## Concept Reinforcement

Determine whether each statement is true or false.

_____ **1.** A number $a$ is divisible by another number $b$ if $b$ is factor of $a$.   [3.2a]

_____ **2.** If a number is not divisible by 6, then it is not divisible by 3.   [3.1b]

_____ **3.** The fraction $\frac{9}{4}$ is equal to the fraction $\frac{13}{6}$.   [3.5c]

_____ **4.** The number 1 is not prime.   [3.2b]

## Guided Solutions

Fill in each blank with the number that creates a correct statement or solution.

**5.** $\dfrac{25}{\square} = 1$   [3.3b]

**6.** $\dfrac{\square}{9} = 0$   [3.3b]

**7.** $\dfrac{8}{\square} = 8$   [3.3b]

**8.** $\dfrac{6}{13} = \dfrac{\square}{39}$   [3.5a]

**9.** Simplify: $\dfrac{70}{225}$.   [3.5b]

$$\frac{70}{225} = \frac{2 \cdot \square \cdot 7}{\square \cdot 3 \cdot 5 \cdot \square} \qquad \text{Factoring the numerator}$$
$$\text{Factoring the denominator}$$

$$= \frac{5}{5} \cdot \frac{\square \cdot 7}{3 \cdot \square \cdot 5} \qquad \text{Factoring the fraction}$$

$$= \square \cdot \frac{\square}{45} \qquad \frac{5}{5} = 1$$

$$= \frac{\square}{\square} \qquad \text{Removing a factor equal to 1}$$

## Mixed Review

To answer Exercises 10–14, consider the following numbers.   [3.1b]

| | | | |
|---|---|---|---|
| 84 | 132 | 594 | 350 |
| 300 | 500 | 120 | 14,850 |
| 17,576 | 180 | 1125 | 504 |
| 224 | 351 | 495 | 1632 |

**10.** Which of the above are divisible by 2 but not by 10?

**11.** Which of the above are divisible by 4 but not by 8?

**12.** Which of the above are divisible by 4 but not by 6?

**13.** Which of the above are divisible by 3 but not by 9?

**14.** Which of the above are divisible by 4, 5, and 6?

Determine whether each number is prime, composite, or neither.   [3.2b]

**15.** 61
**16.** 2
**17.** 91
**18.** 1

Find all the factors of each composite number. Then find the prime factorization of the number. [3.2a], [3.2c]

**19.** 160

**20.** 222

**21.** 98

**22.** 315

What part of each object or set of objects is shaded? [3.3a]

**23.**

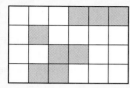

**24.**

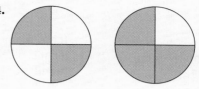

Multiply. [3.4a], [3.4b]

**25.** $7 \cdot \dfrac{1}{9}$

**26.** $\dfrac{4}{15} \cdot \dfrac{2}{3}$

**27.** $\dfrac{5}{11}(-8)$

Simplify. [3.3b], [3.5b]

**28.** $\dfrac{24}{60}$

**29.** $\dfrac{220n}{60n}$

**30.** $\dfrac{17x}{17x}$

**31.** $\dfrac{0}{-23}$

**32.** $\dfrac{54}{186}$

**33.** $\dfrac{-36}{20}$

**34.** $\dfrac{75}{630}$

**35.** $\dfrac{315}{435}$

**36.** $\dfrac{14}{0}$

Use $=$ or $\neq$ for $\square$ to write a true sentence. [3.5c]

**37.** $\dfrac{3}{-7} \ \square \ \dfrac{-39}{91}$

**38.** $\dfrac{19}{3} \ \square \ \dfrac{95}{18}$

**39.** *Veterinary Care.* Of every 500 households that own pet birds, 60 obtain veterinary care for their birds. What is the ratio of households that own birds that seek veterinary care for their birds to all households that own birds? [3.3a], [3.5b]

**Source:** American Veterinary Medical Association

**40.** *Area of an Ice-Skating Rink.* The length of a rectangular ice-skating rink in the atrium of a shopping mall is $\dfrac{7}{100}$ mi. The width is $\dfrac{3}{100}$ mi. What is the area of the rink? [3.4c]

# Understanding Through Discussion and Writing

**41.** Explain a method for finding a composite number that contains exactly two factors other than itself and 1. [3.2b]

**42.** Which of the years from 2000 to 2020, if any, also happen to be prime numbers? Explain at least two ways in which you might go about solving this problem. [3.2b]

**43.** Explain in your own words when it *is* possible to "cancel" and when it *is not* possible to "cancel." [3.5b]

**44.** Can fraction notation be simplified if the numerator and the denominator are two different prime numbers? Why or why not? [3.5b]

# 3.6 Multiplying, Simplifying, and More with Area

## a Simplifying When Multiplying

It is often possible to simplify after we multiply. To make such simplifying easier, it is usually best not to carry out the products in the numerator and the denominator immediately, but to factor and simplify first. Consider the product

$$\frac{5}{6} \cdot \frac{14}{15}.$$

We proceed as follows:

$$\frac{5}{6} \cdot \frac{14}{15} = \frac{5 \cdot 14}{6 \cdot 15}$$    We write the products in the numerator and the denominator, but we do not yet carry out the multiplication.

$$= \frac{5 \cdot 2 \cdot 7}{2 \cdot 3 \cdot 5 \cdot 3}$$    Factoring and identifying common factors

$$= \frac{5 \cdot 2}{5 \cdot 2} \cdot \frac{7}{3 \cdot 3}$$    Factoring the fraction

$$= 1 \cdot \frac{7}{3 \cdot 3}$$

$$= \frac{7}{3 \cdot 3}$$    Removing a factor equal to 1: $\frac{5 \cdot 2}{5 \cdot 2} = 1$

$$= \frac{7}{9}.$$

To multiply and simplify:

a) Write the products in the numerator and the denominator, but do not carry out the multiplication.

b) Factor the numerator and the denominator.

c) Factor the fraction to remove a factor equal to 1, if possible.

d) Carry out the remaining products.

**EXAMPLES** Multiply and simplify.

**1.** $\dfrac{2}{3} \cdot \dfrac{5}{4} = \dfrac{2 \cdot 5}{3 \cdot 4}$    Note that 2 is a common factor of 2 and 4.

$$= \frac{2 \cdot 5}{3 \cdot 2 \cdot 2}$$    Try to go directly to this step.

$$= \frac{2}{2} \cdot \frac{5}{3 \cdot 2}$$

$$= 1 \cdot \frac{5}{3 \cdot 2} = \frac{5}{6}$$    Removing a factor equal to 1: $\frac{2}{2} = 1$

**2.** $\dfrac{6}{7} \cdot \dfrac{-5}{3} = \dfrac{3 \cdot 2 \cdot (-5)}{7 \cdot 3}$    Note that 3 is a common factor of 6 and 3.

$$= \frac{3}{3} \cdot \frac{2(-5)}{7} = \frac{-10}{7}, \text{ or } -\frac{10}{7}$$    Removing a factor equal to 1: $\frac{3}{3} = 1$

**3.** $\dfrac{10}{21} \cdot \dfrac{14a}{15} = \dfrac{5 \cdot 2 \cdot 7 \cdot 2 \cdot a}{7 \cdot 3 \cdot 5 \cdot 3}$   Note that 5 is a common factor of 10 and 15.
Note that 7 is a common factor of 21 and 14a.

$\qquad = \dfrac{5 \cdot 7}{5 \cdot 7} \cdot \dfrac{2 \cdot 2 \cdot a}{3 \cdot 3}$

$\qquad = \dfrac{4a}{9}$ } Removing a factor equal to 1: $\dfrac{5 \cdot 7}{5 \cdot 7} = 1$

**4.** $32 \cdot \dfrac{7}{8} = \dfrac{32}{1} \cdot \dfrac{7}{8} = \dfrac{8 \cdot 4 \cdot 7}{8 \cdot 1}$   Note that 8 is a common factor of 32 and 8.

$\qquad = \dfrac{8}{8} \cdot \dfrac{4 \cdot 7}{1} = 28$   Removing a factor equal to 1: $\dfrac{8}{8} = 1$

-------------------------------------- *Caution!* --------------------------------------

Canceling can be used as follows for these examples.

**1.** $\dfrac{2}{3} \cdot \dfrac{5}{4} = \dfrac{2 \cdot 5}{3 \cdot 2 \cdot 2} = \dfrac{5}{6}$   Removing a factor equal to 1: $\dfrac{2}{2} = 1$

**2.** $\dfrac{6}{7} \cdot \dfrac{-5}{3} = \dfrac{3 \cdot 2(-5)}{7 \cdot 3} = \dfrac{-10}{7}$   Removing a factor equal to 1: $\dfrac{3}{3} = 1$

**3.** $\dfrac{10}{21} \cdot \dfrac{14a}{15} = \dfrac{5 \cdot 2 \cdot 7 \cdot 2 \cdot a}{7 \cdot 3 \cdot 5 \cdot 3} = \dfrac{4a}{9}$   Removing a factor equal to 1: $\dfrac{5 \cdot 7}{5 \cdot 7} = 1$

**4.** $32 \cdot \dfrac{7}{8} = \dfrac{8 \cdot 4 \cdot 7}{8 \cdot 1} = 28$   Removing a factor equal to 1: $\dfrac{8}{8} = 1$

**Remember, only factors can be canceled!**

-------------------------------------------------------------

Do Exercises 1–4.

> Multiply and simplify.
>
> **1.** $\dfrac{2}{3} \cdot \dfrac{7}{8}$   **2.** $\dfrac{4}{5} \cdot \dfrac{-5}{12}$
>
> **3.** $16 \cdot \dfrac{3}{8}$   **4.** $\dfrac{5}{2x} \cdot 4$

## **b** Solving Problems

Recall that problems involving repeated addition and those with keywords "of," "twice," and "product" translate to multiplication.

**EXAMPLE 5** *Landscaping.* Celina's Landscaping uses $\frac{2}{3}$ lb of peat moss when planting a rosebush. How much will be needed to plant 21 rosebushes?

**1. Familiarize.** We first make a drawing or at least visualize the situation. We let $n =$ the number of pounds of peat moss needed.

... 21 rosebushes

$\frac{2}{3}$ pound of peat moss for each rosebush

**2. Translate.** The problem translates to the following equation:

$$n = 21 \cdot \dfrac{2}{3}.$$

**3. Solve.** To solve the equation, we carry out the multiplication:

$n = 21 \cdot \dfrac{2}{3} = \dfrac{21}{1} \cdot \dfrac{2}{3} = \dfrac{21 \cdot 2}{1 \cdot 3}$   Multiplying

$\qquad = \dfrac{3 \cdot 7 \cdot 2}{1 \cdot 3} = \dfrac{3}{3} \cdot \dfrac{7 \cdot 2}{1} = 14.$   Removing the factor $\frac{3}{3}$ and simplifying

*Answers*

**1.** $\dfrac{7}{12}$   **2.** $-\dfrac{1}{3}$   **3.** 6   **4.** $\dfrac{10}{x}$

**4. Check.** We check by repeating the calculation. (This is left to the student.) We can also ask if the answer seems reasonable. We are putting less than a pound of peat moss on each bush, so the answer should be less than 21. Since 14 is less than 21, we have a partial check.

A second partial check can be performed using the units:

$$21 \text{ bushes} \cdot \frac{2}{3} \text{ pounds per bush}$$

$$= 21 \cdot \frac{2}{3} \cdot \text{bushes} \cdot \frac{\text{pounds}}{\text{bush}}$$

$$= 14 \text{ pounds.}$$

Since the resulting unit is pounds, we have another partial check.

**5. State.** Celina's Landscaping will need 14 lb of peat moss to plant 21 rosebushes.

Do Exercise 5.

> **5. Truffles.** Chocolate Delight sells $\frac{4}{5}$-lb boxes of truffles. How many pounds of truffles will be needed to fill 85 boxes?

### Area

We multiply to find the area of a triangle. Consider a triangle with a base of length $b$ and a height of $h$. A rectangle can be formed by splitting and inverting a copy of this triangle.

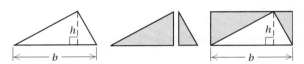

The rectangle's area, $b \cdot h$, is exactly twice the area of the triangle. We have the following result.

---

**AREA OF A TRIANGLE**

The **area $A$ of a triangle** is half the length of the base $b$ times the height $h$:

$$A = \frac{1}{2} \cdot b \cdot h.$$

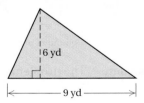

---

**EXAMPLE 6** Find the area of this triangle.

$$A = \frac{1}{2} \cdot b \cdot h$$

$$= \frac{1}{2} \cdot 9 \text{ yd} \cdot 6 \text{ yd}$$

$$= \frac{9 \cdot 6}{2} \text{ yd}^2$$

$$= 27 \text{ yd}^2, \text{ or } 27 \text{ square yards}$$

------ *Caution!* ------

Use square units for units of area.

-----------------------------------------

*Answer*

**5.** 68 pounds

**EXAMPLE 7** Find the area of this triangle.

$$A = \frac{1}{2} \cdot b \cdot h$$

$$= \frac{1}{2} \cdot \frac{10}{3} \text{ cm} \cdot 4 \text{ cm}$$

$$= \frac{1 \cdot 10 \cdot 4}{2 \cdot 3} \text{ cm}^2$$

$$= \frac{1 \cdot 2 \cdot 5 \cdot 4}{2 \cdot 3} \text{ cm}^2 \qquad \text{Removing a factor equal to 1: } \frac{2}{2} = 1$$

$$= \frac{20}{3} \text{ cm}^2$$

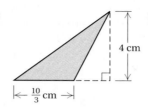

4 cm

$\frac{10}{3}$ cm

Do Exercises 6 and 7.

Find the area.

**6.**

12 m

16 m

**7.**

$\frac{12}{5}$ cm

11 cm

**EXAMPLE 8** Find the area of this kite.

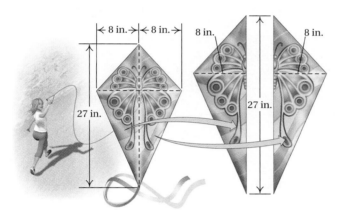

8 in. | 8 in. | 8 in. | 8 in.

27 in.

27 in.

1. **Familiarize.** We look for figures with areas we can calculate using area formulas that we already know. We let $K =$ the kite's area.

2. **Translate.** The kite consists of two triangles, each with a base of 27 in. and a height of 8 in. We can apply the formula $A = \frac{1}{2} \cdot b \cdot h$ for the area of a triangle and then multiply by 2.

*Rephrase:* Kite's area is twice area of long triangle

*Translate:* $K = 2 \cdot \frac{1}{2}(27 \text{ in.}) \cdot (8 \text{ in.})$

3. **Solve.** We have

$$K = 2 \cdot \frac{1}{2} \cdot (27 \text{ in.}) \cdot (8 \text{ in.})$$

$$= 1 \cdot 27 \text{ in.} \cdot 8 \text{ in.} = 216 \text{ in}^2.$$

4. **Check.** We can check by repeating the calculations. The unit, in$^2$, is appropriate for area.

5. **State.** The area of the kite is 216 in$^2$.

Do Exercise 8.

**8.** Find the area. (*Hint*: The figure is made up of a rectangle and a triangle.)

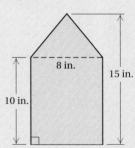

8 in.

15 in.

10 in.

*Answers*

**6.** 96 m$^2$    **7.** $\frac{66}{5}$ cm$^2$

**8.** Rectangle: $(10 \text{ in.}) \cdot (8 \text{ in.}) = 80 \text{ in}^2$;

triangle: $\frac{1}{2}(5 \text{ in.}) \cdot (8 \text{ in.}) = 20 \text{ in}^2$;

$80 \text{ in}^2 + 20 \text{ in}^2 = 100 \text{ in}^2$

**3.6**   **Exercise Set**

For Extra Help

*MyMathLab*

*Math XL*
PRACTICE

WATCH

DOWNLOAD

READ

REVIEW

**a**   Multiply. Don't forget to simplify, if possible.

**1.** $\dfrac{2}{3} \cdot \dfrac{1}{2}$

**2.** $\dfrac{4}{5} \cdot \dfrac{1}{4}$

**3.** $\dfrac{7}{8} \cdot \dfrac{-1}{7}$

**4.** $\dfrac{5}{6} \cdot \dfrac{-1}{5}$

**5.** $\dfrac{2}{3} \cdot \dfrac{6}{7}$

**6.** $\dfrac{2}{5} \cdot \dfrac{3}{10}$

**7.** $\dfrac{2}{9} \cdot \dfrac{3}{10}$

**8.** $\dfrac{3}{5} \cdot \dfrac{10}{9}$

**9.** $\dfrac{9}{-5} \cdot \dfrac{12}{8}$

**10.** $\dfrac{16}{-15} \cdot \dfrac{5}{4}$

**11.** $\dfrac{5x}{9} \cdot \dfrac{4}{5}$

**12.** $\dfrac{25}{4a} \cdot \dfrac{4}{3}$

**13.** $9 \cdot \dfrac{1}{9}$

**14.** $4 \cdot \dfrac{1}{4}$

**15.** $\dfrac{7}{10} \cdot \dfrac{10}{7}$

**16.** $\dfrac{8}{9} \cdot \dfrac{9}{8}$

**17.** $\dfrac{1}{4} \cdot 12$

**18.** $\dfrac{1}{6} \cdot 12$

**19.** $21 \cdot \dfrac{1}{3}$

**20.** $18 \cdot \dfrac{1}{2}$

**21.** $-16\left(-\dfrac{3}{4}\right)$

**22.** $-24\left(-\dfrac{5}{6}\right)$

**23.** $\dfrac{3}{8} \cdot 8a$

**24.** $\dfrac{2}{9} \cdot 9x$

**25.** $\left(-\dfrac{3}{8}\right)\left(-\dfrac{8}{3}\right)$

**26.** $\left(-\dfrac{7}{9}\right)\left(-\dfrac{9}{7}\right)$

**27.** $\dfrac{a}{b} \cdot \dfrac{b}{a}$

**28.** $\dfrac{n}{m} \cdot \dfrac{m}{n}$

**29.** $\dfrac{4}{10} \cdot \dfrac{5}{10}$

**30.** $\dfrac{11}{24} \cdot \dfrac{3}{5}$

**31.** $\dfrac{8}{10} \cdot \dfrac{45}{100}$

**32.** $\dfrac{3}{10} \cdot \dfrac{8}{10}$

**33.** $\dfrac{1}{6} \cdot 360n$

**34.** $\dfrac{1}{3} \cdot 12y$

**35.** $20\left(\dfrac{1}{-6}\right)$

**36.** $35\left(\dfrac{1}{-10}\right)$

**37.** $-8x \cdot \dfrac{1}{-8x}$

**38.** $-5a \cdot \dfrac{1}{-5a}$

**39.** $\dfrac{2x}{9} \cdot \dfrac{27}{2x}$

**40.** $\dfrac{10a}{3} \cdot \dfrac{3}{5a}$

**41.** $\dfrac{7}{10} \cdot \dfrac{34}{150}$

**42.** $\dfrac{15}{22} \cdot \dfrac{4}{7}$

**43.** $\dfrac{36}{85} \cdot \dfrac{25}{-99}$

**44.** $\dfrac{-70}{45} \cdot \dfrac{50}{49}$

**45.** $\dfrac{-98}{99} \cdot \dfrac{27a}{175a}$

**46.** $\dfrac{70}{-49} \cdot \dfrac{63}{300x}$

**47.** $\dfrac{110}{33} \cdot \dfrac{-24}{25x}$

**48.** $\dfrac{-19}{130} \cdot \dfrac{65}{38x}$

**49.** $\left(-\dfrac{11}{24}\right)\dfrac{3}{5}$

**50.** $\left(-\dfrac{15}{22}\right)\dfrac{4}{7}$

**51.** $\dfrac{10a}{21} \cdot \dfrac{3}{8b}$

**52.** $\dfrac{17}{21y} \cdot \dfrac{3x}{5}$

 Solve.

The *pitch* of a screw is the distance between its threads. With each complete rotation, the screw goes in or out a distance equal to its pitch. Use this information to answer Exercises 53 and 54.

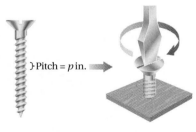

}Pitch = $p$ in.    Each rotation moves the screw in or out $p$ in.

**53.** The pitch of a screw is $\frac{1}{16}$ in. How far will it go into a piece of oak when it is turned 10 complete rotations clockwise?

**54.** The pitch of a screw is $\frac{3}{32}$ in. How far will it come out of a piece of plywood when it is turned 10 complete rotations counterclockwise?

**55.** *Level of Education and Median Income.* The median yearly income of someone with an associate's degree is approximately $\frac{2}{3}$ of the median income of someone with a bachelor's degree. If the median income for those with bachelor's degrees is $72,420, what is the median income of those with associate's degrees?

Source: U.S. Census Bureau

**56.** After Jack completes 60 hr of teacher training in college, he can earn $75 for working a full day as a substitute teacher. How much will he receive for working $\frac{3}{5}$ of a day?

**57.** *Mailing-List Addresses.* Analysts have determined that $\frac{1}{4}$ of the addresses on a mailing list will change in one year. A business has a mailing list of 2500 people. After one year, how many addresses on that list will be incorrect?

**58.** *Shy People.* Sociologists have determined that $\frac{2}{5}$ of the people in the world are shy. A sales manager is interviewing 650 people for an aggressive sales position. How many of these people might be shy?

**59.** A recipe for piecrust calls for $\frac{2}{3}$ cup of flour. A chef is making $\frac{1}{2}$ of the recipe. How much flour should the chef use?

**60.** Of the students in the freshman class, $\frac{4}{5}$ have digital cameras; $\frac{1}{4}$ of these students also join the college photography club. What fraction of the students in the freshman class join the photography club?

**61.** *Assessed Value.* A house worth $154,000 is assessed for $\frac{3}{4}$ of its value. What is the assessed value of the house?

**62.** Roxanne's tuition was $4600. A loan was obtained for $\frac{3}{4}$ of the tuition. How much was the loan?

**63.** *Map Scaling.* On a map, 1 in. represents 240 mi. How much does $\frac{2}{3}$ in. represent?

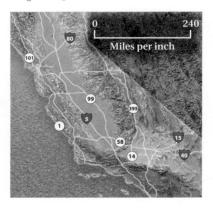

**64.** *Map Scaling.* On a map, 1 in. represents 120 mi. How much does $\frac{3}{4}$ in. represent?

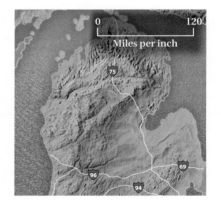

**65.** *Household Budgets.*   A family has an annual income of $42,000. Of this, $\frac{1}{5}$ is spent for food, $\frac{1}{4}$ for housing, $\frac{1}{10}$ for clothing, $\frac{1}{14}$ for savings, $\frac{1}{5}$ for taxes, and the rest for other expenses. How much is spent for each?

**66.** *Household Budgets.*   A family has an annual income of $28,140. Of this, $\frac{1}{5}$ is spent for food, $\frac{1}{4}$ for housing, $\frac{1}{10}$ for clothing, $\frac{1}{14}$ for savings, $\frac{1}{5}$ for taxes, and the rest for other expenses. How much is spent for each?

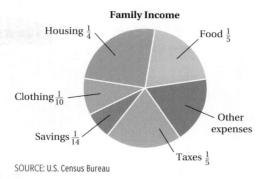

**Family Income**

Housing $\frac{1}{4}$

Food $\frac{1}{5}$

Clothing $\frac{1}{10}$

Savings $\frac{1}{14}$

Other expenses

Taxes $\frac{1}{5}$

SOURCE: U.S. Census Bureau

Find the area.

**67.**

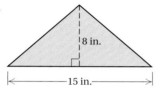

8 in.

15 in.

**68.**

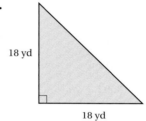

18 yd

18 yd

**69.**

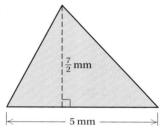

$\frac{7}{2}$ mm

5 mm

**70.**

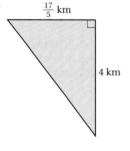

$\frac{17}{5}$ km

4 km

**71.**

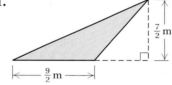

$\frac{7}{2}$ m

$\frac{9}{2}$ m

**72.**

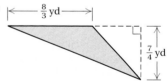

$\frac{8}{3}$ yd

$\frac{7}{4}$ yd

**73.**

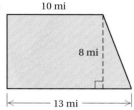

10 mi

8 mi

13 mi

**74.**

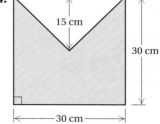

15 cm

30 cm

30 cm

**75.** *Jewelry.* Christie has designed a pair of kite-shaped earrings, as shown below. Determine the surface area of the front of one earring.

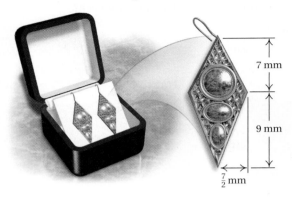

7 mm

9 mm

$\frac{7}{2}$ mm

**76.** *Construction.* Find the total area of the sides and ends of the town office building shown. Do not subtract for any windows, doors, or steps.

25 ft

11 ft

75 ft

50 ft

Dumont Town Offices

## Skill Maintenance

Solve. [1.7b]

**77.** $48 \cdot t = 1680$

**78.** $74 \cdot x = 6290$

**79.** $3125 = 25 \cdot t$

**80.** $2880 = 24 \cdot y$

**81.** $t + 28 = 5017$

**82.** $456 + x = 9002$

**83.** $8797 = y + 2299$

**84.** $10{,}000 = 3593 + m$

## Synthesis

Simplify. Use the list of prime numbers on p. 162.

**85.** $\dfrac{201}{535} \cdot \dfrac{4601}{6499}$

**86.** $\dfrac{667}{899} \cdot \dfrac{558}{621}$

**87.** *College Profile.* Of students entering a college, $\frac{7}{8}$ have completed high school and $\frac{2}{3}$ are older than 20. If $\frac{1}{7}$ of all students are left-handed, what fraction of students entering the college are left-handed high school graduates over the age of 20?

**88.** *College Profile.* Refer to the information in Exercise 87. If 480 students are entering the college, how many of them are left-handed high school graduates 20 years old or younger?

**89.** *Manufacturing.* A candy box is triangular at each end, as shown below. Find the surface area of the box.

30 mm

140 mm

30 mm

26 mm

30 mm

**90.** *Painting.* Shoreline Painting needs to determine the surface area of an octagonal steeple. Find the total area, if the dimensions are as shown below.

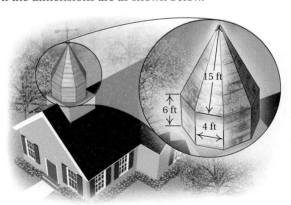

15 ft

6 ft

4 ft

# 3.7 Reciprocals and Division

## a Reciprocals

Each of the following products is 1:

$$8 \cdot \frac{1}{8} = \frac{8}{8} = 1; \qquad \frac{-2}{3} \cdot \frac{3}{-2} = \frac{-6}{-6} = 1.$$

> **OBJECTIVES**
>
> **a** Find the reciprocal of a number.
>
> **b** Divide and simplify using fraction notation.

> **RECIPROCALS**
>
> If the product of two numbers is 1, we say that they are **reciprocals** of each other.* To find the reciprocal of a fraction, interchange the numerator and the denominator.
>
> Number: $\frac{3}{4}$ ⟶ Reciprocal: $\frac{4}{3}$

**SKILL TO REVIEW**
Objective 2.5a: Divide integers.

Divide.

1. $\dfrac{26}{-2}$     2. $\dfrac{-100}{-25}$

**EXAMPLES** Find the reciprocal.

**1.** The reciprocal of $\frac{4}{5}$ is $\frac{5}{4}$.     Note that $\frac{4}{5} \cdot \frac{5}{4} = \frac{20}{20} = 1.$

**2.** The reciprocal of $\frac{a}{b}$ is $\frac{b}{a}$.     Note that $\frac{a}{b} \cdot \frac{b}{a} = \frac{ab}{ba} = 1.$

**3.** The reciprocal of 8 is $\frac{1}{8}$.     Think of 8 as $\frac{8}{1}$: $\frac{8}{1} \cdot \frac{1}{8} = \frac{8}{8} = 1.$

**4.** The reciprocal of $\frac{1}{3}$ is 3.     Note that $\frac{1}{3} \cdot 3 = \frac{3}{3} = 1.$

**5.** The reciprocal of $-\frac{5}{9}$ is $-\frac{9}{5}$.     Negative numbers have negative reciprocals: $\left(-\frac{5}{9}\right)\left(-\frac{9}{5}\right) = \left(\frac{45}{45}\right) = 1.$

> Do Margin Exercises 1–5.

Find the reciprocal.

1. $\dfrac{2}{5}$     2. $\dfrac{-6}{x}$

3. 9     4. $\dfrac{1}{5}$

5. $-\dfrac{3}{10}$

Does 0 have a reciprocal? If it did, it would have to be a number $x$ such that $0 \cdot x = 1$. But 0 times any number is 0. Thus, we have the following.

> **0 HAS NO RECIPROCAL**
>
> The number 0, or $\dfrac{0}{n}$, has no reciprocal. $\left(\text{Recall that } \dfrac{n}{0}, \text{ is not defined.}\right)$

## b Division

Consider the division $\frac{3}{4} \div \frac{1}{8}$. We are asking how many $\frac{1}{8}$'s are in $\frac{3}{4}$. We can answer this by looking at the figure at right. We see that there are six $\frac{1}{8}$'s in $\frac{3}{4}$. Thus,

$$\frac{3}{4} \div \frac{1}{8} = 6.$$

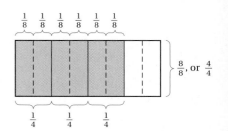

*Answers*

*Skill to Review:*
1. $-13$   2. 4

*Margin Exercises:*
1. $\dfrac{5}{2}$   2. $\dfrac{x}{-6}$   3. $\dfrac{1}{9}$   4. 5   5. $-\dfrac{10}{3}$

---

*A reciprocal is also called a multiplicative inverse.

**Multiplying and Dividing Fractions** On a fraction calculator, multiplication or division of fractions is entered the way it is written. To perform the division $\frac{4}{9} \div \frac{6}{5}$ with such a calculator, the following keystrokes can be used:

$$\boxed{4}\ \boxed{a^b/_c}\ \boxed{9}\ \boxed{\div}$$
$$\boxed{6}\ \boxed{a^b/_c}\ \boxed{5}\ \boxed{=}.$$

The simplified fraction notation is $\frac{10}{27}$. On calculators without an $\boxed{a^b/_c}$ key, parentheses must be used around the second fraction when dividing. The following keystrokes can be used to perform the division and convert the result to fraction notation:

$$\boxed{4}\ \boxed{\div}\ \boxed{9}\ \boxed{\div}\ \boxed{(}\ \boxed{6}\ \boxed{\div}$$
$$\boxed{5}\ \boxed{)}\ \boxed{MATH}\ \boxed{1}\ \boxed{ENTER}.$$

Again, the result is $\frac{10}{27}$.

**Exercises:** Use a calculator to perform the following operations.

1. $\frac{5}{8} \cdot \frac{4}{15}$     2. $\frac{10}{12} \div \frac{3}{8}$

3. $\frac{-6}{7} \div \frac{2}{3}$     4. $\frac{-9}{10} \div \frac{-3}{5}$

We can check this by multiplying:

$$6 \cdot \frac{1}{8} = \frac{6}{1} \cdot \frac{1}{8} = \frac{6}{8} = \frac{2 \cdot 3}{2 \cdot 4} = \frac{2}{2} \cdot \frac{3}{4} = \frac{3}{4}.$$

Here is a faster way to do this division:

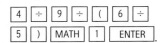

$$\frac{3}{4} \div \frac{1}{8} = \frac{3}{4} \cdot \frac{8}{1} = \frac{3 \cdot 8}{4 \cdot 1} = \frac{24}{4} = 6.$$     Multiplying by the reciprocal of the divisor

---

**DIVISION OF FRACTIONS**

To divide by a fraction, multiply by its reciprocal:

$$\frac{a}{b} \div \frac{c}{d} = \frac{a}{b} \cdot \frac{d}{c}.$$     Multiply by the reciprocal of the divisor.

---

**EXAMPLES** Divide and simplify.

6. $\dfrac{5}{6} \div \dfrac{2}{3} = \dfrac{5}{6} \cdot \dfrac{3}{2}$     Multiplying by the reciprocal of the divisor

$$= \frac{5 \cdot 3}{3 \cdot 2 \cdot 2}$$     Factoring and identifying a common factor

$$= \frac{3}{3} \cdot \frac{5}{2 \cdot 2}$$     Removing a factor equal to 1: $\frac{3}{3} = 1$

$$= \frac{5}{4}$$

7. $\dfrac{-3}{5} \div \dfrac{1}{2} = \dfrac{-3}{5} \cdot 2$     The reciprocal of $\frac{1}{2}$ is 2.

$$= \frac{-3 \cdot 2}{5} = \frac{-6}{5}$$

8. $\dfrac{2a}{5} \div 7 = \dfrac{2a}{5} \cdot \dfrac{1}{7}$     The reciprocal of 7 is $\frac{1}{7}$.

$$= \frac{2a \cdot 1}{5 \cdot 7} = \frac{2a}{35}$$

9. $\dfrac{\dfrac{7}{10}}{-\dfrac{14}{15}} = \dfrac{7}{10} \div \left(-\dfrac{14}{15}\right)$     The fraction bar indicates division.

$$= \frac{7}{10} \cdot \left(-\frac{15}{14}\right)$$     Multiplying by the reciprocal of the divisor

$$= \frac{7 \cdot 5(-3)}{2 \cdot 5 \cdot 7 \cdot 2}$$     Factoring and identifying common factors

$$= \frac{7 \cdot 5}{7 \cdot 5} \cdot \frac{-3}{4}$$     Removing a factor equal to 1: $\frac{7 \cdot 5}{7 \cdot 5} = 1$

$$= -\frac{3}{4}$$

---------------------------------- *Caution!* ----------------------------------

Canceling can be used as follows for Examples 6 and 9.

**6.** $\dfrac{5}{6} \div \dfrac{2}{3} = \dfrac{5}{6} \cdot \dfrac{3}{2} = \dfrac{5 \cdot 3}{6 \cdot 2} = \dfrac{5 \cdot \cancel{3}}{\cancel{3} \cdot 2 \cdot 2} = \dfrac{5}{2 \cdot 2} = \dfrac{5}{4}$

$\phantom{xxxxxxxxx}$ Removing a factor equal to 1: $\frac{3}{3} = 1$

**9.** $\dfrac{7}{10} \div \left(-\dfrac{14}{15}\right) = \dfrac{7}{10} \cdot \left(-\dfrac{15}{14}\right) = \dfrac{7 \cdot \cancel{5}(-3)}{2 \cdot \cancel{5} \cdot 7 \cdot 2} = \dfrac{-3}{4}$, or $-\dfrac{3}{4}$

$\phantom{xxxxxxxxx}$ Removing a factor equal to 1: $\frac{7 \cdot 5}{7 \cdot 5} = 1$

**Remember, if you can't factor, you can't cancel!**

--------------------------------------------------------------------------------

> Do Exercises 6–10.

Why do we multiply by a reciprocal when dividing? To see why, consider $\frac{2}{3} \div \frac{7}{5}$. We will multiply by 1 to find an equivalent expression. To write 1, we use $(5/7)/(5/7)$, since $\frac{5}{7}$ is the reciprocal of $\frac{7}{5}$.

$\dfrac{2}{3} \div \dfrac{7}{5} = \dfrac{\dfrac{2}{3}}{\dfrac{7}{5}}$    Writing fraction notation for the division

$= \dfrac{\dfrac{2}{3}}{\dfrac{7}{5}} \cdot 1$    Multiplying by 1

$= \dfrac{\dfrac{2}{3}}{\dfrac{7}{5}} \cdot \dfrac{\dfrac{5}{7}}{\dfrac{5}{7}}$    Multiplying by 1; $\frac{5}{7}$ is the reciprocal of $\frac{7}{5}$ and $\frac{5/7}{5/7} = 1$

$= \dfrac{\dfrac{2}{3} \cdot \dfrac{5}{7}}{\dfrac{7}{5} \cdot \dfrac{5}{7}}$    Multiplying the numerators and the denominators

$= \dfrac{\dfrac{2}{3} \cdot \dfrac{5}{7}}{1} = \dfrac{2}{3} \cdot \dfrac{5}{7} = \dfrac{10}{21}$    After we multiplied, we got 1 for the denominator. The numerator (in color) shows the multiplication by the reciprocal of $\frac{7}{5}$.

Thus,

$\dfrac{2}{3} \div \dfrac{7}{5} = \dfrac{2}{3} \cdot \dfrac{5}{7} = \dfrac{10}{21}.$

Expressions of the form $\dfrac{\dfrac{a}{b}}{\dfrac{c}{d}}$ are examples of *complex fractions*:

$\dfrac{\dfrac{a}{b}}{\dfrac{c}{d}} = \dfrac{a}{b} \div \dfrac{c}{d}.$

> Do Exercise 11.

Divide and simplify.

**6.** $\dfrac{6}{7} \div \dfrac{3}{4}$

**7.** $\left(-\dfrac{2}{3}\right) \div \dfrac{1}{4}$

**8.** $\dfrac{4}{5} \div 8$

**9.** $60 \div \dfrac{3a}{5}$

**10.** $\dfrac{\dfrac{-6}{7}}{\dfrac{3}{5}}$

**11.** To remember *why* fractions are divided as they are, multiply by 1 to perform the following division, using the reciprocal of $\frac{4}{5}$ to write 1.

$\dfrac{\dfrac{6}{7}}{\dfrac{4}{5}}$

***Answers***

**6.** $\dfrac{8}{7}$ **7.** $-\dfrac{8}{3}$ **8.** $\dfrac{1}{10}$ **9.** $\dfrac{100}{a}$

**10.** $\dfrac{-10}{7}$, or $-\dfrac{10}{7}$ **11.** $\dfrac{15}{14}$

**a** Find the reciprocal.

**1.** $\dfrac{7}{3}$

**2.** $\dfrac{6}{5}$

**3.** 9

**4.** 3

**5.** $\dfrac{1}{7}$

**6.** $\dfrac{1}{4}$

**7.** $-\dfrac{8}{9}$

**8.** $-\dfrac{12}{5}$

**9.** $\dfrac{a}{c}$

**10.** $\dfrac{x}{y}$

**11.** $\dfrac{-3n}{m}$

**12.** $\dfrac{8t}{-7r}$

**13.** $\dfrac{8}{-15}$

**14.** $\dfrac{-6}{25}$

**15.** $7m$

**16.** $5n$

**17.** $\dfrac{1}{4a}$

**18.** $\dfrac{1}{9t}$

**19.** $-\dfrac{1}{3z}$

**20.** $-\dfrac{1}{2x}$

**b** Divide. Don't forget to simplify when possible. Assume that all variables are nonzero.

**21.** $\dfrac{3}{7} \div \dfrac{3}{4}$

**22.** $\dfrac{2}{3} \div \dfrac{3}{4}$

**23.** $\dfrac{3}{5} \div \dfrac{9}{4}$

**24.** $\dfrac{6}{7} \div \dfrac{3}{5}$

**25.** $\dfrac{4}{3} \div \dfrac{1}{3}$

**26.** $\dfrac{10}{9} \div \dfrac{1}{2}$

**27.** $\left(-\dfrac{1}{3}\right) \div \dfrac{1}{6}$

**28.** $\left(-\dfrac{1}{4}\right) \div \dfrac{1}{5}$

**29.** $\left(-\dfrac{10}{21}\right) \div \left(-\dfrac{2}{15}\right)$

**30.** $-\dfrac{15}{28} \div \left(-\dfrac{9}{20}\right)$

**31.** $\dfrac{3}{8} \div 3$

**32.** $\dfrac{5}{6} \div 5$

**33.** $\dfrac{12}{7} \div 16$

**34.** $\dfrac{18}{5} \div 27$

**35.** $(-12) \div \dfrac{3}{2}$

**36.** $(-24) \div \dfrac{3}{8}$

**37.** $\dfrac{x}{8} \div \dfrac{1}{4}$

**38.** $\dfrac{3}{4} \div \dfrac{2}{y}$

**39.** $\dfrac{2}{3} \div (6x)$

**40.** $\dfrac{12}{5} \div (4x)$

**41.** $28 \div \dfrac{4}{5a}$

**42.** $40 \div \dfrac{2}{3m}$

**43.** $\left(-\dfrac{5}{8}\right) \div \left(-\dfrac{5}{8}\right)$

**44.** $\left(-\dfrac{2}{5}\right) \div \left(-\dfrac{2}{5}\right)$

**45.** $\dfrac{-8}{15} \div \dfrac{4}{5}$

**46.** $\dfrac{6}{-13} \div \dfrac{3}{26}$

**47.** $\dfrac{77}{64} \div \dfrac{49}{18}$

**48.** $\dfrac{81}{42} \div \dfrac{33}{56}$

**49.** $120a \div \dfrac{45}{14}$

**50.** $360n \div \dfrac{27n}{8}$

**51.** $\dfrac{\frac{2}{5}}{\frac{3}{7}}$

**52.** $\dfrac{\frac{5}{6}}{\frac{2}{7}}$

**53.** $\dfrac{-\frac{7}{20}}{-\frac{8}{5}}$

**54.** $\dfrac{-\frac{8}{21}}{-\frac{6}{5}}$

**55.** $\dfrac{-\frac{15}{8}}{\frac{9}{10}}$

**56.** $\dfrac{-\frac{27}{10}}{\frac{21}{20}}$

## Skill Maintenance

In each of Exercises 57–64, fill in the blank with the correct term from the given list. Some of the choices may not be used and some may be used more than once.

**57.** The equation $14 + (2 + 30) = (14 + 2) + 30$ illustrates the _____ law of addition.   [1.2a]

**58.** In the product $10 \cdot \frac{3}{4}$, the numbers 10 and $\frac{3}{4}$ are called _____ .   [3.2a]

**59.** A natural number that has exactly two different factors, itself and 1, is called a(n) _____ number.   [3.2b]

**60.** In the fraction $\frac{4}{17}$, we call 17 the _____ .   [3.3a]

**61.** Since $a + 0 = a$ for any number $a$, the number 0 is the _____ identity.   [2.2a]

**62.** The product of 6 and $\frac{1}{6}$ is 1; we say that 6 and $\frac{1}{6}$ are _____ of each other.   [3.7a]

**63.** The sum of 6 and $-6$ is 0; we say that 6 and $-6$ are _____ of each other.   [2.1d]

**64.** A sentence with $=$ is called a(n) _____ .   [1.7a]

associative

commutative

additive

multiplicative

reciprocals

factors

prime

composite

numerator

denominator

equation

expression

opposites

variables

## Synthesis

Simplify.

**65.** $\left( \dfrac{4}{15} \div \dfrac{2}{25} \right)^2$

**66.** $\left( \dfrac{9}{10} \div \dfrac{12}{25} \right)^2$

**67.** $\left( \dfrac{9}{10} \div \dfrac{2}{5} \div \dfrac{3}{8} \right)^2$

**68.** $\dfrac{\left( -\frac{3}{7} \right)^2 \div \frac{12}{5}}{\left( \frac{-2}{9} \right)\left( \frac{9}{2} \right)}$

**69.** $\left( \dfrac{14}{15} \div \dfrac{49}{65} \cdot \dfrac{77}{260} \right)^2$

**70.** $\left( \dfrac{10}{9} \right)^2 \div \dfrac{35}{27} \cdot \dfrac{49}{44}$

Simplify. Use the list of prime numbers on p. 162.

**71.** ▦ $\dfrac{711}{1957} \div \dfrac{10{,}033}{13{,}081}$

**72.** ▦ $\dfrac{8633}{7387} \div \dfrac{485}{581}$

**73.** ▦ $\dfrac{451}{289} \div \dfrac{123}{340}$

**74.** ▦ $\dfrac{530}{490} \div \dfrac{1060}{980}$

# 3.8

## Solving Equations: The Multiplication Principle

## OBJECTIVES

**a** Use the multiplication principle to solve equations.

**b** Solve problems by using the multiplication principle.

**SKILL TO REVIEW**

Objective 2.8b: Use the division principle to solve equations.

Solve.

**1.** $-8t = 32$ **2.** $-x = -9$

With fraction notation, we can solve equations like $a \cdot x = b$ by using multiplication.

### **a** The Multiplication Principle

To divide by a fraction, we multiply by the reciprocal of that fraction. This suggests that we restate the division principle in its more common form—the multiplication principle.

> **THE MULTIPLICATION PRINCIPLE**
>
> For any numbers $a$, $b$, and $c$, with $c \neq 0$,
>
> $$a = b \quad \text{is equivalent to} \quad a \cdot c = b \cdot c.$$

**EXAMPLE 1** Solve: $\frac{3}{4}x = 15$.

We multiply both sides of the equation by the reciprocal of $\frac{3}{4}$.

$$\frac{3}{4}x = 15$$

$$\frac{4}{3} \cdot \frac{3}{4}x = \frac{4}{3} \cdot 15 \qquad \text{Using the multiplication principle; note that } \frac{4}{3} \text{ is the reciprocal of } \frac{3}{4}.$$

$$\left(\frac{4}{3} \cdot \frac{3}{4}\right)x = \frac{4 \cdot 15}{3} \qquad \text{Using an associative law; try to do this mentally.}$$

$$1x = 20 \qquad \text{Multiplying; note that } \frac{4 \cdot 15}{3} = \frac{4 \cdot \cancel{3} \cdot 5}{\cancel{3}}.$$

$$x = 20 \qquad \text{Remember that } 1x \text{ is } x.$$

To confirm that 20 is the solution, we perform a check.

Check:

$$\frac{\frac{3}{4}x = 15}{\begin{array}{c|c} \frac{3}{4} \cdot 20 \;\; ? \;\; 15 & \\ \hline \frac{3 \cdot \cancel{4} \cdot 5}{\cancel{4}} & \qquad \text{Removing a factor equal to 1: } \frac{4}{4} = 1 \\ 3 \cdot 5 \;\big|\; 15 \quad \text{TRUE} & \end{array}}$$

The solution is 20.

Solve.

**1.** $\frac{2}{3}x = 8$ **2.** $\frac{2}{7}a = -6$

Do Margin Exercises 1 and 2.

In an expression like $\frac{3}{4}x$, the constant factor—in this case, $\frac{3}{4}$—is called the **coefficient**. In Example 1, we multiplied on both sides by $\frac{4}{3}$, the reciprocal of the coefficient of $x$. Note that using the multiplication principle to multiply by $\frac{4}{3}$ on both sides is the same as using the division principle to divide by $\frac{3}{4}$ on both sides.

*Answers*

*Skill to Review:*
**1.** $-4$ **2.** 9

*Margin Exercises:*
**1.** 12 **2.** $-21$

**EXAMPLE 2**  Solve: $5a = -\dfrac{7}{3}$.

We have

$$5a = -\frac{7}{3}$$

$$\frac{1}{5} \cdot 5a = \frac{1}{5} \cdot \left(-\frac{7}{3}\right) \qquad \text{Multiplying both sides by } \tfrac{1}{5}, \text{ the reciprocal of 5}$$

$$a = -\frac{1 \cdot 7}{5 \cdot 3} = -\frac{7}{15}.$$

We leave the check to the student. The solution is $-\dfrac{7}{15}$.

**EXAMPLE 3**  Solve: $\dfrac{10}{3} = -\dfrac{4}{9}x$.

We have

$$\frac{10}{3} = -\frac{4}{9}x$$

$$-\frac{9}{4} \cdot \frac{10}{3} = -\frac{9}{4} \cdot \left(-\frac{4}{9}\right)x \qquad \text{Multiplying both sides by } -\tfrac{9}{4}, \text{ the reciprocal of } -\tfrac{4}{9}$$

$$-\frac{3 \cdot 3 \cdot 2 \cdot 5}{2 \cdot 2 \cdot 3} = x$$

$$-\frac{15}{2} = x. \qquad \text{Removing a factor equal to 1: } \frac{3 \cdot 2}{2 \cdot 3} = 1$$

We leave the check to the student. The solution is $-\dfrac{15}{2}$.

Do Exercises 3 and 4.

Solve.

**3.** $-\dfrac{9}{8} = 4x$  **4.** $-\dfrac{6}{7}a = \dfrac{9}{14}$

## (b) Applications and Problem Solving

**EXAMPLE 4**  *Doses of an Antibiotic.*  How many doses, each containing $\frac{15}{4}$ milliliters (mL), can be obtained from a bottle of a children's antibiotic that contains 60 mL?

1. **Familiarize.**  We make a drawing and let $n =$ the number of doses in all.

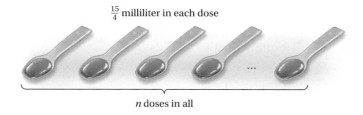

$\frac{15}{4}$ milliliter in each dose

$n$ doses in all

2. **Translate.**  The problem can be translated to the equation

$$\frac{15}{4} \cdot n = 60.$$

*Answers*

3. $-\dfrac{9}{32}$  4. $-\dfrac{3}{4}$

**5.** Each loop in a spring uses $\frac{21}{8}$ in. of wire. How many loops can be made from 210 in. of wire?

**6.** For a party, Jana made an 8-foot submarine sandwich. If one serving is $\frac{2}{3}$ ft, how many servings does Jana's sub contain?

**3. Solve.** To solve the equation, we use the multiplication principle:

$$\frac{4}{15} \cdot \frac{15}{4} \cdot n = \frac{4}{15} \cdot 60 \qquad \text{Multiplying both sides by } \frac{4}{15}$$

$$1n = \frac{4 \cdot 60}{15}$$

$$n = \frac{2 \cdot 2 \cdot 2 \cdot 2 \cdot 3 \cdot 5}{1 \cdot 3 \cdot 5} = \frac{3 \cdot 5}{3 \cdot 5} \cdot \frac{2 \cdot 2 \cdot 2 \cdot 2}{1} = 16$$

**4. Check.** We check by multiplying the number of doses by the size of the dose: $16 \cdot \frac{15}{4} = 60$. Note too that

$$\text{doses} \cdot \frac{\text{mL}}{\text{dose}} = \text{mL},$$

so the units also check.

**5. State.** There are 16 doses in a 60-mL bottle of the antibiotic.

> Do Exercises 5 and 6.

**EXAMPLE 5** *Brick Walkway.* Logan's Landscape specializes in brick and stone walkways. After they complete 63 ft of a brick walkway between two buildings on campus, $\frac{7}{8}$ of the walkway is complete. What is the total length of this walkway?

1. **Familiarize.** We ask: "63 ft is $\frac{7}{8}$ of what length?" We make a drawing or at least visualize the problem. We let $w = $ the length of the walkway.

2. **Translate.** We translate to an equation:

$$\underbrace{63 \text{ ft}}_{\downarrow} \quad \text{is} \downarrow \quad \underset{\downarrow}{\frac{7}{8}} \quad \text{of} \downarrow \quad \underbrace{\text{length of walkway}}$$

$$63 \quad = \quad \frac{7}{8} \quad \cdot \quad w.$$

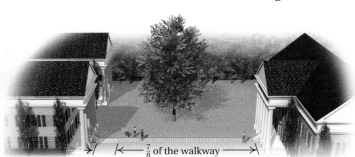

$\frac{7}{8}$ of the walkway
63 ft

3. **Solve.** We multiply on both sides by $\frac{8}{7}$:

$$\frac{8}{7} \cdot 63 = \frac{8}{7} \cdot \frac{7}{8} \cdot w$$

$$\frac{8 \cdot 7 \cdot 9}{7 \cdot 1} = 1w$$

$$72 = w.$$

4. **Check.** We determine whether $\frac{7}{8}$ of 72 is 63:

$$\frac{7}{8} \cdot 72 = \frac{7 \cdot 72}{8 \cdot 1} = \frac{7 \cdot 8 \cdot 9}{8 \cdot 1} = 63.$$

The answer, 72, checks.

5. **State.** The total length of the walkway is 72 ft.

> Do Exercise 7.

**7. Sales Trip.** John Penna sells soybean seeds to seed companies. After he had driven 210 mi, $\frac{5}{6}$ of his sales trip was completed. How long was the total trip?

$\frac{5}{6}$ of the trip
210 mi

*Answers*

**5.** 80 loops **6.** 12 servings **7.** 252 mi

# Translating for Success

1. *Valentine Boxes.* Jane's Fudge Shop is preparing Valentine boxes. How many pounds of fudge will be needed to fill 80 boxes if each box contains $\frac{5}{16}$ lb?

2. *Gallons of Gasoline.* On the third day of a business trip, a sales representative used $\frac{4}{5}$ of a tank of gasoline. If the tank is a 20-gal tank, how many gallons were used on the third day?

3. *Purchasing a Shirt.* Tom received $36 for his birthday. If he spends $\frac{3}{4}$ of the gift on a new shirt, what is the cost of the shirt?

4. *Checkbook Balance.* The balance in Sam's checking account is $1456. He writes a check for $28 and makes a deposit of $52. What is the new balance?

5. *Valentine Boxes.* Jane's Fudge Shop prepared 80 lb of fudge for Valentine boxes. If each box contains $\frac{5}{16}$ lb, how many boxes can be filled?

---

The goal of these matching questions is to practice step (2), *Translate*, of the five-step problem-solving process. Translate each problem to an equation and select a correct translation from equations A–O.

A. $x = \frac{3}{4} \cdot 36$

B. $28 \cdot x = 52$

C. $x = 80 \cdot \frac{5}{16}$

D. $x = 1456 \div 28$

E. $\frac{5}{4} \cdot x = 20$

F. $20 = \frac{4}{5} \cdot x$

G. $x = 12 \cdot 28$

H. $x = \frac{4}{5} \cdot 20$

I. $\frac{3}{4} \cdot x = 36$

J. $x = 1456 - 52 - 28$

K. $x \div 28 = 1456$

L. $x = 52 - 28$

M. $x = 52 \cdot 28$

N. $x = 1456 - 28 + 52$

O. $\frac{5}{16} \cdot x = 80$

*Answers on page A-6*

---

6. *Gasoline Tank.* A gasoline tank contains 20 gal when it is $\frac{4}{5}$ full. How many gallons can it hold when full?

7. *Knitting a Scarf.* It takes Rachel 36 hr to knit a scarf. She can knit only $\frac{3}{4}$ hr per day because she is taking 16 hr of college classes. How many days will it take her to knit the scarf?

8. *Bicycle Trip.* On a recent 52-mi bicycle trip, David stopped to make a cell-phone call after completing 28 mi. How many more miles did he bicycle after the call?

9. *Crème de Menthe Thins.* Andes Candies L.P. makes Crème de Menthe Thins. How many 28-piece packages can be filled with 1456 pieces?

10. *Cereal Donations.* The Williams family donates 28 boxes of cereal weekly to the local Family in Crisis Center. How many boxes does this family donate in one year?

## Exercise Set

**a**  Use the multiplication principle to solve each equation. Don't forget to check!

**1.** $\dfrac{4}{5}x = 12$

**2.** $\dfrac{4}{3}x = 20$

**3.** $\dfrac{7}{3}a = 21$

**4.** $\dfrac{4}{5}a = 24$

**5.** $\dfrac{2}{9}y = -10$

**6.** $\dfrac{3}{8}y = -21$

**7.** $6t = \dfrac{12}{17}$

**8.** $3t = \dfrac{15}{14}$

**9.** $\dfrac{1}{4}x = \dfrac{3}{5}$

**10.** $\dfrac{1}{6}x = \dfrac{2}{7}$

**11.** $\dfrac{3}{2}t = -\dfrac{8}{7}$

**12.** $\dfrac{4}{3}t = -\dfrac{5}{2}$

**13.** $\dfrac{4}{5} = -10a$

**14.** $\dfrac{6}{5} = -12a$

**15.** $x \cdot \dfrac{9}{5} = \dfrac{3}{10}$

**16.** $x \cdot \dfrac{10}{3} = \dfrac{8}{15}$

**17.** $-\dfrac{1}{10}x = 8$

**18.** $-\dfrac{1}{11}x = -5$

**19.** $a \cdot \dfrac{9}{7} = -\dfrac{3}{14}$

**20.** $a\left(-\dfrac{9}{4}\right) = -\dfrac{3}{10}$

**21.** $-x = \dfrac{7}{13}$

**22.** $-x = \dfrac{7}{11}$

**23.** $-x = -\dfrac{27}{31}$

**24.** $-x = -\dfrac{35}{39}$

**25.** $7t = 6$

**26.** $-6t = 1$

**27.** $-24 = -10a$

**28.** $-18 = -20a$

**29.** $-\dfrac{14}{9} = \dfrac{10}{3}\,t$

**30.** $-\dfrac{15}{7} = \dfrac{3}{2}\,t$

**31.** $n \cdot \dfrac{4}{15} = \dfrac{12}{25}$

**32.** $n \cdot \dfrac{5}{16} = \dfrac{15}{14}$

**33.** $-\dfrac{7}{20}\,x = -\dfrac{21}{10}$

**34.** $-\dfrac{7}{15}\,x = -\dfrac{21}{10}$

**35.** $-\dfrac{25}{17} = -\dfrac{35}{34}\,a$

**36.** $-\dfrac{49}{45} = -\dfrac{28}{27}\,a$

 **b**    Solve.

**37.** *Extension Cords.* An electrical supplier sells rolls of SJO 14-3 cable to a company that makes extension cords. It takes $\frac{7}{3}$ ft of cable to make each cord. How many extension cords can be made with a roll of cable containing 2240 ft of cable?

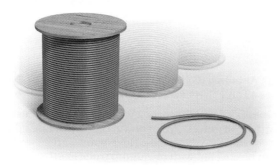

**38.** Benny uses $\frac{2}{5}$ g (gram) of toothpaste each time he brushes his teeth. If Benny buys a 30-g tube, how many times will he be able to brush his teeth?

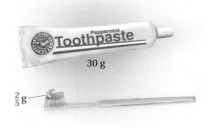

**39.** *Gasoline Tanker.* A tanker that delivers gasoline to gas stations had 1400 gal of gasoline when it was $\frac{7}{9}$ full. How much can the tanker hold when it is full?

**40.** How many $\frac{2}{3}$-cup cereal bowls can be filled from 10 cups of cornflakes?

**41.** *Honey.* A worker bee will produce $\frac{1}{12}$ tsp (teaspoon) of honey in her lifetime. How many worker bees does it take to produce $\frac{3}{4}$ tsp of honey?

Source: www.pbs.org/wgbh/nova/bees/buzz.html

**42.** *Herbal Tea.* At Perfect Tea, Donna fills each bag of Sereni-Tea with $\frac{3}{5}$ g (gram) of chamomile. If she begins with 51 g of chamomile, how many tea bags can she fill?

**43.** *Packaging.* The South Shore Co-op prepackages cheddar cheese in $\frac{3}{4}$-lb packages. How many packages can be made from a 15-lb wheel of cheese?

**44.** *Meal Planning.* Ian purchased 6 lb of cold cuts for a luncheon. If Ian is to allow $\frac{3}{8}$ lb per person, how many people can attend the luncheon?

**45.** *Art Supplies.* The Ferristown School District purchased $\frac{3}{4}$ T (ton) of clay. The clay is to be shared equally among the district's 6 art departments. How much will each art department receive?

**46.** *Gardening.* The Bingham community garden is to be split into 16 equally sized plots. If the garden occupies $\frac{3}{4}$ acre of land, how large will each plot be?

**47.** Yoshi Teramoto sells hardware tools. After driving 180 kilometers (km), he has completed $\frac{5}{8}$ of a sales trip. How long is the total trip? How many kilometers are left to drive?

**48.** A piece of coaxial cable $\frac{4}{5}$ meter (m) long is to be cut into 8 pieces of the same length. What is the length of each piece?

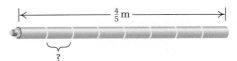

Large quantities of soil, gravel, or mulch are normally sold by the *yard* (yd). Although technically the unit for this type of volume should be *cubic yard* (yd³), in practice only the word *yard* is used. Exercises 49 and 50 make use of this terminology.

**49.** *Gardening.* Green Season Gardening uses about $\frac{2}{3}$ yd of bark mulch per customer every spring. How many customers can they accommodate with one 30-yd batch of bark mulch?

**50.** *Landscaping.* To cover a walkway at a summer cottage with fresh gravel, $\frac{3}{4}$ yd of gravel is needed. Eric's dump truck has a 6-yd capacity. How many cottage walkways can be covered with one dump truck load?

**51.** *Sewing.* A pair of basketball shorts requires $\frac{3}{4}$ yd of nylon. How many pairs of shorts can be made from 24 yd of the fabric?

**52.** *Sewing.* A child's baseball shirt requires $\frac{5}{6}$ yd of cotton fabric. How many shirts can be made from 25 yd of the fabric?

*Pitch of a Screw.* Refer to Exercises 53 and 54 in Exercise Set 3.6.

**53.** After a screw has been turned 8 complete rotations, it is extended $\frac{1}{2}$ in. into a piece of wallboard. What is the pitch of the screw?

**54.** The pitch of a screw is $\frac{3}{32}$ in. How many complete rotations are necessary to drive the screw $\frac{3}{4}$ in. into a piece of pine wood?

## Skill Maintenance

Simplify.

**55.** $-23 + 49$   [2.2a]

**56.** $-69 + 27$   [2.2a]

**57.** $-38 - 29$   [2.3a]

**58.** $-47 - 18$   [2.3a]

**59.** $36 \div (-3)^2 \times (7 - 2)$   [2.5b]

**60.** $(-37 - 12 + 1) \div (-2)^3$   [2.5b]

Form an equivalent expression by combining like terms.   [2.7a]

**61.** $13x + 4x$

**62.** $9a - 5a$

**63.** $2a + 3 + 5a$

**64.** $3x - 7 + x$

## Synthesis

Solve.

**65.** $2x - 7x = -\dfrac{10}{9}$

**66.** $\left(-\dfrac{4}{7}\right)^2 = \left(\dfrac{2^3 - 9}{3}\right)^3 x$

Solve using the five-step problem-solving approach.

**67.** A package of coffee beans weighed $\frac{21}{32}$ lb when it was $\frac{3}{4}$ full. How much could the package hold when completely filled?

**68.** After swimming $\frac{2}{7}$ mi, Katie had swum $\frac{3}{4}$ of the race. How long a race was Katie competing in?

**69.** A block of Swiss cheese is 12 in. long. How many slices will it yield if half of the brick is cut by a slicer set for $\frac{3}{32}$-in. slices and half is cut by a slicer set for $\frac{5}{32}$-in. slices?

**70.** If $\frac{1}{3}$ of a number is $\frac{1}{4}$, what is $\frac{1}{2}$ of the number?

**71.** See Exercise 50. If each cottage requires $\frac{3}{5}$ yd of gravel and Eric charges \$85 for the gravel for each cottage, how much will Eric receive for a full load?

**72.** ▦ See Exercise 49. If each customer at Green Season Gardening uses $\frac{3}{4}$ yd of mulch and Green Season charges each customer \$45 for mulch, how much will Green Season receive from a 25-yd load of mulch?

## Key Terms and Properties

multiple, p. 154
divisible, p. 154
factor, p. 161
factorization, p. 161
prime number, p. 162

composite number, p. 162
prime factorization, p. 163
fraction notation, p. 167
numerator, p. 167
denominator, p. 167

equivalent fractions, p. 183
common denominator, p. 186
cross products, p. 187
reciprocals, p. 201
coefficient, p. 206

*Area of a Triangle*, p. 194
*The Multiplication Principle*, p. 206

## Concept Reinforcement

Determine whether each statement is true or false.

_____ **1.** For any nonzero integer $n$, $\dfrac{n}{n} > \dfrac{0}{n}$.   [3.3b]

_____ **2.** A number is divisible by 10 if its ones digit is 0 or 5.   [3.1b]

_____ **3.** If a number is divisible by 9, then it is also divisible by 3.   [3.1b]

_____ **4.** The fraction $\dfrac{13}{6}$ is larger than the fraction $\dfrac{11}{6}$.   [3.5c]

## Important Concepts

**Objective 3.2a**   Find the factors of a number.

**Example**   Find the factors of 84.

We find as many "two-factor" factorizations as we can.

$1 \cdot 84 \qquad 4 \cdot 21$

$2 \cdot 42 \qquad 6 \cdot 14$

$3 \cdot 28 \qquad 7 \cdot 12 \leftarrow$ Since 8, 9, 10, and 11 are not factors, we are finished.

The factors are 1, 2, 3, 4, 6, 7, 12, 14, 21, 28, 42, and 84.

**Practice Exercise**

**1.** Find the factors of 104.

**Objective 3.2c**   Find the prime factorization of a composite number.

**Example**   Find the prime factorization of 84.

$$
\begin{array}{c}
7 \\
3\overline{)21} \\
2\overline{)42} \\
2\overline{)84}
\end{array}
$$

Thus, $84 = 2 \cdot 2 \cdot 3 \cdot 7$.

**Practice Exercise**

**2.** Find the prime factorization of 104.

**Objective 3.3b** Simplify fraction notation like $n/n$ to 1, $0/n$ to 0, and $n/1$ to $n$.

**Example** Simplify $\frac{6}{6}, \frac{0}{6}$, and $\frac{6}{1}$.

$$\frac{6}{6} = 1, \qquad \frac{0}{6} = 0, \qquad \frac{6}{1} = 6$$

**Practice Exercise**

3. Simplify $\frac{0}{18}, \frac{18}{18}$, and $\frac{18}{1}$.

---

**Objective 3.5a** Multiply by 1 to find an equivalent expression using a different denominator.

**Example** Find a number equivalent to $\frac{2}{7}$ with a denominator of 63.

Since $7 \cdot 9 = 63$, we multiply by $\frac{9}{9}$:

$$\frac{2}{7} = \frac{2}{7} \cdot \frac{9}{9} = \frac{2 \cdot 9}{7 \cdot 9} = \frac{18}{63}.$$

**Practice Exercise**

4. Find a number equivalent to $\frac{7}{12}$ with a denominator of 96.

---

**Objective 3.5b** Simplify fraction notation.

**Example** Simplify: $\frac{315}{1650}$.

Using the test for divisibility by 5, we see that both the numerator and the denominator are divisible by 5:

$$\frac{315}{1650} = \frac{5 \cdot 63}{5 \cdot 330} = \frac{5}{5} \cdot \frac{63}{330} = 1 \cdot \frac{63}{330}$$

$$= \frac{63}{330} = \frac{3 \cdot 21}{3 \cdot 110} = \frac{3}{3} \cdot \frac{21}{110} = 1 \cdot \frac{21}{110} = \frac{21}{110}.$$

**Practice Exercise**

5. Simplify: $\frac{100}{280}$.

---

**Objective 3.5c** Test to determine whether two fractions are equivalent.

**Example** Use = or ≠ for ☐ to write a true sentence:

$$\frac{10}{54} \ \square \ \frac{15}{81}.$$

We find the cross products. We multiply 10 and 81: $10 \cdot 81 = 810$. Then we multiply 54 and 15: $54 \cdot 15 = 810$. Because the cross products are the same, we have

$$\frac{10}{54} = \frac{15}{81}.$$

If the cross products had been different, the fractions would not be equal.

**Practice Exercise**

6. Use = or ≠ for ☐ to write a true sentence:

$$\frac{8}{48} \ \square \ \frac{6}{44}.$$

---

**Objective 3.6a** Multiply and simplify using fraction notation.

**Example** Multiply and simplify: $\frac{7}{16} \cdot \frac{40}{49}$.

$$\frac{7}{16} \cdot \frac{40}{49} = \frac{7 \cdot 40}{16 \cdot 49} = \frac{7 \cdot 2 \cdot 2 \cdot 2 \cdot 5}{2 \cdot 2 \cdot 2 \cdot 2 \cdot 7 \cdot 7}$$

$$= \frac{2 \cdot 2 \cdot 2 \cdot 7}{2 \cdot 2 \cdot 2 \cdot 7} \cdot \frac{5}{2 \cdot 7} = 1 \cdot \frac{5}{14} = \frac{5}{14}$$

**Practice Exercise**

7. Multiply and simplify: $\frac{80}{3} \cdot \frac{21}{72}$.

**Objective 3.7b**  Divide and simplify using fraction notation.

**Example**  Divide and simplify: $\frac{9}{20} \div \frac{18}{25}$.

$$\frac{9}{20} \div \frac{18}{25} = \frac{9}{20} \cdot \frac{25}{18} = \frac{9 \cdot 25}{20 \cdot 18} = \frac{3 \cdot 3 \cdot 5 \cdot 5}{2 \cdot 2 \cdot 5 \cdot 2 \cdot 3 \cdot 3}$$

$$= \frac{3 \cdot 3 \cdot 5}{3 \cdot 3 \cdot 5} \cdot \frac{5}{2 \cdot 2 \cdot 2} = 1 \cdot \frac{5}{8} = \frac{5}{8}$$

**Practice Exercise**

**8.** Divide and simplify: $\frac{9}{4} \div \frac{45}{14}$.

---

**Objective 3.8b**  Solve problems by using the multiplication principle.

**Example**  A rental car had 18 gal of gasoline when its gas tank was $\frac{6}{7}$ full. How much could the tank hold when full? The equation that corresponds to the situation is

$$\frac{6}{7} \cdot g = 18.$$

We multiply both sides by $\frac{7}{6}$:

$$\frac{7}{6} \cdot \frac{6}{7} \cdot g = \frac{7}{6} \cdot 18$$

$$g = \frac{7}{6} \cdot \frac{18}{1} = \frac{7 \cdot 3 \cdot 6}{6 \cdot 1} = 21.$$

The rental car can hold 21 gal of gasoline.

**Practice Exercise**

**9.** A flower vase has $\frac{7}{4}$ cups of water in it when it is $\frac{3}{4}$ full. How much can it hold when full?

---

## Review Exercises

**1.** Multiply by 1, 2, 3, and so on, to find ten multiples of 8. [3.1a]

Use the tests for divisibility to answer Exercises 2–6. [3.1b]

**2.** Determine whether 3920 is divisible by 6.

**3.** Determine whether 68,537 is divisible by 3.

**4.** Determine whether 673 is divisible by 5.

**5.** Determine whether 4936 is divisible by 2.

**6.** Determine whether 5238 is divisible by 9.

Find all the factors of each number. [3.2a]

**7.** 60　　　　　　　　　　**8.** 176

Classify each number as prime, composite, or neither. [3.2b]

**9.** 37　　　**10.** 1　　　**11.** 91

Find the prime factorization of each number. [3.2c]

**12.** 70　　　　　　　　**13.** 72

**14.** 45　　　　　　　　**15.** 150

**16.** 648　　　　　　　　**17.** 1200

**18.** Identify the numerator and the denominator of $\frac{9}{7}$.
[3.3a]

What part is shaded? [3.3a]

**19.** 　　　　**20.**

**21.** For a committee in the United States Senate that consists of 3 Democrats and 5 Republicans, what is the ratio of   [3.3a]

   **a)** Democrats to Republicans?
   **b)** Republicans to Democrats?
   **c)** Democrats to the total number of members of the committee?

Simplify, if possible. Assume that all variables are nonzero.

**22.** $\dfrac{0}{6}$   [3.3b]   **23.** $\dfrac{74}{74}$   [3.3b]   **24.** $\dfrac{48}{1}$   [3.3b]

**25.** $\dfrac{7x}{7x}$   [3.3b]   **26.** $-\dfrac{10}{15}$   [3.5b]   **27.** $\dfrac{7}{28}$   [3.5b]

**28.** $\dfrac{-42}{42}$   [3.5b]   **29.** $\dfrac{9m}{12m}$   [3.5b]   **30.** $\dfrac{-12}{-30}$

                                                                      [3.5b]

**31.** $\dfrac{-27}{0}$   [3.3b]   **32.** $\dfrac{140}{490}$   [3.5b]   **33.** $\dfrac{288}{2025}$

                                                                      [3.5b]

Find an equivalent expression for each number, using the denominator indicated. Use multiplication by 1.   [3.5a]

**34.** $\dfrac{5}{7} = \dfrac{?}{21}$   **35.** $\dfrac{-6}{11} = \dfrac{?}{55}$

**36.** Simplify, if possible, the fractions on this circle graph.   [3.5b]

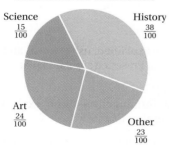

**Museums in the United States**

Science $\frac{15}{100}$

History $\frac{38}{100}$

Art $\frac{24}{100}$

Other $\frac{23}{100}$

Use $=$ or $\neq$ for $\square$ to write a true sentence.   [3.5c]

**37.** $\dfrac{3}{5} \ \square \ \dfrac{4}{6}$   **38.** $\dfrac{4}{-7} \ \square \ \dfrac{-8}{14}$

**39.** $\dfrac{4}{5} \ \square \ \dfrac{5}{6}$   **40.** $\dfrac{4}{3} \ \square \ \dfrac{28}{21}$

Find the reciprocal of each number.   [3.7a]

**41.** $\dfrac{2}{13}$   **42.** $-7$

**43.** $\dfrac{1}{8}$   **44.** $\dfrac{3x}{5y}$

Perform the indicated operation and, if possible, simplify.

**45.** $\dfrac{2}{9} \cdot \dfrac{7}{5}$   [3.4b]   **46.** $\dfrac{3}{x} \cdot \dfrac{y}{7}$   [3.4b]

**47.** $\dfrac{3}{4} \cdot \dfrac{8}{9}$   [3.6a]   **48.** $-10 \cdot \dfrac{7}{5}$   [3.6a]

**49.** $\dfrac{11}{3} \cdot \dfrac{30}{77}$   [3.6a]   **50.** $\dfrac{4a}{7} \cdot \dfrac{7}{4a}$   [3.6a]

**51.** $\dfrac{6}{5} \cdot 20x$   [3.6a]   **52.** $\dfrac{3}{14} \div \dfrac{6}{7}$   [3.7b]

**53.** $20 \div \dfrac{3}{4}$   [3.7b]   **54.** $-\dfrac{5}{36} \div \left(-\dfrac{25}{12}\right)$

                                                                          [3.7b]

**55.** $21 \div \dfrac{7}{2a}$   [3.7b]   **56.** $-\dfrac{23}{25} \div \dfrac{23}{25}$   [3.7b]

**57.** $\dfrac{\frac{21}{30}}{\frac{14}{15}}$   [3.7b]   **58.** $\dfrac{-\frac{2}{3}}{-\frac{3}{2}}$   [3.7b]

Solve.   [3.8a]

**59.** $\dfrac{2}{3}x = 160$   **60.** $\dfrac{3}{8} = -\dfrac{5}{4}t$

**61.** $-\dfrac{1}{7}n = -4$   **62.** $y \cdot \dfrac{1}{2} = \dfrac{1}{3}$

Find the area. [3.6b]

**63.**

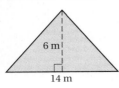

6 m
14 m

**64.**

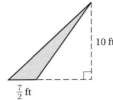

10 ft

$\frac{7}{2}$ ft

Solve. [3.8b]

**65.** A road crew repaves $\frac{1}{12}$ mi of road each day. How long will it take the crew to repave a $\frac{3}{4}$-mi stretch of road?

**66.** *Cotton Production.* In 2008, the United States accounted for approximately $\frac{3}{25}$ of the world production of cotton. The total world cotton production was about 113,000,000 bales. How much cotton did the United States produce?

**Source:** Agricultural Marketing Resource Center

**67.** After driving 600 km, the Buxton family has completed $\frac{3}{5}$ of their vacation. How long is the total trip?

**68.** Molly is making a pepper steak recipe that calls for $\frac{2}{3}$ cup of green bell peppers. How much would be needed to make $\frac{1}{2}$ recipe?

**69.** The Winchester swim team has 4 swimmers in a $\frac{2}{3}$-mi relay race. How far will each person swim?

**70.** A book bag requires $\frac{4}{5}$ yd of fabric. How many bags can be made from 48 yd?

**71.** Solve: $\frac{2}{13} \cdot x = \frac{1}{2}$. [3.8a]

**A.** $\frac{1}{13}$ **B.** 13 **C.** $\frac{4}{13}$ **D.** $\frac{13}{4}$

**72.** Multiply and simplify: $\frac{15}{26} \cdot \frac{13}{90}$. [3.6a]

**A.** $\frac{195}{234}$ **B.** $\frac{1}{12}$ **C.** $\frac{3}{36}$ **D.** $\frac{13}{156}$

## Synthesis

**73.** Simplify: $\frac{15x}{14z} \cdot \frac{17yz}{35xy} \div \left( -\frac{3}{7} \right)^2$. [3.6a], [3.7b]

**74.** What digit(s) could be inserted in the ones place to make 574 _____ divisible by 6? [3.1b]

**75.** In the division below, find $a$ and $b$. [3.7b]

$$\frac{19}{24} \div \frac{a}{b} = \frac{187,853}{268,224}$$

**76.** A prime number that remains a prime number when its digits are reversed is called a **palindrome prime**. For example, 17 is a palindrome prime because both 17 and 71 are primes. Which of the following numbers are palindrome primes? [3.2b]

13, 91, 16, 11, 15, 24, 29, 101, 201, 37

# Understanding Through Discussion and Writing

**1.** A student incorrectly insists that $\frac{2}{5} \div \frac{3}{4}$ is $\frac{15}{8}$. What mistake is he probably making? [3.7b]

**2.** Use the number 9432 to explain why the test for divisibility by 9 works. [3.1b]

**3.** A student claims that "taking $\frac{1}{2}$ of a number is the same as dividing by $\frac{1}{2}$." Explain the error in this reasoning. [3.7b]

**4.** On p. 178 we explained, using words and pictures, why $\frac{2}{5} \cdot \frac{3}{4}$ equals $\frac{6}{20}$. Present a similar explanation of why $\frac{2}{3} \cdot \frac{4}{7}$ equals $\frac{8}{21}$. [3.4a]

**5.** Without performing the division, explain why $5 \div \frac{1}{7}$ is a greater number than $5 \div \frac{2}{3}$. [3. 5c], [3.7b]

**6.** If a fraction's numerator and denominator have no factors (other than 1) in common, can the fraction be simplified? Why or why not? [3.5b]

Test   For Extra Help

Step-by-step test solutions are found on the Chapter Test Prep Videos available via the Video Resources on DVD, in **MyMathLab**  , and on You Tube (search "BittingerPrealgebra" and click on "Channels").

**1.** Determine whether 5682 is divisible by 3. Do not use long division.

**2.** Determine whether 7018 is divisible by 5. Do not use long division.

**3.** Find all the factors of 90.

**4.** Determine whether 93 is prime, composite, or neither.

Find the prime factorization of each number.

**5.** 36

**6.** 60

**7.** Identify the numerator and the denominator of $\frac{4}{9}$.

**8.** What part is shaded?

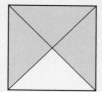

**9.** What part of the set is shaded?

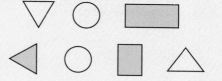

**10.** *Cholesterol.*   Morrison's cholesterol test showed that his total cholesterol level was 180 mg/dL, his HDL level was 47 mg/dL, and his LDL level was 93 mg/dL.

**a)** What was the ratio of total cholesterol to HDL cholesterol?
**b)** What was the ratio of HDL cholesterol to LDL cholesterol?

Simplify, if possible. Assume that all variables are nonzero.

**11.** $\dfrac{32}{1}$

**12.** $\dfrac{-12}{-12}$

**13.** $\dfrac{0}{16}$

**14.** $\dfrac{-8}{24}$

**15.** $\dfrac{42}{7}$

**16.** $\dfrac{9x}{45x}$

**17.** $\dfrac{-62}{0}$

**18.** $\dfrac{72}{108}$

Use = or ≠ for ☐ to write a true sentence.

**19.** $\dfrac{3}{4}$ ☐ $\dfrac{6}{8}$

**20.** $\dfrac{5}{4}$ ☐ $\dfrac{9}{7}$

**21.** Find an equivalent expression for $\dfrac{3}{8}$ with a denominator of 40.

Find the reciprocal.

**22.** $\dfrac{a}{42}$

**23.** −9

Perform the indicated operation. Simplify, if possible.

**24.** $\dfrac{2}{3} \cdot \dfrac{15}{4}$

**25.** $\dfrac{2}{11} \div \dfrac{3}{4}$

**26.** $3 \cdot \dfrac{x}{8}$

**27.** $\dfrac{\frac{4}{7}}{-\frac{8}{3}}$

**28.** $12 \div \dfrac{2}{3}$

**29.** $\dfrac{22c}{15} \cdot \dfrac{5}{33c}$

Solve.

**30.** A $\frac{3}{4}$-lb slab of cheese is shared equally by 5 people. How much does each person receive?

**31.** Monroe weighs $\frac{5}{7}$ of his dad's weight. If his dad weighs 175 lb, how much does Monroe weigh?

**32.** $\dfrac{7}{8} \cdot x = 56$

**33.** $\dfrac{7}{10} = \dfrac{-2}{5} \cdot t$

**34.** Find the area.

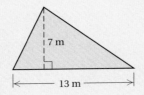

**35.** In which figure does the shaded part represent $\frac{7}{6}$ of the figure?

A.

B.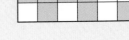

C.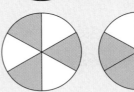

D.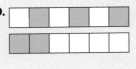

# Synthesis

**36.** Grandma Shelby left $\frac{2}{3}$ of her $\frac{7}{8}$-acre apple farm to Karl. Karl gave $\frac{1}{4}$ of his share to his oldest daughter, Shannon. How much land did Shannon receive?

**37.** Simplify: $\left(-\dfrac{3}{8}\right)^2 \div \dfrac{6}{7} \cdot \dfrac{2}{9} \div (-5)$.

# Cumulative Review

**1.** Write a word name: 2,056,783.

Add.

**2.**
```
  2 7 4 3
+ 8 2 3 9
```

**3.** $-29 + (-14)$

**4.** $-45 + 12$

Subtract.

**5.**
```
  6 3 2 4
- 4 1 9 5
```

**6.** $27 - 50$

**7.** $-12 - (-4)$

Multiply and, if possible, simplify.

**8.**
```
  7 3 5
×   2 3
```

**9.** $-52 \cdot 6$

**10.** $\frac{6}{7} \cdot (-35x)$

**11.** $\frac{2}{9} \cdot \frac{21}{10}$

Divide and, if possible, simplify.

**12.** $1\,3 \overline{)\,3\,0\,5\,8}$

**13.** $-85 \div 5$

**14.** $-16 \div \frac{4}{7}$

**15.** $\frac{3}{7} \div \frac{9}{14}$

**16.** Round 4509 to the nearest ten.

**17.** Estimate the product by rounding to the nearest hundred. Show your work.
```
    9 2 1
×   4 5 3
```

**18.** Find the absolute value: $|-479|$.

**19.** Simplify: $10^2 \div 5(-2) - 8(2 - 8)$.

**20.** Determine whether 98 is prime, composite, or neither.

**21.** Evaluate $a - b^2$ for $a = -5$ and $b = 4$.

Solve.

**22.** $a + 24 = 49$

**23.** $7x = 49$

**24.** $\frac{2}{9} \cdot a = -10$

**25.** $48 = -4t$

**26.** $2x + 13 = 3$

**27.** $-x = -10$

**37.** A 48-oz coffee pot is poured into 6 mugs. How much will each mug hold if the coffee is poured out evenly?

**38.** A 1996 truck that gets 17 miles per gallon is traded in toward a 2009 van that gets 25 miles per gallon. How many more miles per gallon does the newer vehicle get?

Combine like terms.

**28.** $8 - 4x - 13 + 9x$

**29.** $-12x + 7y + 15x$

**39.** There are 7000 students at La Poloma College, and $\frac{5}{8}$ of them live in dorms. How many live in dorms?

Simplify, if possible.

**30.** $\dfrac{97}{97}$

**31.** $\dfrac{0}{81}$

**40.** A thermos of iced tea had 3 qt of tea when it was $\frac{3}{5}$ full. How much tea could it hold when full?

**32.** $\dfrac{63x}{1}$

**33.** $\dfrac{-10}{54}$

**41.** Tony has jogged $\frac{2}{3}$ of a course that is $\frac{9}{10}$ of a mile long. How far has Tony gone?

Find the reciprocal.

**34.** $\dfrac{2}{5}$

**35.** $57$

## Synthesis

**42.** Evaluate $\dfrac{ab}{c}$ for $a = -\dfrac{2}{5}$, $b = \dfrac{10}{13}$, and $c = \dfrac{26}{27}$.

**43.** Evaluate $-|xy|^2$ for $x = -\dfrac{3}{5}$ and $y = \dfrac{1}{2}$.

**36.** Find an equivalent expression for $\frac{3}{10}$ with a denominator of 70. Use multiplying by 1.

**44.** Wayne and Patty each earn $85 a day, while Janet earns $90 a day. They decide to pool their earnings from three days and spend $\frac{2}{5}$ of that on entertainment and save the rest. How much will Wayne, Patty, and Janet end up saving?

# Fraction Notation: Addition, Subtraction, and Mixed Numerals

## Real-World Application

Black bears typically have two cubs. In January 2007 in northern New Hampshire, a black bear sow gave birth to a litter of 5 cubs. This is so rare that Tom Sears, a wildlife photographer, spent 28 hr per week for six weeks watching for the perfect opportunity to photograph this family of six. At the time of this photo, an observer estimated that the cubs weighed $7\frac{1}{2}$ lb, 8 lb, $9\frac{1}{2}$ lb, $10\frac{5}{8}$ lb, and $11\frac{3}{4}$ lb. What was the average weight of the cubs?

*Source:* Andrew Timmins, New Hampshire Fish and Game Department, *Northcountry News,* Warren, NH; Tom Sears, photographer

***This problem appears as Exercise 57 in Section 4.8.***

# 4.1

## Least Common Multiples

### OBJECTIVE

a) Find the least common multiple, or LCM, of two or more numbers.

**SKILL TO REVIEW**

Objective 3.1a: Find some multiples of a number.

Multiply by 1, 2, 3, and so on, to find six multiples of each number.

**1.** 8          **2.** 25

**1.** Find the LCM of 9 and 15 by examining lists of multiples.

**2.** Find the LCM of 8 and 14 by examining lists of multiples.

In order to add or subtract fractions, all fractions must share a common denominator. Before discussing common denominators, we look at how to find the **least common multiple** of two or more numbers.

### a) Finding Least Common Multiples

> **LEAST COMMON MULTIPLE, LCM**
>
> The **least common multiple**, or LCM, of two natural numbers is the smallest number that is a multiple of both numbers.

**EXAMPLE 1** Find the LCM of 20 and 30.

First, we list some multiples of 20 by multiplying 20 by 1, 2, 3, and so on:

$$20, 40, 60, 80, 100, 120, 140, 160, 180, 200, 220, 240, \ldots$$

Then we list some multiples of 30 by multiplying 30 by 1, 2, 3, and so on:

$$30, 60, 90, 120, 150, 180, 210, 240, \ldots$$

Now we determine the smallest number *common* to both lists. The LCM of 20 and 30 is 60.

Do Margin Exercises 1 and 2.

Next we develop two more efficient methods for finding LCMs. You may choose to learn only one method. (Consult your instructor.) However, if you intend to study algebra, you should definitely learn method 2.

### Method 1: Finding LCMs Using One List of Multiples

The first method works especially well when the numbers are relatively small.

> *Method 1.* To find the LCM of a set of numbers using a list of multiples:
>
> **1.** Determine whether the largest number is a multiple of the others. If it is, it is the LCM. That is, if the largest number has the others as factors, the LCM is that number.
>
> **2.** If not, check multiples of the largest number until you get one that is a multiple of each of the others.

**EXAMPLE 2** Find the LCM of 8 and 10.

**1.** 10 is the larger number, but it is not a multiple of 8.

**2.** Check multiples of 10:

$$2 \cdot 10 = 20, \quad \text{Not a multiple of 8}$$
$$3 \cdot 10 = 30, \quad \text{Not a multiple of 8}$$
$$4 \cdot 10 = 40. \quad \text{A multiple of both 8 and 10}$$

The LCM = 40.

*Answers*

*Skill to Review:*
**1.** 8, 16, 24, 32, 40, 48
**2.** 25, 50, 75, 100, 125, 150

*Margin Exercises:*
**1.** 45    **2.** 56

**EXAMPLE 3** Find the LCM of 4 and 14.

1. 14 is the larger number, but it is not a multiple of 4.
2. Check multiples of 14:

   $2 \cdot 14 = 28.$     A multiple of 4

The LCM = 28.

**EXAMPLE 4** Find the LCM of 8 and 32.

1. 32 is the larger number and 32 is a multiple of 8, so it is the LCM.

The LCM = 32.

> Do Exercises 3–6.

Find each LCM using one list of multiples.

**3.** 6, 9          **4.** 12, 15

**5.** 9, 36          **6.** 3, 20

To find the least common multiple of three numbers, we find the LCM of two of the numbers and then find the LCM of that number and the third number.

**EXAMPLE 5** Find the LCM of 4, 10, and 15. We can start by finding the LCM of any two of the numbers. Let's use 10 and 15:

1. 15 is the larger number, but it is not a multiple of 10.
2. Check multiples of 15:

   $2 \cdot 15 = 30.$     A multiple of 10. The LCM of 10 and 15 is 30.

Note now that any multiple of 30 will automatically be a multiple of 10 and 15. Thus, to find the LCM of 10, 15, and 4, we need only find the LCM of 30 and 4:

1. 30 is not a multiple of 4.
2. Check multiples of 30:

   $2 \cdot 30 = 60.$     Since it is a multiple of 4, we know that 60 is the LCM of 30 and 4.

The LCM of 4, 10, and 15 is 60.

> Do Exercises 7 and 8.

Find each LCM using lists of multiples.

**7.** 30, 40, 50          **8.** 10, 20, 40

## Method 2: Finding LCMs Using Prime Factorizations

A second method for finding LCMs uses prime factorizations and is usually the best method when the numbers are large. Consider again 20 and 30. Their prime factorizations are $20 = 2 \cdot 2 \cdot 5$ and $30 = 2 \cdot 3 \cdot 5$. Let's look at these prime factorizations in order to find the LCM. Any multiple of 20 will have to have *two* 2's as factors and *one* 5 as a factor. Any multiple of 30 will need to have *one* 2, *one* 3, and *one* 5 as factors. The smallest number satisfying these conditions is

            ┌─── Two 2's, one 5; $2 \cdot 2 \cdot 3 \cdot 5$ is a multiple of 20.
        ↓ ↓     ↓
    $2 \cdot 2 \cdot 3 \cdot 5.$
        ↑ ↑ ↑
            └─── One 2, one 3, one 5; $2 \cdot 2 \cdot 3 \cdot 5$ is a multiple of 30.

*Answers*

**3.** 18   **4.** 60   **5.** 36
**6.** 60   **7.** 600   **8.** 40

Don't hesitate to write out any missing steps that you'd like to see included. For instance, in Example 5, it is stated (in red) that 60 is a multiple of 4. To solidify your understanding, you may want to write in "60 = 4 · 15."

Thus, the LCM of 20 and 30 is 2 · 2 · 3 · 5, or 60. It has all the factors of 20 and all the factors of 30, but the factors are not repeated when they are common to both numbers.

Note that each prime factor is used the greatest number of times that it occurs in either of the individual factorizations.

*Method 2.* To find the LCM of two numbers using prime factorizations:

1. Write the prime factorization of each number.
2. Select one of the factorizations and see whether it contains the other.

   a) If it does, it is the LCM.

   b) If it does not, multiply that factorization by those prime factors of the other number that it lacks. The final product is the LCM.

3. As a check, make sure that the LCM includes each factor the greatest number of times that it occurs in either factorization.

**EXAMPLE 6**   Find the LCM of 18 and 21.

1. We begin by writing the prime factorization of each number:

   $18 = 2 \cdot 3 \cdot 3$   and   $21 = 3 \cdot 7$.

2. We select the factorization of 18: $2 \cdot 3 \cdot 3$.

   a) We note that $2 \cdot 3 \cdot 3$ does not contain the other factorization, $3 \cdot 7$.

   b) To find the LCM of 18 and 21, we multiply $2 \cdot 3 \cdot 3$ by the factor of 21 that it lacks, 7:

   $$\text{LCM} = 2 \cdot 3 \cdot 3 \cdot 7.$$

   18 is a factor.
   21 is a factor.

3. The greatest number of times that 2 occurs as a factor of 18 or 21 is **one** time; the greatest number of times that 3 occurs as a factor of 18 or 21 is **two** times; and the greatest number of times that 7 occurs as a factor of 18 or 21 is **one** time. To check, note that the LCM has exactly **one** 2, **two** 3's, and **one** 7. The LCM is $2 \cdot 3 \cdot 3 \cdot 7$, or 126.

**EXAMPLE 7**   Find the LCM of 7 and 21.

1. Because 7 is prime, we think of $7 = 7$ as a "factorization":

   $7 = 7$   and   $21 = 3 \cdot 7$.

2. One factorization, $3 \cdot 7$, contains the other. Thus, the LCM is $3 \cdot 7$, or 21.

**EXAMPLE 8** Find the LCM of 24 and 36.

1. We write the prime factorization of each number:

$$24 = 2 \cdot 2 \cdot 2 \cdot 3 \quad \text{and} \quad 36 = 2 \cdot 2 \cdot 3 \cdot 3.$$

2. We select the factorization of 24.

   a) The factorization of 24 does not contain the factorization of 36.

   b) To find the LCM of 24 and 36, we multiply the factorization of 24 by any prime factors of 36 that it lacks. We need another factor of 3.

$$\text{LCM} = 2 \cdot 2 \cdot 2 \cdot 3 \cdot 3.$$

   24 is a factor.
   36 is a factor.

3. Note that the LCM includes 2 and 3 the greatest number of times that each appears as a factor of either 24 or 36. The LCM is $2 \cdot 2 \cdot 2 \cdot 3 \cdot 3$, or 72.

Do Exercises 9–12.

Exponential notation is often helpful when writing least common multiples. Let's reconsider Example 8 using exponents. The largest exponents indicate the greatest number of times that 2 and 3 occur as factors.

$$24 = 2 \cdot 2 \cdot 2 \cdot 3 = 2^3 \cdot 3^1 \qquad 2^3 \text{ is the greatest power of } 2$$
$$36 = 2 \cdot 2 \cdot 3 \cdot 3 = 2^2 \cdot 3^2 \qquad 3^2 \text{ is the greatest power of } 3$$
$$\text{LCM} = 2 \cdot 2 \cdot 2 \cdot 3 \cdot 3 = 2^3 \cdot 3^2, \text{ or } 72.$$

Note that the greatest power of each factor is used to construct the LCM.

Lining up the different prime numbers in the factorizations can help us construct the LCM. This method also works well when finding the LCM of more than two numbers.

**EXAMPLE 9** Find the LCM of 27, 90, and 84.

We find the prime factorization of each number and write the factorizations in exponential notation.

$$27 = 3 \cdot 3 \cdot 3 = 3^3$$
$$90 = 2 \cdot 3 \cdot 3 \cdot 5 = 2 \cdot 3^2 \cdot 5$$
$$84 = 2 \cdot 2 \cdot 3 \cdot 7 = 2^2 \cdot 3 \cdot 7$$

No one factorization contains the others. The prime numbers 2, 3, 5, and 7 appear as factors.

We write the factorizations, lining up all the powers of 2, the powers of 3, and so on.

$$27 = \qquad 3^3$$
$$90 = 2 \quad \cdot 3^2 \cdot 5$$
$$84 = 2^2 \cdot 3 \cdot \quad 7$$

The LCM is formed by choosing the greatest power of each factor:

$$2^2 \cdot 3^3 \cdot 5 \cdot 7 = 3780.$$

The LCM of 27, 90, and 84 is 3780.

Do Exercises 13 and 14.

Use prime factorizations to find the LCM.

**9.** 8, 10  **10.** 5, 30

**11.** 18, 40  **12.** 12, 48

Find the LCM.

**13.** 8, 18, 30  **14.** 10, 20, 25

*Answers*

**9.** 40  **10.** 30  **11.** 360
**12.** 48  **13.** 360  **14.** 100

**EXAMPLE 10**  Find the LCM of 8 and 25.

We write the prime factorization of each number in exponential notation.

$$8 = 2 \cdot 2 \cdot 2 = 2^3$$
$$25 = 5 \cdot 5 = 5^2$$

The prime numbers 2 and 5 appear as factors. We write the factorizations as

$$8 = 2^3$$
$$25 = \quad 5^2.$$

Note that the two numbers, 8 and 25, have no common prime factor. When this is the case, the LCM is just the product of the two numbers. Thus, the LCM is $2^3 \cdot 5^2 = 8 \cdot 25 = 200$.

Do Exercises 15 and 16.

The same method works perfectly with variables.

**EXAMPLE 11**  Find the LCM of $7a^2b$ and $ab^3$.

We have the following factorizations:

$$7a^2b = 7 \cdot a \cdot a \cdot b \quad \text{and} \quad ab^3 = a \cdot b \cdot b \cdot b.$$

No one factorization contains the other.

Consider the factorization of $7a^2b$, which is $7 \cdot a \cdot a \cdot b$. Since $ab^3$ contains two more factors of $b$, we multiply the factorization of $7a^2b$ by $b \cdot b$.

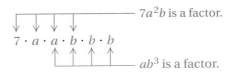

As a second approach, we find the greatest power of each factor using exponential notation.

$$7a^2b = 7 \cdot a^2 \cdot b$$
$$ab^3 = \quad a \cdot b^3$$

The LCM is $7 \cdot a \cdot a \cdot b \cdot b \cdot b$, or $7a^2b^3$.

Do Exercises 17 and 18.

**EXAMPLE 12**  Find the LCM of $12x^2y^3z$ and $18x^4z^3$.

We write the factorizations using exponential notation:

$$12x^2y^3z = 2^2 \cdot 3 \cdot x^2 \cdot y^3 \cdot z \quad \text{and} \quad 18x^4z^3 = 2 \cdot 3^2 \cdot x^4 \cdot z^3.$$

No one factorization contains the other.

We form the LCM using the greatest power of each factor.

$$12x^2y^3z = 2^2 \cdot 3 \quad \cdot x^2 \cdot y^3 \cdot z$$
$$18x^4z^3 = 2 \quad \cdot 3^2 \cdot x^4 \cdot \quad z^3$$

The LCM is $2^2 \cdot 3^2 \cdot x^4 \cdot y^3 \cdot z^3$, or $36x^4y^3z^3$.

Do Exercise 19.

Find the LCM.
**15.** 4, 9

**16.** 5, 6, 7

Find the LCM.
**17.** $xy, yz$

**18.** $5a^2, a^3b$

**19.** Find the LCM of $8a^3b^2$ and $10a^2c^4$.

*Answers*

**15.** 36  **16.** 210  **17.** $xyz$
**18.** $5a^3b$  **19.** $40a^3b^2c^4$

**a**    Find the LCM of each set of numbers or expressions.

**1.** 2, 4            **2.** 3, 15            **3.** 10, 25            **4.** 10, 15            **5.** 20, 40

**6.** 8, 12            **7.** 18, 27            **8.** 9, 11            **9.** 30, 50            **10.** 8, 36

**11.** 30, 40            **12.** 21, 27            **13.** 18, 24            **14.** 12, 18            **15.** 60, 70

**16.** 35, 45            **17.** 16, 36            **18.** 24, 32            **19.** 18, 20            **20.** 36, 48

**21.** 2, 3, 7            **22.** 2, 5, 9            **23.** 3, 6, 15            **24.** 6, 12, 18            **25.** 24, 36, 12

**26.** 8, 16, 22            **27.** 5, 12, 15            **28.** 12, 18, 40            **29.** 9, 12, 6            **30.** 8, 16, 12

**31.** 180, 100, 450            **32.** 18, 30, 50, 48            **33.** 8, 48            **34.** 16, 32            **35.** 10, 21

**36.** 14, 15            **37.** 75, 100            **38.** 81, 90            **39.** 12, 15, 60            **40.** 24, 36, 72

**41.** $ab, bc$            **42.** $7x, xy$            **43.** $3x, 9x^2$            **44.** $10x^4, 5x^3$            **45.** $4x^3, x^2y$

**46.** $6ab^2, 9a^3b$            **47.** $6r^3st^4, 8rs^2t$            **48.** $3m^2n^4p^5, 9mn^2p^4$            **49.** $a^3b, b^2c, ac^2$            **50.** $x^2z^3, x^3y, y^2z$

*Applications of LCMs: Planet Orbits.* Jupiter, Saturn, and Uranus all revolve around the sun. Jupiter takes 12 yr, Saturn 30 yr, and Uranus 84 yr to make a complete revolution. On a certain night, you look at Jupiter, Saturn, and Uranus and wonder how many years it will take before they have the same position again. (*Hint*: The number of years is the LCM of 12, 30, and 84.)
Source: *The Handy Science Answer Book*

**51.** How often will Jupiter and Saturn appear in the same direction in the night sky as seen from the earth?

**52.** How often will Jupiter and Uranus appear in the same direction in the night sky as seen from the earth?

**53.** How often will Saturn and Uranus appear in the same direction in the night sky as seen from the earth?

**54.** How often will Jupiter, Saturn, and Uranus appear in the same direction in the night sky as seen from the earth?

## Skill Maintenance

**55.** *Tornadoes.* The United States reports more tornadoes per year than any other country in the world. In 2009, 1156 tornadoes were recorded. This was a decrease of 534 tornadoes from 2008. How many tornadoes occurred in 2008? [1.8a]

Sources: National Climatic Data Center, *State of the Climate: Tornadoes*, Annual 2008 and 2009.

**56.** *Population of Africa.* The population of South America is approximately $\frac{2}{5}$ of the population of Africa. In 2010, the population of South America was about 390,000,000. What was the population of Africa in 2010? [3.8b]

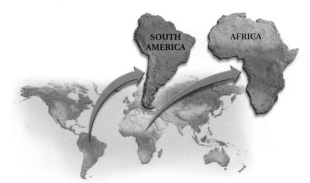

Perform the indicated operation and, if possible, simplify.

**57.** $-38 + 52$ [2.2a]

**58.** $-18 \div \left(\dfrac{2}{3}\right)$ [3.7b]

**59.** $23 \cdot 345$ [1.4a]

**60.** $\dfrac{4}{5} \cdot \dfrac{10}{12}$ [3.6a]

**61.** $\dfrac{4}{5} \div \left(-\dfrac{7}{10}\right)$ [3.7b]

**62.** $382 - 549$ [2.3a]

## Synthesis

Use a calculator and the multiples method to find the LCM of each pair of numbers.

**63.** 288; 324

**64.** 2700; 7800

**65.** 7719; 18,011

**66.** 17,385; 24,339

**67.** The tables at a flea market are either 6 ft long or 8 ft long. If one row is all 6-ft tables and one row is all 8-ft tables, what is the shortest common row length?

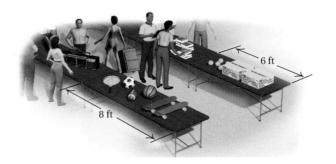

*African Artistry.* In southern Africa, the design of every woven handbag, or *gipatsi* (plural *sipatsi*), is created by repeating two or more geometric patterns. Each pattern encircles the bag, sharing the strands of fabric with any pattern above or below. The length, or period, of each pattern is the number of strands required to construct the pattern. For a gipatsi to be considered beautiful, each individual pattern must fit a whole number of times around the bag.

Source: Gerdes, Paulus. *Women, Art and Geometry in Southern Africa.* Asmara, Eritrea: Africa World Press, Inc., p. 5.

**68.** A weaver is using two patterns to create a gipatsi. Pattern A is 10 strands long, and pattern B is 3 strands long. What is the smallest number of strands that can be used to complete the gipatsi?

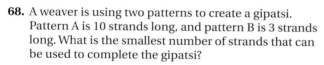

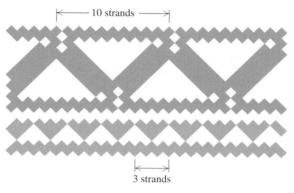

**69.** A weaver is using a four-strand pattern, a six-strand pattern, and an eight-strand pattern. What is the smallest number of strands that can be used to complete the gipatsi?

**70.** *Prescriptions.* Prescriptions for a 30-day supply of simvastatin and a 14-day supply of pain medication are filled at a pharmacy. Assuming the prescriptions are refilled regularly, how long will it be until they are both refilled on the same day?

**71.** Consider $a^3b^2$ and $a^2b^5$. Determine whether each of the following is the LCM of $a^3b^2$ and $a^2b^5$. Tell why or why not.
 **a)** $a^3b^3$
 **b)** $a^2b^5$
 **c)** $a^3b^5$

**72.** Use Example 9 to help find the LCM of 27, 90, 84, 210, 108, and 50.

**73.** Use Examples 6 and 8 to help find the LCM of 18, 21, 24, 36, 63, 56, and 20.

**74.** Find three different pairs of numbers for which 56 is the LCM. Do not use 56 itself in any of the pairs.

**75.** Find three different pairs of numbers for which 54 is the LCM. Do not use 54 itself in any of the pairs.

# 4.2

## Addition, Order, and Applications

**SKILL TO REVIEW**
Objective 3.2c: Find the prime factorization of a composite number.

Find the prime factorization of each number.

**1.** 96          **2.** 1400

**1.** Find $\frac{1}{5} + \frac{3}{5}$.

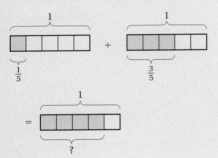

Add and, if possible, simplify.

**2.** $\frac{5}{13} + \frac{9}{13}$          **3.** $\frac{1}{3} + \frac{2}{3}$

**4.** $\frac{-5}{12} + \left(\frac{-1}{12}\right)$          **5.** $\frac{3}{x} + \frac{-7}{x}$

*Answers*

See next page.

### a  Like Denominators

Addition using fraction notation corresponds to combining or putting like things together, just as when we combined like terms. For example,

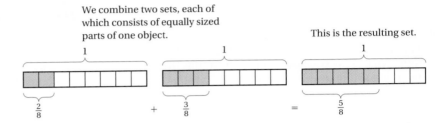

2 eighths + 3 eighths = 5 eighths,

or     $2 \cdot \frac{1}{8} + 3 \cdot \frac{1}{8} = 5 \cdot \frac{1}{8}$,     or     $\frac{2}{8} + \frac{3}{8} = \frac{5}{8}$.

Do Margin Exercise 1.

To add when denominators are the same,

**a)** add the numerators,

**b)** keep the denominator, and          $\frac{2}{6} + \frac{5}{6} = \frac{2 + 5}{6} = \frac{7}{6}$

**c)** simplify, if possible.

**EXAMPLES**  Add and, if possible, simplify.

**1.** $\frac{2}{4} + \frac{1}{4} = \frac{2 + 1}{4} = \frac{3}{4}$          No simplifying is possible.

**2.** $\frac{3}{12} + \frac{5}{12} = \frac{3 + 5}{12} = \frac{8}{12}$          Adding numerators; the denominator remains unchanged.

$= \frac{4}{4} \cdot \frac{2}{3} = \frac{2}{3}$          Simplifying by removing a factor equal to 1: $\frac{4}{4} = 1$

**3.** $\frac{-11}{6} + \frac{3}{6} = \frac{-11 + 3}{6} = \frac{-8}{6}$

$= \frac{2}{2} \cdot \frac{-4}{3} = \frac{-4}{3}$, or $-\frac{4}{3}$          Removing a factor equal to 1: $\frac{2}{2} = 1$

**4.** $-\frac{2}{a} + \left(-\frac{3}{a}\right) = \frac{-2}{a} + \frac{-3}{a}$          Recall that $-\frac{m}{n} = \frac{-m}{n}$.

$= \frac{-2 + (-3)}{a} = \frac{-5}{a}$, or $-\frac{5}{a}$

Do Exercises 2–5.

We may need to add fractions when combining like terms.

**EXAMPLE 5**  Simplify by combining like terms: $\frac{2}{7}x + \frac{3}{7}x$.

$$\frac{2}{7}x + \frac{3}{7}x = \left(\frac{2}{7} + \frac{3}{7}\right)x \quad \text{Try to do this step mentally.}$$
$$= \frac{5}{7}x$$

Do Exercises 6 and 7.

Simplify by combining like terms.

**6.** $\frac{3}{10}a + \frac{1}{10}a$  **7.** $-\frac{3}{4}x + \frac{1}{4}x$

## b Different Denominators

We cannot add $\frac{1}{2} + \frac{1}{3}$ by simply adding numerators. However, by rewriting $\frac{1}{2}$ as $\frac{1}{2} \cdot \frac{3}{3} = \frac{3}{6}$ and $\frac{1}{3}$ as $\frac{1}{3} \cdot \frac{2}{2} = \frac{2}{6}$, we can determine the sum.

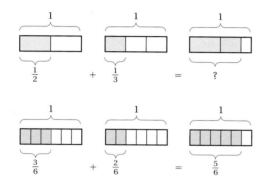

When denominators are different, we can find a common denominator by multiplying by 1. Consider the addition $\frac{3}{4} + \frac{1}{6}$ using two different denominators.

**A.** We use 24 as a common denominator:

$$\frac{3}{4} + \frac{1}{6} = \frac{3}{4} \cdot \frac{6}{6} + \frac{1}{6} \cdot \frac{4}{4}$$
$$= \frac{18}{24} + \frac{4}{24} = \frac{22}{24} = \frac{11}{12}.$$

**B.** We use 12 as a common denominator:

$$\frac{3}{4} + \frac{1}{6} = \frac{3}{4} \cdot \frac{3}{3} + \frac{1}{6} \cdot \frac{2}{2}$$
$$= \frac{9}{12} + \frac{2}{12} = \frac{11}{12}.$$

We had to simplify at the end of (A), but not in (B). In (B), we used the *least* common multiple of the denominators, 12, as the common denominator. That number is called the **least common denominator**, or **LCD**.

To add when denominators are different:

**a)** Find the least common multiple of the denominators. That number is the least common denominator, LCD.

**b)** Multiply by 1, writing 1 in the form of $n/n$, to find an equivalent sum in which the LCD appears in each fraction.

**c)** Add the numerators, keeping the same denominator.

**d)** Simplify, if possible.

*Answers*

*Skill to Review:*
**1.** $2 \cdot 2 \cdot 2 \cdot 2 \cdot 2 \cdot 3$
**2.** $2 \cdot 2 \cdot 2 \cdot 5 \cdot 5 \cdot 7$

*Margin Exercises:*
**1.** $\frac{4}{5}$  **2.** $\frac{14}{13}$  **3.** 1  **4.** $-\frac{1}{2}$  **5.** $-\frac{4}{x}$
**6.** $\frac{2}{5}a$  **7.** $-\frac{1}{2}x$

**EXAMPLE 6** Add: $\frac{1}{8} + \frac{3}{4}$.

a) Since 4 is a factor of 8, the LCM of 4 and 8 is 8. Thus, the LCD is 8.

b) We need to find a fraction equivalent to $\frac{3}{4}$ with a denominator of 8:

$$\frac{1}{8} + \frac{3}{4} = \frac{1}{8} + \frac{3}{4} \cdot \frac{2}{2}.$$  *Think:* $4 \times \square = 8$. The answer is 2, so we multiply by 1, using $\frac{2}{2}$.

c) We add: $\frac{1}{8} + \frac{6}{8} = \frac{7}{8}$.   Adding numerators

d) No simplifying is possible. The sum is $\frac{7}{8}$.

In Examples 7–10, we follow the same steps without labeling them.

**EXAMPLE 7** Add: $\frac{5}{6} + \frac{1}{9}$.

The LCD is 18.   $6 = 2 \cdot 3$ and $9 = 3 \cdot 3$, so the LCM of 6 and 9 is $2 \cdot 3 \cdot 3$, or 18.

$$\frac{5}{6} + \frac{1}{9} = \frac{5}{6} \cdot \frac{3}{3} + \frac{1}{9} \cdot \frac{2}{2}$$  *Think:* $9 \times \square = 18$. The answer is 2, so we multiply by 1, using $\frac{2}{2}$.
*Think:* $6 \times \square = 18$. The answer is 3, so we multiply by 1, using $\frac{3}{3}$.

$$= \frac{15}{18} + \frac{2}{18} = \frac{17}{18}$$

Do Exercises 8 and 9.

Add using the least common denominator.

**8.** $\frac{2}{3} + \frac{1}{6}$

**9.** $\frac{3}{8} + \frac{5}{6}$

**EXAMPLE 8** Add: $\frac{3}{-5} + \frac{11}{10}$.

$$\frac{3}{-5} + \frac{11}{10} = \frac{-3}{5} + \frac{11}{10}$$  Recall that $\frac{m}{-n} = \frac{-m}{n}$. We generally avoid negative signs in the denominator. The LCD is 10.

$$= \frac{-3}{5} \cdot \frac{2}{2} + \frac{11}{10}$$

$$= \frac{-6}{10} + \frac{11}{10}$$   We may still have to simplify, but simplifying is almost always easier if the LCD has been used.

$$= \frac{5}{10} = \frac{1}{2}$$

Do Exercise 10.

**10.** Add: $\frac{1}{-6} + \frac{7}{18}$.

**EXAMPLE 9** Add: $\frac{5}{8} + 2$.

$$\frac{5}{8} + 2 = \frac{5}{8} + \frac{2}{1}$$   Rewriting 2 in fraction notation

$$= \frac{5}{8} + \frac{2}{1} \cdot \frac{8}{8}$$   The LCD is 8.

$$= \frac{5}{8} + \frac{16}{8} = \frac{21}{8}$$

**11.** Add: $7 + \frac{3}{5}$.

Do Exercise 11.

*Answers*

**8.** $\frac{5}{6}$   **9.** $\frac{29}{24}$   **10.** $\frac{2}{9}$   **11.** $\frac{38}{5}$

**EXAMPLE 10**   Add: $\dfrac{9}{70} + \dfrac{11}{21} + \dfrac{-4}{15}$.

We need to determine the LCM of 70, 21, and 15:

$$70 = 2 \cdot 5 \cdot 7, \qquad 21 = 3 \cdot 7, \qquad 15 = 3 \cdot 5.$$

The LCM is $2 \cdot 3 \cdot 5 \cdot 7$, or 210.

$$\frac{9}{70} + \frac{11}{21} + \frac{-4}{15} = \frac{9}{70} \cdot \frac{3}{3} + \frac{11}{21} \cdot \frac{2 \cdot 5}{2 \cdot 5} + \frac{-4}{15} \cdot \frac{7 \cdot 2}{7 \cdot 2}$$

> In each case, we multiply by 1 to obtain the LCD. To form 1, look at the prime factorization of the LCD and use the factor(s) missing from each denominator.

$$= \frac{9 \cdot 3}{70 \cdot 3} + \frac{11 \cdot 10}{21 \cdot 10} + \frac{-4 \cdot 14}{15 \cdot 14}$$

$$= \frac{27}{210} + \frac{110}{210} + \frac{-56}{210}$$

$$= \frac{137 + (-56)}{210} = \frac{81}{210}$$

$$= \frac{3 \cdot 3 \cdot 3 \cdot 3}{2 \cdot 3 \cdot 5 \cdot 7} = \frac{3}{3} \cdot \frac{3 \cdot 3 \cdot 3}{2 \cdot 5 \cdot 7} = \frac{27}{70}$$

Do Exercises 12 and 13.

Add.

**12.** $\dfrac{4}{10} + \dfrac{1}{100} + \dfrac{3}{1000}$

**13.** $\dfrac{7}{10} + \dfrac{-2}{21} + \dfrac{1}{7}$

## c Order

When two fractions share a common denominator, the larger number can be found by comparing numerators. For example, 4 is greater than 3, so $\frac{4}{5}$ is greater than $\frac{3}{5}$.

$$\frac{4}{5} > \frac{3}{5}$$

Similarly, because $-6$ is less than $-2$, we have

$$\frac{-6}{7} < \frac{-2}{7}, \quad \text{or} \quad -\frac{6}{7} < -\frac{2}{7}.$$

Do Exercises 14–16.

Use $<$ or $>$ for $\square$ to form a true sentence.

**14.** $\dfrac{3}{8} \ \square \ \dfrac{5}{8}$

**15.** $\dfrac{7}{10} \ \square \ \dfrac{6}{10}$ 　　 **16.** $\dfrac{-2}{9} \ \square \ \dfrac{-5}{9}$

**EXAMPLE 11**   Use $<$ or $>$ for $\square$ to form a true sentence:

$$\frac{5}{8} \ \square \ \frac{2}{3}.$$

You can confirm that the LCD is 24. We multiply by 1 to find two fractions equivalent to $\frac{5}{8}$ and $\frac{2}{3}$ with denominators the same:

$$\frac{5}{8} \cdot \frac{3}{3} = \frac{15}{24}; \qquad \frac{2}{3} \cdot \frac{8}{8} = \frac{16}{24}. \qquad \frac{5}{8} \ \square \ \frac{2}{3} \text{ is equivalent to } \frac{15}{24} \ \square \ \frac{16}{24}.$$

Since $15 < 16$, it follows that $\frac{15}{24} < \frac{16}{24}$. Thus,

$$\frac{5}{8} < \frac{2}{3}.$$

Use < or > for ☐ to form a true sentence.

**17.** $\dfrac{2}{3}$ ☐ $\dfrac{3}{4}$

**18.** $\dfrac{-3}{4}$ ☐ $\dfrac{-8}{12}$

**19.** $\dfrac{7}{8}$ ☐ $\dfrac{5}{6}$

**EXAMPLE 12**  Use < or > for ☐ to form a true sentence:

$$-\frac{89}{100} \ \square \ -\frac{9}{10}.$$

We rewrite $-\frac{9}{10}$ with a denominator of 100: $-\frac{9}{10} \cdot \frac{10}{10} = -\frac{90}{100}$. We then have

$$\frac{-89}{100} \ \square \ \frac{-90}{100}. \qquad \text{Recall that } -\frac{m}{n} = \frac{-m}{n}.$$

Since $-89 > -90$, it follows that $-\frac{89}{100} > -\frac{90}{100}$, so

$$-\frac{89}{100} > -\frac{9}{10}.$$

Do Exercises 17–19.

## (d) Applications and Problem Solving

**EXAMPLE 13**  *Subflooring.*  A contractor uses two layers of subflooring under a ceramic tile floor. First, she installs a $\frac{3}{4}$-in. layer of oriented strand board (OSB). Then a $\frac{1}{2}$-in. sheet of cement board is mortared to the OSB. The mortar is $\frac{1}{8}$-in. thick. What is the total thickness of the two installed subfloors?

1. **Familiarize.**  We first make a drawing. We let $T =$ the total thickness of the subfloors.

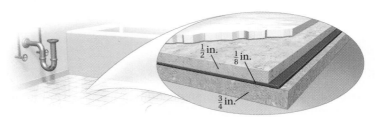

2. **Translate.**  The problem can be translated to an equation as follows.

| OSB | plus | mortar | plus | cement board | is | total thickness |
|---|---|---|---|---|---|---|
| ↓ | ↓ | ↓ | ↓ | ↓ | ↓ | ↓ |
| $\frac{3}{4}$ | $+$ | $\frac{1}{8}$ | $+$ | $\frac{1}{2}$ | $=$ | $T$ |

3. **Solve.**  To solve the equation, we carry out the addition.

$$\frac{3}{4} + \frac{1}{8} + \frac{1}{2} = T \qquad \text{The LCM of the denominators is 8.}$$

$$\frac{3}{4} \cdot \frac{2}{2} + \frac{1}{8} + \frac{1}{2} \cdot \frac{4}{4} = T \qquad \text{Multiplying by 1 to obtain the LCD}$$

$$\frac{6}{8} + \frac{1}{8} + \frac{4}{8} = T$$

$$\frac{11}{8} = T$$

4. **Check.**  We check by repeating the calculation.
5. **State.**  The total thickness of the installed subfloors is $\frac{11}{8}$ in.

Do Exercise 20.

**20. Fruit Salad.**  A caterer prepares a mixed berry salad with $\frac{7}{8}$ qt of strawberries, $\frac{3}{4}$ qt of raspberries, and $\frac{5}{16}$ qt of blueberries. What is the total amount of berries in the salad?

*Answers*

17. <   18. <   19. >   20. $\dfrac{31}{16}$ qt

**a** , **b** Add and, if possible, simplify.

1. $\dfrac{4}{9} + \dfrac{1}{9}$

2. $\dfrac{3}{11} + \dfrac{5}{11}$

3. $\dfrac{4}{7} + \dfrac{3}{7}$

4. $\dfrac{7}{8} + \dfrac{1}{8}$

5. $\dfrac{7}{10} + \dfrac{3}{-10}$

6. $\dfrac{1}{-6} + \dfrac{5}{6}$

7. $\dfrac{9}{a} + \dfrac{4}{a}$

8. $\dfrac{2}{t} + \dfrac{3}{t}$

9. $\dfrac{-1}{4} + \dfrac{-1}{4}$

10. $\dfrac{7}{12} + \dfrac{-5}{12}$

11. $\dfrac{2}{9}x + \dfrac{5}{9}x$

12. $\dfrac{3}{11}a + \dfrac{2}{11}a$

13. $\dfrac{3}{32}t + \dfrac{13}{32}t$

14. $\dfrac{3}{25}x + \dfrac{12}{25}x$

15. $-\dfrac{2}{x} + \left(-\dfrac{7}{x}\right)$

16. $-\dfrac{7}{a} + \dfrac{5}{a}$

17. $\dfrac{1}{8} + \dfrac{1}{6}$

18. $\dfrac{1}{9} + \dfrac{1}{6}$

19. $\dfrac{-4}{5} + \dfrac{7}{10}$

20. $\dfrac{-3}{4} + \dfrac{-1}{12}$

21. $\dfrac{7}{12} + \dfrac{3}{8}$

22. $\dfrac{7}{8} + \dfrac{1}{16}$

23. $\dfrac{3}{20} + 4$

24. $\dfrac{2}{15} + 3$

25. $\dfrac{5}{-8} + \dfrac{5}{6}$

26. $\dfrac{5}{-6} + \dfrac{7}{9}$

27. $\dfrac{3}{10}x + \dfrac{7}{100}x$

28. $\dfrac{9}{20}a + \dfrac{3}{40}a$

29. $\dfrac{5}{12} + \dfrac{8}{15}$

30. $\dfrac{3}{16} + \dfrac{1}{12}$

31. $\dfrac{7}{8} + \dfrac{0}{1}$

32. $\dfrac{0}{6} + \dfrac{5}{3}$

33. $\dfrac{-7}{10} + \dfrac{-29}{100}$

34. $\dfrac{-3}{10} + \dfrac{-27}{100}$

35. $-\dfrac{1}{10}x + \dfrac{1}{15}x$

36. $-\dfrac{1}{6}x + \dfrac{1}{4}x$

37. $-5t + \dfrac{2}{7}t$

38. $-4x + \dfrac{3}{5}x$

39. $-\dfrac{5}{12} + \dfrac{7}{-24}$

40. $-\dfrac{1}{18} + \dfrac{5}{-12}$

**41.** $\dfrac{3}{16} + \dfrac{5}{16} + \dfrac{4}{16}$

**42.** $\dfrac{3}{8} + \dfrac{1}{8} + \dfrac{2}{8}$

**43.** $\dfrac{4}{10} + \dfrac{3}{100} + \dfrac{7}{1000}$

**44.** $\dfrac{7}{10} + \dfrac{2}{100} + \dfrac{9}{1000}$

**45.** $\dfrac{3}{10} + \dfrac{5}{12} + \dfrac{8}{15}$

**46.** $\dfrac{1}{2} + \dfrac{3}{8} + \dfrac{1}{4}$

**47.** $\dfrac{5}{6} + \dfrac{25}{52} + \dfrac{7}{4}$

**48.** $\dfrac{15}{24} + \dfrac{7}{36} + \dfrac{91}{48}$

**49.** $\dfrac{2}{9} + \dfrac{7}{10} + \dfrac{-4}{15}$

**50.** $\dfrac{5}{12} + \dfrac{-3}{8} + \dfrac{1}{10}$

**51.** $-\dfrac{3}{4} + \dfrac{1}{5} + \dfrac{-7}{10}$

**52.** $\dfrac{1}{3} + \dfrac{-7}{9} + \dfrac{-1}{2}$

---

**c**    Use $<$ or $>$ for $\square$ to form a true sentence.

**53.** $\dfrac{3}{8} \,\square\, \dfrac{2}{8}$

**54.** $\dfrac{7}{9} \,\square\, \dfrac{5}{9}$

**55.** $\dfrac{2}{3} \,\square\, \dfrac{5}{6}$

**56.** $\dfrac{11}{18} \,\square\, \dfrac{5}{9}$

**57.** $\dfrac{-2}{7} \,\square\, \dfrac{-5}{7}$

**58.** $\dfrac{-4}{5} \,\square\, \dfrac{-3}{5}$

**59.** $\dfrac{9}{15} \,\square\, \dfrac{7}{10}$

**60.** $\dfrac{5}{14} \,\square\, \dfrac{8}{21}$

**61.** $\dfrac{3}{4} \,\square\, -\dfrac{1}{5}$

**62.** $\dfrac{3}{8} \,\square\, -\dfrac{13}{16}$

**63.** $\dfrac{-7}{20} \,\square\, \dfrac{-6}{15}$

**64.** $\dfrac{-7}{12} \,\square\, \dfrac{-9}{16}$

---

Arrange each group of fractions from smallest to largest.

**65.** $\dfrac{3}{10},\ \dfrac{5}{12},\ \dfrac{4}{15}$

**66.** $\dfrac{5}{6},\ \dfrac{19}{21},\ \dfrac{11}{14}$

---

**d**    Solve.

**67.** *Riding a Segway®.* Tate rode a Segway Personal Transporter $\frac{5}{6}$ mi to the library, then $\frac{3}{4}$ mi to class, and then $\frac{3}{2}$ mi to his part-time job. How far did he ride his Segway?

**68.** *Caffeine.* To cut back on their caffeine intake, Michelle and Gerry mix caffeinated and decaffeinated coffee beans before grinding for a customized mix. They mix $\frac{3}{16}$ lb of decaffeinated beans with $\frac{5}{8}$ lb of caffeinated beans. What is the total amount of coffee beans in the mixture?

---

**69.** For a community project, an earth science class dedicated one hour per day for three days to the state highway beautification project. The students collected trash along a $\frac{4}{5}$-mi stretch of highway the first day, $\frac{5}{8}$ mi the second day, and $\frac{1}{2}$ mi the third day. How many miles along the highway did they clean?

**70.** Alyse bought $\frac{1}{3}$ lb of orange pekoe tea and $\frac{1}{2}$ lb of English cinnamon tea. How many pounds of tea did she buy?

**71.** *Meteorology.* On Monday, April 15, it rained $\frac{1}{2}$ in. in the morning and $\frac{3}{8}$ in. in the afternoon. How much did it rain altogether?

**72.** *Nursing.* Janine took $\frac{1}{5}$ g of ibuprofen before lunch and $\frac{1}{2}$ g after lunch. How much did she take altogether?

**73.** A park naturalist hikes $\frac{3}{5}$ mi to a lookout, another $\frac{3}{10}$ mi to an osprey's nest, and finally $\frac{3}{4}$ mi to a campsite. How far does the naturalist hike?

**74.** A triathlete runs $\frac{7}{8}$ mi, canoes $\frac{1}{3}$ mi, and swims $\frac{1}{6}$ mi. How many miles does the triathlete cover?

**75.** *Baking.* A baker used $\frac{1}{2}$ lb of flour for rolls, $\frac{1}{4}$ lb for donuts, and $\frac{1}{3}$ lb for cookies. How much flour was used?

**76.** *Baking.* A recipe for muffins calls for $\frac{1}{2}$ qt (quart) of buttermilk, $\frac{1}{3}$ qt of skim milk, and $\frac{1}{16}$ qt of oil. How many quarts of liquid ingredients does the recipe call for?

**77.** *Iced Brownies.* The campus culinary arts department is preparing brownies for the international student reception. Students in the catering program are icing the $\frac{11}{16}$-in. $\left(\frac{11"}{16}\right)$ butterscotch brownies with a $\frac{5}{32}$-in. $\left(\frac{5"}{32}\right)$ layer of icing. What is the thickness of the iced brownie?

**78.** *Carpentry.* To cut expenses, a carpenter sometimes glues two kinds of plywood together. He glues a $\frac{1}{4}$-in. $\left(\frac{1"}{4}\right)$ piece of walnut plywood to a $\frac{3}{8}$-in. $\left(\frac{3"}{8}\right)$ piece of less expensive plywood. What is the total thickness of these pieces?

**79.** *Punch Recipe.* A recipe for strawberry punch calls for $\frac{1}{5}$ qt of ginger ale and $\frac{3}{5}$ qt of strawberry soda. How much liquid is needed? If the recipe is doubled, how much liquid is needed? If the recipe is halved, how much liquid is needed?

**80.** *Concrete Mix.* A cubic meter of concrete mix contains 420 kg (kilograms) of cement, 150 kg of stone, and 120 kg of sand. What is the total weight of a cubic meter of the mix? What fractional part is cement? stone? sand? Add these fractional amounts. What is the result?

## Skill Maintenance

Subtract. [2.3a]

**81.** $-7 - 6$

**82.** $-5 - (-9)$

**83.** $9 - 17$

**84.** $-8 - 23$

Evaluate. [2.6a]

**85.** $\dfrac{x - y}{3}$, for $x = 7$ and $y = -3$

**86.** $3(x + y)$ and $3x + 3y$, for $x = 5$ and $y = 9$

*Presidential Elections.* The results of the five closest presidential elections are listed in the following table. Use these data for Exercises 87–92. [1.8a]

| DATE | PRESIDENT | ELECTORAL VOTES | POPULAR VOTES |
|------|-----------|-----------------|---------------|
| 1916 | Woodrow Wilson (D) | 277 | 9,126,300 |
| | Charles E. Hughes (R) | 254 | 8,546,789 |
| 1960 | John F. Kennedy (D) | 303 | 34,226,731 |
| | Richard M. Nixon (R) | 219 | 34,108,157 |
| 1968 | Richard M. Nixon (R) | 301 | 31,785,480 |
| | Hubert H. Humphrey (D) | 191 | 31,275,166 |
| | George C. Wallace (3rd Party) | 46 | 9,906,473 |
| 1976 | Jimmy Carter (D) | 297 | 40,830,763 |
| | Gerald R. Ford (R) | 240 | 39,147,793 |
| 2000 | George W. Bush (R) | 271 | 50,459,211 |
| | Albert A. Gore (D) | 266 | 51,003,894 |

SOURCE: *The World Almanac*, 2008, p. 544

**87.** How many more popular votes did Albert A. Gore have than George W. Bush in the 2000 presidential election?

**88.** How many more electoral votes did George W. Bush have than Albert A. Gore in the 2000 presidential election?

**89.** In the 1960 presidential election, how many more electoral votes did John F. Kennedy have than Richard M. Nixon?

**90.** In the 1968 presidential election, how many more popular votes did Richard M. Nixon have than Hubert H. Humphrey?

**91.** How much greater was the total of the popular vote counts in the 2000 presidential election than in the 1976 election?

**92.** How much greater was the total of the popular vote counts in the 2000 presidential election than in the 1960 election?

## Synthesis

Add and, if possible, simplify.

**93.** $\dfrac{3}{10}t + \dfrac{2}{7} + \dfrac{2}{15}t + \dfrac{3}{5}$

**94.** $\dfrac{2}{9} + \dfrac{4}{21}x + \dfrac{4}{15} + \dfrac{3}{14}x$

**95.** $5t^2 + \dfrac{6}{a}t + 2t^2 + \dfrac{3}{a}t$

Use $<$, $>$, or $=$ for $\square$ to form a true sentence.

**96.** $\boxdot\ \dfrac{10}{97} + \dfrac{67}{137}\ \square\ \dfrac{8123}{13,289}$

**97.** $\boxdot\ \dfrac{12}{169} + \dfrac{53}{103}\ \square\ \dfrac{10,192}{17,407}$

**98.** $\boxdot\ \dfrac{37}{157} + \dfrac{20}{107}\ \square\ \dfrac{6942}{16,799}$

**99.** A guitarist's band is booked for Friday and Saturday nights at a local club. The guitarist's group is a trio on Friday and expands to a quintet on Saturday. Thus, the guitarist is paid one-third of one-half the weekend's pay for Friday and one-fifth of one-half the weekend's pay for Saturday. What fractional part of the total pay did the guitarist receive for the weekend's work? If the band was paid $1200, how much did the guitarist receive?

**100.** $\boxdot$ Consider only the numbers 2, 3, 4, and 5. Assume each is placed in a blank in the following.

$$\dfrac{\square}{\square} + \dfrac{\square}{\square} = ?$$

What placement of the numbers in the blanks yields the largest sum?

**101.** $\boxdot$ In the sum below, $a$ and $b$ are digits (so $1b$ is a two-digit number and $35a$ is a three-digit number). Find $a$ and $b$. (*Hint*: $a < 4$ and $b > 6$.)

$$\dfrac{a}{17} + \dfrac{1b}{23} = \dfrac{35a}{391}$$

**102.** $\boxdot$ Use a standard calculator. Arrange the following in order from smallest to largest.

$$\dfrac{3}{4}, \dfrac{17}{21}, \dfrac{13}{15}, \dfrac{7}{9}, \dfrac{15}{17}, \dfrac{13}{12}, \dfrac{19}{22}$$

# 4.3

## Subtraction, Equations, and Applications

### a) Subtraction

#### Like Denominators

Let's consider the difference $\frac{4}{8} - \frac{3}{8}$.

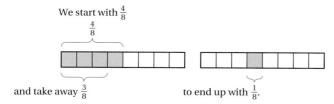

We start with $\frac{4}{8}$

$\frac{4}{8}$

and take away $\frac{3}{8}$   to end up with $\frac{1}{8}$.

We start with 4 eighths and take away 3 eighths:

4 eighths − 3 eighths = 1 eighth,

or $\quad 4 \cdot \frac{1}{8} - 3 \cdot \frac{1}{8} = \frac{1}{8}, \quad$ or $\quad \frac{4}{8} - \frac{3}{8} = \frac{1}{8}.$

---

To subtract when denominators are the same,

**a)** subtract the numerators,

**b)** keep the denominator, and

**c)** simplify, if possible.

$$\frac{7}{10} - \frac{4}{10} = \frac{7 - 4}{10} = \frac{3}{10}$$

---

**EXAMPLES**   Subtract and, if possible, simplify.

**1.** $\dfrac{8}{13} - \dfrac{3}{13} = \dfrac{8 - 3}{13} = \dfrac{5}{13}$

**2.** $\dfrac{3}{35} - \dfrac{13}{35} = \dfrac{3 - 13}{35} = \dfrac{-10}{35} = \dfrac{5}{5} \cdot \dfrac{-2}{7} = \dfrac{-2}{7}, \text{ or } -\dfrac{2}{7}$   Removing a factor equal to 1: $\frac{5}{5} = 1$

**3.** $\dfrac{13}{2a} - \dfrac{5}{2a} = \dfrac{13 - 5}{2a} = \dfrac{8}{2a} = \dfrac{2}{2} \cdot \dfrac{4}{a} = \dfrac{4}{a}$   Removing a factor equal to 1: $\frac{2}{2} = 1$

**4.** $-\dfrac{7}{t} - \dfrac{2}{t} = \dfrac{-7 - 2}{t} = \dfrac{-9}{t}, \text{ or } -\dfrac{9}{t}$

Do Margin Exercises 1–4.

#### Different Denominators

---

To subtract when denominators are different:

**a)** Find the least common multiple of the denominators. That number is the least common denominator, LCD.

**b)** Multiply by 1, using an appropriate notation, $n/n$, to express each fraction in an equivalent form that contains the LCD.

**c)** Subtract the numerators, keeping the same denominator.

**d)** Simplify, if possible.

---

### OBJECTIVES

**a** Subtract using fraction notation.

**b** Solve equations of the type $x + a = b$ and $a + x = b$, where $a$ and $b$ may be fractions.

**c** Solve applied problems involving subtraction with fraction notation.

**SKILL TO REVIEW**
Objective 2.8a: Use the addition principle to solve equations.

Solve.

**1.** $-2 + x = -8$     **2.** $y - 3 = -3$

Subtract and, if possible, simplify.

**1.** $\dfrac{7}{8} - \dfrac{3}{8}$     **2.** $\dfrac{5}{9a} - \dfrac{1}{9a}$

**3.** $\dfrac{7}{10} - \dfrac{13}{10}$     **4.** $-\dfrac{2}{x} - \dfrac{4}{x}$

*Answers*

*Skill to Review:*
1. $-6$   2. $0$

*Margin Exercises:*
1. $\dfrac{1}{2}$   2. $\dfrac{4}{9a}$   3. $-\dfrac{3}{5}$   4. $-\dfrac{6}{x}$

**EXAMPLE 5**  Subtract: $\dfrac{2}{5} - \dfrac{3}{8}$.

**a)**  The LCM of 5 and 8 is 40, so the LCD is 40.

**b)**  We need to find numbers equivalent to $\dfrac{2}{5}$ and $\dfrac{3}{8}$ with denominators of 40:

$$\frac{2}{5} - \frac{3}{8} = \frac{2}{5} \cdot \frac{8}{8} - \frac{3}{8} \cdot \frac{5}{5}.$$

*Think*: $8 \times \square = 40$. The answer is 5, so we multiply by 1, using $\frac{5}{5}$.

*Think*: $5 \times \square = 40$. The answer is 8, so we multiply by 1, using $\frac{8}{8}$.

**c)**  We subtract:  $\dfrac{16}{40} - \dfrac{15}{40} = \dfrac{16 - 15}{40} = \dfrac{1}{40}.$

**d)**  Since $\dfrac{1}{40}$ cannot be simplified, we are finished. The answer is $\dfrac{1}{40}$.

> Do Exercise 5.

**5.** Subtract: $\dfrac{3}{4} - \dfrac{2}{3}$.

**EXAMPLE 6**  Subtract: $\dfrac{7}{12} - \dfrac{5}{6}$.

Since 12 is a multiple of 6, the LCM of 6 and 12 is 12. The LCD is 12.

$$\frac{7}{12} - \frac{5}{6} = \frac{7}{12} - \frac{5}{6} \cdot \frac{2}{2}$$  *Think*: $6 \times \square = 12$. The answer is 2, so we multiply by 1, using $\frac{2}{2}$.

$$= \frac{7}{12} - \frac{10}{12}$$

$$= \frac{7 - 10}{12} = \frac{-3}{12}$$  $7 - 10 = 7 + (-10) = -3$

$$= \frac{3}{3} \cdot \frac{-1}{4} = \frac{-1}{4}, \text{ or } -\frac{1}{4}$$  Simplifying by removing a factor equal to 1: $\frac{3}{3} = 1$

**EXAMPLE 7**  Subtract: $\dfrac{17}{24} - \dfrac{4}{15}$.

We need to find the LCM of 24 and 15:

$$\left.\begin{array}{l} 24 = 2 \cdot 2 \cdot 2 \cdot 3, \\ 15 = 3 \cdot 5. \end{array}\right\} \quad \text{The LCM is } 2 \cdot 2 \cdot 2 \cdot 3 \cdot 5, \text{ or } 120.$$

$$\frac{17}{24} - \frac{4}{15} = \frac{17}{24} \cdot \frac{5}{5} - \frac{4}{15} \cdot \frac{8}{8}$$

Multiplying by 1 to obtain the LCD. To form 1, use the factors of the LCM that each denominator lacks. Note that $2 \cdot 2 \cdot 2 = 8$.

$$= \frac{85}{120} - \frac{32}{120} = \frac{85 - 32}{120} = \frac{53}{120}$$

> Do Exercises 6–9.

**Subtract.**

**6.** $\dfrac{5}{6} - \dfrac{2}{3}$

**7.** $\dfrac{2}{5} - \dfrac{7}{10}$

**8.** $\dfrac{2}{3} - \dfrac{5}{6}$

**9.** $\dfrac{11}{28} - \dfrac{5}{16}$

**EXAMPLE 8**  Simplify by combining like terms: $\dfrac{7}{8}x - \dfrac{3}{4}x$.

$$\frac{7}{8}x - \frac{3}{4}x = \left(\frac{7}{8} - \frac{3}{4}\right)x$$  Try to do this step mentally.

$$= \left(\frac{7}{8} - \frac{6}{8}\right)x = \frac{1}{8}x$$  Multiplying $\frac{3}{4}$ by $\frac{2}{2}$ and subtracting

> Do Exercise 10.

**10.** Simplify: $\dfrac{9}{10}x - \dfrac{3}{5}x$.

*Answers*

**5.** $\dfrac{1}{12}$   **6.** $\dfrac{1}{6}$   **7.** $-\dfrac{3}{10}$   **8.** $-\dfrac{1}{6}$

**9.** $\dfrac{9}{112}$   **10.** $\dfrac{3}{10}x$

## b Solving Equations

Section 2.8 introduced the addition principle as one way to form equivalent equations. We can use that principle here to solve equations containing fractions.

**EXAMPLE 9** Solve: $x - \dfrac{1}{3} = -\dfrac{2}{7}$.

$$x - \frac{1}{3} = -\frac{2}{7}$$

$$x - \frac{1}{3} + \frac{1}{3} = -\frac{2}{7} + \frac{1}{3} \qquad \text{Using the addition principle:} \atop \text{adding } \frac{1}{3} \text{ to both sides}$$

$$x + 0 = -\frac{2}{7} + \frac{1}{3} \qquad \text{Adding } \frac{1}{3} \text{ "undid" the subtraction of } \frac{1}{3} \atop \text{on the left-hand side of the equation.}$$

$$x = -\frac{2}{7} \cdot \frac{3}{3} + \frac{1}{3} \cdot \frac{7}{7} \qquad \text{Multiplying by 1 to obtain the} \atop \text{LCD, 21}$$

$$x = -\frac{6}{21} + \frac{7}{21} = \frac{1}{21} \qquad \text{The solution appears to be } \frac{1}{21}.$$

Check:
$$\begin{array}{c|c} x - \dfrac{1}{3} = -\dfrac{2}{7} \\ \hline \dfrac{1}{21} - \dfrac{1}{3} \;?\; -\dfrac{2}{7} \\ \dfrac{1}{21} - \dfrac{1}{3} \cdot \dfrac{7}{7} \\ \dfrac{1}{21} - \dfrac{7}{21} \\ \dfrac{-6}{21} \\ -\dfrac{2}{7} \cdot \dfrac{3}{3} \;\Big|\; -\dfrac{2}{7} \quad \text{TRUE} \end{array}$$

Our answer checks. The solution is $\frac{1}{21}$.

Recall that we can also create equivalent equations if we subtract the same number on both sides of an equation.

**EXAMPLE 10** Solve: $x + \dfrac{1}{4} = \dfrac{3}{5}$.

$$x + \frac{1}{4} - \frac{1}{4} = \frac{3}{5} - \frac{1}{4} \qquad \text{Using the addition principle:} \atop \text{adding } -\frac{1}{4} \text{ to, or subtracting } \frac{1}{4} \text{ from,} \atop \text{both sides}$$

$$x + 0 = \frac{3}{5} \cdot \frac{4}{4} - \frac{1}{4} \cdot \frac{5}{5} \qquad \text{The LCD is 20. We multiply by 1 to} \atop \text{get the LCD.}$$

$$x = \frac{12}{20} - \frac{5}{20} = \frac{7}{20}$$

The solution is $\frac{7}{20}$. We leave the check to the student.

Do Exercises 11–13.

**STUDY TIPS**

**PLAN FOR YOUR ABSENCES**

Occasionally you know in advance that you will be forced to miss an upcoming class. It is usually best to alert your instructor to this situation as soon as possible. He or she may permit you to make up a missed quiz or test *if you provide enough advance notice.* Make an effort to find out what assignment will be given in your absence and try your best to learn the material on your own. Sometimes it is possible to attend another section of the course or watch a video lecture for the one class that you must miss.

Solve.

**11.** $x - \dfrac{2}{5} = \dfrac{1}{5}$

**12.** $x + \dfrac{2}{3} = \dfrac{5}{6}$

**13.** $\dfrac{3}{5} + t = -\dfrac{7}{8}$

*Answers*

**11.** $\dfrac{3}{5}$  **12.** $\dfrac{1}{6}$  **13.** $-\dfrac{59}{40}$

## (c) Applications and Problem Solving

**EXAMPLE 11** *Fraction of the Moon Illuminated.* From anywhere on earth, the moon appears to be a circular disk. At midnight on October 15, 2010 (Eastern Daylight Time), $\frac{3}{5}$ of the moon appeared illuminated. By October 18, 2010, the illuminated portion had increased to $\frac{17}{20}$. How much more of the moon appeared illuminated on October 18 than on October 15?

**Source:** Astronomical Applications Department, U.S. Naval Observatory, Washington DC 20392

1. **Familiarize.** We let $m =$ the additional part of the moon that appeared illuminated.

2. **Translate.** We translate to an equation:

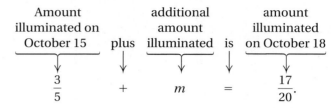

$$\frac{3}{5} + m = \frac{17}{20}.$$

3. **Solve.** To solve the equation, we subtract $\frac{3}{5}$ on both sides:

$$\frac{3}{5} + m = \frac{17}{20}$$

$$\frac{3}{5} + m - \frac{3}{5} = \frac{17}{20} - \frac{3}{5} \qquad \text{Subtracting } \frac{3}{5} \text{ on both sides}$$

$$m + 0 = \frac{17}{20} - \frac{3}{5} \cdot \frac{4}{4} \qquad \begin{array}{l}\text{The LCD is 20. We multiply by 1} \\ \text{to obtain the LCD.}\end{array}$$

$$m = \frac{17}{20} - \frac{12}{20} = \frac{5}{20} = \frac{5 \cdot 1}{5 \cdot 4} = \frac{5}{5} \cdot \frac{1}{4} = \frac{1}{4}.$$

4. **Check.** To check, we add:

$$\frac{3}{5} + \frac{1}{4} = \frac{3}{5} \cdot \frac{4}{4} + \frac{1}{4} \cdot \frac{5}{5} = \frac{12}{20} + \frac{5}{20} = \frac{17}{20}.$$

5. **State.** On October 18, $\frac{1}{4}$ more of the moon was illuminated than on October 15.

**14. Fraction of the Moon Illuminated.** At midnight on August 26, 2010 (Eastern Daylight Time), $\frac{19}{20}$ of the moon appeared illuminated. By August 28, 2010, the illuminated part had decreased to $\frac{16}{25}$. How much less of the moon appeared illuminated on August 28 than on August 26?

**Source:** Astronomical Applications Department, U.S. Naval Observatory, Washington DC 20392

*Answer*

14. $\frac{31}{100}$

Do Exercise 14.

---

### Calculator Corner

**Adding and Subtracting Fractions** To calculate $\frac{2}{15} + \frac{5}{12}$, the following keystrokes can be used (note that the $\boxed{a^b/_c}$ key usually doubles as the $\boxed{^d/_c}$ key): $\boxed{2}$ $\boxed{a^b/_c}$ $\boxed{1}$ $\boxed{5}$ $\boxed{+}$ $\boxed{5}$ $\boxed{a^b/_c}$ $\boxed{1}$ $\boxed{2}$ $\boxed{=}$. The display that appears, $\boxed{11 \lrcorner 20}$ or $\boxed{11/20}$, represents the fraction $\frac{11}{20}$. Fractions with numerators greater than denominators may be given in mixed numeral notation (see Section 4.5). To convert to fraction notation, press $\boxed{\text{2nd}}$ $\boxed{a^b/_c}$ $\boxed{=}$. Some graphing calculators do not have an $\boxed{a^b/_c}$ key. To perform the above addition on such a calculator, we use the $\boxed{\text{MATH}}$ key as follows:

$$\boxed{2} \boxed{÷} \boxed{1} \boxed{5} \boxed{+} \boxed{5} \boxed{÷} \boxed{1} \boxed{2} \boxed{\text{MATH}} \boxed{1} \boxed{\text{ENTER}}.$$

**Exercises:** Calculate.

1. $\dfrac{3}{8} + \dfrac{1}{4}$

2. $\dfrac{5}{12} + \dfrac{3}{10}$

3. $\dfrac{8}{7} + \dfrac{-1}{3}$

4. $\dfrac{19}{20} - \dfrac{17}{35}$

5. $\dfrac{-9}{30} - \dfrac{1}{25}$

6. $\dfrac{7}{23} - \left(-\dfrac{9}{29}\right)$

# Translating for Success

1. *Bubble Wrap.* One-Stop Postal Center orders bubble wrap in 64-yd rolls. On average, $\frac{3}{4}$ yd is used per small package. How many small packages can be prepared with 2 rolls of bubble wrap?

2. *Distance from College.* The post office is $\frac{7}{9}$ mi from the community college. The medical clinic is $\frac{2}{5}$ as far from the college as the post office is. How far is the clinic from the college?

3. *Swimming.* Andrew swims $\frac{7}{9}$ mi every day. One day, he swims $\frac{2}{5}$ mi by 11:00 A.M. How much farther must Andrew swim to reach his daily goal?

4. *Tuition.* The average tuition at Waterside University is $12,000. If a loan is obtained for $\frac{1}{3}$ of the tuition, how much is the loan?

5. *Thermos Bottle Capacity.* A thermos bottle holds $\frac{11}{12}$ gal. How much is in the bottle when it is $\frac{4}{7}$ full?

The goal of these matching questions is to practice step (2), *Translate,* of the five-step problem-solving process. Translate each word problem to an equation and select a correct translation from equations A–O.

**A.** $\frac{3}{4} \cdot 64 = x$

**B.** $\frac{1}{3} \cdot 12{,}000 = x$

**C.** $\frac{1}{3} + \frac{2}{5} = x$

**D.** $\frac{2}{5} + x = \frac{7}{9}$

**E.** $\frac{2}{5} \cdot \frac{7}{9} = x$

**F.** $\frac{3}{4} \cdot x = 64$

**G.** $\frac{4}{7} = x + \frac{11}{12}$

**H.** $\frac{2}{5} = x + \frac{7}{9}$

**I.** $\frac{4}{7} \cdot \frac{11}{12} = x$

**J.** $\frac{3}{4} \cdot x = 128$

**K.** $\frac{1}{3} \cdot x = 12{,}000$

**L.** $\frac{1}{3} + \frac{2}{5} + x = 1$

**M.** $\frac{2}{5} = \frac{7}{9}x$

**N.** $\frac{4}{7} + x = \frac{11}{12}$

**O.** $\frac{1}{3} + x = \frac{2}{5}$

*Answers on page A-7*

6. *Cutting Rope.* A piece of rope $\frac{11}{12}$ yd long is cut into two pieces. One piece is $\frac{4}{7}$ yd long. How long is the other piece?

7. *Planting Corn.* Each year, Prairie State Farm plants 64 acres of corn. With good weather, $\frac{3}{4}$ of the planting can be completed by April 20. How many acres can be planted by April 20 with good weather?

8. *Painting Trim.* A painter used $\frac{1}{3}$ gal of white paint for the trim in the library and $\frac{2}{5}$ gal for the trim in the family room. How much paint was used for the trim in the two rooms?

9. *Lottery Winnings.* Sally won $12,000 in a state lottery and decides to give the net amount after taxes to three charities. One received $\frac{1}{3}$ of the money, and a second received $\frac{2}{5}$. What fractional part did the third charity receive?

10. *Reading Assignment.* When Lowell had read 64 pages of his political science assignment, he had completed $\frac{3}{4}$ of his required reading. How many total pages were assigned?

**a** Subtract and, if possible, simplify.

**1.** $\dfrac{5}{6} - \dfrac{1}{6}$

**2.** $\dfrac{7}{5} - \dfrac{2}{5}$

**3.** $\dfrac{9}{16} - \dfrac{13}{16}$

**4.** $\dfrac{5}{12} - \dfrac{7}{12}$

**5.** $\dfrac{8}{a} - \dfrac{6}{a}$

**6.** $\dfrac{4}{t} - \dfrac{9}{t}$

**7.** $-\dfrac{3}{8} - \dfrac{1}{8}$

**8.** $-\dfrac{3}{10} - \dfrac{1}{10}$

**9.** $\dfrac{3}{5a} - \dfrac{7}{5a}$

**10.** $\dfrac{2}{7t} - \dfrac{10}{7t}$

**11.** $\dfrac{10}{3t} - \dfrac{4}{3t}$

**12.** $\dfrac{9}{2a} - \dfrac{5}{2a}$

**13.** $\dfrac{7}{8} - \dfrac{1}{16}$

**14.** $\dfrac{4}{3} - \dfrac{5}{6}$

**15.** $\dfrac{7}{15} - \dfrac{4}{5}$

**16.** $\dfrac{3}{28} - \dfrac{3}{4}$

**17.** $\dfrac{3}{4} - \dfrac{1}{20}$

**18.** $\dfrac{3}{4} - \dfrac{4}{16}$

**19.** $\dfrac{2}{15} - \dfrac{5}{12}$

**20.** $\dfrac{11}{16} - \dfrac{9}{10}$

**21.** $\dfrac{7}{10} - \dfrac{23}{100}$

**22.** $\dfrac{9}{10} - \dfrac{3}{100}$

**23.** $\dfrac{7}{15} - \dfrac{3}{25}$

**24.** $\dfrac{18}{25} - \dfrac{4}{35}$

**25.** $\dfrac{-41}{100} - \dfrac{3}{10}$

**26.** $\dfrac{-13}{100} - \dfrac{7}{20}$

**27.** $\dfrac{2}{3} - \dfrac{1}{8}$

**28.** $\dfrac{3}{4} - \dfrac{1}{2}$

**29.** $-\dfrac{3}{10} - \dfrac{7}{25}$

**30.** $-\dfrac{5}{18} - \dfrac{2}{27}$

**31.** $\dfrac{3}{8} - \dfrac{5}{12}$

**32.** $\dfrac{2}{9} - \dfrac{7}{12}$

**33.** $\dfrac{-5}{18} - \dfrac{7}{24}$

**34.** $\dfrac{-7}{25} - \dfrac{2}{15}$

**35.** $\dfrac{13}{90} - \dfrac{17}{120}$

**36.** $\dfrac{8}{25} - \dfrac{29}{150}$

**37.** $\dfrac{2}{3}x - \dfrac{4}{9}x$

**38.** $\dfrac{7}{4}x - \dfrac{5}{12}x$

**39.** $\dfrac{2}{5}a - \dfrac{3}{4}a$

**40.** $\dfrac{4}{7}a - \dfrac{1}{3}a$

**b** Solve.

**41.** $x - \dfrac{4}{9} = \dfrac{3}{9}$

**42.** $x - \dfrac{3}{11} = \dfrac{7}{11}$

**43.** $a + \dfrac{2}{11} = \dfrac{6}{11}$

**44.** $a + \dfrac{4}{15} = \dfrac{13}{15}$

**45.** $y + \dfrac{1}{30} = \dfrac{1}{10}$

**46.** $y + \dfrac{1}{3} = \dfrac{5}{6}$

**47.** $a - \dfrac{3}{8} = \dfrac{3}{4}$

**48.** $x - \dfrac{3}{10} = \dfrac{2}{5}$

**49.** $\dfrac{2}{3} + x = \dfrac{4}{5}$

**50.** $\dfrac{4}{5} + x = \dfrac{6}{7}$

**51.** $\dfrac{3}{8} + a = \dfrac{1}{12}$

**52.** $\dfrac{5}{6} + a = \dfrac{2}{9}$

**53.** $n - \dfrac{3}{10} = -\dfrac{1}{6}$

**54.** $n - \dfrac{3}{4} = -\dfrac{5}{12}$

**55.** $x + \dfrac{3}{4} = -\dfrac{1}{2}$

**56.** $x + \dfrac{5}{6} = -\dfrac{11}{12}$

**c** Solve.

**57.** For a research paper, Kaitlyn spent $\dfrac{3}{4}$ hr searching the Internet on google.com and $\dfrac{1}{3}$ hr on chacha.com. How many more hours did she spend on google.com than on chacha.com?

**58.** As part of a fitness program, Deb swims $\dfrac{1}{2}$ mi every day. One day she had already swum $\dfrac{1}{5}$ mi. How much farther did Deb need to swim?

**59.** *Tire Tread.*  A long-life tire has a tread depth of $\dfrac{3}{8}$ in. instead of a more typical $\dfrac{11}{32}$-in. depth. How much deeper is the long-life tread?

**Source:** *Popular Science*

**60.** A baker has a dispenser containing $\dfrac{15}{16}$ cup of icing and puts $\dfrac{1}{12}$ cup on a cinnamon roll. How much icing remains in the dispenser?

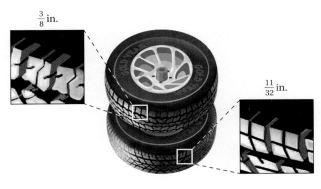

$\dfrac{3}{8}$ in.

$\dfrac{11}{32}$ in.

**61.** From a $\dfrac{4}{5}$-lb wheel of cheese, a $\dfrac{1}{4}$-lb piece was served. How much cheese remained on the wheel?

**62.** Jovan has an $\dfrac{11}{10}$-lb mixture of cashews and peanuts that includes $\dfrac{3}{5}$ lb of cashews. How many pounds of peanuts are in the mixture?

**63.** Jorge's $\dfrac{3}{4}$-hr drive to a job was part city and part country driving. If $\dfrac{2}{5}$ hr was city driving, how much time was spent on country driving?

**64.** Addie spent $\dfrac{3}{4}$ hr listening to Green Day and Coldplay. She spent $\dfrac{2}{3}$ hr listening to Green Day. How many hours were spent listening to Coldplay?

**65. *Woodworking.*** Natalie is replacing a $\frac{3}{4}$-in.-thick shelf in her bookcase. If her replacement board is $\frac{15}{16}$ in. thick, how much must it be planed down before the repair can be completed?

**66. *Furniture Cleaner.*** A $\frac{2}{3}$-cup mixture of lemon juice and olive oil makes an excellent cleaner for wood furniture. If the mixture contains $\frac{1}{4}$ cup of lemon juice, how much olive oil is in the cleaner?

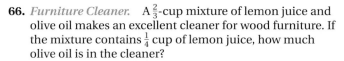

$\frac{1}{4}$ cup    +    ?    =    $\frac{2}{3}$ cup

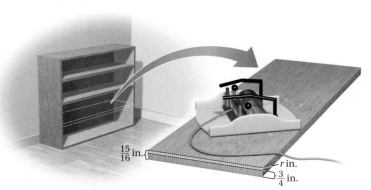

$\frac{15}{16}$ in.

$r$ in.

$\frac{3}{4}$ in.

**67.** Blake used $\frac{1}{3}$ cup of maple syrup in preparing the batter for a batch of maple oatbran muffins. Sheila pointed out that the recipe actually calls for $\frac{5}{8}$ cup of syrup. How much more syrup should Blake add to the batter?

**68.** Amber added $\frac{1}{3}$ qt of two-cycle oil to a fuel mixture for her lawn mower. She then noticed that the owner's manual indicates $\frac{1}{2}$ qt should have been added. How much more two-cycle oil should Amber add to the mixture?

## Skill Maintenance

Divide and, if possible, simplify.   [3.7b]

**69.** $\dfrac{3}{7} \div \dfrac{9}{4}$

**70.** $\dfrac{9}{10} \div \dfrac{3}{5}$

**71.** $7 \div \dfrac{1}{3}$

**72.** $\dfrac{1}{4} \div 8$

Solve.

**73. *Crayons.*** The Crayola 64 box of crayons with a built-in sharpener celebrated its 50th birthday in 2008. Since 1958, approximately 200 million Crayola 64 boxes have been sold. How many crayons have been sold altogether in the 64 box?   [1.8a]
**Source:** Crayola LCC

**74.** A batch of fudge requires $\frac{3}{4}$ cup of sugar. How much sugar is needed to make 12 batches?   [3.6b]

**75.** $3x - 8 = 25$   [2.8d]

**76.** $5x + 9 = 24$   [2.8d]

## Synthesis

Simplify.

**77.** $\dfrac{7}{8} - \dfrac{3}{4} - \dfrac{1}{16}$

**78.** $\dfrac{9}{10} - \dfrac{1}{2} - \dfrac{2}{15}$

**79.** $\dfrac{2}{5} - \dfrac{1}{6}\,(-3)^2$

**80.** $\dfrac{7}{8} - \dfrac{1}{10}\left(-\dfrac{5}{6}\right)^2$

**81.** $-4 \cdot \dfrac{3}{7} - \dfrac{1}{7} \cdot \dfrac{4}{5}$

**82.** $\left(\dfrac{5}{6}\right)^2 + \left(\dfrac{3}{4}\right)^2$

**83.** $\left(-\dfrac{2}{5}\right)^3 - \left(-\dfrac{3}{10}\right)^3$

**84.** $\dfrac{3}{17} - \dfrac{2}{19} - \left(\dfrac{3}{17} - \dfrac{2}{19}\right)$

**85.** The circle graph below shows how long shoppers stay when visiting a mall. What portion of shoppers stay for 0–2 hr?

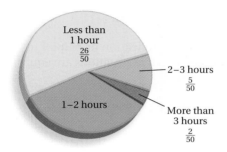

**86.** A new Chevrolet Aveo costs $12,600. Pam will pay $\frac{1}{2}$ of the cost, Sam will pay $\frac{1}{4}$ of the cost, Jan will pay $\frac{1}{6}$ of the cost, and Nan will pay the rest.

   **a)** How much will Nan pay?
   **b)** What fractional part will Nan pay?

**87.** Mark Romano owns $\frac{7}{12}$ of Romano-Chrenka Chevrolet and Lisa Romano owns $\frac{1}{6}$. If Paul and Ella Chrenka own the remaining share of the dealership equally, what fractional piece does Paul own?

**88.** Mazzi's meat slicer cut 8 slices of turkey and 3 slices of Vermont cheddar. If each turkey slice was $\frac{1}{16}$ in. thick and each cheddar slice was $\frac{5}{32}$ in. thick, how tall was the pile of meat and cheese?

**89.** As part of a rehabilitation program, an athlete must swim and then walk a total of $\frac{9}{10}$ km each day. If one lap in the swimming pool is $\frac{3}{80}$ km, how far must the athlete walk after swimming 10 laps?

**90.** A mountain climber, beginning at sea level, climbs $\frac{3}{5}$ km, descends $\frac{1}{4}$ km, climbs $\frac{1}{3}$ km, and then descends $\frac{1}{7}$ km. At what elevation does the climber finish?

**91.** ▦ Solve: $1x + \dfrac{16}{323} = \dfrac{10}{187}$.

**92.** ▦ Determine what whole number $a$ must be in order for the following to be true:

$$\dfrac{10 + a}{23} = \dfrac{330}{391} - \dfrac{a}{17}.$$

**93.** *Microsoft Interview.* The following is a question taken from an employment interview with Microsoft. Try to answer it. "Given a gold bar that can be cut exactly twice and a contractor who must be paid one-seventh of a gold bar every day for seven days, how should the bar be cut?"

**Source:** *Fortune Magazine,* January 22, 2001

# Solving Equations: Using the Principles Together

**a** Solve equations that involve fractions and require use of both the addition principle and the multiplication principle.

**b** Solve equations by using the multiplication principle to clear fractions.

**SKILL TO REVIEW**

Objective 3.8a: Use the multiplication principle to solve equations.

Solve.

**1.** $\dfrac{2}{3}x = \dfrac{5}{6}$    **2.** $18 = -\dfrac{2}{3}t$

We have used the multiplication principle to solve equations like

$$\frac{2}{3}x = \frac{5}{6} \quad \text{and} \quad 7 = \frac{5}{4}t,$$

and we have used the addition principle to solve equations like

$$\frac{4}{5} + x = \frac{1}{2} \quad \text{and} \quad \frac{7}{3} = t - \frac{2}{9}.$$

We are now ready to solve equations in which both principles are required.

## **a** Using the Principles Together

Recall that we use the addition and multiplication principles to write equivalent equations. In the following steps, all five equations are equivalent:

| | |
|---|---|
| $5x - 2 = 43$ | We first isolate $5x$. |
| $5x - 2 + 2 = 43 + 2$ | Using the addition principle |
| $5x = 45$ | We now isolate $x$. |
| $\dfrac{1}{5} \cdot 5x = \dfrac{1}{5} \cdot 45$ | Using the multiplication principle |
| $x = 9.$ | The solution of $x = 9$ is the solution of $5x - 2 = 43$. |

As a check, note that $5 \cdot 9 - 2 = 45 - 2 = 43$, as desired. The solution is 9.

**EXAMPLE 1**   Solve: $\dfrac{3}{4}x - \dfrac{1}{8} = \dfrac{1}{2}$.

We first isolate $\frac{3}{4}x$ by adding $\frac{1}{8}$ to both sides:

$$\frac{3}{4}x - \frac{1}{8} = \frac{1}{2}$$

$$\frac{3}{4}x - \frac{1}{8} + \frac{1}{8} = \frac{1}{2} + \frac{1}{8} \qquad \text{Using the addition principle}$$

$$\frac{3}{4}x + 0 = \frac{4}{8} + \frac{1}{8} \qquad \text{Writing with a common denominator}$$

$$\frac{3}{4}x = \frac{5}{8}.$$

Next, we isolate $x$ by multiplying both sides by $\frac{4}{3}$:

$$\frac{3}{4}x = \frac{5}{8} \qquad \text{Note that the reciprocal of } \tfrac{3}{4} \text{ is } \tfrac{4}{3}.$$

$$\frac{4}{3} \cdot \frac{3}{4}x = \frac{4}{3} \cdot \frac{5}{8} \qquad \text{Using the multiplication principle}$$

$$1x = \frac{20}{24}, \text{ or } \frac{5}{6}. \qquad \text{Simplifying; the solution appears to be } \tfrac{5}{6}.$$

*Answers*

*Skill to Review:*

**1.** $\dfrac{5}{4}$    **2.** $-27$

Check:

$$\frac{3}{4}x - \frac{1}{8} = \frac{1}{2}$$

$$\frac{3}{4} \cdot \frac{5}{6} - \frac{1}{8} \; ? \; \frac{1}{2}$$

$$\frac{\cancel{3} \cdot 5}{4 \cdot 2 \cdot \cancel{3}} - \frac{1}{8}$$     Removing a factor equal to 1: $\frac{3}{3} = 1$

$$\frac{5}{8} - \frac{1}{8}$$

$$\frac{1}{2} \quad \bigg| \quad \frac{1}{2} \quad \text{TRUE}$$

The solution is $\frac{5}{6}$.

Do Exercises 1 and 2.

Solve.

**1.** $\dfrac{3}{5}t - \dfrac{8}{15} = \dfrac{2}{15}$

**2.** $\dfrac{1}{2}x - \dfrac{1}{5} = \dfrac{7}{10}$

**EXAMPLE 2**   Solve: $5 + \dfrac{9}{2}t = -\dfrac{7}{2}$.

We first isolate $\frac{9}{2}t$ by subtracting 5 from both sides:

$$5 + \frac{9}{2}t = -\frac{7}{2}$$

$$5 + \frac{9}{2}t - 5 = -\frac{7}{2} - 5 \qquad \text{Subtracting 5 from both sides}$$

$$\frac{9}{2}t = -\frac{7}{2} - \frac{10}{2} \qquad \text{Writing 5 as } \tfrac{10}{2} \text{ to use the LCD}$$

$$\frac{9}{2}t = -\frac{17}{2} \qquad \text{Note that the reciprocal of } \tfrac{9}{2} \text{ is } \tfrac{2}{9}.$$

$$\frac{2}{9} \cdot \frac{9}{2}t = \frac{2}{9} \cdot \left(-\frac{17}{2}\right) \qquad \text{Multiplying both sides by } \tfrac{2}{9}$$

$$1t = -\frac{2 \cdot 17}{9 \cdot 2} \qquad \text{Removing a factor equal to 1: } \tfrac{2}{2} = 1$$

$$t = -\frac{17}{9}.$$

Check:

$$5 + \frac{9}{2}t = -\frac{7}{2}$$

$$5 + \frac{9}{2}\left(-\frac{17}{\cancel{9}}\right) \; ? \; -\frac{7}{2} \qquad \text{Removing a factor equal to 1: } \tfrac{9}{9} = 1$$

$$5 + \left(-\frac{17}{2}\right)$$

$$\frac{10}{2} + \left(\frac{-17}{2}\right)$$

$$\frac{10 - 17}{2}$$

$$\frac{-7}{2} \quad \bigg| \quad -\frac{7}{2} \quad \text{TRUE}$$

The solution is $-\frac{17}{9}$.

Do Exercises 3 and 4.

Solve.

**3.** $3 + \dfrac{14}{5}t = -\dfrac{21}{5}$

**4.** $2x + 4 = \dfrac{1}{2}$

Sometimes the variable appears on the right side of the equation. The strategy for solving the equation remains the same.

**Answers**

1. $\dfrac{10}{9}$    2. $\dfrac{9}{5}$    3. $-\dfrac{18}{7}$    4. $-\dfrac{7}{4}$

**EXAMPLE 3** Solve: $20 = 6 - \frac{2}{3}x$.

Our plan is to first use the addition principle to isolate $-\frac{2}{3}x$ and then use the multiplication principle to isolate $x$.

$$20 = 6 - \frac{2}{3}x$$

$$20 - 6 = 6 - \frac{2}{3}x - 6 \qquad \text{Subtracting 6 (or adding } -6) \text{ on both sides}$$

$$14 = -\frac{2}{3}x$$

$$\left(-\frac{3}{2}\right)14 = \left(-\frac{3}{2}\right)\left(-\frac{2}{3}x\right) \qquad \text{Multiplying both sides by } -\frac{3}{2}$$

$$-\frac{3 \cdot 14}{2} = 1x$$

$$-\frac{3 \cdot 7 \cdot 2}{2} = 1x \qquad \text{Removing a factor equal to 1: } \frac{2}{2} = 1$$

$$-21 = x$$

Check: 
$$\begin{array}{c|c} 20 = 6 - \dfrac{2}{3}x \\ \hline 20 \ ? \ 6 - \dfrac{2}{3}(-21) \\ 6 + \dfrac{42}{3} \\ 20 \ \bigg| \ 6 + 14 \qquad \text{TRUE} \end{array}$$

The solution is $-21$.

**5.** Solve: $9 - \frac{3}{4}x = 21$.

Do Exercise 5.

## **b** Clearing Fractions

We now look at an alternative approach for solving Examples 1–3. Key to this approach is using the multiplication principle in the *first* step to produce an equivalent equation that is "cleared of fractions."

To "clear fractions," we identify the LCM of the denominators and use the multiplication principle. Because the LCM is a common multiple of the denominators, when both sides of the equation are multiplied by the LCM, the resulting terms can all be simplified. An equivalent equation can then be written without using fractions. We demonstrate this approach by solving Examples 1 and 2 by clearing fractions.

-------------------------------------- *Caution!* --------------------------------

We can "clear fractions" in equations, not in expressions. Do not multiply to clear fractions when simplifying an expression.

-------------------------------------------------------------------------------

Either of the methods discussed in this section can be used to solve equations that contain fractions, but it is important for students planning to continue in algebra to thoroughly understand *both* methods.

*Answer*

**5.** $-16$

**EXAMPLE 4**  Solve Example 1 by clearing fractions:

$$\frac{3}{4}x - \frac{1}{8} = \frac{1}{2}.$$

The LCM of the denominators is 8, so we begin by multiplying both sides of the equation by 8:

$$\frac{3}{4}x - \frac{1}{8} = \frac{1}{2}$$

$$8\left(\frac{3}{4}x - \frac{1}{8}\right) = 8 \cdot \frac{1}{2} \qquad \text{Multiplying both sides by 8. We use parentheses when we are multiplying more than one term.}$$

---

*Caution!*

$$\frac{8 \cdot 3}{4}x - 8 \cdot \frac{1}{8} = \frac{8}{2} \qquad$$ Use the distributive law carefully! Here we multiply every term inside the parentheses by 8.

---

$$\frac{4 \cdot 2 \cdot 3}{4}x - 1 = 4 \qquad \text{Factoring and simplifying}$$

$$6x - 1 = 4 \qquad \text{The equation is now cleared of fractions. This is the advantage of using this method.}$$

$$6x - 1 + 1 = 4 + 1 \qquad \text{Adding 1 to both sides}$$

$$6x = 5$$

$$\frac{6x}{6} = \frac{5}{6} \qquad \text{Dividing both sides by 6 (or multiplying both sides by } \frac{1}{6})$$

$$x = \frac{5}{6}. \qquad \text{Simplifying}$$

Since $\frac{5}{6}$ was the solution in Example 1, we have a check. The solution is $\frac{5}{6}$.

**EXAMPLE 5**  Solve Example 2 by clearing fractions:

$$5 + \frac{9}{2}t = -\frac{7}{2}.$$

The LCM of the denominators is 2, so we begin by multiplying both sides of the equation by 2:

$$2\left(5 + \frac{9}{2}t\right) = 2\left(-\frac{7}{2}\right) \qquad \text{Using the multiplication principle}$$

$$2 \cdot 5 + \frac{2 \cdot 9}{2}t = -\frac{2 \cdot 7}{2} \qquad \text{Using the distributive law; multiplying every term by 2}$$

$$10 + 9t = -7 \qquad \text{Simplifying and removing a factor equal to 1: } \frac{2}{2} = 1. \text{ The equation is now cleared of fractions.}$$

$$10 + 9t - 10 = -7 - 10 \qquad \text{Subtracting 10 from both sides}$$

$$9t = -17 \qquad \text{Simplifying}$$

$$\frac{9t}{9} = -\frac{17}{9} \qquad \text{Dividing both sides by 9 (or multiplying both sides by } \frac{1}{9})$$

$$t = -\frac{17}{9}. \qquad \text{Simplifying}$$

Since the solution in Example 2 is also $-\frac{17}{9}$, we have a check. The solution is $-\frac{17}{9}$.

Do Exercises 6 and 7.

**6.** Solve Example 3 by clearing fractions:

$$20 = 6 - \frac{2}{3}x.$$

**7.** Solve Margin Exercise 1 by clearing fractions:

$$\frac{3}{5}t - \frac{8}{15} = \frac{2}{15}.$$

*Answers*

**6.** $-21$   **7.** $\frac{10}{9}$

**a** Solve using the addition principle and/or the multiplication principle. Don't forget to check!

**1.** $6x - 3 = 15$

**2.** $7x - 6 = 22$

**3.** $5x + 7 = 10$

**4.** $19 = 2x + 4$

**5.** $8 = 3x + 11$

**6.** $2a + 9 = -7$

**7.** $\frac{2}{3}y - 8 = 1$

**8.** $\frac{3}{10}y - 7 = 3$

**9.** $\frac{3}{2}t - \frac{1}{4} = \frac{1}{2}$

**10.** $\frac{1}{4}t + \frac{1}{8} = \frac{1}{2}$

**11.** $\frac{1}{5}x + \frac{3}{10} = \frac{3}{5}$

**12.** $\frac{4}{3}x - \frac{2}{15} = \frac{2}{15}$

**13.** $5 - \frac{3}{4}x = 6$

**14.** $3 - \frac{2}{5}x = 6$

**15.** $-1 + \frac{2}{5}t = -\frac{4}{5}$

**16.** $-2 + \frac{1}{6}t = -\frac{7}{4}$

**17.** $12 = 8 + \frac{7}{2}t$

**18.** $7 = 5 + \frac{3}{2}t$

**19.** $-11 = \frac{2}{3}x - 7$

**20.** $-10 = \frac{2}{5}x - 4$

**21.** $7 = a + \frac{14}{5}$

**22.** $9 = a + \frac{47}{10}$

**23.** $\frac{2}{5}t - 1 = \frac{7}{5}$

**24.** $-\frac{53}{4} = \frac{3}{2}a + 2$

**25.** $\frac{39}{8} = \frac{11}{4} - \frac{1}{2}x$

**26.** $\frac{7}{2} = \frac{13}{2} - \frac{1}{7}y$

**27.** $-\frac{13}{3}s + \frac{11}{2} = \frac{35}{4}$

**28.** $-\frac{11}{5}t + \frac{36}{5} = \frac{7}{2}$

**29.** $\frac{1}{2}x - \frac{1}{4} = \frac{1}{2}$

**30.** $\frac{1}{3}x - \frac{1}{6} = \frac{2}{3}$

**31.** $7 = \frac{4}{9}t + 5$

**32.** $5 = \frac{4}{7}t + 3$

**33.** $-3 = \frac{3}{4}t - \frac{1}{2}$

**34.** $-2 = \frac{4}{3}t - \frac{5}{6}$

**35.** $\frac{4}{3} - \frac{5}{6}x = \frac{3}{2}$

**36.** $\frac{3}{2} - \frac{5}{3}x = \frac{5}{6}$

**37.** $-\frac{3}{4} = -\frac{5}{6} - \frac{1}{2}x$

**38.** $-\frac{1}{4} = -\frac{2}{3} - \frac{1}{6}x$

**39.** $\frac{4}{3} - \frac{1}{5}t = \frac{3}{4}$

**40.** $\frac{2}{5} - \frac{3}{4}t = \frac{4}{3}$

## Skill Maintenance

Divide.    [2.5a]

**41.** $39 \div (-3)$

**42.** $56 \div (-7)$

**43.** $(-72) \div (-4)$

**44.** $(-81) \div (-3)$

Solve.    [2.3b]

**45.** Jeremy withdraws $200 from his bank's ATM (automated teller machine), makes a $90 deposit, and then withdraws another $40. How much has Jeremy's account balance changed?

**46.** Animal Instinct, a pet supply store, makes a profit of $850 on Friday and $375 on Saturday, but suffers a loss of $45 on Sunday. Find the total profit or loss for the three days.

Divide and simplify.    [3.7b]

**47.** $\frac{10}{7} \div (2m)$

**48.** $45n \div \frac{9}{4}$

## Synthesis

Solve.

**49.** ▦ $\frac{553}{2451}a - \frac{13}{57} = \frac{29}{43}$

**50.** ▦ $\frac{1081}{3599}x - \frac{17}{61} = \frac{19}{59}$

**51.** ▦ $\frac{11}{17} = \frac{13}{41} - \frac{23}{29}t$

**52.** $-\frac{a}{5} + \frac{31}{4} = \frac{16}{3}$

**53.** $\frac{47}{5} - \frac{a}{4} = \frac{44}{7}$

**54.** $\frac{49}{8} + \frac{2x}{9} = 4$

**55.** The perimeter of the figure shown is 15 cm. Solve for $x$.

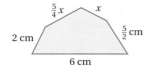

**56.** The perimeter of the figure is 15 cm. Solve for $n$.

# 4.5

# Mixed Numerals

## OBJECTIVES

**a)** Convert between mixed numerals and fraction notation.

**b)** Divide, writing the quotient as a mixed numeral.

---

**SKILL TO REVIEW**
Objective 1.5a: Divide whole numbers.

Divide.

**1.** $735 \div 16$

**2.** $2\,3\,\overline{)\,6\,0\,2\,3}$

---

Convert to a mixed numeral.

**1.** $1 + \dfrac{2}{3} = \Box\dfrac{\Box}{\Box}$

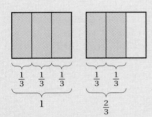

**2.** $2 + \dfrac{3}{4} = \Box\dfrac{\Box}{\Box}$

**3.** $12 + \dfrac{2}{7}$

## a Mixed Numerals

The following figure illustrates the use of a **mixed numeral**. The bolt shown is $2\frac{3}{8}$ in. long. The length is given as a whole-number part, 2, and a fractional part less than 1, $\frac{3}{8}$. We can represent the measurement as $\frac{19}{8}$, but $2\frac{3}{8}$ makes the length easier to visualize and is thus more descriptive.

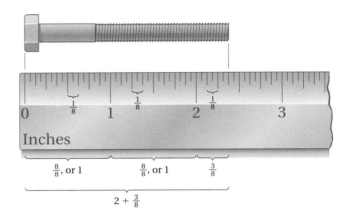

A mixed numeral $2\frac{3}{8}$ represents a sum:

$$2\frac{3}{8} \quad \text{means} \quad 2 + \frac{3}{8}.$$

This is a whole number.     This is a fraction less than 1.

**EXAMPLES**   Convert to a mixed numeral.

**1.** $7 + \dfrac{2}{5} = 7\dfrac{2}{5}$         **2.** $4 + \dfrac{3}{10} = 4\dfrac{3}{10}$

> Do Margin Exercises 1–3.

   The notation $2\frac{3}{4}$ has a plus sign left out. To aid in understanding, we sometimes write the missing plus sign: $2 + \frac{3}{4}$. Similarly, the notation $-5\frac{2}{3}$ has a minus sign left out, since $-5\frac{2}{3} = -\left(5 + \frac{2}{3}\right) = -5 - \frac{2}{3}$.

   Mixed numerals can be displayed on the number line, as shown here.

---

---

**EXAMPLES** Convert to fraction notation.

**3.** $2\dfrac{3}{4} = 2 + \dfrac{3}{4}$  Inserting the missing plus sign

$= \dfrac{2}{1} + \dfrac{3}{4}$  $2 = \dfrac{2}{1}$

$= \dfrac{2}{1} \cdot \dfrac{4}{4} + \dfrac{3}{4}$  Finding a common denominator

$= \dfrac{8}{4} + \dfrac{3}{4}$

$= \dfrac{11}{4}$  Adding

**4.** $4\dfrac{3}{10} = 4 + \dfrac{3}{10} = \dfrac{4}{1} + \dfrac{3}{10} = \dfrac{4}{1} \cdot \dfrac{10}{10} + \dfrac{3}{10} = \dfrac{40}{10} + \dfrac{3}{10} = \dfrac{43}{10}$

Do Exercises 4 and 5.

Using Example 4, we can develop a faster way to convert.

To convert from a mixed numeral like $4\dfrac{3}{10}$ to fraction notation:

(a) Multiply the whole number by the denominator: $4 \cdot 10 = 40$.
(b) Add the result to the numerator: $40 + 3 = 43$.
(c) Keep the denominator.

$$\overset{+}{\underset{\times}{4\,\tfrac{3}{10}}} = \dfrac{43}{10}$$

**EXAMPLES** Convert to fraction notation.

**5.** $6\dfrac{2}{3} = \dfrac{20}{3}$  $6 \cdot 3 = 18, 18 + 2 = 20$

**6.** $8\dfrac{2}{9} = \dfrac{74}{9}$  $8 \cdot 9 = 72, 72 + 2 = 74$

**7.** $10\dfrac{7}{8} = \dfrac{87}{8}$  $10 \cdot 8 = 80, 80 + 7 = 87$

Do Exercises 6-9.

To find the opposite of the number in Example 5, we can write either $-6\dfrac{2}{3}$ or $-\dfrac{20}{3}$. Thus, to convert a negative mixed numeral to fraction notation, we remove the negative sign for purposes of computation and then include it in the answer.

**EXAMPLES** Convert to fraction notation.

**8.** $-5\dfrac{1}{3} = -\left(5 + \dfrac{1}{3}\right) = -\dfrac{16}{3}$  $3 \cdot 5 = 15;\ 15 + 1 = 16;$ include the negative sign

**9.** $-7\dfrac{5}{6} = -\left(7 + \dfrac{5}{6}\right) = -\dfrac{47}{6}$  $6 \cdot 7 = 42;\ 42 + 5 = 47$

Do Exercises 10 and 11.

Convert to fraction notation.

**4.** $4\dfrac{2}{5}$  **5.** $6\dfrac{1}{10}$

Convert to fraction notation. Use the faster method.

**6.** $4\dfrac{5}{6}$  **7.** $9\dfrac{1}{4}$

**8.** $20\dfrac{2}{3}$  **9.** $1\dfrac{9}{13}$

Convert to fraction notation.

**10.** $-6\dfrac{2}{5}$  **11.** $-7\dfrac{2}{9}$

*Answers*

**4.** $\dfrac{22}{5}$  **5.** $\dfrac{61}{10}$  **6.** $\dfrac{29}{6}$  **7.** $\dfrac{37}{4}$
**8.** $\dfrac{62}{3}$  **9.** $\dfrac{22}{13}$  **10.** $-\dfrac{32}{5}$  **11.** $-\dfrac{65}{9}$

## Writing Mixed Numerals

We can find a mixed numeral for $\frac{5}{3}$ as follows:

$$\frac{5}{3} = \frac{3}{3} + \frac{2}{3} = 1 + \frac{2}{3} = 1\frac{2}{3}.$$

In terms of objects, we can think of $\frac{5}{3}$ as $\frac{3}{3}$, or 1, plus $\frac{2}{3}$, as shown below.

$$\frac{5}{3} = \qquad \frac{3}{3}, \text{ or } 1 \qquad + \qquad \frac{2}{3}$$

Fraction symbols like $\frac{5}{3}$ also indicate division; $\frac{5}{3}$ means $5 \div 3$. Let's divide the numerator by the denominator.

$$3\overline{)5} \quad \begin{array}{r} 1 \\ \hline 3 \\ \hline 2 \end{array} \leftarrow 2 \div 3 = \frac{2}{3}$$

Thus, $\frac{5}{3} = 1\frac{2}{3}$.

> To convert from fraction notation to a mixed numeral, divide.
>
> $$\frac{13}{5} \qquad 5\overline{)13} \begin{array}{r} 2 \\ \hline 10 \\ \hline 3 \end{array} \qquad 2\frac{3}{5}$$
>
> The divisor — The quotient — The remainder

**EXAMPLES** Convert to a mixed numeral.

**10.** $\dfrac{69}{10}$ $\qquad 10\overline{)69}\begin{array}{r} 6 \\ \hline 60 \\ \hline 9 \end{array}$ $\qquad \dfrac{69}{10} = 6\dfrac{9}{10}$

**11.** $\dfrac{122}{8}$ $\qquad 8\overline{)122}\begin{array}{r} 15 \\ \hline 8 \\ \hline 42 \\ 40 \\ \hline 2 \end{array}$ $\qquad \dfrac{122}{8} = 15\dfrac{2}{8} = 15\dfrac{1}{4}$

> Simplify the fraction part of a mixed numeral, if possible.

Do Exercises 12–15.

A fraction larger than 1, such as $\frac{27}{8}$, is sometimes referred to as an "improper" fraction. However, the use of notation such as $\frac{27}{8}$, $\frac{11}{9}$, and $\frac{89}{10}$ is quite "proper" and very common in algebra.

Convert to a mixed numeral.

**12.** $\dfrac{7}{3}$

**13.** $\dfrac{11}{10}$

**14.** $\dfrac{110}{6}$

**15.** $\dfrac{231}{18}$

*Answers*

**12.** $2\frac{1}{3}$    **13.** $1\frac{1}{10}$    **14.** $18\frac{1}{3}$    **15.** $12\frac{5}{6}$

The same procedure also works with negative numbers. Of course, the result will be a negative mixed numeral.

**EXAMPLE 12**  Convert $\dfrac{-9}{4}$ to a mixed numeral.

Since $\quad 4\overline{)9},\quad$ we have $\quad \dfrac{9}{4} = 2\dfrac{1}{4}.\quad$ Thus, $\quad \dfrac{-9}{4} = -2\dfrac{1}{4}.$

$$4\overline{)9}\atop{\dfrac{8}{1}}$$ with quotient $2$

Do Exercises 16 and 17.

Do Exercises 16 and 17.

**b** **Finding Quotients and Averages**

It is quite common when performing long division to express the quotient as a mixed numeral. As in Examples 10–12, the remainder becomes the numerator of the fraction part of the mixed numeral.

**EXAMPLE 13**  Divide. Write a mixed numeral for the quotient.

$$7\,\overline{)\,6\ 3\ 4\ 1}$$

We first divide as usual.

$$\begin{array}{r} 9\ 0\ 5 \\ 7\,\overline{)\,6\ 3\ 4\ 1} \\ \underline{6\ 3\ 0\ 0} \\ 4\ 1 \\ \underline{3\ 5} \\ 6 \end{array} \qquad \dfrac{6341}{7} = 905\dfrac{6}{7}$$

The answer is 905 R 6, or $905\dfrac{6}{7}$. Using fraction notation, we write $\dfrac{6341}{7} = 905\dfrac{6}{7}$.

Do Exercises 18 and 19.

Do Exercises 18 and 19.

**EXAMPLE 14**  *Nutrition.*  Each of the following five items from fast food restaurants contains fewer than 10 g of fat. This list gives the total fat, in grams, contained in each item. How much fat is contained, on average, in these items?

**Source:** www.lowfatlifestyle.com

| | |
|---|---|
| Burger King Chicken Tenders | 9 g |
| Wendy's Chili, small | 7 g |
| Subway's 6" Sweet Onion Chicken Teriyaki | 5 g |
| McDonald's Chef Salad | 8 g |
| Chick-fil-A's Chargrilled Chicken Cool Wrap | 7 g |

Recall from Section 1.9 that to find the *average* of a set of values, we add the values and divide that sum by the number of values being added.

$$\text{Average fat grams} = \dfrac{9\,\text{g} + 7\,\text{g} + 5\,\text{g} + 8\,\text{g} + 7\,\text{g}}{5} = \dfrac{36\,\text{g}}{5} = 7\dfrac{1}{5}\,\text{g}$$

On average, these foods contain $7\dfrac{1}{5}$ g of fat per item.

Do Exercise 20.

Do Exercise 20.

---

Convert to a mixed numeral.

**16.** $\dfrac{-12}{5}$    **17.** $-\dfrac{134}{12}$

Divide. Write a mixed numeral for the answer.

**18.** $6\,\overline{)\,4\ 8\ 4\ 6}$

**19.** $4\ 5\,\overline{)\,6\ 0\ 5\ 3}$

**20.** Over the last 4 yr, Roland Thompson's raspberry patch has yielded 48, 35, 65, and 75 qt of berries. Find the average yield for the four years.

*Answers*

**16.** $-2\dfrac{2}{5}$   **17.** $-11\dfrac{1}{6}$   **18.** $807\dfrac{2}{3}$

**19.** $134\dfrac{23}{45}$   **20.** $55\dfrac{3}{4}$ qt

**4.5** **Exercise Set**

For Extra Help

**MyMathLab**

Math XL
PRACTICE

WATCH

DOWNLOAD

READ

REVIEW

**a** Convert to fraction notation.

**1.** $7\frac{2}{3}$

**2.** $6\frac{2}{5}$

**3.** $6\frac{1}{4}$

**4.** $8\frac{1}{2}$

**5.** $-20\frac{1}{8}$

**6.** $-10\frac{1}{3}$

**7.** $5\frac{1}{10}$

**8.** $8\frac{1}{10}$

**9.** $20\frac{3}{5}$

**10.** $30\frac{4}{5}$

**11.** $-33\frac{1}{3}$

**12.** $-66\frac{2}{3}$

**13.** $1\frac{5}{8}$

**14.** $1\frac{3}{5}$

**15.** $-12\frac{3}{4}$

**16.** $-15\frac{2}{3}$

**17.** $5\frac{7}{10}$

**18.** $7\frac{3}{100}$

**19.** $-5\frac{7}{100}$

**20.** $-6\frac{4}{15}$

Convert to a mixed numeral.

**21.** $\frac{16}{3}$

**22.** $\frac{19}{8}$

**23.** $\frac{45}{6}$

**24.** $\frac{30}{9}$

**25.** $\frac{57}{10}$

**26.** $\frac{-89}{10}$

**27.** $\frac{65}{9}$

**28.** $\frac{65}{8}$

**29.** $\frac{-33}{6}$

**30.** $\frac{-50}{8}$

**31.** $\frac{46}{4}$

**32.** $\frac{39}{9}$

**33.** $\frac{-12}{8}$

**34.** $-\frac{57}{6}$

**35.** $\frac{307}{5}$

**36.** $\frac{227}{4}$

**37.** $-\frac{413}{50}$

**38.** $\frac{467}{100}$

**b** Divide. Write a mixed numeral for the answer.

**39.** $8\overline{)869}$

**40.** $3\overline{)2126}$

**41.** $7\overline{)6345}$

**42.** $9\overline{)9110}$

**43.** $21\overline{)852}$

**44.** $85\overline{)7670}$

**45.** $-302 \div 15$

**46.** $-475 \div 13$

**47.** $471 \div (-21)$

**48.** $545 \div (-25)$

*Charities.* One way to rate the efficiency of charitable organizations is to calculate how much of the donations is spent on fundraising. The table below lists some charitable organizations and how much of every $100 donated is spent on fundraising.
**Source:** www.charitynavigator.org

**49.** What is the average fundraising expense per $100 of the humanitarian organizations in the list?

| CHARITY | FUNDRAISING EXPENSES PER $100 | TYPE |
|---|---|---|
| Books for Africa | $1 | Humanitarian |
| Food for the Poor | $2 | Humanitarian |
| American Museum of Natural History | $3 | Museum |
| American Red Cross | $4 | Humanitarian |
| Boston Children's Museum | $6 | Museum |
| Philadelphia Museum of Art | $7 | Museum |
| Doctors Without Borders, USA | $11 | Humanitarian |
| Please Touch Museum | $12 | Museum |
| Oxfam America | $14 | Humanitarian |
| Salt Lake Art Center | $16 | Museum |
| Allentown Art Museum | $31 | Museum |

**50.** What is the average fundraising expense per $100 of the museums in the list?

**51.** What is the average fundraising expense per $100 of the first 5 organizations in the list?

**52.** What is the average fundraising expense per $100 of the last 5 organizations in the list?

## Skill Maintenance

Multiply and simplify. [3.6a]

**53.** $\dfrac{7}{9} \cdot \dfrac{24}{21}$

**54.** $\dfrac{5}{12} \cdot \dfrac{9}{10}$

**55.** $\dfrac{7}{10} \cdot \dfrac{5}{14}$

**56.** $\dfrac{21}{35} \cdot \dfrac{25}{12}$

**57.** $-\dfrac{17}{25} \cdot \dfrac{15}{34}$

**58.** $\dfrac{7}{20} \cdot \left(-\dfrac{45}{49}\right)$

## Synthesis

Write a mixed numeral for each number or sum listed.

**59.**  $\dfrac{128{,}236}{541}$

**60.** 🖩 $\dfrac{103{,}676}{349}$

**61.** $\dfrac{56}{7} + \dfrac{2}{3}$

**62.** $\dfrac{72}{12} + \dfrac{5}{6}$

**63.** $\dfrac{12}{5} + \dfrac{19}{15}$

**64.** There are $\frac{366}{7}$ weeks in a leap year.

**65.** There are $\frac{365}{7}$ weeks in a year.

**66.** *Athletics.* At a track and field meet, the hammer that is thrown has a wire length ranging from 3 ft $10\frac{1}{4}$ in. to 3 ft $11\frac{3}{4}$ in., a $4\frac{1}{8}$-in. grip, and a 16-lb ball with a diameter of $4\frac{3}{8}$ in. to $5\frac{1}{8}$ in. Give specifications for the wire length and diameter of an "average" hammer.

# Mid-Chapter Review

## Concept Reinforcement

Determine whether each statement is true or false.

_____ **1.** If $\dfrac{a}{b} > \dfrac{c}{b}, b \neq 0$, then $a > c$.   [4.2c]

_____ **2.** The mixed numeral $1\frac{1}{2}$ and the fraction $\frac{6}{4}$ represent the same number.   [4.5a]

_____ **3.** The least common multiple of two natural numbers is the smallest number that is a factor of both.   [4.1a]

_____ **4.** To add fractions when denominators are the same, we keep the numerator and add the denominators.   [4.2a]

## Guided Solutions

Fill in each blank with the number that creates a correct solution.

**5.** Subtract: $\dfrac{11}{42} - \dfrac{3}{35}$.   [4.3a]

$$\dfrac{11}{42} - \dfrac{3}{35} = \dfrac{11}{2 \cdot \Box \cdot 7} - \dfrac{3}{\Box \cdot 7} \qquad \text{Factoring the denominators}$$

$$= \dfrac{11}{2 \cdot \Box \cdot 7} \cdot \left(\dfrac{\Box}{\Box}\right) - \dfrac{3}{\Box \cdot 7} \cdot \left(\dfrac{\Box \cdot \Box}{\Box \cdot \Box}\right) \qquad \text{Multiplying by 1 to get the LCD}$$

$$= \dfrac{11 \cdot \Box}{2 \cdot 3 \cdot 7 \cdot \Box} - \dfrac{3 \cdot \Box \cdot \Box}{5 \cdot 7 \cdot \Box \cdot \Box} \qquad \text{Multiplying}$$

$$= \dfrac{\Box}{2 \cdot 3 \cdot 5 \cdot 7} - \dfrac{\Box}{2 \cdot 3 \cdot 5 \cdot 7} \qquad \text{Simplifying}$$

$$= \dfrac{\Box - \Box}{2 \cdot 3 \cdot 5 \cdot 7} = \dfrac{\Box}{\Box} \qquad \text{Subtracting and simplifying}$$

**6.** Solve: $x + \dfrac{1}{8} = \dfrac{2}{3}$.   [4.3b]

$$x + \dfrac{1}{8} = \dfrac{2}{3}$$

$$x + \dfrac{1}{8} - \Box = \dfrac{2}{3} - \Box \qquad \text{Subtracting on both sides}$$

$$x + \Box = \dfrac{2}{3} \cdot \dfrac{\Box}{\Box} - \dfrac{1}{8} \cdot \dfrac{\Box}{\Box} \qquad \text{Multiplying by 1 to get the LCD}$$

$$x = \dfrac{\Box}{\Box} - \dfrac{\Box}{\Box} \qquad \text{Simplifying and multiplying}$$

$$x = \dfrac{\Box}{\Box} \qquad \text{Subtracting}$$

The solution is $\dfrac{\Box}{\Box}$.

## Mixed Review

7. Match each set of numbers in the first column with its least common multiple in the second column by drawing connecting lines.    [4.1a]

| | |
|---|---|
| 45 and 50 | 120 |
| 50 and 80 | 720 |
| 30 and 24 | 400 |
| 18, 24, and 80 | 450 |
| 30, 45, and 50 | |

Calculate and simplify.    [4.2b], [4.3a]

8. $\dfrac{1}{5} + \dfrac{7}{45}$

9. $\dfrac{5}{6} + \dfrac{2}{3} + \dfrac{7}{12}$

10. $\dfrac{2}{9} - \dfrac{1}{6}$

11. $\dfrac{1}{15} - \dfrac{5}{18}$

12. $\dfrac{19}{48} - \dfrac{11}{30}$

13. $-\dfrac{3}{8}x + \dfrac{1}{12}x$

14. $\dfrac{-3}{40} + \dfrac{-5}{24}$

15. $\dfrac{8}{65} - \dfrac{2}{35}$

Solve.

16. Miguel jogs for $\frac{4}{5}$ mi, rests, and then jogs for another $\frac{2}{3}$ mi. How far does he jog in all?    [4.2d]

17. One weekend, Kirby spent $\frac{39}{5}$ hr playing two iPod games—Brain Challenge: Cerebral Burn and Scrabble: Go for a Triple Word Score. She spent $\frac{11}{4}$ hr playing Scrabble. How many hours did she spend playing Brain Challenge?    [4.3c]

18. Arrange in order from smallest to largest: $\dfrac{4}{9}, \dfrac{3}{10}, \dfrac{2}{7}$, and $\dfrac{1}{5}$.    [4.2c]

19. Solve: $\dfrac{2}{5} + x = \dfrac{9}{16}$.    [4.3b]

20. Solve: $\dfrac{3}{4}x + 1 = \dfrac{1}{3}$.    [4.4a], [4.4b]

21. Divide: $15\overline{)263}$. Write a mixed numeral for the answer.    [4.5b]

22. Fraction notation for $9\frac{3}{8}$ is which of the following?    [4.5a]

    **A.** $\dfrac{27}{8}$    **B.** $\dfrac{93}{8}$    **C.** $\dfrac{75}{8}$    **D.** $\dfrac{80}{3}$

23. Mixed numeral notation for $-\frac{39}{4}$ is which of the following?    [4.5a]

    **A.** $-35\frac{1}{4}$    **B.** $-\dfrac{4}{39}$    **C.** $-9\frac{3}{4}$    **D.** $-36\frac{3}{4}$

## Understanding Through Discussion and Writing

24. Is the LCM of two numbers always larger than either number? Why or why not?    [4.1a]

25. Explain the role of multiplication when adding using fraction notation with different denominators.    [4.2b]

26. A student made the following error:

    $$\dfrac{8}{5} - \dfrac{8}{2} = \dfrac{8}{3}.$$

    Find at least two ways to convince him of the mistake.    [4.3a]

27. Are the numbers $2\frac{1}{3}$ and $2 \cdot \frac{1}{3}$ equal? Why or why not?    [4.5a]

# 4.6

## Addition and Subtraction of Mixed Numerals; Applications

### SKILL TO REVIEW

Objective 3.5b: Simplify fraction notation.

Simplify.

**1.** $\dfrac{18}{32}$     **2.** $\dfrac{78}{117}$

Add.

**1.**     $\begin{array}{r} 4 \\ + 5\frac{1}{3} \\ \hline \end{array}$     **2.**     $\begin{array}{r} 2\frac{3}{10} \\ + 7\frac{1}{10} \\ \hline \end{array}$

**3.** Add.

$\begin{array}{r} 8\frac{4}{5} \\ + 3\frac{7}{10} \\ \hline \end{array}$

**4.** Add.

$\begin{array}{r} 5\frac{3}{4} \\ + 8\frac{5}{6} \\ \hline \end{array}$

### **a** Addition Using Mixed Numerals

To find the sum $1\frac{5}{8} + 3\frac{1}{8}$, we first add the fractions. Then we add the whole numbers and, if possible, simplify the fraction part.

$$1\frac{5}{8} = \qquad 1\frac{5}{8}$$
$$+\ 3\frac{1}{8} = \qquad +\ 3\frac{1}{8}$$
$$\overline{\phantom{+\ 3}\frac{6}{8}} \qquad \overline{4\frac{6}{8}} = 4\frac{3}{4}$$

↑ Add the fractions.  ↑ Add the whole numbers.  └── Simplifying: $\frac{6}{8} = \frac{3}{4}$

Do Margin Exercises 1 and 2.

The fraction part of a mixed numeral should always be less than 1.

**EXAMPLE 1**   Add: $5\frac{2}{3} + 3\frac{5}{6}$. Write a mixed numeral for the answer.

The LCD is 6.

$$5\frac{2}{3} \cdot \frac{2}{2} = \quad 5\frac{4}{6}$$
$$+\ 3\frac{5}{6} \quad = +\ 3\frac{5}{6}$$
$$\overline{\qquad\qquad 8\frac{9}{6}} = 8 + \frac{9}{6}$$
$$= 8 + 1\frac{1}{2}$$
$$= 9\frac{1}{2}$$

To find a mixed numeral for $\frac{9}{6}$, we divide:

$$\begin{array}{r} 1 \\ 6\overline{)9} \\ 6 \\ \hline 3 \end{array} \qquad \frac{9}{6} = 1\frac{3}{6} = 1\frac{1}{2}$$

$\frac{19}{2}$ is also a correct answer, but it is not a mixed numeral.

Do Exercise 3.

**EXAMPLE 2**   Add: $10\frac{5}{6} + 7\frac{3}{8}$.

The LCD is 24.

$$10\frac{5}{6} \cdot \frac{4}{4} = \quad 10\frac{20}{24}$$
$$+\ 7\frac{3}{8} \cdot \frac{3}{3} = +\ 7\frac{9}{24}$$
$$\overline{\qquad\qquad 17\frac{29}{24}} = 17 + \frac{29}{24}$$
$$= 17 + 1\frac{5}{24} \qquad \text{Writing } \frac{29}{24} \text{ as a mixed numeral, } 1\frac{5}{24}$$
$$= 18\frac{5}{24}$$

Do Exercise 4.

**Answers**

Skill to Review:

**1.** $\dfrac{9}{16}$   **2.** $\dfrac{2}{3}$

Margin Exercises:

**1.** $9\frac{1}{3}$   **2.** $9\frac{2}{5}$   **3.** $12\frac{1}{2}$   **4.** $14\frac{7}{12}$

## b Subtraction Using Mixed Numerals

**EXAMPLE 3**  Subtract: $7\frac{3}{4} - 2\frac{1}{4}$.

$$\begin{array}{r} 7\frac{3}{4} = \\ -\ 2\frac{1}{4} = \\ \hline \frac{2}{4} \end{array}$$

$$\begin{array}{r} 7\frac{3}{4} \\ -\ 2\frac{1}{4} \\ \hline 5\frac{2}{4} = 5\frac{1}{2} \end{array} \leftarrow \text{Simplifying: } \frac{2}{4} = \frac{1}{2}$$

Subtract the fractions.    Subtract the whole numbers.

**EXAMPLE 4**  Subtract: $9\frac{4}{5} - 3\frac{1}{2}$.

The LCD is 10.

$$\begin{array}{r} 9\frac{4}{5} \cdot \frac{2}{2} = \ \ 9\frac{8}{10} \\ -\ 3\frac{1}{2} \cdot \frac{5}{5} = -\ 3\frac{5}{10} \\ \hline 6\frac{3}{10} \end{array}$$

Subtract.

**5.**  $10\frac{7}{8}$
$-\ 9\frac{3}{8}$

**6.**  $8\frac{2}{3}$
$-\ 5\frac{1}{2}$

Do Exercises 5 and 6.

**EXAMPLE 5**  Subtract: $13 - 9\frac{3}{8}$.

$$\begin{array}{r} 13 \ = \ \ 12\frac{8}{8} \\ -\ 9\frac{3}{8} = -\ 9\frac{3}{8} \\ \hline 3\frac{5}{8} \end{array}$$

We "borrow" 1, or $\frac{8}{8}$, from 13:
$13 = 12 + 1 = 12 + \frac{8}{8} = 12\frac{8}{8}$.

**EXAMPLE 6**  Subtract: $7\frac{1}{15} - 2\frac{1}{6}$.

The LCD is 30.

$$\left.\begin{array}{r} 7\frac{1}{15} \cdot \frac{2}{2} = \ \ 7\frac{2}{30} \\ -\ 2\frac{1}{6} \cdot \frac{5}{5} = -\ 2\frac{5}{30} \end{array}\right\}$$

To subtract $\frac{5}{30}$ from $\frac{2}{30}$,
we borrow 1, or $\frac{30}{30}$, from 7:
$7\frac{2}{30} = 6 + 1 + \frac{2}{30} = 6 + \frac{30}{30} + \frac{2}{30} = 6\frac{32}{30}$.

We can write this as

$$\begin{array}{r} 7\frac{2}{30} = \ \ 6\frac{32}{30} \\ -\ 2\frac{5}{30} = -\ 2\frac{5}{30} \\ \hline 4\frac{27}{30} = 4\frac{9}{10}. \end{array} \quad \text{Simplifying: } \frac{27}{30} = \frac{9}{10}$$

Subtract.

**7.**  $5$
$-\ 1\frac{1}{3}$

**8.**  $8\frac{1}{9}$
$-\ 4\frac{5}{6}$

Do Exercises 7 and 8.

***Answers***

**5.** $1\frac{1}{2}$   **6.** $3\frac{1}{6}$   **7.** $3\frac{2}{3}$   **8.** $3\frac{5}{18}$

To combine like terms, we use the distributive law and add or subtract.

**EXAMPLE 7**   Combine like terms:  **(a)** $9\frac{3}{4}x - 4\frac{1}{2}x$;  **(b)** $4\frac{5}{6}t + 2\frac{7}{9}t$.

**a)** $9\frac{3}{4}x - 4\frac{1}{2}x = \left(9\frac{3}{4} - 4\frac{1}{2}\right)x$     Using the distributive law; this is often done mentally.

$$= \left(9\frac{3}{4} - 4\frac{2}{4}\right)x \qquad \text{The LCD is 4.}$$

$$= 5\frac{1}{4}x \qquad \text{Subtracting}$$

**b)** $4\frac{5}{6}t + 2\frac{7}{9}t = \left(4\frac{5}{6} + 2\frac{7}{9}\right)t$     This step is often performed mentally.

$$= \left(4\frac{15}{18} + 2\frac{14}{18}\right)t \qquad \text{The LCD is 18.}$$

$$= 6\frac{29}{18}t = 7\frac{11}{18}t$$

> Do Exercises 9–11.

Do Exercises 9–11.

## (c) Applications and Problem Solving

**EXAMPLE 8**  *Widening a Driveway.*  Sherry and Woody are widening their existing $17\frac{1}{4}$-ft driveway by adding $5\frac{9}{10}$ ft on one side. What is the width of the new driveway?

1. **Familiarize.**  We let $w$ = the width of the new driveway, in feet.

2. **Translate.**  We translate as follows.

$$\underbrace{\text{Existing width}} + \underbrace{\text{additional width}} = \underbrace{\text{new width}}$$

$$17\frac{1}{4} \quad + \quad 5\frac{9}{10} \quad = \quad w.$$

3. **Solve.**  The translation tells us what to do. We add. The LCD is 20.

$$17\frac{1}{4} = \quad 17\frac{1}{4} \cdot \frac{5}{5} = \quad 17\frac{5}{20}$$

$$+ \; 5\frac{9}{10} = + \; 5\frac{9}{10} \cdot \frac{2}{2} = + \; 5\frac{18}{20}$$

$$\overline{\qquad\qquad\qquad\qquad 22\frac{23}{20} = 23\frac{3}{20}}$$

Thus, $w = 23\frac{3}{20}$.

---

**Combine like terms.**

**9.** $7\frac{1}{12}t + 1\frac{5}{12}t$

**10.** $7\frac{11}{12}x - 5\frac{2}{3}x$

**11.** $5\frac{11}{15}x + 8\frac{3}{10}x$

---

*Answers*

**9.** $8\frac{1}{2}t$    **10.** $2\frac{1}{4}x$    **11.** $14\frac{1}{30}x$

**4. Check.** We check by repeating the calculation. We also note that the answer is larger than either of the widths, which means that the answer is reasonable.

**5. State.** The width of the new driveway is $23\frac{3}{20}$ ft.

Do Exercise 12.

**EXAMPLE 9** *Masonry.* Baytown Village Stone Creations is making a custom stone bench as shown below. The recommended height for the bench is 18 in. The thickness of the stone bench is $3\frac{3}{8}$ in. Each of the two supporting legs is made up of three stacked stones. Two of the stones measure $3\frac{1}{2}$ in. and $5\frac{1}{4}$ in. How much must the third stone measure?

18 in.

$3\frac{3}{8}$ in.

$3\frac{1}{2}$ in.

$5\frac{1}{4}$ in.

?

**1. Familiarize.** Let $x$ = the unknown measurement. It is part of a height totaling 18 in.

**2. Translate.** We rephrase and translate as follows.

*Rephrase:*  The total height  is  the sum of four measurements

*Translate:*  18  =  $3\frac{3}{8} + 3\frac{1}{2} + 5\frac{1}{4} + x.$

**3. Solve.** We solve for $x$.

$$18 = 3\frac{3}{8} + 3\frac{1}{2} + 5\frac{1}{4} + x$$

$$18 = 3\frac{3}{8} + 3\frac{4}{8} + 5\frac{2}{8} + x \qquad \text{Writing the fractions with a common denominator, 8}$$

$$18 = 11\frac{9}{8} + x \qquad \text{Adding: } 3 + 3 + 5 = 11 \text{ and } \frac{3}{8} + \frac{4}{8} + \frac{2}{8} = \frac{9}{8}$$

$$18 = 12\frac{1}{8} + x \qquad 11\frac{9}{8} = 12\frac{1}{8}$$

$$18 - 12\frac{1}{8} = 12\frac{1}{8} + x - 12\frac{1}{8} \qquad \text{Subtracting } 12\frac{1}{8} \text{ from both sides}$$

$$5\frac{7}{8} = x \qquad 18 - 12\frac{1}{8} = 17\frac{8}{8} - 12\frac{1}{8} = 5\frac{7}{8}$$

**4. Check.** We check by adding the heights of the four stones:

$$3\frac{3}{8} + 3\frac{1}{2} + 5\frac{1}{4} + 5\frac{7}{8} = 3\frac{3}{8} + 3\frac{4}{8} + 5\frac{2}{8} + 5\frac{7}{8} = 16\frac{16}{8} = 16 + 2 = 18.$$

**5. State.** The third stone measures $5\frac{7}{8}$ in.

Do Exercise 13.

**12. Travel Distance.** On a two-day business trip, Paul drove $213\frac{7}{10}$ mi the first day and $107\frac{5}{8}$ mi the second day. What was the total distance that Paul drove?

**13.** There are $20\frac{1}{3}$ gal of water in a rainbarrel; $5\frac{3}{4}$ gal are poured out and $8\frac{2}{3}$ gal are returned after a heavy rainfall. How many gallons of water are then in the barrel?

*Answers*

**12.** $321\frac{13}{40}$ mi  **13.** $23\frac{1}{4}$ gal

## d Negative Mixed Numerals

Consider the numbers $5\frac{3}{4}$ and $-5\frac{3}{4}$ on the number line.

$$-5\tfrac{3}{4} \qquad\qquad\qquad\qquad 5\tfrac{3}{4}$$

$$\xleftarrow{\hspace{1em}} \;{-7}\;{-6}\;{-5}\;{-4}\;{-3}\;{-2}\;{-1}\;\;0\;\;1\;\;2\;\;3\;\;4\;\;5\;\;6\;\;7\; \xrightarrow{\hspace{1em}}$$

Note that just as $5\frac{3}{4}$ means $5 + \frac{3}{4}$, we can regard $-5\frac{3}{4}$ as $-5 - \frac{3}{4}$.

Consider the subtraction $4 - 4\frac{1}{2}$. We know that if we have \$4 and make a \$$4\frac{1}{2}$ purchase, we will owe half a dollar. Thus,

$$4 - 4\frac{1}{2} = -\frac{1}{2}.$$

The correct answer, $-\frac{1}{2}$, can be obtained by rewriting the subtraction as addition (see Section 2.3):

$$4 - 4\frac{1}{2} = 4 + \left(-4\frac{1}{2}\right).$$

Because $-4\frac{1}{2}$ has the greater absolute value, the answer will be negative. The difference in absolute value is $4\frac{1}{2} - 4 = \frac{1}{2}$, so

$$4 - 4\frac{1}{2} = -\frac{1}{2}.$$

Another way to see this is to convert to fraction notation and subtract:

$$4 - 4\frac{1}{2} = \frac{8}{2} - \frac{9}{2} = \frac{8}{2} + \left(-\frac{9}{2}\right) = -\frac{1}{2}.$$

**EXAMPLE 10**  Subtract: $3\frac{2}{7} - 4\frac{2}{5}$.

Since $4\frac{2}{5}$ is greater than $3\frac{2}{7}$, the answer will be negative. We can also see this by rewriting the subtraction as $3\frac{2}{7} + \left(-4\frac{2}{5}\right)$. The difference in absolute values is

$$
\begin{array}{rcccc}
4\dfrac{2}{5} = & 4\;\dfrac{2}{5}\cdot\dfrac{7}{7} & = & 4\dfrac{14}{35}\\[2mm]
-\,3\dfrac{2}{7} = & -\,3\;\dfrac{2}{7}\cdot\dfrac{5}{5} & = & -\,3\dfrac{10}{35}\\[2mm]
\hline
& & & 1\dfrac{4}{35}.
\end{array}
$$

> Because $-4\frac{2}{5}$ has the larger absolute value, we make the answer negative.

Thus, $3\frac{2}{7} - 4\frac{2}{5} = -1\frac{4}{35}.$

Do Exercises 14–16.

Subtract.

**14.** $7 - 7\frac{3}{4}$

**15.** $5\frac{1}{2} - 9\frac{3}{4}$

**16.** $4\frac{2}{3} - 7\frac{1}{6}$

**EXAMPLE 11**  Subtract: $-6\frac{4}{5} - \left(-9\frac{3}{10}\right)$.

We write the subtraction as addition:

$$-6\frac{4}{5} - \left(-9\frac{3}{10}\right) = -6\frac{4}{5} + 9\frac{3}{10}.$$  Instead of subtracting, we add the opposite.

Since $9\frac{3}{10}$ has the greater absolute value, the answer will be positive. The difference in absolute values is

$$
\begin{array}{llll}
9\dfrac{3}{10} = & 9\dfrac{3}{10} & = & 9\dfrac{3}{10} = & 8\dfrac{13}{10} \\[2mm]
-6\dfrac{4}{5} = & -6\dfrac{4}{5}\cdot\dfrac{2}{2} & = & -6\dfrac{8}{10} = & -6\dfrac{8}{10} \\[2mm]
& & & & 2\dfrac{5}{10} = 2\dfrac{1}{2}.
\end{array}
$$

Thus, $-6\frac{4}{5} - \left(-9\frac{3}{10}\right) = 2\frac{1}{2}$.

We can check by converting to fraction notation and redoing the calculation:

$$-6\frac{4}{5} - \left(-9\frac{3}{10}\right) = -\frac{34}{5} - \left(-\frac{93}{10}\right)$$

$$= -\frac{34}{5} + \frac{93}{10}$$  Adding the opposite

$$= -\frac{68}{10} + \frac{93}{10}$$  Writing with a common denominator

$$= \frac{25}{10} = \frac{5}{2} = 2\frac{1}{2}.$$  $-68 + 93 = 25$

Do Exercises 17 and 18.

In Section 2.2, we saw that to add two negative numbers we add absolute values and make the answer negative.

**EXAMPLE 12**  Add: $-4\frac{1}{6} + \left(-5\frac{2}{9}\right)$.

We add the absolute values and make the answer negative.

$$
\begin{array}{llll}
4\dfrac{1}{6} = & 4\dfrac{1}{6}\cdot\dfrac{3}{3} & = & 4\dfrac{3}{18} & \text{Adding absolute values}\\[2mm]
+5\dfrac{2}{9} = & +5\dfrac{2}{9}\cdot\dfrac{2}{2} & = & +5\dfrac{4}{18} \\[2mm]
& & & 9\dfrac{7}{18}
\end{array}
$$

Thus, $-4\frac{1}{6} + \left(-5\frac{2}{9}\right) = -9\frac{7}{18}$.

Do Exercise 19.

Subtract.

**17.** $-7\frac{1}{3} - \left(-5\frac{1}{2}\right)$

**18.** $-4\frac{1}{10} - \left(-7\frac{2}{5}\right)$

**19.** Add: $-7\frac{1}{10} + \left(-6\frac{2}{15}\right)$.

**Answers**

**17.** $-1\frac{5}{6}$   **18.** $3\frac{3}{10}$   **19.** $-13\frac{7}{30}$

**a** , **b**    Perform the indicated operation. Write a mixed numeral for each answer.

**1.**   $\begin{array}{r} 6 \\ + 5\frac{2}{5} \\ \hline \end{array}$

**2.**   $\begin{array}{r} 3 \\ + 6\frac{5}{7} \\ \hline \end{array}$

**3.**   $\begin{array}{r} 2\frac{7}{8} \\ + 6\frac{5}{8} \\ \hline \end{array}$

**4.**   $\begin{array}{r} 2\frac{5}{6} \\ + 5\frac{5}{6} \\ \hline \end{array}$

**5.**   $\begin{array}{r} 4\frac{1}{4} \\ + 1\frac{1}{12} \\ \hline \end{array}$

**6.**   $\begin{array}{r} 4\frac{1}{12} \\ + 5\frac{1}{6} \\ \hline \end{array}$

**7.**   $\begin{array}{r} 7\frac{3}{4} \\ + 5\frac{5}{6} \\ \hline \end{array}$

**8.**   $\begin{array}{r} 4\frac{3}{8} \\ + 6\frac{5}{12} \\ \hline \end{array}$

**9.**   $\begin{array}{r} 3\frac{2}{5} \\ + 8\frac{7}{10} \\ \hline \end{array}$

**10.**   $\begin{array}{r} 5\frac{1}{2} \\ + 3\frac{7}{10} \\ \hline \end{array}$

**11.**   $\begin{array}{r} 6\frac{3}{8} \\ + 10\frac{5}{6} \\ \hline \end{array}$

**12.**   $\begin{array}{r} \frac{5}{8} \\ + 1\frac{5}{6} \\ \hline \end{array}$

**13.**   $\begin{array}{r} 18\frac{4}{5} \\ + 2\frac{7}{10} \\ \hline \end{array}$

**14.**   $\begin{array}{r} 15\frac{5}{8} \\ + 11\frac{3}{4} \\ \hline \end{array}$

**15.**   $\begin{array}{r} 14\frac{5}{8} \\ + 13\frac{1}{4} \\ \hline \end{array}$

**16.**   $\begin{array}{r} 16\frac{1}{4} \\ + 15\frac{7}{8} \\ \hline \end{array}$

**17.**   $\begin{array}{r} 8\frac{9}{10} \\ - 1\frac{7}{10} \\ \hline \end{array}$

**18.**   $\begin{array}{r} 6\frac{7}{8} \\ - 5\frac{3}{8} \\ \hline \end{array}$

**19.**   $\begin{array}{r} 9\frac{3}{5} \\ - 3\frac{1}{2} \\ \hline \end{array}$

**20.**   $\begin{array}{r} 8\frac{2}{3} \\ - 7\frac{1}{2} \\ \hline \end{array}$

**21.**   $\begin{array}{r} 4\frac{1}{5} \\ - 2\frac{3}{5} \\ \hline \end{array}$

**22.**   $\begin{array}{r} 5\frac{1}{8} \\ - 2\frac{3}{8} \\ \hline \end{array}$

**23.**   $\begin{array}{r} 19 \\ - 5\frac{3}{4} \\ \hline \end{array}$

**24.**   $\begin{array}{r} 17 \\ - 3\frac{7}{8} \\ \hline \end{array}$

**25.**   $\begin{array}{r} 34 \\ - 18\frac{5}{8} \\ \hline \end{array}$

**26.**   $\begin{array}{r} 23 \\ - 19\frac{3}{4} \\ \hline \end{array}$

**27.**   $\begin{array}{r} 21\frac{1}{6} \\ - 13\frac{3}{4} \\ \hline \end{array}$

**28.**   $\begin{array}{r} 42\frac{1}{10} \\ - 23\frac{7}{12} \\ \hline \end{array}$

**29.**   $\begin{array}{r} 25\frac{1}{9} \\ - 13\frac{5}{6} \\ \hline \end{array}$

**30.**   $\begin{array}{r} 23\frac{5}{16} \\ - 14\frac{7}{12} \\ \hline \end{array}$

Combine like terms.

**31.** $1\frac{1}{8}t + 7\frac{5}{8}t$

**32.** $6\frac{1}{4}x + 8\frac{3}{4}x$

**33.** $9\frac{1}{2}x - 7\frac{1}{2}x$

**34.** $7\frac{3}{4}x - 2\frac{1}{4}x$

**35.** $5\frac{9}{10}y + 2\frac{2}{5}y$

**36.** $9\frac{3}{4}t + 2\frac{3}{8}t$

**37.** $37\frac{5}{9}t - 25\frac{2}{3}t$

**38.** $23\frac{1}{6}t - 19\frac{2}{3}t$

**39.** $2\frac{5}{6}x + 7\frac{3}{8}x$

**40.** $7\frac{3}{20}t + 1\frac{2}{15}t$

**41.** $11a - 8\frac{2}{3}a$

**42.** $6a - 3\frac{7}{10}a$

 **Solve.**

**43.** *Plumbing.* A plumber uses pipes of lengths $10\frac{5}{16}$ in. and $8\frac{3}{4}$ in. when installing a sink. How much pipe is used?

**44.** *Writing Supplies.* The standard pencil is $6\frac{7}{8}$ in. wood and $\frac{1}{2}$ in. eraser. What is the total length of the standard pencil?

**Source:** Eberhard Faber American

**45.** For a family barbecue, Cayla bought packages of hamburger weighing $1\frac{2}{3}$ lb and $5\frac{3}{4}$ lb. What was the total weight of the meat?

**46.** Marsha's Butcher Shop sold packages of sliced turkey breast weighing $1\frac{1}{3}$ lb and $4\frac{3}{5}$ lb. What was the total weight of the meat?

**47.** *NCAA Football Goalposts.* In college football, the distance between goalposts was reduced from $23\frac{1}{3}$ ft to $18\frac{1}{2}$ ft. By how much was it reduced?

**Source:** NCAA

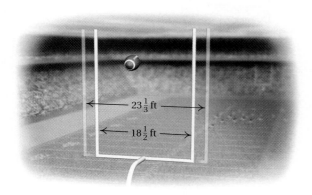

**48.** *Nail Length.* A 30d nail is $4\frac{1}{2}$ in. long. A 5d nail is $1\frac{3}{4}$ in. long. How much longer is the 30d nail than the 5d nail? (The "d" stands for "penny," which was used years ago in England to specify the number of pennies needed to buy 100 nails. Today, "penny" is used only to indicate the length of the nail.)

**Source:** Thomas J. Glover, *Pocket Ref*, 2nd ed. (Littleton, CO: Sequoia Publishing, Inc.), p. 280.

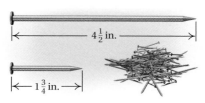

**49.** Tara is 66 in. tall and her son, Tom, is $59\frac{7}{12}$ in. tall. How much taller is Tara?

**50.** Nicholas is $73\frac{2}{3}$ in. tall and his daughter, Kendra, is $71\frac{5}{16}$ in. tall. How much shorter is Kendra?

**51.** *Interior Design.* Sue, an interior designer, worked $10\frac{1}{2}$ hr over a three-day period. If Sue worked $2\frac{1}{2}$ hr on the first day and $4\frac{1}{5}$ hr on the second, how many hours did she work on the third day?

**52.** *Painting.* Geri had $3\frac{1}{2}$ gal of paint. It took $2\frac{3}{4}$ gal to paint the family room. She estimated that it would take $2\frac{1}{4}$ gal to paint the living room. How much more paint was needed?

**53.** Creative Glass sells a framed beveled mirror as shown below. Its dimensions are $30\frac{1}{2}$ in. wide by $36\frac{5}{8}$ in. high. What is the perimeter of (total distance around) the framed mirror?

$36\frac{5}{8}$ in.

$30\frac{1}{2}$ in.

**54.** One standard book size is $8\frac{1}{2}$ in. by $9\frac{3}{4}$ in. What is the perimeter of (total distance around) the front cover of such a book?

$8\frac{1}{2}$ in.

$9\frac{3}{4}$ in.

**55.** *Planting Flowers.* A landscaper planted $4\frac{1}{2}$ flats of impatiens, $6\frac{2}{3}$ flats of snapdragons, and $3\frac{3}{8}$ flats of phlox. How many flats did she plant altogether?

**56.** *Carpentry.* When cutting wood with a saw, a carpenter must take into account the thickness of the saw blade. Suppose that from a piece of wood 36 in. long, a carpenter cuts a $15\frac{3}{4}$-in. length with a saw blade that is $\frac{1}{8}$ in. thick. How long is the piece that remains?

Find the perimeter of (distance around) each figure.

**57.**

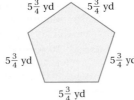

$5\frac{3}{4}$ yd  $5\frac{3}{4}$ yd

$5\frac{3}{4}$ yd  $5\frac{3}{4}$ yd

$5\frac{3}{4}$ yd

**58.**

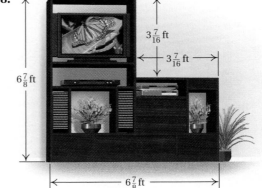

$3\frac{7}{16}$ ft

$3\frac{7}{16}$ ft

$6\frac{7}{8}$ ft

$6\frac{7}{8}$ ft

**59.**

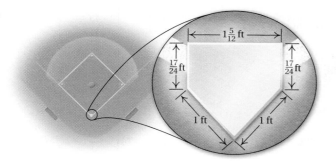

**60.**

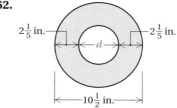

Find the length $d$ in each figure.

**61.**

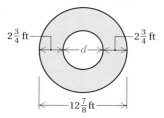

**62.**

$2\frac{1}{5}$ in. —— $2\frac{1}{5}$ in.

$d$

$10\frac{1}{2}$ in.

**63.** *Sewing from a Pattern.* Regan plans to make a dress in size 8. Using 45-in. fabric, she needs $1\frac{3}{8}$ yd for the dress, $\frac{5}{8}$ yd of contrasting fabric for the band at the bottom, and $3\frac{3}{8}$ yd for the jacket. How many yards in all of 45-in. fabric are needed to make the outfit?

**64.** *Sewing from a Pattern.* Gloria wants to make an outfit in size 12. Using 45-in. fabric, she needs $2\frac{3}{4}$ yd for the dress and $3\frac{1}{2}$ yd for the jacket. How many yards in all of 45-in. fabric are needed to make the outfit?

**65.** *Window Dimensions.* The Sanchez family is replacing a window on their second floor. The original window measures $4\frac{5}{6}$ ft $\times$ $8\frac{1}{4}$ ft. The new window is $2\frac{1}{3}$ ft wider. What are the dimensions of the new window?

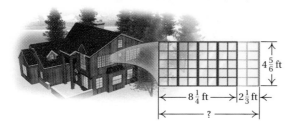

**66.** Find the smallest length of a bolt, not including the head, that will pass through a piece of tubing with an outside diameter of $\frac{1}{2}$ in., a washer $\frac{1}{16}$ in. thick, a piece of tubing with a $\frac{3}{4}$-in. outside diameter, another washer, and a nut $\frac{3}{16}$ in. thick.

**d** Perform the indicated operation. Write a mixed numeral for each answer.

**67.** $9 - 9\frac{2}{5}$

**68.** $8 - 8\frac{3}{7}$

**69.** $3\frac{1}{2} - 6\frac{3}{4}$

**70.** $5\frac{1}{2} - 7\frac{3}{4}$

**71.** $3\frac{4}{5} + \left(-7\frac{2}{3}\right)$

**72.** $2\frac{3}{7} + \left(-5\frac{1}{2}\right)$

**73.** $-3\frac{1}{5} - 4\frac{2}{5}$

**74.** $-5\frac{3}{8} - 4\frac{1}{8}$

**75.** $-4\frac{1}{12} + 6\frac{2}{3}$

**76.** $-2\frac{3}{4} + 5\frac{3}{8}$

**77.** $-6\frac{1}{9} - \left(-4\frac{2}{9}\right)$

**78.** $-2\frac{3}{5} - \left(-1\frac{1}{5}\right)$

## Skill Maintenance

Solve.

**79.** Rick's Market sells Swiss cheese in $\frac{3}{4}$-lb packages. How many packages can be made from a 12-lb slab of cheese? [3.8b]

**80.** Holstein's Dairy produced 4578 oz of milk one morning. How many 16-oz cartons could be filled? How much milk was left over? [1.8a]

Determine whether the first number is divisible by the second. [3.1a], [3.1b]

**81.** 9993 by 3

**82.** 9993 by 9

**83.** 2345 by 9

**84.** 2345 by 5

**85.** 2335 by 10

**86.** 7764 by 6

**87.** 18,888 by 2

**88.** 18,888 by 6

**89.** Multiply and simplify: $\frac{15}{9} \cdot \frac{18}{39}$. [3.6a]

**90.** Divide and simplify: $\frac{12}{25} \div \frac{24}{5}$. [3.7b]

## Synthesis

Calculate each of the following. Write the result as a mixed numeral.

**91.** ▦ $3289\frac{1047}{1189} + 5278\frac{32}{41}$

**92.** ▦ $4230\frac{19}{73} - 5848\frac{17}{29}$

Solve.

**93.** $35\frac{2}{3} + n = 46\frac{1}{4}$

**94.** $42\frac{7}{9} = x - 13\frac{2}{5}$

**95.** $-15\frac{7}{8} = 12\frac{1}{2} + t$

**96.** A post for a pier is 29 ft long. Half of the post extends above the water's surface and $8\frac{3}{4}$ ft of the post is buried in mud. How deep is the water at that location?

**97.** An algebra text is $1\frac{1}{8}$ in. thick, $9\frac{3}{4}$ in. long, and $8\frac{1}{2}$ in. wide. If the front, back, and spine of the book were unfolded, they would form a rectangle. What would the perimeter of that rectangle be?

# 4.7

# Multiplication and Division of Mixed Numerals; Applications

Carrying out addition and subtraction with mixed numerals is usually easier if the numbers are left as mixed numerals. With multiplication and division, however, it is easier to convert the numbers to fraction notation first. If the result is a fraction with the numerator greater than the denominator, the answer should be written as a mixed numeral.

## a Multiplication

> **MULTIPLICATION USING MIXED NUMERALS**
>
> To multiply using mixed numerals, first convert to fraction notation and multiply. Then convert the answer to a mixed numeral, if appropriate.

**EXAMPLE 1** Multiply: $6 \cdot 2\frac{1}{2}$.

$$6 \cdot 2\frac{1}{2} = \frac{6}{1} \cdot \frac{5}{2} = \frac{6 \cdot 5}{1 \cdot 2} = \frac{2 \cdot 3 \cdot 5}{2 \cdot 1} = 15$$

Removing a factor equal to 1: $\frac{2}{2} = 1$

Here we write fraction notation.

**EXAMPLE 2** Multiply: $3\frac{1}{2} \cdot \frac{3}{4}$.

$$3\frac{1}{2} \cdot \frac{3}{4} = \frac{7}{2} \cdot \frac{3}{4} = \frac{21}{8} = 2\frac{5}{8}$$

Recall that common denominators are *not* required when multiplying fractions.

> Do Margin Exercises 1 and 2.

**EXAMPLE 3** Multiply: $-10 \cdot 5\frac{2}{3}$.

$$-10 \cdot 5\frac{2}{3} = -\frac{10}{1} \cdot \frac{17}{3} = -\frac{170}{3} = -56\frac{2}{3}$$

**EXAMPLE 4** Multiply: $2\frac{1}{4} \cdot 5\frac{2}{3}$.

$$2\frac{1}{4} \cdot 5\frac{2}{3} = \frac{9}{4} \cdot \frac{17}{3} = \frac{9 \cdot 17}{4 \cdot 3} = \frac{3 \cdot 3 \cdot 17}{2 \cdot 2 \cdot 3} = \frac{51}{4} = 12\frac{3}{4}$$

-------------------------------- *Caution!* --------------------------------

$2\frac{1}{4} \cdot 5\frac{2}{3} \neq 10\frac{2}{12}$. A common error is to multiply the whole numbers and then the fractions. The correct answer, $12\frac{3}{4}$, is found only after converting to fraction notation.

> Do Exercises 3 and 4.

## OBJECTIVES

**a** Multiply using mixed numerals.

**b** Divide using mixed numerals.

**c** Evaluate expressions using mixed numerals.

**d** Solve applied problems involving multiplication and division with mixed numerals.

**SKILL TO REVIEW**
Objective 3.7b: Divide and simplify using fraction notation.

Divide and simplify.

**1.** $85 \div \frac{17}{5}$    **2.** $\frac{7}{65} \div \frac{21}{25}$

Multiply.

**1.** $8 \cdot 3\frac{1}{2}$    **2.** $5\frac{1}{2} \cdot \frac{3}{7}$

Multiply.

**3.** $-2 \cdot 6\frac{2}{5}$    **4.** $3\frac{1}{3} \cdot 2\frac{1}{2}$

*Answers*
*Skill to Review:*
**1.** 25    **2.** $\frac{5}{39}$

*Margin Exercises:*
**1.** 28    **2.** $2\frac{5}{14}$    **3.** $-12\frac{4}{5}$    **4.** $8\frac{1}{3}$

## b Division

The division $1\frac{1}{2} \div \frac{1}{6}$ is shown here. This division means "How many $\frac{1}{6}$'s are in $1\frac{1}{2}$?" We see that the answer is 9.

$\frac{1}{6}$ goes into $1\frac{1}{2}$ nine times.

When we divide using mixed numerals, we convert to fraction notation first.

$$1\frac{1}{2} \div \frac{1}{6} = \frac{3}{2} \div \frac{1}{6} \qquad \text{We convert } 1\frac{1}{2} \text{ to fraction notation.}$$

$$= \frac{3}{2} \cdot \frac{6}{1} \qquad \text{To divide by } \frac{1}{6}, \text{ multiply by the reciprocal, } \frac{6}{1}.$$

$$= \frac{3 \cdot 6}{2 \cdot 1} = \frac{3 \cdot 3 \cdot 2}{2 \cdot 1} = \frac{3 \cdot 3}{1} \cdot \frac{2}{2} = \frac{3 \cdot 3}{1} \cdot 1 = 9$$

> **DIVISION USING MIXED NUMERALS**
>
> To divide using mixed numerals, first write fraction notation and divide. Then convert the answer to a mixed numeral, if appropriate.

**EXAMPLE 5** Divide: $32 \div 3\frac{1}{5}$.

$$32 \div 3\frac{1}{5} = \frac{32}{1} \div \frac{16}{5} \qquad \text{Converting to fraction notation}$$

$$= \frac{32}{1} \cdot \frac{5}{16} = \frac{32 \cdot 5}{1 \cdot 16} = \frac{2 \cdot 16 \cdot 5}{1 \cdot 16} = 10 \qquad \begin{array}{l}\text{Removing a factor} \\ \text{equal to 1: } \frac{16}{16} = 1\end{array}$$

$\uparrow$ Remember to multiply by the reciprocal of the divisor.

-------------------------------------------------- *Caution!* --------------------------------------------------

The reciprocal of $3\frac{1}{5}$ is neither $5\frac{1}{3}$ nor $3\frac{5}{1}$!

-----------------------------------------------------------------------------------------------------

**5.** Divide: $63 \div 5\frac{1}{4}$.

Do Exercise 5.

**EXAMPLE 6** Divide: $2\frac{1}{3} \div 1\frac{3}{4}$.

$$2\frac{1}{3} \div 1\frac{3}{4} = \frac{7}{3} \div \frac{7}{4} = \frac{7}{3} \cdot \frac{4}{7} = \frac{7 \cdot 4}{7 \cdot 3} = \frac{4}{3} = 1\frac{1}{3} \qquad \begin{array}{l}\text{Removing a factor} \\ \text{equal to 1: } \frac{7}{7} = 1\end{array}$$

Divide.

**6.** $2\frac{1}{4} \div 1\frac{1}{5}$

**7.** $1\frac{3}{4} \div \left(-2\frac{1}{2}\right)$

**EXAMPLE 7** Divide: $-1\frac{3}{5} \div \left(-3\frac{1}{3}\right)$.

$$-1\frac{3}{5} \div \left(-3\frac{1}{3}\right) = -\frac{8}{5} \div \left(-\frac{10}{3}\right) = \frac{8}{5} \cdot \frac{3}{10} \qquad \begin{array}{l}\text{The product or quotient of} \\ \text{two negatives is positive.}\end{array}$$

$$= \frac{2 \cdot 4 \cdot 3}{5 \cdot 2 \cdot 5} = \frac{12}{25} \qquad \begin{array}{l}\text{Removing a factor} \\ \text{equal to 1: } \frac{2}{2} = 1\end{array}$$

Do Exercises 6 and 7.

**Answers**

**5.** 12   **6.** $1\frac{7}{8}$   **7.** $-\frac{7}{10}$

## (c) Evaluating Expressions

Mixed numerals can appear in algebraic expressions.

**EXAMPLE 8**  A train traveling $r$ miles per hour for $t$ hours travels a total of $rt$ miles. (*Remember*: Distance = Rate · Time.)

a) Find the distance traveled by a 60-mph train in $2\frac{3}{4}$ hr.

b) Find the distance traveled if the speed of the train is $26\frac{1}{2}$ mph and the time is $2\frac{2}{3}$ hr.

a) We evaluate $rt$ for $r = 60$ and $t = 2\frac{3}{4}$:

$$rt = 60 \cdot 2\frac{3}{4}$$

$$= \frac{60}{1} \cdot \frac{11}{4}$$

$$= \frac{15 \cdot \cancel{4} \cdot 11}{1 \cdot \cancel{4}} = 165. \qquad \begin{array}{l}\text{Removing a factor} \\ \text{equal to 1: } \frac{4}{4} = 1\end{array}$$

In $2\frac{3}{4}$ hr, a 60-mph train travels 165 mi.

b) We evaluate $rt$ for $r = 26\frac{1}{2}$ and $t = 2\frac{2}{3}$:

$$rt = 26\frac{1}{2} \cdot 2\frac{2}{3}$$

$$= \frac{53}{2} \cdot \frac{8}{3} = \frac{53 \cdot 2 \cdot 4}{2 \cdot 3} \qquad \begin{array}{l}\text{Removing a factor} \\ \text{equal to 1: } \frac{2}{2} = 1\end{array}$$

$$= \frac{212}{3} = 70\frac{2}{3}.$$

In $2\frac{2}{3}$ hr, a $26\frac{1}{2}$-mph train travels $70\frac{2}{3}$ mi.

**EXAMPLE 9**  Evaluate $x + yz$ for $x = 7\frac{1}{3}$, $y = \frac{1}{3}$, and $z = 5$.

We substitute and follow the rules for order of operations:

$$x + yz = 7\frac{1}{3} + \frac{1}{3} \cdot 5$$

$$= 7\frac{1}{3} + \frac{1}{3} \cdot \frac{5}{1} \qquad \text{Multiply first; then add.}$$

$$= 7\frac{1}{3} + \frac{5}{3}$$

$$\left.\begin{array}{l}= 7\frac{1}{3} + 1\frac{2}{3} \\ \\ = 8\frac{3}{3} = 9.\end{array}\right\} \quad \text{Adding mixed numerals}$$

Do Exercises 8–10.

Evaluate.

**8.** $rt$, for $r = 78$ and $t = 2\frac{1}{4}$

**9.** $7xy$, for $x = 9\frac{2}{5}$ and $y = 2\frac{3}{7}$

**10.** $x - y \div z$, for $x = 5\frac{7}{8}$, $y = \frac{1}{4}$, and $z = 2$

**Answers**

**8.** $175\frac{1}{2}$  **9.** $159\frac{4}{5}$  **10.** $5\frac{3}{4}$

## (d) Applications and Problem Solving

**EXAMPLE 10** *Average Speed in Indianapolis 500.* Arie Luyendyk won the Indianapolis 500 in 1990 with the highest average speed of about 186 mph. This record high is about $2\frac{12}{25}$ times the average speed of the first winner, Ray Harroun, in 1911. What was the average speed of the winner in the first Indianapolis 500?

Source: Indianapolis Motor Speedway

1. **Familiarize.** We ask the question "186 is $2\frac{12}{25}$ times what number?" We let $s$ = the average speed in 1911. Then the average speed in 1990 was $2\frac{12}{25} \cdot s$.

2. **Translate.** The problem can be translated to an equation as follows:

$$\underbrace{\text{Average speed in 1990}}_{\downarrow} \text{ is } 2\frac{12}{25} \text{ times } \underbrace{\text{average speed in 1911}}_{\downarrow}$$

$$186 = 2\frac{12}{25} \cdot s.$$

3. **Solve.** To solve the equation, we multiply on both sides.

$$186 = \frac{62}{25} \cdot s \qquad \text{Converting } 2\frac{12}{25} \text{ to fraction notation}$$

$$\frac{25}{62} \cdot 186 = \frac{25}{62} \cdot \frac{62}{25} s \qquad \text{Using the multiplication principle}$$

$$\frac{25 \cdot 186}{62} = 1 \cdot s \qquad \text{Multiplying}$$

$$\frac{25 \cdot 3 \cdot \cancel{62}}{\cancel{62} \cdot 1} = 1 \cdot s \qquad \begin{array}{l}\text{Since 62 is in the denominator, we check to}\\\text{see if it is a factor of 186. Since it is, we can}\\\text{remove a factor equal to 1: } \frac{62}{62} = 1.\end{array}$$

$$\frac{25 \cdot 3}{1} = 1 \cdot s$$

$$75 = s \qquad \text{Simplifying}$$

4. **Check.** If the average speed in 1911 was about 75 mph, we find the average speed in 1990 by multiplying 75 by $2\frac{12}{25}$:

$$2\frac{12}{25} \cdot 75 = \frac{62}{25} \cdot 75 = \frac{62 \cdot 75}{25 \cdot 1} = \frac{62 \cdot \cancel{25} \cdot 3}{\cancel{25} \cdot 1} = 62 \cdot 3 = 186.$$

The answer checks.

5. **State.** The average speed of the winner in the first Indianapolis 500 was about 75 mph.

Do Exercises 11 and 12.

Solve.

11. Kyle's pickup truck travels on an interstate highway at 65 mph for $3\frac{1}{2}$ hr. How far does it travel?

12. Holly's minivan traveled 302 mi on $15\frac{1}{10}$ gal of gas. How many miles per gallon did it get?

*Answers*

11. $227\frac{1}{2}$ mi   12. 20 mpg

**EXAMPLE 11** *Koi Pond.* Colleen designed a koi fish pond for her backyard. Using the dimensions shown in the diagram below, determine the area of Colleen's backyard before the pond was constructed and the area of the yard remaining after the pond was completed.

Sources: en.wikipedia.org; pondliner.com

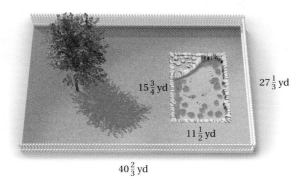

1. **Familiarize.** From the diagram, we know that the dimensions of the backyard are $40\frac{2}{3}$ yd by $27\frac{1}{3}$ yd and the dimensions of the pond are $15\frac{3}{4}$ yd by $11\frac{1}{2}$ yd. We let $B =$ the area of the yard before the construction, $P =$ the area of the pond, and $R =$ the area of the yard that remains after the pond is complete.

2. **Translate.** This is a multistep problem. To find $R$, the area of the yard remaining, we first find $B$, the area of the yard before construction, and $P$, the area of the pond. Then $R$ is the area of the yard $B$ minus the area of the pond $P$:

$$R = B - P.$$

We find each area using the formula for the area of a rectangle, $A = l \times w$.

3. **Solve.** We carry out the calculations as follows:

$$B = \text{length} \times \text{width}$$
$$= 40\frac{2}{3} \cdot 27\frac{1}{3}$$
$$= \frac{122}{3} \cdot \frac{82}{3}$$
$$= \frac{10{,}004}{9} = 1111\frac{5}{9} \text{ sq yd};$$

$$P = \text{length} \times \text{width}$$
$$= 15\frac{3}{4} \cdot 11\frac{1}{2}$$
$$= \frac{63}{4} \cdot \frac{23}{2}$$
$$= \frac{1449}{8} = 181\frac{1}{8} \text{ sq yd.}$$

Then

$$R = B - P$$
$$= 1111\frac{5}{9} - 181\frac{1}{8}$$
$$= 1111\frac{40}{72} - 181\frac{9}{72}$$
$$= 930\frac{31}{72} \text{ sq yd.}$$

**13.** A room measures $22\frac{1}{2}$ ft by $15\frac{1}{2}$ ft. A 9-ft by 12-ft rug is placed in the center of the room. How much floor area is not covered by the rug?

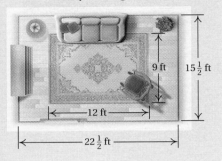

9 ft   12 ft   $15\frac{1}{2}$ ft   $22\frac{1}{2}$ ft

**4. Check.**   We can perform a check by repeating the calculations or we can round the measurements to the nearest yard and estimate a solution.

$$B \approx 41 \times 27 \approx 1107;$$
$$P \approx 16 \times 12 \approx 192;$$
$$R \approx 1107 - 192 \approx 915 \text{ sq yd}$$

Since 915 sq yd is close to $930\frac{31}{72}$ sq yd, the answer seems reasonable.

**5. State.**   The area of the yard before construction of the pond was $1111\frac{5}{9}$ sq yd. The area of the yard remaining after the pond was completed was $930\frac{31}{72}$ sq yd.

Do Exercise 13.

---

## Calculator Corner

**Operations on Fractions and Mixed Numerals**   Fraction calculators can add, subtract, multiply, and divide fractions and mixed numerals. The $\boxed{a^b/c}$ key is used to enter fractions and mixed numerals. To find $\frac{3}{4} + \frac{1}{2}$, for example, we press $\boxed{3}$ $\boxed{a^b/c}$ $\boxed{4}$ $\boxed{+}$ $\boxed{1}$ $\boxed{a^b/c}$ $\boxed{2}$ $\boxed{=}$. Note that 3/4 and 1/2 appear on the display as $\boxed{3 \lrcorner 4}$ and $\boxed{1 \lrcorner 2}$, respectively. The result is given as the mixed numeral $1\frac{1}{4}$ and is displayed as $\boxed{1 \lrcorner 1 \lrcorner 4}$. Fraction results that are greater than 1 are always displayed as mixed numerals. To express this result as a fraction, we press $\boxed{\text{SHIFT}}$ $\boxed{d/c}$. We get $\boxed{5 \lrcorner 4}$, or 5/4.

To find $3\frac{2}{3} \cdot 4\frac{1}{5}$, we press $\boxed{3}$ $\boxed{a^b/c}$ $\boxed{2}$ $\boxed{a^b/c}$ $\boxed{3}$ $\boxed{\times}$ $\boxed{4}$ $\boxed{a^b/c}$ $\boxed{1}$ $\boxed{a^b/c}$ $\boxed{5}$ $\boxed{=}$. The calculator displays $\boxed{15 \lrcorner 2 \lrcorner 5}$, so the product is $15\frac{2}{5}$.

Some calculators are capable of displaying mixed numerals in the way in which we write them, as shown below.

On calculators without an $\boxed{a^b/c}$ key, mixed numerals can be entered as sums. For example, $1\frac{1}{3}$ is entered by pressing $\boxed{1}$ $\boxed{+}$ $\boxed{1}$ $\boxed{\div}$ $\boxed{3}$. When subtracting a mixed numeral written as a sum, parentheses are essential. To be safe, you can enclose every mixed numeral in parentheses. Thus, to find $6 - 4\frac{5}{8}$, we press $\boxed{6}$ $\boxed{-}$ $\boxed{(}$ $\boxed{4}$ $\boxed{+}$ $\boxed{5}$ $\boxed{\div}$ $\boxed{8}$ $\boxed{)}$ $\boxed{\text{ENTER}}$. The answer is written as a decimal number. Convert to a mixed numeral as described in Section 4.5. The difference is $1\frac{3}{8}$. Consult your owner's manual or an instructor if you need help with your calculator.

**Exercises:**   Perform each calculation. Give the answer as a mixed numeral, when appropriate.

**1.** $4\frac{1}{3} + 5\frac{4}{5}$

**2.** $9\frac{2}{7} - 8\frac{1}{4}$

**3.** $7\frac{2}{9} - 5\frac{1}{7}$

**4.** $8\frac{17}{19} - 9\frac{2}{11}$

**5.** $2\frac{1}{3} \cdot 4\frac{3}{5}$

**6.** $10\frac{7}{10} \div 3\frac{5}{6}$

**7.** $-7\frac{2}{9} \div 4\frac{1}{5}$

**8.** $-7\frac{9}{16} \cdot 3\frac{4}{7}$

---

*Answer*

**13.** $240\frac{3}{4}$ ft$^2$

# Translating for Success

**1.** *Raffle Tickets.* At the Happy Hollow Camp Fall Festival, Rico and Becca, together, spent $270 on raffle tickets that sell for $ $\frac{9}{20}$ each. How many tickets did they buy?

**2.** *Irrigation Pipe.* Jed uses two pipes, one of which measures $5\frac{1}{3}$ ft, to repair the irrigation system in the Buxtons' lawn. The total length of the two pipes is $8\frac{7}{12}$ ft. How long is the other pipe?

**3.** *Vacation Days.* Together, Helmut and Claire have 36 vacation days a year. Helmut has 22 vacation days per year. How many does Claire have?

**4.** *Enrollment in Japanese Classes.* Last year at Lakeside Community College, 225 students enrolled in basic mathematics. This number is $4\frac{1}{2}$ times as many as the number who enrolled in Japanese. How many enrolled in Japanese?

**5.** *Bicycling.* Cole rode his bicycle $5\frac{1}{3}$ mi on Saturday and $8\frac{7}{12}$ mi on Sunday. How far did he ride on the weekend?

---

The goal of these matching questions is to practice step (2), *Translate*, of the five-step problem-solving process. Translate each word problem to an equation and select a correct translation from equations A–O.

**A.** $13\frac{11}{12} = x + 5\frac{1}{3}$

**B.** $\frac{3}{4} \cdot x = 1\frac{2}{3}$

**C.** $\frac{20}{9} \cdot 270 = x$

**D.** $225 = 4\frac{1}{2} \cdot x$

**E.** $98 \div 2\frac{1}{3} = x$

**F.** $22 + x = 36$

**G.** $x = 4\frac{1}{2} \cdot 225$

**H.** $x = 5\frac{1}{3} + 8\frac{7}{12}$

**I.** $22 \cdot x = 36$

**J.** $x = \frac{3}{4} \cdot 1\frac{2}{3}$

**K.** $5\frac{1}{3} + x = 8\frac{7}{12}$

**L.** $\frac{9}{20} \cdot 270 = x$

**M.** $1\frac{2}{3} + \frac{3}{4} = x$

**N.** $98 - 2\frac{1}{3} = x$

**O.** $\frac{9}{20} \cdot x = 270$

*Answers on page A-8*

---

**6.** *Deli Order.* For a promotional open house for contractors last year, the Bayside Builders Association ordered 225 turkey sandwiches. Because of increased registrations this year, $4\frac{1}{2}$ times as many sandwiches are needed. How many sandwiches are ordered?

**7.** *Dog Ownership.* In Sam's community, $\frac{9}{20}$ of the households own at least one dog. There are 270 households. How many own at least one dog?

**8.** *Magic Tricks.* Samantha has 98 ft of rope and needs to cut it into $2\frac{1}{3}$-ft pieces to be used in a magic trick. How many pieces can be cut from the rope?

**9.** *Painting.* Laura needs $1\frac{2}{3}$ gal of paint to paint the ceiling of the exercise room and $\frac{3}{4}$ gal of the same paint for the bathroom. How much paint does Laura need?

**10.** *Chocolate Fudge Bars.* A recipe for chocolate fudge bars that serves 16 includes $1\frac{2}{3}$ cups of sugar. How much sugar is needed for $\frac{3}{4}$ of this recipe?

**a**   Multiply. Write a mixed numeral for each answer.

**1.** $8 \cdot 2\frac{5}{6}$

**2.** $10 \cdot 3\frac{3}{4}$

**3.** $6\frac{2}{3} \cdot \frac{1}{4}$

**4.** $-\frac{1}{3} \cdot 5\frac{2}{5}$

**5.** $20\left(-2\frac{5}{6}\right)$

**6.** $6\frac{3}{8} \cdot 4\frac{1}{3}$

**7.** $3\frac{1}{2} \cdot 4\frac{2}{3}$

**8.** $4\frac{1}{5} \cdot 5\frac{1}{4}$

**9.** $-2\frac{3}{10} \cdot 4\frac{2}{5}$

**10.** $4\frac{7}{10} \cdot 5\frac{3}{10}$

**11.** $\left(-6\frac{3}{10}\right)\left(-5\frac{7}{10}\right)$

**12.** $-20\frac{1}{2} \cdot \left(-10\frac{1}{5}\right)$

**b**   Divide. Write a mixed numeral for each answer whenever possible.

**13.** $20 \div 3\frac{1}{5}$

**14.** $18 \div 2\frac{1}{4}$

**15.** $8\frac{2}{5} \div 7$

**16.** $3\frac{3}{8} \div 3$

**17.** $6\frac{1}{4} \div 3\frac{3}{4}$

**18.** $5\frac{4}{5} \div 2\frac{1}{2}$

**19.** $-1\frac{7}{8} \div 1\frac{2}{3}$

**20.** $-4\frac{3}{8} \div 2\frac{5}{6}$

**21.** $5\frac{1}{10} \div 4\frac{3}{10}$

**22.** $4\frac{1}{10} \div 2\frac{1}{10}$

**23.** $-20\frac{1}{4} \div (-90)$

**24.** $-12\frac{1}{2} \div (-50)$

**c**   Evaluate.

**25.** $lw$, for $l = 2\frac{3}{5}$ and $w = 9$

**26.** $mv$, for $m = 7$ and $v = 3\frac{2}{5}$

**27.** $rs$, for $r = 5$ and $s = 3\frac{1}{7}$

**28.** $rt$, for $r = 5\frac{2}{3}$ and $t = 4\frac{1}{5}$

**29.** $mt$, for $m = 6\dfrac{2}{9}$ and $t = -4\dfrac{3}{8}$

**30.** $M \div NP$, for $M = 2\dfrac{1}{4}$, $N = -5$, and $P = 2\dfrac{1}{3}$

**31.** $R \cdot S \div T$, for $R = 4\dfrac{2}{3}$, $S = 1\dfrac{3}{7}$, and $T = -5$

**32.** $a - bc$, for $a = 18$, $b = 2\dfrac{1}{5}$, and $c = 3\dfrac{3}{4}$

**33.** $r + ps$, for $r = 5\dfrac{1}{2}$, $p = 3$, and $s = 2\dfrac{1}{4}$

**34.** $s + rt$, for $s = 3\dfrac{1}{2}$, $r = 5\dfrac{1}{2}$, and $t = 7\dfrac{1}{2}$

**35.** $m + n \div p$, for $m = 7\dfrac{2}{5}$, $n = 4\dfrac{1}{2}$, and $p = 6$

**36.** $x - y \div z$, for $x = 9$, $y = 2\dfrac{1}{2}$, and $z = 3\dfrac{3}{4}$

 Solve.

**37.** *Beagles.*  There are about 155,000 Labrador retrievers registered with The American Kennel Club. This is $3\dfrac{4}{9}$ times the number of beagles registered. How many beagles are registered?

**Source:** The American Kennel Club

**38.** *Landscaping.*  Emily seeds lawns for Sam's Superior Lawn Care. When she walks at a rapid pace, the wheel on the broadcast spreader completes $150\dfrac{2}{3}$ revolutions per minute. How many revolutions does the wheel complete in 15 min?

**39.** *Sodium Consumption.*  The average American woman consumes $1\dfrac{1}{3}$ tsp of sodium each day. How much sodium does the average American woman consume in 30 days?

**Source:** *Nutrition Action Health Letter,* March 1994

**40.** *Exercise.*  At one point during a spinning class, Kea's bicycle wheel was completing $76\dfrac{2}{3}$ revolutions per minute. How many revolutions did the wheel complete in 6 min?

**41.** *Landscape Design.* A sidewalk alongside a garden at the conservatory is to be $14\frac{2}{5}$ yd long. Rectangular stone tiles that are each $1\frac{1}{8}$ yd long are used to form the sidewalk. How many tiles are used?

**42.** *Aeronautics.* Most space shuttles orbit the earth once every $1\frac{1}{2}$ hr. How many orbits are made every 24 hr?

**43.** *Population.* The population of Alabama is $6\frac{4}{5}$ times that of Alaska. The population of Alabama is approximately 4,700,000. What is the population of Alaska?
**Source:** U.S. Census Bureau

**44.** *Population.* The population of Iowa is $2\frac{1}{3}$ times the population of Hawaii. The population of Hawaii is approximately 1,300,000. What is the population of Iowa?
**Source:** U.S. Census Bureau

**45.** *Temperature.* Fahrenheit temperature can be obtained from Celsius (Centigrade) temperature by multiplying by $1\frac{4}{5}$ and adding 32°. What Fahrenheit temperature corresponds to a Celsius temperature of 20°?

**46.** *Temperature.* Fahrenheit temperature can be obtained from Celsius (Centigrade) temperature by multiplying by $1\frac{4}{5}$ and adding 32°. What Fahrenheit temperature corresponds to the Celsius temperature of boiling water, 100°?

**47.** *Newspaper Circulation.* In 2008, the daily circulation of *USA Today* was about $3\frac{1}{5}$ times the daily circulation of the *Washington Post*. The circulation of the *Washington Post* was about 665,000. What was the daily circulation of *USA Today*?
**Source:** *Editor and Publisher International*

**48.** *Earned Doctorates.* In 2006, the number of doctorates earned in English and American Languages and Literature was $1\frac{7}{12}$ times the number of doctorates earned in Foreign Languages and Literature. There were approximately 600 doctorates earned in Foreign Languages and Literature. How many doctorates were earned in English and American Languages and Literature?
**Source:** Association of Departments of Foreign Languages

**49.** *Art.* Cecilia hired an artist to paint a mural on the wall in her twin sons' bedroom. The dimensions of the mural are $6\frac{2}{3}$ ft by $9\frac{3}{8}$ ft. What is the area of the mural?

**50.** *Weight of Water.* The weight of water is $62\frac{1}{2}$ lb per cubic foot. What is the weight of $2\frac{1}{4}$ cubic feet of water?

**51.** *Population.* The population of India is about $3\frac{3}{4}$ times the population of the United States. In 2008, the population of India was approximately 1,149,000,000. What was the population of the United States in 2008?

Source: Population Division/International Programs Center, U.S. Census Bureau, U.S. Dept. of Commerce

**52.** *Population.* The population of Cleveland is about $1\frac{1}{3}$ times the population of Cincinnati. In 2007, the population of Cincinnati was approximately 332,400. What was the population of Cleveland in 2007?

Source: U.S. Census Bureau

**53.** *Doubling a Recipe.* The chef of a 5-star hotel is doubling a recipe for chocolate cake. The cake requires $2\frac{3}{4}$ cups of flour and $1\frac{1}{3}$ cups of sugar. How much flour and sugar will she need?

**54.** *Half of a Recipe.* A caterer is following a salad dressing recipe that calls for $1\frac{7}{8}$ cups of mayonnaise and $1\frac{1}{6}$ cups of sugar. How much mayonnaise and sugar will he need if he prepares $\frac{1}{2}$ of the amount of salad dressing?

**55.** *Weight of Water.* The weight of water is $62\frac{1}{2}$ lb per cubic foot. How many cubic feet would be occupied by 25,000 lb of water?

**56.** *Weight of Water.* The weight of water is $8\frac{1}{3}$ lb per gallon. Harry rolls his lawn with an 800-lb capacity roller. Express the water capacity of the roller in gallons.

**57.** *Servings of Salmon.* A serving of filleted fish is generally considered to be about $\frac{1}{3}$ lb. How many servings can be prepared from $5\frac{1}{2}$ lb of salmon fillet?

**58.** *Servings of Tuna.* A serving of fish steak (cross section) is generally $\frac{1}{2}$ lb. How many servings can be prepared from a cleaned $18\frac{3}{4}$-lb tuna?

**59.** A car traveled 213 mi on $14\frac{2}{10}$ gal of gas. How many miles per gallon did it get?

**60.** A car traveled 385 mi on $15\frac{4}{10}$ gal of gas. How many miles per gallon did it get?

**61.** *Word Processing.* Kelly wants to create a table using Microsoft® Word software. She needs to have two columns each $1\frac{1}{2}$ in. wide and five columns each $\frac{3}{4}$ in. wide. Will this table fit on a piece of standard paper that is $8\frac{1}{2}$ in. wide? If so, how wide will each margin be if her margins on each side are to be of equal width?

**62.** *Construction.* A rectangular lot has dimensions of $302\frac{1}{2}$ ft by $205\frac{1}{4}$ ft. A building with dimensions of 100 ft by $25\frac{1}{2}$ ft is built on the lot. How much area is left over?

**63.** *Landscaping.* The previous owners of Ashley's new home had a large L-shaped vegetable garden consisting of a rectangle that was $15\frac{1}{2}$ ft by 20 ft adjacent to one that was $10\frac{1}{2}$ ft by $12\frac{1}{2}$ ft. Ashley wants to cover the garden with sod. What is the total area of the sod she must purchase?

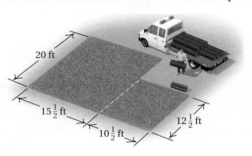

**64.** *Home Furnishings.* An L-shaped sunroom consists of a rectangle that is $9\frac{1}{2}$ ft by 12 ft adjacent to one that is $9\frac{1}{2}$ ft by 8 ft. What is the total area of a carpet that covers the floor?

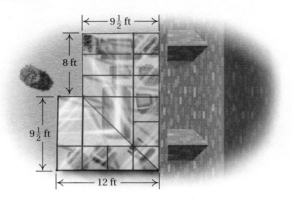

Find the area of each shaded region.

**65.**

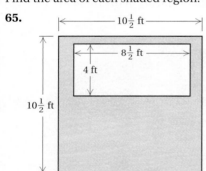

**66.**

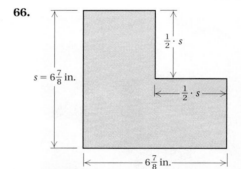

**67.**

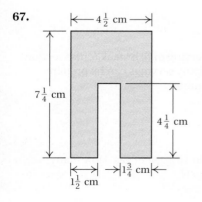

**68.**

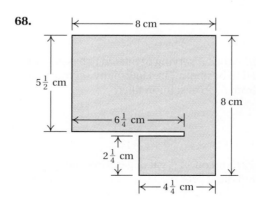

## Skill Maintenance

In each of Exercises 69–76, fill in the blank with the correct term from the given list. Some of the choices may not be used and some may be used more than once.

69. The set $\{\ldots, -3, -2, -1, 0, 1, 2, 3, \ldots\}$ is the set of
   _____ . [2.1a]

70. When denominators are the same, we say that fractions have a(n)
   _____ denominator. [4.2a]

71. The numbers 81, 95, and 248 are examples of _____ numbers. [3.2b]

72. The number 22,223,133 is _____ by 9 because the sum of its digits is _____ by 9. [3.1b]

73. To add fractions with different denominators, we must first find the _____ of the denominators. [4.2b]

74. In the equation $2 + 3 = 5$, the numbers 2 and 3 are called _____ . [1.2a]

75. In the expression $\dfrac{c}{d}$, we call $c$ the _____ . [3.3a]

76. The number 0 has no _____ . [3.7a]

identity
reciprocal
least common multiple
irrational numbers
integers
numerator
denominator
equal
common
prime
composite
product
divisible
digits
factors
addends

## Synthesis

Simplify. Write each answer as a mixed numeral whenever possible.

77. $-8 \div \dfrac{1}{2} + \dfrac{3}{4} + \left(-5 - \dfrac{5}{8}\right)^2$

78. $\left(\dfrac{5}{9} - \dfrac{1}{4}\right)(-12) + \left(-4 - \dfrac{3}{4}\right)^2$

79. $\dfrac{1}{3} \div \left(\dfrac{1}{2} - \dfrac{1}{5}\right) \times \dfrac{1}{4} + \dfrac{1}{6}$

80. $\dfrac{7}{8} - 1\dfrac{1}{8} \times \dfrac{2}{3} + \dfrac{9}{10} \div \dfrac{3}{5}$

81. Find $r$ if
$$\dfrac{1}{r} = \dfrac{1}{40} + \dfrac{1}{60} + \dfrac{1}{80}.$$

82. *Heights.* Find the average height of the following NBA players:

| | |
|---|---|
| LeBron James | 6 ft 8 in. |
| Dwayne Wade | 6 ft 4 in. |
| Dwight Howard | 6 ft 11 in. |
| Dirk Nowitzki | 7 ft 0 in. |
| Al Jefferson | 6 ft 10 in. |

83. *Water Consumption.* According to the U.S. Department of Energy, washing one load of clothes uses $2\frac{2}{3}$ times the amount of hot water required for the average shower. If the average shower uses 12 gal of hot water, how much hot water will two showers and two loads of wash require?

# 4.8

## Order of Operations and Complex Fractions

## OBJECTIVES

**a** Simplify expressions containing fraction notation using the rules for order of operations.

**b** Simplify complex fractions.

**SKILL TO REVIEW**

Objective 1.9c: Simplify expressions using the rules for order of operations.

Simplify.

**1.** $22 - 3 \cdot 4$

**2.** $6(4 + 1)^2 - 3^4 \div 3$

Simplify.

**1.** $\dfrac{2}{5} \cdot \dfrac{5}{8} + \dfrac{1}{4}$

**2.** $\dfrac{1}{3} \cdot \dfrac{3}{4} \div \dfrac{5}{8} - \dfrac{1}{10}$

**3.** Simplify: $\dfrac{3}{4} \cdot 16 + 8\dfrac{2}{3}$.

### a Order of Operations

Like expressions containing integers, expressions containing fraction notation follow the rules for order of operations.

> **RULES FOR ORDER OF OPERATIONS**
>
> 1. Do all calculations within parentheses ( ), brackets [ ], braces { }, absolute-value symbols, numerators, or denominators.
> 2. Evaluate all exponential expressions.
> 3. Do all multiplications and divisions in order from left to right.
> 4. Do all additions and subtractions in order from left to right.

**EXAMPLE 1** Simplify: $\dfrac{1}{6} + \dfrac{2}{3} \div \dfrac{1}{2} \cdot \dfrac{5}{8}$.

$$\dfrac{1}{6} + \dfrac{2}{3} \div \dfrac{1}{2} \cdot \dfrac{5}{8} = \dfrac{1}{6} + \dfrac{2}{3} \cdot \dfrac{2}{1} \cdot \dfrac{5}{8}$$

Doing the division first by multiplying by the reciprocal of $\frac{1}{2}$

$$= \dfrac{1}{6} + \dfrac{2 \cdot 2 \cdot 5}{3 \cdot 1 \cdot 8}$$

Doing the multiplications in order from left to right

$$= \dfrac{1}{6} + \dfrac{2 \cdot 2 \cdot 5}{3 \cdot 1 \cdot 2 \cdot 2 \cdot 2}$$

Factoring

$$= \dfrac{1}{6} + \dfrac{5}{6}$$

Removing a factor equal to 1: $\dfrac{2 \cdot 2}{2 \cdot 2} = 1$; simplifying

$$= \dfrac{6}{6}, \quad \text{or} \quad 1$$

Doing the addition

Do Margin Exercises 1 and 2.

**EXAMPLE 2** Simplify: $\dfrac{2}{3} \cdot 24 - 11\dfrac{1}{2}$.

$$\dfrac{2}{3} \cdot 24 - 11\dfrac{1}{2} = \dfrac{2 \cdot 24}{3 \cdot 1} - 11\dfrac{1}{2}$$

Doing the multiplication first

$$= \dfrac{2 \cdot 3 \cdot 8}{3 \cdot 1} - 11\dfrac{1}{2}$$

Factoring

$$= 2 \cdot 8 - 11\dfrac{1}{2}$$

Removing a factor equal to 1: $\dfrac{3}{3} = 1$

$$= 16 - 11\dfrac{1}{2}$$

Multiplying

$$= 15\dfrac{2}{2} - 11\dfrac{1}{2}$$

$$= 4\dfrac{1}{2}, \quad \text{or} \quad \dfrac{9}{2}$$

Subtracting

Do Exercise 3.

*Answers*

*Skill to Review:*
**1.** 10   **2.** 123

*Margin Exercises:*
**1.** $\dfrac{1}{2}$   **2.** $\dfrac{3}{10}$   **3.** $20\dfrac{2}{3}$, or $\dfrac{62}{3}$

**EXAMPLE 3** Simplify.

**a)** $-\dfrac{1}{2} - \dfrac{2}{3}\left(-\dfrac{1}{2}\right)^2$

**b)** $1 - \dfrac{2}{3}\left(\dfrac{1}{3} - \dfrac{1}{2}\right)$

**a)** The parentheses here are not grouping symbols, so we begin by evaluating the exponential expression.

$$-\frac{1}{2} - \frac{2}{3}\left(-\frac{1}{2}\right)^2 = -\frac{1}{2} - \frac{2}{3} \cdot \frac{1}{4} \qquad \left(-\frac{1}{2}\right)^2 = \left(-\frac{1}{2}\right) \cdot \left(-\frac{1}{2}\right) = \frac{1}{4}$$

$$= -\frac{1}{2} - \frac{2 \cdot 1}{3 \cdot 2 \cdot 2} \qquad \text{Multiplying and factoring}$$

$$= -\frac{1}{2} - \frac{1}{6} \qquad \text{Simplifying}$$

$$= -\frac{3}{6} - \frac{1}{6} \qquad \text{Writing with the LCD, 6, in order to subtract}$$

$$= -\frac{4}{6} = -\frac{2}{3} \qquad \text{Subtracting and simplifying}$$

**b)** We first subtract within the parentheses.

$$1 - \frac{2}{3}\left(\frac{1}{3} - \frac{1}{2}\right) = 1 - \frac{2}{3}\left(\frac{2}{6} - \frac{3}{6}\right) \qquad \text{Writing with the LCD, 6, in order to subtract}$$

$$= 1 - \frac{2}{3}\left(-\frac{1}{6}\right) \qquad \begin{array}{l}\text{Subtracting within the parentheses:} \\ \dfrac{2}{6} - \dfrac{3}{6} = \dfrac{2}{6} + \left(-\dfrac{3}{6}\right) = -\dfrac{1}{6}\end{array}$$

$$= 1 - \left(-\frac{2 \cdot 1}{3 \cdot 2 \cdot 3}\right) \qquad \begin{array}{l}\text{There are no exponential expres-} \\ \text{sions, so we multiply and factor.}\end{array}$$

$$= 1 - \left(-\frac{1}{9}\right) \qquad \text{Simplifying}$$

$$= \frac{9}{9} + \frac{1}{9} = \frac{10}{9} \qquad \begin{array}{l}\text{Writing with the LCD, 9, and} \\ \text{adding the opposite of } -\dfrac{1}{9}\end{array}$$

Do Exercises 4–6.

Simplify.

**4.** $\left(\dfrac{3}{4}\right)^2 - \dfrac{1}{2} \div \left(-\dfrac{4}{5}\right)$

**5.** $1 - \left(\dfrac{2}{3} - \dfrac{3}{4}\right)^2$

**6.** $\left(\dfrac{2}{3} + \dfrac{3}{4}\right) \div 2\dfrac{1}{3} - \left(\dfrac{1}{2}\right)^3$

## b Complex Fractions

A **complex fraction** is a fraction in which the numerator and/or denominator contain one or more fractions. The following are some examples of complex fractions.

$$\dfrac{\frac{2}{3}}{-7} \leftarrow \begin{array}{l}\text{The numerator} \\ \text{contains a fraction.}\end{array} \qquad \dfrac{-\frac{1}{5}}{-\frac{9}{10}} \begin{array}{l}\leftarrow \text{The numerator} \\ \phantom{\leftarrow} \text{contains a fraction.} \\ \leftarrow \text{The denominator} \\ \phantom{\leftarrow} \text{contains a fraction.}\end{array}$$

Since a fraction bar represents division, complex fractions can be rewritten using the division symbol $\div$.

*Answers*

**4.** $1\dfrac{3}{16}$    **5.** $\dfrac{143}{144}$    **6.** $\dfrac{27}{56}$

**EXAMPLE 4**  Simplify.

**a)** $\dfrac{\frac{2}{3}}{-7}$    **b)** $\dfrac{-\frac{1}{5}}{-\frac{9}{10}}$

**a)**

$$\dfrac{\frac{2}{3}}{-7} = \frac{2}{3} \div (-7) \quad \text{Rewriting using a division symbol}$$

$$= \frac{2}{3} \cdot \left(\frac{1}{-7}\right) \quad \text{Multiplying by the reciprocal of the divisor}$$

$$= \frac{2 \cdot 1}{3(-7)} \quad \text{Multiplying numerators and multiplying denominators}$$

$$= -\frac{2}{21} \quad \text{This expression cannot be simplified.}$$

**b)**

$$\dfrac{-\frac{1}{5}}{-\frac{9}{10}} = -\frac{1}{5} \div \left(-\frac{9}{10}\right) \quad \text{Rewriting using a division symbol}$$

$$= -\frac{1}{5} \cdot \left(-\frac{10}{9}\right) \quad \text{Multiplying by the reciprocal of the divisor; the product will be positive.}$$

$$= \frac{1 \cdot 2 \cdot \cancel{5}}{\cancel{5} \cdot 3 \cdot 3} \quad \text{Multiplying numerators and multiplying denominators; factoring}$$

$$= \frac{2}{9} \quad \text{Removing a factor equal to 1 and simplifying}$$

Do Exercises 7 and 8.

Complex fractions may contain variables.

**EXAMPLE 5**  Simplify: $\dfrac{\frac{x}{24}}{\frac{15}{28}}$.

$$\dfrac{\frac{x}{24}}{\frac{15}{28}} = \frac{x}{24} \div \frac{15}{28} \quad \text{Rewriting using a division symbol}$$

$$= \frac{x}{24} \cdot \frac{28}{15} \quad \text{Multiplying by the reciprocal of the divisor}$$

$$= \frac{x \cdot 2 \cdot 2 \cdot 7}{2 \cdot 2 \cdot 2 \cdot 3 \cdot 3 \cdot 5} \quad \text{Multiplying numerators and multiplying denominators; factoring}$$

$$= \frac{x \cdot 7}{2 \cdot 3 \cdot 3 \cdot 5} \quad \text{Removing a factor equal to 1: } \frac{2 \cdot 2}{2 \cdot 2} = 1$$

$$= \frac{7x}{90} \quad \text{Multiplying}$$

Do Exercises 9 and 10.

Simplify.

**7.** $\dfrac{10}{-\frac{5}{8}}$    **8.** $\dfrac{\frac{7}{5}}{-\frac{10}{7}}$

Simplify.

**9.** $\dfrac{-\frac{x}{20}}{\frac{3}{10}}$

**10.** $\dfrac{-\frac{7}{12}}{-\frac{x}{18}}$

*Answers*

**7.** $-16$    **8.** $-\dfrac{49}{50}$    **9.** $-\dfrac{x}{6}$    **10.** $\dfrac{21}{2x}$

When the numerator or denominator of a complex fraction consists of more than one term, first simplify the numerator and/or denominator separately.

**EXAMPLE 6** Simplify: $\dfrac{\dfrac{1}{2} - \dfrac{2}{3}}{1\dfrac{7}{8}}$.

$$\dfrac{\dfrac{1}{2} - \dfrac{2}{3}}{1\dfrac{7}{8}} = \dfrac{\dfrac{3}{6} - \dfrac{4}{6}}{\dfrac{15}{8}}$$ 
Writing the fractions in the numerator with a common denominator

Writing the mixed numeral in the denominator as a fraction

$$= \dfrac{-\dfrac{1}{6}}{\dfrac{15}{8}}$$ 
Subtracting in the numerator of the complex fraction

$$= -\dfrac{1}{6} \div \dfrac{15}{8}$$ 
Rewriting using a division symbol

$$= -\dfrac{1}{6} \cdot \dfrac{8}{15}$$ 
Multiplying by the reciprocal of the divisor

$$= -\dfrac{1 \cdot 2 \cdot 2 \cdot 2}{2 \cdot 3 \cdot 3 \cdot 5}$$ 
Multiplying numerators and multiplying denominators; factoring

$$= -\dfrac{4}{45}$$ 
Removing a factor equal to 1: $\dfrac{2}{2} = 1$

Do Exercises 11 and 12.

**EXAMPLE 7** *Harvesting Walnut Trees.* A woodland owner decided to harvest five walnut trees in order to improve the growing conditions of the remaining trees. The logs she sold measured $7\frac{5}{8}$ ft, $8\frac{1}{4}$ ft, $8\frac{3}{4}$ ft, $9\frac{1}{8}$ ft, and $10\frac{1}{2}$ ft. What is the average length of the logs?

Recall that to compute an average, we add the numbers and then divide the sum by the number of addends. We have

$$\dfrac{7\frac{5}{8} + 8\frac{1}{4} + 8\frac{3}{4} + 9\frac{1}{8} + 10\frac{1}{2}}{5} = \dfrac{7\frac{5}{8} + 8\frac{2}{8} + 8\frac{6}{8} + 9\frac{1}{8} + 10\frac{4}{8}}{5}$$

$$= \dfrac{42\frac{18}{8}}{5} = \dfrac{42\frac{9}{4}}{5} = \dfrac{\frac{177}{4}}{5}$$ 
Adding, simplifying, and converting to fraction notation

$$= \dfrac{177}{4} \div 5$$ 
Rewriting using a division symbol

$$= \dfrac{177}{4} \cdot \dfrac{1}{5}$$ 
Multiplying by the reciprocal of 5

$$= \dfrac{177}{20} = 8\dfrac{17}{20}.$$ 
Converting to a mixed numeral

The average length of the logs is $8\frac{17}{20}$ ft.

Do Exercises 13–15.

Simplify.

**11.** $\dfrac{\dfrac{7}{12} + \dfrac{5}{6}}{\dfrac{4}{9}}$

**12.** $\dfrac{-\dfrac{3}{5}}{\dfrac{2}{3} - \dfrac{7}{10}}$

**13.** Rachel has triplets. Their birth weights are $3\frac{1}{2}$ lb, $2\frac{3}{4}$ lb, and $3\frac{1}{8}$ lb. What is the average weight of her babies?

**14.** Find the average of
$$\dfrac{1}{2}, \dfrac{1}{3}, \text{ and } \dfrac{5}{6}.$$

**15.** Find the average of $\dfrac{3}{4}$ and $\dfrac{4}{5}$.

*Answers*

**11.** $3\dfrac{3}{16}$   **12.** $18$   **13.** $3\dfrac{1}{8}$ lb   **14.** $\dfrac{5}{9}$

**15.** $\dfrac{31}{40}$

**a** Simplify.

**1.** $\dfrac{1}{8} + \dfrac{1}{4} \cdot \dfrac{2}{3}$

**2.** $\dfrac{2}{5} - \dfrac{4}{5} \div \dfrac{2}{3}$

**3.** $-\dfrac{1}{6} - 3\left(-\dfrac{5}{9}\right)$

**4.** $1 + \dfrac{2}{3}\left(-\dfrac{6}{25}\right)$

**5.** $\dfrac{9}{10} - \left(\dfrac{2}{5} - \dfrac{3}{8}\right)$

**6.** $-\dfrac{5}{9} + \left(\dfrac{1}{3} - \dfrac{5}{6}\right)$

**7.** $\dfrac{5}{8} \div \dfrac{1}{4} - \dfrac{2}{3} \cdot \dfrac{4}{5}$

**8.** $\dfrac{4}{7} \cdot \dfrac{7}{15} + \dfrac{2}{3} \div 8$

**9.** $\dfrac{7}{8} \div \dfrac{1}{2} \cdot \dfrac{1}{4}$

**10.** $\dfrac{7}{10} \cdot \dfrac{4}{5} \div \dfrac{2}{3}$

**11.** $\dfrac{3}{4} - \dfrac{2}{3} \cdot \left(\dfrac{1}{2} + \dfrac{2}{5}\right)$

**12.** $\dfrac{3}{4} \div \dfrac{1}{2} \cdot \left(\dfrac{8}{9} - \dfrac{2}{3}\right)$

**13.** $\dfrac{4}{5} \div \left(\dfrac{2}{9} \cdot \dfrac{1}{2}\right) \cdot \left(-\dfrac{5}{6}\right)$

**14.** $-\dfrac{4}{9} \cdot \left(\dfrac{3}{8} \div \dfrac{1}{2}\right) \div \left(-\dfrac{2}{3}\right)$

**15.** $\left(\dfrac{2}{3}\right)^2 - \dfrac{1}{3} \cdot 1\dfrac{1}{4}$

**16.** $-1\dfrac{3}{5} - \dfrac{9}{10} + \left(-\dfrac{1}{2}\right)^2$

**17.** $-\dfrac{12}{25}\left(\dfrac{3}{4} - \dfrac{1}{2}\right)^2$

**18.** $\dfrac{2}{3}\left(\dfrac{1}{3} - \dfrac{1}{5}\right)^2$

**19.** $-\dfrac{3}{4} \div \left(\dfrac{2}{3} - \dfrac{1}{6}\right) + \dfrac{1}{2}$

**20.** $-6 \div \left(\dfrac{1}{2} - \dfrac{2}{3}\right) + \dfrac{1}{3}$

**21.** $\left(-\dfrac{3}{2}\right)^2 - 2\left(\dfrac{1}{4} - \dfrac{3}{2}\right)$

**22.** $\left(-\dfrac{1}{2}\right)^3 - \dfrac{3}{4}\left(\dfrac{1}{3} - \dfrac{1}{6}\right)$

**23.** $\dfrac{1}{2} - \left(\dfrac{1}{2}\right)^2 + \left(\dfrac{1}{2}\right)^3$

**24.** $1 + \dfrac{1}{4} + \left(\dfrac{1}{4}\right)^2 - \left(\dfrac{1}{4}\right)^3$

**25.** $\left(\dfrac{3}{5} - \dfrac{1}{2}\right) \div \left(\dfrac{3}{4} - \dfrac{3}{10}\right)$

**26.** $\left(\dfrac{2}{3} + \dfrac{3}{4}\right) \div \left(\dfrac{5}{6} - \dfrac{1}{3}\right)$

**b** Simplify.

**27.** $\dfrac{\frac{3}{8}}{\frac{11}{8}}$

**28.** $\dfrac{-\frac{1}{8}}{\frac{3}{4}}$

**29.** $\dfrac{-4}{\frac{6}{7}}$

**30.** $\dfrac{-\frac{3}{8}}{-12}$

**31.** $\dfrac{\frac{1}{40}}{-\frac{1}{50}}$

**32.** $\dfrac{\frac{7}{9}}{\frac{3}{9}}$

**33.** $\dfrac{-\frac{1}{10}}{-10}$

**34.** $\dfrac{28}{-\frac{7}{4}}$

**35.** $\dfrac{\frac{5}{18}}{-1\frac{2}{3}}$

**36.** $\dfrac{-2\frac{1}{5}}{\frac{7}{10}}$

**37.** $\dfrac{\dfrac{x}{28}}{\dfrac{5}{8}}$

**38.** $\dfrac{\dfrac{x}{15}}{\dfrac{3}{4}}$

**39.** $\dfrac{\dfrac{n}{14}}{-\dfrac{2}{3}}$

**40.** $\dfrac{-\dfrac{t}{10}}{\dfrac{4}{45}}$

**41.** $\dfrac{-\dfrac{3}{35}}{-\dfrac{x}{10}}$

**42.** $\dfrac{\dfrac{7}{40}}{-\dfrac{6}{x}}$

**43.** $\dfrac{-\dfrac{5}{8}}{\left(\dfrac{3}{2}\right)^2}$

**44.** $\dfrac{\left(-\dfrac{2}{3}\right)^2}{\left(\dfrac{9}{10}\right)^2}$

**45.** $\dfrac{\dfrac{1}{6}-\dfrac{5}{9}}{\dfrac{2}{3}}$

**46.** $\dfrac{\dfrac{7}{12}}{\dfrac{1}{4}-\dfrac{5}{8}}$

**47.** $\dfrac{\dfrac{1}{4}-\dfrac{3}{8}}{\dfrac{1}{2}-\dfrac{7}{8}}$

**48.** $\dfrac{\dfrac{1}{2}-\dfrac{3}{5}}{\dfrac{2}{5}-\dfrac{1}{2}}$

**49.** Find the average of $\dfrac{2}{3}$ and $\dfrac{7}{8}$.

**50.** Find the average of $\dfrac{1}{4}$ and $\dfrac{1}{5}$.

**51.** Find the average of $\dfrac{1}{6}, \dfrac{1}{8}$, and $\dfrac{3}{4}$.

**52.** Find the average of $\dfrac{4}{5}, \dfrac{1}{2}$, and $\dfrac{1}{10}$.

**53.** Find the average of $3\dfrac{1}{2}$ and $9\dfrac{3}{8}$.

**54.** Find the average of $10\dfrac{2}{3}$ and $24\dfrac{5}{6}$.

**55.** *Hiking the Appalachian Trail.* Ellen camped and hiked for three consecutive days along a section of the Appalachian Trail. The distances she hiked on the three days were $15\dfrac{5}{32}$ mi, $20\dfrac{3}{16}$ mi, and $12\dfrac{7}{8}$ mi. Find the average of these distances.

**56.** *Vertical Leaps.* Eight-year-old Zachary registered vertical leaps of $12\dfrac{3}{4}$ in., $13\dfrac{3}{4}$ in., $13\dfrac{1}{2}$ in., and 14 in. Find his average vertical leap.

**57.** *Black Bear Cubs.* Black bears typically have two cubs. In January 2007 in northern New Hampshire, a black bear sow gave birth to a litter of 5 cubs. This is so rare that Tom Sears, a wildlife photographer, spent 28 hr per week for six weeks watching for the perfect opportunity to photograph this family of six. At the time of this photo, an observer estimated that the cubs weighed $7\frac{1}{2}$ lb, 8 lb, $9\frac{1}{2}$ lb, $10\frac{5}{8}$ lb, and $11\frac{3}{4}$ lb. What was the average weight of the cubs?

Source: Andrew Timmins, New Hampshire Fish and Game Department, *Northcountry News*, Warren, NH; Tom Sears, photographer

**58.** *Acceleration.* The results of a road acceleration test for five cars are given in the graph below. The test measures the time in seconds required to go from 0 mph to 60 mph. What was the average time?

**Acceleration: 0 mph to 60 mph**

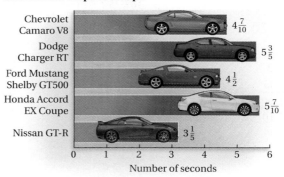

SOURCE: www.zeroto60times.com

## Skill Maintenance

Divide and simplify.  [3.7b]

**59.** $\dfrac{4}{5} \div \left(-\dfrac{3}{10}\right)$

**60.** $\left(-\dfrac{3}{10}\right) \div \left(-\dfrac{4}{5}\right)$

**61.** Classify the given numbers as prime, composite, or neither.

$$1, 5, 7, 9, 14, 23, 43 \quad [3.2b]$$

**62.** Divide: $7865 \div 132$.  [1.5a]

Solve.

**63.** *Luncheon Servings.* Ian purchased 6 lb of cold cuts for a luncheon. If he has allowed $\frac{3}{8}$ lb per person, how many people did he invite to the luncheon?  [3.8b]

**64.** *Cholesterol.* A 3-oz serving of crabmeat contains 85 milligrams (mg) of cholesterol. A 3-oz serving of shrimp contains 128 mg of cholesterol. How much more cholesterol is in the shrimp?  [1.8a]

## Synthesis

Simplify.

**65.** $\left(1\frac{1}{2} - 1\frac{1}{3}\right)^2 \cdot 144 - \dfrac{9}{10} \div 4\frac{1}{5}$

**66.** 🔲 $\left(3\frac{1}{2} - 2\frac{1}{3}\right)^3 - 30 \cdot 2\frac{1}{2} \div (-2)^5$

**67.** $\dfrac{\dfrac{2}{3}x - \dfrac{1}{2}x}{\left(1\frac{1}{4} + \dfrac{1}{2}\right)^2}$

**68.** $\dfrac{\dfrac{x}{24}}{\dfrac{x}{48}}$

Estimate each of the following as $0$, $\frac{1}{2}$, or $1$.

**69.** $\dfrac{2}{99}$

**70.** $\dfrac{19}{20}$

**71.** $\dfrac{13}{27}$

**72.** $\dfrac{101}{100}$

**73.** $\dfrac{215}{429}$

**74.** $\dfrac{1}{1000}$

# Summary and Review

## Key Terms

least common multiple, p. 224          mixed numeral, p. 256          complex fraction, p. 289
least common denominator, p. 233

## Concept Reinforcement

Determine whether each statement is true or false.

_____ **1.** The mixed numeral $5\frac{2}{3}$ can be represented by the sum $5 \cdot \frac{3}{3} + \frac{2}{3}$.   [4.5a]

_____ **2.** The least common multiple of two numbers is always larger than or equal to the larger number.   [4.1a]

_____ **3.** To clear fractions in an equation, multiply both sides by the LCM of all denominators in the equation.   [4.4b]

_____ **4.** The product of any two mixed numerals is greater than 1.   [4.7a]

## Important Concepts

**Objective 4.1a**   Find the least common multiple, or LCM, of two or more numbers.

**Example**   Find the LCM of 105 and 90.

$$105 = 3 \cdot 5 \cdot 7,$$
$$90 = 2 \cdot 3 \cdot 3 \cdot 5;$$
$$\text{LCM} = 2 \cdot 3 \cdot 3 \cdot 5 \cdot 7 = 630$$

**Practice Exercise**

**1.** Find the LCM of 52 and 78.

---

**Objective 4.2b**   Add using fraction notation when denominators are different.

**Example**   Add: $\frac{5}{24} + \frac{7}{45}$.

$$\frac{5}{24} + \frac{7}{45} = \frac{5}{2 \cdot 2 \cdot 2 \cdot 3} + \frac{7}{3 \cdot 3 \cdot 5}$$

$$= \frac{5}{2 \cdot 2 \cdot 2 \cdot 3} \cdot \frac{3 \cdot 5}{3 \cdot 5} + \frac{7}{3 \cdot 3 \cdot 5} \cdot \frac{2 \cdot 2 \cdot 2}{2 \cdot 2 \cdot 2}$$

$$= \frac{75}{360} + \frac{56}{360} = \frac{131}{360}$$

**Practice Exercise**

**2.** Add: $\frac{19}{60} + \frac{11}{36}$.

---

**Objective 4.2c**   Use $<$ or $>$ to form a true statement with fraction notation.

**Example**   Use $<$ or $>$ for $\square$ to write a true sentence:

$$\frac{5}{12} \,\square\, \frac{9}{16}.$$

The LCD is 48. Thus, we have

$$\frac{5}{12} \cdot \frac{4}{4} \,\square\, \frac{9}{16} \cdot \frac{3}{3}, \quad \text{or} \quad \frac{20}{48} \,\square\, \frac{27}{48}.$$

Since $20 < 27$, $\frac{20}{48} < \frac{27}{48}$ and thus $\frac{5}{12} < \frac{9}{16}$.

**Practice Exercise**

**3.** Use $<$ or $>$ for $\square$ to write a true sentence:

$$\frac{3}{13} \,\square\, \frac{5}{12}.$$

**Objective 4.3a** Subtract using fraction notation.

**Example** Subtract: $\dfrac{7}{12} - \dfrac{11}{60}$.

$$\dfrac{7}{12} - \dfrac{11}{60} = \dfrac{7}{12} \cdot \dfrac{5}{5} - \dfrac{11}{60} = \dfrac{35}{60} - \dfrac{11}{60}$$

$$= \dfrac{35 - 11}{60} = \dfrac{24}{60} = \dfrac{2 \cdot \cancel{12}}{5 \cdot \cancel{12}} = \dfrac{2}{5}$$

**Practice Exercise**

4. Subtract: $\dfrac{29}{35} - \dfrac{5}{7}$.

---

**Objective 4.4a** Solve equations that involve fractions and require use of both the addition principle and the multiplication principle.

**Example** Solve: $\dfrac{1}{2}x + \dfrac{1}{6} = \dfrac{5}{8}$.

$$\dfrac{1}{2}x + \dfrac{1}{6} = \dfrac{5}{8}$$

$$\dfrac{1}{2}x + \dfrac{1}{6} - \dfrac{1}{6} = \dfrac{5}{8} - \dfrac{1}{6}$$

$$\dfrac{1}{2}x = \dfrac{5}{8} \cdot \dfrac{3}{3} - \dfrac{1}{6} \cdot \dfrac{4}{4}$$

$$\dfrac{1}{2}x = \dfrac{15}{24} - \dfrac{4}{24}$$

$$\dfrac{1}{2}x = \dfrac{11}{24}$$

$$\dfrac{2}{1}\left(\dfrac{1}{2}x\right) = \dfrac{2}{1}\left(\dfrac{11}{24}\right)$$

$$1 \cdot x = \dfrac{\cancel{2} \cdot 11}{1 \cdot \cancel{2} \cdot 12}$$

$$x = \dfrac{11}{12}$$

The solution is $\dfrac{11}{12}$.

**Practice Exercise**

5. Solve: $\dfrac{2}{9} + \dfrac{2}{3}x = \dfrac{1}{6}$.

---

**Objective 4.5a** Convert between mixed numerals and fraction notation.

**Example** Convert $2\dfrac{5}{13}$ to fraction notation: $2\dfrac{5}{13} = \dfrac{31}{13}$.

**Example** Convert $\dfrac{40}{9}$ to a mixed numeral: $\dfrac{40}{9} = 4\dfrac{4}{9}$.

**Practice Exercises**

6. Convert $8\dfrac{2}{3}$ to fraction notation.

7. Convert $\dfrac{47}{6}$ to a mixed numeral.

---

**Objective 4.6b** Subtract using mixed numerals.

**Example** Subtract: $3\dfrac{3}{8} - 1\dfrac{4}{5}$.

$$3\dfrac{3}{8} = \quad 3\dfrac{15}{40} = \quad 2\dfrac{55}{40}$$

$$\underline{-1\dfrac{4}{5} = \quad -1\dfrac{32}{40} = \quad -1\dfrac{32}{40}}$$

$$1\dfrac{23}{40}$$

**Practice Exercise**

8. Subtract: $10\dfrac{5}{7} - 2\dfrac{3}{4}$.

**Objective 4.7a** Multiply using mixed numerals.

**Example** Multiply: $7\frac{1}{4} \cdot 5\frac{3}{10}$. Write a mixed numeral for the answer.

$$7\frac{1}{4} \cdot 5\frac{3}{10} = \frac{29}{4} \cdot \frac{53}{10}$$

$$= \frac{1537}{40} = 38\frac{17}{40}$$

**Practice Exercise**

9. Multiply: $4\frac{1}{5} \cdot 3\frac{7}{15}$.

**Objective 4.7d** Solve applied problems involving multiplication and division with mixed numerals.

**Example** The population of New York is $3\frac{1}{3}$ times that of Missouri. The population of New York is approximately 19,000,000. What is the population of Missouri?

Translate:

$$\underbrace{\text{Population of New York}}_{19,000,000} \quad \underset{=}{\text{is}} \quad \underset{3\frac{1}{3}}{3\frac{1}{3}} \quad \underset{\cdot}{\text{times}} \quad \underbrace{\text{population of Missouri}}_{x.}$$

Solve:

$$19,000,000 = \frac{10}{3} \cdot x$$

$$\frac{19,000,000}{\frac{10}{3}} = \frac{\frac{10}{3} \cdot x}{\frac{10}{3}}$$

$$19,000,000 \cdot \frac{3}{10} = x$$

$$5,700,000 = x.$$

The population of Missouri is about 5,700,000.

**Practice Exercise**

10. The population of Louisiana is $2\frac{1}{2}$ times the population of West Virginia. The population of West Virginia is approximately 1,800,000. What is the population of Louisiana?

**Objective 4.8a** Simplify expressions containing fraction notation using the rules for order of operations.

**Example** Simplify: $\left(\frac{4}{5}\right)^2 - \frac{1}{5} \cdot 2\frac{1}{8}$.

$$\left(\frac{4}{5}\right)^2 - \frac{1}{5} \cdot 2\frac{1}{8} = \frac{16}{25} - \frac{1}{5} \cdot \frac{17}{8}$$

$$= \frac{16}{25} - \frac{17}{40}$$

$$= \frac{16}{25} \cdot \frac{8}{8} - \frac{17}{40} \cdot \frac{5}{5}$$

$$= \frac{128}{200} - \frac{85}{200} = \frac{43}{200}$$

**Practice Exercise**

11. Simplify: $\frac{3}{2} \cdot 1\frac{1}{3} \div \left(\frac{2}{3}\right)^2$.

# Review Exercises

Find the LCM.  [4.1a]

**1.** 12 and 18

**2.** 18 and 45

**3.** 3, 6, and 30

**4.** 26, 36, and 54

Perform the indicated operation and, if possible, simplify.
[4.2a, b], [4.3a]

**5.** $\dfrac{2}{9} + \dfrac{5}{9}$

**6.** $\dfrac{7}{x} + \dfrac{2}{x}$

**7.** $-\dfrac{6}{5} + \dfrac{11}{15}$

**8.** $\dfrac{5}{16} + \dfrac{3}{24}$

**9.** $\dfrac{7}{9} - \dfrac{5}{9}$

**10.** $\dfrac{1}{4} - \dfrac{3}{8}$

**11.** $\dfrac{10}{27} - \dfrac{2}{9}$

**12.** $\dfrac{5}{6} - \dfrac{7}{9}$

Use $<$ or $>$ for $\square$ to form a true sentence.  [4.2c]

**13.** $\dfrac{4}{7} \,\square\, \dfrac{5}{9}$

**14.** $-\dfrac{8}{9} \,\square\, -\dfrac{11}{13}$

Solve.  [4.3b], [4.4a]

**15.** $x + \dfrac{2}{5} = \dfrac{7}{8}$

**16.** $7a - 3 = 25$

**17.** $5 + \dfrac{16}{3}x = \dfrac{5}{9}$

**18.** $\dfrac{22}{5} = \dfrac{16}{5} + \dfrac{5}{2}x$

Solve by using the multiplication principle to clear fractions.
[4.4b]

**19.** $\dfrac{5}{3}x + \dfrac{5}{6} = \dfrac{3}{2}$

Convert to fraction notation.  [4.5a]

**20.** $7\dfrac{1}{2}$

**21.** $8\dfrac{3}{8}$

**22.** $4\dfrac{1}{3}$

**23.** $-1\dfrac{5}{7}$

Convert to a mixed numeral.  [4.5a]

**24.** $\dfrac{7}{3}$

**25.** $\dfrac{-27}{4}$

**26.** $\dfrac{63}{5}$

**27.** $\dfrac{7}{2}$

**28.** Divide. Write a mixed numeral for the answer.
$$7896 \div (-9) \quad [4.5b]$$

**29.** Gina's golf scores were 80, 82, and 85. What was her average score?  [4.5b]

Perform the indicated operation. Write a mixed numeral for each answer.  [4.6a, b, d]

**30.** $\quad 7\dfrac{3}{5}$
$$+\ 2\dfrac{4}{5}$$

**31.** $\quad 6\dfrac{1}{3}$
$$+\ 5\dfrac{2}{5}$$

**32.** $-3\dfrac{5}{6} + \left(-5\dfrac{1}{6}\right)$

**33.** $-2\dfrac{3}{4} + 4\dfrac{1}{2}$

**34.** $\quad 14$
$$-\ 6\dfrac{2}{9}$$

**35.** $\quad 9\dfrac{3}{5}$
$$-\ 4\dfrac{13}{15}$$

**36.** $4\dfrac{5}{8} - 9\dfrac{3}{4}$

**37.** $-7\dfrac{1}{2} - 6\dfrac{3}{4}$

Combine like terms.  [4.2b], [4.6b]

**38.** $\dfrac{4}{9}x + \dfrac{1}{3}x$

**39.** $8\dfrac{3}{10}a - 5\dfrac{1}{8}a$

Perform the indicated operation. Write a mixed numeral or integer for each answer, unless the answer is less than 1.  [4.7a, b]

**40.** $6 \cdot 2\frac{2}{3}$

**41.** $-5\frac{1}{4} \cdot \frac{2}{3}$

**42.** $2\frac{1}{5} \cdot 1\frac{1}{10}$

**43.** $2\frac{2}{5} \cdot 2\frac{1}{2}$

**44.** $-54 \div 2\frac{1}{4}$

**45.** $2\frac{2}{5} \div \left(-1\frac{7}{10}\right)$

**46.** $3\frac{1}{4} \div 26$

**47.** $4\frac{1}{5} \div 4\frac{2}{3}$

Evaluate.  [4.7c]

**48.** $5x - y$, for $x = 3\frac{1}{5}$ and $y = 2\frac{2}{7}$

**49.** $2a \div b$, for $a = 5\frac{2}{11}$ and $b = 3\frac{4}{5}$

Solve.  [4.2d], [4.3c], [4.6c], [4.7d]

**50.** *Sewing.*  Kim wants to make slacks and a jacket. She needs $1\frac{5}{8}$ yd of 60-in. fabric for the slacks and $2\frac{5}{8}$ yd for the jacket. How many yards in all does Kim need to make the outfit?

**51.** The San Diaz drama club had $\frac{3}{8}$ of a vegetarian pizza, $1\frac{1}{2}$ cheese pizzas, and $1\frac{1}{4}$ pepperoni pizzas remaining after a cast party. How many pizzas remained altogether?

**52.** *Turkey Servings.*  Turkey contains $1\frac{1}{3}$ servings per pound. How many pounds are needed for 32 servings?

**53.** What is the sum of the areas in the figure below?

**54.** In the figure above, how much larger is the area of rectangle $A$ than the area of rectangle $B$?

**55.** *Painting a Border.*  Katie hired an artist to paint a decorative border around the top of her son's bedroom. The artist charges $20 per foot. The room measures $11\frac{3}{4}$ ft $\times$ $9\frac{1}{2}$ ft. What is Katie's cost for the project?

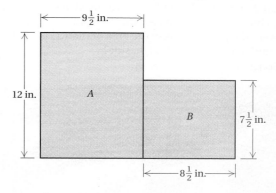

$9\frac{1}{2}$ ft

$11\frac{3}{4}$ ft

**56.** *Cake Recipe.*  A wedding-cake recipe requires 12 cups of shortening. Being calorie-conscious, the wedding couple decides to reduce the shortening by $3\frac{5}{8}$ cups and replace it with prune purée. How many cups of shortening are used in their new recipe?

**57.** *Humane Society Pie Sale.*  Green River's Humane Society recently hosted its annual ice cream social. Each of the 83 pies donated was cut into 6 pieces. At the end of the evening, 382 pieces of pie had been sold. How many pies were sold? How many were left over? Express your answers in mixed numerals.

**58.** *Building a Ziggurat.* The dimensions of each square brick that King Nebuchadnezzar used over 2500 yr ago to build ziggurats was $13\frac{1}{4}$ in. $\times$ $13\frac{1}{4}$ in. $\times$ $3\frac{1}{4}$ in. What were the perimeter and the area of the $13\frac{1}{4}$ in. $\times$ $13\frac{1}{4}$ in. side? of the $13\frac{1}{4}$ in. $\times$ $3\frac{1}{4}$ in. side?

**Source:** www.eartharchitecture.org

Simplify each expression using the rules for order of operations.  [4.8a]

**59.** $\dfrac{1}{2} + \dfrac{1}{8} \div \dfrac{1}{4}$

**60.** $\dfrac{4}{5} - \dfrac{1}{2} \cdot \left( \dfrac{1}{4} - \dfrac{3}{5} \right)$

**61.** $20\dfrac{3}{4} - 1\dfrac{1}{2} \times 12 + \left( \dfrac{1}{2} \right)^2$

**62.** Find the average of $\dfrac{1}{2}, \dfrac{1}{4}, \dfrac{1}{3}$, and $\dfrac{1}{5}$.  [4.8b]

Simplify.  [4.8b]

**63.** $\dfrac{\dfrac{1}{3}}{\dfrac{5}{8} - 1}$

**64.** $\dfrac{-\dfrac{x}{7}}{\dfrac{3}{14}}$

**65.** Simplify: $\dfrac{1}{4} + \dfrac{2}{5} \div 5^2$.  [4.8a]

  **A.** $\dfrac{133}{500}$        **B.** $\dfrac{3}{500}$

  **C.** $\dfrac{117}{500}$        **D.** $\dfrac{5}{2}$

**66.** Solve: $x + \dfrac{2}{3} = 5$.  [4.3b]

  **A.** $\dfrac{15}{2}$        **B.** $5\dfrac{2}{3}$

  **C.** $\dfrac{10}{3}$        **D.** $4\dfrac{1}{3}$

## Synthesis

**67.** Find $r$ if

$$\dfrac{1}{r} = \dfrac{1}{100} + \dfrac{1}{150} + \dfrac{1}{200}.$$  [4.2b], [4.4b]

**68.** Place the numbers 3, 4, 5, and 6 in the boxes in order to make a true equation:  [4.5a]

$$\dfrac{\square}{\square} + \dfrac{\square}{\square} = 3\dfrac{1}{4}.$$

**69.** Find the largest integer for which each fraction is greater than 1.  [4.2c]

  **a)** $\dfrac{7}{\square}$        **b)** $\dfrac{11}{\square}$

  **c)** $\dfrac{\square}{-27}$        **d)** $\dfrac{\square}{-\frac{1}{2}}$

# Understanding Through Discussion and Writing

**1.** Is the sum of two mixed numerals always a mixed numeral? Why or why not?  [4.6a]

**2.** Write a problem for a classmate to solve. Design the problem so that its solution is found by performing the multiplication $4\frac{1}{2} \cdot 33\frac{1}{3}$.  [4.7d]

**3.** A student insists that $3\frac{2}{5} \cdot 1\frac{3}{7} = 3\frac{6}{35}$. What mistake is he making and how should he have proceeded?  [4.7a]

**4.** Discuss the role of least common multiples in adding and subtracting with fraction notation.  [4.2b], [4.3a]

**5.** Find a real-world situation that fits this equation:

$$2 \cdot 15\frac{3}{4} + 2 \cdot 28\frac{5}{8} = 88\frac{3}{4}.$$  [4.6c], [4.7d]

**6.** A student insists that $5 \cdot 3\frac{2}{7} = (5 \cdot 3) \cdot \left(5 \cdot \frac{2}{7}\right)$. What mistake is she making and how should she have proceeded?  [4.8a]

Test

For Extra Help

CHAPTER
Test Prep
VIDEOS

Step-by-step test solutions are found on the Chapter Test Prep Videos available via the Video Resources on DVD, in **MyMathLab** , and on You Tube (search "BittingerPrealgebra" and click on "Channels").

**1.** Find the LCM of 12 and 16.

Perform the indicated operation and, if possible, simplify.

**2.** $\dfrac{1}{2} + \dfrac{5}{2}$

**3.** $-\dfrac{7}{8} + \dfrac{2}{3}$

**4.** $\dfrac{5}{t} - \dfrac{3}{t}$

**5.** $\dfrac{5}{6} - \dfrac{3}{4}$

**6.** $\dfrac{5}{8} - \dfrac{17}{24}$

Solve.

**7.** $x + \dfrac{2}{3} = \dfrac{11}{12}$

**8.** $-5x - 3 = 9$

**9.** $\dfrac{3}{4} = \dfrac{1}{2} + \dfrac{5}{3}x$

**10.** Use $<$ or $>$ for $\square$ to form a true sentence.

$$\dfrac{6}{7} \ \square \ \dfrac{21}{25}$$

Convert to fraction notation.

**11.** $3\dfrac{1}{2}$

**12.** $-9\dfrac{3}{8}$

**13.** Convert to a mixed numeral:

$$-\dfrac{74}{9}.$$

**14.** Divide. Write a mixed numeral for the answer.

$$1\,1\,\overline{)\,1\,7\,8\,9}$$

Perform the indicated operation. Write a mixed numeral for each answer.

**15.** $\begin{array}{r} 6\dfrac{2}{5} \\ + \ 7\dfrac{4}{5} \\ \hline \end{array}$

**16.** $\begin{array}{r} 3\dfrac{1}{4} \\ + \ 9\dfrac{1}{6} \\ \hline \end{array}$

**17.** $\begin{array}{r} 10\dfrac{1}{6} \\ - \ 5\dfrac{7}{8} \\ \hline \end{array}$

**18.** $14 + \left(-5\dfrac{3}{7}\right)$

**19.** $3\dfrac{4}{5} - 9\dfrac{1}{2}$

Combine like terms.

**20.** $\dfrac{3}{8}x - \dfrac{1}{2}x$

**21.** $5\dfrac{2}{11}a - 3\dfrac{1}{5}a$

Perform the indicated operation.

**22.** $9 \cdot 4\dfrac{1}{3}$

**23.** $6\dfrac{3}{4} \cdot \left(-2\dfrac{2}{3}\right)$

**24.** $33 \div 5\dfrac{1}{2}$

**25.** $2\dfrac{1}{3} \div 1\dfrac{1}{6}$

Evaluate.

**26.** $\frac{2}{3}ab$, for $a = 7$ and $b = 4\frac{1}{5}$

**27.** $4 + mn$, for $m = 7\frac{2}{5}$ and $n = 3\frac{1}{4}$

Solve.

**28.** *Weightlifting.*   In 2002, Hossein Rezazadeh of Iran completed a clean and jerk of 263 kg. This amount was about $2\frac{1}{2}$ times his body weight. How much did Rezazadeh weigh?
**Source:** *The Guinness Book of Records,* 2005

**29.** *Book Order.*   An order of books for a math course weighs 220 lb. Each book weighs $2\frac{3}{4}$ lb. How many books are in the order?

**30.** *Carpentry.*   The following diagram shows a middle drawer support guide for a cabinet drawer. Find each of the following.

   **a)** The short length $a$ across the top
   **b)** The length $b$ across the bottom

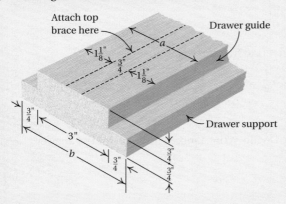

**31.** *Carpentry.*   In carpentry, some pieces of plywood that are called "$\frac{3}{4}$-inch" plywood are actually $\frac{11}{16}$ in. thick. How much thinner is such a piece than its name indicates?

**32.** *Women's Dunks.*   The first three women in the history of college basketball able to dunk a basketball are listed below. Their names, heights, and universities are

   Michelle Snow, $6\frac{5}{12}$ ft, Tennessee;

   Charlotte Smith, $5\frac{11}{12}$ ft, North Carolina;

   Georgeann Wells, $6\frac{7}{12}$ ft, West Virginia.

Find the average height of these women.
**Source:** *USA Today,* 11/30/00, p. 3C

Simplify.

**33.** $\frac{2}{3} + 1\frac{1}{3} \cdot 2\frac{1}{8}$

**34.** $-1\frac{1}{2} - \frac{1}{2}\left(\frac{1}{2} \div \frac{1}{4}\right) + \left(\frac{1}{2}\right)^2$

**35.** $\dfrac{\dfrac{1}{3} - \dfrac{7}{9}}{\dfrac{1}{2} + \dfrac{1}{6}}$

**36.** Find the LCM of 12, 36, and 60.
   **A.** 6           **B.** 12
   **C.** 60          **D.** 180

# Synthesis

**37.** The students in a math class can be organized into study groups of 8 each so that no students are left out. The same class of students can also be organized into groups of 6 so that no students are left out.

   **a)** Find some class sizes for which this will work.
   **b)** Find the smallest such class size.

**38.** Rebecca walks 17 laps at her health club. Trent walks 17 laps at his health club. If the track at Rebecca's health club is $\frac{1}{7}$ mi long and the track at Trent's is $\frac{1}{8}$ mi long, who walks farther? How much farther?

# Cumulative Review

Solve.

**1.** *Cross-Country Skiing.* During a three-day holiday weekend trip, David and Sally Jean cross-country skied $3\frac{2}{3}$ mi on Friday, $6\frac{1}{8}$ mi on Saturday, and $4\frac{3}{4}$ mi on Sunday.

**a)** Find the total number of miles that they skied.

**b)** Find the average number of miles that they skied per day. Express your answer as a mixed numeral.

**2.** How many people can receive equal $16 shares from a total of $496?

**3.** A recipe calls for $\frac{4}{5}$ tsp of salt. How much salt should be used for $\frac{1}{2}$ recipe? for 5 recipes?

**4.** How many pieces, each $2\frac{3}{8}$ ft long, can be cut from a piece of wire 38 ft long?

**5.** An emergency food pantry fund contains $423. From this fund, $148 and $167 are withdrawn for expenses. How much is left in the fund?

**6.** In a walkathon, Jermaine walked $\frac{9}{10}$ mi and Oleta walked $\frac{3}{4}$ mi. What was the total distance that they walked?

What part is shaded?

**7.**

**8.**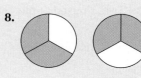

Calculate and simplify.

**9.**
$$\begin{array}{r} 3\,7\,0\,4 \\ +\,5\,2\,7\,8 \end{array}$$

**10.**
$$\begin{array}{r} 7\,6\,0\,5 \\ -\,3\,0\,8\,7 \end{array}$$

**11.**
$$\begin{array}{r} 2\,7\,8 \\ \times\quad 1\,8 \end{array}$$

**12.** $29(-5)$

**13.** $\dfrac{3}{8} + \dfrac{1}{24}$

**14.**
$$\begin{array}{r} 2\frac{3}{4} \\ +\,5\frac{1}{2} \end{array}$$

**15.** $\dfrac{4}{t} - \dfrac{9}{t}$

**16.**
$$\begin{array}{r} 2\frac{1}{3} \\ -\,1\frac{1}{6} \end{array}$$

**17.** $\dfrac{9}{10} \cdot \dfrac{5}{3}$

**18.** $18\left(-\dfrac{5}{6}\right)$

**19.** $-9 - (-25)$

**20.** $2\dfrac{1}{5} \div \dfrac{3}{10}$

Divide. Write the answer in the form 34 R 7.

**21.** $6 \overline{)\ 4\ 2\ 9\ 0}$          **22.** $4\ 5 \overline{)\ 2\ 5\ 3\ 1}$

**23.** Write a mixed numeral for the answer in Exercise 22.

**24.** In the number 2753, what digit names tens?

**25.** *Room Carpeting.* The Chandlers are carpeting an L-shaped family room consisting of one rectangle that is $8\frac{1}{2}$ ft by 11 ft and another rectangle that is $6\frac{1}{2}$ ft by $7\frac{1}{2}$ ft.
   **a)** Find the area of the carpet.
   **b)** Find the perimeter of the carpet.

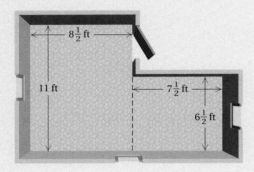

**26.** Round 38,478 to the nearest hundred.

**27.** Find the LCM of 18 and 24.

**28.** Simplify:
$$\left(\frac{1}{2} + \frac{2}{5}\right)^2 \div 3 + 6 \times \left(2 + \frac{1}{4}\right).$$

Use $<$, $>$, or $=$ for $\square$ to write a true sentence.

**29.** $\dfrac{4}{5} \ \square \ \dfrac{4}{6}$     **30.** $\dfrac{3}{13} \ \square \ \dfrac{9}{39}$     **31.** $-\dfrac{5}{12} \ \square \ -\dfrac{3}{7}$

**32.** Evaluate $\dfrac{t + p}{3}$ for $t = -4$ and $p = 16$.

Simplify.

**33.** $\dfrac{36}{45}$       **34.** $\dfrac{0}{27}$       **35.** $\dfrac{-320}{10}$

**36.** Convert to fraction notation: $4\dfrac{5}{8}$.

**37.** Convert to a mixed numeral: $-\dfrac{17}{3}$.

Solve.

**38.** $x + 24 = 117$      **39.** $x + \dfrac{7}{9} = \dfrac{4}{3}$

**40.** $\dfrac{7}{9} \cdot t = -\dfrac{4}{3}$      **41.** $\dfrac{5}{7} = \dfrac{1}{3} + 4a$

**42.** *Matching.* Match each item in the first column with the appropriate item in the second column by drawing connecting lines. There can be more than one correct correspondence for an item.

| | |
|---|---|
| Factors of 68 | 12, 54, 72, 300 |
| Factorization of 68 | 2, 3, 17, 19, 23, 31, 47, 101 |
| Prime factorization of 68 | $2 \cdot 2 \cdot 17$ |
| Numbers divisible by 6 | $2 \cdot 34$ |
| Numbers divisible by 8 | 8, 16, 24, 32, 40, 48, 64, 864 |
| Numbers divisible by 5 | 1, 2, 4, 17, 34, 68 |
| Prime numbers | 70, 95, 215 |

## Synthesis

**43.** Find the smallest prime number that is larger than 2000.

**44.** Solve: $7x - \dfrac{2}{3}(x - 6) = 6\dfrac{5}{7}$.

# Decimal Notation

## Real-World Application

The numbers of overnight camping stays in Park Service campgrounds in the National Park System from 2004 to 2008 were 5.4 million, 5.2 million, 5.0 million, 5.1 million, and 5.0 million, respectively. Find the average number of stays per year during this period.

*Source:* U.S. National Park Service

*This problem appears as Exercise 67 in Section 5.4.*

**SKILL TO REVIEW**

Objective 1.6a: Round to the nearest ten, hundred, or thousand.

Round 4735 to the nearest

**1.** Ten.  **2.** Hundred.

The set of **rational numbers** consists of the **integers**

$$\ldots, \ -3, \ -2, \ -1, \ 0, \ 1, \ 2, \ 3, \ldots$$

and fractions like

$$\frac{1}{2}, \ \frac{2}{3}, \ \frac{-7}{8}, \ \frac{17}{-10}, \text{ and so on.}$$

We used fraction notation for rational numbers in Chapters 3 and 4. Here in Chapter 5, we will use *decimal notation* to represent the set of rational numbers. For example, $\frac{3}{4}$ will be written as 0.75, and $9\frac{1}{2}$ will be written as 9.5. A number written in decimal notation is often simply referred to as a *decimal*.

The word *decimal* comes from the Latin word *decima*, meaning a *tenth part*. Since our usual counting system is based on tens, decimal notation is a natural extension of an already familiar system.

## a Decimal Notation and Word Names

One model of the Magellan GPS navigation system sells for $249.98. The dot in $249.98 is called a **decimal point**. Since $0.98, or 98¢, is $\frac{98}{100}$ of a dollar, it follows that

$$\$249.98 = 249 + \frac{98}{100} \text{ dollars.}$$

Also, since $0.98, or 98¢, has the same value as

9 dimes + 8 cents

and 1 dime is $\frac{1}{10}$ of a dollar and 1 cent is $\frac{1}{100}$ of a dollar, we can write

$$249.98 = 2 \cdot 100 + 4 \cdot 10 + 9 \cdot 1 + 9 \cdot \frac{1}{10} + 8 \cdot \frac{1}{100}.$$

This is an extension of the expanded notation for whole numbers that we used in Chapter 1. The place values are 100, 10, 1, $\frac{1}{10}$, $\frac{1}{100}$, and so on. We can see this on a **place-value chart**. The value of each place is $\frac{1}{10}$ as large as that of the one to its left.

Let's consider decimal notation using a place-value chart to represent 26.2922 min, the men's 10,000-meter run record held by Kenenisa Bekele from Ethiopia.

| PLACE-VALUE CHART | | | | | | | |
|---|---|---|---|---|---|---|---|
| Hundreds | Tens | Ones | Tenths | Hundredths | Thousandths | Ten-Thousandths | Hundred-Thousandths |
| 100 | 10 | 1 | $\frac{1}{10}$ | $\frac{1}{100}$ | $\frac{1}{1000}$ | $\frac{1}{10,000}$ | $\frac{1}{100,000}$ |
| | 2 | 6 . | 2 | 9 | 2 | 2 | |

*Answers*

*Skill to Review:*

1. 4740  2. 4700

The decimal notation 26.2922 means

2 tens + 6 ones + 2 tenths + 9 hundredths + 2 thousandths + 2 ten-thousandths

or $\quad 20 + 6 + \dfrac{2}{10} + \dfrac{9}{100} + \dfrac{2}{1000} + \dfrac{2}{10,000}.$

Using 10,000 as the least common denominator, we have

$$26.2922 = 26 + \frac{2000}{10,000} + \frac{900}{10,000} + \frac{20}{10,000} + \frac{2}{10,000} = 26\frac{2922}{10,000}.$$

We read both 26.2922 and $26\frac{2922}{10,000}$ as

"Twenty-six *and* two thousand nine hundred twenty-two ten-thousandths."

We read the decimal point as "and." Note that the place values to the right of the decimal point always end in *th*. We can also read 26.2922 as "Two six *point* two nine two two" or "Twenty-six point two nine two two."

**STUDY TIPS**

**QUIZ-TEST FOLLOW-UP**

You may have just completed a chapter quiz or test. Immediately after doing so, write out a step-by-step solution of each question that you missed. Visit your instructor or tutor for help with problems that are still giving you trouble. When the week of the final examination arrives, you will be glad to have the excellent study guide these corrected tests provide.

To write a word name from decimal notation,

a) write a word name for the whole number (the number named to the left of the decimal point),

    397.685    → Three hundred ninety-seven

b) write the word "and" for the decimal point, and

    397.685     Three hundred ninety-seven and

c) write a word name for the number named to the right of the decimal point, followed by the place value of the last digit.

    397.685     Three hundred ninety-seven and six hundred eighty-five *thousandths*

**EXAMPLE 1** *Birth Rate.* There were 14.3 births per 1000 Americans in a recent year. Write a word name for 14.3.
**Source:** U.S. Centers for Disease Control and Prevention

    Fourteen and three tenths

Write a word name for each number.

3. 245.89

4. 34.0064

5. 31,079.756

Write in words, as on a check.

6. $4217.56          7. $13.98

**EXAMPLE 2**  Write a word name for the number in this sentence: The world record in the women's pole vault is 5.01 m, set by Yelena Isinbayeva of Russia.
Source: International Association of Athletics Federations

Five and one hundredth

**EXAMPLE 3**  *Indianapolis 500.*  In the 2009 Indianapolis 500-mile race, winner Helio Castroneves had a 1.9819-second margin of victory over second-place finisher Dan Wheldon. Write a word name for 1.9819.
Source: Indianapolis Motor Speedway, historian Donald Davidson

One and nine thousand, eight hundred nineteen ten-thousandths

**EXAMPLE 4**  Write a word name for the number in this sentence: The current one-mile land speed record of 763.035 mph is held by Andy Green.
Source: www.castrol.com

Seven hundred sixty-three and thirty-five thousandths

Do Exercises 1–5.

Decimal notation is also used with money. It is common on a check to write "and ninety-five cents" as "and $\frac{95}{100}$ dollars."

**EXAMPLE 5**  Write $5876.95 in words, as on a check.

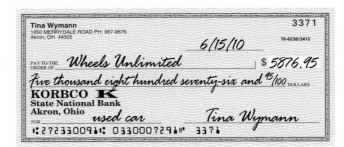

Five thousand, eight hundred seventy-six and $\frac{95}{100}$ dollars

Do Exercises 6 and 7.

## b Converting Between Decimal Notation and Fraction Notation

Since 9.875 means nine and eight hundred seventy-five thousandths, we can write

$$9.875 = 9 + \frac{875}{1000} = \frac{9000}{1000} + \frac{875}{1000} = \frac{9875}{1000}.$$

Decimal notation ———————— Fraction notation

9.875            $\frac{9875}{1000}$

3 decimal places        3 zeros

To convert from decimal to fraction notation,

a) count the number of decimal places,

4.98

↑———— 2 places

b) move the decimal point that many places to the right, and

4.98.  Move 2 places.

c) write the answer over a denominator of 1 followed by that number of zeros.

$\frac{498}{100}$  2 zeros

For a number like 0.876, we write a 0 to call attention to the presence of the decimal point.

**EXAMPLE 6**  Write fraction notation for 0.876. Do not simplify.

$$0.876 \qquad 0.876. \qquad 0.876 = \frac{876}{1000}$$

3 places        3 zeros

Decimals greater than 1 or less than −1 can be written either as fractions or as mixed numerals.

**EXAMPLE 7**  Write 56.23 as a fraction and as a mixed numeral.

$$56.23 \qquad 56.23. \qquad 56.23 = \frac{5623}{100}, \qquad \text{and} \qquad 56.23 = 56\frac{23}{100}$$

2 places        2 zeros

As a check, note that both 56.23 and $56\frac{23}{100}$ are read as "fifty-six and twenty-three hundredths."

**EXAMPLE 8**  Write −2.6073 as a fraction and as a mixed numeral.

$$-2.6073. = -\frac{26,073}{10,000} \qquad \text{and} \qquad -2.6073 = -2\frac{6073}{10,000}$$

4 places        4 zeros

Do Exercises 8–10.

To write $\frac{5328}{10}$ as a decimal, we can first divide to find an equivalent mixed numeral.

$$\frac{5328}{10} = 532\frac{8}{10}$$

Next note that

$$532\frac{8}{10} = 532 + \frac{8}{10}$$
$$= 532.8.$$

```
        5 3 2
1 0 ) 5 3 2 8
      5 0
      ─────
        3 2
        3 0
        ─────
          2 8
          2 0
          ─────
            8
```

Write fraction notation. Do not simplify.

**8.** 0.5491

Write as a fraction and as a mixed numeral.

**9.** 75.069        **10.** −312.9

**Answers**

**8.** $\frac{5491}{10,000}$  **9.** $\frac{75,069}{1000}; 75\frac{69}{1000}$

**10.** $-\frac{3129}{10}; -312\frac{9}{10}$

Thus, if fraction notation has a denominator that is a power of ten, such as 10, 100, 1000, and so on, we convert to decimal notation by reversing the procedure that we just used in Examples 6–8.

> To convert from fraction notation to decimal notation when the denominator is 10, 100, 1000, and so on,
>
> a) count the number of zeros and
>
> $$\frac{8679}{1000}$$
>
> — 3 zeros
>
> b) move the decimal point that number of places to the left. Leave off the denominator.
>
> 8.679.
>
> Move 3 places.
>
> $$\frac{8679}{1000} = 8.679$$

**EXAMPLE 9** Write decimal notation for $\frac{47}{10}$.

$$\frac{47}{10} \qquad 4.7. \qquad \frac{47}{10} = 4.7 \qquad \text{The decimal point is moved to the left.}$$

1 zero    1 place

**EXAMPLE 10** Write decimal notation for $\frac{123{,}067}{10{,}000}$.

$$\frac{123{,}067}{10{,}000} \qquad 12.3067. \qquad \frac{123{,}067}{10{,}000} = 12.3067$$

4 zeros    4 places

To move the decimal point to the left, we may need to write extra 0's.

**EXAMPLE 11** Write decimal notation for $-\frac{9}{100}$.

$$-\frac{9}{100} \qquad -0.09. \qquad -\frac{9}{100} = -0.09$$

2 zeros    2 places

Do Exercises 11–14.

For denominators other than 10, 100, and so on, we will usually perform long division. We convert such fractions to decimal notation in Section 5.5.

If a mixed numeral has a fraction part with a denominator that is a power of ten, such as 10, 100, or 1000, and so on, we first write the mixed numeral as a sum of a whole number and a fraction. Then we convert to decimal notation.

**EXAMPLE 12** Write decimal notation for $23\frac{59}{100}$.

$$23\frac{59}{100} = 23 + \frac{59}{100} = 23 \text{ and } \frac{59}{100} = 23.59$$

Do Exercises 15 and 16.

Write decimal notation for each number.

**11.** $\frac{743}{100}$

**12.** $-\frac{73}{1000}$

**13.** $\frac{67{,}089}{10{,}000}$

**14.** $-\frac{9}{10}$

Write decimal notation for each number.

**15.** $-7\frac{3}{100}$

**16.** $23\frac{47}{1000}$

*Answers*

**11.** 7.43   **12.** −0.073   **13.** 6.7089
**14.** −0.9   **15.** −7.03   **16.** 23.047

## (c) Order

To understand how to compare numbers in decimal notation, consider 0.85 and 0.9. First note that $0.9 = 0.90$ because $\frac{9}{10} = \frac{90}{100}$. Since $0.85 = \frac{85}{100}$, it follows that $\frac{85}{100} < \frac{90}{100}$ and $0.85 < 0.9$. This leads us to a quick way to compare two numbers in decimal notation.

> To compare two positive numbers in decimal notation, start at the left and compare corresponding digits, moving from left to right. When two digits differ, the number with the larger digit is the larger of the two numbers. Extra zeros can be written to the right of the last decimal place.

**EXAMPLE 13**  Which is larger: 2.109 or 2.1?

| 2.109 | 2.109 | 2.109 | 2.109 |
|---|---|---|---|
| ↕ The same | ↕ The same | ↕ The same | ↕ Different; |
| 2.1 | 2.1 | 2.10 | 2.100  9 is larger than 0. |

Thus, 2.109 is larger than 2.1. In symbols, $2.109 > 2.1$.

**EXAMPLE 14**  Which is larger: 0.09 or 0.108?

| 0.09 | 0.09 |
|---|---|
| ↕ The same | ↕ Different; 1 is larger than 0. |
| 0.108 | 0.108 |

Thus, 0.108 is larger than 0.09. In symbols, $0.108 > 0.09$.

Do Exercises 17–20.

As before, we can use the number line to visualize order. We illustrate Examples 13 and 14 below. Larger numbers are always to the right.

Note from the number line that $-2 < -1$. Similarly, $-1.57 < -1.52$.

> To compare two negative numbers in decimal notation, start at the left and compare corresponding digits, moving from left to right. When two digits differ, the number with the smaller digit is the larger of the two numbers.

**EXAMPLE 15**  Which is larger: −3.8 or −3.82?

| −3.8 | −3.80 |
|---|---|
| ↕↕ The same | ↕ Different; 0 is smaller than 2. |
| −3.82 | −3.82 |

Thus, −3.8 is larger than −3.82. In symbols, $-3.8 > -3.82$ (see the number line above).

Do Exercises 21–24.

---

**Which number is larger?**

**17.** 2.04,  2.039

**18.** 0.06,  0.008

**19.** 0.5,  0.58

**20.** 1,  0.9999

**Which number is larger?**

**21.** 0.8989,  0.09898

**22.** 21.006,  21.05

**23.** −34.01,  −34.008

**24.** −9.12,  −8.98

*Answers*

**17.** 2.04  **18.** 0.06  **19.** 0.58  **20.** 1
**21.** 0.8989  **22.** 21.05  **23.** −34.008
**24.** −8.98

## (d) Rounding

We round decimals in much the same way that we round whole numbers. To see how, we use the number line.

**EXAMPLE 16** Round 0.37 to the nearest tenth.

Here is part of the number line, magnified.

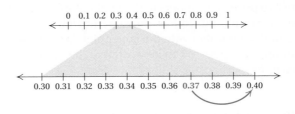

We see that 0.37 is closer to 0.40 than to 0.30. Thus, 0.37 rounded to the nearest tenth is 0.4.

> To round to a certain place:
>
> **a)** Locate the digit in that place.
>
> **b)** Consider the next digit to the right.
>
> **c)** If the digit to the right is 5 or greater, add one to the original digit. If the digit to the right is 4 or less, the original digit does not change. In either case, drop all numbers to the right of the original digit.

**EXAMPLE 17** Round 72.3846 to the nearest hundredth.

**a)** Locate the digit in the hundredths place, 8.

$$7\ 2 . 3\ 8\ 4\ 6$$

**b)** Consider the next digit to the right, 4.

**c)** Since that digit, 4, is less than 5, the original digit does not change. The rounded number is 72.38.

------------------------------ *Caution!* ------------------------------

72.39 is not a correct answer to Example 17. It is *incorrect* to round sequentially from right to left as follows: 72.3846, 72.385, 72.39.

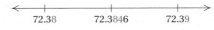

72.3846 is closer to 72.38 than to 72.39.

------------------------------------------------------------------------

**EXAMPLE 18** Round −0.064 to the nearest tenth.

**a)** Locate the digit in the tenths place, 0.

$$-0.0\ 6\ 4$$

**b)** Consider the next digit to the right, 6.

**c)** Since that digit, 6, is greater than 5, round from −0.064 to −0.1.

The answer is −0.1. Since −0.1 < −0.064, we actually rounded *down*.

Do Exercises 25–43.

---

Round to the nearest tenth.

**25.** 2.76     **26.** 13.85

**27.** −234.448     **28.** 7.009

Round to the nearest hundredth.

**29.** 0.6362     **30.** −7.8348

**31.** 34.67514     **32.** −0.02521

Round to the nearest thousandth.

**33.** 0.94347     **34.** −8.00382

**35.** −43.111943     **36.** 37.400526

Round 7459.3548 to the nearest

**37.** Thousandth.

**38.** Hundredth.

**39.** Tenth.     **40.** One.

**41.** Ten. (*Caution:* "Tens" are not "tenths.")

**42.** Hundred.     **43.** Thousand.

*Answers*

**25.** 2.8     **26.** 13.9     **27.** −234.4     **28.** 7.0
**29.** 0.64     **30.** −7.83     **31.** 34.68
**32.** −0.03     **33.** 0.943     **34.** −8.004
**35.** −43.112     **36.** 37.401     **37.** 7459.355
**38.** 7459.35     **39.** 7459.4     **40.** 7459
**41.** 7460     **42.** 7500     **43.** 7000

**5.1** **Exercise Set**

For Extra Help

*MyMathLab*

Math XL
PRACTICE

WATCH

DOWNLOAD

READ

REVIEW

**a** Write a word name for the number in each sentence.

**1.** *Stock Price.* Google's stock was priced at $486.34 per share.
**Source:** New York Stock Exchange

**2.** *Soft-Drink Consumption.* The average American consumed 51.5 gal of carbonated soft drinks in a recent year.
**Source:** U.S. Department of Agriculture

**3.** *Currency Conversion.* One Chinese yuan was worth about $0.146 in U.S. currency recently.
**Source:** Interactive Data

**4.** *Currency Conversion.* One U.S. dollar was worth 0.6714 euros recently.
**Source:** Interactive Data

**5.** *Game System.* The Nintendo Wii game system sold for $249.89 recently.
**Source:** Best Buy

**6.** *Cell-Phone Plan.* The Sprint Talk/Message cell-phone plan with unlimited anytime minutes was priced at $89.99 per month recently.
**Source:** Sprint

**7.** One gallon of paint is equal to 3.7854 liters of paint.

**8.** *Water Weight.* One gallon of water weighs 8.35 lb.

**9.** *Theater Admissions.* In 2008, the country with the largest per capita movie theater admissions was Iceland, with each person attending a movie, on average, 5.4 times a year.
**Source:** Cinema Intelligence Service

**10.** *Letter Dimensions.* The maximum height of a first-class letter is 6.125 inches.
**Source:** United States Postal Service

Write in words, as on a check.
**11.** $524.95         **12.** $149.99         **13.** $36.72         **14.** $0.67

**b** Write each number as a fraction and, if possible, as a mixed numeral. Do not simplify.

**15.** 7.3         **16.** 4.9         **17.** 21.67         **18.** −57.32         **19.** −2.703

**20.** 0.079         **21.** 0.0109         **22.** 1.0008         **23.** −4.0003         **24.** −9.012

**25.** −0.0207         **26.** −0.00104         **27.** 70.00105         **28.** 60.0403

Write decimal notation for each number.

**29.** $\dfrac{3}{10}$          **30.** $\dfrac{73}{10}$          **31.** $-\dfrac{59}{100}$          **32.** $-\dfrac{67}{100}$          **33.** $\dfrac{3798}{1000}$

**34.** $\dfrac{780}{1000}$          **35.** $\dfrac{78}{10,000}$          **36.** $\dfrac{56,788}{100,000}$          **37.** $\dfrac{-18}{100,000}$          **38.** $\dfrac{-2347}{100}$

**39.** $\dfrac{486,197}{1,000,000}$          **40.** $\dfrac{8,953,074}{1,000,000}$          **41.** $7\dfrac{13}{1000}$          **42.** $4\dfrac{909}{1000}$          **43.** $-8\dfrac{431}{1000}$

**44.** $-49\dfrac{32}{1000}$          **45.** $2\dfrac{1739}{10,000}$          **46.** $9243\dfrac{1}{10}$          **47.** $8\dfrac{953,073}{1,000,000}$          **48.** $2256\dfrac{3059}{10,000}$

**c**    Which number is larger?

**49.** 0.06,   0.58          **50.** 0.008,   0.8          **51.** 0.403,   0.410          **52.** 42.06,   42.1

**53.** −5.046,   −5.043          **54.** −324.19,   −325.19          **55.** 234.07,   235.07          **56.** 0.99999,   1

**57.** 0.007,   $\dfrac{7}{100}$          **58.** $\dfrac{73}{10}$,   0.73          **59.** −0.872,   −0.873          **60.** −0.8437,   −0.84384

**d**    Round to the nearest tenth.

**61.** 0.23          **62.** 0.85          **63.** −0.372          **64.** −0.261

**65.** 2.951          **66.** 7.532          **67.** −327.2347          **68.** −8.7493

Round to the nearest hundredth.

**69.** 0.893          **70.** 0.675          **71.** −0.6666          **72.** −7.5252

**73.** 0.9952          **74.** 207.9976          **75.** −0.03488          **76.** −9.27481

Round to the nearest thousandth.

**77.** 0.5724          **78.** 0.6666          **79.** 17.0015          **80.** 123.4562

**81.** −20.20202          **82.** −0.10346          **83.** 9.98487          **84.** 67.100602

Round 809.47321 to the nearest

**85.** Tenth.          **86.** Thousandth.          **87.** Hundredth.          **88.** One.

## Skill Maintenance

Add or subtract, as indicated.

**89.**
$$\begin{array}{r} 6\ 8\ 1 \\ +\ 1\ 4\ 9 \\ \hline \end{array}$$ [1.2a]

**90.** $\dfrac{681}{1000} + \dfrac{149}{1000}$ [4.2a]

**91.**
$$\begin{array}{r} 2\ 6\ 7 \\ -\ \ \ 8\ 5 \\ \hline \end{array}$$ [1.3a]

**92.** $\dfrac{267}{100} - \dfrac{85}{100}$ [4.3a]

**93.** $\dfrac{37}{55} - \dfrac{49}{55}$ [4.3a]

**94.** $-\dfrac{29}{34} + \dfrac{14}{34}$ [4.2a]

**95.**
$$\begin{array}{r} 3\ 4,9\ 0\ 3 \\ -\ \ \ 1,9\ 4\ 5 \\ \hline \end{array}$$ [1.3a]

**96.**
$$\begin{array}{r} 4\ 9\ 3\ 7 \\ +\ 5\ 7\ 8\ 9 \\ \hline \end{array}$$ [1.2a]

## Synthesis

**97.** Arrange the following numbers in order from smallest to largest.

$$-0.989,\ -0.898,\ -1.009,\ -1.09,\ -0.098$$

**98.** Arrange the following numbers in order from smallest to largest.

$$-2.018,\ -2.1,\ -2.109,\ -2.0119,\ -2.108,$$
$$-2.000001$$

*Truncating.* There are other methods of rounding decimal notation. A computer often uses a method called **truncating**. To truncate we drop off decimal places right of the rounding place, which is the same as changing all digits to the right of the rounding place to zeros. For example, rounding 6.78093456285102 to the ninth decimal place, using truncating, gives us 6.780934562. Use truncating to round each of the following to the fifth decimal place, that is, the hundred-thousandth.

**99.** 6.78346123

**100.** 6.783461902

**101.** 99.999999999

**102.** 0.030303030303

*Temperature Change.* The graph below is based on the average U.S. temperature from 1951 through 1980. Each bar indicates, in Celsius degrees, how much above or below average the temperature was for the year. Use the graph for Exercises 103–106.

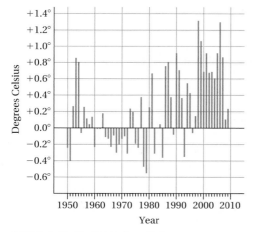

**Annual U.S. Average Temperature Change
(Degrees above or below the 1951–1980 mean)**

SOURCE: Goddard Institute for Space Studies

**103.** For what year(s) was the yearly temperature more than 1.2 degrees above average?

**104.** What was the last year in which the yearly temperature was more than 0.2 degree below average?

**105.** What was the last year in which the yearly temperature was below average?

**106.** For what year(s) was the yearly temperature less than 0.4 degree below average?

# 5.2

# Addition and Subtraction of Decimals

## OBJECTIVES

**a** Add using decimal notation.

**b** Subtract using decimal notation.

**c** Add and subtract negative decimals.

**d** Combine like terms with decimal coefficients.

### SKILLS TO REVIEW
Objective 1.2a: Add whole numbers.
Objective 1.3a: Subtract whole numbers.

Add or subtract.

1. 
$$\begin{array}{r} 3\ 8\ 7 \\ +\ 2\ 5\ 4 \end{array}$$

2. 
$$\begin{array}{r} 4\ 0\ 2\ 3 \\ -\ 1\ 6\ 6\ 7 \end{array}$$

Add.

1. 
$$\begin{array}{r} 0.8\ 4\ 7 \\ +\ 1\ 0.0\ 7 \end{array}$$

2. 
$$\begin{array}{r} 2.1 \\ 0.7\ 3 \\ +\ 3\ 1.3\ 6\ 8 \end{array}$$

Add.

3. $0.02 + 4.3 + 0.649$

4. $0.12 + 3.006 + 0.4357$

5. $0.4591 + 12.9 + 8.70894$

## a Addition

Adding with decimal notation is similar to adding whole numbers. First, we line up the decimal points so that we can add corresponding place-value digits. Then we add digits from the right. For example, we add the thousandths, then the hundredths, and so on, carrying if necessary. If desired, we can write extra zeros to the right of the decimal point so that the number of places is the same in all of the addends.

**EXAMPLE 1**  Add: $56.314 + 17.78$.

$$\begin{array}{r} 5\ 6\ .\ 3\ 1\ 4 \\ +\ 1\ 7\ .\ 7\ 8\ 0 \end{array}$$  Lining up the decimal points in order to add
Writing an extra zero to the far right of the decimal point

$$\begin{array}{r} 5\ 6\ .\ 3\ 1\ 4 \\ +\ 1\ 7\ .\ 7\ 8\ 0 \\ \hline 4 \end{array}$$  Adding thousandths

$$\begin{array}{r} 5\ 6\ .\ 3\ 1\ 4 \\ +\ 1\ 7\ .\ 7\ 8\ 0 \\ \hline 9\ 4 \end{array}$$  Adding hundredths

$$\begin{array}{r} \overset{1}{5}\ 6\ .\ 3\ 1\ 4 \\ +\ 1\ 7\ .\ 7\ 8\ 0 \\ \hline .\ 0\ 9\ 4 \end{array}$$  Adding tenths
We get 10 tenths = 1 one + 0 tenths, so we carry the 1 to the ones column.
Writing a decimal point in the answer

$$\begin{array}{r} \overset{1}{5}\ \overset{1}{6}\ .\ 3\ 1\ 4 \\ +\ 1\ 7\ .\ 7\ 8\ 0 \\ \hline 4\ .\ 0\ 9\ 4 \end{array}$$  Adding ones
We get 14 ones = 1 ten + 4 ones, so we carry the 1 to the tens column.

$$\begin{array}{r} \overset{1}{5}\ \overset{1}{6}\ .\ 3\ 1\ 4 \\ +\ 1\ 7\ .\ 7\ 8\ 0 \\ \hline 7\ 4\ .\ 0\ 9\ 4 \end{array}$$  Adding tens

Do Margin Exercises 1 and 2.

**EXAMPLE 2**  Add: $3.42 + 0.237 + 14.1$.

$$\begin{array}{r} 3.4\ 2\ 0 \\ 0.2\ 3\ 7 \\ +\ 1\ 4.1\ 0\ 0 \\ \hline 1\ 7.7\ 5\ 7 \end{array}$$  Lining up the decimal points and writing extra zeros

Adding

Do Exercises 3–5.

*Answers*

*Skills to Review:*
1. 641   2. 2356

*Margin Exercises:*
1. 10.917   2. 34.198   3. 4.969
4. 3.5617   5. 22.06804

Now we consider the addition 3456 + 19.347. Keep in mind that any whole number has an "unwritten" decimal point at the right that can be followed by zeros. For example, 3456 can also be written 3456.000. When adding, we can always write in the decimal point and extra zeros if desired.

**EXAMPLE 3**  Add: 3456 + 19.347.

$$
\begin{array}{r}
\overset{1}{3\ 4\ 5\ 6}.0\ 0\ 0 \\
+\quad\ \ 1\ 9.3\ 4\ 7 \\
\hline
3\ 4\ 7\ 5.3\ 4\ 7
\end{array}
$$

Writing in the decimal point and extra zeros
Lining up the decimal points
Adding

Do Exercises 6 and 7.

Do Exercises 6 and 7.

Add.
**6.** 789 + 123.67

**7.** 45.78 + 160 + 1.993

## **b**  Subtraction

Subtracting with decimal notation is similar to subtracting whole numbers. First, we line up the decimal points so that we can subtract corresponding place-value digits. Then we subtract digits from the right. In the example below, we subtract first the thousandths, then the hundredths, the tenths, and so on, borrowing if necessary.

**EXAMPLE 4**  Subtract: 56.314 − 17.78.

$$
\begin{array}{r}
5\ 6.3\ 1\ 4 \\
-\ 1\ 7.7\ 8\ 0 \\
\end{array}
$$

Lining up the decimal points in order to subtract
Writing an extra 0

$$
\begin{array}{r}
5\ 6.3\ 1\ 4 \\
-\ 1\ 7.7\ 8\ 0 \\
\hline
4
\end{array}
$$

Subtracting thousandths

$$
\begin{array}{r}
5\ 6.3\ \overset{2\ 11}{\cancel{1}\ 4} \\
-\ 1\ 7.7\ 8\ 0 \\
\hline
3\ 4
\end{array}
$$

Borrowing a tenth to subtract hundredths
Subtracting hundredths

$$
\begin{array}{r}
5\ \overset{12}{6}.\overset{5\ \ 2}{\cancel{3}\ \cancel{1}}\ 4 \\
-\ 1\ 7.7\ 8\ 0 \\
\hline
.5\ 3\ 4
\end{array}
$$

Borrowing a one to subtract tenths
Subtracting tenths; writing a decimal point

$$
\begin{array}{r}
\overset{15\ 12}{\cancel{5}}\ \overset{4\ 5\ 2}{\cancel{6}}.\cancel{3}\ \cancel{1}\ 4 \\
-\ 1\ 7.7\ 8\ 0 \\
\hline
8.5\ 3\ 4
\end{array}
$$

Borrowing a ten to subtract ones
Subtracting ones

$$
\begin{array}{r}
\overset{15\ 12}{\cancel{5}}\ \overset{4\ 5\ 2\ 11}{\cancel{6}}.\cancel{3}\ \cancel{1}\ 4 \\
-\ 1\ 7.7\ 8\ 0 \\
\hline
3\ 8.5\ 3\ 4
\end{array}
$$

Subtracting tens

Check by adding:

$$
\begin{array}{r}
\overset{1\ 1\ 1}{3\ 8.5\ 3\ 4} \\
+\ 1\ 7.7\ 8\ 0 \\
\hline
5\ 6.3\ 1\ 4 \longleftarrow
\end{array}
$$

The answer checks because this is the top number in the subtraction.

Do Exercises 8 and 9.

Do Exercises 8 and 9.

### STUDY TIPS

**NEATNESS COUNTS**
When working with decimals, make an extra effort to write neatly. Lining up decimal places and clearly distinguishing decimal points from commas will prevent mistakes that could result from sloppiness.

Subtract.
**8.** 37.428 − 26.674

**9.**
$$
\begin{array}{r}
0.3\ 4\ 7 \\
-\ 0.0\ 0\ 8
\end{array}
$$

***Answers***
**6.** 912.67  **7.** 207.773  **8.** 10.754
**9.** 0.339

Subtract.

**10.** $2.9 - 0.36$

**11.** $0.43 - 0.18762$

**12.** $5.27 - 0.00008$

---

**EXAMPLE 5**  Subtract: $23.08 - 5.0053$.

$$
\begin{array}{r}
\overset{1}{\phantom{0}}\overset{13}{\phantom{0}}\quad\overset{7}{\phantom{0}}\overset{9}{\phantom{0}}\overset{10}{\phantom{0}} \\
2\ \cancel{3}.0\ \cancel{8}\ \cancel{0}\ \cancel{0} \\
-\quad 5.0\ 0\ 5\ 3 \\
\hline
1\ 8.0\ 7\ 4\ 7
\end{array}
$$

Writing two extra zeros to the right of the last digit

Subtracting

Do Exercises 10-12.

When subtraction involves an integer, the "unwritten" decimal point can be written in. Extra zeros can then be written in to the right of the decimal point.

**EXAMPLE 6**  Subtract: $456 - 2.467$.

$$
\begin{array}{r}
\quad\overset{5}{\phantom{0}}\ \overset{9}{\phantom{0}}\ \overset{9}{\phantom{0}}\ \overset{10}{\phantom{0}} \\
4\ 5\ \cancel{6}.\cancel{0}\ \cancel{0}\ \cancel{0} \\
-\quad\quad 2.4\ 6\ 7 \\
\hline
4\ 5\ 3.5\ 3\ 3
\end{array}
$$

Writing in the decimal point and extra zeros

Subtracting

Do Exercises 13 and 14.

Subtract.

**13.** $1277 - 82.78$

**14.** $5 - 0.0089$

---

## (c) Adding and Subtracting with Negatives

Negative decimals are added or subtracted just like integers.

---

To add a negative number and a positive number:

**a)** Determine the sign of the number with the greater absolute value.

**b)** Subtract the smaller absolute value from the larger one.

**c)** The answer is the difference from part (b) with the sign from part (a).

---

**EXAMPLE 7**  Add: $-13.82 + 4.69$.

**a)** Since $|-13.82| > |4.69|$, the sign of the number with the greater absolute value is negative.

**b)**
$$
\begin{array}{r}
\quad\overset{7}{\phantom{0}}\ \overset{12}{\phantom{0}} \\
1\ 3.\cancel{8}\ \cancel{2} \\
-\quad 4.6\ 9 \\
\hline
9.1\ 3
\end{array}
$$
Finding the difference of the absolute values

**c)** The answer is negative: $-13.82 + 4.69 = -9.13$.

Do Exercises 15 and 16.

Add.

**15.** $7.42 + (-9.38)$

**16.** $-4.201 + 7.36$

---

To add two negative numbers:

**a)** Add the absolute values.

**b)** Make the answer negative.

---

**17.** Add: $-7.49 + (-5.8)$.

**EXAMPLE 8**  Add: $-2.306 + (-3.125)$.

**a)**
$$
\begin{array}{r}
\overset{1}{\phantom{0}} \\
2.3\ 0\ 6 \\
+\ 3.1\ 2\ 5 \\
\hline
5.4\ 3\ 1
\end{array}
$$
$|-2.306| = 2.306$ and $|-3.125| = 3.125$

Adding the absolute values

**b)** $-2.306 + (-3.125) = -5.431$    The answer is negative.

Do Exercise 17.

---

**Answers**

**10.** 2.54    **11.** 0.24238    **12.** 5.26992
**13.** 1194.22    **14.** 4.9911    **15.** −1.96
**16.** 3.159    **17.** −13.29

To subtract, we add the opposite of the number being subtracted.

**EXAMPLE 9** Subtract: $-3.1 - 4.8$.

$$-3.1 - 4.8 = -3.1 + (-4.8) \qquad \text{Adding the opposite of 4.8}$$
$$= -7.9 \qquad \text{The sum of two negative numbers is negative.}$$

**EXAMPLE 10** Subtract: $-7.9 - (-8.5)$.

$$-7.9 - (-8.5) = -7.9 + 8.5 \qquad \text{Adding the opposite of } -8.5$$
$$= 0.6 \qquad \text{Subtracting absolute values. The answer is positive since 8.5 has the larger absolute value.}$$

Do Exercises 18–21.

## (d) Combining Like Terms

Recall that like, or similar, terms have exactly the same variable factors. To combine like terms, we add or subtract coefficients to form an equivalent expression.

**EXAMPLE 11** Combine like terms: $3.2x + 4.6x$.

These are the coefficients.

$$3.2x + 4.6x = (3.2 + 4.6)x \qquad \text{Using the distributive law— try to do this step mentally.}$$
$$= 7.8x \qquad \text{Adding}$$

A similar procedure is used when subtracting like terms.

**EXAMPLE 12** Combine like terms: $4.13a - 7.56a$.

$$4.13a - 7.56a = (4.13 - 7.56)a \qquad \text{Using the distributive law}$$
$$= (4.13 + (-7.56))a \qquad \text{Adding the opposite of 7.56}$$
$$= -3.43a \qquad \text{Subtracting absolute values. The coefficient is negative since } |-7.56| > |4.13|.$$

When more than one pair of like terms is present, we can rearrange the terms and then simplify.

**EXAMPLE 13** Combine like terms: $5.7x - 3.9y - 2.4x + 4.5y$.

$$5.7x - 3.9y - 2.4x + 4.5y$$
$$= 5.7x + (-3.9y) + (-2.4x) + 4.5y \qquad \text{Rewriting as addition}$$
$$= 5.7x + (-2.4x) + 4.5y + (-3.9y) \qquad \text{Using the commutative law to rearrange}$$
$$= 3.3x + 0.6y \qquad \text{Combining like terms}$$

With practice, you will be able to perform many of the above steps mentally.

Do Exercises 22–24.

Subtract.

**18.** $9.25 - 13.41$

**19.** $-5.72 - 4.19$

**20.** $9.8 - (-2.6)$

**21.** $-5.9 - (-3.2)$

**Calculator Corner**

**Addition and Subtraction with Decimal Notation**

To use a calculator to add and subtract with decimal notation, we use the $\boxed{\cdot}$, $\boxed{+}$, $\boxed{-}$, and $\boxed{=}$ keys. To find $47.046 - 28.193$, for example, we press $\boxed{4}$ $\boxed{7}$ $\boxed{\cdot}$ $\boxed{0}$ $\boxed{4}$ $\boxed{6}$ $\boxed{-}$ $\boxed{2}$ $\boxed{8}$ $\boxed{\cdot}$ $\boxed{1}$ $\boxed{9}$ $\boxed{3}$ $\boxed{=}$. The display reads $\boxed{18.853}$, so $47.046 - 28.193 = 18.853$.

**Exercises:** Use a calculator to perform the indicated operations.

**1.**  $274.159$
   $+\ \ 43.486$

**2.** $3.4 + 45 + 0.68$

**3.**  $52.34$
   $-\ \ 18.51$

**4.** $6.09 - 5.1$

**5.** $246 + (-3.07)$

**6.** $-12.7 - (-1.008)$

Combine like terms.

**22.** $5.8x - 2.1x$

**23.** $-5.9a + 7.6a$

**24.** $-4.8y + 7.5 + 2.1y - 2.1$

*Answers*

**18.** $-4.16$   **19.** $-9.91$   **20.** $12.4$
**21.** $-2.7$   **22.** $3.7x$   **23.** $1.7a$
**24.** $-2.7y + 5.4$

## 5.2

# Exercise Set

For Extra Help

**MyMathLab**

Math XL
PRACTICE

WATCH

DOWNLOAD

READ

REVIEW

**a** Add.

**1.**    4 2 6.2 5
    +    3 8.1 2

**2.**    6 4 1.8 2 3
    +    1 4.9 1 5

**3.**    6 5 9.4 0 3
    + 9 1 6.6 1 2

**4.**    8 7 5.7 9 5
    + 3 2 4.8 6 2

**5.**        9.1 0 4
    + 1 2 3.4 5 6

**6.**        3.4 0 9
    + 8 1.0 0 1

**7.** 2.006 + 5.817

**8.** 0.8096 + 0.7856

**9.** 20.7 + 30.0124

**10.** 0.263 + 0.8

**11.** 1.06 + 9

**12.** 12 + 18.08

**13.** 0.34 + 3.5 + 0.127 + 768

**14.** 2.3 + 0.729 + 23

**15.** 17 + 3.24 + 0.256 + 0.3689

**16.**    4 7.8
    2 1 9.8 5 2
    4 3.5 9
    + 6 6 6.7 1 3

**17.**        2.7 0 3
    7 8.3 3
    2 8.0 0 0 9
    + 1 1 8.4 3 4 1

**18.**        1 3.7 2
        9.1 1 2
    6 5 4 2.7 9 0 8
    +        2 3.9 0 1

**b** Subtract.

**19.**    4 7.5 9 6
    −    6.2 1 5

**20.**    1 1.3 4 5
    −    2.1 0 5

**21.**    5 1.3 1
    −    2.2 9

**22.**    3 7.4 5
    −    6.3 2

**23.**
$$\begin{array}{r} 3.6 \\ -\ 0.0\ 3\ 6 \\ \hline \end{array}$$

**24.**
$$\begin{array}{r} 2\ 8.0 \\ -\ \ \ 0.2\ 8 \\ \hline \end{array}$$

**25.**
$$\begin{array}{r} 9\ 2.3\ 4\ 1 \\ -\ \ \ 6.4\ 2 \\ \hline \end{array}$$

**26.**
$$\begin{array}{r} 0.3\ 4\ 6 \\ -\ 0.0\ 3\ 4\ 6 \\ \hline \end{array}$$

**27.**
$$\begin{array}{r} 3.0\ 0\ 7\ 4 \\ -\ 1.3\ 4\ 0\ 8 \\ \hline \end{array}$$

**28.**
$$\begin{array}{r} 3\ 2.7\ 9\ 7\ 8 \\ -\ \ \ 0.0\ 5\ 9\ 2 \\ \hline \end{array}$$

**29.**
$$\begin{array}{r} 6.0\ 7 \\ -\ 2.0\ 0\ 7\ 8 \\ \hline \end{array}$$

**30.**
$$\begin{array}{r} 1.0 \\ -\ 0.9\ 9\ 9\ 9 \\ \hline \end{array}$$

**31.** $30.24 - 0.241$

**32.** $100.12 - 0.112$

**33.** $34.07 - 30.7$

**34.** $36.2 - 16.28$

**35.** $8.45 - 7.405$

**36.** $3.801 - 2.81$

**37.** $6.003 - 2.3$

**38.** $1 - 0.0098$

**39.** $2 - 1.0908$

**40.** $100 - 0.34$

**41.** $624 - 18.79$

**42.** $7.48 - 2.6$

**43.** $57.803 - 4.6$

**44.** $25.008 - 12.4$

**45.** $263.7 - 102.08$

**46.** $19 - 1.198$

**47.** $45 - 0.999$

**48.** $10.05 - 0.392$

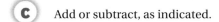 Add or subtract, as indicated.

**49.** $-5.02 + 1.73$

**50.** $-4.31 + 7.66$

**51.** $12.9 - 15.4$

**52.** $27.2 - 31.9$

**53.** $-2.9 + (-4.3)$

**54.** $-7.49 - 1.82$

**55.** $-4.301 + 7.68$

**56.** $-5.952 + 7.98$

**57.** $-12.9 - 3.7$

**58.** $-8.7 - 12.4$

**59.** $-2.1 - (-4.6)$

**60.** $-4.3 - (-2.5)$

**61.** $14.301 + (-17.82)$

**62.** $13.45 + (-18.701)$

**63.** $7.201 - (-2.4)$

**64.** $2.901 - (-5.7)$

**65.** $96.9 + (-21.4)$

**66.** $43.2 + (-10.9)$

**67.** $-3 - (-12.7)$

**68.** $-4.5 - (-7)$

**69.** $-4.9 - 5.392$

**70.** $89.3 - 100$

**71.** $14.7 - 15$

**72.** $-7.201 - 1.9$

**d**   Combine like terms.

**73.** $1.8x + 3.9x$

**74.** $7.9x + 1.3x$

**75.** $17.59a - 12.73a$

**76.** $23.28a - 15.79a$

**77.** $15.2t + 7.9 + 5.9t$

**78.** $29.5t - 4.8 + 7.6t$

**79.** $5.217x - 8.134x$

**80.** $6.317t - 9.429t$

**81.** $4.906y - 7.1 + 3.2y$

**82.** $9.108y + 4.2 + 3.7y$

**83.** $4.8x + 1.9y - 5.7x + 1.2y$

**84.** $3.2r - 4.1t - 5.6t + 1.9r$

**85.** $4.9 - 3.9t - 6 - 4.5t$

**86.** $5 + 9.7x - 7.2 - 12.8x$

## Skill Maintenance

Multiply.  [3.4b]

**87.** $\dfrac{3}{5} \cdot \dfrac{4}{7}$

**88.** $\dfrac{2}{9} \cdot \dfrac{7}{5}$

**89.** $\dfrac{3}{10} \cdot \dfrac{21}{100}$

Evaluate.  [2.6a]

**90.** $8 - 2x^2$, for $x = 3$

**91.** $5 - 3x^2$, for $x = -2$

**92.** $7 + 2x^2 \div 3$, for $x = 6$

## Synthesis

Combine like terms.

**93.** ▦ $-3.928 - 4.39a + 7.4b - 8.073 + 2.0001a - 9.931b - 9.8799a + 12.897b$

**94.** ▦ $79.02x + 0.0093y - 53.14z - 0.02001y - 37.987z - 97.203x - 0.00987y$

**95.** ▦ $39.123a - 42.458b - 72.457a + 31.462b - 59.491 + 37.927a$

**96.** Ryan presses the wrong key when using a calculator and adds 235.7 instead of subtracting it. The incorrect answer is 817.2. What is the correct answer?

**97.** Alicia presses the wrong key when using a calculator and subtracts 349.2 instead of adding it. The incorrect answer is −836.9. What is the correct answer?

**98.** ▦ Find the errors, if any, in the balances in this checkbook.

| 20___ | | RECORD ALL CHARGES OR CREDITS THAT AFFECT YOUR ACCOUNT | | | | | | |
|---|---|---|---|---|---|---|---|---|
| DATE | CHECK NUMBER | TRANSACTION DESCRIPTION | √ T | (−) PAYMENT/ DEBIT | (+ OR −) OTHER | (+) DEPOSIT/ CREDIT | BALANCE FORWARD | |
| | | | | | | | 2767 | 73 |
| 8/16 | 432 | Burch Laundry | | 23 56 | | | 2744 | 16 |
| 8/19 | 433 | Rogers TV | | 20 49 | | | 2764 | 65 |
| 8/20 | | Deposit | | | | 85 00 | 2848 | 65 |
| 8/21 | 434 | Galaxy Records | | 48 60 | | | 2801 | 05 |
| 8/22 | 435 | Electric Works | | 267 95 | | | 2533 | 09 |
| | | | | | | | | |

Find *a*.

**99.**
```
  9 3.a 4 3
- 8 7.9 6 9
───────────
    5.2 7 4
```

**100.**
```
  4 8 1.a 2 4
-   7 2.9 7 8
─────────────
  4 0 8.3 4 6
```

# 5.3

# Multiplication of Decimals

## OBJECTIVES

**a** Multiply using decimal notation.

**b** Convert from notation like 45.7 million to standard notation and convert between dollars and cents.

**c** Evaluate algebraic expressions using decimal notation.

**SKILL TO REVIEW**

Objective 1.4a: Multiply whole numbers.

Multiply.

**1.** 
$$\begin{array}{r} 4\ 2 \\ \times\ 6\ 3 \end{array}$$

**2.** 
$$\begin{array}{r} 7\ 1\ 6 \\ \times\ \ \ 5\ 8 \end{array}$$

## STUDY TIPS

### DON'T GET STUCK ON A QUESTION

Unless you know the material "cold," don't be surprised if a quiz or test includes a question for which you feel unprepared. Should this happen, do not get rattled—simply skip the question and continue with the quiz or test, returning to the trouble spot after the other questions have been answered.

## a Multiplication

To develop an understanding of decimal multiplication, consider $2.3 \times 1.12$:

$$2.3 \times 1.12 = \frac{23}{10} \times \frac{112}{100} = \frac{2576}{1000} = 2.576.$$

Note that the number of decimal places in the product is the sum of the numbers of decimal places in the factors.

$$\begin{array}{r} 1.1\ 2 \quad (2\ \text{decimal places}) \\ \times\ \ \ 2.3 \quad (1\ \text{decimal place}) \\ \hline 2.5\ 7\ 6 \quad (3\ \text{decimal places}) \end{array}$$

Now consider $0.02 \times 3.412$:

$$0.02 \times 3.412 = \frac{2}{100} \times \frac{3412}{1000} = \frac{6824}{100,000} = 0.06824.$$

Again, note that the number of decimal places in the product is the sum of the numbers of decimal places in the factors.

$$\begin{array}{r} 3.4\ 1\ 2 \quad (3\ \text{decimal places}) \\ \times\ \ \ \ \ 0.0\ 2 \quad (2\ \text{decimal places}) \\ \hline 0.0\ 6\ 8\ 2\ 4 \quad (5\ \text{decimal places}) \end{array}$$ The 0 after the decimal point is necessary.

---

To multiply using decimal notation:  $0.8 \times 0.43$

**a)** Ignore the decimal points, for the moment, and multiply as though both factors were integers.

$$\begin{array}{r} \overset{2}{0.4}\ 3 \\ \times\ \ \ 0.8 \\ \hline 3\ 4\ 4 \end{array}$$ Ignore the decimal points for now.

**b)** Place the decimal point in the result. The number of decimal places in the product is the sum of the numbers of places in the factors.

$$\begin{array}{r} 0.4\ 3 \quad (2\ \text{decimal places}) \\ \times\ \ \ 0.8 \quad (1\ \text{decimal place}) \\ \hline 0.3\ 4\ 4 \quad (3\ \text{decimal places}) \end{array}$$

Count the number of decimal places by starting at the far right and moving the decimal point to the left.

---

**EXAMPLE 1** Multiply: $8.3 \times 74.6$.

**a)** Ignore the decimal points and multiply as if both factors were integers.

$$\begin{array}{r} \overset{3}{\phantom{0}}\overset{4}{\phantom{0}} \\ \overset{1}{\phantom{0}}\overset{1}{\phantom{0}} \\ 7\ 4.6 \\ \times\ \ \ \ \ 8.3 \\ \hline 2\ 2\ 3\ 8 \\ 5\ 9\ 6\ 8\ 0 \\ \hline 6\ 1\ 9\ 1\ 8 \end{array}$$ We are not yet finished.

*Answers*

*Skill to Review:*

**1.** 2646  **2.** 41,528

**b)** Place the decimal point in the result. The number of decimal places in the product is the sum, $1 + 1$, of the numbers of decimal places in the factors.

$$
\begin{array}{r}
7\ 4.6 \quad \text{(1 decimal place)} \\
\times \quad\ \ 8.3 \quad \text{(1 decimal place)} \\
\hline
2\ 2\ 3\ 8 \\
5\ 9\ 6\ 8\ 0 \\
\hline
6\ 1\ 9.1\ 8 \quad \text{(2 decimal places)}
\end{array}
$$

Do Exercise 1.

**EXAMPLE 2** Multiply: $0.0032 \times 2148$.

$$
\begin{array}{r}
2\ 1\ 4\ 8 \quad \text{(0 decimal places)} \\
\times\ 0.0\ 0\ 3\ 2 \quad \text{(4 decimal places)} \\
\hline
4\ 2\ 9\ 6 \\
6\ 4\ 4\ 4\ 0 \\
\hline
6.8\ 7\ 3\ 6 \quad \text{(4 decimal places)}
\end{array}
$$

**EXAMPLE 3** Multiply: $-0.14 \times 0.867$.

Multiplying the absolute values, we have

$$
\begin{array}{r}
0.8\ 6\ 7 \quad \text{(3 decimal places)} \\
\times \quad\ \ 0.1\ 4 \quad \text{(2 decimal places)} \\
\hline
3\ 4\ 6\ 8 \\
8\ 6\ 7\ 0 \\
\hline
0.1\ 2\ 1\ 3\ 8 \quad \text{(5 decimal places)}
\end{array}
$$

Since the product of a negative number and a positive number is negative, the answer is $-0.12138$.

We may need to write extra zeros when locating the decimal point.

**EXAMPLE 4** Multiply: $0.018 \times 2.7$.

$$
\begin{array}{r}
0.0\ 1\ 8 \quad \text{(3 decimal places)} \\
\times \quad\ \ 2.7 \quad \text{(1 decimal place)} \\
\hline
1\ 2\ 6 \\
3\ 6\ 0 \\
\hline
0.0\ 4\ 8\ 6 \quad \text{(4 decimal places)} \quad \text{We write an extra zero.}
\end{array}
$$

Do Exercises 2-4.

## Multiplying by 0.1, 0.01, 0.001, and So On

Suppose that a product involves multiplying by a tenth, hundredth, thousandth, or ten-thousandth. We can see a pattern in the following products.

$$
\begin{array}{cccc}
4\ 5.6 & 4\ 5.6 & 4\ 5.6 & 4\ 5.6 \\
\times\ \ 0.1 & \times\ 0.0\ 1 & \times\ 0.0\ 0\ 1 & \times\ 0.0\ 0\ 0\ 1 \\
\hline
4.5\ 6 & 0.4\ 5\ 6 & 0.0\ 4\ 5\ 6 & 0.0\ 0\ 4\ 5\ 6
\end{array}
$$

Note in each case that the product is *smaller* than 45.6. That is, the decimal point in each product is farther to the left than it is in 45.6. Note also that each product can be obtained from 45.6 by moving the decimal point.

---

**1.** Multiply:

$$
\begin{array}{r}
7\ 6.3 \\
\times \quad\ \ 8.2 \\
\end{array}
$$

Multiply.

**2.**
$$
\begin{array}{r}
4\ 2\ 1\ 3 \\
\times\ 0.0\ 0\ 5\ 1 \\
\end{array}
$$

**3.** $2.3 \times 0.0041$

**4.** $5.2014 \times (-2.41)$

*Answers*

**1.** 625.66   **2.** 21.4863   **3.** 0.00943
**4.** $-12.535374$

5.3 Multiplication of Decimals   **325**

To multiply any number by 0.1, 0.01, 0.001, and so on:

a) count the number of decimal places in the tenth, hundredth, thousandth, and so on, and

$0.001 \times 34.45678$

3 places

b) move the decimal point in the other number that many places to the left. Use zeros as placeholders when necessary.

$0.001 \times 34.45678 = 0.034.45678$

Move 3 places to the left.

$0.001 \times 34.45678 = 0.03445678$

---

**EXAMPLES**   Multiply.

**5.** $0.1 \times 45 = 4.5$   Moving the decimal point one place to the left

**6.** $0.01 \times 243.7 = 2.437$   Moving the decimal point two places to the left

**7.** $0.001 \times (-8.2) = -0.0082$   Moving the decimal point three places to the left. This requires writing two extra zeros.

**8.** $0.0001 \times 536.9 = 0.05369$   Moving the decimal point four places to the left. This requires writing one extra zero.

Do Exercises 5–8.

---

Multiply.

**5.** $0.1 \times 746$

**6.** $0.001 \times 732.4$

**7.** $(-0.01) \times 6.2$

**8.** $0.0001 \times 723.6$

---

### Multiplying by 10, 100, 1000, and So On

Next we consider multiplying by 10, 100, 1000, and so on. We see a pattern in the following.

$$
\begin{array}{r}
5.2\ 3\ 7 \\
\times\quad\ 1\ 0 \\
\hline
0\ 0\ 0\ 0 \\
5\ 2\ 3\ 7 \\
\hline
5\ 2.3\ 7\ 0
\end{array}
\qquad
\begin{array}{r}
5.2\ 3\ 7 \\
\times\quad\ 1\ 0\ 0 \\
\hline
0\ 0\ 0\ 0 \\
0\ 0\ 0\ 0 \\
5\ 2\ 3\ 7 \\
\hline
5\ 2\ 3.7\ 0\ 0
\end{array}
\qquad
\begin{array}{r}
5.2\ 3\ 7 \\
\times\quad\ 1\ 0\ 0\ 0 \\
\hline
0\ 0\ 0\ 0 \\
0\ 0\ 0\ 0 \\
0\ 0\ 0\ 0 \\
5\ 2\ 3\ 7 \\
\hline
5\ 2\ 3\ 7.0\ 0\ 0
\end{array}
$$

Note in each case that the product is *larger* than 5.237. That is, the decimal point in each product is farther to the right than the decimal point in 5.237. Also, each product can be obtained from 5.237 by moving the decimal point.

---

To multiply any number by 10, 100, 1000, and so on:

a) count the number of zeros and

$1000 \times 34.45678$

3 zeros

b) move the decimal point in the other number that many places to the right. Use zeros as placeholders when necessary.

$1000 \times 34.45678 = 34.456.78$

Move 3 places to the right.

$1000 \times 34.45678 = 34{,}456.78$

---

---

*Answers*

**5.** 74.6   **6.** 0.7324
**7.** −0.062   **8.** 0.07236

**EXAMPLES** Multiply.

**9.** $10 \times 32.98 = 329.8$     Moving the decimal point one place to the right

**10.** $100 \times 4.7 = 470$     Moving the decimal point two places to the right. The 0 in 470 is a placeholder.

**11.** $1000 \times (-2.4167) = -2416.7$     Moving the decimal point three places to the right

**12.** $10,000 \times 7.52 = 75,200$     Moving the decimal point four places to the right and using two zeros as placeholders

> Do Exercises 9–12.

Multiply.

**9.** $10 \times 53.917$

**10.** $100 \times (-62.417)$

**11.** $1000 \times 83.9$

**12.** $10,000 \times 57.04$

## b   Applications Using Multiplication with Decimal Notation

### Naming Large Numbers

We often see notation like the following in newspapers and magazines and on television and the Internet.

- U.S. venture-capital firms invested $2.6 billion in alternative-energy startup enterprises during the first three quarters of 2007.
  Source: National Venture Capital Association

- Each day about 39.7 billion person-to-person e-mails, 17.1 billion automated e-mails, and 40.5 billion pieces of spam are sent worldwide.
  Source: IDC

- The largest building in the world is the Pentagon, which has 3.7 million square feet of floor space.

To understand such notation, consider the information in the following table.

**NAMING LARGE NUMBERS**

1 hundred $= 100 = 10 \cdot 10 = 10^2$
         ⟶ 2 zeros

1 thousand $= 1000 = 10 \cdot 10 \cdot 10 = 10^3$
         ⟶ 3 zeros

1 million $= 1,000,000 = 10 \cdot 10 \cdot 10 \cdot 10 \cdot 10 \cdot 10 = 10^6$
         ⟶ 6 zeros

1 billion $= 1,000,000,000 = 10^9$
         ⟶ 9 zeros

1 trillion $= 1,000,000,000,000 = 10^{12}$
         ⟶ 12 zeros

Convert the number in each sentence to standard notation.

**13.** The largest building in the world is the Pentagon, which has 3.7 million square feet of floor space.

**14.** Each day about 39.7 billion personal e-mails are sent.

**15.** Losses due to identity fraud totaled $49.3 billion in a recent year.

To convert a large number to standard notation, we proceed as follows.

**EXAMPLE 13** U.S. movie theaters sold 1.42 billion tickets in 2009. Convert 1.42 billion to standard notation.

**Source:** Nash Information Services

$$1.42 \text{ billion} = 1.42 \times 1 \text{ billion}$$
$$= 1.42 \times 1,\underbrace{000,000,000}_{\text{9 zeros}}$$
$$= 1,420,000,000 \quad \text{Moving the decimal point 9 places to the right}$$

Do Exercises 13–15.

## Money Conversion

Converting from dollars to cents is like multiplying by 100. To see why, consider $19.43.

$$\$19.43 = 19.43 \times \$1 \qquad \text{We think of \$19.43 as } 19.43 \times 1 \text{ dollar, or } 19.43 \times \$1.$$
$$= 19.43 \times 100\cent \qquad \text{Substituting } 100\cent \text{ for } \$1: \$1 = 100\cent$$
$$= 1943\cent \qquad \text{Multiplying}$$

> **DOLLARS TO CENTS**
>
> To convert from dollars to cents, move the decimal point two places to the right and change the $ sign in front to a ¢ sign at the end.

**EXAMPLES** Convert from dollars to cents.

**14.** $189.64 = 18,964¢

**15.** $0.75 = 75¢

Do Exercises 16 and 17.

Convert from dollars to cents.
**16.** $15.69      **17.** $0.17

Converting from cents to dollars is like multiplying by 0.01. To see why, consider 65¢.

$$65\cent = 65 \times 1\cent \qquad \text{We think of } 65\cent \text{ as } 65 \times 1 \text{ cent, or } 65 \times 1\cent.$$
$$= 65 \times \$0.01 \qquad \text{Substituting } \$0.01 \text{ for } 1\cent: 1\cent = \$0.01$$
$$= \$0.65 \qquad \text{Multiplying}$$

> **CENTS TO DOLLARS**
>
> To convert from cents to dollars, move the decimal point two places to the left and change the ¢ sign at the end to a $ sign in front.

Convert from cents to dollars.
**18.** 35¢      **19.** 577¢

**EXAMPLES** Convert from cents to dollars.

**16.** 395¢ = $3.95

**17.** 8503¢ = $85.03

Do Exercises 18 and 19.

*Answers*

**13.** 3,700,000   **14.** 39,700,000,000
**15.** $49,300,000,000   **16.** 1569¢
**17.** 17¢   **18.** $0.35   **19.** $5.77

## (c) Evaluating

Algebraic expressions are often evaluated using numbers written in decimal notation.

**EXAMPLE 18**   Evaluate $Prt$ for $P = 780$, $r = 0.12$, and $t = 0.5$.

We will see in Chapter 8 that this product could be used to determine the interest paid on $780, borrowed at 12 percent simple interest, for half a year. We substitute as follows.

$$Prt = 780 \cdot 0.12 \cdot 0.5 = 780 \cdot 0.06 = 46.8 \qquad \text{This would represent } \$46.80.$$

Do Exercise 20.

**20.** Evaluate $lwh$ for $l = 3.2$, $w = 2.6$, and $h = 0.8$. (This is the formula for the volume of a rectangular box.)

**EXAMPLE 19**   Find the perimeter of a stamp that is 3.25 cm long and 2.5 cm wide.

Recall that the perimeter, $P$, of a rectangle of length $l$ and width $w$ is given by the formula

$$P = 2l + 2w.$$

Thus, we evaluate $2l + 2w$ for $l = 3.25$ and $w = 2.5$.

$$2l + 2w = 2 \cdot 3.25 + 2 \cdot 2.5$$
$$= 6.5 + 5.0 \qquad \text{Remember the rules for order of operations.}$$
$$= 11.5$$

The perimeter is 11.5 cm.

Do Exercise 21.

**21.** Find the area of the stamp in Example 19.

**EXAMPLE 20**   *Multiple Births.*   The expression $2.67t + 65.02$ can be used to predict the number of twin births, in thousands, in the United States $t$ years after 1980. Predict the number of twin births in 2015.

Source: Based on information from the Centers for Disease Control

2015 is 35 years after 1980, so we evaluate $2.67t + 65.02$ for $t = 35$.

$$2.67t + 65.02 = 2.67 \cdot 35 + 65.02$$
$$= 93.45 + 65.02$$
$$= 158.47$$

In 2015, there will be approximately 158.47 thousand, or 158,470, twin births.

Do Exercise 22.

**22.** Evaluate $6.28rh + 3.14r^2$ for $r = 1.5$ and $h = 5.1$. (This is the formula for the area of an open can.)

*Answers*

**20.** 6.656    **21.** 8.125 sq cm    **22.** 55.107

**a** Multiply.

**1.**
$$
\begin{array}{r}
6.8 \\
\times\quad 7 \\
\hline
\end{array}
$$

**2.**
$$
\begin{array}{r}
5.7 \\
\times\ 0.9 \\
\hline
\end{array}
$$

**3.**
$$
\begin{array}{r}
0.8\,4 \\
\times\qquad 8 \\
\hline
\end{array}
$$

**4.**
$$
\begin{array}{r}
7.3 \\
\times\ 0.6 \\
\hline
\end{array}
$$

**5.**
$$
\begin{array}{r}
6.3 \\
\times\ 0.0\,4 \\
\hline
\end{array}
$$

**6.**
$$
\begin{array}{r}
7.8 \\
\times\ 0.0\,9 \\
\hline
\end{array}
$$

**7.**
$$
\begin{array}{r}
8\,7 \\
\times\ 0.0\,0\,6 \\
\hline
\end{array}
$$

**8.**
$$
\begin{array}{r}
2\,5.9 \\
\times\ 0.0\,0\,5 \\
\hline
\end{array}
$$

**9.** $10 \times 42.63$

**10.** $100 \times 2.8793$

**11.** $-1000 \times 783.686852$

**12.** $-0.34 \times 1000$

**13.** $-7.8 \times 100$

**14.** $0.00238 \times (-10)$

**15.** $0.1 \times 79.18$

**16.** $0.01 \times 789.235$

**17.** $0.001 \times 97.68$

**18.** $8976.23 \times 0.001$

**19.** $28.7 \times (-0.01)$

**20.** $0.0325 \times (-0.1)$

**21.**
$$
\begin{array}{r}
2.7\,3 \\
\times\quad 1\,6 \\
\hline
\end{array}
$$

**22.**
$$
\begin{array}{r}
8.2\,7 \\
\times\quad 5.4 \\
\hline
\end{array}
$$

**23.**
$$
\begin{array}{r}
0.9\,8\,4 \\
\times\ 0.0\,3\,1 \\
\hline
\end{array}
$$

**24.**
$$
\begin{array}{r}
7.4\,8\,9 \\
\times\qquad 1.7 \\
\hline
\end{array}
$$

**25.** $(-37.4)(-2.4)$

**26.** $569(-1.05)$

**27.** $749(-0.43)$

**28.** $(-0.876)(-0.0204)$

**29.**
$$
\begin{array}{r}
0.8\,7 \\
\times\quad 6\,4 \\
\hline
\end{array}
$$

**30.**
$$
\begin{array}{r}
7.2\,5 \\
\times\quad 6\,0 \\
\hline
\end{array}
$$

**31.**
$$
\begin{array}{r}
4\,6.5\,0 \\
\times\qquad 7\,5 \\
\hline
\end{array}
$$

**32.**
$$
\begin{array}{r}
8.2\,4 \\
\times\ 7\,0\,3 \\
\hline
\end{array}
$$

**33.** $(-0.231)(-0.5)$

**34.** $(-12.3)(-1.08)$

**35.** $9.42 \times (-1000)$

**36.** $-7.6 \times 1000$

**37.** $-95.3 \times (-0.0001)$

**38.** $-4.23 \times (-0.001)$

**b** Convert from dollars to cents.

**39.** $57.06     **40.** $49.85     **41.** $0.95     **42.** $0.49     **43.** $0.01     **44.** $0.09

Convert from cents to dollars.

**45.** 72¢     **46.** 52¢     **47.** 2¢     **48.** 5¢     **49.** 6399¢     **50.** 5238¢

**51.** *Identity Fraud.* About 11.1 million adults were victims of identity fraud in the United States in 2009. Convert 11.1 million to standard notation.
**Source:** Javelin Strategy & Research

**52.** *Low-Cost Laptops.* It is estimated that 9.2 million low-cost laptop computers will be shipped worldwide in 2012. Convert 9.2 million to standard notation.
**Source:** IDC

**53.** *Text Messages.* In December 2009, 152.7 billion text messages were sent in the United States. Convert 152.7 billion to standard notation.
**Source:** CTIA

**54.** *Cell-Phone Ads.* Global spending on advertisements appearing on cell phones is projected to be $16.2 billion in 2011. Convert 16.2 billion to standard notation.
**Source:** eMarketer

**55.** A century is approximately 3.156 billion seconds. Convert 3.156 billion to standard notation.

**56.** The total surface area of earth is 196.8 million square miles. Convert 196.8 million to standard notation.

 **c** Evaluate.

**57.** $P + Prt$, for $P = 10,000$, $r = 0.04$, and $t = 2.5$
(*A formula for adding interest*)

**58.** $6.28r(h + r)$, for $r = 10$ and $h = 17.2$
(*Surface area of a cylinder*)

**59.** $vt + 0.5at^2$, for $v = 10$, $t = 1.5$, and $a = 9.8$
(*A physics formula*)

**60.** $4lh + 2h^2$, for $l = 3.5$ and $h = 1.2$
(*Surface area of a rectangular prism*)

Find **(a)** the perimeter and **(b)** the area of a rectangular room with the given dimensions.

**61.** 12.5 ft long, 9.5 ft wide

**62.** 10.25 ft long, 8 ft wide

**63.** 8.4 m wide, 10.5 m long

**64.** 8.2 yd long, 6.4 yd wide

*Nursing.* The expression $0.0375t + 2.2$ can be used to predict the number of registered nurses, in millions, in the United States $t$ years after 2000. Predict the number of registered nurses in the United States in the year indicated.
**Source:** Based on information from the Bureau of Labor Statistics, U.S. Dept. of Labor

**65.** 2015

**66.** 2020

## Skill Maintenance

Divide.

**67.** $-162 \div 6$   [2.5a]

**68.** $-216 \div (-6)$   [2.5a]

**69.** $-1035 \div (-15)$   [2.5a]

**70.** $-423 \div 3$   [2.5a]

**71.** $-525 \div 25$   [2.5a]

**72.** $675 \div (-25)$   [2.5a]

**73.** $18\overline{)22{,}626}$   [1.5a]

**74.** $17\overline{)20{,}006}$   [1.5a]

## Synthesis

Consider the following names for large numbers in addition to those already discussed in this section:

1 quadrillion $= 1{,}000{,}000{,}000{,}000{,}000 = 10^{15}$;

1 quintillion $= 1{,}000{,}000{,}000{,}000{,}000{,}000 = 10^{18}$;

1 sextillion $= 1{,}000{,}000{,}000{,}000{,}000{,}000{,}000 = 10^{21}$;

1 septillion $= 1{,}000{,}000{,}000{,}000{,}000{,}000{,}000{,}000 = 10^{24}$.

Find each of the following. Express the answer with a name that is a power of 10.

**75.** (1 trillion) $\cdot$ (1 billion)

**76.** (1 million) $\cdot$ (1 billion)

**77.** (1 trillion) $\cdot$ (1 trillion)

**78.** Is a billion millions the same as a million billions? Explain.

**79.** In Great Britain, France, and Germany, a billion means a million millions. Write standard notation for the British number 6.6 billion.

**80.** One light-year (LY) is $9.46 \times 10^{12}$ km. The star Regulus is 85 LY from earth. How many billions of kilometers (km) from earth is Regulus?

Source: *The Cambridge Factfinder*, 4th ed

▦ Evaluate using a calculator.

**81.** $d + vt + at^2$, for $d = 79.2$, $v = 3.029$, $t = 7.355$, and $a = 4.9$ (*A physics formula for distance traveled*)

**82.** $3.14r^2 + 6.28rh$, for $r = 5.756$ in. and $h = 9.047$ in. (*Surface area of a toy silo, including bottom*)

**83.** $0.5(b_1 + b_2)h$, for $b_1 = 9.7$ cm, $b_2 = 13.4$ cm, and $h = 6.32$ cm (*A geometry formula for the area of a trapezoid*)

**84.** $0.5bh$, for $b = 12.59$ cm and $h = 13.72$ cm (*A formula for the area of a triangle*)

**85.** ▦ *Electric Bills.* Recently, electric bills from the Central Vermont Public Service Corporation consisted of a "customer charge" of $0.374 per day plus an "energy charge" of $0.1174 per kilowatt-hour (kWh) for the first 250 kWh used and $0.09079 per kilowatt-hour for each kilowatt-hour in excess of 250. From April 20 to May 20, the Coy-Bergers used 480 kWh of electricity. What was their bill for the period?

| **C V P S C** **Central Vermont** **Public Service** **Corporation** | Coy-Berger R.R. 1 Braintree, VT 05060-9601 | Account No. **54879582** Meter No. **3621458** |
|---|---|---|

**ELECTRIC SERVICE**

| | | |
|---|---|---|
| Current Meter Reading, 05/20 (Actual) | | 22571 |
| Previous Meter Reading, 04/20 (Actual) | − | 22091 |
| Amount of electricity Used in 30 Days | kWh | 480 |

Cost of Electricity Used for 30 Days Ending 05/20

| | | |
|---|---|---|
| Customer Charge . . . . . . . . . | $0.374 per day | $ |
| Energy Charge . . . . . . . | kWh × $0.11740 | $ |
| | kWh × $0.09079 | |

# 5.4

# Division of Decimals

## a Division

### Whole-Number Divisors

Now that we have studied multiplication of decimals, we can develop a procedure for division. The following divisions are justified by the multiplication in each *check*:

This is the dividend. ⟶ 651
This is the divisor. ⟶ $\dfrac{651}{7} = 93$     *Check*: $7 \cdot 93 = 651$

— This is the quotient.

$$\dfrac{65.1}{7} = 9.3 \qquad \text{\textit{Check}:} \ 7 \cdot 9.3 = 65.1$$

$$\dfrac{6.51}{7} = 0.93 \qquad \text{\textit{Check}:} \ 7 \cdot 0.93 = 6.51$$

$$\dfrac{0.651}{7} = 0.093 \qquad \text{\textit{Check}:} \ 7 \cdot 0.093 = 0.651$$

Note that the number of decimal places in each quotient is the same as the number of decimal places in the dividend.

---

To perform long division by a whole number:

a) place the decimal point directly above the decimal point in the dividend and

b) divide as though dividing whole numbers.

```
         0.8 4  ← Quotient
    7 ) 5.8 8   ← Dividend
        5 6 0
          2 8
          2 8
            0  ← Remainder
```
↑ Divisor

---

**EXAMPLE 1**  Divide: $82.08 \div 24$.

We have

— Place the decimal point.

```
         3.4 2
    2 4 ) 8 2.0 8
          7 2 0 0
          1 0 0 8
            9 6 0
              4 8
              4 8
                0
```
Divide as though dividing whole numbers.

Since $(3.42)(24) = 82.08$, the answer checks.

Do Margin Exercises 1–3.

## OBJECTIVES

a  Divide using decimal notation.

b  Simplify expressions using the rules for order of operations.

**SKILL TO REVIEW**
Objective 1.5a: Divide whole numbers.

Divide.

**1.** $5 \overline{\smash{)}2\,4\,5}$

**2.** $2\,3 \overline{\smash{)}1\,9\,7\,8}$

Divide.

**1.** $9 \overline{\smash{)}5.4}$

**2.** $1\,5 \overline{\smash{)}2\,2.5}$

**3.** $8\,2 \overline{\smash{)}3\,8.5\,4}$

***Answers***

*Skill to Review:*
**1.** 49   **2.** 86

*Margin Exercises:*
**1.** 0.6   **2.** 1.5   **3.** 0.47

We can think of a whole-number dividend as having a decimal point at the end with as many 0's as we wish after the decimal point. For example, $12 = 12. = 12.0 = 12.00 = 12.000$, and so on. We can also add 0's after the last digit in the decimal portion of a number: $3.6 = 3.60 = 3.600$, and so on.

**EXAMPLE 2** Divide: $30 \div 8$.

$$
\begin{array}{r}
3.\phantom{0} \\
8\overline{)\phantom{0}3\ 0.} \\
2\ 4 \\
\hline
6
\end{array}
$$
Place the decimal point and divide to find how many ones.

$$
\begin{array}{r}
3.\phantom{0} \\
8\overline{)\phantom{0}3\ 0.0} \\
2\ 4\ \downarrow \\
\hline
6\ 0
\end{array}
$$
Write an extra zero to the far right of the decimal point. This does not change the number. We can bring down one digit at a time.

$$
\begin{array}{r}
3.7 \\
8\overline{)\phantom{0}3\ 0.0} \\
2\ 4 \\
\hline
6\ 0 \\
5\ 6 \\
\hline
4
\end{array}
$$
Divide to find how many tenths.

$$
\begin{array}{r}
3.7\phantom{0} \\
8\overline{)\phantom{0}3\ 0.0\ 0} \\
2\ 4 \\
\hline
6\ 0 \\
5\ 6\ \downarrow \\
\hline
4\ 0
\end{array}
$$
Write another zero.

$$
\begin{array}{r}
3.7\ 5 \\
8\overline{)\phantom{0}3\ 0.0\ 0} \\
2\ 4 \\
\hline
6\ 0 \\
5\ 6 \\
\hline
4\ 0 \\
4\ 0 \\
\hline
0
\end{array}
$$
Divide to find how many hundredths.

Check:
$$
\begin{array}{r}
\overset{6\phantom{.}4}{3.7\ 5} \\
\times \phantom{00}8 \\
\hline
3\ 0.0\ 0
\end{array}
$$

**EXAMPLE 3** Divide: $-4.5 \div 25$.

We first consider $4.5 \div 25$.

$$
\begin{array}{r}
0.1\ 8 \\
25\overline{)\phantom{0}4.5\ 0} \\
2\ 5 \\
\hline
2\ 0\ 0 \\
2\ 0\ 0 \\
\hline
0
\end{array}
$$
$\longleftarrow$ Since the remainder is 0, we are finished.

Since a negative number divided by a positive number is negative, the answer is $-0.18$. To check, note that $(-0.18)25 = -4.5$.

Do Exercises 4–6.

Divide.

**4.** $25\overline{)8}$

**5.** $-23 \div 4$

**6.** $86\overline{)21.5}$

**Answers**

**4.** 0.32   **5.** $-5.75$   **6.** 0.25

## Divisors That Are Not Whole Numbers

Consider the division

$$0.2\,4\,\overline{)\,8.2\,0\,8}$$

We write the division as $\dfrac{8.208}{0.24}$. Then we multiply by 1 to change to a whole-number divisor:

The division $0.24\overline{)8.208}$ is the same as $24\overline{)820.8}$.

$$\frac{8.208}{0.24} = \frac{8.208}{0.24} \times \frac{100}{100} = \frac{820.8}{24}.$$

The divisor is now a whole number.

To divide when the divisor is not a whole number:

**a)** move the decimal point (multiply by 10, 100, and so on) to make the divisor a whole number,

$$0.2\,4\,\overline{)\,8.2\,0\,8}$$
Move 2 places to the right.

**b)** move the decimal point in the dividend the same number of places (multiply the same way), and

$$0.2\,4\,\overline{)\,8.2\,0\,8}$$
Move 2 places to the right.

**c)** place the decimal point for the answer directly above the new decimal point in the dividend and divide as though dividing whole numbers.

$$
\begin{array}{r}
3\,4.2 \\
0.2\,4\,\overline{)\,8.2\,0_\wedge 8} \\
7\,2 \\
\hline
1\,0\,0 \\
9\,6 \\
\hline
4\,8 \\
4\,8 \\
\hline
0
\end{array}
$$

(The new decimal point in the dividend is indicated by a caret.)

---

**EXAMPLE 4** Divide: $5.848 \div 8.6$.

$$8.6\,\overline{)\,5.8\,4\,8}$$

Multiply the divisor by 10. (Move the decimal point 1 place.) Multiply the same way in the dividend. (Move 1 place.)

$$
\begin{array}{r}
0.6\,8 \\
8.6\,\overline{)\,5.8_\wedge 4\,8} \\
5\,1\,6 \\
\hline
6\,8\,8 \\
6\,8\,8 \\
\hline
0
\end{array}
$$

Place a decimal point above the new decimal point and then divide.

*Note:* $\dfrac{5.848}{8.6} = \dfrac{5.848}{8.6} \cdot \dfrac{10}{10} = \dfrac{58.48}{86}.$

**Check:** $(0.68)(8.6) = 5.848$

Do Exercises 7–9.

Do Exercises 7–9.

---

### Calculator Corner

**Finding Remainders**

We can find a whole-number remainder by multiplying the decimal portion of a quotient by the divisor. For example, to find the quotient and the whole-number remainder for $567 \div 13$, we can use a calculator to find that

$$567 \div 13 \approx 43.61538462$$

To isolate the portion to the right of the decimal point, we can subtract 43. When the decimal part of the quotient is multiplied by the divisor, we have

$$0.61538462 \times 13 = 8.00000006.$$

The rounding error in the result may vary, depending on the calculator used. We see that $567 \div 13 = 43\,R\,8$.

**Exercises:** Find the quotient and the whole-number remainder for each of the following.

**1.** $478 \div 17$

**2.** $815 \div 7$

**3.** $824 \div 11$

**4.** $7888 \div 19$

---

**7. a)** Complete.

$$\frac{3.75}{0.25} = \frac{3.75}{0.25} \times \frac{100}{100}$$
$$= \frac{(\quad)}{25}$$

**b)** Divide.

$$0.2\,5\,\overline{)\,3.7\,5}$$

Divide.

**8.** $4.067 \div (-0.83)$

**9.** $-44.8 \div (-3.5)$

---

***Answers***

**7. (a)** 375; **(b)** 15    **8.** $-4.9$    **9.** 12.8

**EXAMPLE 5** Divide: $-12 \div (-0.64)$.

Note first that a negative number divided by a negative number is positive. To find the quotient, we consider $12 \div 0.64$.

$$0.6\,4 \overline{\smash{\big)}\, 1\,2.}$$

Place a decimal point at the end of the whole number.

$$0.6\,4 \overline{\smash{\big)}\, 1\,2.0\,0}$$

Multiply the divisor by 100. (Move the decimal point 2 places.) Multiply the same way in the dividend. (Move 2 places after adding extra zeros.)

$$
\begin{array}{r}
1\,8.7\,5 \\
0.6\,4 \,\overline{\smash{\big)}\, 1\,2.0\,0{\scriptstyle\wedge}0\,0} \\
\underline{6\,4\,0} \\
5\,6\,0 \\
\underline{5\,1\,2} \\
4\,8\,0 \\
\underline{4\,4\,8} \\
3\,2\,0 \\
\underline{3\,2\,0} \\
0
\end{array}
$$

Place a decimal point above the new decimal point and then divide.

We have $-12 \div (-0.64) = 18.75$.

Do Exercises 10 and 11.

### Dividing by 10, 100, 1000, 0.1, 0.01, 0.001, and So On

We can divide quickly by a thousandth, hundredth, tenth, ten, hundred, and so on. Consider $43.9 \div 0.001$.

$$
\begin{array}{r}
4\,3\,9\,0\,0. \\
0.0\,0\,1 \,\overline{\smash{\big)}\, 4\,3.9\,0\,0{\scriptstyle\wedge}}
\end{array}
$$

We move the decimal point in both the divisor and the dividend 3 places to the right. Note that the divisor is now 1.

---

To divide by a number like 10, 100, 1000, 0.1, 0.01, or 0.001:

a) move the decimal point in the divisor until the divisor is 1 and

b) move the decimal point in the dividend the same number of places and the same direction.

---

**EXAMPLE 6** Divide: $\dfrac{0.0732}{10}$.

$$\frac{0.0732}{10} = \frac{0.0732}{10.} = \frac{0.00732}{1.0} = 0.00732$$

1 zero    1 place to the left to change 10 to 1

The answer is $0.00732$.

---

Divide.

**10.** $1.6 \overline{\smash{\big)}\, 2.5}$

**11.** $-9 \div 0.03$

*Answers*

**10.** 1.5625    **11.** $-300$

**EXAMPLE 7** Divide −213.75 by 100.

$$\frac{-213.75}{100} = \frac{-213.75}{100.} = \frac{-2.1375}{1.0} = -2.1375$$

2 zeros    2 places to the left to change 100 to 1

The answer is −2.1375.

**EXAMPLE 8** Divide: $\frac{67.8}{0.1}$.

$$\frac{67.8}{0.1} = \frac{67.8}{0.1} = \frac{678}{1.} = 678$$

1 place    1 place to the right to change 0.1 to 1

The answer is 678.

**EXAMPLE 9** Divide: $\frac{-23.738}{0.001}$.

$$\frac{-23.738}{0.001} = \frac{-23.738}{0.001} = \frac{-23,738}{1.} = -23,738$$

3 places    3 places to the right to change 0.001 to 1

The answer is −23,738.

Do Exercises 12–15.

Divide.

**12.** $\dfrac{12.78}{0.01}$

**13.** $\dfrac{0.176}{100}$

**14.** $\dfrac{98.47}{1000}$

**15.** $\dfrac{-6.7832}{0.1}$

## b  Order of Operations: Decimal Notation

The rules for order of operations apply when simplifying expressions involving decimal notation.

---

**RULES FOR ORDER OF OPERATIONS**

1. Do all calculations within grouping symbols first.
2. Evaluate all exponential expressions.
3. Do all multiplications and divisions in order from left to right.
4. Do all additions and subtractions in order from left to right.

---

**EXAMPLE 10** Simplify: $2.5 \times 25 \div 25{,}000 \times 250$.

There are no grouping symbols or parentheses, so we multiply and divide from left to right:

$$\begin{aligned}
2.5 \times 25 \div 25{,}000 \times 250 &= 62.5 \div 25{,}000 \times 250 \\
&= 0.0025 \times 250 \\
&= 0.625.
\end{aligned}$$

Doing all multiplications and divisions in order from left to right

**EXAMPLE 11**  Simplify: $(5 - 0.06) \div 2 + 3.42 \times 0.1$.

$(5 - 0.06) \div 2 + 3.42 \times 0.1 = 4.94 \div 2 + 3.42 \times 0.1$  Working inside the parentheses

$= 2.47 + 0.342$  Multiplying and dividing in order from left to right

$= 2.812$

**EXAMPLE 12**  Simplify: $13 - [5.4(1.3^2 + 0.21) \div 0.6]$.

$13 - [5.4(1.3^2 + 0.21) \div 0.6]$

$= 13 - [5.4(1.69 + 0.21) \div 0.6]$ ⎫ Working in the innermost
$= 13 - [5.4 \times 1.9 \div 0.6]$ ⎭ parentheses first

$= 13 - [10.26 \div 0.6]$  Multiplying

$= 13 - 17.1$  Dividing

$= -4.1$

Do Exercises 16–18.

Do Exercises 16–18.

**EXAMPLE 13**  *Mountains in Peru.*  The figure below shows a range of very high mountains in Peru, together with their altitudes, given both in feet and in meters. Find the average height of these mountains, in meters.

Source: *National Geographic*, July 1968, p. 130

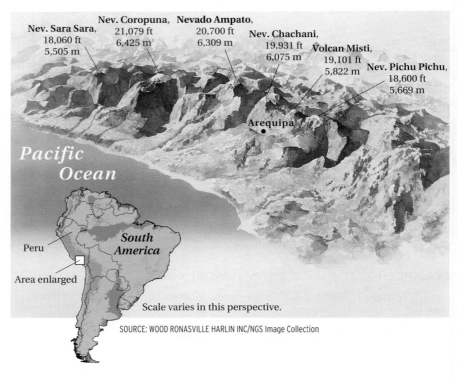

Nev. Sara Sara, 18,060 ft 5,505 m

Nev. Coropuna, 21,079 ft 6,425 m

Nevado Ampato, 20,700 ft 6,309 m

Nev. Chachani, 19,931 ft 6,075 m

Volcan Misti, 19,101 ft 5,822 m

Nev. Pichu Pichu, 18,600 ft 5,669 m

Arequipa

*Pacific Ocean*

Peru

*South America*

Area enlarged

Scale varies in this perspective.

SOURCE: WOOD RONASVILLE HARLIN INC/NGS Image Collection

The **average** of a set of numbers is the sum of the numbers divided by the number of addends. The average of the heights, in meters, is given by

$(5505 + 6425 + 6309 + 6075 + 5822 + 5669) \div 6 = 35,805 \div 6 = 5967.5$.

Thus, the average height of these mountains is 5967.5 m.

Do Exercise 19.

Simplify.

**16.** $625 \div 62.5 \times 25 \div 6250$

**17.** $0.25 \cdot (1 + 0.08) - 0.0274$

**18.** $[(19.7 - 17.2)^2 + 3] \div (-1.25)$

**19. Mountains in Peru.** Refer to the figure in Example 13. Find the average height of the mountains, in feet.

*Answers*

**16.** 0.04  **17.** 0.2426
**18.** −7.4  **19.** 19,578.5 ft

**a** Divide.

1. $2 \overline{\smash{)}\,5.9\,8}$

2. $5 \overline{\smash{)}\,13.5}$

3. $4 \overline{\smash{)}\,9\,5.1\,2}$

4. $8 \overline{\smash{)}\,2\,5.9\,2}$

5. $1\,2 \overline{\smash{)}\,8\,4.9\,6}$

6. $2\,3 \overline{\smash{)}\,2\,5.0\,7}$

7. $1\,5 \overline{\smash{)}\,1\,8}$

8. $3\,0 \overline{\smash{)}\,5\,4}$

9. $5.4 \div (-6)$

10. $3.6 \div (-4)$

11. $-30 \div 0.005$

12. $-100 \div 0.0002$

13. $0.0\,6 \overline{\smash{)}\,8.4}$

14. $0.0\,4 \overline{\smash{)}\,1.6\,8}$

15. $2.6 \overline{\smash{)}\,1\,0\,4}$

16. $3.2 \overline{\smash{)}\,1\,9\,2}$

17. $1.8 \div (-12)$

18. $6 \div (-15)$

19. $8.5 \overline{\smash{)}\,2\,7.2}$

20. $6.2 \overline{\smash{)}\,4\,6.5}$

21. $-31.59 \div 8.1$

22. $-39.06 \div 4.2$

23. $-5 \div (-8)$

24. $-7 \div (-8)$

25. $0.4\,7 \overline{\smash{)}\,0.1\,2\,2\,2}$

26. $0.5\,4 \overline{\smash{)}\,0.2\,7}$

27. $0.0\,3\,2 \overline{\smash{)}\,0.0\,7\,4\,8\,8}$

28. $0.0\,1\,7 \overline{\smash{)}\,1.5\,8\,1}$

29. $-24.969 \div 82$

30. $-25.221 \div 42$

**31.** $\dfrac{213.4567}{100}$     **32.** $\dfrac{769.3265}{1000}$     **33.** $\dfrac{-23.59}{10}$     **34.** $\dfrac{-83.57}{10}$     **35.** $\dfrac{1.0237}{0.001}$     **36.** $\dfrac{3.4029}{0.001}$

**37.** $\dfrac{-92.36}{0.01}$     **38.** $\dfrac{-56.78}{0.001}$     **39.** $\dfrac{0.8172}{10}$     **40.** $\dfrac{0.5678}{1000}$     **41.** $\dfrac{0.97}{0.1}$     **42.** $\dfrac{0.97}{0.001}$

**43.** $\dfrac{52.7}{-1000}$     **44.** $\dfrac{8.9}{-100}$     **45.** $\dfrac{75.3}{-0.001}$     **46.** $\dfrac{63.47}{-0.1}$     **47.** $\dfrac{-75.3}{1000}$     **48.** $\dfrac{23,001}{100}$

**b**    Simplify.

**49.** $14 \times (82.6 + 67.9)$         **50.** $(26.2 - 14.8) \times 12$         **51.** $0.003 + 3.03 \div (-0.01)$

**52.** $42 \times (10.6 + 0.024)$         **53.** $(4.9 - 18.6) \times 13$         **54.** $4.2 \times 5.7 + 0.7 \div 3.5$

**55.** $210.3 - 4.24 \times 1.01$         **56.** $-7.32 + 0.04 \div 0.1^2$         **57.** $0.04 \times 0.1 \div 0.4 \times 50$

**58.** $30 \div 0.2 \times 0.4 \div 10$         **59.** $12 \div (-0.03) - 12 \times 0.03^2$         **60.** $(5 - 0.04)^2 \div 4 + 8.7 \times 0.4$

**61.** $(4 - 2.5)^2 \div 100 + 0.1 \times 6.5$         **62.** $4 \div 0.4 - 0.1 \times 5 + 0.1^2$

**63.** $6 \times 0.9 - 0.1 \div 4 + 0.2^3$         **64.** $5.5^2 \times [(6 - 7.8) \div 0.06 + 0.12]$

**65.** $12^2 \div (12 + 2.4) - [(2 - 2.4) \div 0.8]$         **66.** $0.01 \times \{[(4 - 0.25) \div 2.5] - (4.5 - 4.025)\}$

**67.** *Camping in National Parks.* The graph below shows the numbers of overnight camping stays in Park Service campgrounds in the National Park System from 2004 to 2008. Find the average number of stays per year during this period.

**Camping in Park Service Campgrounds**

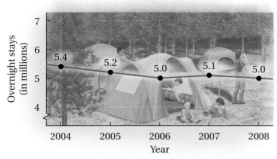

SOURCE: U.S. National Park Service

**68.** *Unclaimed Prizes.* Each year millions of lottery prize dollars go unclaimed. The table below shows the five states with the most unclaimed prize money for a recent year. Find the average amount that went unclaimed in these states.

| STATE | UNCLAIMED PRIZES (in millions) |
|---|---|
| New York | $73.6 |
| Texas | 58.9 |
| Florida | 48.4 |
| Ohio | 33.4 |
| California | 29.4 |

SOURCE: Individual state lotteries

**69.** *Gasoline Consumption.* The information in the graph below shows the average number of miles per gallon for new U.S. passenger cars. Find the average for the entire period.

**New U.S. Passenger Car Fuel Consumption Rates**

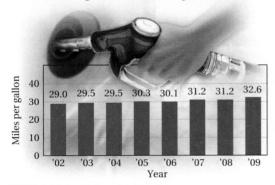

SOURCE: U.S. Federal Highway Administration

**70.** *Life Expectancy.* The information in the following bar graph shows the five highest life expectancies in the world. Determine the average life expectancy for these countries.

**Life Expectancy at Birth 2009**

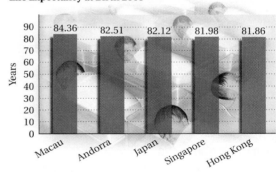

SOURCE: CIA World Factbook 2010

The following table lists the lengths of the longest railway tunnels in the world. Use the table for Exercises 71 and 72.

| TUNNEL | Gotthard Base Switzerland | Brenner Basis Austria | Sei-Kan Japan | Eurotunnel England-France | Lötschberg Switzerland |
|---|---|---|---|---|---|
| LENGTH, IN MILES | 35.5 | 34.4 | 33.5 | 31.3 | 21.5 |
| LENGTH, IN KILOMETERS | 57.1 | 55.4 | 53.9 | 50.5 | 34.6 |

SOURCE: The World's Longest Tunnel Page

**71.** Find the average length of the tunnels, in miles.

**72.** Find the average length of the tunnels, in kilometers.

## Skill Maintenance

Simplify to form an equivalent expression.   [3.5b]

**73.** $\dfrac{33}{44}$

**74.** $\dfrac{49}{56}$

**75.** $-\dfrac{27}{18}$

**76.** $-\dfrac{18}{60}$

**77.** $\dfrac{9a}{27}$

**78.** $\dfrac{12x}{30}$

**79.** $\dfrac{4r}{20r}$

**80.** $\dfrac{10t}{15t}$

## Synthesis

Calculate each of the following.

**81.** ▦ $7.434 \div (-1.2) \times 9.5 + 1.47^2$

**82.** ▦ $-9.46 \times 2.1^2 \div 3.5 + 4.36$

**83.** ▦ $9.0534 - 2.041^2 \times 0.731 \div 1.043^2$

**84.** ▦ $23.042(7 - 4.037 \times 1.46 - 0.932^2)$

Solve.

**85.** $439.57 \times 0.01 \div 1000 \cdot x = 4.3957$

**86.** $5.2738 \div 0.01 \times 1000 \div t = 52.738$

**87.** $0.0329 \div 0.001 \times 10^4 \div x = 3290$

**88.** $-4.302 \times 0.1^2 \div 0.001 \cdot t = -430.2$

**89.** ▦ *Television Ratings.*   A television rating point represents 1,150,000 households. The 2010 Super Bowl was viewed in approximately 78.2 million households, a record for the National Football League. How many rating points did the game receive? Round to the nearest tenth.

**Source:** Based on information from www.huffingtonpost.com

**90.** *Size of Country.*   The world's largest country, Russia, has an area of approximately 6.6 million square miles. The smallest country, Vatican City, has an area of approximately 0.2 square mile. How many times larger is Russia?

**Source:** U.S. Bureau of the Census, International Data Base

▦ *Electric Bills.*   Recently, electric bills from the Central Vermont Public Service Corporation consisted of a "customer charge" of $0.374 per day plus an "energy charge" of $0.1174 per kilowatt-hour (kWh) for the first 250 kWh used and $0.09079 per kilowatt-hour for each kilowatt-hour in excess of 250.

**91.** From August 20 to September 20, the Kaufmans' bill was $59.10. How many kilowatt-hours of electricity did they use? (Round to the nearest kilowatt-hour.)

**92.** From July 20 to August 20, the McGuires' bill was $70. How many kilowatt-hours of electricity did they use? (Round to the nearest kilowatt-hour.)

# Mid-Chapter Review

## Concept Reinforcement

Determine whether each statement is true or false.

_____ **1.** In the number 308.00567, the digit 6 names the tens place.  [5.1a]

_____ **2.** When writing a word name for decimal notation, we write the word "and" for the decimal point.  [5.1a]

_____ **3.** On the number line, $-2.3$ is to the left of $-2.2$.  [5.1c]

## Guided Solutions

Fill in each blank with the number that creates a correct statement or solution.

**4.** Evaluate $P(1 + r)$ for $P = 5000$ and $r = 0.045$.  [5.3c]

$$P(1 + r) = \square(1 + \square) \qquad \text{Substituting}$$
$$= 5000(\square) \qquad \text{Adding within parentheses}$$
$$= \square \qquad \text{Multiplying}$$

**5.** Simplify: $5.6 + 4.3 \times (6.5 - 0.25)^2$.  [5.4b]

$$5.6 + 4.3 \times (6.5 - 0.25)^2 = 5.6 + 4.3 \times (\square)^2 \qquad \text{Carrying out the operation inside parentheses}$$
$$= 5.6 + 4.3 \times \square \qquad \text{Evaluating the exponential expression}$$
$$= 5.6 + \square \qquad \text{Multiplying}$$
$$= \square \qquad \text{Adding}$$

## Mixed Review

**6.** Usain Bolt of Jamaica set a world record of 9.69 sec in the men's 100-m race at the 2008 Summer Olympics. Write a word name for 9.69.  [5.1a]

**Source:** beijing2008.cn

**7.** After removing coins from their pockets at security checkpoints in U.S. airports, travelers forgot to reclaim coins totaling about $1.05 million from 2005 to 2007. Convert 1.05 million to standard notation.  [5.3b]

**Source:** Transportation Security Administration

Write each number as a fraction and, if possible, as a mixed numeral.  [5.1b]

**8.** 4.53

**9.** 0.287

Which number is larger?  [5.1c]

**10.** 0.07, 0.13

**11.** $-5.2, -5.09$

Write decimal notation.  [5.1b]

**12.** $\dfrac{7}{10}$

**13.** $\dfrac{639}{100}$

**14.** $-35\dfrac{67}{100}$

**15.** $8\dfrac{2}{1000}$

Round 28.4615 to the nearest  [5.1d]

**16.** Thousandth.

**17.** Hundredth.

**18.** Tenth.

**19.** One.

Add. [5.2a], [5.2c]

**20.**
```
  4 7.6 3 8
+   2.4 5 7
```

**21.**
```
    1 5.6
  2 3 4.7 2 9
      3.0 8
+ 9 6 1.4 5 3
```

**22.** $-10.5 + 0.27$

**23.** $16 + 0.34 + 1.9$

Subtract. [5.2b], [5.2c]

**24.**
```
  3 2 1.5 7
−     4 9.3 8
```

**25.**
```
    5.6
−   0.0 0 7
```

**26.** $34.3 - 18.75$

**27.** $-6.9 - 13$

Multiply. [5.3a]

**28.**
```
    4.6
×   0.9
```

**29.**
```
    1 5.3
×     6.0 7
```

**30.** $100 \times 81.236$

**31.** $0.1 \times (-0.483)$

Divide. [5.4a]

**32.** $-20.24 \div (-4)$

**33.** $21.76 \div 6.8$

**34.** $76.3 \div 0.1$

**35.** $914.036 \div 1000$

**36.** Convert $20.45 to cents. [5.3b]

**37.** Convert 147¢ to dollars. [5.3b]

**38.** Combine like terms: $3.08x - 7.1 - 4.3x$. [5.2d]

**39.** Evaluate $2(l + w)$ for $l = 1.3$ and $w = 0.8$. [5.3c]

Simplify. [5.4b]

**40.** $6.594 + 0.5318 \div 0.01$

**41.** $7.3 \times 4.6 - 0.8 \div 3.2$

# Understanding Through Discussion and Writing

**42.** A classmate rounds 236.448 to the nearest one and gets 237. Explain the possible error. [5.1d]

**43.** Explain the error in the following:

Subtract.

$$73.089 - 5.0061 = 2.3028 \quad \text{[5.2b]}$$

**44.** Explain why $10 \div 0.2 = 100 \div 2$. [5.4a]

**45.** Kayla made these two computational mistakes:

$$0.247 \div 0.1 = 0.0247; \quad 0.247 \div 10 = 2.47.$$

In each case, how could you convince her that a mistake has been made? [5.4a]

# 5.5

# Using Fraction Notation with Decimal Notation

Now that we know how to divide using decimal notation, we can express *any* fraction as a decimal. This means that any *rational* number (ratio of integers) can be written as a decimal.

## a  Using Division to Find Decimal Notation

Recall that $\frac{a}{b}$ means $a \div b$. This gives us one way of converting fraction notation to decimal notation.

**EXAMPLE 1**  Find decimal notation for $\frac{3}{20}$.

Because $\frac{3}{20}$ means $3 \div 20$, we can perform long division.

$$
\begin{array}{r}
0.1\,5 \\
20\,\overline{)\,3.0\,0} \\
\underline{2\,0} \\
1\,0\,0 \\
\underline{1\,0\,0} \\
0
\end{array}
$$

> We are finished when the remainder is 0.

We have $\frac{3}{20} = 0.15$.

**EXAMPLE 2**  Find decimal notation for $\frac{-7}{8}$.

Since $\frac{-7}{8}$ means $-7 \div 8$ and a negative number divided by a positive number is negative, we know that the decimal will be negative. We divide 7 by 8 and make the result negative.

$$
\begin{array}{r}
0.8\,7\,5 \\
8\,\overline{)\,7.0\,0\,0} \\
\underline{6\,4} \\
6\,0 \\
\underline{5\,6} \\
4\,0 \\
\underline{4\,0} \\
0
\end{array}
$$

Thus, $\frac{-7}{8} = -0.875$.

> Do Margin Exercises 1 and 2.

When division with decimals ends, or *terminates*, as in Examples 1 and 2, the result is called a **terminating decimal**. If the division does not terminate, the result will be a **repeating decimal**.

**SKILL TO REVIEW**
Objective 5.4a:  Divide using decimal notation.

Divide.

**1.** $3 \div 4$  **2.** $25 \div 8$

Find decimal notation.

**1.** $\dfrac{2}{5}$  **2.** $\dfrac{-5}{8}$

***Answers***

*Skill to Review:*
**1.** 0.75  **2.** 3.125

*Margin Exercises:*
**1.** 0.4  **2.** −0.625

**EXAMPLE 3** Find decimal notation for $\frac{5}{6}$.

Since $\frac{5}{6}$ means $5 \div 6$, we have

$$
\begin{array}{r}
0.8\ 3\ 3 \\
6\ )\ \overline{5.0\ 0\ 0} \\
\underline{4\ 8} \\
2\ 0 \\
\underline{1\ 8} \\
2\ 0 \\
\underline{1\ 8} \\
2
\end{array}
$$

Since 2 keeps reappearing as a remainder, the digits repeat and will continue to do so; therefore,

$$\frac{5}{6} = 0.83333\dots.$$ The dots indicate an endless sequence of digits in the quotient.

When there is a repeating pattern, the dots are often replaced by a bar to indicate the repeating part, in this case, only the 3.

$$\frac{5}{6} = 0.8\overline{3}$$

Do Exercises 3 and 4.

**EXAMPLE 4** Find decimal notation for $-\frac{4}{11}$.

Since $-\frac{4}{11}$ is negative, we divide 4 by 11 and make the result negative.

$$
\begin{array}{r}
0.3\ 6\ 3\ 6 \\
1\ 1\ )\ \overline{4.0\ 0\ 0\ 0} \\
\underline{3\ 3} \\
7\ 0 \\
\underline{6\ 6} \\
4\ 0 \\
\underline{3\ 3} \\
7\ 0 \\
\underline{6\ 6} \\
4
\end{array}
$$

Since 7 and 4 keep repeating as remainders, the sequence of digits "36" repeats in the quotient, and

$$\frac{4}{11} = 0.363636\dots, \quad \text{or} \quad 0.\overline{36}.$$

Thus, $-\frac{4}{11} = -0.\overline{36}$.

Do Exercises 5 and 6.

When a fraction is written in simplified form, we can tell from the denominator whether its decimal notation will repeat or terminate.

> For a fraction in simplified form,
> - if the denominator has a prime factor other than 2 or 5, the decimal notation repeats.
> - if the denominator has no prime factor other than 2 or 5, the decimal notation terminates.

---

Find decimal notation for each number.

**3.** $\frac{1}{6}$   **4.** $\frac{2}{3}$

---

Find decimal notation for each number.

**5.** $\frac{5}{11}$   **6.** $-\frac{12}{11}$

---

***Answers***

**3.** $0.1\overline{6}$   **4.** $0.\overline{6}$   **5.** $0.\overline{45}$
**6.** $-1.\overline{09}$

**EXAMPLE 5** Find decimal notation for $\frac{3}{7}$.

Because 7 is not a product of 2's and/or 5's, we expect a repeating decimal.

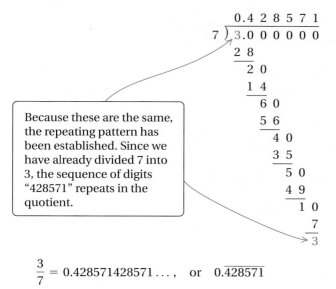

Because these are the same, the repeating pattern has been established. Since we have already divided 7 into 3, the sequence of digits "428571" repeats in the quotient.

$$\frac{3}{7} = 0.428571428571\ldots, \quad \text{or} \quad 0.\overline{428571}$$

Do Exercise 7.

**Calculator Corner**

**Recognizing Repeating Decimals** Short repeating patterns can usually be recognized on a calculator's display. For example, to convert $\frac{1}{11}$ to decimal notation, press [1] [÷] [1] [1] [=]. The result for a 10-digit display is [.0909090909], so the decimal notation is $0.\overline{09}$.

For $-\frac{12}{11}$, the same calculator displays [−1.090909091]. Note that the last digit has been rounded from 0 to 1. The decimal notation for $-\frac{12}{11}$ is $-1.\overline{09}$.

**Exercises:** Convert to decimal notation.

1. $-\dfrac{1}{6}$     2. $\dfrac{7}{11}$

3. $\dfrac{19}{3}$     4. $-\dfrac{514}{9}$

## b  Rounding Repeating Decimals

The repeating part of a decimal can be so long that it will not fit on a calculator. For example, when $\frac{5}{97}$ is written as a decimal, its repeating part is 96 digits long! Most calculators round repeating decimals to 9 or 10 decimal places.

In applied problems, repeating decimals are rounded to get approximate answers. To round a repeating decimal, we can extend the decimal notation at least one place past the rounding digit and then round as before.

**EXAMPLES** Round each to the nearest tenth, hundredth, and thousandth.

|  | *Nearest tenth* | *Nearest hundredth* | *Nearest thousandth* |
|---|---|---|---|
| **6.** $0.8\overline{3} = 0.83333\ldots$ | 0.8 | 0.83 | 0.833 |
| **7.** $3.\overline{09} = 3.090909\ldots$ | 3.1 | 3.09 | 3.091 |
| **8.** $-4.1\overline{763} = -4.1763763\ldots$ | −4.2 | −4.18 | −4.176 |

Do Exercises 8–11.

**EXAMPLE 9** *Gas Mileage.* A car travels 457 mi on 16.4 gal of gasoline. The ratio of number of miles driven to amount of gasoline used is *gas mileage*. Find the gas mileage and convert the ratio to decimal notation rounded to the nearest tenth.

$$\frac{\text{Miles driven}}{\text{Gasoline used}} = \frac{457}{16.4} \approx 27.86 \qquad \text{Dividing to 2 decimal places}$$
$$\approx 27.9 \qquad \text{Rounding to 1 decimal place}$$

The gas mileage is 27.9 miles to the gallon.

Do Exercise 12.

**7.** Find decimal notation for $\frac{5}{7}$.

Round each to the nearest tenth, hundredth, and thousandth.

**8.** $0.\overline{6}$     **9.** $0.6\overline{08}$

**10.** $-7.3\overline{49}$     **11.** $2.6\overline{891}$

**12. Gas Mileage.** A car travels 380 mi on 15.7 gal of gasoline. Find the gasoline mileage and convert the ratio to decimal notation rounded to the nearest tenth.

*Answers*

**7.** $0.\overline{714285}$   **8.** 0.7; 0.67; 0.667
**9.** 0.6; 0.61; 0.608   **10.** −7.3; −7.35; −7.349
**11.** 2.7; 2.69; 2.689   **12.** 24.2 miles per gallon

Multiply by a form of 1 to find decimal notation for each number.

**13.** $\dfrac{4}{5}$     **14.** $-\dfrac{9}{20}$

**15.** $\dfrac{7}{200}$     **16.** $\dfrac{33}{25}$

## c   More with Conversions

Fractions like $\frac{3}{10}$ or $-\frac{71}{1000}$ can be converted to decimal notation without using long division. When a denominator is a factor of 10, 100, and so on, we can convert to decimal notation by finding (perhaps mentally) an equivalent fraction in which the denominator is a power of 10.

**EXAMPLE 10**   Find decimal notation for $-\frac{7}{500}$.

Since $500 \cdot 2 = 1000$, and 1000 is a power of 10, we use $\frac{2}{2}$ as an expression for 1.

$$-\frac{7}{500} = -\frac{7}{500} \cdot \frac{2}{2} = -\frac{14}{1000} = -0.014$$   *Think*: $1000 \div 500 = 2$, and $7 \cdot 2 = 14$.

**EXAMPLE 11**   Find decimal notation for $\frac{9}{25}$.

$$\frac{9}{25} = \frac{9}{25} \cdot \frac{4}{4} = \frac{36}{100} = 0.36$$   Using $\frac{4}{4}$ for 1 to get a denominator of 100

As a check, we can divide.

$$
\begin{array}{r}
0.3\ 6 \\
2\ 5 \overline{)\ 9.0\ 0} \\
7\ 5 \\
\hline
1\ 5\ 0 \\
1\ 5\ 0 \\
\hline
0
\end{array}
$$
Note that multiplication by 1 is much faster.

**EXAMPLE 12**   Find decimal notation for $\frac{7}{4}$.

$$\frac{7}{4} = \frac{7}{4} \cdot \frac{25}{25} = \frac{175}{100} = 1.75$$   Using $\frac{25}{25}$ for 1 to get a denominator of 100. You might also note that 7 quarters is \$1.75.

Do Exercises 13–16.

## d   Calculations with Fraction and Decimal Notation Together

In certain kinds of calculations, fraction and decimal notation might occur together. In such cases, there are at least three ways in which we can proceed.

**EXAMPLE 13**   Calculate: $\frac{2}{3} \times 0.576$.

METHOD 1:   Perhaps the quickest method is to treat 0.576 as $\frac{0.576}{1}$. Then we multiply 0.576 by 2 and divide the result by 3.

$$\frac{2}{3} \times 0.576 = \frac{2}{3} \times \frac{0.576}{1}$$

$$= \frac{2 \times 0.576}{3} = \frac{1.152}{3}$$

$$= 0.384$$

$$
\begin{array}{r}
0.3\ 8\ 4 \\
3 \overline{)\ 1.1\ 5\ 2} \\
9 \\
\hline
2\ 5 \\
2\ 4 \\
\hline
1\ 2 \\
1\ 2 \\
\hline
0
\end{array}
$$

METHOD 2: A second way to do this calculation is to convert the fraction notation to decimal notation so that both numbers are in decimal notation. Since $\frac{2}{3}$ converts to repeating decimal notation, it is first rounded to some chosen decimal place. We choose three decimal places because 0.576 has three decimal places. Then, using decimal notation, we multiply.

$$\frac{2}{3} \times 0.576 = 0.\overline{6} \times 0.576 \approx 0.667 \times 0.576 = 0.384192$$

METHOD 3: A third method is to convert the decimal notation to fraction notation so that both numbers are in fraction notation. The answer can be left in fraction notation and simplified, or we can convert back to decimal notation and, if appropriate, round.

$$\frac{2}{3} \times 0.576 = \frac{2}{3} \cdot \frac{576}{1000} = \frac{2 \cdot 576}{3 \cdot 1000}$$

$$= \frac{2 \cdot 2 \cdot 2 \cdot 2 \cdot 2 \cdot 2 \cdot 3 \cdot 3}{2 \cdot 2 \cdot 2 \cdot 3 \cdot 5 \cdot 5 \cdot 5} \quad \text{Factoring}$$

$$= \frac{2 \cdot 2 \cdot 2 \cdot 3}{2 \cdot 2 \cdot 2 \cdot 3} \cdot \frac{2 \cdot 2 \cdot 2 \cdot 2 \cdot 3}{5 \cdot 5 \cdot 5} \quad \begin{array}{l}\text{Removing a factor equal}\\ \text{to 1: } \frac{2 \cdot 2 \cdot 2 \cdot 3}{2 \cdot 2 \cdot 2 \cdot 3} = 1\end{array}$$

$$= \frac{2 \cdot 2 \cdot 2 \cdot 2 \cdot 3}{5 \cdot 5 \cdot 5} = \frac{48}{125}, \text{ or } 0.384$$

Note that we get an exact answer with methods 1 and 3, but method 2 gives an approximation since we rounded decimal notation for $\frac{2}{3}$.

Do Exercises 17 and 18.

**EXAMPLE 14** *Boating.* A triangular sail from a single-sail day cruiser is 3.4 m wide and 4.2 m tall. Find the area of the sail.

1. **Familiarize.** We make a drawing and recall that the formula for the area, $A$, of a triangle with base $b$ and height $h$ is $A = \frac{1}{2}bh$.

2. **Translate.** We substitute 3.4 for $b$ and 4.2 for $h$.

$$A = \frac{1}{2}bh = \frac{1}{2}(3.4)(4.2) \quad \text{Evaluating}$$

3. **Solve.** We simplify as follows.

$$A = \frac{1}{2}(3.4)(4.2)$$

$$= \frac{3.4}{2}(4.2) \quad \text{Multiplying } \frac{1}{2} \text{ and } \frac{3.4}{1}$$

$$= 1.7(4.2) \quad \text{Dividing}$$

$$= 7.14 \quad \text{Multiplying}$$

4. **Check.** To check, we repeat the calculations using the commutative law. We also rewrite $\frac{1}{2}$ as 0.5.

$$\frac{1}{2}(3.4)(4.2) = 0.5(4.2)(3.4) = (2.1)(3.4) = 7.14$$

Our answer checks.

5. **State.** The area of the sail is 7.14 m² (square meters).

Do Exercise 19.

Calculate.

**17.** $\frac{5}{6} \times 0.864$

**18.** $\frac{1}{3} \times 0.384 + \frac{5}{8} \times 0.6784$

**19.** Find the area of a triangular window that is 3.25 ft wide and 2.6 ft tall.

2.6 ft

3.25 ft

**Answers**

**17.** 0.72  **18.** 0.552  **19.** 4.225 ft²

**a** , **c**  Find decimal notation for each number.

1. $\dfrac{3}{8}$

2. $\dfrac{3}{5}$

3. $\dfrac{-1}{2}$

4. $\dfrac{-1}{4}$

5. $\dfrac{3}{25}$

6. $\dfrac{7}{20}$

7. $\dfrac{9}{40}$

8. $\dfrac{3}{40}$

9. $\dfrac{13}{25}$

10. $\dfrac{17}{25}$

11. $\dfrac{-17}{20}$

12. $\dfrac{-13}{20}$

13. $-\dfrac{9}{16}$

14. $-\dfrac{5}{16}$

15. $\dfrac{7}{5}$

16. $\dfrac{3}{2}$

17. $\dfrac{28}{25}$

18. $\dfrac{31}{20}$

19. $\dfrac{11}{-8}$

20. $\dfrac{17}{-10}$

21. $-\dfrac{39}{40}$

22. $-\dfrac{17}{40}$

23. $\dfrac{121}{200}$

24. $\dfrac{32}{125}$

25. $\dfrac{8}{15}$

26. $\dfrac{7}{9}$

27. $\dfrac{1}{3}$

28. $\dfrac{1}{9}$

29. $\dfrac{-4}{3}$

30. $\dfrac{-8}{9}$

31. $\dfrac{7}{6}$

32. $\dfrac{7}{11}$

33. $-\dfrac{14}{11}$

34. $-\dfrac{7}{11}$

35. $\dfrac{-5}{12}$

36. $\dfrac{-11}{12}$

**37.** $\dfrac{127}{500}$  **38.** $\dfrac{83}{500}$  **39.** $\dfrac{4}{33}$  **40.** $\dfrac{5}{33}$

**41.** $\dfrac{-12}{55}$  **42.** $\dfrac{-5}{22}$  **43.** $\dfrac{4}{7}$  **44.** $\dfrac{2}{7}$

**b**  For Exercises 45–56, round the decimal notation for each number to the nearest tenth, hundredth, and thousandth.

**45.** $\dfrac{4}{11}$  **46.** $\dfrac{3}{11}$  **47.** $-\dfrac{5}{3}$  **48.** $-\dfrac{19}{16}$

**49.** $\dfrac{-8}{17}$  **50.** $\dfrac{-7}{13}$  **51.** $\dfrac{7}{12}$  **52.** $\dfrac{2}{15}$

**53.** $\dfrac{29}{-150}$  **54.** $\dfrac{37}{-150}$  **55.** $\dfrac{7}{-9}$  **56.** $\dfrac{5}{-13}$

Round each to the nearest tenth, hundredth, and thousandth.

**57.** $0.\overline{18}$  **58.** $0.\overline{83}$  **59.** $0.2\overline{7}$  **60.** $3.5\overline{4}$

**61.** For this set of people, what is the ratio, in decimal notation rounded to the nearest thousandth, where appropriate, of
  **a)** women to the total number of people?
  **b)** women to men?
  **c)** men to the total number of people?
  **d)** men to women?

**62.** For this set of pennies and quarters, what is the ratio, in decimal notation rounded to the nearest thousandth, where appropriate, of
  **a)** pennies to quarters?
  **b)** quarters to pennies?
  **c)** pennies to total number of coins?
  **d)** total number of coins to pennies?

*Gas Mileage.* In each of Exercises 63–66, find the gas mileage rounded to the nearest tenth.

**63.** 285 mi; 18 gal

**64.** 396 mi; 17 gal

**65.** 324.8 mi; 18.2 gal

**66.** 264.8 mi; 12.7 gal

**(d)** Calculate and write the result as a decimal.

**67.** $\dfrac{7}{8} \times 12.64$

**68.** $\dfrac{4}{5} \times 384.8$

**69.** $6.84 \div 2\dfrac{1}{2}$

**70.** $8\dfrac{1}{2} \div 2.125$

**71.** $\dfrac{47}{9}(-79.95)$

**72.** $\dfrac{7}{11}(-2.7873)$

**73.** $\dfrac{1}{2} - 0.5$

**74.** $3\dfrac{1}{8} - 2.75$

**75.** $\left(\dfrac{1}{6}\right)0.0765 + \left(\dfrac{3}{4}\right)0.1124$

**76.** $\left(\dfrac{2}{5}\right)6384.1 - \left(\dfrac{5}{8}\right)156.56$

**77.** $\dfrac{3}{4} \times 2.56 - \dfrac{7}{8} \times 3.94$

**78.** $\dfrac{2}{5} \times 3.91 - \dfrac{7}{10} \times 4.15$

**79.** $5.2 \times 1\dfrac{7}{8} \div 0.4$

**80.** $4\dfrac{3}{4} \times 0.5 \div 0.1$

Solve.

**81.** Find the area of a triangular shawl that is 1.8 m long and 1.2 m wide.

**82.** Find the area of a triangular sign that is 1.5 m wide and 1.5 m tall.

**83.** Find the area of a triangular stamp that is 3.4 cm wide and 3.4 cm tall.

**84.** Find the area of a triangular reflector that is 7.4 cm wide and 9.1 cm tall.

**85.** Find the area of the kite shown on the left.

**86.** Find the area of the kite shown on the right.

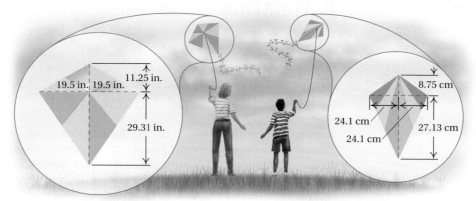

## Skill Maintenance

**87.** Round 3572 to the nearest ten.  [1.6a]

**88.** Round 3572 to the nearest thousand.  [1.6a]

**89.** Round 78,951 to the nearest hundred.  [1.6a]

**90.** Round 19,829,996 to the nearest ten.  [1.6a]

**91.** Simplify: $\dfrac{95}{-1}$.  [3.3b]

**92.** Solve: $5x - 9 = 7x + 11$.  [4.4a]

**93.** Simplify: $9 - 4 + 2 \div (-1) \cdot 6$.  [2.5b]

**94.** Simplify: $\dfrac{-9}{-9}$.  [3.3b]

## Synthesis

▦ Find decimal notation. Save the answers for Exercise 100.

**95.** $\dfrac{1}{7}$   **96.** $\dfrac{2}{7}$   **97.** $\dfrac{3}{7}$   **98.** $\dfrac{4}{7}$   **99.** $\dfrac{5}{7}$

**100.** ▦ From the pattern of answers to Exercises 95–99, predict the decimal notation for $\frac{6}{7}$. Check your answer on a calculator.

Find decimal notation. Save the answers for Exercise 104.

**101.** $\dfrac{1}{9}$   **102.** $\dfrac{1}{99}$   **103.** $\dfrac{1}{999}$

**104.** ▦ From the pattern of Exercises 101–103, predict the decimal notation for $\frac{1}{9999}$. Check your answer on a calculator.

The formula $A = \pi r^2$ is used to find the area, $A$, of a circle with radius $r$. For Exercises 105 and 106, find the area of a circle with the given radius, using $\frac{22}{7}$ for $\pi$. For Exercises 107 and 108, use 3.14 for $\pi$ or a calculator with a $\pi$ key.

**105.** $r = 2.1$ cm   **106.** $r = 1.4$ cm   **107.** ▦ $r = \dfrac{3}{4}$ ft   **108.** ▦ $r = 4\dfrac{1}{2}$ yd

# 5.6

## Estimating

### OBJECTIVE

**a** Estimate sums, differences, products, and quotients.

**SKILL TO REVIEW**

Objective 1.6b: Estimate sums, differences, products, and quotients by rounding.

Estimate by first rounding to the nearest ten.

1.  4 6 7
   − 2 8 4

2.     5 4
   × 2 9

1. Estimate by rounding to the nearest ten the total cost of one grill and one camera. Which of the following is an appropriate estimate?

   **a)** $43     **b)** $400
   **c)** $410    **d)** $430

2. About how much more does the TV cost than the grill? Estimate by rounding to the nearest ten. Which of the following is an appropriate estimate?

   **a)** $100    **b)** $150
   **c)** $160    **d)** $300

### **a** Estimating Sums, Differences, Products, and Quotients

Estimating has many uses. It can be done before a problem is even attempted and it can be done afterward as a check, even when we are using a calculator. In many situations, an estimate is all we need. We usually estimate by rounding the numbers so that there are one or two nonzero digits. Consider the following prices for Examples 1–3.

**8.2 Megapixel Digital Camera**

$289.⁹⁵

Four Burner Gas Grill

$139.⁹⁷

19" Flat-Panel HDTV
$449.99

**EXAMPLE 1** Estimate by rounding to the nearest ten the total cost of one grill and one TV.

We are estimating the sum

$289.95 + $449.99 = Total cost.

The estimate found by rounding the addends to the nearest ten is

$290 + $450 = $740.    (Estimated total cost)

Do Margin Exercise 1.

**EXAMPLE 2** About how much more does the TV cost than the camera? Estimate by rounding to the nearest ten.

We are estimating the difference

$449.99 − $139.97 = Price difference.

The estimate found by rounding each price to the nearest ten is

$450 − $140 = $310.    (Estimated price difference)

Do Exercise 2.

*Answers*

*Skill to Review:*
1. 190    2. 1500

*Margin Exercises:*
1. (d)    2. (c)

**EXAMPLE 3** Estimate the total cost of 4 cameras.

We are estimating the product

$$4 \times \$139.97 = \text{Total cost.}$$

The estimate is found by rounding 139.97 to the nearest ten:

$$4 \times \$140 = \$560.$$

Do Exercise 3.

**EXAMPLE 4** About how many Blu-ray discs at $29.99 each can be purchased for $154?

We estimate the quotient

$$\$154 \div \$29.99.$$

Since we want a whole-number estimate, we choose our rounding appropriately. Rounding $29.99 to the nearest one, we get $30. Since $154 is close to $150, which is a multiple of 30, we estimate

$$\$150 \div \$30,$$

so the answer is 5.

Do Exercise 4.

When estimating, we usually look for numbers that are easy to work with. For example, if multiplying, we might round 0.43 to 0.5 and 8.9 to 10, because 0.5 and 10 are convenient numbers to multiply.

**EXAMPLE 5** Estimate: $4.8 \times 52$. Do not find the actual product. Which of the following is an appropriate estimate?

a) 25    b) 250    c) 2500    d) 360

We round 4.8 to the nearest one and 52 to the nearest ten:

$$5 \times 50 = 250. \quad \text{(Estimated product)}$$

Thus, an approximate estimate is (b).

Other estimates that we might have used in Example 5 are

$$5 \times 52 = 260 \quad \text{or} \quad 4.8 \times 50 = 240.$$

The estimate in Example 5, $5 \times 50 = 250$, is the easiest to do because the factors have the fewest nonzero digits. You could probably do it mentally. In general, we try to round so that a computation has as few nonzero digits as possible while still keeping the estimated value close to the original value.

Do Exercises 5–10.

---

**3.** Estimate the total cost of 6 TVs. Which of the following is an appropriate estimate?
  **a)** $450       **b)** $2700
  **c)** $4500      **d)** $27,000

**4.** About how many Blu-ray discs can be purchased for $485? Choose an appropriate estimate from the following.
  **a)** 16       **b)** 21
  **c)** 25       **d)** 30

Estimate each product. Do not find the actual product. Which of the following is an appropriate estimate?

**5.** $2.4 \times 8$
  **a)** 16       **b)** 34
  **c)** 125      **d)** 5

**6.** $24 \times 0.6$
  **a)** 200      **b)** 5
  **c)** 110      **d)** 20

**7.** $0.86 \times 0.432$
  **a)** 0.04     **b)** 0.4
  **c)** 1.1      **d)** 4

**8.** $0.82 \times 0.1$
  **a)** 800      **b)** 8
  **c)** 0.08     **d)** 80

**9.** $0.12 \times 18.248$
  **a)** 180      **b)** 1.8
  **c)** 0.018    **d)** 18

**10.** $24.234 \times 5.2$
  **a)** 200      **b)** 120
  **c)** 12.5     **d)** 234

*Answers*
**3.** (b)  **4.** (a)  **5.** (a)  **6.** (d)  **7.** (b)
**8.** (c)  **9.** (b)  **10.** (b)

**EXAMPLE 6** Estimate: 82.08 ÷ 24. Which of the following is an appropriate estimate?

**a)** 400          **b)** 16          **c)** 40          **d)** 4

This is about 80 ÷ 20, so the answer is about 4. Thus, an appropriate estimate is (d).

**EXAMPLE 7** Estimate: 94.18 ÷ 3.2. Which of the following is an appropriate estimate?

**a)** 30          **b)** 300          **c)** 3          **d)** 60

This is about 90 ÷ 3, so the answer is about 30. Thus, an appropriate estimate is (a).

**EXAMPLE 8** Estimate: 0.0156 ÷ 1.3. Which of the following is an appropriate estimate?

**a)** 0.2          **b)** 0.002          **c)** 0.02          **d)** 20

This is about 0.02 ÷ 1, so the answer is about 0.02. Thus, an appropriate estimate is (c).

Do Exercises 11–13.

In some cases, it is easier to estimate a quotient by checking products rather than by rounding the divisor and the dividend.

**EXAMPLE 9** Estimate: 0.0074 ÷ 0.23. Which of the following is an appropriate estimate?

**a)** 0.3          **b)** 0.03          **c)** 300          **d)** 3

We estimate 3 for a quotient. We check by multiplying.

$$0.23 \times 3 = 0.69$$

We make the estimate smaller. We estimate 0.3 and check by multiplying.

$$0.23 \times 0.3 = 0.069$$

We make the estimate smaller. We estimate 0.03 and check by multiplying.

$$0.23 \times 0.03 = 0.0069$$

This is about 0.0074, so the quotient is about 0.03. Thus, an appropriate estimate is (b).

Do Exercise 14.

---

Estimate each quotient. Which of the following is an appropriate estimate?

**11.** 59.78 ÷ 29.1
   **a)** 200          **b)** 20
   **c)** 2            **d)** 0.2

**12.** 82.08 ÷ 2.4
   **a)** 40           **b)** 4.0
   **c)** 400          **d)** 0.4

**13.** 0.1768 ÷ 0.08
   **a)** 8            **b)** 10
   **c)** 2            **d)** 20

---

## STUDY TIPS

### PUT MATH TO USE

One excellent way to study math is to use it in your everyday life. The concepts of this section can be easily reinforced if you use estimating when you next go shopping.

---

**14.** Estimate: 0.0069 ÷ 0.15. Which of the following is an appropriate estimate?
   **a)** 0.5          **b)** 50
   **c)** 0.05         **d)** 0.004

---

*Answers*

**11.** (c)   **12.** (a)   **13.** (c)   **14.** (c)

**a** Consider the following prices for Exercises 1–8. Estimate the sums, differences, products, or quotients involved in these problems. Indicate which of the choices is an appropriate estimate.

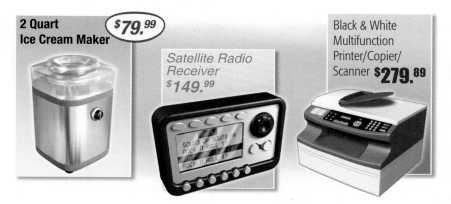

**2 Quart Ice Cream Maker** $79.⁹⁹

Satellite Radio Receiver $149.⁹⁹

Black & White Multifunction Printer/Copier/Scanner $279.⁸⁹

1. Estimate the total cost of one printer and one satellite radio.
   a) $43   b) $4300   c) $360   d) $430

2. Estimate the total cost of one satellite radio and one ice cream maker.
   a) $230   b) $23   c) $2300   d) $400

3. About how much more does the printer cost than the satellite radio?
   a) $1300   b) $200   c) $130   d) $13

4. About how much more does the satellite radio cost than the ice cream maker?
   a) $7000   b) $70   c) $130   d) $700

5. Estimate the total cost of 6 ice cream makers.
   a) $480   b) $48   c) $240   d) $4800

6. Estimate the total cost of 4 printers.
   a) $1200   b) $1120   c) $11,200   d) $600

7. About how many ice cream makers can be purchased for $830?
   a) 120   b) 100   c) 10   d) 1000

8. About how many printers can be purchased for $5627?
   a) 200   b) 20   c) 1800   d) 2000

Estimate by rounding as directed.

9. $0.02 + 1.31 + 0.34$; nearest tenth

10. $0.88 + 2.07 + 1.54$; nearest one

11. $6.03 + 0.007 + 0.214$; nearest one

12. $1.11 + 8.888 + 99.94$; nearest one

13. $52.367 + 1.307 + 7.324$; nearest one

14. $12.9882 + 1.2115$; nearest tenth

**15.** $2.678 - 0.445$;
nearest tenth

**16.** $12.9882 - 1.0115$;
nearest one

**17.** $198.67432 - 24.5007$;
nearest ten

Estimate. Choose a rounding digit that gives one or two nonzero digits. Indicate which of the choices is an appropriate estimate.

**18.** $234.12321 - 200.3223$
  **a)** 600     **b)** 60
  **c)** 300     **d)** 30

**19.** $49 \times 7.89$
  **a)** 400     **b)** 40
  **c)** 4       **d)** 0.4

**20.** $7.4 \times 8.9$
  **a)** 95      **b)** 63
  **c)** 124     **d)** 6

**21.** $98.4 \times 0.083$
  **a)** 80      **b)** 12
  **c)** 8       **d)** 0.8

**22.** $78 \times 5.3$
  **a)** 400     **b)** 800
  **c)** 40      **d)** 8

**23.** $3.6 \div 4$
  **a)** 10      **b)** 1
  **c)** 0.1     **d)** 0.01

**24.** $0.0713 \div 1.94$
  **a)** 3.5     **b)** 0.35
  **c)** 0.035   **d)** 35

**25.** $74.68 \div 24.7$
  **a)** 9       **b)** 3
  **c)** 12      **d)** 120

**26.** $914 \div 0.921$
  **a)** 10      **b)** 100
  **c)** 1000    **d)** 1

**27.** *Fence Posts.* A zoo plans to construct a fence around its proposed animals of the Great Plains exhibit. The perimeter of the area to be fenced is 1760 ft. Estimate the number of wooden fence posts needed if the posts are placed 8.625 ft apart.

**28.** *Ticketmaster.* Recently, Ticketmaster stock sold for $22.25 per share. Estimate how many shares can be purchased for $4400.

**29.** *Day-Care Supplies.* Helen wants to buy 12 boxes of crayons at $1.89 per box for the day care center that she runs. Estimate the total cost of the crayons.

**30.** *Batteries.* Oscar buys 6 packages of AAA batteries at $5.29 per package. Estimate the total cost of the purchase.

## Skill Maintenance

In each of Exercises 31–38, fill in the blank with the correct term from the given list. Some of the choices may not be used and some may be used more than once.

**31.** The decimal $0.57\overline{3}$ is an example of a(n) _____ decimal. [5.5a]

**32.** The least common _____ of two natural numbers is the smallest number that is a multiple of both. [4.1a]

**33.** The sentence $5(3 + 8) = 5 \cdot 3 + 5 \cdot 8$ illustrates the _____ law. [1.4a]

**34.** A(n) _____ of an equation is a replacement for the variable that makes the equation true. [1.7a]

**35.** The number 1 is the _____ identity. [1.4a]

**36.** The sentence $13 + 7 = 7 + 13$ illustrates the _____ law of addition. [1.2a]

**37.** The least common _____ of two or more fractions is the least common _____ of their denominators. [4.2b]

**38.** The number 3728 is _____ by 2 if the last digit is even. [3.1b]

additive

multiplicative

numerator

denominator

commutative

associative

distributive

solution

divisible

terminating

repeating

multiple

factor

## Synthesis

The following were done on a calculator. Estimate to determine whether the decimal point was placed correctly.

**39.** $178.9462 \times 61.78 = 11{,}055.29624$

**40.** $14{,}973.35 \div 298.75 = 501.2$

**41.** $19.7236 - 1.4738 \times 4.1097 = 1.366672414$

**42.** $28.46901 \div 4.9187 - 2.5081 = 3.279813473$

**43.** ▦ Use one of $+$, $-$, $\times$, and $\div$ in each blank to make a true sentence.
  **a)** $(0.37 \,\boxed{\phantom{x}}\, 18.78) \,\boxed{\phantom{x}}\, 2^{13} = 156{,}876.8$
  **b)** $2.56 \,\boxed{\phantom{x}}\, 6.4 \,\boxed{\phantom{x}}\, 51.2 \,\boxed{\phantom{x}}\, 17.4 = 312.84$

# 5.7

## Solving Equations

## OBJECTIVES

**a** Solve equations containing decimals and one variable term.

**b** Solve equations containing decimals and two or more variable terms.

---

**SKILL TO REVIEW**

Objective 2.8d: Solve equations that require use of both the addition principle and the division principle.

Solve.

1. $5x + 7 = -3$

2. $3x - 5 - x = 4 + x$

---

Solve.

1. $6x + 7.4 = 11$

2. $0.2 - 0.1x = 1.4$

---

Solve.

3. $7.4t + 1.25 = 27.89$

4. $-5.7 + 4.8x = -14.82$

---

In Section 2.8, we used a combination of the addition and division principles to solve equations like $5x + 7 = -3$. We now use those same properties to solve similar equations involving decimals.

### **a** Equations with One Variable Term

Recall that equations like $5x + 7 = -3$ are normally solved by first "undoing" the addition and then "undoing" the multiplication. This reverses the order of operations in which we add last and multiply first.

**EXAMPLE 1** Solve: $0.5x + 5 = 8$.

$$0.5x + 5 = 8$$

$$0.5x + 5 - 5 = 8 - 5 \quad \text{Subtracting 5 from both sides}$$

$$0.5x = 3 \quad \text{Simplifying}$$

$$\frac{0.5x}{0.5} = \frac{3}{0.5} \quad \text{Dividing both sides by 0.5}$$

$$x = 6 \quad \text{Simplifying}$$

Check:

$$\begin{array}{c|c} 0.5x + 5 = 8 \\ \hline 0.5(6) + 5 \; ? \; 8 \\ 3 + 5 \\ 8 \; | \; 8 \quad \text{TRUE} \end{array}$$

The solution is 6.

Do Margin Exercises 1 and 2.

**EXAMPLE 2** Solve: $4.2x + 3.7 = -26.12$.

$$4.2x + 3.7 = -26.12$$

$$4.2x + 3.7 - 3.7 = -26.12 - 3.7 \quad \text{Subtracting 3.7 from both sides}$$

$$4.2x = -29.82 \quad \text{Simplifying}$$

$$\frac{4.2x}{4.2} = \frac{-29.82}{4.2} \quad \text{Dividing both sides by 4.2}$$

$$x = -7.1 \quad \text{Simplifying}$$

Check:

$$\begin{array}{c|c} 4.2x + 3.7 = -26.12 \\ \hline 4.2(-7.1) + 3.7 \; ? \; -26.12 \\ -29.82 + 3.7 \\ -26.12 \; | \; -26.12 \quad \text{TRUE} \end{array}$$

The solution is $-7.1$.

Do Exercises 3 and 4.

---

*Answers*

*Skill to Review:*
1. $-2$   2. 9

*Margin Exercises:*
1. 0.6   2. $-12$   3. 3.6   4. $-1.9$

## b Equations with Two or More Variable Terms

Some equations have variable terms on both sides. To solve such an equation, we use the addition principle to get all variable terms on one side of the equation and all constant terms on the other side.

**EXAMPLE 3**   Solve: $10x - 7 = 2x + 13$.

We begin by subtracting $2x$ from (or adding $-2x$ to) each side. This will group all variable terms on one side of the equation.

$$10x - 7 - 2x = 2x + 13 - 2x \qquad \text{Adding } -2x \text{ to both sides}$$
$$8x - 7 = 13 \qquad \text{Combining like terms}$$

We use the addition principle to isolate all constant terms on one side.

$$8x - 7 = 13$$
$$8x - 7 + 7 = 13 + 7 \qquad \text{Adding 7 to both sides}$$
$$8x = 20 \qquad \text{Simplifying (combining like terms)}$$
$$\frac{8x}{8} = \frac{20}{8} \qquad \text{Dividing both sides by 8}$$
$$x = 2.5$$

Check:

$$
\begin{array}{c|c}
\multicolumn{2}{c}{10x - 7 = 2x + 13} \\
\hline
10(2.5) - 7 \;?\; & 2(2.5) + 13 \\
25 - 7 & 5 + 13 \\
18 & 18 \qquad \text{TRUE}
\end{array}
$$

The solution is 2.5.

Sometimes it may be easier to combine all variable terms on the right side and all constant terms on the left side.

**EXAMPLE 4**   Solve: $11 - 3t = 7t + 8$.

We can combine all variable terms on the right side by adding $3t$ to both sides.

$$11 - 3t = 7t + 8$$
$$11 - 3t + 3t = 7t + 8 + 3t \qquad \text{Adding } 3t \text{ to both sides}$$
$$11 = 10t + 8 \qquad \text{Combining like terms}$$
$$11 - 8 = 10t + 8 - 8 \qquad \text{Subtracting 8 from both sides}$$
$$3 = 10t$$
$$\frac{3}{10} = \frac{10t}{10} \qquad \text{Dividing both sides by 10}$$
$$0.3 = t$$

Check:

$$
\begin{array}{c|c}
\multicolumn{2}{c}{11 - 3t = 7t + 8} \\
\hline
11 - 3(0.3) \;?\; & 7(0.3) + 8 \\
11 - 0.9 & 2.1 + 8 \\
10.1 & 10.1 \qquad \text{TRUE}
\end{array}
$$

The solution is 0.3.

### STUDY TIPS

**DOUBLE-CHECK THE NUMBERS**

Solving problems is challenging enough, without miscopying information. Always double-check that you have accurately transferred numbers from the correct exercise in the exercise set.

Solve.

**5.** $10t - 3 = 4t + 18$

**6.** $8 + 4x = 9x - 3$

**7.** $2.1x - 45.3 = 17.3x + 23.1$

Note that in Example 4 the variable appears on the right side of the last equation. It does not matter whether the variable is isolated on the right or left side. What is important is that you have a clear direction to your work as you proceed from step to step.

Do Exercises 5–7.

**EXAMPLE 5** Solve: $5(x + 1) = 7x + 12$.

$$5(x + 1) = 7x + 12$$

$$5 \cdot x + 5 \cdot 1 = 7x + 12 \qquad \text{Using the distributive law to remove parentheses}$$

$$5x + 5 = 7x + 12 \qquad \text{Simplifying}$$

$$5x + 5 - 7x = 7x + 12 - 7x \qquad \text{Subtracting } 7x \text{ from both sides}$$

$$-2x + 5 = 12 \qquad \text{Simplifying}$$

$$-2x + 5 - 5 = 12 - 5 \qquad \text{Subtracting 5 from both sides}$$

$$-2x = 7$$

$$\frac{-2x}{-2} = \frac{7}{-2} \qquad \text{Dividing both sides by } -2$$

$$x = -3.5$$

Check: 

$$\begin{array}{c|c} \multicolumn{2}{c}{5(x + 1) = 7x + 12} \\ \hline 5(-3.5 + 1) \ ? \ 7(-3.5) + 12 \\ 5(-2.5) \ \bigm| \ -24.5 + 12 \\ -12.5 \ \bigm| \ -12.5 \end{array}$$

Check in the original equation.

TRUE

The solution is $-3.5$.

**8.** Solve: $3(x + 5) = 20 - x$.

Do Exercise 8.

**EXAMPLE 6** Solve: $9(x - 3) + 7 = 5x - 47$.

We use the distributive law and combine like terms before using the addition and multiplication principles.

$$9(x - 3) + 7 = 5x - 47$$

$$9x - 27 + 7 = 5x - 47 \qquad \text{Using the distributive law}$$

$$9x - 20 = 5x - 47 \qquad \text{Simplifying}$$

$$9x - 20 - 5x = 5x - 47 - 5x \qquad \text{Subtracting } 5x \text{ from both sides}$$

$$4x - 20 = -47 \qquad \text{Simplifying}$$

$$4x - 20 + 20 = -47 + 20 \qquad \text{Adding 20 to both sides}$$

$$4x = -27 \qquad \text{Simplifying}$$

$$\frac{4x}{4} = -\frac{27}{4} \qquad \text{Dividing both sides by 4}$$

$$x = -6.75$$

Check: 

$$\begin{array}{c|c} \multicolumn{2}{c}{9(x - 3) + 7 = 5x - 47} \\ \hline 9(-6.75 - 3) + 7 \ ? \ 5(-6.75) - 47 \\ 9(-9.75) + 7 \ \bigm| \ -33.75 - 47 \\ -87.75 + 7 \ \bigm| \ -80.75 \\ -80.75 \ \bigm| \ -80.75 \end{array}$$

TRUE

**9.** Solve: $8(x - 2) - 15 = 4x + 2$.

The solution is $-6.75$.

Do Exercise 9.

*Answers*

**5.** 3.5   **6.** 2.2   **7.** −4.5
**8.** 1.25   **9.** 8.25

**a**    Solve. Remember to check.

**1.** $5x = 27$

**2.** $36 \cdot y = 14.76$

**3.** $x + 15.7 = 3.1$

**4.** $x + 13.9 = 4.2$

**5.** $5x - 8 = 22$

**6.** $4x - 7 = 13$

**7.** $6.9x - 8.4 = 4.02$

**8.** $7.1x - 9.3 = 8.45$

**9.** $21.6 + 4.1t = 6.43$

**10.** $12.4 + 3.7t = 2.04$

**11.** $-26.25 = 7.5x + 9$

**12.** $-43.72 = 8.7x + 5$

**13.** $-4.2x + 3.04 = -4.1$

**14.** $-2.9x - 2.24 = -17.9$

**15.** $-3.05 = 7.24 - 3.5t$

**16.** $-4.62 = 5.68 - 2.5t$

**17.** $3 - 1.2y = -2.4$

**18.** $6 - 3.5y = 12.3$

**b**    Solve. Remember to check.

**19.** $9x - 2 = 5x + 34$

**20.** $8x - 5 = 6x + 9$

**21.** $2x + 6 = 7x - 10$

**22.** $3x + 4 = 11x - 6$

**23.** $5y - 3 = 4 + 9y$

**24.** $6y - 5 = 8 + 10y$

**25.** $5.9x + 67 = 7.6x + 16$

**26.** $2.1x + 42 = 5.2x - 20$

**27.** $7.8a + 2 = 2.4a + 19.28$

**28.** $7.5a - 5.16 = 3.1a + 12$

**29.** $6(x + 2) = 4x + 30$

**30.** $5(x + 3) = 3x + 23$

**31.** $5(x + 3) = 15x - 6$

**32.** $2(x + 3) = 4x - 11$

**33.** $7a - 9 = 15(a - 3)$

**34.** $2a - 7 = 12(a - 3)$

**35.** $1.5(y - 6) = 1.3 - y$

**36.** $2.3(5 - y) = 0.2y + 1.7$

**37.** $2.9(x + 8.1) = 7.8x - 3.95$

**38.** $2(x + 7.3) = 6x - 0.83$

**39.** $-6.21 - 4.3t = 9.8(t + 2.1)$

**40.** $-7.37 - 3.2t = 4.9(t + 6.1)$

**41.** $4(x - 2) - 9 = 2x + 9$

**42.** $9(x - 4) + 13 = 4x - 23$

**43.** $2(4y - 1.8) + 0.4 = 8(2y - 0.4)$

**44.** $3(1.2y + 5) = 2(2.5y - 7) + 1$

**45.** $43(7 - 2x) + 34 = 50(x - 4.1) + 744$

**46.** $34(5 - 3.5x) = 12(3x - 8) + 653.5$

## Skill Maintenance

Find the area of each figure. [3.6b]

**47.**
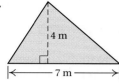

**48.**
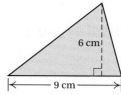

**49.**
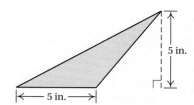

**50.**
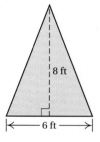

**51.**
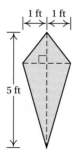

**52.**
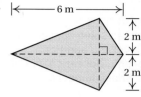

**53.** Subtract: $\dfrac{3}{25} - \dfrac{7}{10}$. [4.3a]

**54.** Simplify: $\dfrac{0}{-18}$. [3.3b]

**55.** Add: $-17 + 24 + (-9)$. [2.2a]

**56.** Solve: $3x - 10 = 14$. [2.8d]

## Synthesis

Solve.

**57.** ▦ $7.035(4.91x - 8.21) + 17.401 = 23.902x - 7.372815$

**58.** ▦ $8.701(3.4 - 5.1x) - 89.321 = 5.401x + 74.65787$

**59.** $5(x - 4.2) + 3[2x - 5(x + 7)] = 39 + 2(7.5 - 6x) + 3x$

**60.** $14(2.5x - 3) + 9x + 5 = 4(3.25 - x) + 2[5x - 3(x + 1)]$

**61.** ▦ $3.5(4.8x - 2.9) + 4.5 = 9.4x - 3.4(x - 1.9)$

**62.** ▦ $4.19 - 1.8(4.5x - 6.4) = 3.1(9.8 + x)$

# 5.8

## Applications and Problem Solving

# OBJECTIVES

**a** Translate key phrases to algebraic expressions.

**b** Solve applied problems involving decimals.

**SKILL TO REVIEW**

Objective 1.8a: Solve applied problems involving addition, subtraction, multiplication, or division of whole numbers.

Solve.

1. A piece of pecan pie has 502 calories. This is 186 calories more than a piece of pumpkin pie contains. How many calories does a piece of pumpkin pie have?

### a Translating to Algebraic Expressions

To translate problems to equations, we need to be able to translate phrases to algebraic expressions. Certain key words in phrases help direct the translation.

| KEY WORDS | SAMPLE PHRASE OR SENTENCE | TRANSLATION |
|---|---|---|
| **Addition (+)** | | |
| added to | 350 lb was added to the car's weight. | $w + 350$ |
| sum of | The sum of a number and 10 | $n + 10$ |
| plus | 13 plus some number | $13 + x$ |
| more than | 176 more than last year's enrollment | $r + 176$ |
| increased by | The original estimate, increased by 50 | $y + 50$ |
| **Subtraction (−)** | | |
| subtracted from | 2 oz was subtracted from the bag's weight. | $w - 2$ |
| difference of | The difference of two prices | $p - q$ |
| minus | A construction crew of size $c$, minus 4 workers | $c - 4$ |
| less than | 18 less than the number of volunteers last month | $v - 18$ |
| decreased by | An essay's grade, decreased by 5 points | $g - 5$ |
| **Multiplication (·)** | | |
| multiplied by | The number of mints in a box, multiplied by 5 | $5 \cdot m$ |
| product of | The product of two numbers | $a \cdot b$ |
| times | 10 times Jon's age | $10j$ |
| twice | Twice the wholesale price | $2w$ |
| of | $\frac{1}{2}$ of the number of pages assigned | $\frac{1}{2}p$ |
| **Division (÷)** | | |
| divided by | A 16-oz bag of almonds, divided by 5 | $16 \div 5$ |
| quotient of | The quotient of 60 and 3 | $60 \div 3$ |
| divided into | 6 divided into the cost of the meal | $c \div 6$ |
| ratio of | The ratio of 456 to the number of gallons of gasoline | $456/g$ |

It is helpful to choose a descriptive variable to represent the unknown. For example, $w$ suggests weight and $g$ suggests the number of gallons of gasoline.

*Answer*

*Skills to Review:*
1. 316 calories

The following tips are helpful in translating phrases to algebraic expressions.

> ### TIPS FOR TRANSLATING PHRASES TO ALGEBRAIC EXPRESSIONS
> - Use a specific number in the statement before translating using the variable.
> - Write down what each variable represents.
> - Check the translation with another number to see if it matches the phrase.
> - Be especially careful with order when subtracting and dividing.

**EXAMPLE 1**  Translate each phrase to an algebraic expression.

**a)** A number added to 5

**b)** A number subtracted from 5

**a)** Let $n$ represent the number. Then the phrase "a number added to 5" can be translated $5 + n$, or $n + 5$.

**b)** Let $n$ represent the number. Then the phrase "a number subtracted from 5" can be translated $5 - n$.

Do Exercises 1 and 2.

Translate to an algebraic expression.

1. Twenty more than a number

2. Twenty less than a number

------- *Caution!* -------

Note that in Example 1, there are two correct translations for part (a) and only one correct translation for part (b). Recall that numbers can be added in either order, so $5 + n$ and $n + 5$ represent the same value. However, order is important in subtraction, so $5 - n$ and $n - 5$ do NOT represent the same value.

**EXAMPLE 2**  Translate each phrase to an algebraic expression.

**a)** Than's age increased by six

**b)** Half of some number

**c)** Twice the cost

**d)** Seven more than twice the weight

**e)** Fifteen divided by a number

**f)** A number divided by fifteen

**g)** Six less than the product of two numbers

**h)** Nine times the sum of two numbers

| *Phrase* | *Variable(s)* | *Algebraic Expression* |
|---|---|---|
| **a)** Than's age increased by six | Let $a$ = Than's age. | $a + 6$, or $6 + a$ |
| **b)** Half of some number | Let $n$ = the number. | $\frac{1}{2}n$, or $\frac{n}{2}$, or $n \div 2$ |
| **c)** Twice the cost | Let $c$ = the cost. | $2c$ |
| **d)** Seven more than twice the weight | Let $w$ = the weight. | $2w + 7$, or $7 + 2w$ |
| **e)** Fifteen divided by a number | Let $x$ = the number. | $15 \div x$, or $\frac{15}{x}$ |

*Answers*

1. Let $n$ = the number; $n + 20$, or $20 + n$
2. Let $n$ = the number; $n - 20$

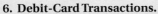

Translate to an algebraic expression.

**3.** The height of a tree, decreased by 50

**4.** Four less than ten times a number

**5.** Fourteen more than the product of the hourly rate and the number of hours worked

**6. Debit-Card Transactions.**
Debit-card transactions in 2000 totaled 9.8 billion. The estimated number of transactions for 2010 is 42.5 billion. How many more debit-card transactions were there in 2010 than in 2000?

Source: *The Nilson Report*

| | | |
|---|---|---|
| **f)** A number divided by fifteen | Let $x$ = the number. | $x \div 15$, or $\dfrac{x}{15}$ |
| **g)** Six less than the product of two numbers | Let $m$ and $n$ = the numbers. | $mn - 6$ |
| **h)** Nine times the sum of two numbers | Let $a$ and $b$ = the numbers. | $9(a + b)$ |

Do Exercises 3–5.

## (b) Solving Applied Problems

**EXAMPLE 3** *Canals.* The Panama Canal in Panama is 50.7 mi long. The Suez Canal in Egypt is 119.9 mi long. How much longer is the Suez Canal?

Panama Canal      Suez Canal

1. **Familiarize.** We let $l$ = the distance in miles that the length of the longer canal differs from the length of the shorter canal.

2. **Translate.** We translate as follows, using the given information:

| Length of Panama Canal, the shorter canal | plus | additional length | is | length of Suez Canal, the longer canal |
|:---:|:---:|:---:|:---:|:---:|
| 50.7 mi | + | $l$ | = | 119.9 mi. |

3. **Solve.** We solve the equation by subtracting 50.7 mi on both sides:

$$50.7 + l = 119.9$$
$$50.7 + l - 50.7 = 119.9 - 50.7$$
$$l = 69.2.$$

4. **Check.** We can check by adding.

$$\begin{array}{r} 5\ 0.7 \\ +\ \ 6\ 9.2 \\ \hline 1\ 1\ 9.9 \end{array}$$

The answer checks.

5. **State.** The Suez Canal is 69.2 mi longer than the Panama Canal.

Do Exercise 6.

**EXAMPLE 4** *Medicine.* A 100-unit syringe is often used by diabetics and nurses to administer insulin. Each unit on the syringe represents 0.01 cc (cubic centimeter). Each day, Wendy averages 42 units of insulin. How many cc's will Wendy use in a typical week?

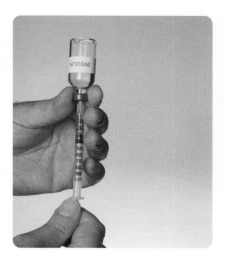

1. **Familiarize.** We let $a$ = the amount of insulin used.

2. **Translate.** We translate as follows.

$$\underbrace{\text{Amount used each day}}_{0.42 \text{ cc}} \quad \underbrace{\text{times}}_{\times} \quad \underbrace{\text{number of days in a week}}_{7} \quad \underbrace{\text{is}}_{=} \quad \underbrace{\text{total amount injected}}_{a}$$

3. **Solve.** To solve the equation, we carry out the multiplication.

$$\begin{array}{r} 0.4\ 2 \\ \times \quad\quad 7 \\ \hline 2.9\ 4 \end{array}$$

Thus, $a$ = 2.94 cc.

4. **Check.** We can obtain a partial check by rounding and estimating:

$$0.42 \text{ cc} \times 7 \approx 0.4 \text{ cc} \times 7 = 2.8 \text{ cc} \approx 2.94 \text{ cc}.$$

5. **State.** Wendy uses 2.94 cc's of insulin in a typical week.

Do Exercise 7.

Do Exercise 7.

## Multistep Problems

**EXAMPLE 5** *Student Loans.* Upon graduation from college, Eric must repay a loan that totals $12,267. The loan is to be paid back over 5 yr in equal monthly payments. Find the amount of each payment.

1. **Familiarize.** We let $m$ = the size of each monthly payment. To find the number of payments, we determine how many months there are in 5 yr. (We could also find the amount paid per year and divide that number by 12.)

2. **Translate.** To find the amount of the monthly payment, we note that the amount owed is split up, or *divided*, into payments of equal size. To find the number of payments, we determine that in 5 yr there are

$$5 \cdot 12 = 60 \text{ months}. \quad \text{There are 12 months in a year.}$$

We have

$$\underbrace{\text{Amount of payment}}_{m} \quad \underbrace{\text{is}}_{=} \quad \underbrace{\text{total amount owed}}_{12,267} \quad \underbrace{\text{divided by}}_{\div} \quad \underbrace{\text{number of payments}}_{60}.$$

3. **Solve.** To solve, we carry out the division.

$$\begin{array}{r} 2\ 0\ 4.4\ 5 \\ 6\,0\ )\overline{1\,2,2\,6\,7.0\,0} \\ \underline{1\,2\,0} \\ 2\,6\,7 \\ \underline{2\,4\,0} \\ 2\,7\,0 \\ \underline{2\,4\,0} \\ 3\,0\,0 \\ \underline{3\,0\,0} \\ 0 \end{array}$$

$m = \$204.45$

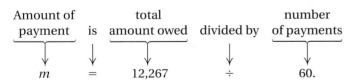

7. **Mileage Rates.** The Internal Revenue Service allowed a tax deduction of 16.5¢ per mile driven for medical or moving purposes in 2010. What deduction, in dollars, would be allowed for driving 1862 mi during a move?

Source: Internal Revenue Service

*Answer*

7. $307.23

4. **Check.** To check, we solve the problem using an alternative method. We first find how much is repaid each year:

$$\$12,267 \div 5 = \$2453.40.$$

We then divide this amount by 12:

$$\$2453.40 \div 12 = \$204.45.$$

5. **State.** Eric's monthly payments will be $204.45.

Do Exercise 8.

**EXAMPLE 6** *Gas Mileage.* Ava filled her gas tank and noted that the odometer read 67,507.8. After the next filling, the odometer read 68,006.1. It took 16.5 gal to fill the tank. How many miles per gallon did Ava get?

1. **Familiarize.** We first make a drawing.

This is a two-step problem. First, we find the number of miles that have been driven between fillups. We let $n$ = the number of miles driven and $m$ = the number of miles per gallon.

2., 3. **Translate** and **Solve.** We translate and solve as follows:

First reading plus number of miles driven is second reading

$$67{,}507.8 \quad + \quad n \quad = \quad 68{,}006.1.$$

To solve the equation, we subtract 67,507.8 on both sides:

$$n = 68{,}006.1 - 67{,}507.8$$
$$= 498.3$$

$$\begin{array}{r} 6\,8{,}0\,0\,6.1 \\ -\ 6\,7{,}5\,0\,7.8 \\ \hline 4\,9\,8.3 \end{array}$$

Next, we divide the total number of miles driven by the number of gallons. This gives us $m$. The division that corresponds to the situation is

$$498.3 \div 16.5 = m. \qquad m = \text{the number of miles per gallon}$$

$$\begin{array}{r} 3\,0.2 \\ 1\,6.5\,)\,\overline{4\,9\,8.3_\wedge 0} \\ \underline{4\,9\,5} \\ 3\,3\,0 \\ \underline{3\,3\,0} \\ 0 \end{array} \qquad m = 30.2$$

4. **Check.** To check, we first multiply the number of miles per gallon times the number of gallons to find the number of miles driven:

$$16.5 \times 30.2 = 498.3.$$

Then we add 498.3 to 67,507.8 to find the new odometer reading:

$$67{,}507.8 + 498.3 = 68{,}006.1. \qquad \text{The mileage 30.2 checks.}$$

5. **State.** Ava got 30.2 miles per gallon.

Do Exercise 9.

Example 7 involves a formula giving the area of a circle.

In any circle, a **diameter** is a segment that passes through the center of the circle with endpoints on the circle. A **radius** is a segment with one endpoint on the center and the other endpoint on the circle. The area, $A$, of a circle with radius of length $r$ is given by

$$A = \pi \cdot r^2,$$

where $\pi \approx 3.14$.
  The length $r$ of a radius of a circle is half the length $d$ of a diameter:

$$r = \frac{1}{2}d.$$

**EXAMPLE 7**  The Northfield Tap and Die Company stamps 6-cm-wide discs out of metal squares that are 6 cm by 6 cm. How much metal remains after the disc has been punched out?

1. **Familiarize.**  We make, and label, a drawing. We let $a = $ the amount of metal remaining, in square centimeters, and list the relevant formulas.

    For a square with sides of length $s$, $Area = s^2$.
    For a circle with radius of length $r$, $Area = \pi \cdot r^2$, where $\pi \approx 3.14$.

    The radius $r$ of a circle is half of its diameter:  $r = \frac{1}{2}d$.

    We also list the known information about the square and the disc:

    Square:   The side $s = 6$ cm.
    Disc:   The diameter $d = 6$ cm.

        The radius $r = \frac{1}{2}(6\text{ cm}) = 3$ cm.

The shaded area is the amount of metal remaining.

2. **Translate.**  To find the amount left over, we subtract the area of the disc from the area of the square.

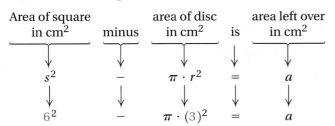

3. **Solve.**  We simplify as follows.

$$6^2 - 3.14(3)^2 = a$$
$$36 - 3.14 \cdot 9 = a$$
$$36 - 28.26 = a$$
$$7.74 = a$$

4. **Check.**  We can repeat our calculation as a check. Note that 7.74 is less than the area of the disc, which in turn is less than the area of the square. This agrees with the impression given by our drawing.

5. **State.**  The amount of material left over is 7.74 cm$^2$.

**10.** Suppose that an 8-in.-wide disc is punched out of an 8-in. by 8-in. sheet of metal. How much material is left over?

*Answer*
**10.** 13.76 in$^2$

Do Exercise 10.

**EXAMPLE 8** *Photo Books.* Apple sells iPhoto books that users of the iPhoto application can create. The price of a large, softcover, 20-page book is $19.99. Additional pages are 69 cents per page. Marta has $35 to spend on a book. What is the greatest number of pages she can put in the book?
Source: www.apple.com

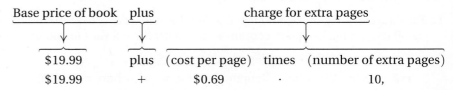

1. **Familiarize.** Suppose that Marta put 30 pages in the book. She would have to pay an additional price per page for 30 − 20, or 10, pages. The price would be

| Base price of book | plus | charge for extra pages | | |
|---|---|---|---|---|
| ↓ | ↓ | | ↓ | |
| $19.99 | plus | (cost per page) | times | (number of extra pages) |
| $19.99 | + | $0.69 | · | 10, |

which is $19.99 + $6.90, or $26.89. This familiarizes us with the way in which price is calculated. Note that we convert 69 cents to $0.69 so that only one unit, dollars, is used. Note also that Marta can put more than 30 pages in the book. To see just how many pages she can add, we could make and check more guesses, but this would be very time-consuming. Instead, we let $p$ = the number of extra pages Marta adds to the 20-page book. Note that the total number of pages in the book is then $20 + p$.

2. **Translate.** The problem can be rephrased and translated as follows.

| Base price of book | plus | cost per page | times | number of extra pages | is | cost of book |
|---|---|---|---|---|---|---|
| ↓ | ↓ | ↓ | ↓ | ↓ | ↓ | ↓ |
| $19.99 | + | $0.69 | · | $p$ | = | $35 |

3. **Solve.** We solve the equation.

$$19.99 + 0.69p = 35$$
$$0.69p = 15.01 \qquad \text{Subtracting 19.99 from both sides}$$
$$p = \frac{15.01}{0.69} \qquad \text{Dividing both sides by 0.69}$$
$$p \approx 21.75 \qquad \text{Rounding}$$

4. **Check.** We check in the original problem. Since Marta cannot pay for parts of a page, we must round the answer *down* to 21 in order to keep the cost under $35. This makes the total length of the book 20 + 21, or 41, pages. If Marta adds 21 pages to the book, the additional cost will be 21 times $0.69, or $14.49. If we add $14.49 to the base price of $19.99, we get $34.48, which is just under the $35 Marta has to spend.

5. **State.** With $35, Marta can make a 41-page book.

**11. Bike Rentals.** Mike's Bikes rents mountain bikes. The shop charges $4.00 insurance for each rental plus $6.00 per hour. For how many hours can a person rent a bike with $25.00?

Do Exercise 11.

*Answer*
11. 3.5 hr

## Problems with More than One Unknown

**EXAMPLE 9** *Multi-sport Recreation.* Around the Bend Expeditions offers a 2-day canyon trip combining biking and canoeing. Those participating will ride mountain bikes for 12 miles longer than the distance paddled. The total length of the trip is 43 miles. How long is the biking portion of the trip, and how long is the canoeing portion of the trip?

1. **Familiarize.** We first list the quantities we are asked to find:

   Length of biking portion, in miles

   Length of canoeing portion, in miles

   We will want to represent both quantities using only one variable. To do so, we use the second sentence in the problem: *Those participating will ride mountain bikes for 12 miles longer than the distance paddled.* Since the biking portion of the trip is described in terms of the canoeing portion, we let $x = $ the length of the canoeing portion, in miles. Then the length of the biking portion, also in miles, is $x + 12$.

   Length of biking portion, in miles:     $x + 12$

   Length of canoeing portion, in miles:     $x$

2. **Translate.** We use the total length of the trip to translate to an equation. Note that all lengths are in miles.

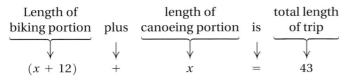

$$(x + 12) + x = 43$$

3. **Solve.** We solve the equation.

   $$(x + 12) + x = 43$$

   | | |
   |---|---|
   | $x + 12 + x = 43$ | Removing parentheses |
   | $2x + 12 = 43$ | Combining like terms |
   | $2x = 31$ | Subtracting 12 from both sides |
   | $x = 15.5$ | Dividing both sides by 2 |

   Recall that we are looking for two quantities. We will use the value of $x$ to find both of them. We return to the list of unknowns in the *Familiarize* step.

   Length of biking portion, in miles:     $x + 12 = 15.5 + 12 = 27.5$

   Length of canoeing portion, in miles:     $x = 15.5$

4. **Check.** There are two statements in the problem to verify. First since $27.5 - 15.5 = 12$, the biking portion is 12 miles longer than the canoeing portion. Second, since $27.5 + 15.5 = 43$, the total length of the trip is 43 miles.

5. **State.** The biking portion of the trip is 27.5 miles, and the canoeing portion is 15.5 miles.

Do Exercise 12.

----------- *Caution!* -----------

Do not skip the *Familiarize* step, particularly in problems involving more than one unknown. Describing the unknowns using one variable is often the most challenging part of the solution process.

12. Holly and Lorenzo worked together on a project for a psychology class. Holly spent twice as much time on the project as Lorenzo did. They spent a total of 10.5 hours on the project. How much time did each person work on the project?

*Answer*

12. Holly: 7 hr; Lorenzo: 3.5 hr

# Translating for Success

1. *Gas Mileage.* Art filled his SUV's gas tank and noted that the odometer read 38,271.8 mi. At the next filling, the odometer read 38,677.9 mi. It took 28.4 gal to fill the tank. How many miles per gallon did the SUV get?

2. *Dimensions of a Parking Lot.* Seals' parking lot is a rectangle that measures 85.2 ft by 52.3 ft. What is the area of the parking lot?

3. *Game Snacks.* Three students pay $18.40 for snacks at a football game. What is each person's share?

4. *Electrical Wiring.* An electrician needs 1314 ft of wiring cut into $2\frac{1}{2}$-ft pieces. How many pieces will she have?

5. *College Tuition.* Wayne needs $4638 for the fall semester's tuition. On the day of registration, he has only $3092. How much does he need to borrow?

---

The goal of these matching questions is to practice step (2), *Translate*, of the five-step problem-solving process. Translate each word problem to an equation and select a correct translation from equations A–O.

**A.** $2\frac{1}{2} \cdot n = 1314$

**B.** $18.4 \times 3.26 = n$

**C.** $n = 85.2 \times 52.3$

**D.** $19 - (-4) = n$

**E.** $3 \times 18.40 = n$

**F.** $2\frac{1}{2} \cdot 1314 = n$

**G.** $3092 + n = 4638$

**H.** $18.4 \cdot n = 3.26$

**I.** $\dfrac{406.1}{28.4} = n$

**J.** $52.3 \cdot n = 85.2$

**K.** $n = 19 + (-4)$

**L.** $52.3 + n = 85.2$

**M.** $3092 + 4638 = n$

**N.** $3 \cdot n = 18.40$

**O.** $85.2 + 52.3 = n$

*Answers on page A-10*

---

6. *Cost of Gasoline.* What is the cost, in dollars, of 18.4 gal of gasoline at $3.26 per gallon?

7. *Temperature.* At noon, the temperature in Pierre was 19°F. At midnight, the temperature had fallen to −4°F. By how many degrees had the temperature fallen?

8. *Acres Planted.* This season Sam planted 85.2 acres of corn and 52.3 acres of soybeans. Find the total number of acres that he planted.

9. *Amount Inherited.* Tara inherited $2\frac{1}{2}$ times as much as her cousin. Her cousin received $1314. How much did Tara receive?

10. *Travel Funds.* The athletic department needs travel funds of $4638 for the tennis team and $3092 for the basketball team. What is the total amount needed for travel?

## 5.8 Exercise Set

For Extra Help

**MyMathLab**

 PRACTICE

 WATCH

 DOWNLOAD

 READ

REVIEW

**a**    Translate to an algebraic expression. Choice of variables used may vary.

**1.** Five more than Ron's age

**2.** The product of four and $t$

**3.** 6 more than $b$

**4.** 7 more than Jen's weight

**5.** 9 less than $c$

**6.** 4 less than $d$

**7.** A number decreased by 16

**8.** A number increased by 20

**9.** 8 times Nate's speed

**10.** The ratio of a number and 100

**11.** $x$ divided by 17

**12.** 100 divided by the hourly rate

**13.** 20 added to half of a number

**14.** 18 subtracted from twice a number

**15.** 20 less than 4 times a number

**16.** 35 more than 10 times a number

**17.** The sum of the box's length and width

**18.** The Cessna's speed minus the wind speed

**19.** 10 more than the product of the rate and the time

**20.** 50 less than the product of the length and width

**21.** The sum of 10 times a number and the number

**22.** The difference of a number and twice the number

**23.** 5 times the difference of two numbers

**24.** One fourth of the sum of two numbers

**b**    Solve using the five-step problem-solving procedure.

**25.** *Record Movie Revenue.*    The movie *Avatar* took in $2.63 billion in its lifetime. This is $0.78 billion more than *Titanic* took in. How much did *Titanic* take in during its lifetime?
Source: www.the-numbers.com

**26.** *Counterfeit Money.*    The amount of counterfeit money passed in the United States is on the rise. About $64.4 million was passed in 2008. This amount was $25.2 million more than entered circulation in 1999. How much counterfeit money was passed in 1999?
Source: U.S. Secret Service

*Hurricane Damage.* The amounts of damage caused by the five most costly Atlantic hurricanes in the United States are shown in the table below. Use this table to do Exercises 27 and 28.

**Most Costly Hurricanes**

| RANK | HURRICANE | YEAR | COST IN 2007 DOLLARS (in billions) |
|------|-----------|------|-----------------------------------|
| 1 | Katrina | 2005 | $81.2 |
| 2 | Andrew | 1992 | 38.1 |
| 3 | Wilma | 2005 | 30.4 |
| 4 | Ivan | 2004 | 18.1 |
| 5 | Charley | 2004 | 16.2 |

SOURCE: National Hurricane Center

**27.** How much more costly was Hurricane Katrina than Hurricane Andrew?

**28.** What was the total cost of the two hurricanes that occurred in 2005?

**29.** *Gasoline Cost.* What is the cost, in dollars, of 20.4 gal of gasoline at 224.9 cents per gallon? (224.9 cents = $2.249) Round the answer to the nearest cent.

**30.** *Gasoline Cost.* What is the cost, in dollars, of 15.3 gal of gasoline at 289.9 cents per gallon? (289.9 cents = $2.899) Round the answer to the nearest cent.

**31.** *Cost of Bottled Water.* The cost of a year's supply of a popular brand of bottled water, based on the recommended consumption of 64 oz per day, at the supermarket price of $3.99 for a six-pack of half-liter bottles is $918.82. This is $918.31 more than the cost of drinking the same amount of tap water for a year. What is the cost of drinking tap water for a year?

**Source:** American Water Works Association

**32.** *Body Temperature.* Normal body temperature is 98.6°F. During an illness, a patient's temperature rose 4.2°. What was the new temperature?

**33.** *Lottery Winnings.* The largest lotto jackpot ever won in California totaled $193,000,000 and was shared equally by 3 winners. How much was each winner's share? Round to the nearest cent.

**Source:** California State Lottery

**34.** *Lunch Costs.* A group of 4 students pays $47.84 for lunch and splits the cost equally. What is each person's share?

**35.** *Stamp.* Find the area and the perimeter of the stamp shown here.

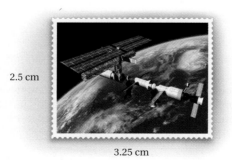

2.5 cm

3.25 cm

**36.** *Pole Vault Pit.* Find the area and the perimeter of the landing area of the pole vault pit shown here.

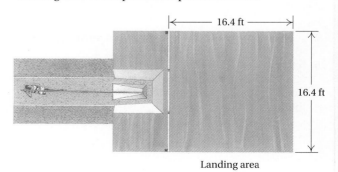

16.4 ft

16.4 ft

Landing area

**37.** *Odometer Reading.* The Binford family's odometer reads 22,456.8 at the beginning of a trip. The family's online driving directions tell them that they will be driving 234.7 mi. What will the odometer read at the end of the trip?

**38.** *Miles Driven.* Petra bought gasoline when the odometer read 14,296.3. At the next gasoline purchase, the odometer read 14,515.8. How many miles had been driven?

**39.** *Gas Mileage.* Peggy filled her van's gas tank and noted that the odometer read 26,342.8. After the next filling, the odometer read 26,736.7. It took 19.5 gal to fill the tank. How many miles per gallon did the van get?

**40.** *Gas Mileage.* Henry filled his Honda's gas tank and noted that the odometer read 18,943.2. After the next filling, the odometer read 19,306.2. It took 13.2 gal to fill the tank. How many miles per gallon did the car get?

**41.** Andrew bought a DVD of the movie *Horton Hears a Who* for his nephew for $23.99 plus $1.68 sales tax. He paid for it with a $50 bill. How much change did he receive?

**42.** Claire bought a copy of the book *Make Way for Ducklings* for her daughter for $16.95 plus $0.85 sales tax. She paid for it with a $20 bill. How much change did she receive?

**43.** *Nursing.* Phil injects a total of 38 units of insulin, each day for a week (see Example 4). How many cc's of insulin does he use in a week?

**44.** *Chemistry.* The water in a filled tank weighs 748.45 lb. One cubic foot of water weighs 62.5 lb. How many cubic feet of water does the tank hold?

**45.** *Egg Costs.* A restaurant owner bought 20 dozen eggs for $25.80. Find the cost of each egg to the nearest tenth of a cent (thousandth of a dollar).

**46.** *Weight Loss.* A person weighing 170 lb burns 8.6 calories per minute while mowing a lawn. One must burn about 3500 calories in order to lose 1 lb. How many pounds would be lost by mowing for 2 hr? Round to the nearest tenth.

**47.** *Travel Poster.* Sam is decorating his dorm room with a travel poster. The dimensions of the poster are as shown. How much area is not devoted to the painting?

15.7 in.
22.2 in.
27.4 in.
19.3 in.

**48.** *Earth Day Poster.* An Earth Day poster that is 61.8 cm by 73.2 cm includes a 2-cm border. What is the area of the poster inside the border?

61.8 cm
73.2 cm

**49.** *Study Cards.* An instructor allows her students to bring one 7.6-cm by 12.7-cm index card to the final exam, with notes of any sort written on the card. If both sides of the card are used, how much area is available for notes?

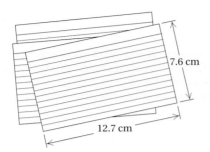

7.6 cm

12.7 cm

**50.** *Stamps.* Find the total area of the stamps shown.

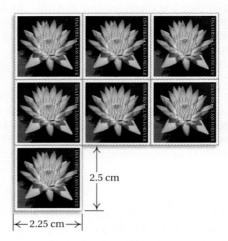

2.5 cm

2.25 cm

Find the length *d* in each figure.

**51.**

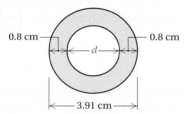

0.8 cm — — 0.8 cm

*d*

3.91 cm

**52.**

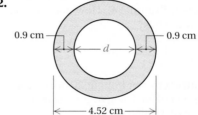

0.9 cm — — 0.9 cm

*d*

4.52 cm

**53.** *Carpentry.* A round, 6-ft-wide hot tub is being built into a 12-ft by 30-ft rectangular deck. How much decking is needed for the surface of the deck?

12 ft    6 ft    30 ft

**54.** A 4-ft by 4-ft tablecloth is cut from a round tablecloth that is 6 ft wide. Find the area of the cloth left over.

**55.** *Web Space.* Penn State charges $8 a month for full access plus 3 cents per gigabyte for Web space. Nikki's bill for one month was $11.75. How many gigabytes did she pay for?
Source: aset.its.psu.edu

**56.** *Service Calls.* JoJo's Service Center charges $30 for a house call plus $37.50 for each hour the job takes. For how long has a repairperson worked on a house call if the bill comes to $123.75?

**57.** Frank has been sent to the store with $40 to purchase 6 lb of cheese at $4.79 a pound and as many bottles of seltzer, at $0.64 a bottle, as possible. How many bottles of seltzer should Frank buy?

**58.** Janice has been sent to the store with $30 to purchase 5 pt of strawberries at $2.49 a pint and as many bags of chips, at $1.39 a bag, as possible. How many bags of chips should Janice buy?

**59.** *Hours Worked.* Ben works evenings as a waiter and weekends doing lawn maintenance. Last week, he worked 6 more hours as a waiter than he did doing lawn maintenance. He worked a total of 27 hours. How many hours did he work at each job?

**60.** *Hours Spent Eating.* The U.S. Bureau of Labor Statistics tracks the average time that Americans spend eating as a primary activity, as well as the number of hours spent eating as a secondary activity. In 2007, on average, Americans spent 0.4 hour a day more eating as a secondary activity than they did eating as a primary activity. They spent a total of 2.6 hours a day eating. How many hours did they spend eating as a primary activity and how many hours as a secondary activity?

**61.** *Endangered Species.* Part of the giant panda's habitat is protected by the Chinese government. There are 0.4 million more acres unprotected than there are protected acres. The total size of the giant panda's habitat is 5.4 million acres. How many acres are protected and how many are not protected?
**Source:** www.worldwildlife.org

**62.** *Hours of Daylight.* On December 22, Fairbanks, Alaska, has 16.6 more hours of darkness than it has daylight. How many hours of daylight and how many hours of darkness are there in Fairbanks on that day? [*Hint:* There are 24 hours in a day.]
**Source:** "Alaska's Winter Daylight," by Kimi Ross, at www.bellaonline.com

**63.** *e-Mail.* In 2008, approximately 210 billion e-mail messages were sent each day. The number of spam messages was about five times the number of non-spam messages. How many of each type of message were sent each day in 2008?
**Source:** Radicati Group and ICF International

**64.** *Laptop Computers.* Approximately 160 million laptop computers, including netbooks, were sold in 2009. The number of laptops sold that were not netbooks was four times the number of netbooks sold. How many of each type of computer were sold in 2009?
**Source:** IDC research estimate, quoted in "How Laptops Took over the World," in guardian.co.uk

**65.** *Vacation Spending.* Emily spent three times as much on lodging as she did for food on a recent vacation. She spent a total of $261.20 for food and lodging. How much did she spend for each?

**66.** *Homework.* Ian spends twice as long on his science homework as he does on his history homework. One week, he spent a total of 10.5 hours on science and history homework. How much time did he spend on each subject?

## Skill Maintenance

Simplify, if possible.

**67.** $\dfrac{0}{-13}$ [3.3b]

**68.** $\dfrac{12}{0}$ [3.3b]

**69.** $\dfrac{-76}{-76}$ [3.3b]

**70.** Add: $-\dfrac{4}{5} + \dfrac{7}{10}$. [4.2b]

**71.** Subtract: $\dfrac{8}{11} - \dfrac{4}{3}$. [4.3a]

**72.** Add: $4\dfrac{1}{3} + 2\dfrac{1}{2}$. [4.6a]

**73.** Solve: $4x - 7 = 9x + 13$. [5.7b]

**74.** Solve: $-\dfrac{2}{9}x = 12$. [3.8a]

## Synthesis

**75.** *Evening News.* The graph below shows the average number of viewers of network evening news for 2006–2009. Determine the average yearly decrease in the number of viewers.

**Evening News**

| Year | Number of average viewers (in millions) |
|------|------|
| 2009 | 22.3 |
| 2008 | 22.8 |
| 2007 | 23.1 |
| 2006 | 24.3 |

*Source*: www.stateofthemedia.org

**76.**  A "French Press" coffee pot requires no filters, but costs $34.95. Kenny could buy a plastic drip cone for $4.49, but the cone requires filters which cost $0.04 per pot dripped. How many pots of coffee must Kenny make for the French Press pot to be the more economical purchase?

**77.** If the daily rental for a car is $18.90 plus a certain price per mile, and Lindsey must drive 190 mi in one day and still stay within a $55.00 budget, what is the highest price per mile that Lindsey can afford?

**78.** A 25-ft by 30-ft yard contains an 8-ft-wide, round fountain. How many 1-lb bags of grass seed should be purchased to seed the lawn if 1 lb of seed covers 300 ft²?

**79.** Find the shaded area. What assumptions must you make?

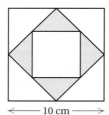

$\longleftarrow$ 10 cm $\longrightarrow$

**80.** You can drive from home to work using either of two routes:

*Route A*: Via interstate highway, 7.6 mi, with a speed limit of 65 mph.

*Route B*: Via a country road, 5.6 mi, with a speed limit of 50 mph.

Assuming you drive at the posted speed limit, how much time can you save by taking the faster route?

## Key Terms

rational numbers, p. 306          terminating decimal, p. 345          repeating decimal, p. 345

## Concept Reinforcement

Determine whether each statement is true or false.

_____ **1.** One thousand billion is one trillion.   [5.3b]

_____ **2.** The number of decimal places in the product of two numbers is the product of the numbers of places in the factors.   [5.3a]

_____ **3.** When we divide a positive number by 0.1, 0.01, 0.001, and so on, the quotient is larger than the dividend.   [5.4a]

_____ **4.** For a fraction with a factor other than 2 or 5 in its denominator, decimal notation terminates.   [5.5a]

_____ **5.** An estimate found by rounding to the nearest ten is usually more accurate than one found by rounding to the nearest hundred.   [5.6a]

## Important Concepts

**Objective 5.1b**   Convert between decimal notation and fraction notation.

**Example**   Write fraction notation for 5.347.

$$5.347 \qquad 5.347. \qquad \frac{5347}{1000}$$

3 decimal places     Move 3 places     3 zeros
                     to the right.

$$5.347 = \frac{5347}{1000}$$

**Example**   Write decimal notation for $\frac{29}{1000}$.

$$\frac{29}{1000} \qquad 0.029.$$

3 zeros     Move 3 places to the left.

$$\frac{29}{1000} = 0.029$$

**Example**   Write decimal notation for $4\frac{63}{100}$.

$$4\frac{63}{100} = 4 + \frac{63}{100} = 4 \text{ and } \frac{63}{100} = 4.63$$

**Practice Exercise**

**1.** Write fraction notation for 0.03.

**Practice Exercise**

**2.** Write decimal notation for $\frac{817}{10}$.

**Practice Exercise**

**3.** Write decimal notation for $42\frac{159}{1000}$.

**Objective 5.1d**   Round decimal notation to the nearest thousandth, hundredth, tenth, one, ten, hundred, or thousand.

**Example**   Round 19.7625 to the nearest hundredth.

Locate the digit in the hundredths place, 6. Consider the next digit to the right, 2. Since that digit, 2, is 4 or lower, the original digit does not change.

19.7625
↓
19.76

**Practice Exercise**

**4.** Round 153.346 to the nearest hundredth.

---

**Objective 5.2a**   Add using decimal notation.

**Example**   Add: 14.26 + 63.589.

$$
\begin{array}{r}
\overset{\scriptstyle 1}{1\ 4.2}\ 6\ 0 \\
+\ 6\ 3.5\ 8\ 9 \\
\hline
7\ 7.8\ 4\ 9
\end{array}
$$   Writing an extra zero

**Practice Exercise**

**5.** Add: 5.54 + 33.071.

---

**Objective 5.2b**   Subtract using decimal notation.

**Example**   Subtract: 67.345 − 24.28.

$$
\begin{array}{r}
6\ 7.\overset{\scriptstyle 2}{3}\ \overset{\scriptstyle 14}{4}\ 5 \\
-\ 2\ 4.2\ 8\ 0 \\
\hline
4\ 3.0\ 6\ 5
\end{array}
$$   Writing an extra zero

**Practice Exercise**

**6.** Subtract: 221.04 − 13.192.

---

**Objective 5.3a**   Multiply using decimal notation.

**Example**   Multiply: $1.8 \times 0.04$.

$$
\begin{array}{rl}
1.8 & \text{(1 decimal place)} \\
\times\ 0.0\ 4 & \text{(2 decimal places)} \\
\hline
0.0\ 7\ 2 & \text{(3 decimal places)}
\end{array}
$$

**Example**   Multiply: $0.001 \times 87.1$.

$0.001 \times 87.1$      $0.087.1$

↑            ↶

3 decimal places     Move 3 places to the left. We write an extra zero.

$0.001 \times 87.1 = 0.0871$

**Example**   Multiply: $63.4 \times 100$.

$63.4 \times 100$      $63.40.$

↑          ↷

2 zeros     Move 2 places to the right. We write an extra zero.

$63.4 \times 100 = 6340$

**Practice Exercise**

**7.** Multiply: $5.46 \times 3.5$.

**Practice Exercise**

**8.** Multiply: $17.6 \times 0.01$.

**Practice Exercise**

**9.** Multiply: $1000 \times 60.437$.

**Objective 5.4a** Divide using decimal notation.

**Example** Divide: $21.35 \div 6.1$.

$$
\begin{array}{r}
3.5 \\
6.1\,\overline{)\,2\,1.3_{\wedge}5} \\
1\,8\,3 \\
\hline
3\,0\,5 \\
3\,0\,5 \\
\hline
0
\end{array}
$$

**Example** Divide: $\dfrac{16.7}{1000}$.

$$\frac{16.7}{1000} = \frac{016.7}{1\,000.} = \frac{0.0167}{1.0} = 0.0167$$

3 zeros    Move 3 places to the left
         to change 1000 to 1.

$$\frac{16.7}{1000} = 0.0167$$

**Example** Divide: $\dfrac{42.93}{0.001}$.

$$\frac{42.93}{0.001} = \frac{42.930}{0.001} = \frac{42{,}930}{1.} = 42{,}930$$

3 decimal    Move 3 places to the right
places       to change 0.001 to 1.

$$\frac{42.93}{0.001} = 42{,}930$$

**Practice Exercise**

**10.** Divide: $26.64 \div 3.6$.

**Practice Exercise**

**11.** Divide: $\dfrac{4.7}{100}$.

**Practice Exercise**

**12.** Divide: $\dfrac{156.9}{0.01}$.

# Review Exercises

Convert the number in each sentence to standard notation. [5.3b]

**1.** Russia has the largest total area of any country in the world, at 6.59 million square miles.

**2.** Americans eat more than 3.1 billion lb of chocolate each year.
**Source:** Chocolate Manufacturers' Association

Write a word name. [5.1a]

**3.** 3.47            **4.** 0.031

**5.** 27.0001         **6.** 0.9

Write in fraction notation and, if possible, as a mixed numeral. [5.1b]

**7.** 0.09            **8.** $-4.561$

**9.** $-0.089$        **10.** 3.0227

Write in decimal notation. [5.1b]

**11.** $\dfrac{34}{1000}$          **12.** $\dfrac{42{,}603}{10{,}000}$

**13.** $27\dfrac{91}{100}$        **14.** $-867\dfrac{6}{1000}$

Which number is larger? [5.1c]

**15.** 0.034, 0.0185       **16.** $-0.91$, $-0.19$

**17.** 0.741, 0.6943      **18.** 1.038, 1.041

Round 17.4287 to the nearest [5.1d]

**19.** Tenth.          **20.** Hundredth.

**21.** Thousandth.       **22.** One.

Perform the indicated operation.

**23.**
$$
\begin{array}{r}
2\ 3\ 6.2\ 3\ 1 \\
2\ 6\ 3.4 \\
+\quad\ \ 0.1\ 9\ 8 \\
\hline
\end{array}
$$
[5.2a]

**24.**
$$
\begin{array}{r}
3\ 7.6\ 4\ 5 \\
-\quad\ \ 8.4\ 9\ 7 \\
\hline
\end{array}
$$
[5.2b]

**25.** $219.3 + 2.8 + 7$  [5.2a]

**26.** $745.0109 - 59.959$  [5.2b]

**27.** $-37.8 + (-19.5)$  [5.2c]

**28.** $-7.52 - (-9.89)$  [5.2c]

**29.**
$$
\begin{array}{r}
4\ 8 \\
\times\ 0.2\ 7 \\
\hline
\end{array}
$$
[5.3a]

**30.** $-3.7(0.29)$  [5.3a]

**31.**
$$
\begin{array}{r}
2\ 4.6\ 8 \\
\times\ 1\ 0\ 0\ 0 \\
\hline
\end{array}
$$
[5.3a]

**32.** $2\ 5\overline{)8\ 0}$  [5.4a]

**33.** $11.52 \div (-7.2)$  [5.4a]

**34.** $\dfrac{276.3}{1000}$  [5.4a]

Combine like terms.  [5.2d]

**35.** $3.7x - 5.2y - 1.5x - 3.9y$

**36.** $7.94 - 3.89a + 4.63 + 1.05a$

**37.** Evaluate: $P - Prt$ for $P = 1000$, $r = 0.05$, and $t = 1.5$. (*A formula for depreciation*)  [5.3c]

**38.** Simplify: $9 - 3.2(-1.5) + 5.2^2$.  [5.4b]

**39.** Convert 1549 cents to dollars.  [5.3b]

**40.** Round $248.\overline{27}$ to the nearest hundredth.  [5.5b]

Multiply by a form of 1 to find decimal notation for each number.  [5.5c]

**41.** $\dfrac{13}{5}$

**42.** $\dfrac{32}{25}$

Use division to find decimal notation for each number. [5.5a]

**43.** $\dfrac{13}{4}$

**44.** $-\dfrac{7}{6}$

**45.** Calculate: $\dfrac{4}{15} \times 79.05$.  [5.5d]

Solve. Remember to check.

**46.** $t - 4.3 = -7.5$  [5.7a]

**47.** $4.1x + 5.6 = -6.7$  [5.7a]

**48.** $6x - 11 = 8x + 4$  [5.7b]

**49.** $3(x + 2) = 5x - 7$  [5.7b]

Solve.  [5.8b]

**50.** In the United States, there are 51.81 telephone poles for every 100 people. In Canada, there are 40.65 poles for every 100 people. How many more telephone poles for every 100 people are there in the United States?

**51.** Stacia, a coronary intensive care nurse, earned $620.74 during a recent 40-hr week. What was her hourly wage? Round to the nearest cent.

**52.** *Landscaping.*   A rectangular yard is 20 ft by 15 ft. The yard is covered with grass except for a circular flower garden with an 8-ft diameter. Find the area of grass in the yard.

**53.** Derek had $6274.35 in his checking account. He used $485.79 to buy a camera with his debit card. How much was left in his account?

**54.** *Credit Card Processing.*   Pay Right charges $150 for its software, $8.95 a month for service, and 21¢ per online transaction. Timeless Treasures paid $178.90 for its first month of service. How many transactions did the company process that month?

**55.** *Recycling.*   Lisa volunteers at a community recycling center. One Saturday, she processed 130 more pounds of newspaper than she did of glass. She processed a total of 261.4 pounds of newspaper and glass. How many pounds of each did she process?

**56.** *Gas Mileage.* Inge wants to estimate gas mileage per gallon. With an odometer reading 36,057.1, she fills up. At 36,217.6 mi, the tank is refilled with 11.1 gal. Find the mileage per gallon. Round to the nearest tenth.

**57.** *Seafood Consumption.* The following graph shows the annual consumption, in pounds, of seafood per person in the United States in recent years.

  **a)** Find the total per capita consumption for the six years.

  **b)** Find the average per capita consumption for the six years.

**Seafood Consumption**

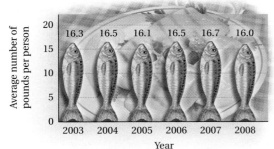

SOURCE: U.S. Department of Agriculture

**58.** A taxi driver charges $7.25 plus 95 cents a mile for out-of-town fares. How far can an out-of-towner travel on $15.23?

**59.** One pound of lean boneless ham contains 4.5 servings. It costs $5.99 per pound. What is the cost per serving? Round to the nearest cent.

**60.** *Construction.* A rectangular room measures 14.5 ft by 16.25 ft. How many feet of crown molding are needed to go around the top of the room? How many square feet of bamboo tiles are needed for the floor of the room?

**61.** Estimate the quotient $82.304 \div 17.287$ by rounding to the nearest ten. [5.6a]

  **A.** 0.4              **B.** 4

  **C.** 40             **D.** 400

**62.** Translate to an algebraic expression: 15 less than twice the price. [5.8a]

  **A.** $2p - 15$         **B.** $15 - 2p$

  **C.** $2 + p - 15$     **D.** $2(p - 15)$

## Synthesis

**63.** 🖩 In each of the following, use $+$, $-$, $\times$, or $\div$ in each blank to make a true sentence. [5.4b]

  **a)** $2.56 - 6.4 \ \square \ 51.2 - 17.4 + 89.7 = 119.66$

  **b)** $(11.12 \ \square \ 0.29)3^4 = 877.23$

**64.** Arrange from smallest to largest:

$$-\frac{2}{3}, \ -\frac{15}{19}, \ -\frac{11}{13}, \ \frac{-5}{7}, \ \frac{-13}{15}, \ \frac{-17}{20}.$$

[5.1c], [5.5a]

**65.** *Automobile Leases.* Quentin can lease a Ford Fusion for $225 a month. He must pay an additional 20 cents per mile for all miles over 10,000 in one year. In 2010, his total bill for leasing the car was $5952. How many miles did he drive the car in 2010? [5.8b]

Source: www.edmunds.com

**66.** 🖩 Sal's sells Sicilian pizza as a 17-in. by 20-in. pie for $15 or as an 18-in.-diameter round pie for $14. Which is a better buy and why? [5.8b]

# Understanding Through Discussion and Writing

**1.** Describe in your own words a procedure for converting from decimal notation to fraction notation. [5.1b]

**2.** A student insists that $346.708 \times 0.1 = 3467.08$. How could you convince him that a mistake had been made without checking on a calculator? [5.3a]

**3.** When is long division *not* the fastest way to convert from fraction notation to decimal notation? [5.5a]

**4.** Consider finding decimal notation for $\frac{44}{125}$. Discuss as many ways as you can for finding such notation and give the answer. [5.5a]

Test

For Extra Help

Step-by-step test solutions are found on the Chapter Test Prep Videos available via the Video Resources on DVD, in **MyMathLab** , and on You Tube (search "BittingerPrealgebra" and click on "Channels").

Convert the number in each sentence to standard notation.

**1.** The United States issued 18.4 million passports in 2007, an all-time high, due in part to new rules requiring a passport for travel to Canada and Mexico.
**Source:** Associated Press

**2.** An ethanol boom helped to deliver a corn crop of a record 13.1 billion bushels in 2007.
**Source:** U.S. Department of Agriculture

Write a word name.

**3.** 2.34

**4.** 105.0005

Write in fraction notation and, if possible, as a mixed numeral.

**5.** $-0.91$

**6.** 2.769

Write in decimal notation.

**7.** $\dfrac{74}{1000}$

**8.** $-\dfrac{37{,}047}{10{,}000}$

**9.** $756\dfrac{9}{100}$

**10.** $91\dfrac{703}{1000}$

Which number is larger?

**11.** 0.07, 0.162

**12.** 0.078, 0.06

**13.** $-0.09$, $-0.9$

Round 5.6783 to the nearest

**14.** One.

**15.** Hundredth.

**16.** Thousandth.

**17.** Tenth.

Perform the indicated operation.

**18.**
$$\begin{array}{r} 4\,0\,2.3 \\ 2.8\,1 \\ 0.1\,0\,9 \\ \hline \end{array}$$

**19.**
$$\begin{array}{r} 0.1\,2\,5 \\ \times\quad 0.2\,4 \\ \hline \end{array}$$

**20.**
$$\begin{array}{r} 2\,1\,3.4\,5 \\ \times\quad 0.0\,0\,1 \\ \hline \end{array}$$

**21.**
$$\begin{array}{r} 5\,2.0\,9\,1 \\ -\quad 7.3\,4\,5 \\ \hline \end{array}$$

**22.** $342.9 + 8.1 + 5.37$

**23.** $-9.5 + 7.3$

**24.** $2 - 0.0054$

**25.** $1000 \times 73.962$

**26.** $4\,)\overline{1\,9}$

**27.** $3.3\,)\overline{1\,0\,0.3\,2}$

**28.** $\dfrac{-346.82}{1000}$

**29.** $\dfrac{346.82}{0.01}$

**30.** Convert $179.82 to cents.

**31.** Combine like terms:
$$4.1x + 5.2 - 3.9y + 5.7x - 9.8.$$

**32.** Evaluate: $2l + 4w + 2h$ for $l = 2.4$, $w = 1.3$, and $h = 0.8$. (*The total girth of a postal package*)

**33.** Simplify: $20 \div 5(-2)^2 - 8.4$.

Multiply by a form of 1 to find decimal notation for each number.

**34.** $\dfrac{8}{5}$

**35.** $\dfrac{21}{4}$

Use division to find decimal notation for each number.

**36.** $-\dfrac{7}{16}$

**37.** $\dfrac{14}{9}$

**38.** Round the answer in Exercise 37 to the nearest hundredth.

Calculate.

**39.** $3 \div (-0.3) \cdot 2 - 1.5^2$

**40.** $(8 - 1.23) \div 4 + 5.6 \times 0.02$

**41.** $\dfrac{3}{8} \times 45.6 - \dfrac{1}{5} \times 36.9$

Solve. Remember to check.

**42.** $17y - 3.12 = -58.2$

**43.** $9t - 4 = 6t + 26$

**44.** $4 + 2(x - 3) = 7x - 9$

**45.** *Cell-Phone Plan.* In 2008, at&t offered its FamilyTalk cell-phone plan with 2 lines and 1400 anytime minutes for a monthly access fee of $89.99. Minutes in excess of 1400 were charged at the rate of $0.40 per minute. One month, Mr. and Mrs. Tews used their cell phones for 1510 min. What was the charge?
Source: at&t

**46.** *Gas Mileage.* Tina wants to estimate the gas mileage in her economy car. At 76,843 mi, she fills the tank with 14.3 gal of gasoline. At 77,310 mi, she fills the tank with 16.5 gal of gasoline. Find the mileage per gallon. Round to the nearest tenth.

**47.** *Checking Account Balance.* Nicholas had a balance of $820 in his checking account before making purchases of $123.89, $56.68, and $46.98 with his debit card. What was the balance after the purchases had been made?

**48.** The office manager for the Drake, Smith, and Hartner law firm buys 7 cases of copy paper at $41.99 per case. What is the total cost?

**49.** *Top American Sky Routes.* The figure below shows the numbers of airline passengers carried on the top five sky routes in the United States from February 2007 to January 2008. Note that these routes go in both directions. Find the average number of passengers on these routes.

**Top American Sky Routes**

| | Number of passengers (in millions) |
|---|---|
| New York City–Chicago | 3.45 |
| Washington–Chicago | 2.75 |
| Atlanta–Orlando | 2.75 |
| Atlanta–New York City | 2.70 |
| Los Angeles–Chicago | 2.67 |

SOURCE: U.S. Bureau of Transportation Statistics

**50.** About how many gallons of gasoline, at $2.749 per gallon, can be bought with $20?

**A.** 1 gallon   **B.** 3 gallons   **C.** 7 gallons   **D.** 12 gallons

# Synthesis

**51.** Use one of the words *sometimes*, *never*, or *always* to complete each of the following.

**a)** The product of two numbers greater than 0 and less than 1 is _____ less than 1.

**b)** The product of two numbers greater than 1 is _____ less than 1.

**c)** The product of a number greater than 1 and a number less than 1 is _____ equal to 1.

**d)** The product of a number greater than 1 and a number less than 1 is _____ equal to 0.

**52.** Silver's Health Club charges a $79 membership fee and $42.50 a month. Allise has a coupon that will allow her to join the club for $299 for six months. How much will Allise save if she uses the coupon?

**53.** *Travel Costs.* Roundtrip airfare between Burlington, VT, and Newark, NJ, often costs $229. One estimate of the true cost of driving is $1.35 per mile. Is it more economical to fly or drive the 600 mi for **(a)** an individual; **(b)** a couple; **(c)** a family of 4?

**Source:** commutesolutions.org

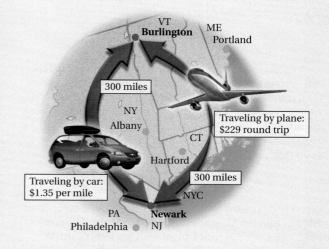

# Cumulative Review

Convert to fraction notation.

**1.** $2\dfrac{2}{9}$                     **2.** 3.051

Find decimal notation.

**3.** $-\dfrac{7}{5}$                     **4.** $\dfrac{6}{11}$

**5.** Determine whether 43 is prime, composite, or neither.

**6.** Determine whether 2,053,752 is divisible by 5.

Calculate.

**7.** $48 + 12 \div 4 - 10 \times 2 + 6892 \div 4$

**8.** $4.7 - \{0.1[1.2(3.95 - 1.65) + 1.5 \div 2.5]\}$

Round to the nearest hundredth.

**9.** 584.973                     **10.** $218.\overline{5}$

**11.** Estimate the product $16.392 \times 9.715$ by rounding to the nearest one.

**12.** Estimate by rounding to the nearest tenth:
$2.714 + 4.562 - 3.31 - 0.0023$.

**13.** Evaluate $a \div 3 \cdot b$ for $a = 18$ and $b = 2$.

**14.** Combine like terms:
$-4p + 9 + 11p - 17$.

Add and simplify.

**15.**    $2\dfrac{1}{4}$
       $+\ 3\dfrac{4}{5}$

**16.**    $\begin{array}{r} 3\,4{,}9\,2\,1 \\ 9\,3{,}0\,9\,2 \\ +\ 1\,1{,}1\,0\,3 \\ \hline \end{array}$

**17.** $\dfrac{1}{6} + \dfrac{2}{3} + \dfrac{8}{9}$                     **18.** $143.9 + 2.053$

Subtract and simplify.

**19.** $723{,}041 - 12{,}904$                     **20.** $5.903 - 19$

**21.** $5\dfrac{1}{7} - 4\dfrac{3}{7}$                     **22.** $\dfrac{9}{10} - \dfrac{10}{11}$

Multiply and simplify.

**23.** $-\dfrac{3}{8} \cdot \dfrac{4}{9}$

**24.**    $\begin{array}{r} 2\,5\,3\,2 \\ \times\ 2\,1\,0\,0 \\ \hline \end{array}$

**25.**    $\begin{array}{r} 2\,3.9 \\ \times\ \ \ 0.2 \\ \hline \end{array}$

**26.**    $\begin{array}{r} 2\,7.9\,4\,3\,1 \\ \times\ \ \ \ \ 0.0\,0\,1 \\ \hline \end{array}$

Divide and simplify.

**27.** $1\,6.5\,)\overline{\,3\,5.0\,1\,3}$                     **28.** $2\,6\,)\overline{\,4\,7{,}9\,1\,8}$

**29.** $13.8621 \div 0.001$

**30.** $-\dfrac{4}{9} \div \left(-\dfrac{8}{15}\right)$

Solve.

**31.** $3(x - 7) = 5x + 2$

**32.** $-75 \cdot x = 2100$

**33.** $2.4x - 7.1 = 2.05$

**34.** $1062 + y = 368{,}313$

**35.** $t + \dfrac{5}{6} = \dfrac{8}{9}$

**36.** $-\dfrac{7}{8} \cdot t = \dfrac{7}{16}$

**37.** *Organ Transplants.* There were 16,646 kidney transplants, 6136 liver transplants, and 2147 heart transplants in the United States in a recent year. How many transplants of these three organs were performed that year?
**Source:** U.S. Department of Health and Human Services

**38.** *Hospital Revenue.* Hospitals received $265 billion in revenue from private health insurance payments in a recent year. This was $88 billion more than was received from Medicare payments. How much revenue was received from Medicare payments?
**Source:** U.S. Census Bureau

**39.** After Lauren made a $450 down payment on a sofa, $\dfrac{3}{10}$ of the total cost was paid. How much did the sofa cost?

**40.** Joshua's tuition was $3600. He obtained a loan for $\dfrac{2}{3}$ of the tuition. How much was the loan?

**41.** The balance in Elliott's checking account is $314.79. After a check is written for $56.02, what is the balance in the account?

**42.** A clerk in Leah's Delicatessen sold $1\dfrac{1}{2}$ lb of ham, $2\dfrac{3}{4}$ lb of turkey, and $2\dfrac{1}{4}$ lb of roast beef. How many pounds of meat were sold altogether?

**43.** A triangular sail has a height of 16 ft and a base of 11 ft. Find its area.

**44.** A 4-in. by 5-in. rectangle is punched from a round piece of steel that is 9 in. wide. How much steel will be left over?

Simplify.

**45.** $\left(\dfrac{3}{4}\right)^2 - \dfrac{1}{8} \cdot \left(3 - 1\dfrac{1}{2}\right)^2$

**46.** $1.2 \times 12.2 \div 0.1 \times 3.6$

## Synthesis

**47.** Using a manufacturer's coupon, Lucy bought 2 cartons of orange juice and received a third carton free. The price of each carton was $3.59. What was the cost per carton with the coupon? Round to the nearest cent.

**48.** A carton of gelatin mix packages weighs $15\dfrac{3}{4}$ lb. Each package weighs $1\dfrac{3}{4}$ oz. How many packages are in the carton? ($1$ lb $= 16$ oz)

# Introduction to Graphing and Statistics

## Real-World Application

A gardener planted six varieties of bearded iris in a new garden on campus. Students from the horticulture department were assigned to record data on the range of heights for each variety. The vertical bar graph on page 405 shows the results of their study. The length of the light green shaded portion and the blossom of each bar illustrates the range of heights. What is the range of heights for the border bearded iris?

*Source: www.irises.org/classification.htm*

***This problem appears as Exercise 5 in Exercise Set 6.2.***

# 6.1

# Tables and Pictographs

## OBJECTIVES

**a** Extract and interpret data from tables.

**b** Extract and interpret data from pictographs.

**SKILL TO REVIEW**

Objective 4.7a: Multiply using mixed numerals.

Multiply.

1. $3\frac{1}{2} \times 800$

2. $1\frac{3}{4} \times 3020$

## a Reading and Interpreting Tables

A **table** is often used to present data in rows and columns.

**EXAMPLE 1** *Population Density.* The following table lists data regarding 11 countries around the world.

| COUNTRY | POPULATION | LAND AREA (in square miles) | POPULATION DENSITY (per square mile) | PERCENT OF POPULATION THAT IS URBAN |
|---|---|---|---|---|
| Australia | 20,434,176 | 2,941,299 | 7 | 88.2 |
| Brazil | 190,010,647 | 3,265,077 | 58 | 84.2 |
| China | 1,321,851,888 | 3,600,947 | 367 | 40.4 |
| Finland | 5,238,460 | 117,558 | 45 | 61.1 |
| France | 63,718,187 | 210,669 | 258 | 76.7 |
| Germany | 82,400,996 | 134,836 | 611 | 75.2 |
| India | 1,129,866,154 | 1,147,955 | 984 | 28.7 |
| Japan | 127,433,494 | 144,689 | 881 | 65.8 |
| Kenya | 36,913,721 | 219,789 | 168 | 20.7 |
| Mexico | 108,700,891 | 742,490 | 146 | 76.0 |
| United States | 301,139,947 | 3,537,439 | 85 | 80.8 |

SOURCE: *The World Almanac*, 2008

Use the table in Example 1 to answer Margin Exercises 1–5.

1. Which country has the smallest amount of land area?

2. Which country has the greatest population density?

3. Which countries have population densities smaller than 100 per square mile?

4. What is the population of Mexico?

5. Find the average of the land areas of the countries listed.

a) Which country has the largest population?

b) Which countries have a population of over 1 billion?

c) What is the population density of Australia?

d) Find the average of the population densities in the countries listed.

Careful examination of the table will give the answers.

a) To determine which country has the largest population, we look down the column headed "Population" and find the largest number. That number is 1,321,851,888. Then we look across that row to find the name of the country, China.

b) To determine which countries have a population of over 1 billion, we look down the column headed "Population" and find any numbers greater than 1 billion. Those numbers are 1,321,851,888 and 1,129,866,154. Then we look across those rows to find the names of the countries, China and India.

c) To determine the population density of Australia, we look down the column headed "Country" and find the name Australia. Then we look across that row to the column headed "Population Density." Using the units given at the top of the column, the population density of Australia is 7 per square mile.

*Answers*

*Skill to Review:*
1. 2800   2. 5285

*Margin Exercises:*
1. Finland   2. India   3. Australia, Brazil, Finland, and the United States
4. 108,700,891   5. About 1,460,250 sq mi

**d)** We determine the average of all the numbers in the column headed "Population Density":

$$\frac{7 + 58 + 367 + 45 + 258 + 611 + 984 + 881 + 168 + 146 + 85}{11}$$

$$= \frac{3610}{11} \approx 328.2 \text{ per sq mi.}$$

Do Margin Exercises 1–5 on the preceding page.

**EXAMPLE 2** *Frosted Flakes Nutrition Facts.* Most foods are required by law to provide factual information regarding nutrition, as shown in the following table of nutrition facts from a box of Frosted Flakes cereal. You as a consumer must be careful in interpreting the data.

### Nutrition Facts

Serving Size  3/4 Cup (30g/1.1oz)
Servings Per Container  About 16

| Amount Per Serving | Cereal | Cereal with 1/2 Cup Vitamins A&D Fat Free Milk |
|---|---|---|
| Calories | 110 | 150 |
| Calories from Fat | 0 | 0 |

| | % Daily Value** | |
|---|---|---|
| Total Fat 0g* | 0% | 0% |
| Saturated Fat 0g | 0% | 0% |
| Trans Fat 0g | | |
| Cholesterol 0mg | 0% | 0% |
| Sodium 140mg | 6% | 9% |
| Potassium 20mg | 1% | 6% |
| Total Carbohydrate 27g | 9% | 11% |
| Dietary Fiber 1g | 3% | 3% |
| Sugars 11g | | |
| Other Carbohydrate 15g | | |
| Protein 1g | | |

| | | |
|---|---|---|
| Vitamin A | 10% | 15% |
| Vitamin C | 10% | 10% |
| Calcium | 0% | 15% |
| Iron | 25% | 25% |
| Vitamin D | 10% | 25% |
| Thiamin | 25% | 30% |
| Riboflavin | 25% | 35% |
| Niacin | 25% | 25% |
| Vitamin B$_6$ | 25% | 25% |
| Folic Acid | 25% | 25% |
| Vitamin B$_{12}$ | 25% | 35% |

\*Amount in cereal. One half cup of fat free milk contributes an additional 40 calories, 65mg sodium, 6g total carbohydrates (6g sugars), and 4g protein.

\*\*Percent Daily Values are based on a 2,000 calorie diet. Your daily values may be higher or lower depending on your calorie needs.

SOURCE: Kelloggs

Suppose your morning bowl of cereal consists of $1\frac{1}{2}$ cups of Frosted Flakes with 1 cup of fat-free milk.

**a)** How many calories have you consumed?

**b)** What percent of the daily value for dietary fiber have you consumed?

**c)** A nutritionist recommends that you look for foods that provide 10% or more of the daily value for vitamin C. Do you get that with your bowl of Frosted Flakes?

**d)** Suppose you are trying to limit your daily caloric intake to 2000 calories. How many bowls of cereal would it take to exceed the 2000 calories?

First, note that $1\frac{1}{2}$ cups of cereal is twice the serving size: $2\left(\frac{3}{4}\right) = \frac{6}{4} = \frac{3}{2} = 1\frac{1}{2}$. Also, 1 cup of milk is twice the serving size of the milk.

**a)** We look at the column marked "with $\frac{1}{2}$ Cup Fat Free Milk" and note that a serving size of $\frac{3}{4}$ cup of cereal with $\frac{1}{2}$ cup of fat-free milk contains 150 calories. Since you are having twice that amount, you are consuming

$$2 \times 150 \text{ calories, or } 300 \text{ calories.}$$

**b)** We read across from "Dietary Fiber" and note that in $\frac{3}{4}$ cup of cereal with $\frac{1}{2}$ cup of fat-free milk, you get 3% of the daily value for dietary fiber. Since you are doubling that, you get 6% of the daily value for dietary fiber.

**c)** We find the row labeled "Vitamin C" on the left and look under the column labeled "with $\frac{1}{2}$ Cup Fat Free Milk." Note that you get 10% of the daily value in $\frac{3}{4}$ cup of cereal with $\frac{1}{2}$ cup of fat-free milk, and since you are doubling that, you are more than satisfying the 10% requirement.

Use the nutrition facts data shown in the figure and the bowl of cereal described in Example 2 to answer Margin Exercises 6–10.

**6.** How many calories from fat are in your bowl of cereal?

**7.** A nutritionist recommends that you look for foods that provide 12% or more of the daily value for iron. Do you get that with your bowl of Frosted Flakes?

**8.** How much sodium have you consumed? (*Hint:* Use the data listed in the first footnote below the table of nutrition facts.)

**9.** What percent of the daily value for sodium have you consumed?

**10.** How much protein have you consumed? (*Hint:* Use the data listed in the first footnote below the table of nutrition facts.)

*Answers*

**6.** 0 calories  **7.** Yes  **8.** 410 mg  **9.** 18%
**10.** 10 g

**d)** From part (a), we know that you are consuming 300 calories per bowl. Dividing 2000 by 300 gives $\frac{2000}{300} \approx 6.67$. Thus, if you eat 7 bowls of cereal in this manner, you will exceed the 2000 calories.

Do Exercises 6–10 on the preceding page.

## b  Reading and Interpreting Pictographs

*Pictographs* (or *picture graphs*) are another way to show information. Instead of actually listing the amounts to be considered, a **pictograph** uses symbols to represent the amounts. In addition, a *key* is given telling what each symbol represents.

**EXAMPLE 3**  *Touchdown Passes.*  The following pictograph shows the career-high number of touchdown passes in a season for seven quarterbacks in the National Football League. Located below the graph is a key that tells you that each ⬤ symbol represents 10 touchdown passes.

**Touchdown Passes (Career high for quarterback)**

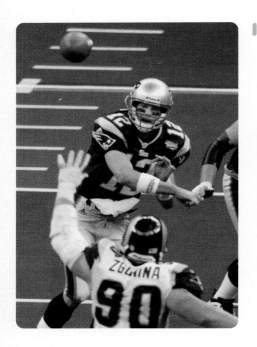

|  | PLAYER | YEAR | TOUCHDOWN PASSES |
|---|---|---|---|
| Retired | John Elway | 1997 | ⬤⬤◖ |
|  | Dan Marino | 1984 | ⬤⬤⬤◖ |
|  | Joe Montana | 1987 | ⬤⬤◖ |
|  | Roger Staubach | 1978 | ⬤⬤◖ |
| Active | Tom Brady | 2007 | ⬤⬤⬤⬤⬤ |
|  | Brett Favre | 2001 and 2003 | ⬤⬤⬤⬤◖ |
|  | Peyton Manning | 2004 | ⬤⬤⬤⬤◖ |

SOURCE: National Football League

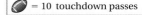 ⬤ = 10 touchdown passes

**a)** Which quarterback has the greatest number of touchdown passes?

**b)** Which quarterback has the least number of touchdown passes?

**c)** How many more touchdown passes does Peyton Manning have in his career high than John Elway?

We can compute the answers by first reading the pictograph.

**a)** The quarterback with the most symbols has the greatest number of touchdown passes: Tom Brady, with 5 (symbols) × 10, or 50, touchdown passes.

**b)** The quarterback with the fewest symbols has the least number of touchdown passes: Roger Staubach, with about $2\frac{1}{2}$ (symbols) × 10, or 25, touchdown passes.

**c)** From the graph, we see that there are about $4\frac{9}{10}$ × 10, or 49, touchdown passes for Peyton Manning, and about $2\frac{7}{10}$ × 10, or 27, for John Elway. Thus, Peyton Manning has 49 − 27, or 22, more touchdown passes in his career high than John Elway.

Do Exercises 11–13.

Note that we must estimate the fractional part of any incomplete symbol. For this reason, answers based on partial symbols may vary slightly.

Use the pictograph in Example 3 to answer Margin Exercises 11–13.

**11.** How many more touchdown passes does Dan Marino have in his career high in a season than Brett Favre?

**12.** How many more touchdown passes does Joe Montana have in his career high in a season than Roger Staubach?

**13.** What is the average number of career-high touchdown passes in a season for the seven quarterbacks?

*Answers*

**11.** 16 touchdown passes   **12.** 6 touchdown passes   **13.** About 37 touchdown passes

**6.1**

# Exercise Set

**For Extra Help**

**MyMathLab**

Math XL
PRACTICE

WATCH

DOWNLOAD

READ

REVIEW

**a** *Veggie Burgers.* Veggie burgers tend to be higher in fiber and lower in calories than conventional burgers. However, they are high in sodium content. Comparative information on veggie burgers from eight companies is listed in the following table. Use the data for Exercises 1–10.

| COMPANY AND PRODUCT | PER PATTY (ABOUT 2.5 OZ) | | | | | |
|---|---|---|---|---|---|---|
| | Cost | Calories | Fat (g) | Protein (g) | Fiber (g) | Sodium (mg) |
| **Amy's** All American The Classic | $1.12 | 120 | 3.0 | 10 | 3 | 390 |
| **Boca** All American Flame Grilled Meatless | 0.96 | 90 | 3.0 | 14 | 3 | 280 |
| **Dr. Praeger's Sensible Foods** California | 1.08 | 110 | 4.5 | 5 | 4 | 250 |
| **Franklin Farms** Portabella Fresh | 1.00 | 100 | 1.5 | 14 | 3 | 390 |
| **Gardenburger** Portabella | 0.98 | 90 | 2.5 | 5 | 5 | 360 |
| **Lightlife** Meatless Light | 1.48 | 120 | 1.5 | 16 | 3 | 500 |
| **MorningStar Farms** Garden | 1.07 | 110 | 3.5 | 10 | 3 | 350 |
| **Veggie Patch** Garlic Portabella | 1.00 | 120 | 6.0 | 11 | 4 | 320 |

SOURCE: *Consumer Reports*, June 2008

1. How many calories are in a Franklin Farms Portabella Fresh veggie burger?

2. Which veggie burger contains the most protein?

3. Which veggie burgers have less than 110 calories?

4. How many grams of fat are in a MorningStar Farms Garden veggie burger?

5. What is the average of the fiber content in these eight veggie burgers?

6. What is the average number of calories in these eight veggie burgers?

7. Which veggie burger is the most expensive? the least expensive?

8. Which veggie burger contains the most sodium? the least sodium?

9. Which veggie burger contains the greatest number of grams of fat? the least number of grams of fat?

10. Which veggie burgers cost less than $1.00?

*Heat Index.* In warm weather, a person can feel hot because of reduced heat loss from the skin caused by higher humidity. The **temperature–humidity index**, or **apparent temperature**, is what the temperature would have to be with no humidity in order to give the same heat effect. The following table lists the apparent temperatures for various actual temperatures and relative humidities. Use this table for Exercises 11–22.

| ACTUAL TEMPERATURE (°F) | RELATIVE HUMIDITY | | | | | | | | | |
|---|---|---|---|---|---|---|---|---|---|---|
| | 10% | 20% | 30% | 40% | 50% | 60% | 70% | 80% | 90% | 100% |
| | APPARENT TEMPERATURE (°F) | | | | | | | | | |
| 75° | 75 | 77 | 79 | 80 | 82 | 84 | 86 | 88 | 90 | 92 |
| 80° | 80 | 82 | 85 | 87 | 90 | 92 | 94 | 97 | 99 | 102 |
| 85° | 85 | 88 | 91 | 94 | 97 | 100 | 103 | 106 | 108 | 111 |
| 90° | 90 | 93 | 97 | 100 | 104 | 107 | 111 | 114 | 118 | 121 |
| 95° | 95 | 99 | 103 | 107 | 111 | 115 | 119 | 123 | 127 | 131 |
| 100° | 100 | 105 | 109 | 114 | 118 | 123 | 127 | 132 | 137 | 141 |
| 105° | 105 | 110 | 115 | 120 | 125 | 131 | 136 | 141 | 146 | 151 |

In Exercises 11–14, find the apparent temperature for the given actual temperature and humidity combinations.

**11.** 80°, 60%

**12.** 90°, 70%

**13.** 85°, 90%

**14.** 95°, 80%

**15.** Which temperature–humidity combinations give an apparent temperature of 100°?

**16.** Which temperature–humidity combinations give an apparent temperature of 111°?

**17.** At a relative humidity of 50%, what actual temperatures give an apparent temperature above 100°?

**18.** At a relative humidity of 90%, what actual temperatures give an apparent temperature above 100°?

**19.** At an actual temperature of 95°, what relative humidities give an apparent temperature above 100°?

**20.** At an actual temperature of 85°, what relative humidities give an apparent temperature above 100°?

**21.** At a relative humidity of 50%, what is the difference in actual temperature required to raise the apparent temperature from 82° to 104°?

**22.** At a relative humidity of 70%, what is the difference in actual temperature required to raise the apparent temperature from 94° to 127°?

*Planets.* Use the following table, which lists information about the planets, for Exercises 23–32.

| PLANET | AVERAGE DISTANCE FROM SUN (in miles) | DIAMETER (in miles) | LENGTH OF PLANET'S DAY IN EARTH TIME (in days) | TIME OF REVOLUTION IN EARTH TIME (in years) |
|---|---|---|---|---|
| Mercury | 35,983,000 | 3,031 | 58.82 | 0.24 |
| Venus | 67,237,700 | 7,520 | 224.59 | 0.62 |
| Earth | 92,955,900 | 7,926 | 1.00 | 1.00 |
| Mars | 141,634,800 | 4,221 | 1.03 | 1.88 |
| Jupiter | 483,612,200 | 88,846 | 0.41 | 11.86 |
| Saturn | 888,184,000 | 74,898 | 0.43 | 29.46 |
| Uranus | 1,782,000,000 | 31,763 | 0.45 | 84.01 |
| Neptune | 2,794,000,000 | 31,329 | 0.66 | 164.78 |

SOURCE: *The Handy Science Answer Book*, Gale Research, Inc.

**23.** Find the average distance from the sun to Jupiter.

**24.** How long is a day on Venus?

**25.** Which planet has a time of revolution of 164.78 yr?

**26.** Which planet has a diameter of 4221 mi?

**27.** Which planets have an average distance from the sun that is greater than 1,000,000 mi?

**28.** Which planets have a diameter that is less than 100,000 mi?

**29.** About how many earth diameters would it take to equal one Jupiter diameter?

**30.** How much longer is the longest time of revolution than the shortest?

**31.** What is the average of the diameters of the planets?

**32.** What is the average of the average distances from the sun of the planets?

**b** *Rhino Population.* The rhinoceros is considered one of the world's most endangered animals. The total number of rhinoceroses worldwide is approximately 20,700. The following pictograph shows the populations of the five remaining rhino species. Located in the graph is a key that tells you that each symbol 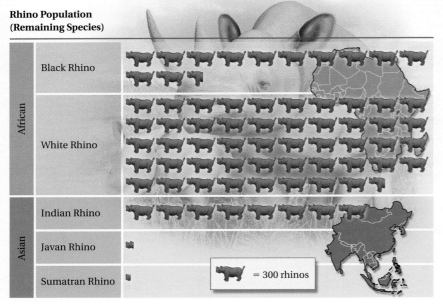 represents 300 rhinos. Use the pictograph for Exercises 33–38.

**33.** Which species has the greatest number of rhinos?

**34.** Which species has the least number of rhinos?

**35.** How many more black rhinos are there than Indian rhinos?

**36.** How many more white rhinos are there than black rhinos?

**Rhino Population (Remaining Species)**

African
- Black Rhino
- White Rhino

Asian
- Indian Rhino
- Javan Rhino
- Sumatran Rhino

= 300 rhinos

SOURCE: World Wildlife Fund, 2008

**37.** What is the average number of rhinos in these five species?

**38.** How does the white rhino population compare with the Indian rhino population?

## Skill Maintenance

Perform the indicated operation and, if possible, simplify.

**39.** $-\dfrac{3}{8} + \dfrac{5}{16}$ [4.2b]

**40.** $-\dfrac{2}{7} - \dfrac{3}{5}$ [4.3a]

**41.** $-\dfrac{8}{9} \div \dfrac{2}{3}$ [3.7b]

Solve.

**42.** $9x - 5 = -23$ [2.8d]

**43.** $3x - 2 = 7x + 10$ [5.7b]

**44.** $-4x = 3x - 7$ [5.7b]

## Synthesis

The chart below shows phone rates for some long-distance plans offered in California. All rates are in cents per minute. Use the chart for Exercises 45 and 46.

**45.** Alan spoke the same number of minutes to a friend in Germany as he did to a friend in South Africa. If his ECG bill was for $18.40, for how many minutes did he speak to each friend?

**46.** Each month, Patti makes 2 hr of long distance calls in-state and 4 hr of state-to-state calls. Which plan will be the least expensive, and how much will her bill be each month?

| PLAN | IN-STATE | STATE-TO-STATE, CONTINENTAL U.S. | GERMANY | SOUTH AFRICA |
|---|---|---|---|---|
| ECG Easy 2.5 | 3.9 | 2.5 | 6 | 17 |
| Pioneer Telephone | 3.4 | 2.7 | 4.5 | 12.5 |
| telna | 3.4 | 4.8 | 3.8 | 7.9 |
| TCI | 2.9 | 3.0 | 3.9 | 9.9 |

# 6.2

# Bar Graphs and Line Graphs

## a Reading and Interpreting Bar Graphs

A **bar graph** is convenient for showing comparisons because you can tell at a glance which amount represents the largest or smallest quantity. This is true of pictographs as well; however, with bar graphs, a *second scale* is usually included so that a more accurate determination of the amount can be made.

**EXAMPLE 1** *Foreign Visits by Presidents.*   Theodore Roosevelt's trip to Panama in 1904 was the first presidential foreign visit. The 42nd and 43rd presidents of the United States made more than twice as many foreign visits while in office as previous presidents. The following horizontal bar graph shows the number of foreign visits by the 34th–43rd presidents.

**OBJECTIVES**

**a** Extract and interpret data from bar graphs.

**b** Draw bar graphs.

**c** Extract and interpret data from line graphs.

**d** Draw line graphs.

**Foreign Visits by U.S. Presidents While in Office**

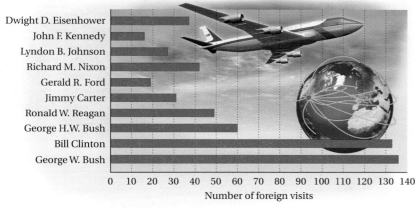

SOURCE: State Department, White House

**a)**  Which president in this list made the fewest foreign visits while in office?

**b)**  How many foreign visits did Ronald Reagan make?

**c)**  Which president made approximately 130 foreign visits?

We look at the graph to answer the questions.

**a)**  The shortest bar is for John F. Kennedy. Thus, John F. Kennedy made the fewest foreign visits.

**b)**  We move to the right along the bar that represents Ronald Reagan. We can read, fairly accurately, that he made approximately 49 foreign visits.

**c)**  We locate the line representing 130 visits and then go up until we reach a bar that ends close to 130. We can go to the left and read the name of the president, Bill Clinton.

> Do Exercises 1–3.

Bar graphs are often drawn vertically, and sometimes a double bar graph is used to make comparisons.

Use the bar graph in Example 1 to answer Margin Exercises 1–3.

**1.** How many foreign visits did George H. W. Bush make?

**2.** Which presidents made fewer than 30 foreign visits?

**3.** Which president made approximately 40 foreign visits?

***Answers***

**1.** 60 foreign visits   **2.** John F. Kennedy, Lyndon B. Johnson, Gerald R. Ford
**3.** Richard M. Nixon

**EXAMPLE 2** *Childhood Cancer.* The following graph indicates the survival rates for various childhood cancers, comparing the rates for 1962 and the present.

**Childhood Cancer Survival Rates**

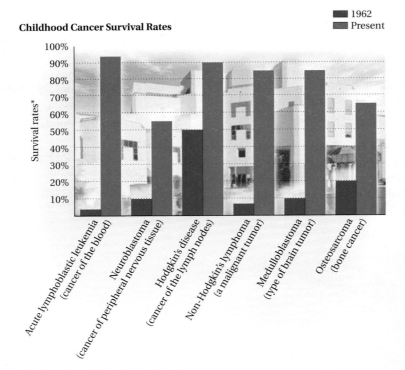

■ 1962
■ Present

SOURCE: St. Jude Children's Research Hospital
*Cancer survival of 5 years or more based on national averages over the past 10 years.

a) What was the survival rate for osteosarcoma in 1962? in the present?

b) For which cancers is the present survival rate more than 85%?

c) How many of the cancers had survival rates lower than 50% in 1962? in the present?

d) The difference in survival rates from 1962 to the present is 75% for which type of cancer?

We look at the graph to answer the questions.

a) We go to the right, across the bottom, to the blue bar (1962) above osteosarcoma. Next, we go to the top of that bar and, from there, back to the left to read 20% on the vertical scale. We follow the same procedure for the red bar (the present) and read 65% on the vertical scale. The survival rate for osteosarcoma in 1962 was 20% and presently is 65%.

b) We move up the vertical scale to 85%. From there, we move to the right across all red bars, noting any that are above 85%. We see that the survival rates for acute lymphoblastic leukemia and Hodgkin's disease are greater than 85%.

c) We move up the vertical scale to 50%. From there, we move to the right across all blue bars, noting any that are below 50%. We see that five of the six blue bars are below 50%. We also see that all six red bars are above 50%. Thus, in 1962 five of the six cancers had survival rates below 50%. Today, none of the six cancers has a survival rate below 50%.

d) Looking at the heights of each set of bars, we see that the survival rate for medulloblastoma has increased from 10% in 1962 to 85% at the present. Thus, the difference in survival rates is 75% for medulloblastoma.

Do Exercises 4–6.

Use the bar graph in Example 2 to answer Margin Exercises 4–6.

**4.** For which cancer(s) is the present survival rate between 50% and 70%?

**5.** What is the present survival rate for acute lymphoblastic leukemia?

**6.** What is the difference in survival rates for non-Hodgkin's lymphoma from 1962 to the present?

*Answers*

**4.** Neuroblastoma and osteosarcoma
**5.** About 94%     **6.** About 85% − 7% = 78%

## ⓑ Drawing Bar Graphs

**EXAMPLE 3** *Population by Age.* Listed below are U.S. population data for selected age groups. Make a vertical bar graph of the data.

| AGE GROUP | PERCENT OF POPULATION |
|---|---|
| 5–17 years | 17% |
| 18 years and older | 76 |
| 10–49 years | 54 |
| 16–64 years | 66 |
| 55 years and older | 25 |
| 65 years and older | 13 |
| 85 years and older | 2 |

SOURCE: U.S. Census Bureau

First, we indicate the age groups in seven equally spaced intervals on the horizontal scale and give the horizontal scale the title "Age category." (See Figure 1 below.)

Next, we scale the vertical axis. To do so, we look over the data and note that it ranges from 2% to 76%. We start the vertical scaling at 0, labeling the marks by 10's from 0 to 80. We give the vertical scale the title "Percent of population."

Finally, we draw vertical bars to show the various percents, as shown in Figure 2. We give the graph the overall title "U.S. Population by Age."

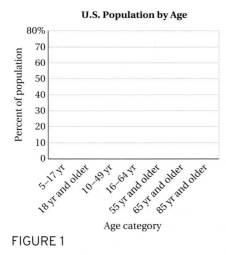

FIGURE 1

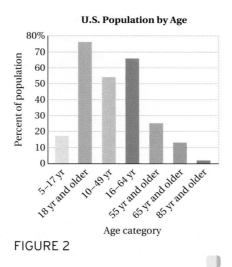

FIGURE 2

Do Exercise 7.

**7. Planetary Moons.** Make a horizontal bar graph to show the number of moons orbiting the various planets.

| PLANET | MOONS |
|---|---|
| Earth | 1 |
| Mars | 2 |
| Jupiter | 63 |
| Saturn | 60 |
| Uranus | 27 |
| Neptune | 13 |

SOURCE: National Aeronautics and Space Administration

Bar graphs that are drawn with unmarked scales may be misleading. This is especially true when the scale does not begin with 0 or when an inconsistent scale is used. The graph on the next page creates such a misleading image.

*Answer*

7.

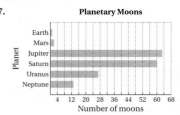

**8.** What is wrong with the following graph?

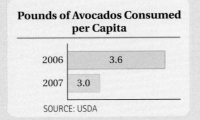

**Pounds of Avocados Consumed per Capita**

2006     3.6

2007     3.0

SOURCE: USDA

Note that the bar representing $1122 is five times as long as the bar representing $988 but 1122 is not five times 988. As drawn, the graph implies that the cost of books and supplies in 2009–2010 was five times that in 2007–2008.

The bars on this graph are misleading because the horizontal scale does not start at 0. Only the top part of each full bar is shown here, magnifying the differences between the years.

Do Exercise 8.

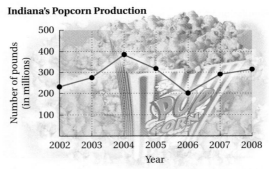

**College Costs More**

It's not just tuition that is increasing for college students. The cost of books and supplies is going up as well.

**Average per-student cost of books and supplies, 4-year public colleges**

2007–2008   $988

2008–2009   $1010

2009–2010   $1122

SOURCE: The College Board, *Trends in College Pricing*

## c   Reading and Interpreting Line Graphs

**Line graphs** are often used to show a change over time as well as to indicate patterns or trends.

**EXAMPLE 4**   *Popcorn.*   Americans consume 16 billion quarts of popcorn annually. Indiana ranks second in the United States in the production of popcorn. The following line graph shows the production of popcorn, in millions of pounds per year, in Indiana from 2002 to 2008.

**Indiana's Popcorn Production**

SOURCE: *The Hoosier Farmer*, Spring 2008; Indiana Corn Growers Association

**a)** For which year was the production of popcorn the highest?

**b)** Between which years did production increase?

**c)** For which years was popcorn production about 315 million lb?

We look at the graph to answer the questions.

**a)** The greatest number of pounds was about 385 million in 2004.

**b)** Reading the graph from left to right, we see that popcorn production increased from 2002 to 2003, from 2003 to 2004, from 2006 to 2007, and from 2007 to 2008.

**c)** We look left to right along a line at 315, as shown at the top of the next page. We see that there are points close to 315 million at 2005 and at 2008.

*Answer*

**8.** The graph is not drawn to scale. The bar for 2006 should be shorter or the bar for 2007 should be longer.

**Indiana's Popcorn Production**

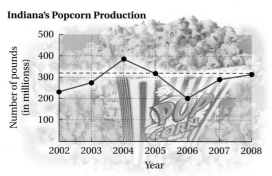

SOURCE: *The Hoosier Farmer, Spring 2008*; Indiana Corn Growers Association

Do Exercises 9–11.

Use the line graph in Example 4 to answer Margin Exercises 9–11.

**9.** For which year was the production of popcorn the lowest?

**10.** Between which years did production decrease?

**11.** For which year was popcorn production about 230 million lb?

**EXAMPLE 5**  *Monthly Loan Payment.*  Suppose you borrow $110,000 at an interest rate of $5\frac{1}{2}\%$ to buy a condominium. The following graph shows the monthly payment required to pay off the loan, depending on the length of the loan.

**$110,000 Loan Repayment**

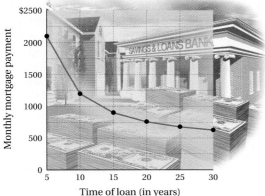

**a)** Estimate the monthly payment for a loan of 15 yr.

**b)** What time period corresponds to a monthly payment of about $760?

**c)** By how much does the monthly payment decrease when the loan period is increased from 10 yr to 20 yr?

We look at the graph to answer the questions.

**a)** We find the time period labeled "15" on the bottom scale and move up from that point to the line. We then go straight across to the left and find that the monthly payment is about $900.

**b)** We locate $760 on the vertical axis. Then we move to the right until we come to the line. The point $760 is on the line at the 20-yr time period.

**c)** The graph shows that the monthly payment for 10 yr is about $1200; for 20 yr, it is about $760. Thus, the monthly payment is decreased by $1200 − $760, or $440. (It should be noted that you will pay back $760 · 20 · 12 − $1200 · 10 · 12, or $38,400, more in interest for a 20-yr loan.)

Do Exercises 12–14.

Use the line graph in Example 5 to answer Margin Exercises 12–14.

**12.** Estimate the monthly payment for a loan of 25 yr.

**13.** What time period corresponds to a monthly payment of about $625?

**14.** By how much does the monthly payment decrease when the loan period is increased from 5 yr to 20 yr?

*Answers*

**9.** 2006    **10.** 2004–2005 and 2005–2006
**11.** 2002    **12.** $675    **13.** 30 yr
**14.** About $1340

## d) Drawing Line Graphs

**EXAMPLE 6** *Temperature in Enclosed Vehicle.* The temperature inside an enclosed vehicle increases rapidly with time. Listed in the table below are the inside temperatures of an enclosed vehicle for specified elapsed times when the outside temperature is 80°F. Make a line graph of the data.

| ELAPSED TIME | TEMPERATURE IN ENCLOSED VEHICLE WITH OUTSIDE TEMPERATURE 80°F |
|---|---|
| 10 min | 99° |
| 20 min | 109° |
| 30 min | 114° |
| 40 min | 118° |
| 50 min | 120° |
| 60 min | 123° |

SOURCES: General Motors; Jan Null, Golden Gate Weather Services

First, we indicate the 10-min elapsed time intervals on the horizontal scale and give the horizontal scale the title "Elapsed time (in minutes)." See the figure on the left below. Next, we scale the vertical axis by 10's beginning with 80 to show the number of degrees and give the vertical scale the title "Temperature (in degrees)." The jagged line at the base of the vertical scale indicates that an unnecessary portion of the scale has been omitted. We also give the graph the overall title "Temperature in Enclosed Vehicle with Outside Temperature 80°F."

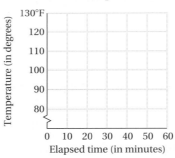

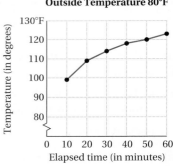

Next, we mark the temperature at the appropriate level above each elapsed time. (See the figure on the right above.) Then we draw line segments connecting the points. The rapid change in temperature can be observed easily from the graph.

Do Exercise 15.

---

**15. Touchdown Passes.** Listed below are the numbers of touchdown passes for Peyton Manning of the Indianapolis Colts for the years 2003 to 2009. Make a line graph of the data.

| YEAR (season) | TOUCHDOWN PASSES |
|---|---|
| 2003 | 29 |
| 2004 | 49 |
| 2005 | 28 |
| 2006 | 31 |
| 2007 | 31 |
| 2008 | 27 |
| 2009 | 33 |

SOURCE: National Football League

*Answer*

15.
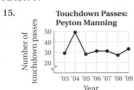

## 6.2

## Exercise Set

For Extra Help

**MyMathLab**

Math XL
PRACTICE

WATCH

DOWNLOAD

READ

REVIEW

**a**  *Bearded Irises.*    A gardener planted six varieties of bearded iris in a new garden on campus. Students from the horticulture department were assigned to record data on the range of heights for each variety. The following vertical bar graph shows the results of their study. The length of the light green shaded portion and the blossom of each bar illustrates the range of heights. For example, the range of heights for the miniature dwarf bearded iris is 2 in. to 9 in.

1. Which variety of iris has a minimum height of 17 in.?

2. What is the range of heights for the standard dwarf bearded iris?

3. Find the average of the maximum heights of all the varieties of irises shown in the graph.

4. Which variety of iris has a maximum height of 28 in.?

5. What is the range of heights for the border bearded iris?

7. Which variety of iris has the smallest range in heights?

9. What is the difference between the maximum heights of the tallest iris and the shortest iris?

**Bearded Irises**

Height (in inches)

Variety of bearded iris

SOURCE: www.irises.org/classification.htm

6. Find the average of the minimum heights of all the varieties of irises shown in the graph.

8. Which irises have a maximum height less than 16 in.?

10. Which irises have a range in heights less than 10 in.?

*Chocolate Desserts.* The following horizontal bar graph shows the average caloric content of various kinds of chocolate desserts and drinks. Use the bar graph for Exercises 11–18.

**11.** Estimate how many calories there are in 1 cup of hot cocoa with skim milk.

**12.** Estimate how many calories there are in a 2-oz candy bar with peanuts.

**13.** Which dessert has the highest caloric content?

**14.** Which item in the bar graph has the lowest caloric content?

**15.** Which dessert contains about 460 calories?

**16.** Which items in the bar graph contain about 300 calories?

**17.** How many more calories are there in 1 cup of hot cocoa made with whole milk than in 1 cup of hot cocoa made with skim milk?

**18.** If Emily drinks a 4-cup chocolate milkshake, how many calories does she consume?

**Chocolate Desserts**

Premium chocolate ice cream, 1 cup
Chocolate cake with fudge frosting, 1 slice
Chocolate milkshake, 1 cup
Candy bar with peanuts, 2 oz.
Hot cocoa with whole milk, 1 cup
Hot cocoa with skim milk, 1 cup
Chocolate syrup, 2 T

0    100   200   300   400   500   600
Calories

*Bachelor's Degrees.* The graph shown at right provides data on the number of bachelor's degrees conferred on men and on women in selected years. Use the bar graph for Exercises 19–22.

**19.** In which years were more bachelor's degrees conferred on men than on women?

**20.** How many more bachelor's degrees were conferred on women in 2007 than in 1970?

**21.** How many more bachelor's degrees were conferred on women than on men in 2000?

**22.** In which years were the numbers of bachelor's degrees conferred on men and on women each greater than 500,000?

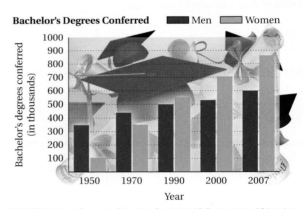

**Bachelor's Degrees Conferred**  ■ Men  □ Women

Bachelor's degrees conferred (in thousands)

1000
900
800
700
600
500
400
300
200
100

1950   1970   1990   2000   2007
Year

SOURCE: National Center for Education Statistics, U.S. Department of Education

**23.** *Cost of Adult Day Care.* The following table lists the average daily cost of adult day care in six cities in the United States, along with the national average. Make a horizontal bar graph to illustrate the data.

| CITY | AVERAGE DAILY COST OF ADULT DAY CARE |
|------|------|
| Chicago | $50 |
| Denver | 61 |
| New York City | 107 |
| Pittsburgh | 53 |
| San Antonio | 32 |
| San Diego | 77 |
| U.S. Average | 64 |

SOURCES: www.consumerhealthratings.com; MetLife Mature Market Institute survey, 2008

Use the data and the bar graph in Exercise 23 to do Exercises 24–27.

**24.** Which city has the highest average daily cost of adult day care?

**25.** In which cities is the average daily cost of adult day care less than $75?

**26.** What is the average daily cost of adult day care in the cities listed in the table? How does it compare with the national average?

**27.** How much higher is the average daily cost of adult day care in New York City than the national average?

**28.** *Commuting Time.* The following table lists the average commuting time in six metropolitan areas with more than 1 million people. Make a vertical bar graph to illustrate the data.

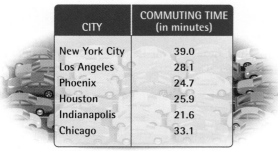

| CITY | COMMUTING TIME (in minutes) |
|------|------|
| New York City | 39.0 |
| Los Angeles | 28.1 |
| Phoenix | 24.7 |
| Houston | 25.9 |
| Indianapolis | 21.6 |
| Chicago | 33.1 |

SOURCE: U.S. Census Bureau

Use the data and the bar graph in Exercise 28 to do Exercises 29–32.

**29.** Which city has the longest commuting time?

**30.** Which city has the shortest commuting time?

**31.** How much longer is the average commuting time in New York City than in Indianapolis?

**32.** What is the average commuting time for the six cities?

**c** *International Tourists to Beijing.* The line graph below illustrates the number of international tourists, in millions, who visited Beijing in recent years. Use the graph for Exercises 33–36.

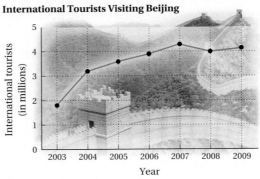

**International Tourists Visiting Beijing**

SOURCE: *Beijing Statistics Year Book*

**33.** Estimate how many international tourists visited Beijing in 2004; in 2007.

**34.** How many more international tourists visited Beijing in 2008 than in 2003?

**35.** Between which years did the number of international tourists decrease?

**36.** Between which two years was the greatest increase in the number of international tourists?

*Prime Time for Bank Crimes.* In 2006, 7272 bank crimes occurred in the United States. The line graph below shows the number of those incidents that occurred by time of day. Use the graph for Exercises 37–40.

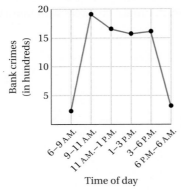

**37.** In which time interval did the most bank crimes occur?

**38.** In which time interval did the smallest number of bank crimes occur?

**39.** Estimate how many more bank crimes occurred between 9 A.M. and 11 A.M. than between 3 P.M. and 6 P.M.

**40.** Estimate how many bank crimes occurred between 11 A.M. and 1 P.M.

**(d)**

**41.** *Longevity Beyond Age 65.* The data in the table below tell us the number of years that a 65-year-old male in the given year can expect to live beyond age 65. Draw a line graph using the horizontal axis to scale "Year."

| YEAR | AVERAGE NUMBER OF YEARS MEN ARE ESTIMATED TO LIVE BEYOND AGE 65 |
|------|-----------------------------------------------------------------|
| 1980 | 14 |
| 1990 | 15 |
| 2000 | 15.9 |
| 2010 | 16.4 |
| 2020 | 16.9 |
| 2030 | 17.5 |

SOURCE: 2000 Social Security Report

**42.** What was the amount of increase in longevity (years beyond 65) between 1980 and 2000?

**43.** What is the expected amount of increase in longevity between 1980 and 2030?

**44.** What is the expected amount of increase in longevity between 2020 and 2030?

**45.** What is the expected amount of increase in longevity between 2000 and 2030?

## Skill Maintenance

**46.** How many 12-oz bottles can be filled from a vat containing 408 oz of catsup?  [1.8a]

**47.** Managers of pizza restaurants know that if 50 pizzas are ordered in an evening, people will request extra cheese on 9 of them. Find the ratio of pizzas ordered with extra cheese to pizzas ordered.  [3.3a]

**48.** A can of Cola-Cola contains 12 fl oz (fluid ounces). How many fluid ounces are in a six-pack?  [1.8a]

**49.** 24 is $\frac{3}{4}$ of what number?  [3.8b]

**50.** $\frac{2}{3}$ of 75 is what number?  [3.4c]

**51.** $\frac{3}{5}$ of 30 is what number?  [3.4c]

**52.** Solve: $-9 = -2x + 3$.  [2.8d]

**53.** Solve: $17 = -3x - 4$.  [2.8d]

## Synthesis

**54.** Draw the bar graph shown at the top right of page 402 using an accurate scale.

**55.** Draw the bar graph in Margin Exercise 8 using an accurate scale.

# Ordered Pairs and Equations in Two Variables

## OBJECTIVES

**a** Plot a point, given its coordinates. Find coordinates, given a point.

**b** Determine the quadrant in which a point lies.

**c** Determine whether an ordered pair is a solution of an equation with two variables.

**SKILL TO REVIEW**
Objective 2.6a: Evaluate an algebraic expression by substitution.

1. Evaluate $2x + 3y$ for $x = 5$ and $y = -3$.

2. Evaluate $x - y$ for $x = 2$ and $y = -10$.

Plot these points on the graph below.

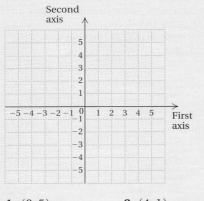

1. $(2, 5)$        2. $(4, 1)$

Bar graphs and line graphs are used to illustrate relationships between the items or quantities listed along the bottom and the side of the graph. The horizontal and vertical sides of a bar graph or line graph are often called the **axes** (pronounced ăk′sēz; singular: **axis**). By using two perpendicular number lines as axes, we can use points to represent solutions of certain equations. First we look at graphing points using such axes.

## a Points and Ordered Pairs

When two number lines are used as axes, a grid can be formed. The grid provides a helpful way of locating any point on the plane. Just as a location in a city might be given as the intersection of an avenue and a side street, a point on a plane might be described as the intersection of a vertical line and a horizontal line. For the point shown in the figure below, these lines pass through 3 on the horizontal axis and 4 on the vertical axis. Thus, the **first coordinate** of this point is 3 and the **second coordinate** is 4. **Ordered pair** notation, (3, 4), provides a quick way of stating this.

------ *Caution!* ------

When writing an ordered pair, you should *always* list the coordinate from the horizontal axis first.

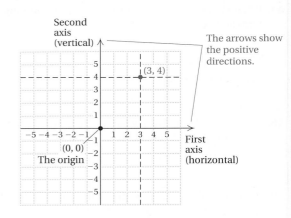

The point (0, 0), where the axes cross each other, is called the **origin**. To graph, or *plot*, the point (3, 4), we can begin at the origin and move horizontally (along the first axis) to the number 3. From there, we move up 4 units vertically and make a "dot."

It is important to always make sure that the first coordinate matches the number that would be below (or above) the point on the horizontal axis. Similarly, the second coordinate should always match the number that would be to the left (or right) of the point on the vertical axis.

Do Margin Exercises 1 and 2.

---

*Answers*

*Skill to Review:*
**1.** 1    **2.** 12

*Margin Exercises:*
**1 and 2.**
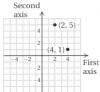

**EXAMPLE 1** Plot the points $(-5, 2)$ and $(2, -5)$.

To plot $(-5, 2)$, we locate $-5$ on the first, or horizontal, axis. From there we go up 2 units and make a dot.

To plot $(2, -5)$, we locate 2 on the first, or horizontal, axis. Then we go down 5 units and make a dot. Note that the order of the numbers within a pair is important: $(2, -5) \neq (-5, 2)$.

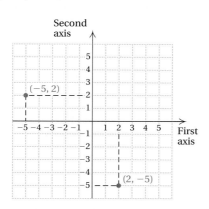

Do Exercises 3–8.

To determine the coordinates of a given point, we first look directly above or below the point to find the point's horizontal coordinate. Then we look to the left (or right) of the point to identify the vertical coordinate.

**EXAMPLE 2** Determine the coordinates of points $A$, $B$, $C$, $D$, $E$, $F$, and $G$.

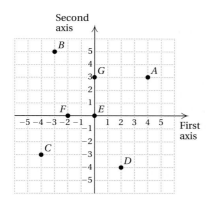

We look below point $A$ to see that its first coordinate is 4. Looking to the left of point $A$, we find that its second coordinate is 3. Thus, the coordinates of point $A$ are $(4, 3)$. The coordinates of the other points are

$B$: $(-3, 5)$;  $C$: $(-4, -3)$;  $D$: $(2, -4)$;
$E$: $(0, 0)$;  $F$: $(-2, 0)$;  $G$: $(0, 3)$.

Do Exercise 9.

---

Plot these points on the graph below.

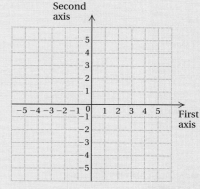

**3.** $(-2, 5)$          **4.** $(-3, -4)$

**5.** $(5, -3)$          **6.** $(-2, -1)$

**7.** $(0, -3)$          **8.** $\left(2\frac{1}{2}, 0\right)$

**9.** Determine the coordinates of points $A$, $B$, $C$, $D$, $E$, $F$, and $G$ on the graph below.

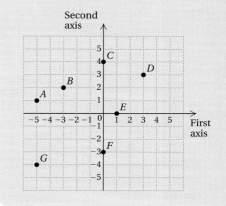

*Answers*

3–8.

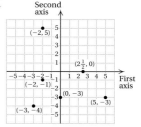

**9.** $A$: $(-5, 1)$; $B$: $(-3, 2)$; $C$: $(0, 4)$; $D$: $(3, 3)$; $E$: $(1, 0)$; $F$: $(0, -3)$; $G$: $(-5, -4)$

## b  Quadrants

The axes divide the plane into four regions, or **quadrants**. For any point in region I (the *first quadrant*), both coordinates are positive. For any point in region II (the *second quadrant*), the first coordinate is negative and the second coordinate is positive. In region III (the *third quadrant*), both coordinates are negative. In region IV (the *fourth quadrant*), the first coordinate is positive and the second coordinate is negative.

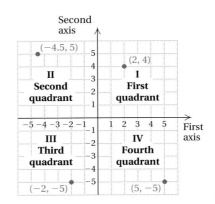

As the figure above illustrates, the point $(2, 4)$ is in the first quadrant, $(-4.5, 5)$ is in the second quadrant, $(-2, -5)$ is in the third quadrant, and $(5, -5)$ is in the fourth quadrant.

Do Exercises 10–15.

**10.** What can you say about the coordinates of a point in the third quadrant?

**11.** What can you say about the coordinates of a point in the fourth quadrant?

In which quadrant is each point located?

**12.** $(5, 3)$

**13.** $(-6, -4)$

**14.** $(10, -14)$

**15.** $\left(-13, 9\frac{1}{2}\right)$

## c  Solutions of Equations

The coordinate system we have just introduced is called the **Cartesian** coordinate system, in honor of the great mathematician and philosopher René Descartes (1596–1650). Legend has it that Descartes hit upon the idea of the coordinate system after watching a fly stop several times on the ceiling over his bed. Descartes used this coordinate system as a method of presenting solutions of equations containing two variables. Equations like $3x + 2y = 8$ have ordered pairs as solutions. In Section 6.4, we will find solutions and graph them. Here we simply practice checking to see if an ordered pair is a solution.

To determine whether an ordered pair is a solution of an equation, we normally substitute the first coordinate for the variable that comes first alphabetically and the second coordinate for the variable that is last alphabetically. The letters $x$ and $y$ are used most often.

**EXAMPLE 3**   Determine whether the ordered pair $(2, 1)$ is a solution of the equation $3x + 2y = 8$.

$$\frac{3x + 2y = 8}{\begin{array}{c|c} 3 \cdot 2 + 2 \cdot 1 \; ? \; 8 \\ 6 + 2 \\ 8 \; \big| \; 8 \end{array}}$$   Substituting 2 for $x$ and 1 for $y$ (alphabetical order of variables)

TRUE

Since the equation becomes true, $(2, 1)$ is a solution.

In a similar manner, we can show that $(0, 4)$ and $(4, -2)$ are also solutions of $3x + 2y = 8$. In fact, there is an infinite number of solutions of $3x + 2y = 8$.

*Answers*

**10.** Both are negative numbers.
**11.** The first, or horizontal, coordinate is positive; the second, or vertical, coordinate is negative.   **12.** I   **13.** III
**14.** IV   **15.** II

**EXAMPLE 4** Determine whether the ordered pair $(-2, 3)$ is a solution of the equation $2t = 4s - 8$.

We substitute:

$$
\begin{array}{c|c}
\multicolumn{2}{c}{2t = 4s - 8} \\
\hline
2 \cdot 3 \; ? \; 4(-2) - 8 \\
6 \;\; \big| \;\; -8 - 8 \\
6 \;\; \big| \;\; -16 \quad \text{FALSE}
\end{array}
$$

Using alphabetical order, substituting $-2$ for $s$ and $3$ for $t$

Since the equation becomes false, $(-2, 3)$ is not a solution.

**Unless stated otherwise, the coordinates of an ordered pair correspond alphabetically to the variables.**

Do Exercises 16 and 17.

**16.** Determine whether $(5, 1)$ is a solution of $y = 2x + 3$.

**17.** Determine whether $(-13.6, 25.4)$ is a solution of $3x + 2y = 10$.

*Answers*
**16.** No   **17.** Yes

## Calculator Corner

**Checking Solutions**   Solutions of equations in two variables can be checked on a calculator. For instance, to demonstrate that $(5.1, -3.65)$ is a solution of $3x + 2y = 8$, we press either

$$\boxed{3}\;\boxed{\times}\;\boxed{5}\;\boxed{\cdot}\;\boxed{1}\;\boxed{+}\;\boxed{2}\;\boxed{\times}\;\boxed{3}\;\boxed{\cdot}\;\boxed{6}\;\boxed{5}\;\boxed{+/-}\;\boxed{=}$$

or

$$\boxed{3}\;\boxed{\times}\;\boxed{5}\;\boxed{\cdot}\;\boxed{1}\;\boxed{+}\;\boxed{2}\;\boxed{\times}\;\boxed{(-)}\;\boxed{3}\;\boxed{\cdot}\;\boxed{6}\;\boxed{5}\;\boxed{\text{ENTER}}.$$

The result, 8, shows that $(5.1, -3.65)$ is a solution.

Most calculators now have memory keys. These keys enable us to store and recall a number as needed. On most calculators, we store a number using a key labeled $\boxed{\text{STO}}$, $\boxed{\text{M}}$, or $\boxed{\text{Min}}$. Once a number has been stored, we can retrieve or recall the number by pressing a key labeled $\boxed{\text{RCL}}$ or $\boxed{\text{MR}}$.

To show that $(7.35, 10.7)$ is a solution of $2t = 4s - 8$, we can first evaluate and store the right side of the equation:

$$\boxed{4}\;\boxed{\times}\;\boxed{7}\;\boxed{\cdot}\;\boxed{3}\;\boxed{5}\;\boxed{-}\;\boxed{8}\;\boxed{=}\;\boxed{\text{STO}}.$$

The result, 21.4, has been stored in the calculator's memory, so we need not worry about writing it down. To complete the check, we clear the calculator and evaluate the left side of the equation:

$$\boxed{2}\;\boxed{\times}\;\boxed{1}\;\boxed{0}\;\boxed{\cdot}\;\boxed{7}\;\boxed{=}.$$

To show that this result matches the number stored earlier, we do not clear the display, but instead subtract the number stored:

$$\boxed{-}\;\boxed{\text{RCL}}\;\boxed{=}.$$

The result, 0, indicates that $2 \times 10.7$ and $4 \times 7.35 - 8$ are equal. A result other than 0 would indicate that the ordered pair in question does not check.

As always, keystrokes may vary, so consult your owner's manual if the above keystrokes do not work for your calculator.

**Exercises:**   Determine whether each point is a solution of the given equation.

**1.** $(7.9, 3.2);\ 5x + 4y = 52.3$

**2.** $(1.9, 2.3);\ 7x - 8y = 5.1$

**3.** $(4.3, 4.75);\ 5y = 6x - 7$

**4.** $(3.8, -4.3);\ 9a = 17 - 4b$

**5.** $(9.4, -3.9);\ 3a - 15 = 29 + 4b$

**6.** $(5.6, 8.8);\ 4y + 23 = 7x + 19$

**7.** $(-2.4, 8.5625);\ 3.5x + 17.4 = 3.2y - 18.4$

**8.** $(1.8, 2.6);\ 9.2x - 15.3 = 4.8y - 13.7$

**6.3** **Exercise Set**

For Extra Help

**MyMathLab**

Math XL
PRACTICE

WATCH

DOWNLOAD

READ

REVIEW

**a** Plot each group of points on the given graph below.

**1.** $(4, 4)$ $(-2, 4)$ $(5, -3)$ $(-5, -5)$ $(0, 4)$ $(0, -4)$ $(3, 0)$ $(-4, 0)$

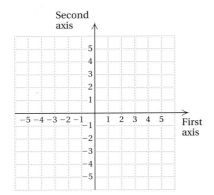

**2.** $(2, 5)$ $(-1, 3)$ $(3, -2)$ $(-2, -4)$ $(0, 4)$ $(0, -5)$ $(5, 0)$ $(-5, 0)$

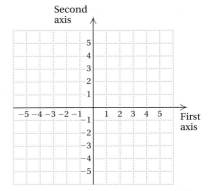

**3.** $(-2, -4)$ $(5, -4)$ $\left(0, 3\frac{1}{2}\right)$ $\left(4, 3\frac{1}{2}\right)$ $(-1, -3)$ $(-1, 5)$ $(4, -1)$ $(-2, 0)$

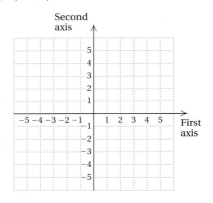

**4.** $(-3, -1)$ $(5, 1)$ $(-1, -5)$ $(0, 0)$ $(0, 1)$ $(-4, 0)$ $\left(2, 3\frac{1}{2}\right)$ $\left(4\frac{1}{2}, -2\right)$

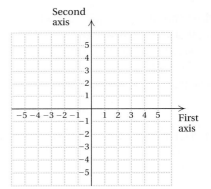

Determine the coordinates of points *A*, *B*, *C*, *D*, *E*, and *F*.

**5.**

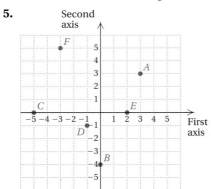

**6.**

**7.**

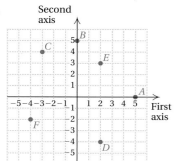

**8.**

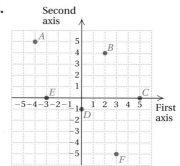

**b** In which quadrant is each point located?

**9.** $(-5, 3)$

**10.** $(-12, 1)$

**11.** $(100, -1)$

**12.** $\left(35\frac{1}{2}, -2\frac{1}{2}\right)$

**13.** $(-6.5, -1.9)$

**14.** $(-3.4, -5.9)$

**15.** $\left(3\frac{7}{10}, 9\frac{1}{11}\right)$

**16.** $(1895, 1492)$

Complete each sentence using the words *positive* or *negative* or the numerals I, II, III, or IV.

**17.** In quadrant IV, first coordinates are always _____ and second coordinates are always _____.

**18.** In quadrant III, first coordinates are always _____ and second coordinates are always _____.

**19.** In quadrant _____, both coordinates are always negative.

**20.** In quadrant _____, both coordinates are always positive.

**21.** In quadrants I and _____, the first coordinate is always _____.

**22.** In quadrants II and _____, the second coordinate is always _____.

**c** Determine whether each ordered pair is a solution of the given equation.

**23.** $(4, 3)$; $y = 2x - 5$

**24.** $(1, 7)$; $y = 2x + 5$

**25.** $(2, -3)$; $3x - y = 4$

**26.** $(-1, 4)$; $2x + y = 6$

**27.** $(-2, -1)$; $3x + 2y = -8$

**28.** $(0, -4)$; $4x + 2y = -9$

**29.** $(5, -4)$; $3x + y = 19$

**30.** $(-1, 7)$; $x - y = -8$

**31.** $\left(2\dfrac{1}{3},6\right);\quad 2y-3x=3$

**32.** $\left(3,1\dfrac{1}{4}\right);\quad 2x-4y=1$

**33.** $(2.4,0.7);\quad y=5x-11.3$

**34.** $(1.8,7.4);\quad y=3x+2$

## Skill Maintenance

Solve.

**35.** $3x-4=17$   [2.8d]

**36.** $5(x-2)=3x-4$   [5.7b]

**37.** $-\dfrac{1}{9}t=\dfrac{2}{3}t$   [5.7b]

**38.** Simplify: $\dfrac{90}{51}$.   [3.5b]

**39.** Combine like terms:   [4.6b]
$$7\dfrac{2}{11}a-5\dfrac{1}{3}a.$$

**40.** Simplify:   [2.7a]
$$3(x-5)+4x-9.$$

## Synthesis

Determine whether each ordered pair is a solution of the given equation.

**41.** ▦ $(-2.37,1.23);\ 5.2x+6.1y=-4.821$

**42.** ▦ $(4.16,-9.35);\ 6.5x-7.2y=-94.35$

In Exercises 43–46, determine the quadrant(s) in which the point could be located.

**43.** The first coordinate is positive.

**44.** The second coordinate is negative.

**45.** The first and second coordinates are equal.

**46.** The first coordinate is the opposite of the second coordinate.

**47.** The points $(-1,1)$, $(4,1)$, and $(4,-5)$ are three vertices of a rectangle. Find the coordinates of the fourth vertex.

**48.** A parallelogram is a four-sided polygon with two pairs of parallel sides (two examples are shown below). Three parallelograms share the vertices $(-2,-3)$, $(-1,2)$, and $(4,-3)$. Find the fourth vertex of each parallelogram.

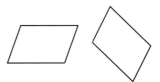

**49.** Find the perimeter of a rectangle with vertices at $(5,3)$, $(5,-2)$, $(-3,-2)$, and $(-3,3)$.

**50.** Find the area of a rectangle with vertices at $(0,9)$, $(0,-4)$, $(5,-4)$, and $(5,9)$.

# 6.4

## Graphing Linear Equations

In Section 6.3, we saw how to determine whether an ordered pair is a solution of an equation in two variables. Once we can find a few ordered pairs that solve an equation, we will be able to graph the equation.

### a  Finding Solutions

To solve an equation with one variable, like $3x + 2 = 8$, we isolate the variable, $x$, on one side of the equation. To solve an equation with two variables, we will first replace one variable with some number choice and then solve the resulting equation.

**EXAMPLE 1**  Find a solution of $x + y = 7$. Let $x = 5$.

If $x$ is 5, then the solution will be an ordered pair $(5, y)$. We find $y$ by substituting 5 for $x$ in $x + y = 7$:

$$5 + y = 7.$$

We solve as follows:

$$5 + y = 7$$
$$5 + y - 5 = 7 - 5 \qquad \text{Subtracting 5 from both sides}$$
$$y = 2.$$

Since $5 + 2 = 7$, the ordered pair $(5, 2)$ is a solution of $x + y = 7$.

> Do Margin Exercise 1.

**EXAMPLE 2**  Complete these solutions of $2x + 3y = 8$: $(\square, 2)$; $(-2, \square)$.

Recall that each solution will be of the form $(x, y)$. Thus, to complete the pair $(\square, 2)$, we replace $y$ with 2 and solve for $x$:

$$2x + 3y = 8$$
$$2x + 3 \cdot 2 = 8 \qquad \text{Substituting 2 for } y$$
$$2x + 6 = 8$$
$$2x + 6 - 6 = 8 - 6 \qquad \text{Subtracting 6 from both sides}$$
$$2x = 2$$
$$\tfrac{1}{2} \cdot 2x = \tfrac{1}{2} \cdot 2 \qquad \text{Multiplying both sides by } \tfrac{1}{2}$$
$$x = 1.$$

Thus, $(1, 2)$ is a solution of $2x + 3y = 8$. Since $2(1) + 3(2) = 8$, the solution checks.

To complete the pair $(-2, \square)$, we replace $x$ with $-2$ and solve for $y$:

$$2x + 3y = 8$$
$$2(-2) + 3y = 8 \qquad \text{Substituting } -2 \text{ for } x$$
$$-4 + 3y = 8$$
$$3y = 12 \qquad \text{Adding 4 to both sides}$$
$$y = 4. \qquad \text{Dividing both sides by 3}$$

Thus, $(-2, 4)$ is also a solution of $2x + 3y = 8$. Since $2(-2) + 3(4) = 8$, the solution checks.

> Do Exercise 2.

---

### OBJECTIVES

**a**  Find solutions of equations in two variables.

**b**  Graph linear equations in two variables.

**c**  Graph equations for horizontal or vertical lines.

**SKILL TO REVIEW**
Objective 2.8d: Solve equations that require use of both the addition principle and the division principle.

Solve.
1. $3x + 2 \cdot 4 = 23$
2. $4(-1) - y = 7$

1. Find a solution of $x - y = 3$. Let $y = 5$.

2. Complete these solutions of $5x + y = 10$: $(1, \square)$; $(\square, -5)$.

***Answers***

*Skill to Review:*
1. 5   2. $-11$

*Margin Exercises:*
1. $(8, 5)$   2. $(1, 5)$; $(3, -5)$

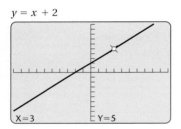
**3.** Find three solutions of $x + 2y = 5$. Answers may vary.

**4.** Find three solutions of $y = -2x + 1$. Answers may vary.

*Answers*

**3.** $(1, 2), (5, 0), (3, 1)$
**4.** $(0, 1), (2, -3), (-2, 5)$

---

**EXAMPLE 3**   Find three solutions of $2x - y = 5$.

We are free to use *any* number as a replacement for either $x$ or $y$. To find one solution, we select 1 as a replacement for $x$. We then solve for $y$:

| | |
|---|---|
| $2x - y = 5$ | We are looking for an ordered pair $(1, \Box)$. |
| $2 \cdot 1 - y = 5$ | Substituting 1 for $x$ |
| $2 - y = 5$ | |
| $-y = 3$ | Subtracting 2 from both sides |
| $-1y = 3$ | Recall that $-a = -1 \cdot a$. |
| $y = -3.$ | Dividing both sides by $-1$ |

Thus, $(1, -3)$ is one solution of $2x - y = 5$.

To find a second solution, we choose to replace $y$ with 0 and solve for $x$:

| | |
|---|---|
| $2x - y = 5$ | We are looking for an ordered pair $(\Box, 0)$. |
| $2x - 0 = 5$ | Substituting 0 for $y$ |
| $2x = 5$ | Simplifying |
| $x = 2.5.$ | Dividing both sides by 2 |

Thus, $(2.5, 0)$ is a second solution of $2x - y = 5$.

To find a third solution, we can replace $x$ with 0 and solve for $y$:

| | |
|---|---|
| $2x - y = 5$ | We are looking for an ordered pair $(0, \Box)$. |
| $2 \cdot 0 - y = 5$ | Substituting 0 for $x$ |
| $0 - y = 5$ | |
| $-y = 5$ | |
| $-1y = 5$ | Try to do this step mentally. |
| $y = -5.$ | Dividing both sides by $-1$ |

The pair $(0, -5)$ is a third solution of $2x - y = 5$.

Note that three different choices for $x$ or $y$ would have given three different solutions. There are an infinite number of ordered pairs that are solutions, so it is unlikely for two students to have solutions that match entirely.

---

To find a solution of an equation with two variables:

1. Choose a replacement for one variable.
2. Solve for the other variable.
3. Write the solution as an ordered pair.

---

Do Exercises 3 and 4.

## b Graphing Equations

Equations like those considered in Examples 1–3 are in the form $Ax + By = C$. All equations that can be written this way are said to be **linear** because the solutions of each equation, when graphed, form a straight line. An equation $Ax + By = C$ is called the **standard form** of a linear equation. When the line representing the solutions is drawn, we say that we have *graphed* the equation.

Since solutions of $Ax + By = C$ are written in the form $(x, y)$, we label the horizontal axis as the $x$-axis and the vertical axis as the $y$-axis.

**EXAMPLE 4**  Graph: $2x - y = 5$.

We first need to calculate several solutions of $2x - y = 5$. In Example 3, we found that $(1, -3)$, $(2.5, 0)$, and $(0, -5)$ are solutions of the equation.

Next, we plot the points. As expected, the points describe a straight line. We draw the line and label it with the equation, as shown on the right below.

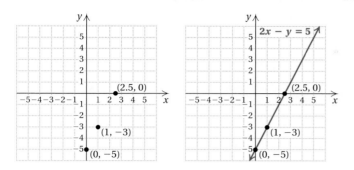

Note that two points are enough to determine a line, but we generally calculate and graph at least three ordered pairs before drawing each line. If the points do not all line up, we know that a mistake has been made.

Do Exercise 5.

Linear equations are not always written in standard form. Equations like $y = 2x$ or $y = x + 2$ are also linear. To find solutions of equations like $y = 2x$, we usually choose values for $x$ and then calculate $y$.

**EXAMPLE 5**  Graph: $y = 2x$.

First, we find some ordered pairs that are solutions. To find three ordered pairs, we can choose any three values for $x$ and then calculate the corresponding values for $y$. One good choice is 0; we also choose $-2$ and 3.

If $x$ is $0$, then $y = 2x = 2 \cdot 0 = 0$.     Thus, $(0, 0)$ is a solution.

If $x$ is $-2$, then $y = 2x = 2(-2) = -4$.     Thus, $(-2, -4)$ is a solution.

If $x$ is $3$, then $y = 2x = 2 \cdot 3 = 6$.     Thus, $(3, 6)$ is a solution.

We can compute additional pairs if we wish and list the ordered pairs that are solutions in a table.

| $x$ | $y$ $y = 2x$ | $(x, y)$ |
|-----|-----|-----|
| 3 | 6 | $(3, 6)$ |
| $-2$ | $-4$ | $(-2, -4)$ |
| 0 | 0 | $(0, 0)$ |
| 1 | 2 | $(1, 2)$ |

Substitute for $x$.
Compute the value of $y$.
Form the ordered pair $(x, y)$.
Plot the points.
Draw and label the graph.

Next, we plot these points. We draw the line, or graph, with a ruler and label it $y = 2x$.

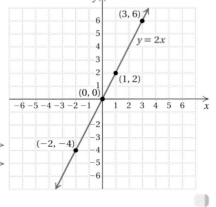

Do Exercises 6 and 7.

**5.** Graph $x + 2y = 5$. Use the results from Margin Exercise 3.

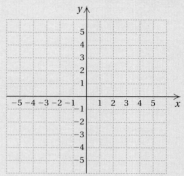

Graph.

**6.** $y = 3x$

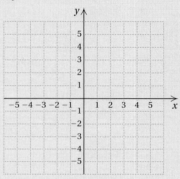

**7.** $y = \frac{1}{2}x$

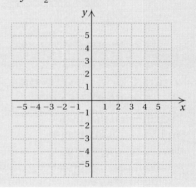

*Answers*

**5.**

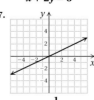

$x + 2y = 5$

**6.**

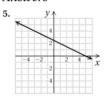

$y = 3x$

**7.**

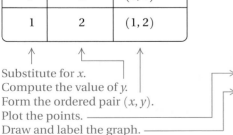

$y = \frac{1}{2}x$

Graph.

**8.** $y = -x$ (or $y = -1 \cdot x$)

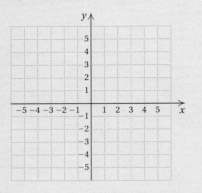

**9.** $y = -2x$

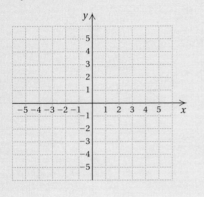

*Answers*

**8.**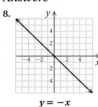
$y = -x$

**9.**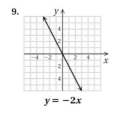
$y = -2x$

**EXAMPLE 6**  Graph: $y = -3x$.

We make a table of solutions. Then we plot the points, draw the line with a ruler, and label the line $y = -3x$.

If $x$ is 0, then $y = -3 \cdot 0 = 0$.
If $x$ is 1, then $y = -3 \cdot 1 = -3$.
If $x$ is $-2$, then $y = -3(-2) = 6$.
If $x$ is 2, then $y = -3 \cdot 2 = -6$.

| $x$ | $y$ $y = -3x$ | $(x, y)$ |
|---|---|---|
| 0 | 0 | $(0, 0)$ |
| 1 | $-3$ | $(1, -3)$ |
| $-2$ | 6 | $(-2, 6)$ |
| 2 | $-6$ | $(2, -6)$ |

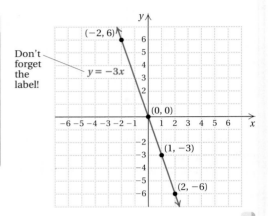

Don't forget the label!

Do Exercises 8 and 9.

**EXAMPLE 7**  Graph: $y = x + 2$.

We make a table of solutions. Then we plot the points, draw the line with a ruler, and label it.

If $x$ is 0, then $y = 0 + 2 = 2$.
If $x$ is 1, then $y = 1 + 2 = 3$.
If $x$ is $-1$, then $y = -1 + 2 = 1$.
If $x$ is 3, then $y = 3 + 2 = 5$.

| $x$ | $y$ $y = x + 2$ | $(x, y)$ |
|---|---|---|
| 0 | 2 | $(0, 2)$ |
| 1 | 3 | $(1, 3)$ |
| $-1$ | 1 | $(-1, 1)$ |
| 3 | 5 | $(3, 5)$ |

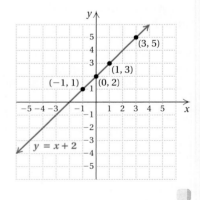

The values of $x$ in these examples were *chosen*. Different choices for $x$ would yield different points, but the same line.

**For linear equations, tables can be formed using any numbers for $x$.**

**EXAMPLE 8** Graph: $y = \frac{2}{3}x + 1$.

We make a table of solutions, plot the points, and draw and label the line. For this table, we selected multiples of 3 as $x$-values to avoid fraction values for $y$.

If $x$ is 6, then $y = \frac{2}{3} \cdot 6 + 1 = 4 + 1 = 5$.
If $x$ is 3, then $y = \frac{2}{3} \cdot 3 + 1 = 2 + 1 = 3$.
If $x$ is 0, then $y = \frac{2}{3} \cdot 0 + 1 = 0 + 1 = 1$.
If $x$ is $-3$, then $y = \frac{2}{3} \cdot (-3) + 1 = -2 + 1 = -1$.

| $x$ | $y = \frac{2}{3}x + 1$ | $(x, y)$ |
|-----|------------------------|----------|
| 6 | 5 | $(6, 5)$ |
| 3 | 3 | $(3, 3)$ |
| 0 | 1 | $(0, 1)$ |
| $-3$ | $-1$ | $(-3, -1)$ |

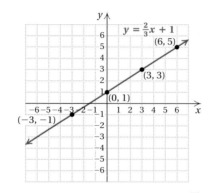

Do Exercises 10–12.

## c Graphing Horizontal or Vertical Lines

Any equation in the form $Ax + By = C$ is linear, provided $A$ and $B$ are not both zero. If $A$ is 0 and $B$ is nonzero, there is no $x$-term and the graph is a horizontal line. If $B$ is 0 and $A$ is nonzero, there is no $y$-term and the graph is a vertical line. In Examples 9 and 10, we consider both of these possibilities.

**EXAMPLE 9** Graph: $y = 3$.

We regard $y = 3$ as $0 \cdot x + y = 3$. No matter what number we choose for $x$, we find that $y$ must be 3 if the equation is to be solved.

Choose any number for $x$. →

| $x$ | $y = 3$ | $(x, y)$ |
|-----|---------|----------|
| $-2$ | 3 | $(-2, 3)$ |
| 0 | 3 | $(0, 3)$ |
| 4 | 3 | $(4, 3)$ |

$y$ must be 3.

All pairs will have 3 as the $y$-coordinate.

When we plot $(-2, 3)$, $(0, 3)$, and $(4, 3)$ and connect the points, we obtain a horizontal line. Any ordered pair of the form $(x, 3)$ is a solution, so the line is 3 units above the $x$-axis, as shown in the graph at the top of the next page.

Graph.

**10.** $y = x + 1$

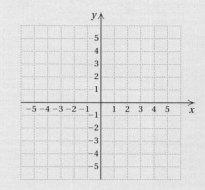

**11.** $y = -2x + 1$

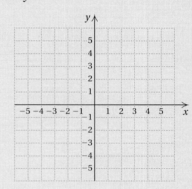

**12.** $y = \frac{3}{5}x$

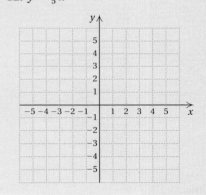

*Answers*

10.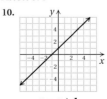
$y = x + 1$

11.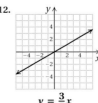
$y = -2x + 1$

12.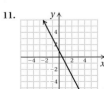
$y = \frac{3}{5}x$

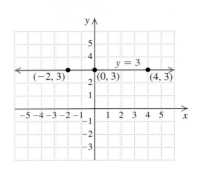

Graph.

**13.** $y = 4$

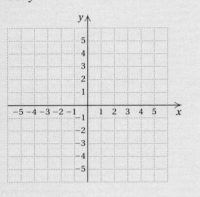

**14.** $x = 5$

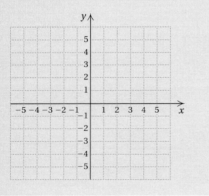

**Answers**

**13.**

$y = 4$

**14.**

$x = 5$

**EXAMPLE 10** Graph: $x = -4$.

We regard $x = -4$ as $x + 0 \cdot y = -4$ and make a table with all $-4$'s in the $x$-column.

| $x$ | | |
|---|---|---|
| $x = -4$ | $y$ | $(x, y)$ |
| $-4$ | $-5$ | $(-4, -5)$ |
| $-4$ | $1$ | $(-4, 1)$ |
| $-4$ | $3$ | $(-4, 3)$ |

$x$ must be $-4$. →

Choose any number for $y$.

All pairs will have $-4$ as the $x$-coordinate.

When we plot $(-4, -5)$, $(-4, 1)$, and $(-4, 3)$ and connect them, we obtain a vertical line. Any ordered pair of the form $(-4, y)$ is a solution, so the line is 4 units left of the $y$-axis.

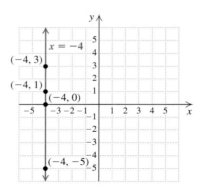

Do Exercises 13 and 14.

## HORIZONTAL AND VERTICAL LINES

The graph of $y = b$ is a horizontal line.
The graph of $x = a$ is a vertical line.

**a** For each equation, use the indicated value to find an ordered pair that is a solution.

**1.** $x + y = 8$; let $x = 5$

**2.** $x + y = 5$; let $x = 4$

**3.** $2x + y = 7$; let $x = 3$

**4.** $x + 2y = 9$; let $y = 4$

**5.** $y = 3x - 1$; let $x = 5$

**6.** $y = 2x + 7$; let $x = 3$

**7.** $x + 3y = 1$; let $x = 10$

**8.** $5x + y = 7$; let $y = -8$

**9.** $2x + 5y = 15$; let $x = 0$

**10.** $5x + 2y = 18$; let $x = 0$

**11.** $3x - 2y = 8$; let $y = -1$

**12.** $2x - 5y = 12$; let $y = -2$

For each equation, complete the given ordered pairs.

**13.** $x + y = 4$; $(\square, 3)$; $(-1, \square)$

**14.** $x - y = 6$; $(\square, 2)$; $(9, \square)$

**15.** $x - y = 4$; $(\square, 3)$; $(10, \square)$

**16.** $x + y = 10$; $(\square, 8)$; $(3, \square)$

**17.** $2x + 3y = 30$; $(0, \square)$; $(\square, 0)$

**18.** $3x + 2y = 24$; $(0, \square)$; $(\square, 0)$

**19.** $3x + 5y = 14$; $(3, \square)$; $(\square, 4)$

**20.** $4x + 3y = 11$; $(5, \square)$; $(\square, 2)$

**21.** $y = 4x$; $(\square, 4)$; $(-2, \square)$

**22.** $y = 6x$; $(\square, 6)$; $(-2, \square)$

**23.** $2x + 5y = 3$; $(0, \square)$; $(\square, 0)$

**24.** $5x + 7y = 9$; $(0, \square)$; $(\square, 0)$

For each equation, find three solutions. Answers may vary.

**25.** $x + y = 9$

**26.** $x + y = 19$

**27.** $y = 4x$

**28.** $y = 5x$

**29.** $3x + y = 13$

**30.** $x + 5y = 12$

**31.** $y = 3x - 1$

**32.** $y = 2x + 5$

**33.** $y = -7x$

**34.** $y = -4x$

**35.** $4 + y = x$

**36.** $3 + y = x$

**37.** $3x + 2y = 12$

**38.** $2x + 3y = 18$

**39.** $y = \frac{1}{3}x + 2$

**40.** $y = \frac{1}{2}x + 5$

**b** Graph each equation.

**41.** $x + y = 6$

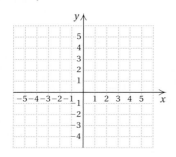

**42.** $x + y = 4$

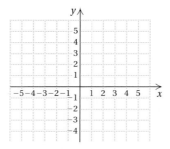

**43.** $x - 1 = y$

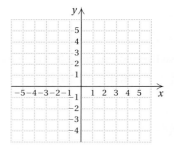

**44.** $x - 2 = y$

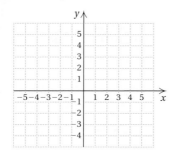

**45.** $y = x - 4$

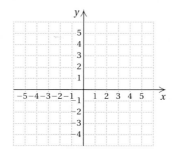

**46.** $y = x - 5$

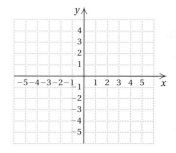

**47.** $y = \dfrac{1}{3}x$

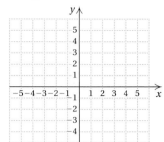

**48.** $y = -\dfrac{1}{3}x$

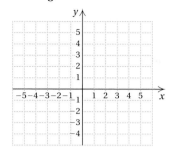

**49.** $y = x$

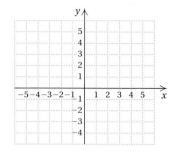

**50.** $y = x - 3$

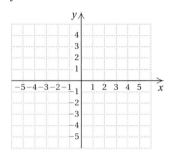

**51.** $y = 2x - 1$

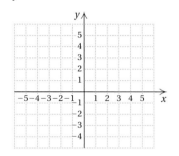

**52.** $y = 2x - 3$

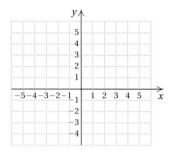

**53.** $y = 2x + 1$

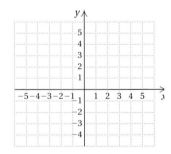

**54.** $y = 3x + 1$

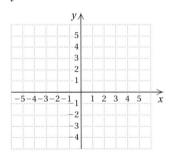

**55.** $y = \dfrac{2}{5}x$

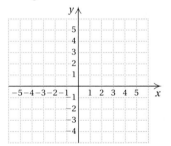

**56.** $y = \dfrac{3}{4}x$

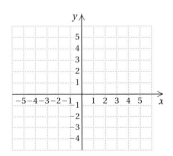

**57.** $y = -x + 4$

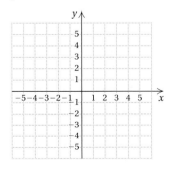

**58.** $y = -x + 5$

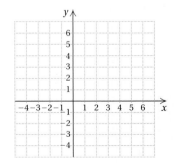

**59.** $y = \dfrac{2}{3}x + 1$

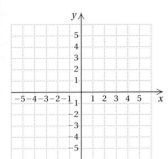

**60.** $y = \dfrac{2}{5}x - 1$

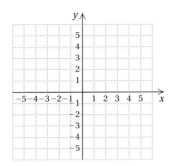

**c** Graph.

**61.** $y = 2$

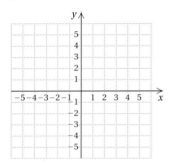

**62.** $y = 1$

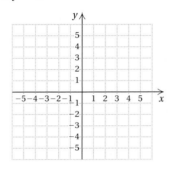

**63.** $x = 2$

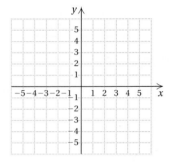

**64.** $x = 3$

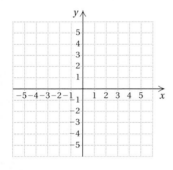

**65.** $x = -3$

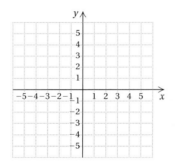

**66.** $x = -1$

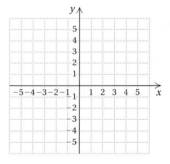

**67.** $y = -4$

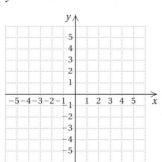

**68.** $y = -2$

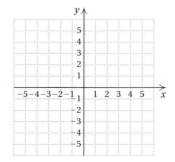

## Skill Maintenance

**69.** The tunes on Miles Davis's classic *Kind of Blue* album are approximately 9 min, $9\frac{1}{2}$ min, $5\frac{1}{2}$ min, $11\frac{1}{2}$ min, and $9\frac{1}{2}$ min long. Find the average length of a tune on that album.   [4.6c], [4.7d]

**70.** The books on Sherry's nightstand are 243, 410, 352, and 274 pages long. What is the average length of a book on the nightstand?   [5.8b]

**71.** A recipe for a batch of chili calls for $\frac{3}{4}$ cup of red wine vinegar. How much vinegar is needed to make $2\frac{1}{2}$ batches of chili?   [4.7d]

Simplify.

**72.** $-\dfrac{49}{77}$   [3.5b]

**73.** $-8 - 5^2 \cdot 2(3 - 4)$   [2.5b]

**74.** $\dfrac{3}{10}\left(-\dfrac{25}{12}\right)$   [3.4b]

## Synthesis

Find three solutions of each equation. Then graph the equation.

**75.** ▦ $21x - 70y = -14$

**76.** ▦ $25x + 80y = 100$

**77.** ▦ $50x + 75y = 180$

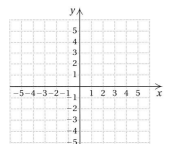

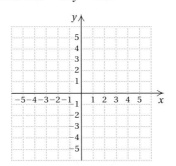

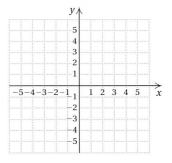

**78.** Use the graph in Example 4 to find three solutions of $2x - y = 5$. Do not use the ordered pairs already listed.

**79.** List all solutions of $x + y = 6$ that use only whole numbers.

**80.** Graph three solutions of $y = |x|$ in the second quadrant and another three solutions in the first quadrant.

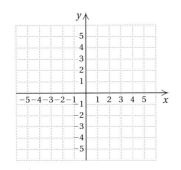

*To the Student and the Instructor*: Exercises marked with a [graphing calculator symbol] symbol are meant to be solved using a graphing calculator.

**81.** [graphing calculator symbol] Use a graphing calculator to graph each of the following.

**(a)** $y = -0.63x + 2.8$

**(b)** $y = 2.3x - 4.1$

# Mid-Chapter Review

## Concept Reinforcement

Determine whether each statement is true or false.

_____ **1.** The key on a pictograph indicates what each symbol represents.   [6.1b]

_____ **2.** The solution of an equation in two variables is written as an ordered pair.   [6.3c]

_____ **3.** The graph of the equation $x = 5$ is a horizontal line.   [6.4c]

## Guided Solutions

Fill in each blank with the number that creates a correct solution.

**4.** Determine whether the ordered pair $(2, -1)$ is a solution of the equation $3x + y = 5$.   [6.3c]

$$
\begin{array}{c|c}
3x + y = 5 & \\
\hline
3 \cdot \square + (\square) \; ? \; 5 & \\
\square + (\square) & \\
\square & 5 \quad \text{TRUE}
\end{array}
$$

Since $3x + y = 5$ becomes true, $(2, -1)$ is a solution.

**5.** Find a solution of $x - y = 6$. Let $x = 1$.   [6.4a]

$$x - y = 6$$
$$\square - y = 6$$
$$-y = \square$$
$$y = \square$$

Thus, $(1, \square)$ is a solution of $x - y = 6$.

## Mixed Review

_Downsizing._   In tight economic times, companies sometimes downsize their products. That is, they charge the same price for a package that contains less product. The following table lists products that have been downsized recently. Use this table for Exercises 6–10.   [6.1a]

| PRODUCT | SIZE (in ounces) OLD | SIZE (in ounces) NEW | PERCENT SMALLER |
|---|---|---|---|
| Breyer's ice cream | 56 | 48 | 14 |
| Hellmann's mayonnaise | 32 | 30 | 6 |
| Hershey's Special Dark chocolate bar | 8 | 6.8 | 15 |
| Iams cat food | 6 | 5.5 | 8 |
| Nabisco Chips Ahoy cookies | 16 | 15.25 | 5 |
| Skippy creamy peanut butter | 18 | 16.3 | 9 |
| Tropicana orange juice | 96 | 89 | 7 |

SOURCE: _Consumer Reports_, October 2008

**6.** How much less ice cream is in the new Breyer's ice cream package than in the old package?

**7.** By what percent has the size of Hellmann's mayonnaise changed in the downsizing process?

**8.** Which product in the table has the greatest percent of decrease?

**9.** How much less orange juice is in the new Tropicana orange juice package than in the old package?

**10.** Which product in the table has the smallest percent of decrease?

In which quadrant is each point located?   [6.3b]

**11.** $(-2, -12)$

**12.** $(-5, 6)$

**13.** $\left(\dfrac{1}{2}, -8\right)$

**14.** Determine whether the ordered pair $(0, -5)$ is a solution of $5x + y = 5$.   [6.3c]

**15.** Find a solution of $y = x + 3$. Let $x = -10$.   [6.4a]

**16.** Complete this solution of $3x - y = 7$: $(-1, \square)$.   [6.4a]

**17.** Find three solutions of $2x + y = 5$. Answers may vary. [6.4a]

Graph each equation.   [6.4b], [6.4c]

**18.** $x + y = 3$

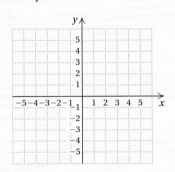

**19.** $y = x - 2$

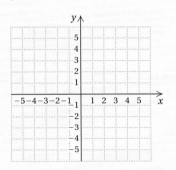

**20.** $y = 4x$

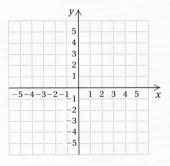

**21.** $y = 2x - 4$

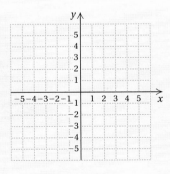

**22.** $y = -3$

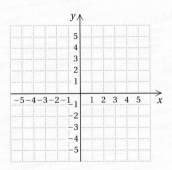

**23.** $x = 1$

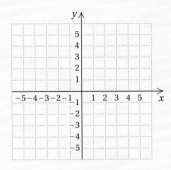

# Understanding Through Discussion and Writing

**24.** Under what conditions will the points $(a, b)$ and $(b, a)$ be in the same quadrant?   [6.3b]

**25.** In which quadrant, if any, is the point $(5, 0)$? Why? [6.3b]

**26.** To graph a linear equation, a student plots three points and discovers that the points do not line up with each other. What should the student do next?   [6.4b]

**27.** What is the greatest number of quadrants that a line can pass through? Why?   [6.3b], [6.4b]

# 6.5

# Means, Medians, and Modes

## OBJECTIVES

**a** Find the mean of a set of numbers and solve applied problems involving means.

**b** Find the median of a set of numbers and solve applied problems involving medians.

**c** Find the mode of a set of numbers and solve applied problems involving modes.

**d** Compare two sets of data using their means.

A **statistic** is a number that is derived from a set of data and used to describe the data. One type of statistic is a *center point* or *measure of central tendency.* Three commonly used center points are the mean, the median, and the mode.

## **a** Means

The most commonly used center point is the *average* of the set of numbers. We have already computed an average in Sections 1.9, 4.8, and 5.4. Although the word "average" is often used in everyday speech, in math we generally use the word *mean* instead.

> **MEAN**
>
> To find the **mean** of a set of numbers, add the numbers and then divide the sum by the number of items of data.

**SKILLS TO REVIEW**
Objectives 1.9c and 5.4b: Simplify expressions using the rules for order of operations.

1. Find the average of 282, 137, 5280, and 193.

2. Find the average of $23.40, $89.15, and $148.17 to the nearest cent.

**EXAMPLE 1** On a 4-day trip, a car was driven the following number of miles each day: 240, 302, 280, 320. What was the mean number of miles per day?

$$\frac{240 + 302 + 280 + 320}{4} = \frac{1142}{4}, \quad \text{or} \quad 285.5$$

The car was driven an average of 285.5 mi per day. Had the car been driven exactly 285.5 mi each day, the same total distance (1142 mi) would have been traveled.

Do Margin Exercises 1–3.

Find the mean.

1. 12, 15, 27    2. 10.5, 9.5, 8.2, 7.2

3. In a five-game series, Antonio scored 26, 21, 13, 14, and 23 points. Find the mean number of points scored per game.

4. **Home-Run Batting Average.** Babe Ruth hit 714 home runs in 22 seasons played in the major leagues. What was his average number of home runs per season? Round to the nearest tenth.

Source: Major League Baseball

**EXAMPLE 2** *Scoring Average.* Michael Jordan is the third highest scorer in the history of the National Basketball Association. He scored 32,292 points in 1072 games. What was the average number of points scored per game? Round to the nearest tenth.

Source: National Basketball Association

We know the total number of points, 32,292, and the number of games, 1072. We divide and round to the nearest tenth:

$$\frac{32,292}{1072} = 30.123134\ldots \approx 30.1.$$

Michael Jordan's average was about 30.1 points per game.

**Answers**

*Skills to Review:*
1. 1473    2. $86.91

*Margin Exercises:*
1. 18    2. 8.85    3. 19.4
4. 32.5 home runs per season

Do Exercise 4.

**EXAMPLE 3** *Grade Point Average.* In most colleges, students are assigned grade point values for grades obtained. The **grade point average**, or **GPA**, is the average of the grade point values for each credit hour taken. At most colleges, grade point values are assigned as follows.

A: 4.0      B: 3.0      C: 2.0      D: 1.0      F: 0.0

Meg earned the following grades for one semester. What was her grade point average?

| COURSE | GRADE | NUMBER OF CREDIT HOURS IN COURSE |
|---|---|---|
| Colonial History | B | 3 |
| Basic Mathematics | A | 4 |
| English Literature | A | 3 |
| French | C | 4 |
| Time Management | D | 1 |

Because some of Meg's courses carried more credit than others, the grades in those courses carry more weight in her GPA. To find the GPA, we first multiply the grade point value by the number of credit hours for the course.

Colonial History      $3.0 \cdot 3 = 9$
Basic Mathematics    $4.0 \cdot 4 = 16$
English Literature    $4.0 \cdot 3 = 12$    Multiplying grade point values (in color) by the number of credits for each course. Meg earned 46 *quality points.*
French               $2.0 \cdot 4 = 8$
Time Management      $1.0 \cdot 1 = \underline{1}$
                                    46 (Total)

The total number of credit hours taken is $3 + 4 + 3 + 4 + 1$, or 15. We divide 46 by 15 and round to the nearest tenth.

$$GPA = \frac{46}{15} \approx 3.1$$

Meg's grade point average was 3.1.

Do Exercise 5.

**5. Grade Point Average.** Alex earned the following grades one semester.

| GRADE | NUMBER OF CREDIT HOURS IN COURSE |
|---|---|
| B | 3 |
| C | 4 |
| C | 4 |
| A | 2 |

What was Alex's grade point average? Assume that the grade point values are 4.0 for an A, 3.0 for a B, and so on. Round to the nearest tenth.

**EXAMPLE 4** *Grading.* To get a B in math, Geraldo must score a mean of 80 on five tests. On the first four tests, his scores were 79, 88, 64, and 78. What is the lowest score that Geraldo can get on the last test and still get a B?

An average of 80 is equivalent to scoring 80 on each test. Thus, Geraldo needs a total score of $5 \cdot 80$, or 400.

The total of the scores on the first four tests is $79 + 88 + 64 + 78 = 309$. Thus, Geraldo needs to get at least $400 - 309$, or 91, in order to get a B. We can check this as follows.

$$\frac{79 + 88 + 64 + 78 + 91}{5} = \frac{400}{5}, \text{ or } 80$$

Do Exercise 6.

**6.** To get an A in math, Rosa must have a mean test grade of at least 90 on four tests. On the first three tests, her scores were 80, 100, and 86. What is the lowest score that Rosa can get on the last test and still get an A?

*Answers*
**5.** 2.5   **6.** 94

## ⓑ Medians

Another measure of central tendency is the *median*. Medians are useful when we wish to de-emphasize unusually extreme numbers. For example, suppose a small class scored as follows on an exam.

| Phil: 78 | Pat: 56 | Matt: 82 |
| Jill: 81 | Olga: 84 | |

Let's first list the scores in order from smallest to largest.

56,  78,  81,  82,  84

↑
Middle score

The middle score—in this case, 81—is called the **median**. Note that because of the extremely low score of 56, the mean of the scores is 76.2. In this example, the median may be a better center point of the scores.

**EXAMPLE 5**   What is the median of this set of numbers?

99,  870,  91,  98,  106,  90,  98

We first rearrange the numbers in order from smallest to largest. Then we locate the middle number, 98.

90,  91,  98,  98,  99,  106,  870

↑
Middle number

The median is 98.

Do Exercises 7–9.

Do Exercises 7–9.

Find the median.

**7.** 17, 13, 18, 14, 19

**8.** 20, 14, 13, 19, 16, 18, 17

**9.** 78, 81, 83, 91, 103, 102, 122, 119, 88

---

**MEDIAN**

Once the data in a set are listed in order, from smallest to largest, the **median** is the middle number if there is an odd number of values. If there is an even number of values, the median is the number that is the average of the two middle numbers.

---

**EXAMPLE 6**   *Salaries.*   The salaries of the six employees (one of whom is the owner) of Top Notch Grill are as follows.

$35,000,  $29,000,  $32,000,  $31,000,  $93,000,  $30,000

What is the median salary at the restaurant?

We rearrange the numbers in order from smallest to largest. The two middle numbers are $31,000 and $32,000. Thus, the median is halfway between $31,000 and $32,000 (the average of $31,000 and $32,000):

$29,000,  $30,000,  $31,000,  $32,000,  $35,000,  $93,000

$$\text{Median} = \frac{\$31{,}000 + \$32{,}000}{2} = \frac{\$63{,}000}{2} = \$31{,}500.$$

The average of the middle numbers is 31,500.

The median salary is $31,500.

Do Exercises 10 and 11.

Find the median.
**10. Salaries of Part-Time Typists.**
$3300, $4000, $3900, $3600, $3800, $3400

**11.** 68, 34, 67, 69, 34, 70

*Answers*
**7.** 17   **8.** 17   **9.** 91
**10.** $3700   **11.** 67.5

In Example 6, the mean salary is $41,666.67, whereas the median salary is $31,500. If you were interviewing for a job at Top Notch Grill, the median would probably give a better indication of what you would likely earn.

## (c) Modes

The final type of center-point statistic is the **mode**.

> **MODE**
>
> The **mode** of a set of data is the number or numbers that occur most often. If each number occurs the same number of times, there is no mode.

**EXAMPLE 7** Find the mode of these data.

    13,  14,  17,  17,  18,  19

The number that occurs most often is 17. Thus, the mode is 17.

A set of data has just one mean and just one median, but it can have more than one mode. It may also have no mode—when all numbers are equally represented. For example, the set of data 5, 7, 11, 13, 19 has no mode.

**EXAMPLE 8** Find the modes of these data.

    33,  34,  34,  34,  35,  36,  37,  37,  37,  38,  39,  40

There are two numbers that occur most often, 34 and 37. Thus, the modes are 34 and 37.

Do Exercises 12–15.

Which statistic is best for a particular situation? If someone is bowling, the *mean* from several games is a good indicator of that person's ability. If someone is buying a home, the *median* price for a neighborhood is often most indicative of what homes sell for there. Finally, if someone is reordering for a clothing store, the *mode* of the sizes sold is probably the most important statistic.

**Calculator Corner**

**Computing Means**  Means can be easily computed on a calculator if we remember the order in which operations are performed. For example, to calculate $\dfrac{85 + 92 + 79}{3}$ on most calculators, we press

    [8][5][+][9][2][+][7][9][=][÷][3][=],

or

    [(][8][5][+][9][2][+][7][9][)][÷][3][=].

The answer is $85.\overline{3}$.

**Exercises:**

1. What would the result have been if we had not used parentheses in the latter sequence of keystrokes?

2. Use a calculator to solve Examples 1–4.

Find any modes that exist.

**12.** 23, 45, 45, 45, 78

**13.** 34, 34, 67, 67, 68, 70

**14.** 13, 24, 27, 28, 67, 89

**15.** In a lab, Gina determined the mass, in grams, of each of five eggs:

    15 g, 19 g, 19 g, 14 g, 18 g.

  **a)** What is the mean?

  **b)** What is the median?

  **c)** What is the mode?

*Answers*

**12.** 45   **13.** 34, 67   **14.** No mode exists.
**15. (a)** 17 g; **(b)** 18 g; **(c)** 19 g

## (d) Comparing Two Sets of Data

We have seen how to calculate means, medians, and modes from data. Sometimes we want to know which of two groups is "better." One way to find out is by comparing the means.

**EXAMPLE 9** *Growth of Wheat.*
A university agriculture department experiments to see which of two kinds of wheat is better. (In this situation, the shorter wheat is considered "better.") The researchers grow both kinds under similar conditions and measure stalk heights, in inches, as follows. Which kind is better?

| WHEAT A STALK HEIGHTS (in inches) | | | | WHEAT B STALK HEIGHTS (in inches) | | | |
|---|---|---|---|---|---|---|---|
| 16.2 | 42.3 | 19.5 | 25.7 | 19.7 | 18.4 | 19.7 | 17.2 |
| 25.6 | 18.0 | 15.6 | 41.7 | 19.7 | 14.6 | 32.0 | 25.7 |
| 22.6 | 26.4 | 18.4 | 12.6 | 14.0 | 21.6 | 42.5 | 32.6 |
| 41.5 | 13.7 | 42.0 | 21.6 | 22.6 | 10.9 | 26.7 | 22.8 |

Note that it is difficult to analyze the data at a glance because the numbers are close together. We need a way to compare the two groups. Let's compute the mean of each set of data.

Wheat A: Average

$$= \frac{\left( \begin{array}{l} 16.2 + 25.6 + 22.6 + 41.5 + 42.3 + 18.0 + 26.4 + 13.7 \\ + 19.5 + 15.6 + 18.4 + 42.0 + 25.7 + 41.7 + 12.6 + 21.6 \end{array} \right)}{16}$$

$$\approx 25.21 \text{ in.}$$

Wheat B: Average

$$= \frac{\left( \begin{array}{l} 19.7 + 19.7 + 14.0 + 22.6 + 18.4 + 14.6 + 21.6 + 10.9 \\ + 19.7 + 32.0 + 42.5 + 26.7 + 17.2 + 25.7 + 32.6 + 22.8 \end{array} \right)}{16}$$

$$\approx 22.54 \text{ in.}$$

We see that the mean stalk height of wheat B is less than that of wheat A. Thus, wheat B is "better."

Do Exercise 16.

**16. Battery Testing.** Two kinds of batteries were tested to see how long, in hours, they kept an emergency light running. On the basis of this test, which battery is better?

| BATTERY A (in hours) | | | |
|---|---|---|---|
| 26.6 | 28.3 | 27.1 | 27.6 |
| 27.6 | 28.0 | 26.8 | 27.4 |
| 28.1 | 28.2 | 26.9 | 27.3 |

| BATTERY B (in hours) | | | |
|---|---|---|---|
| 28.5 | 27.6 | 28.6 | 27.5 |
| 27.7 | 27.9 | 26.9 | 27.8 |
| 28.3 | 27.9 | 28.7 | 27.6 |

*Answer*

**16.** Battery A: average hours ≈ 27.49; battery B: average hours ≈ 27.92; battery B is better

# Translating for Success

1. *Perimeter.* A rectangular garden is 35 ft long and 27 ft wide. How many feet of fencing are needed to surround the garden?

2. *Craft Sale.* Marissa sold crocheted mittens at a craft sale. She sold 35 pairs on Friday and 27 pairs on Saturday. What was the mean number of pairs of mittens sold each day?

3. *Jogging.* David must jog 2 mi a day for training for basketball. If the alley behind his house is $\frac{1}{5}$ mi long, how many times must he run the length of the alley?

4. *Taxi Fare.* A taxi driver charges $2.25 plus 95 cents a mile. How far can Tonya travel on $11.75?

5. *Fudge.* Amber made $3\frac{1}{2}$ lb of fudge and divided it evenly among 3 packages for gifts. How much fudge was in each package?

---

The goal of these matching questions is to practice step (2), *Translate*, of the five-step problem-solving process. Translate each word problem to an equation and select a correct translation from equations A–O.

**A.** $x = 2 \div \frac{1}{5}$

**B.** $x = 4 + (-8)$

**C.** $x = 3 \cdot 3\frac{1}{2}$

**D.** $x = 4 - (-8)$

**E.** $35 + 27 = x$

**F.** $x = 2 \cdot 35 + 2 \cdot 27$

**G.** $x = 3\frac{1}{2} \div 3$

**H.** $x = 248 \div 15\frac{5}{10}$

**I.** $x = 3 + 3\frac{1}{2}$

**J.** $x = 2 \cdot \frac{1}{5}$

**K.** $35 \cdot 27 = x$

**L.** $2.25 + 0.95 + x = 11.75$

**M.** $248 \div x = 15\frac{5}{10}$

**N.** $x = \dfrac{35 + 27}{2}$

**O.** $2.25 + 0.95x = 11.75$

*Answers on page A-13*

---

6. *Temperature.* The weather forecast for the low temperature in Muncy on December 14 was 4°F. The actual low was −8°F. How many degrees lower than the forecast was the actual temperature?

7. *Pizza Recipe.* A recipe for pizza crust calls for $3\frac{1}{2}$ cups of flour. How many cups of flour are needed for 3 pizza crusts?

8. *Total Purchase.* Oliver spent $11.75 at the Country Store. He purchased candy for 95 cents, nuts for $2.25, and coffee. How much did the coffee cost?

9. *Mileage.* A car traveled 248 mi on $15\frac{5}{10}$ gal of gas. How many miles per gallon did it get?

10. *Cell Phone Minutes.* On Friday, Lathan made phone calls of lengths 35 min and 27 min. How many minutes did he use?

**a**, **b**, **c**   For each set of numbers, find the mean, the median, and any modes that exist.

**1.** 17, 19, 29, 18, 14, 29

**2.** 72, 83, 85, 88, 92

**3.** 5, 37, 20, 20, 35, 5, 25

**4.** 13, 32, 25, 27, 13

**5.** 4.3, 7.4, 1.2, 5.7, 8.3

**6.** 13.4, 13.4, 12.6, 42.9

**7.** 234, 228, 234, 229, 234, 278

**8.** $29.95, $28.79, $30.95, $29.95

**9.** *Atlantic Storms and Hurricanes.*   The following bar graph shows the number of Atlantic storms and hurricanes that formed in various months from 1980 to 2009. What is the mean number for the 9 months given? the median? the mode?

**Atlantic Storms and Hurricanes**

Tropical storm and hurricane formation in 1980–2009, by month

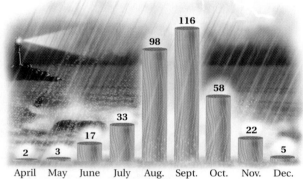

SOURCE: Colorado State University, Department of Atmospheric Science, Phil Klotzbach, Ph.D., Research Scientist, and the National Hurricane Center

**10.** *Tornadoes.*   The following bar graph shows the average number of tornado deaths by month since 1950. What is the mean number of tornado deaths for the 12 months? the median? the mode?

Average Number of Deaths by Tornado by Month

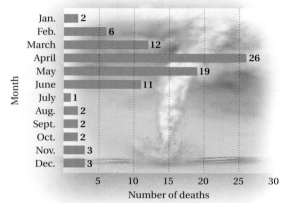

SOURCE: National Weather Service's Storm Prediction Center

**11.** *Gas Mileage.*   The 2009 Kia Sedona gets 253 mi of highway driving on 11 gal of gasoline. What is the average number of miles expected per gallon—that is, what is its gas mileage?
Source: *Motor Trend*, November 2008

**12.** *Gas Mileage.*   The 2009 Chevrolet Cobalt SS gets 330 mi of city driving on 15 gal of gasoline. What is the average number of miles expected per gallon—that is, what is its gas mileage?
Source: *Road & Track*, November 2008

**13.** *Grading.*   To get a B in math, Rich must average at least 80 on five tests. Scores on his first four tests were 80, 74, 81, and 75. What is the lowest score that Rich can get on the last test and still receive a B?

**14.** *Grading.*   To get an A in math, Sybil must average at least 90 on five tests. Scores on her first four tests were 90, 91, 81, and 92. What is the lowest score that Sybil can get on the last test and still receive an A?

**15.** *Length of Pregnancy.*   Marta was pregnant 270 days, 259 days, and 272 days for her first three pregnancies. After her fourth pregnancy, Marta's average pregnancy was exactly the worldwide average of 266 days. How long was her fourth pregnancy?
Source: David Crystal (ed.). *The Cambridge Factfinder* (Cambridge: Cambridge University Press, 2000). p. 90

**16.** *Male Height.*   Jason's brothers are 174 cm, 180 cm, 179 cm, and 172 cm tall. The average male is 176.5 cm tall. How tall is Jason if he and his brothers have an average height of 176.5 cm?

*Grade Point Average.* The tables in Exercises 17 and 18 show the grades of a student for one semester. In each case, find the grade point average. Assume that the grade point values are 4.0 for an A, 3.0 for a B, and so on. Round to the nearest tenth.

**17.**

| COURSE | GRADE | NUMBER OF CREDIT HOURS IN COURSE |
|--------|-------|----------------------------------|
| Chemistry | B | 4 |
| Prealgebra | A | 5 |
| French I | D | 3 |
| Pastels | C | 4 |

**18.**

| COURSE | GRADE | NUMBER OF CREDIT HOURS IN COURSE |
|--------|-------|----------------------------------|
| Botany | A | 5 |
| U.S. History | C | 4 |
| Drawing I | F | 3 |
| Basic Math | B | 5 |

**19.** *Brussels Sprouts.* The following prices per stalk of Brussels sprouts were found at five supermarkets:

$3.99, $4.49, $4.99, $3.99, $3.49.

What was the mean price per stalk? the median price? the mode?

**20.** *Cheddar Cheese Prices.* The following prices per pound of sharp cheddar cheese were found at five supermarkets:

$5.99, $6.79, $5.99, $6.99, $6.79.

What was the mean price per pound? the median price? the mode?

**d** Solve.

**21.** *Light-Bulb Testing.* An experiment was performed to compare the lives of two types of light bulb. Several bulbs of each type were tested and the results are listed in the following table. On the basis of this test, which bulb is better?

| BULB A: HOTLIGHT TIMES (in hours) | | | BULB B: BRIGHTBULB TIMES (in hours) | | |
|------|------|------|------|------|------|
| 983 | 964 | 1214 | 979 | 1083 | 1344 |
| 1417 | 1211 | 1521 | 984 | 1445 | 975 |
| 1084 | 1075 | 892 | 1492 | 1325 | 1283 |
| 1423 | 949 | 1322 | 1325 | 1352 | 1432 |

**22.** *Cola Testing.* An experiment was conducted to determine which of two colas tastes better. Students drank each cola and gave it a rating from 1 to 10, with 10 indicating the best taste. The results are given in the following table. On the basis of this test, which cola tastes better?

| COLA A: VERVCOLA | | | | COLA B: COLA–COLA | | | |
|----|----|----|----|----|----|----|----|
| 6 | 8 | 10 | 7 | 10 | 9 | 9 | 6 |
| 7 | 9 | 9 | 8 | 8 | 8 | 10 | 7 |
| 5 | 10 | 9 | 10 | 8 | 7 | 4 | 3 |
| 9 | 4 | 7 | 6 | 7 | 8 | 10 | 9 |

## Skill Maintenance

Multiply.

**23.** $14 \cdot 14$   [1.4a]

**24.** $\frac{2}{3} \cdot \frac{2}{3}$   [3.4b]

**25.** $1.4 \times 1.4$   [5.3a]

**26.** $1.414 \times 1.414$   [5.3a]

**27.** $12.86 \times 17.5$   [5.3a]

**28.** $222 \times 0.5678$   [5.3a]

**29.** $\frac{4}{5} \cdot \frac{3}{28}$   [3.6a]

**30.** $\frac{28}{45} \cdot \frac{3}{2}$   [3.6a]

Solve.   [5.8b]

**31.** A disc jockey charges a $40 setup fee and $50 an hour. How long can the disc jockey work for $165?

**32.** To rent a floor sander costs $15 an hour plus a $10 supply fee. For how long can the machine be rented if $100 has been budgeted for the sander?

## Synthesis

*Bowling Averages.*   Bowling averages are always computed by rounding down to the nearest integer. For example, suppose a bowler gets a total of 599 for 3 games. To find the average, we divide 599 by 3 and drop the amount to the right of the decimal point.

$$\frac{599}{3} \approx 199.67 \qquad \text{The bowler's average is 199.}$$

**33.** 🖩 If Frances bowls 4176 in 23 games, what is her average?

**34.** 🖩 If Eric bowls 4621 in 27 games, what is his average?

**35.** *Hank Aaron.*   Hank Aaron averaged $34\frac{7}{22}$ home runs per year over a 22-yr career. After 21 yr, Aaron had averaged $35\frac{10}{21}$ home runs per year. How many home runs did Aaron hit in his final year?

**36.** Because of a poor grade on the fifth and final test, Chris's mean test grade fell from 90.5 to 84.0. What did Chris score on the fifth test? Assume that all tests are equally important.

**37.** *Price Negotiations.*   Amy offers $3200 for a used Ford Taurus advertised at $4000. The first offer from Jim, the car's owner, is to "split the difference" and sell the car for $(3200 + 4000) \div 2$, or $3600. Amy's second offer is to split the difference between Jim's offer and her first offer. Jim's second offer is to split the difference between Amy's second offer and his first offer. If this pattern continues and Amy accepts Jim's third (and final) offer, how much will she pay for the car?

**38.** The ordered set of data 18, 21, 24, $a$, 36, 37, $b$ has a median of 30 and a mean of 32. Find $a$ and $b$.

**39.** *Gas Mileage.*   The Honda Insight, a gas/electric hybrid car, averages 61 mpg in city driving and has a 10.5-gal gas tank. How much city driving can be done on $\frac{3}{4}$ of a tank of gas?

**Sources:** Based on information from EPA and Honda Motors

**40.** 🖩 After bowling 15 games, Liz had an average of 207. In her 16th game, Liz bowled a 244 and raised her average to 210. Andrew also had a 207 average after 15 games, but needed to bowl a 255 in his 16th game in order to raise his average to 210. How is this possible?

## (a) Making Predictions

Sometimes we use data to make predictions or estimates of missing data points. One process for doing so is called **interpolation**. Interpolation enables us to estimate missing "in-between values" on the basis of known information.

**EXAMPLE 1** *Monthly Mortgage Payments.* When money is borrowed and then repaid in monthly installments, the payment amount increases as the total number of payments decreases. The table below lists the size of a monthly payment when $110,000 is borrowed (at 9% interest) for various lengths of time. Use interpolation to estimate the monthly payment on a 35-yr loan.

| YEAR | MONTHLY PAYMENT |
|------|-----------------|
| 5    | $2283.42        |
| 10   | 1393.43         |
| 15   | 1115.69         |
| 20   | 989.70          |
| 25   | 923.12          |
| 30   | 885.08          |
| 35   | ?               |
| 40   | 848.50          |

To use interpolation, we first plot the points and look for a trend. It seems reasonable to draw a line between the points corresponding to 30 and 40. We can "zoom-in" to better visualize the situation. To estimate the second coordinate that is paired with 35, we trace a vertical line up from 35 to the graph and then left to the vertical axis. Thus, we estimate the value to be 867. We can also estimate this value by averaging $885.08 and $848.50.

$$\frac{\$885.08 + \$848.50}{2} = \$866.79$$

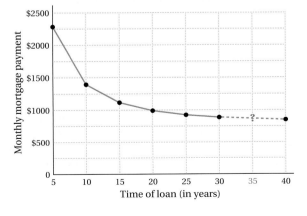

**$110,000 Loan Repayment**

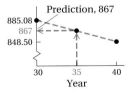

## OBJECTIVES

(a) Make predictions from a set of data using interpolation or extrapolation.

(b) Calculate the probability of an event occurring.

**SKILL TO REVIEW**
Objective 5.5a: Use division to convert fraction notation to decimal notation.

Convert to decimal notation.

**1.** $\dfrac{4}{5}$     **2.** $\dfrac{2}{25}$

## STUDY TIPS

### IF YOU MISS A CLASS

You can get information about a missed class from a classmate or from your instructor. In addition to finding out the assignment, try to determine what problems were reviewed, what new concepts were discussed, and what announcements (regarding quizzes, hand-in assignments, schedule of class meetings, etc.) you may have missed. Your instructor may make this information available online.

*Answers*

*Skill to Review:*
**1.** 0.8     **2.** 0.08

When we estimate in this manner to find an in-between value, we are *interpolating*. Real-world information about the data might tell us that an estimate found in this way is unreliable. For example, data from the stock market might be too erratic for interpolation.

Do Exercise 1.

We often analyze data with the intention of going "beyond" the data. One process for doing so is called **extrapolation**.

**EXAMPLE 2** *Movies Released.* The data in the following table and graphs show the number of movies released over a period of years. Use extrapolation to estimate the number of movies released in 2010.

| YEAR | MOVIES RELEASED |
|------|------|
| 2006 | 808 |
| 2007 | 1018 |
| 2008 | 1044 |
| 2009 | 1160 |

SOURCE: www.the-numbers.com

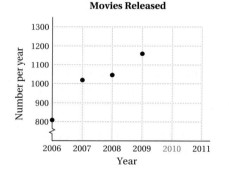

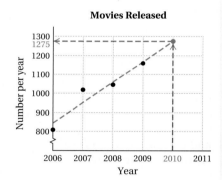

First, we analyze the data and note that they increase from 2006 through 2009. Then we draw a "representative" line through the data and beyond. To estimate a value for 2010, we draw a vertical line up from 2010 until it hits the representative line. We go to the left and read off a value—about 1275. When we estimate in this way to find a "go-beyond value," we are *extrapolating*. Estimates found with this method vary depending on the "representative" line chosen.

Do Exercise 2.

In calculus and statistics, other methods of interpolation and extrapolation are developed. The two basic concepts remain unchanged, but more complicated methods of determining what line "best fits" the given data are used. These methods often involve use of a graphing calculator or computer software.

*Answers*
**1.** About 1.635 Mw  **2.** 94%

## b  Probability

The predictions made in Examples 1 and 2 have a good chance of being reasonably accurate. Using *probability*, we can attach a numerical value to the likelihood that a specific event will occur.

Suppose we were to flip a fair coin. Because the coin is just as likely to land heads as it is to land tails, we say that the *probability* of it landing heads is $\frac{1}{2}$. Similarly, if we roll a fair die (plural: dice), we are as likely to roll a 🎲 as we are to roll a 🎲, 🎲, 🎲, 🎲, or 🎲. Because of this, we say that the probability of rolling a 🎲 is $\frac{1}{6}$.

**EXAMPLE 3**  A die is about to be rolled. Find the probability that a number greater than 4 will be rolled.

Since a 🎲, 🎲, 🎲, 🎲, 🎲, and 🎲 are all equally likely to be rolled, and since two of these possibilities involve numbers greater than 4, we have

$$\begin{array}{l}\text{The probability of rolling} \\ \text{a number greater than 4}\end{array} = \dfrac{2}{6} \begin{array}{l}\leftarrow \text{ Number of ways to roll a 5 or 6} \\ \leftarrow \text{ Number of (equally likely) possible} \\ \quad\text{ outcomes}\end{array}$$

$$= \frac{1}{3}.$$

A coin landing heads and rolling a 🎲 are examples of "events." To find the probability, or likelihood, of an event occurring, we use the following principle.

---

### THE PRIMARY PRINCIPLE OF PROBABILITY

If an event $E$ can occur $m$ ways out of $n$ equally likely possible outcomes, then

$$\text{The probability of } E \text{ occurring } = \frac{m}{n}.$$

---

**EXAMPLE 4**  A spinner such as the one shown below is used by flicking the arrow so that it rotates quickly. Find the probability that the arrow will come to rest on a green space. (Assume that the arrow will not rest on a line.)

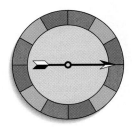

There are 12 sections of equal size, so there are 12 equally likely possible outcomes. Since 4 of the possibilities are green, we have

$$\begin{array}{l}\text{Probability of} \\ \text{landing on green}\end{array} = \dfrac{\text{Number of ways to land on green}}{\text{Number of ways to land on a space}}$$

$$= \frac{4}{12} = \frac{1}{3}.$$

Do Exercise 3.

3. Find the probability of the arrow in Example 4 landing on red or blue.

*Answer*

3. $\frac{2}{3}$

**EXAMPLE 5**  A cloth bag contains 20 equally sized marbles: 5 are red, 7 are blue, and 8 are yellow. A marble is randomly selected. Find the probability that (**a**) a red marble is selected; (**b**) a blue marble is selected; (**c**) a yellow marble is selected.

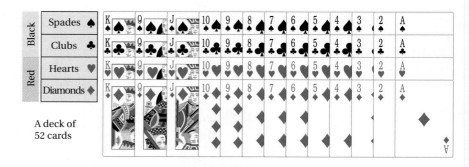

a)  Since all 20 marbles are equally likely to be selected, we have

$$\text{The probability of selecting a red marble} = \frac{\text{Number of ways to select a red marble}}{\text{Number of ways to select any marble}}$$

$$= \frac{5}{20} = \frac{1}{4}, \text{ or } 0.25$$

b)  $$\text{The probability of selecting a blue marble} = \frac{\text{Number of ways to select a blue marble}}{\text{Number of ways to select any marble}}$$

$$= \frac{7}{20}, \text{ or } 0.35$$

c)  $$\text{The probability of selecting a yellow marble} = \frac{\text{Number of ways to select a yellow marble}}{\text{Number of ways to select any marble}}$$

$$= \frac{8}{20} = \frac{2}{5}, \text{ or } 0.4$$

> Do Exercise 4.

**4.** A school play is attended by 500 people: 250 children, 100 seniors, and 150 (nonsenior) adults. After everyone has been seated, one audience member is selected at random. Find the probability of each of the following.

**a)** A child is selected.

**b)** A senior is selected.

**c)** A (nonsenior) adult is selected.

Many probability problems involve a standard deck of 52 playing cards. Such a deck is made up as shown below.

| | | K | Q | J | 10 | 9 | 8 | 7 | 6 | 5 | 4 | 3 | 2 | A |
|---|---|---|---|---|---|---|---|---|---|---|---|---|---|---|
| Black | Spades ♠ | K | Q | J | 10 | 9 | 8 | 7 | 6 | 5 | 4 | 3 | 2 | A |
| | Clubs ♣ | K | Q | J | 10 | 9 | 8 | 7 | 6 | 5 | 4 | 3 | 2 | A |
| Red | Hearts ♥ | K | Q | J | 10 | 9 | 8 | 7 | 6 | 5 | 4 | 3 | 2 | A |
| | Diamonds ♦ | K | Q | J | 10 | 9 | 8 | 7 | 6 | 5 | 4 | 3 | 2 | A |

A deck of 52 cards

**EXAMPLE 6**  A card is randomly selected from a well-shuffled (mixed) deck of cards. Find the probability that (**a**) the card is a jack; (**b**) the card is a club.

a)  $$\text{The probability of selecting a jack} = \frac{\text{Number of ways to select a jack}}{\text{Number of ways to select any card}}$$

$$= \frac{4}{52} = \frac{1}{13}$$

b)  $$\text{The probability of selecting a club} = \frac{\text{Number of ways to select a club}}{\text{Number of ways to select any card}}$$

$$= \frac{13}{52} = \frac{1}{4}$$

**5.** A card is randomly selected from a well-shuffled deck of cards. Find the probability of each of the following.

**a)** The card is a diamond.

**b)** The card is a king or queen.

> Do Exercise 5.

*Answers*

**4.** (a) $\frac{1}{2}$, or 0.5; (b) $\frac{1}{5}$, or 0.2;

(c) $\frac{3}{10}$, or 0.3    **5.** (a) $\frac{1}{4}$, or 0.25; (b) $\frac{2}{13}$

## 6.6

# Exercise Set

**For Extra Help**

**MyMathLab**

Math XL
PRACTICE

WATCH

DOWNLOAD

READ

REVIEW

**a**    Use interpolation or extrapolation to find the missing data values.

**1.** *Study Time and Grades.* A math instructor asked her students to keep track of how much time each spent studying the chapter on decimal notation. They collected the information together with test scores from that chapter's test. The data are given in the following table. Estimate the missing value.

| STUDY TIME (in hours) | TEST GRADE |
|---|---|
| 13 | 80 |
| 15 | 85 |
| 17 | 80 |
| 19 | ? |
| 21 | 86 |
| 23 | 91 |

**2.** *Maximum Heart Rate.* A person's maximum heart rate depends on his or her gender, age, and resting heart rate. The following table relates resting heart rate and maximum heart rate for a 30-yr-old woman. Estimate the missing value.

| RESTING HEART RATE (in beats per minute) | MAXIMUM HEART RATE (in beats per minute) |
|---|---|
| 58 | 173 |
| 65 | 178 |
| 70 | ? |
| 78 | 185 |
| 85 | 188 |

SOURCE: American Heart Association

**3.** *Movie Rental.*

| YEAR | NUMBER OF REDBOX DVD KIOSKS (in thousands) |
|---|---|
| 2006 | 2 |
| 2007 | 6 |
| 2008 | 13 |
| 2009 | 20 |
| 2010 | ? |

SOURCE: Redbox

**4.** *National Park Visits.*

| YEAR | VISITS TO NATIONAL PARKS (in millions) |
|---|---|
| 2005 | 63.5 |
| 2006 | 60.4 |
| 2007 | 62.3 |
| 2008 | 61.2 |
| 2009 | ? |

SOURCE: U.S. National Park Service

**5.** *Major League Baseball Salaries.*

| YEAR | AVERAGE SALARY OF MAJOR LEAGUE BASEBALL PLAYERS (in millions of dollars) |
|---|---|
| 2004 | 2.3 |
| 2005 | 2.5 |
| 2006 | 2.7 |
| 2007 | 2.8 |
| 2008 | 2.9 |
| 2009 | 3.0 |
| 2010 | ? |

SOURCE: Major League Baseball Players Association

**6.** *Cruise Ships.*

| YEAR | NUMBER OF U.S. CRUISE SHIPS |
|---|---|
| 2003 | 134 |
| 2004 | 144 |
| 2005 | 145 |
| 2006 | 151 |
| 2007 | 159 |
| 2008 | ? |

SOURCE: Business Research and Economic Advisors

**7.** *Veterans Benefits.*

| YEAR | U.S. VETERANS BENEFITS (in billions of dollars) |
|------|-------------------------------------------------|
| 2003 | 32 |
| 2004 | 35 |
| 2005 | 37 |
| 2006 | 39 |
| 2007 | ? |
| 2008 | 45 |

SOURCE: *Statistical Abstract of the United States*, 2010.

**8.** *Alternative-Fueled Vehicles.*

| YEAR | ALTERNATIVE-FUELED VEHICLES IN THE U.S. (in thousands) |
|------|--------------------------------------------------------|
| 2003 | 534 |
| 2004 | 565 |
| 2005 | ? |
| 2006 | 635 |
| 2007 | 696 |

SOURCE: U.S. Department of Transportation

**b**    Find each of the following probabilities.

*Rolling a Die.*    In Exercises 9–12, assume that a die is about to be rolled.

**9.** Find the probability that a 🎲 is rolled.

**10.** Find the probability that a 🎲 is rolled.

**11.** Find the probability that an odd number is rolled.

**12.** Find the probability that a number greater than 2 is rolled.

*Playing Cards.*    In Exercises 13–18, assume that one card is randomly selected from a well-shuffled deck (see p. 442).

**13.** Find the probability that the card is the jack of spades.

**14.** Find the probability that the card is a picture card (jack, queen, or king).

**15.** Find the probability that an 8 or a 6 is selected.

**16.** Find the probability that a black 5 is selected.

**17.** Find the probability that a red picture card (jack, queen, or king) is selected.

**18.** Find the probability that a 10 is selected.

*Candy Colors.*    A box of candy was opened by the authors and found to contain the following number of gumdrops.

| | |
|------------|---|
| Strawberry | 7 |
| Lemon | 8 |
| Orange | 9 |
| Cherry | 4 |
| Lime | 5 |
| Grape | 6 |

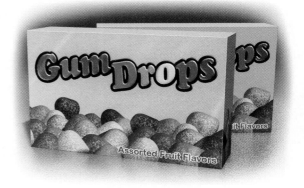

In Exercises 19–22, assume that one of the gumdrops is randomly chosen from the box.

**19.** Find the probability that a cherry gumdrop is selected.

**20.** Find the probability that an orange gumdrop is selected.

**21.** Find the probability that the gumdrop is *not* lime.

**22.** Find the probability that the gumdrop is *not* lemon.

## Skill Maintenance

In each of Exercises 23–30, fill in the blank with the correct term from the given list. Some of the choices may not be used and some may be used more than once.

23. The set $\{1, 2, 3, 4, 5, \ldots\}$ is called the set of _____ numbers.   [1.1b]

24. The number $\frac{2}{5}$ is written in _____ notation, and the equivalent number 0.4 is written in _____ notation.   [3.3a], [5.1a]

25. To find the _____ of a set of numbers, add the numbers and then divide by the number of items of data.   [6.5a]

26. The decimal $0.\overline{1518}$ is an example of a(n) _____ decimal.   [5.5a]

27. Values in between known values can be estimated using _____.   [6.6a]

28. The statement $a(b + c) = ab + ac$ illustrates the _____ law.   [1.4a]

29. The statement $x + t = t + x$ illustrates the _____ law of addition.   [1.2a]

30. Two perpendicular number lines used to graph ordered pairs are called _____.   [6.3a]

decimal

fraction

terminating

repeating

axes

quadrants

commutative

associative

distributive

mean

median

mode

natural

whole

interpolation

extrapolation

## Synthesis

31. A coin is flipped twice. What is the probability that two heads will occur?

32. A coin is flipped twice. What is the probability that one head and one tail will occur?

33. A die is rolled twice. What is the probability that a ⚁ is rolled twice?

34. A day is chosen randomly during a leap year. What is the probability that the day is in July?

35. A die is rolled. What is the probability that a 7 is rolled?

36. A die is rolled. What is the probability that a ⚀, ⚁, ⚂, ⚃, ⚄, or ⚅ is rolled?

# Summary and Review

## Key Terms

table, p. 392
pictograph, p. 394
bar graph, p. 399
line graph, p. 402
axes, p. 410
coordinate, p. 410

ordered pair, p. 410
origin, p. 410
quadrants, p. 412
linear equation, p. 418
statistic, p. 430
mean, p. 430

grade point average, p. 431
median, p. 432
mode, p. 433
interpolation, p. 439
extrapolation, p. 440

## Concept Reinforcement

Determine whether each statement is true or false.

_____ 1. To find the mean of a set of numbers, add the numbers and then multiply by the number of items of data.   [6.5a]

_____ 2. If each number in a set of data occurs the same number of times, there is no mode.   [6.5c]

_____ 3. If there is an odd number of items in an ordered set of data, the middle number is the median.   [6.5b]

## Important Concepts

**Objective 6.1a**   Extract and interpret data from tables.

**Example**   Which oatmeal listed below has the greatest number of calories?

| PRODUCT | Cost | Calories | Fat (g) | Fiber (g) | Sugars (g) |
|---|---|---|---|---|---|
| Quaker Quick-1 Minute | 0.19 | 150 | 3.0 | 4 | 1 |
| Kashi Heart to Heart Golden Brown Maple | 0.44 | 160 | 2.0 | 5 | 12 |
| Quaker Organic Maple & Brown Sugar | 0.54 | 150 | 2.0 | 3 | 12 |
| Nature's Path Organic Maple Nut | 0.47 | 200 | 4.0 | 4 | 12 |

Header note: PER PACKET (instant) OR SERVING (longer–cooking)

SOURCE: *Consumer Reports*, November 2008

We look down the column headed "Calories" and find the largest number, 200. The name of the oatmeal in that row is Nature's Path Organic Maple Nut.

**Practice Exercises**

1. Which oatmeal has the greatest cost per serving? What is  that cost?

2. How many grams of sugar are in the Kashi oatmeal?

**Objective 6.2a**   Extract and interpret data from bar graphs.

**Example**   The horizontal bar graph below shows the building costs of selected stadiums. Estimate how much more Lucas Oil Stadium cost than Invesco Field cost.

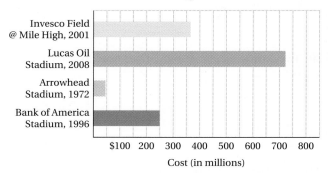

**Cost of Sports Stadiums**

SOURCE: National Football League

We estimate each cost.
*Lucas Oil Stadium:* $720 million; *Invesco Field:* $360 million
*Difference:* $720 million − $360 million = $360 million

**Practice Exercises**

3. Which stadium cost less than $100 million?

4. Estimate how much more Invesco Field cost than Bank of America Stadium cost.

**Objective 6.3c**   Determine whether an ordered pair is a solution of an equation with two variables.

**Example**   Determine whether the ordered pair $(-2, 1)$ is a solution of the equation $x + 5y = 3$.

$$\frac{x + 5y = 3}{-2 + 5 \cdot 1 \overset{?}{\,} 3}$$
$$\begin{array}{c|c} -2 + 5 & \\ 3 & 3 \quad \text{TRUE} \end{array}$$

Since the equation becomes true for $x = -2$ and $y = 1$, $(-2, 1)$ is a solution.

**Practice Exercise**

5. Determine whether the ordered pair $(2, -4)$ is a solution of the equation $3x + y = 10$.

---

**Objective 6.4b**   Graph linear equations in two variables.

**Example**   Graph: $y = 2x - 3$.

We make a table of solutions. Then we plot the points, draw the line with a ruler, and label it.

| $x$ | $y$ | $(x, y)$ |
|-----|-----|----------|
| 0 | $-3$ | $(0, -3)$ |
| 1 | $-1$ | $(1, -1)$ |
| 2 | 1 | $(2, 1)$ |
| 3 | 3 | $(3, 3)$ |

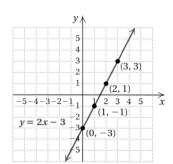

**Practice Exercise**

6. Graph: $y = x - 4$.

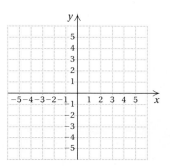

---

**Objectives 6.5a, b, c**   Find the mean, the median, and the mode of a set of numbers.

**Example**   Find the mean, the median, and the mode of this set of numbers:

2.6,   3.5,   61.8,   10.4,   3.5,   21.6,   10.4,   3.5.

*Mean:* We add the numbers and divide by the number of data items:

$$\frac{2.6 + 3.5 + 61.8 + 10.4 + 3.5 + 21.6 + 10.4 + 3.5}{8}$$

$$= \frac{117.3}{8} = 14.6625.$$

The *mean* is 14.6625.

*Median:* We first rearrange the numbers from the smallest to the largest:

2.6,   3.5,   3.5,   3.5,   10.4,   10.4,   21.6,   61.8.

There is an even number of numbers. The median is halfway between the middle two. The average of the middle numbers is

$$\frac{3.5 + 10.4}{2} = \frac{13.9}{2} = 6.95.$$

The *median* is 6.95.

*Mode:* The number that occurs most often is 3.5. The *mode* is 3.5.

**Practice Exercise**

7. Find the mean, the median, and the mode of this set of numbers:

8,   13,   1,   4,   8,   7,   15.

**Objective 6.6b** Calculate the probability of an event occurring.

**Example**  A company randomly selects a month of the year for an annual party. What is the probability that they choose a month whose name begins with J?

There are 12 months in a year, so there are 12 equally likely possible outcomes. There are 3 months whose names begin with J: January, June, and July. We have

$$\text{The probability that the name of the month begins with J} = \frac{\text{Number of months whose names begin with J}}{\text{Number of months in the year}}$$

$$= \frac{3}{12}$$

$$= \frac{1}{4}.$$

**Practice Exercise**

8. A month of the year is randomly selected. What is the probability that the name of the month begins with A?

# Review Exercises

*UPS Mailing Costs.* The table below lists the charges for three types of UPS delivery service of packages of various weights from zip code 46143 to zip code 60614. Use this table for Exercises 1–3.  [6.1a]

| UPS PACKAGE | UPS GROUND | UPS NEXT DAY SAVER DELIVERY BY END OF DAY | UPS NEXT DAY AIR DELIVERY BY 10:30 A.M. |
|---|---|---|---|
| 1 lb | $ 9.13 | $21.29 | $24.30 |
| 2 | 9.25 | 22.90 | 25.21 |
| 3 | 9.32 | 24.03 | 28.06 |
| 4 | 9.52 | 25.75 | 30.05 |
| 5 | 9.88 | 26.61 | 30.37 |
| 6 | 10.11 | 27.79 | 32.79 |
| 7 | 10.52 | 28.17 | 34.24 |
| 8 | 10.85 | 29.51 | 34.78 |

SOURCE: United Parcel Service

1. Find the cost of a 5-lb UPS Next Day Air delivery by 10:30 A.M.

2. Find the cost of a 3-lb UPS Ground delivery.

3. How much would you save by sending the package in Exercise 1 by UPS Next Day Saver delivery?

*PGA Champions by Age.* The pictograph below shows the number of Professional Golf Association (PGA) champions by age. Use this graph for Exercises 4–6.  [6.1b]

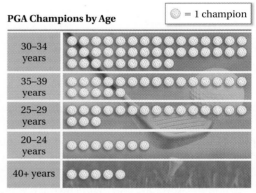

PGA Champions by Age      = 1 champion

SOURCE: Professional Golf Association

4. How many PGA champions were 25–29 years old?

5. Of the age categories listed, which one has the most champions?

6. How many more champions were 30–34 years old than 35–39 years old?

*Melting Snow Runoff.* The amount of snow that accumulates during the winter in California is vital to the state's water supply. The bar graph below shows the state's unimpaired runoff as a percentage of normal for 2005 – 2010. Use this graph for Exercises 7–10. [6.2a]

**California's Snow Runoff**

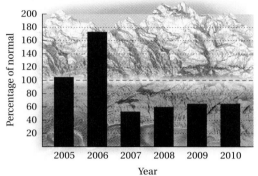

SOURCE: California Department of Water Resources

**7.** Which years had less than normal runoff?

**8.** What was the percentage of runoff in 2008?

**9.** What was the difference in the runoffs for 2005 and 2007?

**10.** What year had over 170% runoff?

*National Park Visits.* Kobuk Valley National Park in Alaska is the least-visited park in the U.S. National Park System. This is largely due to the fact that there are no roads to the park. The line graph below shows the number of visitors to the park for 2002–2009. Use this graph for Exercises 11–15. [6.2c]

**Visitors to Kobuk Valley**

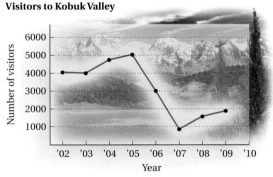

SOURCE: National Park Service

**11.** In which year were there the most visitors to Kobuk Valley?

**12.** In which year were there the fewest visitors to Kobuk Valley?

**13.** In which years were there fewer than 2000 visitors to Kobuk Valley?

**14.** In which year were there 3000 visitors to Kobuk Valley?

**15.** How many more visitors were there to Kobuk Valley in 2005 than there were in 2007?

*First-Class Postage.* The table below lists the cost of first-class postage in various years. Use the table for Exercises 16 and 17.

| YEAR | FIRST-CLASS POSTAGE |
|------|---------------------|
| 1995 | 32¢ |
| 1999 | 33 |
| 2001 | 34 |
| 2002 | 37 |
| 2006 | 39 |
| 2007 | 41 |
| 2008 | 42 |
| 2009 | 44 |

SOURCE: U.S. Postal Service

**16.** Make a vertical bar graph of the data. [6.2b]

**17.** Make a line graph of the data. [6.2d]

Determine the coordinates for each point. [6.3a]

**18.** *A*

**19.** *B*

**20.** *C*

**21.** *D*

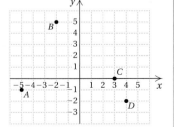

Plot each point on the graph below.    [6.3a]

**22.** $(2, 5)$        **23.** $(0, -3)$        **24.** $(-4, -2)$

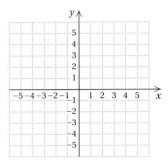

In which quadrant is each point located?    [6.3b]

**25.** $(3, -8)$        **26.** $(-20, -14)$        **27.** $\left( 4\dfrac{9}{10}, 1\dfrac{3}{10} \right)$

**28.** Complete these solutions of $2x + 4y = 10$:
$(1, \square); (\square, -2)$.    [6.4a]

Graph on a plane.    [6.4b, c]

**29.** $y = 2x - 5$

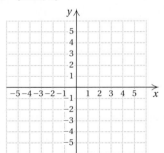

**30.** $y = -\dfrac{3}{4}x$

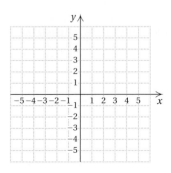

**31.** $x + y = 4$

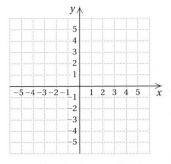

**32.** $x = -5$

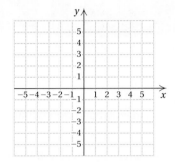

**33.** $y = 6$

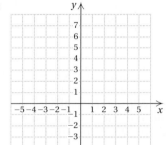

**34.** To get an A in math, Naomi must average at least 90 on four tests. Her scores on the first three tests were 94, 78, and 92. What is the lowest score she can receive on the last test and still get an A?    [6.5a]

**35.** *Gas Mileage.*    A 2009 Pontiac Solstice GXP gets 532 mi of highway driving on 19 gal of gasoline. What is the gas mileage?    [6.5a]
**Source:** *Car and Driver,* December 2008

Find the mean.  [6.5a]

**36.** 26, 34, 43, 51

**37.** 11, 14, 17, 18, 7

**38.** 0.2, 1.7, 1.9, 2.4

**39.** 700, 2700, 3000, 900, 1900

**40.** $2, $14, $17, $17, $21, $29

**41.** 20, 190, 280, 470, 470, 500

Find the median.  [6.5b]

**42.** 26, 34, 43, 51

**43.** 7, 11, 14, 17, 18

**44.** 0.2, 1.7, 1.9, 2.4

**45.** 700, 900, 1900, 2700, 3000

**46.** $2, $17, $21, $29, $14, $17

**47.** 470, 20, 190, 280, 470, 500

Find the mode or modes, if any exist.  [6.5c]

**48.** 26, 34, 43, 26, 51

**49.** 17, 7, 11, 11, 14, 17, 18

**50.** 0.2, 0.2, 1.7, 1.9, 2.4, 0.2

**51.** 700, 700, 800, 2700, 800

**52.** $14, $17, $21, $29, $17, $2

**53.** 20, 20, 20, 20, 20, 500

**54.** One summer, a student worked part time as a veterinary assistant. She earned the following weekly amounts over a six-week period: $360, $192, $240, $216, $420, and $132. What was the mean amount earned per week? the median?  [6.5a, b]

**55.** *Grade Point Average.*  Find the grade point average for one semester given the following grades. Assume the grade point values are 4.0 for A, 3.0 for B, and so on. Round to the nearest tenth.  [6.5a]

| COURSE | GRADE | NUMBER OF CREDIT HOURS IN COURSE |
|---|---|---|
| Basic Math | A | 5 |
| English | B | 3 |
| Computer Applications | C | 4 |
| Russian | B | 3 |
| College Skills | B | 1 |

**56.** *Battery Testing.*  An experiment was performed to compare battery quality. Two kinds of batteries were tested to see how long, in hours, they kept a hand radio running. On the basis of this test, which battery is better?  [6.5d]

| BATTERY A: TIMES (in hours) | | | BATTERY B: TIMES (in hours) | | |
|---|---|---|---|---|---|
| 38.9 | 39.3 | 40.4 | 39.3 | 38.6 | 38.8 |
| 53.1 | 41.7 | 38.0 | 37.4 | 47.6 | 37.9 |
| 36.8 | 47.7 | 48.1 | 46.9 | 37.8 | 38.1 |
| 38.2 | 46.9 | 47.4 | 47.9 | 50.1 | 38.2 |

**57.** Use extrapolation and the graph for Exercises 11–15 to estimate the number of visitors to Kobuk Valley National Park in 2010.  [6.6a]

A deck of 52 playing cards is thoroughly shuffled and a card is randomly selected.   [6.6b]

**58.** Find the probability that the 5 of clubs is selected.

**59.** Find the probability that a red card is selected.

**60.** Find the mode(s), if any exist, of this set of data.

6,  9,  6,  8,  8,  5,  10,  5,  9,  10   [6.5c]

**A.** 8                    **B.** 5, 6, 8, 9, 10
**C.** 9                    **D.** No mode exists.

**61.** What is the mean of this set of data?

$$\frac{1}{2}, \ \frac{1}{3}, \ \frac{1}{4}, \ \frac{1}{5} \quad [6.5a]$$

**A.** $\frac{77}{240}$              **B.** $\frac{1}{3}$

**C.** $\frac{7}{24}$               **D.** $\frac{1}{4}$

## Synthesis

**62.** ▦ Find three solutions and then graph
$34x + 47y = 100$.   [6.4a, b]

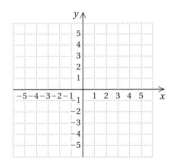

**63.** A typing pool consists of four senior typists who earn $12.35 per hour and nine other typists who earn $11.15 per hour. Find the mean hourly wage. [6.5a]

**64.** The ordered set of data 298, 301, 305, $a$, 323, $b$, 390 has a median of 316 and a mean of 326. Find $a$ and $b$.   [6.5a, b]

Graph on a plane.   [6.4b]

**65.** $1\frac{2}{3}x + \frac{3}{4}y = 2$

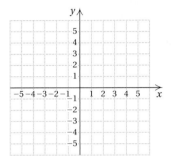

**66.** $\frac{3}{4}x - 2\frac{1}{2}y = 3$

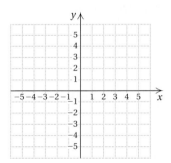

# Understanding Through Discussion and Writing

**1.** Find a real-world situation that fits this equation:

$$T = \frac{20{,}500 + 22{,}800 + 23{,}400 + 26{,}000}{4}. \quad [6.5a]$$

**2.** Can bar graphs always, sometimes, or never be converted to line graphs? Why?   [6.2b, d]

**3.** Is it possible for the mean of a set of numbers to be greater than all but one of the numbers in the set? Why or why not?   [6.5a]

**4.** Is it possible for the median of a set of four numbers to be one of the numbers in the set? Why or why not? [6.5b]

**5.** One way the U.S. Census Bureau reports income is by giving a 3-yr mean of median incomes. Why would this statistic be chosen?   [6.5a, b]

**6.** Would a company considering expansion be more interested in interpolation or extrapolation? Why? [6.6a]

CHAPTER
6

Test    For Extra Help

Step-by-step test solutions are found on the Chapter Test Prep Videos available via the Video Resources on DVD, in *MyMathLab* , and on You[Tube] (search "BittingerPrealgebra" and click on "Channels").

This table lists the number of calories burned during various walking activities. Use it for Exercises 1 and 2.

1. Which activity provides the greatest benefit in burned calories for a person who weighs 132 lb?

2. What activities would allow a person weighing 154 lb to burn at least 250 calories every 30 min?

| | CALORIES BURNED IN 30 MIN | | |
|---|---|---|---|
| WALKING ACTIVITY | 110 lb | 132 lb | 154 lb |
| Walking | | | |
| Fitness (5 mph) | 183 | 213 | 246 |
| Mildly energetic (3.5 mph) | 111 | 132 | 159 |
| Strolling (2 mph) | 69 | 84 | 99 |
| Hiking | | | |
| 3 mph with 20-lb load | 210 | 249 | 285 |
| 3 mph with 10-lb load | 195 | 228 | 264 |
| 3 mph with no load | 183 | 213 | 246 |

*Waste Generated.*    The number of pounds of waste generated per person per year varies greatly among countries around the world. In the pictograph at right, each symbol represents approximately 100 lb of waste. Use the pictograph for Exercises 3–6.

3. In which country does each person generate 1300 lb of waste per year?

4. In which countries does each person generate more than 1500 lb of waste per year?

5. How many pounds of waste per person per year are generated in Canada?

6. How many more pounds of waste per person per year are generated in the United States than in Mexico?

**Amount of Waste Generated (per person per year)**

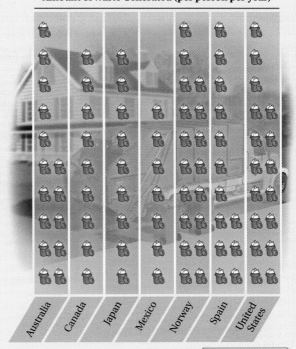

SOURCE: OECD, Key Environmental Indicators 2008    🗑 = 100 pounds

**7.** *Animal Speeds.*   The following table lists maximum speeds of movement for various animals, in miles per hour, compared to the speed of the fastest human. Make a vertical bar graph of the data.

| ANIMAL | SPEED (in miles per hour) |
|---|---|
| Human | 28 |
| Chicken | 9 |
| Elephant | 25 |
| Cheetah | 70 |
| Zebra | 40 |
| Lion | 50 |
| Pronghorn antelope | 61 |
| Elk | 45 |

SOURCE: *Natural History Magazine,* © American Museum of Natural History

Refer to the table in Exercise 7 for Exercises 8–11.

**8.** By how much does the fastest speed exceed the slowest speed?

**9.** How many times faster can a cheetah run than the fastest human?

**10.** Find the mean of all the speeds.

**11.** Find the median of all the speeds.

*Food Dollars Spent Away from Home.*   The line graph below shows the percent of food dollars spent away from home for various years, and projected to 2010. Use the graph for Exercises 12–15.

**Food Dollars Spent Away from Home**

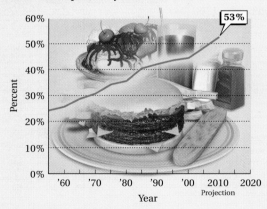

SOURCES: U.S. Bureau of Labor Statistics; National Restaurant Association

**12.** What percent of food dollars will be spent away from home in 2010?

**13.** What percent of food dollars were spent away from home in 1985?

**14.** In what year was the percent of food dollars spent away from home about 30%?

**15.** Use extrapolation to estimate the percent of food dollars spent away from home in 2015.

In which quadrant is each point located?

**16.** $\left(-\frac{1}{2}, 7\right)$

**17.** $(-5, -6)$

Determine the coordinates of each point.

**18.** A

**19.** B

**20.** C

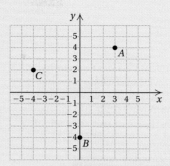

**21.** Complete the following solution of the equation $y - 3x = -10$: $(\square, 2)$.

Graph.

**22.** $y = 2x - 2$

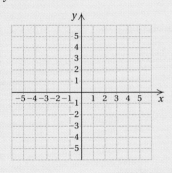

**23.** $y = -\frac{3}{2}x$

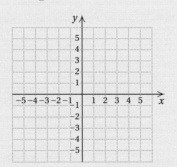

**24.** $x = -2$

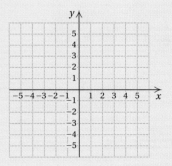

Find the mean.

**25.** 45, 49, 52, 54

**26.** 1, 2, 3, 4, 5

**27.** 3, 17, 17, 18, 18, 20

Find the median and any modes that exist.

**28.** 45, 47, 54, 54

**29.** 1, 2, 3, 4, 5

**30.** 20, 17, 17, 18, 3, 18

**31.** *Gas Mileage.* A 2009 Honda Fit gets 462 mi of highway driving on 14 gal of gasoline. What is the gas mileage?
Source: *Car and Driver*, December 2008

**32.** To get a C in chemistry, Kegan must score an average of 70 on four tests. Scores on his first three tests were 68, 71, and 65. What is the lowest score that Kegan can receive on the last test and still get a C?

**33.** *Chocolate Bars.* An experiment was performed to compare the quality of new Swiss chocolate bars being introduced in the United States. People were asked to taste the candies and rate them on a scale of 1 to 10, with 10 being the best. On the basis of this test, which chocolate bar is better?

| BAR A: SWISS PECAN | | | BAR B: SWISS HAZELNUT | | |
|---|---|---|---|---|---|
| 9 | 10 | 8 | 10 | 6 | 8 |
| 10 | 9 | 7 | 9 | 10 | 10 |
| 6 | 9 | 10 | 8 | 7 | 6 |
| 7 | 8 | 8 | 9 | 10 | 8 |

**34.** *Grade Point Average.* Find the grade point average for one semester given the following grades. Assume the grade point values are 4.0 for A, 3.0 for B, and so on. Round to the nearest tenth.

| COURSE | GRADE | NUMBER OF CREDIT HOURS IN COURSE |
|---|---|---|
| Introductory Algebra | B | 3 |
| English | A | 3 |
| Business | C | 4 |
| Spanish | B | 3 |
| Typing | B | 2 |

**35.** Tomas has 15 cherry jellybeans, 22 vanilla jellybeans, and 18 lime jellybeans. What is the probability that a randomly chosen jellybean will be vanilla?

## Synthesis

Graph.

**36.** $\frac{1}{4}x + 3\frac{1}{2}y = 1$

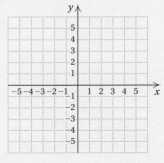

**37.** $\frac{5}{6}x - 2\frac{1}{3}y = 1$

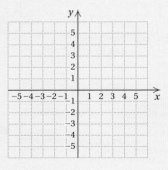

**38.** Find the area of a rectangle whose vertices are $(-3, 1)$, $(5, 1)$, $(5, 8)$, and $(-3, 8)$.

1. *Net Worth.*   In 2009, Warren Buffett of the United States was worth $62.1 billion. Write standard notation for 62.1 billion.

2. *Gas Mileage.*   A 2009 Saturn Outlook gets 312 mi of highway driving on 13 gal of gasoline. What is the gas mileage?
   **Source:** *Motor Trend*, November 2008

3. In 402,513, what does the digit 5 mean?

4. Evaluate:  $3 + 5^3$.

5. Find all the factors of 60.

6. Round 52.045 to the nearest tenth.

7. Convert to fraction notation:  $3\frac{3}{10}$.

8. Convert from cents to dollars:  210¢.

9. Find $-x$ when $x = -9$.

10. Evaluate $2x - y$ for $x = 3$ and $y = 8$.

Perform the indicated operation and, if possible, simplify.

11. $\dfrac{-2}{15} + \dfrac{3}{10}$

12. $-18 + (-21)$

13. $\dfrac{14}{15} - \dfrac{3}{5}$

14. $350 - 24.57$

15. $3\dfrac{3}{7} \cdot 4\dfrac{3}{8}$

16. $(17.4)(-2.43)$

17. $\dfrac{13}{15} \div \dfrac{26}{27}$

18. $-1334.183 \div (-21.4)$

Solve.

19. $x + \dfrac{2}{3} = -\dfrac{1}{5}$

20. $\dfrac{2}{5} \cdot y = \dfrac{3}{10}$

21. $\dfrac{3}{8}x + 2 = 11$

22. $3(x - 5) = 7x + 2$

Solve.

23. *Energy Consumption.*   In a recent year, American utility companies generated 1464 billion kilowatt-hours of electricity using coal, 455 billion using nuclear power, 273 billion using natural gas, 250 billion using hydroelectric plants, 118 billion using petroleum, and 12 billion using geothermal technology and other methods. How many kilowatt-hours of electricity were produced that year?

24. A piece of fabric $1\frac{3}{4}$ yd long is cut into 7 equal strips. What is the length of each strip?

25. A recipe calls for $\frac{3}{4}$ cup of sugar. How much sugar should be used for $\frac{1}{2}$ of the recipe?

**26.** *Peanut Products.* In any given year, the average American eats 2.7 lb of peanut butter, 1.5 lb of salted peanuts, 1.2 lb of peanut candy, 0.7 lb of in-shell peanuts, and 0.1 lb of peanuts in other forms. How many pounds of peanuts and products containing peanuts does the average American eat in one year?

**27.** A landscaper bought 22 evergreen trees for $210. What was the cost of each tree? Round to the nearest cent.

**28.** Find the perimeter and the area of the rectangle.

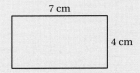

**29.** In which quadrant is the point $(-4, 9)$ located?

**30.** Graph: $y = \dfrac{1}{2}x - 4$.

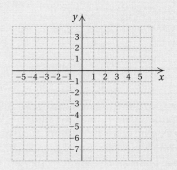

**31.** Combine like terms: $6x + 4y - 8x - 3y$.

**32.** Multiply: $5(2a - 3b + 1)$.

*FedEx.* The following table lists the cost of delivering a package by FedEx Priority Overnight shipping from zip code 46143 to zip code 80403. Use the table for Exercises 33–35.

| WEIGHT (in pounds) | COST |
|:---:|:---:|
| 1 | $38.08 |
| 2 | 41.86 |
| 3 | 46.11 |
| 4 | 50.25 |
| 5 | 54.69 |
| 6 | 58.83 |
| 7 | 62.97 |
| 8 | 66.86 |
| 9 | 71.00 |
| 10 | 74.59 |

SOURCE: Federal Express Corporation

**33.** Find the mean and the median of these costs.

**34.** Make a vertical bar graph of the data.

**35.** Make a line graph of the data.

## Synthesis

**36.** Simplify:

$$\left(\frac{3}{4}\right)^2 - \frac{1}{8} \cdot \left(3 - 1\frac{1}{2}\right)^2.$$

**37.** Add and write the answer as a mixed numeral:

$$-5\frac{42}{100} + \frac{355}{100} + \frac{89}{10} + \frac{17}{1000}.$$

**38.** A square with sides parallel to the axes has the point (2, 3) at its center. Find the coordinates of the square's vertices if each side is 8 units long.

# Ratio and Proportion

# Real-World Application

Of the 60 shark attacks recorded worldwide in 2008, 41 occurred in U.S. waters. What is the ratio of the number of shark attacks in U.S. waters to the number of shark attacks worldwide?

*Source:* University of Florida

***This problem appears as Example 7 in Section 7.1.***

# 7.1

# Introduction to Ratios

## OBJECTIVES

**a** Find fraction notation for ratios.

**b** Simplify ratios.

**SKILL TO REVIEW**

Objective 3.5b: Simplify fraction notation.

Simplify.

1. $\dfrac{16}{64}$    2. $\dfrac{40}{24}$

## a Ratios

> **RATIO**
>
> A **ratio** is the quotient of two quantities.

In the 2009–2010 regular basketball season, the Boston Celtics scored a total of 8136 points and allowed their opponents a total of 7836 points. The *ratio* of points scored to points allowed is given by the fraction notation

$$\dfrac{8136}{7836} \begin{array}{l} \leftarrow \text{Points scored} \\ \leftarrow \text{Points allowed} \end{array}$$

or by the colon notation

$$8136 : 7836.$$

Points scored ⟋    ⟍ Points allowed

We read both forms of notation as "the ratio of 8136 to 7836," listing the numerator first and the denominator second.

> **RATIO NOTATION**
>
> The **ratio** of $a$ to $b$ is given by the fraction notation $\dfrac{a}{b}$, where $a$ is the numerator and $b$ is the denominator, or by the colon notation $a : b$.

**EXAMPLE 1**  Write the ratio of 7 to 8.

The ratio is $\dfrac{7}{8}$,  or  $7 : 8$.

**EXAMPLE 2**  Write the ratio of 31.4 to 100.

The ratio is $\dfrac{31.4}{100}$,  or  $31.4 : 100$.

**EXAMPLE 3**  Write the ratio of $4\frac{2}{3}$ to $5\frac{7}{8}$. You need not simplify.

The ratio is $\dfrac{4\frac{2}{3}}{5\frac{7}{8}}$,  or  $4\frac{2}{3} : 5\frac{7}{8}$.

Do Margin Exercises 1–3.

In most of our work, we will use fraction notation for ratios.

1. Write the ratio of 5 to 11.

2. Write the ratio of 57.3 to 86.1.

3. Write the ratio of $6\dfrac{3}{4}$ to $7\dfrac{2}{5}$.

*Answers*

*Skill to Review:*

1. $\dfrac{1}{4}$    2. $\dfrac{5}{3}$

*Margin Exercises:*

1. $\dfrac{5}{11}$, or $5 : 11$    2. $\dfrac{57.3}{86.1}$, or $57.3 : 86.1$

3. $\dfrac{6\frac{3}{4}}{7\frac{2}{5}}$, or $6\dfrac{3}{4} : 7\dfrac{2}{5}$

**EXAMPLE 4** *Wind Speeds.* The average wind speed in Chicago is 10.4 mph. The average wind speed in Boston is 12.5 mph. Find the ratio of the wind speed in Chicago to the wind speed in Boston.

Source: *The Handy Geography Answer Book*

The ratio is $\dfrac{10.4}{12.5}$.

**EXAMPLE 5** *Record Rainfall.* The greatest rainfall ever recorded in the United States during a 12-month period was 739 inches in Kukui, Maui, Hawaii, from December 1981 to December 1982. What is the ratio of amount of rainfall, in inches, to time, in months? of time, in months, to amount of rainfall, in inches?

Source: *Time Almanac*

The ratio of amount of rainfall, in inches, to time, in months, is

$$\dfrac{739}{12} \cdot \quad \begin{array}{l} \leftarrow \text{Rainfall} \\ \leftarrow \text{Time} \end{array}$$

The ratio of time, in months, to amount of rainfall, in inches, is

$$\dfrac{12}{739} \cdot \quad \begin{array}{l} \leftarrow \text{Time} \\ \leftarrow \text{Rainfall} \end{array}$$

**EXAMPLE 6** Refer to the triangle below.

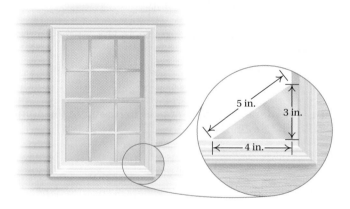

5 in. 3 in. 4 in.

a) What is the ratio of the length of the longest side to the length of the shortest side?

$$\dfrac{5}{3} \quad \begin{array}{l} \leftarrow \text{Longest side} \\ \leftarrow \text{Shortest side} \end{array}$$

b) What is the ratio of the length of the shortest side to the length of the longest side?

$$\dfrac{3}{5} \quad \begin{array}{l} \leftarrow \text{Shortest side} \\ \leftarrow \text{Longest side} \end{array}$$

Do Exercises 4–6; Exercise 6 is on the next page.

**4. Record Snowfall.** The greatest snowfall recorded in North America during a 24-hr period was 76 in. in Silver Lake, Colorado, on April 14–15, 1921. What is the ratio of amount of snowfall, in inches, to time, in hours?

Source: U.S. Army Corps of Engineers

**5. Frozen Fruit Drinks.** A Berries & Kreme Chiller from Krispy Kreme contains 960 calories, while Smoothie King's MangoFest drink contains 258 calories. What is the ratio of the number of calories in the Krispy Kreme drink to the number of calories in the Smoothie King drink? of the number of calories in the Smoothie King drink to the number of calories in the Krispy Kreme drink?

Source: Physicians Committee for Responsible Medicine

*Answers*

4. $\dfrac{76}{24}$   5. $\dfrac{960}{258}; \dfrac{258}{960}$

**6.** In the triangle below, what is the ratio of the length of the shortest side to the length of the longest side?

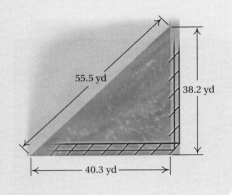

55.5 yd

38.2 yd

40.3 yd

**EXAMPLE 7** *Shark Attacks.* Of the 60 shark attacks recorded worldwide in 2008, 41 occurred in U.S. waters. The bar graph below shows the breakdown by state.

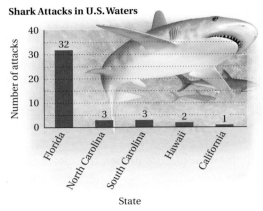

**Shark Attacks in U.S. Waters**

Number of attacks

40 — 30 — 20 — 10 — 0

32

3    3    2    1

Florida    North Carolina    South Carolina    Hawaii    California

State

SOURCE: University of Florida

**a)** What is the ratio of the number of shark attacks in U.S. waters to the number of shark attacks worldwide?

**b)** What is the ratio of the number of shark attacks in Hawaii to the number of shark attacks in South Carolina?

**c)** What is the ratio of the number of shark attacks in Florida to the total number of shark attacks in the other four states?

**a)** The ratio of the number of shark attacks in U.S. waters to the number of shark attacks worldwide is

$\dfrac{41}{60}.$  ← Attacks in U.S. waters
← Attacks worldwide

**b)** The ratio of the number of shark attacks in Hawaii to the number of shark attacks in South Carolina is

$\dfrac{2}{3}.$  ← Attacks in Hawaii
← Attacks in South Carolina

**7. Soap Box Derby.** Of the 538 participants in the 2008 All-American Soap Box Derby, 296 were boys and 242 were girls. What was the ratio of girls to boys? of boys to girls? of boys to total number of participants?
**Source:** All-American Soap Box Derby

**c)** Of the 41 shark attacks in U.S. waters, 32 occurred in Florida. We subtract to determine how many shark attacks took place in the other four states. We have

$$41 - 32 = 9.$$

Thus, the ratio of the number of shark attacks in Florida to the number of shark attacks in the other four states is

$\dfrac{32}{9}.$  ← Attacks in Florida
← Attacks in other four states

Do Exercise 7.

## b Simplifying Notation for Ratios

Sometimes a ratio can be simplified. This provides a means of finding other numbers with the same ratio.

**EXAMPLE 8** Find the ratio of 6 to 8. Then simplify to find two other numbers in the same ratio.

We write the ratio in fraction notation and then simplify:

$$\frac{6}{8} = \frac{2 \cdot 3}{2 \cdot 4} = \frac{2}{2} \cdot \frac{3}{4} = 1 \cdot \frac{3}{4} = \frac{3}{4}.$$

Thus, 3 and 4 have the same ratio as 6 and 8. We can express this by saying "6 is to 8 as 3 is to 4."

Do Exercise 8.

**EXAMPLE 9** Find the ratio of 50 to 10. Then simplify to find two other numbers in the same ratio.

We write the ratio in fraction notation and then simplify. The simplified form must also be in fraction notation.

$$\frac{50}{10} = \frac{10 \cdot 5}{10 \cdot 1} = \frac{10}{10} \cdot \frac{5}{1} = \frac{5}{1}$$

Since this is a ratio, leave the 1 in the denominator.

Thus, 50 is to 10 as 5 is to 1.

**EXAMPLE 10** Find the ratio of 2.4 to 10. Then simplify to find two other numbers in the same ratio.

We first write the ratio in fraction notation. Next, we multiply by 1 to clear the decimal from the numerator. Then we simplify.

$$\frac{2.4}{10} = \frac{2.4}{10} \cdot \frac{10}{10} = \frac{24}{100} = \frac{4 \cdot 6}{4 \cdot 25} = \frac{4}{4} \cdot \frac{6}{25} = \frac{6}{25}$$

Thus, 2.4 is to 10 as 6 is to 25.

Do Exercises 9–11.

**EXAMPLE 11** An HDTV screen that measures 46 in. diagonally has a width of 40 in. and a height of $22\frac{1}{2}$ in. Find the ratio of width to height and simplify.

We first write the ratio. To simplify, we could rewrite $22\frac{1}{2}$ in fraction notation as $\frac{45}{2}$ or in decimal notation as 22.5.

$$\text{The ratio is } \frac{40}{22\frac{1}{2}} = \frac{40}{22.5} = \frac{40}{22.5} \cdot \frac{10}{10} = \frac{400}{225}$$

$$= \frac{25 \cdot 16}{25 \cdot 9} = \frac{25}{25} \cdot \frac{16}{9}$$

$$= \frac{16}{9}.$$

Thus, we can say that the ratio of width to height is 16 to 9, which can also be expressed as 16:9.

Do Exercise 12.

8. Find the ratio of 18 to 27. Then simplify to find two other numbers in the same ratio.

9. Find the ratio of 18 to 3. Then simplify to find two other numbers in the same ratio.

10. Find the ratio of 3.6 to 12. Then simplify to find two other numbers in the same ratio.

11. Find the ratio of 1.2 to 1.5. Then simplify to find two other numbers in the same ratio.

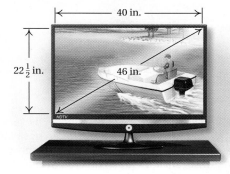

12. An HDTV screen that measures 44 in. diagonally has a width of 38.4 in. and a height of 21.6 in. Find the ratio of height to width and simplify.

***Answers***

8. 18 is to 27 as 2 is to 3.     9. 18 is to 3 as 6 is to 1.
10. 3.6 is to 12 as 3 is to 10.     11. 1.2 is to 1.5 as 4 is to 5.     12. $\frac{9}{16}$

**a**  Find fraction notation for each ratio. You need not simplify.

**1.** 4 to 5

**2.** 3 to 2

**3.** 178 to 572

**4.** 329 to 967

**5.** 0.4 to 12

**6.** 2.3 to 22

**7.** 3.8 to 7.4

**8.** 0.6 to 0.7

**9.** 56.78 to 98.35

**10.** 456.2 to 333.1

**11.** $8\frac{3}{4}$ to $9\frac{5}{6}$

**12.** $10\frac{1}{2}$ to $43\frac{1}{4}$

**13.** *Space Plane.* It is estimated that it will take 4 hr to fly from London to Sydney on the beyond-the-atmosphere space plane being developed in Europe. This 10,600-mi trip currently takes about 21 hr. What is the ratio of the time of the current trip to the time of the trip on the space plane? of the time of the trip on the space plane to the time of the current trip?

**Source:** EADS Atrium

**14.** *Silicon in the Earth's Crust.* Every 100 tons of the earth's crust contains about 28 tons of silicon. What is the ratio of the weight of silicon to the weight of crust? of the weight of crust to the weight of silicon?

**Source:** *The Handy Science Answer Book*

**15.** *Health Insurance Coverage.* Of every 100 people in the United States, 84.7 are covered by health insurance. What is the ratio of all people to those covered by health insurance? of those covered by health insurance to all people?

**Source:** U.S. Census Bureau

**16.** *Price of a Book.* The list price of the book *The Last Lecture* by Randy Pausch is $21.95, but the book recently sold for $12.07 on amazon.com. What is the ratio of the list price to the Amazon price? of the Amazon price to the list price?

**Source:** amazon.com

**17.** *Tax Freedom Day.* Of the 366 days in 2008 (a leap year), the average American worked 113 days to pay his or her federal, state, and local taxes. Find the ratio of the number of days worked to pay taxes in 2008 to the number of days in the year.

**Source:** Tax Foundation

**18.** *Spending Categories.* The average American worked 60 days in 2008 to pay housing and household operating expenses and 35 days to pay for food. Find the ratio of the number of days worked to pay for food to the number of days worked to pay housing and household operating expenses.

**Source:** Tax Foundation

**19.** *Population Estimates.* It is estimated that, of every 1000 people in the United States in 2050, 126 will be from 20 to 29 years old. What is the ratio of all people to those in the 20- to 29-year age group? of those in the 20- to 29-year age group to all people?

Source: U.S. Census Bureau

**20.** *Population Estimates.* It is estimated that, of every 1000 people in the United States in 2050, 118 will be age 75 and older. What is the ratio of all people to those age 75 and older? of those age 75 and older to all people?

Source: U.S. Census Bureau

**21.** *Field Hockey.* A diagram of the playing area for field hockey is shown below. What is the ratio of width to length? of length to width?

Source: *Sports: The Complete Visual Reference*

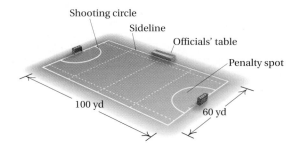

Shooting circle
Sideline
Officials' table
Penalty spot
100 yd
60 yd

**22.** *The Leaning Tower of Pisa.* The Leaning Tower of Pisa was reopened to the public in 2001 following a 10-yr stabilization project. The 184.5-ft tower now leans about 13 ft from its base. What is the ratio of the distance that it leans to its height? of its height to the distance that it leans?

Source: CNN

184.5 ft

13 ft

**b** Find the ratio of the first number to the second and simplify.

**23.** 4 to 6

**24.** 6 to 10

**25.** 18 to 24

**26.** 28 to 36

**27.** 4.8 to 10

**28.** 5.6 to 10

**29.** 2.8 to 3.6

**30.** 4.8 to 6.4

**31.** 20 to 30

**32.** 40 to 60

**33.** 56 to 100

**34.** 42 to 100

**35.** 128 to 256

**36.** 232 to 116

**37.** 0.48 to 0.64

**38.** 0.32 to 0.96

**39.** $3\frac{1}{5}$ to $4\frac{1}{10}$

**40.** $2\frac{1}{4}$ to $5\frac{1}{2}$

**41.** $7\frac{1}{5}$ to $2\frac{3}{10}$

**42.** $6\frac{3}{10}$ to $1\frac{4}{5}$

**43.** In this rectangle, find the ratios of length to width and of width to length.

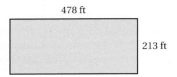

478 ft

213 ft

**44.** In this right triangle, find the ratios of shortest length to longest length and of longest length to shortest length.

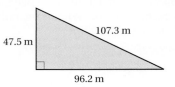

107.3 m

47.5 m

96.2 m

## Skill Maintenance

Use = or ≠ for ☐ to write a true sentence.   [3.5c]

**45.** $\frac{12}{8}$ ☐ $\frac{6}{4}$

**46.** $\frac{4}{7}$ ☐ $\frac{5}{9}$

**47.** $\frac{7}{2}$ ☐ $\frac{31}{9}$

**48.** $\frac{17}{25}$ ☐ $\frac{68}{100}$

Divide. Write decimal notation for the answer.   [5.4a]

**49.** 200 ÷ 4

**50.** 95 ÷ 10

**51.** 232 ÷ 16

**52.** 342 ÷ 2.25

Solve.   [4.6c]

**53.** Rocky is $187\frac{1}{10}$ cm tall and his daughter is $180\frac{3}{4}$ cm tall. How much taller is Rocky?

**54.** Aunt Louise is $168\frac{1}{4}$ cm tall and her son is $150\frac{7}{10}$ cm tall. How much taller is Aunt Louise?

## Synthesis

**55.** ▦ See Exercises 53 and 54. Find the ratio of the parents' total height to their children's total height and simplify.

*Fertilizer.*   Exercises 56 and 57 refer to a common lawn fertilizer known as "5, 10, 15." This mixture contains 5 parts of potassium for every 10 parts of phosphorus and 15 parts of nitrogen. (This is often denoted 5 : 10 : 15.)

**56.** Find and simplify the ratio of potassium to nitrogen and of nitrogen to phosphorus.

**57.** Simplify the ratio 5 : 10 : 15.

# 7.2

# Rates and Unit Prices

## a Rates

A 2010 Kia Sportage EX can travel 414 mi on 18 gal of gasoline. Let's consider the ratio of miles to gallons:

**Source:** Kia Motors America, Inc.

$$\frac{414 \text{ mi}}{18 \text{ gal}} = \frac{414}{18} \frac{\text{miles}}{\text{gallon}} = \frac{23}{1} \frac{\text{miles}}{\text{gallon}}$$

$$= 23 \text{ miles per gallon} = 23 \text{ mpg}.$$

> "per" means "division," or "for each."

The ratio

$$\frac{414 \text{ mi}}{18 \text{ gal}}, \quad \text{or} \quad \frac{414}{18} \frac{\text{mi}}{\text{gal}}, \quad \text{or 23 mpg},$$

is called a **rate**.

> **OBJECTIVES**
>
> **a** Give the ratio of two different measures as a rate.
>
> **b** Find unit prices and use them to compare purchases.

> **RATE**
>
> When a ratio is used to compare two different kinds of measure, we call it a **rate**.

Suppose David says his car travels 392.4 mi on 16.8 gal of gasoline. Is the mpg (mileage) of his car better than that of the Kia Sportage above? To determine this, it helps to convert the ratio to decimal notation and perhaps round. Then we have

$$\frac{392.4 \text{ miles}}{16.8 \text{ gallons}} = \frac{392.4}{16.8} \text{ mpg} \approx 23.357 \text{ mpg}.$$

Since 23.357 > 23, David's car gets better mileage than the Kia Sportage does.

**EXAMPLE 1**  It takes 60 oz of grass seed to seed 3000 sq ft of lawn. What is the rate in ounces per square foot?

$$\frac{60 \text{ oz}}{3000 \text{ sq ft}} = \frac{1}{50} \frac{\text{oz}}{\text{sq ft}}, \quad \text{or} \quad 0.02 \frac{\text{oz}}{\text{sq ft}}$$

**EXAMPLE 2**  Martina bought 5 lb of organic russet potatoes for $4.99. What was the rate in cents per pound?

$$\frac{\$4.99}{5 \text{ lb}} = \frac{499 \text{ cents}}{5 \text{ lb}} = 99.8 \text{¢/lb}$$

Rates use two units. Paying attention to units can be helpful in solving certain rate problems. For example, if we expect an answer to be a rate in dollars per month, we look for a calculation that causes an amount of money (in dollars) to be divided by a length of time (in months). In this manner, the rate, dollars/month, will be formed.

> **STUDY TIPS**
>
> **AIM FOR MASTERY**
>
> In each exercise set, you may encounter one or more exercises that give you trouble. Do not ignore these problems, but instead work to master them. These are the problems (not the "easy" problems) that you will need to revisit as you progress through the course. Consider marking these exercises with a star so that you can easily locate them when studying or reviewing for a quiz or test.

**EXAMPLE 3** A pharmacy student employed as a pharmacist's assistant earned $3930 for working 3 months one summer. What was the rate of pay per month?

The rate of pay is the ratio of money earned to length of time worked, or

$$\frac{\$3930}{3 \text{ mo}} = 1310 \frac{\text{dollars}}{\text{month}}, \text{ or } \$1310 \text{ per month.}$$

**EXAMPLE 4** *Ratio of Strike-outs to Home Runs.* In the 2009 baseball season, Dustin Pedroia of the Boston Red Sox had 45 strikeouts and 15 home runs. What was his strikeout to home-run rate?

**Source:** Major League Baseball

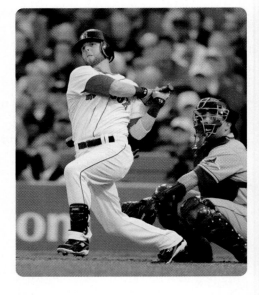

$$\frac{45 \text{ strikeouts}}{15 \text{ home runs}} = \frac{45}{15} \frac{\text{strikeouts}}{\text{home runs}} = \frac{45}{15} \text{ strikeouts per home run}$$

$$= 3 \text{ strikeouts per home run}$$

Do Exercises 1–7.

A ratio of distance traveled to time is called *speed*. What is the rate, or speed, in miles per hour or in kilometers per hour?

**1.** 45 mi, 9 hr

**2.** 120 mi, 10 hr

**3.** 89 km, 13 hr (Round to the nearest hundredth.)

What is the rate, or speed, in feet per second?

**4.** 2200 ft, 2 sec

**5.** 52 ft, 13 sec

**6.** 242 ft, 16 sec

**7. Babe Ruth.** In his baseball career, Babe Ruth had 1330 strikeouts and 714 home runs. What was his home-run to strikeout rate?

**Source:** Major League Baseball

## b Unit Pricing

**UNIT PRICE**

A **unit price**, or **unit rate**, is the ratio of price to the number of units.

**EXAMPLE 5** *Unit Price of Pears.* Ruth bought a 15.2-oz can of pears for $1.30. What is the unit price in cents per ounce?

$$\text{Unit price} = \frac{\text{Price}}{\text{Number of units}}$$

$$= \frac{\$1.30}{15.2 \text{ oz}} = \frac{130 \text{ cents}}{15.2 \text{ oz}} = \frac{130}{15.2} \frac{\text{cents}}{\text{oz}}$$

$$\approx 8.553 \text{ cents per ounce}$$

Do Exercise 8.

**8. Unit Price of Pasta Sauce.**
Gregory bought a 26-oz jar of pasta sauce for $2.79. What is the unit price in cents per ounce?

*Answers*

**1.** 5 mi/hr, or 5 mph   **2.** 12 mi/hr, or 12 mph
**3.** $\frac{89}{13}$ km/h, or 6.85 km/h   **4.** 1100 ft/sec
**5.** 4 ft/sec   **6.** $\frac{121}{8}$ ft/sec, or 15.125 ft/sec
**7.** $\frac{714}{1330}$ home runs per strikeout ≈ 0.537 home run per strikeout   **8.** 10.731¢/oz

Unit prices enable us to do comparison shopping and determine the best buy of a product on the basis of price. It is often helpful to change all prices to cents so that we can compare unit prices more easily.

**EXAMPLE 6** *Unit Price of Salad Dressing.* At the request of his customers, Angelo started bottling and selling the popular basil and lemon salad dressing that he serves in his neighborhood café. The dressing is sold in four sizes of containers as listed in the table below. Compute the unit price of each size of container and determine which size is the best buy on the basis of unit price alone.

| SIZE | PRICE | UNIT PRICE |
|------|-------|------------|
| 10 oz | $2.49 | 24.900¢/oz |
| 16 oz | $3.59 | 22.438¢/oz |
| 20 oz | $4.09 | 20.450¢/oz Lowest unit price |
| 32 oz | $6.79 | 21.219¢/oz |

We compute the unit price of each size and fill in the chart:

10 oz: $\dfrac{\$2.49}{10 \text{ oz}} = \dfrac{249 \text{ cents}}{10 \text{ oz}} = \dfrac{249}{10} \dfrac{\text{cents}}{\text{oz}} = 24.900¢/\text{oz};$

16 oz: $\dfrac{\$3.59}{16 \text{ oz}} = \dfrac{359 \text{ cents}}{16 \text{ oz}} = \dfrac{359}{16} \dfrac{\text{cents}}{\text{oz}} \approx 22.438¢/\text{oz};$

20 oz: $\dfrac{\$4.09}{20 \text{ oz}} = \dfrac{409 \text{ cents}}{20 \text{ oz}} = \dfrac{409}{20} \dfrac{\text{cents}}{\text{oz}} = 20.450¢/\text{oz};$

32 oz: $\dfrac{\$6.79}{32 \text{ oz}} = \dfrac{679 \text{ cents}}{32 \text{ oz}} = \dfrac{679}{32} \dfrac{\text{cents}}{\text{oz}} \approx 21.219¢/\text{oz}.$

On the basis of unit price alone, we see that the 20-oz container is the best buy.

Do Exercise 9.

Although we often think that "bigger is cheaper," this is not always the case, as we see in Example 6. In addition, even when a larger package has a lower unit price than a smaller package, it still might not be the best buy for you. For example, some of the food in a large package could spoil before it is used, or you might not have room to store a large package.

**Calculator Corner**

**Reciprocals of Rates** If, in Example 1, we wished to determine the rate of lawn coverage in sq ft/oz, we could simply find the reciprocal of 0.02 oz/sq ft by pressing ⟨0⟩ ⟨·⟩ ⟨0⟩ ⟨2⟩ ⟨1/x⟩. The rate of lawn coverage is 50 sq ft/oz.

**Exercise:**

1. Use the solution of Example 4 to find Pedroia's rate of home runs per strikeout.

**9. Cost of Mayonnaise.** Complete the following table for Hellmann's mayonnaise sold on an online shopping site. Which size has the lowest unit price?

**Source:** Peapod

| SIZE | PRICE | UNIT PRICE |
|------|-------|------------|
| 8 oz | $2.79 | |
| 10 oz | $3.69 | |
| 30 oz | $5.39 | |

*Answer*

**9.** 34.875¢/oz; 36.900¢/oz; 17.967¢/oz; the 30-oz size has the lowest unit price.

**a**  In Exercises 1–4, find each rate, or speed, as a ratio of distance to time. Round to the nearest hundredth where appropriate.

**1.** 120 km, 3 hr

**2.** 18 mi, 9 hr

**3.** 217 mi, 29 sec

**4.** 443 m, 48 sec

**5.** *Chevrolet Cobalt LS—City Driving.*  A 2010 Chevrolet Cobalt LS will travel 312.5 mi on 12.5 gal of gasoline in city driving. What is the rate in miles per gallon?
**Source:** Chevrolet

**6.** *Mazda 3—City Driving.*  A 2010 Mazda 3 will travel 348 mi on 14.5 gal of gasoline in city driving. What is the rate in miles per gallon?
**Source:** Ford Motor Company

**7.** *Mazda 3—Highway Driving.*  A 2010 Mazda 3 will travel 643.5 mi on 19.5 gal of gasoline in highway driving. What is the rate in miles per gallon?
**Source:** Ford Motor Company

**8.** *Chevrolet Cobalt LS—Highway Driving.*  A 2010 Chevrolet Cobalt LS will travel 499.5 mi on 13.5 gal of gasoline in highway driving. What is the rate in miles per gallon?
**Source:** Chevrolet

**9.** *Population Density of Monaco.*  Monaco is a tiny country on the Mediterranean coast of France. It has an area of 0.75 square mile and a population of 32,796 people. What is the rate of number of people per square mile? The rate per square mile is called the *population density.* Monaco has the highest population density of any country in the world.
**Source:** *The World Factbook*

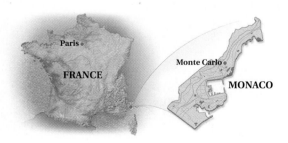

**10.** *Population Density of Australia.*  The continent of Australia, with the island state of Tasmania, has an area of 2,967,893 sq mi and a population of 21,007,310 people. What is the rate of number of people per square mile? The rate per square mile is called the *population density.* Australia has one of the lowest population densities in the world.
**Source:** *The World Factbook*

**11.** A car is driven 500 mi in 20 hr. What is the rate in miles per hour? in hours per mile?

**12.** A student eats 3 hamburgers in 15 min. What is the rate in hamburgers per minute? in minutes per hamburger?

**13.** *Points per Game.* Dwayne Wade of the Miami Heat scored 2045 points in 77 games during the 2009–2010 basketball season. What was the rate in points per game?
**Source:** National Basketball Association

**14.** *Rebounds per Game.* Dwight Howard of the Orlando Magic got 1082 rebounds in 82 games during the 2009–2010 basketball season. What was the rate in rebounds per game?
**Source:** National Basketball Association

**15.** *Lawn Watering.* Watering a lawn adequately requires 623 gal of water for every 1000 ft$^2$. What is the rate in gallons per square foot?

**16.** A car is driven 200 km on 40 L of gasoline. What is the rate in kilometers per liter?

**17.** *Speed of Light.* Light travels 186,000 mi in 1 sec. What is its rate, or speed, in miles per second?
**Source:** *The Handy Science Answer Book*

**18.** *Speed of Sound.* Sound travels 1100 ft in 1 sec. What is its rate, or speed, in feet per second?
**Source:** *The Handy Science Answer Book*

**19.** Impulses in nerve fibers travel 310 km in 2.5 hr. What is the rate, or speed, in kilometers per hour?

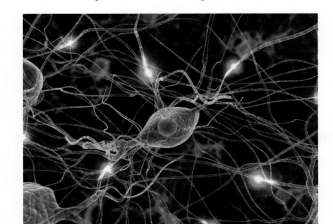

**20.** A black racer snake can travel 4.6 km in 2 hr. What is its rate, or speed, in kilometers per hour?

**21.** *Elephant Heart Rate.* The heart of an elephant, at rest, will beat an average of 1500 beats in 60 min. What is the rate in beats per minute?

**Source:** *The Handy Science Answer Book*

**22.** *Human Heart Rate.* The heart of a human, at rest, will beat an average of 4200 beats in 60 min. What is the rate in beats per minute?

**Source:** *The Handy Science Answer Book*

**b**   Find each unit price in Exercises 23–30. Then, in each exercise, determine which size is the better buy based on unit price alone.

**23.** *Hidden Valley Ranch Dressing.*

| SIZE | PRICE | UNIT PRICE |
|------|-------|------------|
| 16 oz | $4.19 | |
| 20 oz | $5.29 | |

**24.** *Miracle Whip.*

| SIZE | PRICE | UNIT PRICE |
|------|-------|------------|
| 32 oz | $5.29 | |
| 48 oz | $8.29 | |

**25.** *Cascade Powder Detergent.*

| SIZE | PRICE | UNIT PRICE |
|------|-------|------------|
| 45 oz | $3.99 | |
| 75 oz | $5.49 | |

**26.** *Bush's Homestyle Baked Beans.*

| SIZE | PRICE | UNIT PRICE |
|------|-------|------------|
| 16 oz | $1.59 | |
| 28 oz | $2.09 | |

**27.** *Jif Creamy Peanut Butter.*

| SIZE | PRICE | UNIT PRICE |
|------|-------|------------|
| 18 oz | $2.50 | |
| 28 oz | $4.89 | |

**28.** *Hills Brothers Coffee.*

| SIZE | PRICE | UNIT PRICE |
|------|-------|------------|
| 26 oz | $8.99 | |
| 39 oz | $12.99 | |

**29.** *Campbell's Condensed Tomato Soup.*

| SIZE | PRICE | UNIT PRICE |
|------|-------|------------|
| 10.7 oz | $1.39 | |
| 26 oz | $2.69 | |

**30.** *Nabisco Saltines.*

| SIZE | PRICE | UNIT PRICE |
|------|-------|------------|
| 16 oz | $3.59 | |
| 32 oz | $5.19 | |

Find the unit price of each brand in Exercises 31–34. Then, in each exercise, determine which brand is the best buy based on unit price alone.

**31.** *Vanilla Ice Cream.*

| BRAND | SIZE | PRICE |
|-------|------|-------|
| B | 32 oz | $5.99 |
| E | 48 oz | $6.99 |

**32.** *Orange Juice.*

| BRAND | SIZE | PRICE |
|-------|------|-------|
| M | 54 oz | $4.79 |
| T | 59 oz | $5.99 |

**33.** *Tomato Ketchup.*

| BRAND | SIZE | PRICE |
|-------|------|-------|
| A | 24 oz | $2.49 |
| B | 36 oz | $3.29 |
| H | 46 oz | $3.69 |

**34.** *Yellow Mustard.*

| BRAND | SIZE | PRICE |
|-------|------|-------|
| F | 14 oz | $1.29 |
| G | 19 oz | $1.99 |
| P | 20 oz | $2.49 |

## Skill Maintenance

Replace the ☐ with < or > to form a true statement.   [4.2c]

**35.** $\dfrac{3}{11} \; \square \; \dfrac{5}{13}$

**36.** $\dfrac{9}{7} \; \square \; \dfrac{7}{5}$

**37.** $\dfrac{4}{9} \; \square \; \dfrac{3}{7}$

**38.** $\dfrac{3}{19} \; \square \; \dfrac{2}{17}$

**39.** $-\dfrac{3}{10} \; \square \; -\dfrac{2}{7}$

**40.** $-\dfrac{9}{8} \; \square \; -\dfrac{15}{13}$

**41.** There are 20.6 million people in this country who play the piano and 18.9 million who play the guitar. How many more play the piano than the guitar?   [5.8b]

**42.** A serving of fish steak (cross section) is generally $\frac{1}{2}$ lb. How many servings can be prepared from a cleaned $12\frac{3}{4}$-lb salmon?   [4.7d]

## Synthesis

**43.** Recently, some manufacturers have changed the size and price of a product in such a way that the consumer thinks the price of the product has been lowered when, in reality, a higher unit price is being charged. Some aluminum juice cans are now concave (curved in) on the bottom. Suppose the volume of the can in the figure has been reduced from a fluid capacity of 6 oz to 5.5 oz, and the price of each can has been reduced from 65¢ to 60¢. Find the unit price of each container in cents per ounce.

**44.** ▦ Use the formula for the area of a circle, $A = \pi r^2$, to determine which is a better deal: a 14-in. pizza for $10.50 or a 16-in. pizza for $11.95. Use 3.14 for $\pi$.

**45.** ▦ Suppose that, 25 mi from where you're standing, a bolt of lightning splits a tree. How long will it take for you to hear the accompanying crack of thunder? How long will it take for you to see the flash of light? (*Hint*: Use the information in Exercises 17 and 18.)

**46.** ▦ A six-pack of 12-oz cans of ginger ale was recently on sale for $2.99. Find the unit price in ounces per dollar.

**47.** Suppose that a pasta manufacturer shrinks the size of a box from 1 lb to 14 oz, but keeps the price at 85¢ a box. By how much does the unit price change?

# 7.3

## Proandtions

### OBJECTIVES

**a** Determine whether two pairs of numbers are proportional.

**b** Solve proportions.

**SKILL TO REVIEW**
Objective 3.5c: Test to determine whether two fractions are equivalent.

Use = or ≠ for □ to write a true sentence.

**1.** $\frac{18}{12} \square \frac{9}{6}$   **2.** $\frac{4}{7} \square \frac{5}{8}$

### **a** Proportions

When two pairs of numbers, such as 3, 2 and 6, 4, have the same ratio, we say that they are **proportional**. The equation

$$\frac{3}{2} = \frac{6}{4}$$

states that the pairs 3, 2 and 6, 4 are proportional. Such an equation is called a **proportion**. We sometimes read $\frac{3}{2} = \frac{6}{4}$ as "3 is to 2 as 6 is to 4."

Since ratios can be written using fraction notation, we can use the test for equality of fractions discussed in Section 3.5 to determine whether two ratios are the same.

**EXAMPLE 1**   Determine whether 1, 2 and 3, 6 are proportional.

We can use cross products:

$$1 \cdot 6 = 6 \qquad \frac{1}{2} \overset{?}{=} \frac{3}{6} \qquad 2 \cdot 3 = 6.$$

Since the cross products are the same, $6 = 6$, we know that $\frac{1}{2} = \frac{3}{6}$, so the numbers are proportional.

**EXAMPLE 2**   Determine whether 2, 5 and 4, 7 are proportional.

We can use cross products:

$$2 \cdot 7 = 14 \qquad \frac{2}{5} \overset{?}{=} \frac{4}{7} \qquad 5 \cdot 4 = 20.$$

Since the cross products are not the same, $14 \neq 20$, we know that $\frac{2}{5} \neq \frac{4}{7}$, so the numbers are not proportional.

Do Margin Exercises 1–3.

Determine whether the two pairs of numbers are proportional.

**1.** 3, 4 and 6, 8

**2.** 1, 4 and 10, 39

**3.** 1, 2 and 20, 39

**EXAMPLE 3**   Determine whether 3.2, 4.8 and 0.16, 0.24 are proportional.

We can use cross products:

$$3.2 \times 0.24 = 0.768 \qquad \frac{3.2}{4.8} \overset{?}{=} \frac{0.16}{0.24} \qquad 4.8 \times 0.16 = 0.768.$$

Since the cross products are the same, $0.768 = 0.768$, we know that $\frac{3.2}{4.8} = \frac{0.16}{0.24}$, so the numbers are proportional.

Do Exercises 4 and 5.

Determine whether the two pairs of numbers are proportional.

**4.** 6.4, 12.8 and 5.3, 10.6

**5.** 6.8, 7.4 and 3.4, 4.2

*Answers*

*Skill to Review:*
1. =   2. ≠

*Margin Exercises:*
1. Yes   2. No   3. No   4. Yes   5. No

**EXAMPLE 4** Determine whether $4\frac{2}{3}$, $5\frac{1}{2}$ and $8\frac{7}{8}$, $16\frac{1}{3}$ are proportional.

We can use cross products:

$$4\frac{2}{3} \cdot 16\frac{1}{3} = \frac{14}{3} \cdot \frac{49}{3} \qquad \frac{4\frac{2}{3}}{5\frac{1}{2}} \overset{?}{=} \frac{8\frac{7}{8}}{16\frac{1}{3}} \qquad 5\frac{1}{2} \cdot 8\frac{7}{8} = \frac{11}{2} \cdot \frac{71}{8}$$

$$= \frac{686}{9} \qquad\qquad\qquad\qquad\qquad\qquad = \frac{781}{16}$$

$$= 76\frac{2}{9}; \qquad\qquad\qquad\qquad\qquad\qquad = 48\frac{13}{16}.$$

Since the cross products are not the same, $76\frac{2}{9} \neq 48\frac{13}{16}$, we know that the numbers are not proportional.

Do Exercise 6.

**6.** Determine whether $4\frac{2}{3}$, $5\frac{1}{2}$ and 14, $16\frac{1}{2}$ are proportional.

## b Solving Proportions

Often one of the four numbers in a proportion is unknown. Cross products can be used to find the missing number and "solve" the proportion.

**EXAMPLE 5** Solve: $\frac{x}{3} = \frac{4}{6}$.

We form an equivalent equation by equating cross products. Then we solve for $x$.

$$\frac{x}{3} = \frac{4}{6}$$

$$x \cdot 6 = 3 \cdot 4 \qquad \text{Equating cross products (finding cross products and setting them equal)}$$

$$\frac{6x}{6} = \frac{12}{6} \qquad \text{Dividing both sides by 6}$$

$$x = \frac{12}{6} = 2$$

We can check that 2 is the solution by replacing $x$ with 2 and finding cross products:

$$2 \cdot 6 = 12 \qquad \frac{2}{3} \overset{?}{=} \frac{4}{6} \qquad 3 \cdot 4 = 12.$$

Since the cross products are the same, it follows that $\frac{2}{3} = \frac{4}{6}$. Thus, the pairs of numbers 2, 3 and 4, 6 are proportional, and 2 is the solution.

Do Exercise 7.

**7.** Solve: $\frac{x}{63} = \frac{2}{9}$.

**SOLVING PROPORTIONS**

To solve $\frac{a}{b} = \frac{c}{d}$ for a specific variable, equate cross products and then divide on both sides to get that variable alone. (Assume $b, d \neq 0$.)

**EXAMPLE 6** Solve: $\dfrac{x}{7} = \dfrac{5}{3}$. Write a mixed numeral for the answer.

We have

$$\frac{x}{7} = \frac{5}{3}$$

$$x \cdot 3 = 7 \cdot 5 \qquad \text{Equating cross products}$$

$$\frac{3x}{3} = \frac{35}{3} \qquad \text{Dividing both sides by 3}$$

$$x = \frac{35}{3}, \text{ or } 11\frac{2}{3}.$$

The solution is $11\frac{2}{3}$.

Do Exercise 8.

**8.** Solve: $\dfrac{x}{9} = \dfrac{5}{4}$. Write a mixed numeral for the answer.

**EXAMPLE 7** Solve: $\dfrac{7.7}{15.4} = \dfrac{y}{2.2}$.

We have

$$\frac{7.7}{15.4} = \frac{y}{2.2}$$

$$(7.7)(2.2) = 15.4y \qquad \text{Equating cross products}$$

$$\frac{(7.7)(2.2)}{15.4} = \frac{15.4y}{15.4} \qquad \text{Dividing both sides by 15.4}$$

$$\frac{16.94}{15.4} = y \qquad \text{Simplifying}$$

$$1.1 = y. \qquad \text{Dividing:}$$

$$
\begin{array}{r}
1.1 \\
15.4 \,\overline{\smash{\big)}\, 16.9{\wedge}4} \\
\underline{15\ 4}\phantom{0} \\
1\ 5\ 4 \\
\underline{1\ 5\ 4} \\
0
\end{array}
$$

The solution is 1.1.

**EXAMPLE 8** Solve: $\dfrac{8}{x} = \dfrac{5}{3}$. Write decimal notation for the answer.

We have

$$\frac{8}{x} = \frac{5}{3}$$

$$8 \cdot 3 = x \cdot 5 \qquad \text{Equating cross products}$$

$$\frac{8 \cdot 3}{5} = \frac{x \cdot 5}{5} \qquad \text{Dividing both sides by 5}$$

$$\frac{24}{5} = x \qquad \text{Simplifying}$$

$$4.8 = x. \qquad \text{Dividing}$$

The solution is 4.8.

Do Exercises 9 and 10.

Solve: Write decimal notation for the answer.

**9.** $\dfrac{21}{5} = \dfrac{n}{2.5}$

**10.** $\dfrac{6}{x} = \dfrac{25}{11}$

*Answers*

**8.** $11\dfrac{1}{4}$     **9.** 10.5     **10.** 2.64

**EXAMPLE 9**  Solve: $\dfrac{3.4}{4.93} = \dfrac{10}{n}$.

We have

$$\dfrac{3.4}{4.93} = \dfrac{10}{n}$$

$3.4n = 4.93 \times 10$      Equating cross products

$\dfrac{3.4n}{3.4} = \dfrac{49.3}{3.4}$      Dividing by 3.4

$n = \dfrac{49.3}{3.4}$      Simplifying

$n = 14.5.$      Dividing

The solution is 14.5.

**EXAMPLE 10**  Solve: $\dfrac{4\frac{2}{3}}{5\frac{1}{2}} = \dfrac{14}{x}$.

We have

$$\dfrac{4\frac{2}{3}}{5\frac{1}{2}} = \dfrac{14}{x}$$

$4\dfrac{2}{3} \cdot x = 5\dfrac{1}{2} \cdot 14$      Equating cross products

$\dfrac{14}{3} \cdot x = \dfrac{11}{2} \cdot 14$      Converting to fraction notation

$\dfrac{3}{14} \cdot \dfrac{14}{3} \cdot x = \dfrac{3}{14} \cdot \dfrac{11}{2} \cdot 14$      Multiplying both sides by $\dfrac{3}{14}$

$x = \dfrac{3 \cdot 11 \cdot \cancel{14}}{\cancel{14} \cdot 2}$      Multiplying

$x = \dfrac{11 \cdot 3}{2}$      Simplifying by removing a factor of 1: $\dfrac{14}{14} = 1$

$x = \dfrac{33}{2}$, or $16\dfrac{1}{2}$.

The solution is $\dfrac{33}{2}$, or $16\dfrac{1}{2}$.

Do Exercises 11 and 12.

**Calculator Corner**

**Solving Proportions**

We can use a calculator when solving proportions. In Example 7, for instance, after equating cross products and dividing by 15.4 on both sides, we have

$$y = \dfrac{(7.7)(2.2)}{15.4}.$$

To compute $y$ on a calculator, we can press $\boxed{7}\ \boxed{.}\ \boxed{7}\ \boxed{\times}\ \boxed{2}\ \boxed{.}\ \boxed{2}\ \boxed{\div}$ $\boxed{1}\ \boxed{5}\ \boxed{.}\ \boxed{4}\ \boxed{=}$. The result is 1.1, so $y = 1.1$.

**Exercises:**   Solve each proportion.

1. $\dfrac{15.75}{20} = \dfrac{a}{35}$

2. $\dfrac{32}{x} = \dfrac{25}{20}$

3. $\dfrac{t}{57} = \dfrac{17}{64}$

4. $\dfrac{71.2}{a} = \dfrac{42.5}{23.9}$

5. $\dfrac{29.6}{3.15} = \dfrac{x}{4.23}$

6. $\dfrac{a}{3.01} = \dfrac{1.7}{0.043}$

Solve.

**11.** $\dfrac{0.4}{0.9} = \dfrac{4.8}{t}$

**12.** $\dfrac{8\frac{1}{3}}{x} = \dfrac{10\frac{1}{2}}{3\frac{3}{4}}$

*Answers*

**11.** 10.8    **12.** $\dfrac{125}{42}$, or $2\dfrac{41}{42}$

**a**    Determine whether the two pairs of numbers are proportional.

**1.** 5, 6 and 7, 9

**2.** 7, 5 and 6, 4

**3.** 1, 2 and 10, 20

**4.** 7, 3 and 21, 9

**5.** 2.4, 3.6 and 1.8, 2.7

**6.** 4.5, 3.8 and 6.7, 5.2

**7.** $5\frac{1}{3}, 8\frac{1}{4}$ and $2\frac{1}{5}, 9\frac{1}{2}$

**8.** $2\frac{1}{3}, 3\frac{1}{2}$ and 14, 21

**b**    Solve.

**9.** $\dfrac{18}{4} = \dfrac{x}{10}$

**10.** $\dfrac{x}{45} = \dfrac{20}{25}$

**11.** $\dfrac{x}{8} = \dfrac{9}{6}$

**12.** $\dfrac{8}{10} = \dfrac{n}{5}$

**13.** $\dfrac{t}{12} = \dfrac{5}{6}$

**14.** $\dfrac{12}{4} = \dfrac{x}{3}$

**15.** $\dfrac{2}{5} = \dfrac{8}{n}$

**16.** $\dfrac{10}{6} = \dfrac{5}{x}$

**17.** $\dfrac{n}{15} = \dfrac{10}{30}$

**18.** $\dfrac{2}{24} = \dfrac{x}{36}$

**19.** $\dfrac{16}{12} = \dfrac{24}{x}$

**20.** $\dfrac{8}{12} = \dfrac{20}{x}$

**21.** $\dfrac{6}{11} = \dfrac{12}{x}$

**22.** $\dfrac{8}{9} = \dfrac{32}{n}$

**23.** $\dfrac{20}{7} = \dfrac{80}{x}$

**24.** $\dfrac{36}{x} = \dfrac{9}{5}$

**25.** $\dfrac{12}{9} = \dfrac{x}{7}$

**26.** $\dfrac{x}{20} = \dfrac{16}{15}$

**27.** $\dfrac{x}{13} = \dfrac{2}{9}$

**28.** $\dfrac{8}{11} = \dfrac{x}{5}$

**29.** $\dfrac{100}{25} = \dfrac{20}{n}$

**30.** $\dfrac{35}{125} = \dfrac{7}{m}$

**31.** $\dfrac{6}{y} = \dfrac{18}{15}$

**32.** $\dfrac{15}{y} = \dfrac{3}{4}$

**33.** $\dfrac{x}{3} = \dfrac{0}{9}$

**34.** $\dfrac{x}{6} = \dfrac{1}{6}$

**35.** $\dfrac{1}{2} = \dfrac{7}{x}$

**36.** $\dfrac{2}{5} = \dfrac{12}{x}$

**37.** $\dfrac{1.2}{4} = \dfrac{x}{9}$

**38.** $\dfrac{x}{11} = \dfrac{7.1}{2}$

**39.** $\dfrac{8}{2.4} = \dfrac{6}{y}$

**40.** $\dfrac{3}{y} = \dfrac{5}{4.5}$

**41.** $\dfrac{t}{0.16} = \dfrac{0.15}{0.40}$

**42.** $\dfrac{0.12}{0.04} = \dfrac{t}{0.32}$

**43.** $\dfrac{0.5}{n} = \dfrac{2.5}{3.5}$

**44.** $\dfrac{6.3}{0.9} = \dfrac{0.7}{n}$

**45.** $\dfrac{1.28}{3.76} = \dfrac{4.28}{y}$

**46.** $\dfrac{10.4}{12.4} = \dfrac{6.76}{t}$

**47.** $\dfrac{7}{\frac{1}{4}} = \dfrac{28}{x}$

**48.** $\dfrac{5}{\frac{1}{3}} = \dfrac{3}{x}$

**49.** $\dfrac{\frac{1}{5}}{\frac{1}{10}} = \dfrac{\frac{1}{10}}{x}$

**50.** $\dfrac{\frac{1}{4}}{\frac{1}{2}} = \dfrac{\frac{1}{2}}{x}$

**51.** $\dfrac{y}{\frac{3}{5}} = \dfrac{\frac{7}{12}}{\frac{14}{15}}$

**52.** $\dfrac{\frac{5}{8}}{\frac{5}{4}} = \dfrac{y}{\frac{3}{2}}$

**53.** $\dfrac{x}{1\frac{3}{5}} = \dfrac{2}{15}$

**54.** $\dfrac{1}{7} = \dfrac{x}{4\frac{1}{2}}$

**55.** $\dfrac{2\frac{1}{2}}{3\frac{1}{3}} = \dfrac{x}{4\frac{1}{4}}$

**56.** $\dfrac{3\frac{1}{2}}{y} = \dfrac{6\frac{1}{2}}{4\frac{2}{3}}$

**57.** $\dfrac{5\frac{1}{5}}{6\frac{1}{6}} = \dfrac{y}{3\frac{1}{2}}$

**58.** $\dfrac{10\frac{3}{8}}{12\frac{2}{3}} = \dfrac{5\frac{3}{4}}{y}$

## Skill Maintenance

In each of Exercises 59–66, fill in the blank with the correct term from the given list. Some of the choices may not be used.

**59.** A ratio is the _____ of two quantities.   [7.1a]

**60.** A number is divisible by 9 if the _____ of the digits is divisible by 9.   [3.1b]

**61.** To compute a(n) _____ of a set of numbers, add the numbers and then divide by the number of addends.   [6.5a]

**62.** To convert from _____ to _____, move the decimal point two places to the right and change the $ sign in front to the ¢ sign at the end.   [5.3b]

**63.** The numbers −3 and 3 are called _____.   [2.1d]

**64.** The number 0.125 is an example of a(n) _____ decimal.   [5.5a]

**65.** The sentence $\dfrac{2}{5} \cdot \dfrac{4}{9} = \dfrac{4}{9} \cdot \dfrac{2}{5}$ illustrates the _____ law of multiplication.   [1.4a]

**66.** To solve $\dfrac{x}{a} = \dfrac{c}{d}$ for $x$, equate the _____ and divide on both sides to get $x$ alone.   [7.3b]

cross products
cents
dollars
terminating
repeating
sum
difference
opposites
reciprocals
associative
commutative
mean
median
mode
product
quotient

## Synthesis

▦ Solve.

**67.** $\dfrac{1728}{5643} = \dfrac{836.4}{x}$

**68.** $\dfrac{328.56}{627.48} = \dfrac{y}{127.66}$

**69.** $\dfrac{x}{4} = \dfrac{x-1}{6}$

**70.** $\dfrac{x+3}{5} = \dfrac{x}{7}$

**71.** Use a sequence of steps—each of which can be justified—to show that for $a, b, c, d \neq 0$,

$$\dfrac{a}{b} = \dfrac{c}{d} \quad \text{is equivalent to} \quad \dfrac{d}{b} = \dfrac{c}{a}.$$

**72.** *Strikeouts per Home Run.*   Baseball Hall-of-Famer Babe Ruth had 1330 strikeouts and 714 home runs in his career. Hall-of-Famer Mike Schmidt had 1883 strikeouts and 548 home runs in his career. Find the rate of strikeouts per home run for each player. (These rates were considered among the highest in the history of the game and yet each made the Hall of Fame.)

# Mid-Chapter Review

## Concept Reinforcement

Determine whether each statement is true or false.

_____ **1.** A ratio is a quotient of two quantities.   [7.1a]

_____ **2.** A rate is a ratio.   [7.2a]

_____ **3.** The largest size package of an item always has the lowest unit price.   [7.2b]

_____ **4.** If $\dfrac{x}{t} = \dfrac{y}{s}$, then $xy = ts$.   [7.3b]

## Guided Solutions

**5.** What is the rate, or speed, in miles per hour?   [7.2a]

120 mi,  2 hr

$$\frac{120\ \square}{\square\ \text{hr}} = \frac{120\ \square}{\square\ \text{hr}} = \square\ \text{mi/hr}$$

**6.** Solve: $\dfrac{x}{4} = \dfrac{3}{6}$.   [7.3b]

$$\frac{x}{4} = \frac{3}{6}$$

$x \cdot \square = \square \cdot 3$     Equating cross products

$\dfrac{x \cdot 6}{\square} = \dfrac{4 \cdot 3}{\square}$     Dividing on both sides

$x = \square$     Simplifying

## Mixed Review

Find fraction notation for each ratio.   [7.1a]

**7.** 4 to 7

**8.** 313 to 199

**9.** 35 to 17

**10.** 59 to 101

Find the ratio of the first number to the second and simplify.   [7.1b]

**11.** 8 to 12

**12.** 25 to 75

**13.** 32 to 28

**14.** 100 to 76

**15.** 112 to 56

**16.** 15 to 3

**17.** 2.4 to 8.4

**18.** 0.27 to 0.45

Find each rate, or speed, as a ratio of distance to time. Round to the nearest hundredth where appropriate.   [7.2a]

**19.** 243 mi, 4 hr          **20.** 146 km, 3 hr          **21.** 65 m, 5 sec          **22.** 97 ft, 6 sec

**23.** *Record Snowfall.*   The greatest recorded snowfall in a single storm occurred in the Mt. Shasta Ski Bowl in California, when 189 in. fell during a seven-day storm in 1959. What is the rate in inches per day?   [7.2a]
**Source:** U.S. Army Corps of Engineers

**24.** *Free Throws.*   During the 2009–2010 basketball season, Peja Stojakovic of the New Orleans Hornets attempted 107 free throws and made 96 of them. What is the rate in number of free throws made to number of free throws attempted? Round to the nearest thousandth.   [7.2a]
**Source:** National Basketball Association

**25.** Jerome bought an 18-oz jar of grape jelly for $2.09. What is the unit price in cents per ounce?   [7.2b]

**26.** Martha bought 12 oz of deli honey ham for $5.99. What is the unit price in cents per ounce?   [7.2b]

Determine whether the two pairs of numbers are proportional.   [7.3a]

**27.** 3, 7 and 15, 35          **28.** 9, 7 and 7, 5          **29.** 2.4, 1.5 and 3.2, 2.1          **30.** $1\frac{3}{4}, 1\frac{1}{3}$ and $8\frac{3}{4}, 6\frac{2}{3}$

Solve.   [7.3b]

**31.** $\dfrac{9}{15} = \dfrac{x}{20}$          **32.** $\dfrac{x}{24} = \dfrac{30}{18}$          **33.** $\dfrac{12}{y} = \dfrac{20}{15}$          **34.** $\dfrac{2}{7} = \dfrac{10}{y}$

**35.** $\dfrac{y}{1.2} = \dfrac{1.1}{0.6}$          **36.** $\dfrac{0.24}{0.02} = \dfrac{y}{0.36}$          **37.** $\dfrac{\frac{1}{4}}{x} = \dfrac{\frac{1}{8}}{\frac{1}{4}}$          **38.** $\dfrac{1\frac{1}{2}}{3\frac{1}{4}} = \dfrac{7\frac{1}{2}}{x}$

# Understanding Through Discussion and Writing

**39.** Can every ratio be written as the ratio of some number to 1? Why or why not?   [7.1a]

**40.** What can be concluded about a rectangle's width if the ratio of length to perimeter is 1 to 3? Make some sketches and explain your reasoning.   [7.1a, b]

**41.** Instead of equating cross products, a student solves $\frac{x}{7} = \frac{5}{3}$ by multiplying on both sides by the least common denominator, 21. Is his approach a good one? Why or why not?   [7.3b]

**42.** An instructor predicts that a student's test grade will be proportional to the amount of time the student spends studying. What is meant by this? Write an example of a proportion that involves the grades of two students and their study times.   [7.3b]

## a Applications and Problem Solving

Proportions have applications in such diverse fields as business, chemistry, health sciences, and home economics, as well as in many areas of daily life. Proportions are useful in making predictions.

**EXAMPLE 1** *Predicting Total Distance.* Donna drives her delivery van 800 mi in 3 days. At this rate, how far will she drive in 15 days?

1. **Familiarize.** We let $d$ = the distance traveled in 15 days.

2. **Translate.** We translate to a proportion. We make each side the ratio of distance to time, with distance in the numerator and time in the denominator.

$$\text{Distance in 15 days} \longrightarrow \frac{d}{15} = \frac{800}{3} \longleftarrow \text{Distance in 3 days}$$
$$\text{Time} \longrightarrow \qquad \qquad \longleftarrow \text{Time}$$

It may help to verbalize the proportion above as "the unknown distance $d$ is to 15 days as the known distance 800 mi is to 3 days."

3. **Solve.** Next, we solve the proportion:

$$3 \cdot d = 15 \cdot 800 \qquad \text{Equating cross products}$$
$$\frac{3 \cdot d}{3} = \frac{15 \cdot 800}{3} \qquad \text{Dividing both sides by 3}$$
$$d = \frac{15 \cdot 800}{3}$$
$$d = 4000. \qquad \text{Multiplying and dividing}$$

4. **Check.** We substitute into the proportion and check cross products:

$$\frac{4000}{15} = \frac{800}{3};$$
$$4000 \cdot 3 = 12{,}000; \qquad 15 \cdot 800 = 12{,}000.$$

The cross products are the same.

5. **State.** Donna will drive 4000 mi in 15 days.

Do Exercise 1.

Problems involving proportions can be translated in more than one way. For Example 1, any one of the following is also a correct translation:

$$\frac{15}{d} = \frac{3}{800}, \qquad \frac{15}{3} = \frac{d}{800}, \qquad \frac{800}{d} = \frac{3}{15}.$$

Equating the cross products in each proportion gives us the equation $3 \cdot d = 15 \cdot 800$, which is the equation we obtained in Example 1.

**1. Burning Calories.** The readout on Mary's treadmill indicates that she burns 108 calories when she walks for 24 min. How many calories will she burn if she walks at the same rate for 30 min?

*Answer*

**1.** 135 calories

**2. Determining Paint Needs.**
Lowell and Chris run a summer painting company to pay for their college expenses. They can paint 1600 ft² of clapboard with 4 gal of paint. How much paint would be needed for a building with 6000 ft² of clapboard?

**EXAMPLE 2** *Recommended Dosage.* To control a fever, a doctor suggests that a child who weighs 28 kg be given 320 mg of a liquid pain reliever. If the dosage is proportional to the child's weight, how much of the medication is recommended for a child who weighs 35 kg?

1. **Familiarize.** We let $t$ = the number of milligrams of the liquid pain reliever.

2. **Translate.** We translate to a proportion, keeping the amount of medication in the numerators.

$$\text{Medication suggested} \rightarrow \frac{320}{28} = \frac{t}{35} \leftarrow \text{Medication suggested}$$
$$\text{Child's weight} \rightarrow \qquad\qquad \leftarrow \text{Child's weight}$$

3. **Solve.** Next, we solve the proportion:

$$320 \cdot 35 = 28 \cdot t \qquad \text{Equating cross products}$$

$$\frac{320 \cdot 35}{28} = \frac{28 \cdot t}{28} \qquad \text{Dividing both sides by 28}$$

$$\frac{320 \cdot 35}{28} = t$$

$$400 = t. \qquad \text{Multiplying and dividing}$$

4. **Check.** We substitute into the proportion and check cross products:

$$\frac{320}{28} = \frac{400}{35};$$

$$320 \cdot 35 = 11{,}200; \qquad 28 \cdot 400 = 11{,}200.$$

The cross products are the same.

5. **State.** The dosage for a child who weighs 35 kg is 400 mg.

Do Exercise 2.

**EXAMPLE 3** *Purchasing Tickets.* Carey bought 8 tickets to an international food festival for $52. How many tickets could she purchase with $90?

1. **Familiarize.** We let $n$ = the number of tickets that can be purchased with $90.

2. **Translate.** We translate to a proportion, keeping the number of tickets in the numerators.

$$\text{Tickets} \rightarrow \frac{8}{52} = \frac{n}{90} \leftarrow \text{Tickets}$$
$$\text{Cost} \rightarrow \qquad\qquad \leftarrow \text{Cost}$$

*Answer*
2. 15 gal

**3. Solve.** Next, we solve the proportion:

$$52 \cdot n = 8 \cdot 90 \qquad \text{Equating cross products}$$

$$\frac{52 \cdot n}{52} = \frac{8 \cdot 90}{52} \qquad \text{Dividing both sides by 52}$$

$$n = \frac{8 \cdot 90}{52}$$

$$n \approx 13.8. \qquad \text{Multiplying and dividing}$$

Because it is impossible to buy a fractional part of a ticket, we must round our answer *down* to 13.

**4. Check.** As a check, we use a different approach: We find the cost per ticket and then divide $90 by that price. Since $52 \div 8 = 6.50$ and $90 \div 6.50 \approx 13.8$, we have a check.

**5. State.** Carey could purchase 13 tickets with $90.

Do Exercise 3.

**3. Purchasing Shirts.** If 2 shirts can be bought for $47, how many shirts can be bought with $200?

**EXAMPLE 4** *Waist-to-Hip Ratio.* To reduce the risk of heart disease, it is recommended that a man's waist-to-hip ratio be 0.9 or lower. Mac's hip measurement is 40 in. To meet the recommendation, what should his waist measurement be?

**Source:** Mayo Clinic

Waist measurement is the smallest measurement below the ribs but above the navel.

Hip measurement is the largest measurement around the widest part of the buttocks.

**1. Familiarize.** Note that $0.9 = \frac{9}{10}$. We let $w$ = Mac's waist measurement.

**2. Translate.** We translate to a proportion as follows:

$$\text{Waist measurement} \rightarrow \frac{w}{40} = \frac{9}{10} \begin{array}{l} \leftarrow \text{Recommended} \\ \leftarrow \text{waist-to-hip ratio} \end{array}$$
$$\text{Hip measurement} \rightarrow$$

**3. Solve.** Next, we solve the proportion:

$$10 \cdot w = 40 \cdot 9 \qquad \text{Equating cross products}$$

$$\frac{10 \cdot w}{10} = \frac{40 \cdot 9}{10} \qquad \text{Dividing both sides by 10}$$

$$w = \frac{40 \cdot 9}{10}$$

$$w = 36. \qquad \text{Multiplying and dividing}$$

*Answer*

**3.** 8 shirts

4. **Check.** As a check, we divide 36 by 40: $36 \div 40 = 0.9$. This is the desired ratio.

5. **State.** Mac's recommended waist measurement is 36 in. or less.

Do Exercise 4.

**EXAMPLE 5** *Construction Plans.* Architects make blueprints of projects to be constructed. These are scale drawings in which lengths are in proportion to actual sizes. The Hennesseys are adding a rectangular deck to their house. The architectural blueprints are rendered such that $\frac{3}{4}$ in. on the drawing is actually 2.25 ft on the deck. The width of the deck on the drawing is 4.3 in. How wide is the deck in reality?

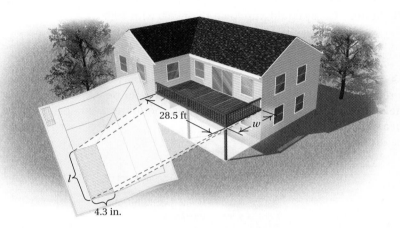

1. **Familiarize.** We let $w$ = the width of the deck.

2. **Translate.** Then we translate to a proportion, using 0.75 for $\frac{3}{4}$ in.

$$\text{Measure on drawing} \rightarrow \frac{0.75}{2.25} = \frac{4.3}{w} \leftarrow \text{Width on drawing}$$
$$\text{Measure on deck} \rightarrow \qquad\qquad \leftarrow \text{Width on deck}$$

3. **Solve.** Next, we solve the proportion:

$$0.75 \times w = 2.25 \times 4.3 \qquad \text{Equating cross products}$$

$$\frac{0.75 \times w}{0.75} = \frac{2.25 \times 4.3}{0.75} \qquad \text{Dividing both sides by 0.75}$$

$$w = \frac{2.25 \times 4.3}{0.75}$$

$$w = 12.9.$$

4. **Check.** We substitute into the proportion and check cross products:

$$\frac{0.75}{2.25} = \frac{4.3}{12.9};$$

$$0.75 \times 12.9 = 9.675; \qquad 2.25 \times 4.3 = 9.675.$$

The cross products are the same.

5. **State.** The width of the deck is 12.9 ft.

Do Exercise 5.

*Answers*

4. 34 in. or less   5. 9.5 in.

**EXAMPLE 6** *Estimating a Wildlife Population.* To determine the number of fish in a lake, a conservationist catches 225 fish, tags them, and throws them back into the lake. Later, 108 fish are caught, and it is found that 15 of them are tagged. Estimate how many fish are in the lake.

1. **Familiarize.** Our strategy is to form two different ratios which can be used to represent the ratio of tagged fish to all fish in the lake. One way to write such a ratio is simply

$$\frac{\text{Number of tagged fish}}{\text{Number of fish in lake}}.$$

A second way to represent this ratio assumes that the tagged fish become uniformly distributed throughout the lake. We then form the ratio

$$\frac{\text{Number of tagged fish caught}}{\text{Total number of fish caught}}.$$

We let $F =$ the total number of fish in the lake.

2. **Translate.** We translate to a proportion as follows:

$$\begin{array}{ll} \text{Fish tagged originally} \longrightarrow & \dfrac{225}{F} = \dfrac{15}{108}. \longleftarrow \text{Tagged fish caught later} \\ \text{Fish in lake} \longrightarrow & \phantom{\dfrac{225}{F} = \dfrac{15}{108}.} \longleftarrow \text{Fish caught later} \end{array}$$

3. **Solve.** Next, we solve the proportion:

$$225 \cdot 108 = F \cdot 15 \qquad \text{Equating cross products}$$

$$\frac{225 \cdot 108}{15} = \frac{F \cdot 15}{15} \qquad \text{Dividing both sides by 15}$$

$$\frac{225 \cdot 108}{15} = F$$

$$1620 = F. \qquad \text{Multiplying and dividing}$$

4. **Check.** We substitute into the proportion and check cross products:

$$\frac{225}{1620} = \frac{15}{108};$$

$$225 \cdot 108 = 24{,}300; \qquad 1620 \cdot 15 = 24{,}300.$$

The cross products are the same.

5. **State.** We estimate that there are 1620 fish in the lake.

Do Exercise 6.

6. **Estimating a Deer Population.** To determine the number of deer in a forest, a conservationist catches 153 deer, tags them, and releases them. Later, 62 deer are caught, and it is found that 18 of them are tagged. Estimate how many deer are in the forest.

*Answer*

6. 527 deer

# Translating for Success

1. **Calories in Cereal.** There are 140 calories in a $1\frac{1}{2}$-cup serving of Brand A cereal. How many calories are there in 6 cups of the cereal?

2. **Calories in Cereal.** There are 140 calories in 6 cups of Brand B cereal. How many calories are there in a $1\frac{1}{2}$-cup serving of the cereal?

3. **Gallons of Gasoline.** Cody's SUV traveled 310 mi on 15.5 gal of gasoline. At this rate, how many gallons would be needed to travel 465 mi?

4. **Gallons of Gasoline.** Elizabeth's new fuel-efficient car traveled 465 mi on 15.5 gal of gasoline. At this rate, how many gallons will be needed to travel 310 mi?

5. **Perimeter.** What is the perimeter of a rectangular field that measures 83.7 m by 62.4 m?

The goal of these matching questions is to practice step (2), *Translate*, of the five-step problem-solving process. Translate each word problem to an equation and select a correct translation from equations A–O.

**A.** $\dfrac{310}{15.5} = \dfrac{465}{x}$

**B.** $180 = 1\frac{1}{2} \cdot x$

**C.** $x = 71\frac{1}{8} - 76\frac{1}{2}$

**D.** $71\frac{1}{8} \cdot x = 74$

**E.** $74 \cdot 71\frac{1}{8} = x$

**F.** $x = 83.7 + 62.4$

**G.** $71\frac{1}{8} + x = 76\frac{1}{2}$

**H.** $x = 1\frac{2}{3} \cdot 180$

**I.** $\dfrac{140}{6} = \dfrac{x}{1\frac{1}{2}}$

**J.** $x = 2(83.7 + 62.4)$

**K.** $\dfrac{465}{15.5} = \dfrac{310}{x}$

**L.** $x = 83.7 \cdot 62.4$

**M.** $x = 180 \div 1\frac{2}{3}$

**N.** $\dfrac{140}{1\frac{1}{2}} = \dfrac{x}{6}$

**O.** $x = 1\frac{2}{3} \div 180$

*Answers on page A-15*

6. **Electric Bill.** Last month Todd's electric bills for his two rentals were \$83.70 and \$62.40. What was the total electric bill for the two properties?

7. **Package Tape.** A postal service center uses rolls of package tape that each contain 180 ft of tape. If it takes an average of $1\frac{2}{3}$ ft of tape per package, how many packages can be taped with one roll?

8. **Online Price.** Jane spent \$180 for an area rug in a department store. Later, she saw the same rug for sale online and realized she had paid $1\frac{1}{2}$ times the online price. What was the online price?

9. **Heights of Sons.** Henry's three sons play basketball on three different college teams. Jeff's, Jason's, and Jared's heights are 74 in., $71\frac{1}{8}$ in., and $76\frac{1}{2}$ in., respectively. How much taller is Jared than Jason?

10. **Total Investment.** An investor bought 74 shares of stock at \$71 per share. What was the total investment?

**a**    Solve.

**1.** *Study Time and Test Grades.*    An English instructor asserted that students' test grades are directly proportional to the amount of time spent studying. Lisa studies for 9 hr for a particular test and gets a score of 75. At this rate, for how many hours would she have had to study to get a score of 92?

**2.** *Study Time and Test Grades.*    A mathematics instructor asserted that students' test grades are directly proportional to the amount of time spent studying. Brent studies for 15 hr for a final exam and gets a score of 75. At this rate, what score would he have received if he had studied for 18 hr?

**3.** *Lefties.*    In a class of 40 students, on average, 6 will be left-handed. If a class includes 9 "lefties," how many students would you estimate are in the class?

**4.** *Sugaring.*    When 20 gal of maple sap are boiled down, the result is $\frac{1}{2}$ gal of maple syrup. How much sap is needed to produce 9 gal of syrup?

**Source:** University of Maine

**5.** *Gas Mileage.*    A 2010 Ford Mustang with a 5.4 L engine will travel 341 mi on 15.5 gal of gasoline in highway driving.

   **a)** How many gallons of gasoline will it take to drive 2690 mi from Boston to Phoenix?

   **b)** How far can the car be driven on 140 gal of gasoline?

**Source:** Ford Motor Company

**6.** *Gas Mileage.*    A 2010 Volkswagen Beetle convertible will travel 462 mi on 16.5 gal of gasoline in highway driving.

   **a)** How many gallons of gasoline will it take to drive 1650 mi from Pittsburgh to Albuquerque?

   **b)** How far can the car be driven on 130 gal of gasoline?

**Source:** Volkswagen of America, Inc.

**7.** *Overweight Americans.*    A recent study determined that of every 100 Americans, 66 are overweight or obese. It is estimated that the U.S. population will be about 322 million in 2015. At the given rate, how many Americans would be considered overweight or obese in 2015?

**Source:** U.S. Centers for Disease Control and Prevention

**8.** *Prevalence of Diabetes.*    A recent study determined that of every 1000 Americans in the 65- to 74-year age group, 185 have been diagnosed with diabetes. It is estimated that there will be about 26.6 million Americans in this age group in 2015. At the given rate, how many in this age group will be diagnosed with diabetes in 2015?

**Source:** U.S. Centers for Disease Control and Prevention

**9.** *Quality Control.* A quality-control inspector examined 100 lightbulbs and found 7 of them to be defective. At this rate, how many defective bulbs will there be in a lot of 2500?

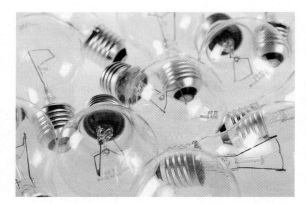

**11.** *Painting.* Fred uses 3 gal of paint to cover 1275 ft$^2$ of siding. How much siding can Fred paint with 7 gal of paint?

**13.** *Publishing.* Every 6 pages of an author's manuscript corresponds to 5 published pages. How many published pages will a 540-page manuscript become?

**15.** *Currency Exchange.* On 15 April 2010, 1 U.S. dollar was worth about 0.647374 British pound.
  **a)** How much would 50 U.S. dollars be worth in British pounds?
  **b)** How much would a car that costs 8640 British pounds cost in U.S. dollars?

**17.** *Currency Exchange.* On 15 April 2010, 1 U.S. dollar was worth about 93.588 Japanese yen.
  **a)** How much would 125 U.S. dollars be worth in Japanese yen?
  **b)** While traveling in Japan, Dana bought a vase for 5120 Japanese yen. How much did it cost in U.S. dollars?

**10.** *Grading.* A high school math teacher must grade 32 reports on famous mathematicians. She can grade 4 reports in 90 min. At this rate, how long will it take her to grade all 32 reports?

**12.** *Waterproofing.* Bonnie can waterproof 450 ft$^2$ of decking with 2 gal of sealant. How many gallons should Bonnie buy for a 1200-ft$^2$ deck?

**14.** *Turkey Servings.* An 8-lb turkey breast contains 36 servings of meat. How many pounds of turkey breast would be needed for 54 servings?

**16.** *Currency Exchange.* On 15 April 2010, 1 U.S. dollar was worth about 1.05531 Swiss francs.
  **a)** How much would 360 U.S. dollars be worth in Swiss francs?
  **b)** How much would a pair of jeans that costs 80 Swiss francs cost in U.S. dollars?

**18.** *Currency Exchange.* On 15 April 2010, 1 U.S. dollar was worth about 12.1484 Mexican pesos.
  **a)** How much would 150 U.S. dollars be worth in Mexican pesos?
  **b)** While traveling in Mexico, Jake bought a watch that cost 3600 Mexican pesos. How much did it cost in U.S. dollars?

**19.** *Mileage.* Jean bought a new car. In the first 8 months, it was driven 9000 mi. At this rate, how many miles will the car be driven in 1 yr?

**20.** *Coffee Production.* Coffee beans from 18 trees are required to produce enough coffee each year for a person who drinks 2 cups of coffee per day. Jared brews 15 cups of coffee each day for himself and his coworkers. How many coffee trees are required for this?

**21.** *Cap'n Crunch's Peanut Butter Crunch® Cereal.* The nutritional chart on the side of a box of Quaker Cap'n Crunch's Peanut Butter Crunch® cereal states that there are 110 calories in a $\frac{3}{4}$-cup serving. How many calories are there in 6 cups of the cereal?

**22.** *Rice Krispies® Cereal.* The nutritional chart on the side of a box of Kellogg's Rice Krispies® cereal states that there are 130 calories in a $1\frac{1}{4}$-cup serving. How many calories are there in 5 cups of the cereal?

**Nutrition Facts**
Serving Size 3/4 Cup (27g)

| Amount Per Serving | Cereal Alone | with 1/2 Cup Vitamin A&D Fortified Skim Milk |
|---|---|---|
| **Calories** | 110 | 150 |
| Calories from Fat | 25 | 25 |
| | % Daily Value | |
| **Total Fat** 2.5g | **4%** | **4%** |
| Saturated Fat 1g | **5%** | **6%** |
| Trans Fat 0g | | |
| Polyunsaturated Fat 0.5g | | |
| Monounsaturated Fat 1g | | |
| **Cholesterol** 0mg | **0%** | **1%** |
| **Sodium** 200mg | **8%** | **10%** |
| **Potassium** 65mg | **2%** | **7%** |
| **Total Carbohydrate** 21g | **7%** | **9%** |
| Dietary Fiber 1g | **3%** | **3%** |
| Sugars 9g | | |
| Other Carbohydrate 11g | | |
| **Protein** 2g | | |

**Nutrition Facts**
Serving Size 1$\frac{1}{4}$ Cups (33g/1.2oz )

| Amount Per Serving | Cereal | Cereal with 1/2 Cup Vitamins A&D Fat Free Milk |
|---|---|---|
| **Calories** | 130 | 170 |
| Calories from Fat | 0 | 0 |
| | % Daily Value | |
| **Total Fat** 0g | **0%** | **0%** |
| Saturated Fat 0g | **0%** | **0%** |
| *Trans* Fat 0g | | |
| **Cholesterol** 0mg | **0%** | **0%** |
| **Sodium** 220mg | **9%** | **12%** |
| **Potassium** 30mg | **1%** | **7%** |
| **Total Carbohydrate** 29g | **10%** | **11%** |
| Dietary Fiber 0g | **0%** | **0%** |
| Sugars 4g | | |
| Other Carbohydrate 25g | | |
| **Protein** 2g | | |

**23.** *Metallurgy.* In a metal alloy, the ratio of zinc to copper is 3 to 13. If there are 520 lb of copper, how many pounds of zinc are there?

**24.** *Class Size.* A college advertises that its student-to-faculty ratio is 14 to 1. If 56 students register for Introductory Spanish, how many sections of the course would you expect to see offered?

**25.** *Painting.* Helen can paint 950 ft$^2$ with 2 gal of paint. How many 1-gal cans does she need in order to paint a 30,000-ft$^2$ wall?

**26.** *Snow to Water.* Under typical conditions, $1\frac{1}{2}$ ft of snow will melt to 2 in. of water. To how many inches of water will $5\frac{1}{2}$ ft of snow melt?

**27.** *Grass-Seed Coverage.* It takes 60 oz of grass seed to seed 3000 ft² of lawn. At this rate, how much would be needed for 5000 ft² of lawn?

**28.** *Grass-Seed Coverage.* In Exercise 27, how much seed would be needed for 7000 ft² of lawn?

**29.** *Estimating a Deer Population.* To determine the number of deer in a game preserve, a forest ranger catches 318 deer, tags them, and releases them. Later, 168 deer are caught, and it is found that 56 of them are tagged. Estimate how many deer are in the game preserve.

**30.** *Estimating a Trout Population.* To determine the number of trout in a lake, a conservationist catches 112 trout, tags them, and throws them back into the lake. Later, 82 trout are caught, and it is found that 32 of them are tagged. Estimate how many trout there are in the lake.

**31.** *Map Scaling.* On a road atlas map, 1 in. represents 16.6 mi. If two cities are 3.5 in. apart on the map, how far apart are they in reality?

**32.** *Map Scaling.* On a map, $\frac{1}{4}$ in. represents 50 mi. If two cities are $3\frac{1}{4}$ in. apart on the map, how far apart are they in reality?

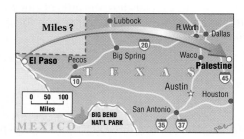

**33. Gasoline Mileage.** Nancy's van traveled 84 mi on 6.5 gal of gasoline. At this rate, how many gallons would be needed to travel 126 mi?

**34. Bicycling.** Roy bicycled 234 mi in 14 days. At this rate, how far would Roy bicycle in 42 days?

**35. Home Runs.** After playing 8 games in the 2010 Major League Baseball season, Ryan Howard of the Philadelphia Phillies had hit 3 home runs.

**a)** At this rate, how many games would it take him to hit 50 home runs?

**b)** At this rate, how many home runs would Howard hit in the 162-game baseball season?

**Source:** Major League Baseball

**36. Hits.** After playing 10 games in the 2010 Major League Baseball season, Ichiro Suzuki of the Seattle Mariners had 10 hits.

**a)** At this rate, how many games would it take him to get 200 hits?

**b)** At this rate, how many hits would Suzuki get in the 162-game baseball season?

**Source:** Major League Baseball

## Skill Maintenance

Determine whether each number is prime, composite, or neither.   [3.2b]

**37.** 1            **38.** 83            **39.** 28            **40.** 93            **41.** 47

Find the prime factorization of each number.   [3.2c]

**42.** 808            **43.** 56            **44.** 866            **45.** 93            **46.** 2020

## Synthesis

**47.** ▦ Carney College is expanding from 850 to 1050 students. To avoid any rise in the student-to-faculty ratio, the faculty of 69 professors must also be increased. How many new faculty positions should be created?

**48.** ▦ In recognition of her outstanding work, Sheri's salary has been increased from $26,000 to $29,380. Tim is earning $23,000 and is requesting a proportional raise. How much more should he ask for?

**49. Baseball Statistics.** Cy Young, one of the greatest baseball pitchers of all time, gave up an average of 2.63 earned runs every 9 innings. Young pitched 7356 innings, more than anyone else in the history of baseball. How many earned runs did he give up?

**50.** ▦ **Real-Estate Values.** According to Coldwell Banker Real Estate Corporation, a home selling for $189,000 in Austin, Texas, would sell for $437,850 in Denver, Colorado. How much would a $350,000 home in Denver sell for in Austin? Round to the nearest $1000.

**Source:** Coldwell Banker Real Estate Corporation

**51.** ▦ The ratio 1 : 3 : 2 is used to estimate the relative costs of a CD player, receiver, and speakers when shopping for a sound system. That is, the receiver should cost three times the amount spent on the CD player and the speakers should cost twice the amount spent on the CD player. If you had $900 to spend, how would you allocate the money, using this ratio?

# 7.5

# Geometric Applications

## OBJECTIVES

**a** Find lengths of sides of similar triangles using proportions.

**b** Use proportions to find lengths in pairs of figures that differ only in size.

## a Proportions and Similar Triangles

Look at the pair of triangles below. Note that they appear to have the same shape, but their sizes are different. These are examples of **similar triangles**. You can imagine using a magnifying glass to enlarge the smaller triangle and get the larger. This process works because the corresponding sides of each triangle have the same ratio. That is, the following proportion is true.

$$\frac{a}{d} = \frac{b}{e} = \frac{c}{f}$$

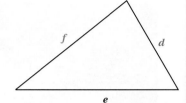

### SIMILAR TRIANGLES

**Similar triangles** have the same shape. The lengths of their corresponding sides have the same ratio—that is, they are proportional.

**EXAMPLE 1** The triangles below are similar triangles. Find the missing length $x$.

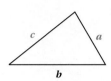

The ratio of $x$ to 9 is the same as the ratio of 24 to 8 or 21 to 7. We get the proportions

$$\frac{x}{9} = \frac{24}{8} \quad \text{and} \quad \frac{x}{9} = \frac{21}{7}.$$

> For both proportions, the sides of the larger triangle are in the numerators, and the sides of the smaller triangle are in the denominators.

We can solve either one of these proportions. We use the first one:

$$\frac{x}{9} = \frac{24}{8}$$

$x \cdot 8 = 9 \cdot 24$     Equating cross products

$\dfrac{x \cdot 8}{8} = \dfrac{9 \cdot 24}{8}$     Dividing both sides by 8

$x = 27.$     Simplifying

The missing length $x$ is 27. Other proportions could also be used.

Do Exercise 1.

**1.** This pair of triangles is similar. Find the missing length $x$.

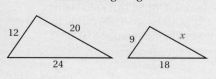

*Answer*

**1.** 15

Similar triangles and proportions can often be used to find lengths that would ordinarily be difficult to measure. For example, we could find the height of a flagpole without climbing it or the distance across a river without crossing it.

**EXAMPLE 2** How tall is a flagpole that casts a 56-ft shadow at the same time that a 6-ft man casts a 5-ft shadow?

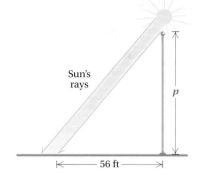

If we use the sun's rays to represent the third side of the triangle in our drawing of the situation, we see that we have similar triangles. Let $p =$ the height of the flagpole. The ratio of 6 to $p$ is the same as the ratio of 5 to 56. Thus, we have the proportion

Height of man $\longrightarrow \dfrac{6}{p} = \dfrac{5}{56}.$ $\longleftarrow$ Length of shadow of man
Height of pole $\longrightarrow$ $\longleftarrow$ Length of shadow of pole

*Solve:* $6 \cdot 56 = 5 \cdot p$     Equating cross products

$\dfrac{6 \cdot 56}{5} = \dfrac{5 \cdot p}{5}$     Dividing both sides by 5

$\dfrac{6 \cdot 56}{5} = p$     Simplifying

$67.2 = p.$

The height of the flagpole is 67.2 ft.

> Do Exercise 2.

**EXAMPLE 3** *Rafters of a House.* Carpenters use similar triangles to determine the lengths of rafters for a house. They first choose the pitch of the roof, or the ratio of the rise over the run. Then, using a triangle with that ratio, they calculate the length of the rafter needed for the house. Loren is constructing rafters for a roof with a 6/12 pitch on a house that is 30 ft wide. Using a rafter guide, Loren knows that the rafter length corresponding to the 6/12 pitch is 13.4. Find the length $x$ of the rafter of the house to the nearest tenth of a foot.

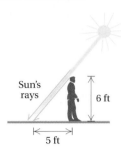

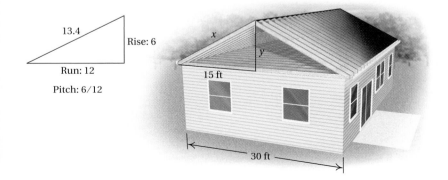

**2.** How tall is a flagpole that casts a 45-ft shadow at the same time that a 5.5-ft woman casts a 10-ft shadow?

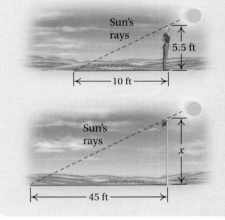

*Answer*

2. 24.75 ft

We have the proportion

Length of rafter
in 6/12 triangle → $\dfrac{13.4}{x} = \dfrac{12}{15}$. ← Run in 6/12 triangle
Length of rafter → ← Run in similar
on the house    triangle on the house

*Solve:* $13.4 \cdot 15 = x \cdot 12$     Equating cross products

$\dfrac{13.4 \cdot 15}{12} = \dfrac{x \cdot 12}{12}$     Dividing both sides by 12

$\dfrac{13.4 \cdot 15}{12} = x$

$16.8 \text{ ft} \approx x.$     Rounding to the nearest tenth of a foot

The length $x$ of the rafter of the house is about 16.8 ft.

Do Exercise 3.

**3. Rafters of a House.** Referring to Example 3, find the length $y$ of the rise of the rafter of the house to the nearest tenth of a foot.

## b Proportions and Other Geometric Shapes

When one geometric figure is a magnification of another, the figures are similar. Thus, the corresponding lengths are proportional.

**EXAMPLE 4** The sides in the photographs below are proportional. Find the width of the larger photograph.

We let $x =$ the width of the photograph. Then we translate to a proportion.

Larger width → $\dfrac{x}{2.5} = \dfrac{10.5}{3.5}$ ← Larger length
Smaller width →      ← Smaller length

*Solve:* $3.5 \times x = 2.5 \times 10.5$     Equating cross products

$\dfrac{3.5 \times x}{3.5} = \dfrac{2.5 \times 10.5}{3.5}$     Dividing both sides by 3.5

$x = \dfrac{2.5 \times 10.5}{3.5}$     Simplifying

$x = 7.5.$

Thus, the width of the larger photograph is 7.5 cm.

Do Exercise 4.

**4.** The sides in the photographs below are proportional. Find the width of the larger photograph.

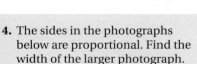
6 cm
←10 cm→

$x$
← 35 cm →

*Answers*
**3.** 7.5 ft    **4.** 21 cm

**EXAMPLE 5**   A scale model of an addition to an athletic facility is 12 cm wide at the base and rises to a height of 15 cm. If the actual base is to be 52 ft, what will the actual height of the addition be?

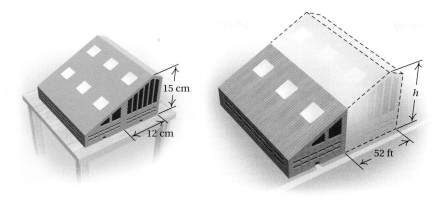

We let $h$ = the height of the addition. Then we translate to a proportion.

$$\text{Width in model} \rightarrow \frac{12}{52} = \frac{15}{h} \begin{matrix} \leftarrow \text{Height in model} \\ \leftarrow \text{Actual height} \end{matrix}$$
$$\text{Actual width} \rightarrow$$

*Solve:* $12 \cdot h = 52 \cdot 15$     Equating cross products

$\dfrac{12 \cdot h}{12} = \dfrac{52 \cdot 15}{12}$     Dividing both sides by 12

$h = \dfrac{52 \cdot 15}{12} = 65.$

Thus, the height of the addition will be 65 ft.

Do Exercise 5.

**5.** Refer to the figures in Example 5. If a skylight on the model is 3 cm wide, how wide will the actual skylight be?

---

## STUDY TIPS

### DON'T BE OVERWHELMED

Don't let the idea of starting your homework or studying for a test overwhelm you. Begin by choosing a few homework problems to do or focus on a single topic to review for a test. Often, getting started is the most difficult part of an assignment. Once you have cleared this mental hurdle, your task will seem less overwhelming.

*Answer*

**5.** 13 ft

**a** The triangles in each exercise are similar. Find the missing lengths.

**1.**

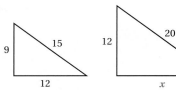

**2.**

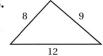

**3.**

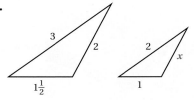

**4.**

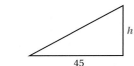

**5.**

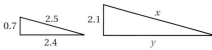

**6.**

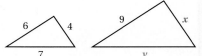

**7.**

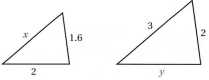

**8.**

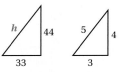

**9.** When a tree 8 m tall casts a shadow 5 m long, how long a shadow is cast by a person 2 m tall?

**10.** How tall is a flagpole that casts a 42-ft shadow at the same time that a $5\frac{1}{2}$-ft woman casts a 7-ft shadow?

**11.** How tall is a tree that casts a 27-ft shadow at the same time that a 4-ft fence post casts a 3-ft shadow?

**12.** How tall is a tree that casts a 32-ft shadow at the same time that an 8-ft light pole casts a 9-ft shadow?

**13.** Find the height $h$ of the wall.

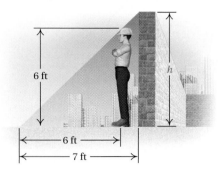

**14.** Find the length $L$ of the lake. Assume that the ratio of $L$ to 120 yd is the same as the ratio of 720 yd to 30 yd.

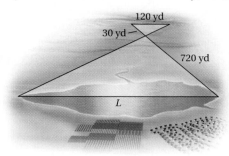

**15.** Find the distance across the river. Assume that the ratio of $d$ to 25 ft is the same as the ratio of 40 ft to 10 ft.

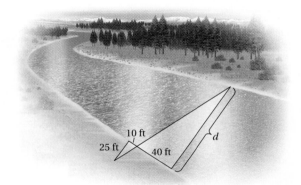

**16.** To measure the height of a hill, a string is stretched from level ground to the top of the hill. A 3-ft stick is placed under the string, touching it at point $P$, a distance of 5 ft from point $G$, where the string touches the ground. The string is then detached and found to be 120 ft long. How high is the hill?

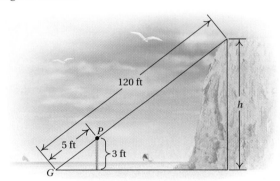

**b** In each of Exercises 17–26, the sides in each pair of figures are proportional. Find the missing lengths.

**17.**

**18.**

**19.**

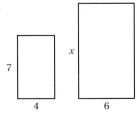

**20.**

**21.**

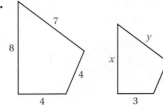

**22.**

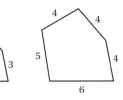

**23.**

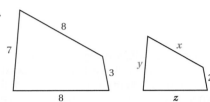

**24.**

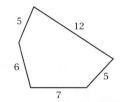

**25.**

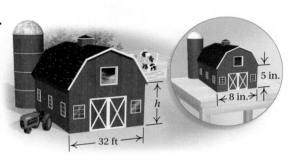

**26.**

**27.** A scale model of an addition to an athletic facility is 15 cm wide at the base and rises to a height of 19 cm. If the actual base is to be 120 ft, what will the actual height of the addition be?

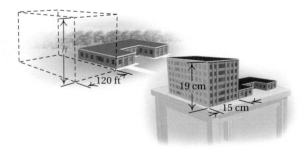

**28.** Refer to the figures in Exercise 27. If a skylight on the model is 3 cm wide, how wide will the actual skylight be?

## Skill Maintenance

**29.** *Expense Needs.* Hassam has $34.97 to spend for a book at $49.95, a calculator at $14.88, and a sweatshirt at $29.95. How much more money does Hassam need to make these purchases?  [5.8b]

**30.** Divide: $80.892 \div 8.4$.  [5.4a]

Multiply.  [5.3a]

**31.** $-8.4 \times 80.892$

**32.** $0.01 \times 274.568$

**33.** $100 \times 274.568$

**34.** $-0.002 \times (-274.568)$

Find decimal notation and round to the nearest thousandth, if appropriate.  [5.5a, b]

**35.** $\dfrac{17}{20}$

**36.** $-\dfrac{73}{40}$

**37.** $-\dfrac{10}{11}$

**38.** $\dfrac{43}{51}$

## Synthesis

*Hockey Goals.* An official hockey goal is 6 ft wide. To make scoring more difficult, goalies often locate themselves far in front of the goal to "cut down the angle." In Exercises 39 and 40, suppose that a slapshot from point $A$ is attempted and that the goalie is 2.7 ft wide. Determine how far from the goal the goalie should be located if point $A$ is the given distance from the goal. (*Hint*: First find how far the goalie should be from point $A$.)

**39.**  25 ft

**40.** 35 ft

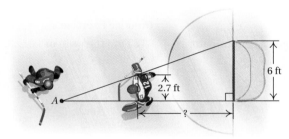

**41.** A miniature basketball hoop is built for the model referred to in Exercise 27. An actual hoop is 10 ft high. How high should the model hoop be?

 Solve. Round the answer to the nearest thousandth.

**42.** $\dfrac{8664.3}{10{,}344.8} = \dfrac{x}{9776.2}$

**43.** $\dfrac{12.0078}{56.0115} = \dfrac{789.23}{y}$

 The triangles in each exercise are similar triangles. Find the lengths not given.

**44.**

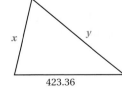

**45.**

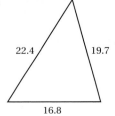

# Summary and Review

## Key Terms

ratio, p. 460
rate, p. 467

unit price, p. 468
unit rate, p. 468

proportion, p. 474
similar triangles, p. 494

## Concept Reinforcement

Determine whether each statement is true or false.

_____ 1. When we simplify a ratio like $\frac{8}{12}$, we find two other numbers in the same ratio.   [7.1b]

_____ 2. The proportion $\frac{a}{b} = \frac{c}{d}$ can also be written as $\frac{c}{a} = \frac{d}{b}$.   [7.3a]

_____ 3. Similar triangles must be the same size.   [7.5a]

_____ 4. Lengths of corresponding sides of similar figures have the same ratio.   [7.5b]

## Important Concepts

**Objective 7.1a**   Find fraction notation for ratios.

**Example**   Find the ratio of 7 to 18.

   Write a fraction with a numerator of 7 and a denominator of 18: $\frac{7}{18}$.

**Practice Exercise**

  **1.** Find the ratio of 17 to 3.

**Objective 7.1b**   Simplify ratios.

**Example**   Simplify the ratio 8 to 2.5.

   We first write the ratio in fraction notation. Next, we multiply by 1 to clear the decimal from the denominator. Then we simplify.

$$\frac{8}{2.5} = \frac{8}{2.5} \cdot \frac{10}{10} = \frac{80}{25} = \frac{5 \cdot 16}{5 \cdot 5}$$

$$= \frac{5}{5} \cdot \frac{16}{5} = \frac{16}{5}$$

**Practice Exercise**

  **2.** Simplify the ratio 3.2 to 2.8.

**Objective 7.2a**   Give the ratio of two different measures as a rate.

**Example**   A driver travels 156 mi on 6.5 gal of gas. What is the rate in miles per gallon?

$$\frac{156 \text{ mi}}{6.5 \text{ gal}} = \frac{156}{6.5} \frac{\text{mi}}{\text{gal}} = 24 \frac{\text{mi}}{\text{gal}}, \text{ or } 24 \text{ mpg}$$

**Practice Exercise**

  **3.** A student earned $120 for working 16 hr. What was the rate of pay per hour?

**Objective 7.2b** Find unit prices and use them to compare purchases.

**Example** A 16-oz can of Garcia diced tomatoes costs $1.00. A 20-oz can of Del's diced tomatoes costs $1.23. Which has the lower unit price?

We find each unit price:

For Garcia: $\dfrac{\$1.00}{16\text{ oz}} = \dfrac{100\text{ cents}}{16\text{ oz}} = 6.25\,\dfrac{\text{cents}}{\text{oz}};$

For Del's: $\dfrac{\$1.23}{20\text{ oz}} = \dfrac{123\text{ cents}}{20\text{ oz}} = 6.15\,\dfrac{\text{cents}}{\text{oz}}.$

Thus, Del's has the lower unit price.

**Practice Exercise**

4. A 28-oz jar of Brand A spaghetti sauce costs $2.79. A 32-oz jar of Brand B spaghetti sauce costs $3.29. Find the unit price of each brand and determine which is a better buy based on unit price alone.

---

**Objective 7.3a** Determine whether two pairs of numbers are proportional.

**Example** Determine whether 3, 4 and 7, 9 are proportional.

We have

$$3 \cdot 9 = 27 \qquad \overset{?}{\dfrac{3}{4} = \dfrac{7}{9}} \qquad 4 \cdot 7 = 28.$$

Since the cross products are not the same ($27 \neq 28$), $\dfrac{3}{4} \neq \dfrac{7}{9}$ and the numbers are not proportional.

**Practice Exercise**

5. Determine whether 7, 9 and 21, 27 are proportional.

---

**Objective 7.3b** Solve proportions.

**Example** Solve: $\dfrac{3}{4} = \dfrac{y}{7}$.

$$\dfrac{3}{4} = \dfrac{y}{7}$$

$3 \cdot 7 = 4 \cdot y$     Equating cross products

$\dfrac{3 \cdot 7}{4} = \dfrac{4 \cdot y}{4}$     Dividing both sides by 4

$\dfrac{21}{4} = y$

The solution is $\dfrac{21}{4}$.

**Practice Exercise**

6. Solve: $\dfrac{9}{x} = \dfrac{8}{3}$.

---

**Objective 7.4a** Solve applied problems involving proportions.

**Example** Martina bought 3 tickets to a campus theater production for $16.50. How much would 8 tickets cost?

We translate to a proportion.

$$\text{Tickets} \longrightarrow \dfrac{3}{16.50} = \dfrac{8}{c} \longleftarrow \text{Tickets}$$
$$\text{Cost} \longrightarrow \qquad\qquad \longleftarrow \text{Cost}$$

$3 \cdot c = 16.50 \cdot 8$     Equating cross products

$\dfrac{3 \cdot c}{3} = \dfrac{16.50 \cdot 8}{3}$     Dividing both sides by 3

$c = 44$

Eight tickets would cost $44.

**Practice Exercise**

7. On a map, $\frac{1}{2}$ in. represents 50 mi. If two cities are $1\frac{3}{4}$ in. apart on the map, how far apart are they in reality?

**Objective 7.5a** Find lengths of sides of similar triangles using proportions.

**Example** The triangles below are similar. Find the missing length $x$.

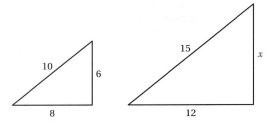

The ratio of 6 to $x$ is the same as the ratio of 8 to 12 (and also as the ratio of 10 to 15). We write and solve a proportion:

$$\frac{6}{x} = \frac{8}{12}$$

$$6 \cdot 12 = x \cdot 8$$

$$\frac{6 \cdot 12}{8} = \frac{x \cdot 8}{8}$$

$$9 = x.$$

The missing length is 9. (We could also have used other proportions, including $\dfrac{6}{x} = \dfrac{10}{15}$, to find $x$.)

**Practice Exercise**

8. The triangles below are similar. Find the missing length $y$.

 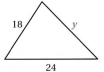

## Review Exercises

Write fraction notation for each ratio. Do not simplify. [7.1a]

1. 47 to 84

2. 46 to 1.27

3. 83 to 100

4. 0.72 to 197

5. At Preston Seafood Market, 12,480 lb of tuna and 16,640 lb of salmon were sold one year. [7.1a, b]
   a) Write fraction notation for the ratio of tuna sold to salmon sold.
   b) Write fraction notation for the ratio of salmon sold to the total number of pounds of both kinds of fish sold.

Find the ratio of the first number to the second number and simplify. [7.1b]

6. 9 to 12

7. 3.6 to 6.4

8. *Gas Mileage.* The Chrysler PT Cruiser will travel 377 mi on 14.5 gal of gasoline in highway driving. What is the rate in miles per gallon? [7.2a]
   Source: Chrysler Motor Corporation

9. *Flywheel Revolutions.* A certain flywheel makes 472,500 revolutions in 75 min. What is the rate of spin in revolutions per minute? [7.2a]

10. A lawn requires 319 gal of water for every 500 ft². What is the rate in gallons per square foot? [7.2a]

11. *Calcium Supplement.* The price for a particular calcium supplement is $18.99 for 300 tablets. Find the unit price in cents per tablet. [7.2b]

12. Nola bought a 24-oz loaf of 12-grain bread for $2.69. Find the unit price in cents per ounce. [7.2b]

**13.** *Vegetable Oil.* Find the unit price. Then determine which size has the lowest unit price. [7.2b]

| PACKAGE | PRICE | UNIT PRICE |
|---|---|---|
| 32 oz | $4.79 | |
| 48 oz | $5.99 | |
| 64 oz | $9.99 | |

Determine whether the two pairs of numbers are proportional. [7.3a]

**14.** $9, 15$ and $36, 60$

**15.** $24, 37$ and $40, 46.25$

Solve. [7.3b]

**16.** $\dfrac{8}{9} = \dfrac{x}{36}$

**17.** $\dfrac{6}{x} = \dfrac{48}{56}$

**18.** $\dfrac{120}{\frac{3}{7}} = \dfrac{7}{x}$

**19.** $\dfrac{4.5}{120} = \dfrac{0.9}{x}$

Solve. [7.4a]

**20.** *Quality Control.* A factory manufacturing computer circuits found 3 defective circuits in a lot of 65 circuits. At this rate, how many defective circuits can be expected in a lot of 585 circuits?

**21.** *Exchanging Money.* On 15 April 2010, 1 U.S. dollar was worth about 0.9968 Canadian dollar.
   **a)** How much would 250 U.S. dollars be worth in Canada?
   **b)** While traveling in Canada, Jamal saw a sweatshirt that cost 50 Canadian dollars. How much would it cost in U.S. dollars?

**22.** A train travels 448 mi in 7 hr. At this rate, how far will it travel in 13 hr?

**23.** Fifteen acres of land is required to produce 54 bushels of tomatoes. At this rate, how many acres are required to produce 97.2 bushels of tomatoes?

**24.** *Trash Production.* A study shows that 5 people generate 23 lb of trash each day. The population of Austin, Texas, is 743,074. How many pounds of trash are produced in Austin in one day?
Sources: U.S. Environmental Protection Agency; U.S. Census Bureau

**25.** *Snow to Water.* Under typical conditions, $1\frac{1}{2}$ ft of snow will melt to 2 in. of water. To how many inches of water will $4\frac{1}{2}$ ft of snow melt?

**26.** *Lawyers in Chicago.* In Illinois, there are about 4.8 lawyers for every 1000 people. The population of Chicago is 2,842,518. How many lawyers would you expect there to be in Chicago?
Sources: American Bar Association; U.S. Census Bureau

Each pair of triangles in Exercises 27 and 28 is similar. Find the missing length(s). [7.5a]

**27.**

**28.**

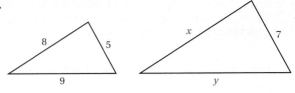

**29.** How tall is a billboard that casts a 25-ft shadow at the same time that an 8-ft sapling casts a 5-ft shadow?   [7.5a]

**30.** The lengths in the figures below are proportional. Find the missing lengths.   [7.5b]

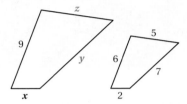

**31.** *Turkey Servings.*   A 25-lb turkey serves 18 people. Find the rate in servings per pound.   [7.2a]
  **A.** 0.36 serving/lb      **B.** 0.72 serving/lb
  **C.** 0.98 serving/lb      **D.** 1.39 servings/lb

**32.** If 3 dozen eggs cost $5.04, how much will 5 dozen eggs cost?   [7.4a]
  **A.** $6.72      **B.** $6.96
  **C.** $8.40      **D.** $10.08

## Synthesis

**33.** *Paper Towels.*   Find the unit price and determine which item would be the best buy based on unit price alone.   [7.2b]

| PACKAGE | PRICE | UNIT PRICE PER SHEET |
|---|---|---|
| 8 rolls, 60 (2 ply) sheets per roll | $6.38 | |
| 15 rolls, 60 (2 ply) sheets per roll | $13.99 | |
| 6 big rolls, 165 (2 ply) sheets per roll | $10.99 | |

**34.** It takes Yancy Martinez 10 min to type two-thirds of a page of his term paper. At this rate, how long will it take him to type a 7-page term paper?   [7.4a]

**35.** Sari is making beaded bracelets from purple beads and lavender beads. If the ratio of purple to lavender beads is 3 to 5 and each bracelet contains 40 beads, how many bracelets can Sari make from 60 purple beads? How many lavender beads will she need?   [7.4a]

**36.** A medical insurance claims service analyst can type 8500 keystrokes per hour. At this rate, how many minutes would 34,000 keystrokes require?   [7.4a]
Source: www.or.regence.com

**37.** Shine-and-Glo Painters uses 2 gal of finishing paint for every 3 gal of primer. Each gallon of finishing paint covers 450 ft$^2$. If a surface of 4950 ft$^2$ needs both primer and finishing paint, how many gallons of each should be purchased?   [7.4a]

# Understanding Through Discussion and Writing

**1.** If you were a college president, which would you prefer: a low or high faculty-to-student ratio? Why?   [7.1a]

**2.** Can unit prices be used to solve proportions that involve money? Explain why or why not.   [7.2b], [7.4a]

**3.** Write a proportion problem for a classmate to solve. Design the problem so that the solution is "Leslie would need 16 gal of gasoline in order to travel 368 mi."   [7.4a]

**4.** Is it possible for two triangles to have two pairs of sides that are proportional without the triangles being similar? Why or why not?   [7.5a]

Test

**For Extra Help**

Step-by-step test solutions are found on the Chapter Test Prep Videos available via the Video Resources on DVD, in *MyMathLab*, and on YouTube (search "BittingerPrealgebra" and click on "Channels").

Write fraction notation for each ratio. Do not simplify.

**1.** 85 to 97

**2.** 0.34 to 124

Find the ratio of the first number to the second number and simplify.

**3.** 18 to 20

**4.** 0.75 to 0.96

**5.** What is the rate in feet per second?

10 feet,   16 seconds

**6.** *Ham Servings.*   A 12-lb shankless ham contains 16 servings. What is the rate in servings per pound?

**7.** *Gas Mileage.*   The 2010 Chevrolet Malibu will travel 319 mi on 14.5 gal of gasoline in city driving. What is the rate in miles per gallon?

**Source:** General Motors Corporation

**8.** *Bagged Salad Greens.*   Ron bought a 16-oz bag of salad greens for $2.49. Find the unit price in cents per ounce.

**9.** The table below lists prices for concentrated liquid laundry detergent. Find the unit price of each size in cents per ounce. Then determine which has the lowest unit price.

| SIZE | PRICE | UNIT PRICE |
|---|---|---|
| 40 oz | $6.59 | |
| 50 oz | $6.99 | |
| 100 oz | $11.49 | |
| 150 oz | $24.99 | |

**10.** Simplify the ratio of length to width in this rectangle.

0.15

0.32

Determine whether the two pairs of numbers are proportional.

**11.** 7, 8  and  63, 72

**12.** 1.3, 3.4  and  5.6, 15.2

Solve.

**13.** $\dfrac{9}{4} = \dfrac{27}{x}$

**14.** $\dfrac{150}{2.5} = \dfrac{x}{6}$

**15.** $\dfrac{x}{100} = \dfrac{27}{64}$

**16.** $\dfrac{68}{y} = \dfrac{17}{25}$

Solve.

**17.** *Distance Traveled.*   An ocean liner traveled 432 km in 12 hr. At this rate, how far would the boat travel in 42 hr?

**18.** *Time Loss.*   A watch loses 2 min in 10 hr. At this rate, how much will it lose in 24 hr?

**19.** *Map Scaling.* On a map, 3 in. represents 225 mi. If two cities are 7 in. apart on the map, how far are they apart in reality?

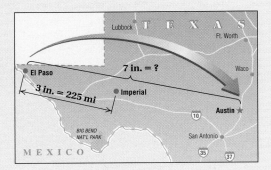

**20.** *Tower Height.* A birdhouse built on a pole that is 3 m tall casts a shadow 5 m long. At the same time, the shadow of a tower is 110 m long. How tall is the tower?

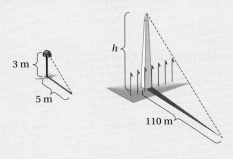

**21.** *Exchanging Money.* On 15 April 2010, 1 U.S. dollar was worth about 7.761 Hong Kong dollars.

**a)** How much would 450 U.S. dollars be worth in Hong Kong dollars?

**b)** While traveling in Hong Kong, Mitchell saw a DVD player that cost 795 Hong Kong dollars. How much would it cost in U.S. dollars?

**22.** *Thanksgiving Dinner.* A traditional turkey dinner for 8 people cost about $33.81 in a recent year. How much would it cost to serve a turkey dinner for 14 people?

**Source:** American Farm Bureau Federation

The sides in each pair of figures are proportional. Find the missing lengths.

**23.**

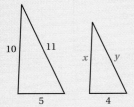

**24.**

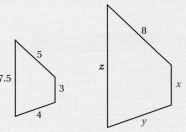

**25.** Lucita walks $4\frac{1}{2}$ mi in $1\frac{1}{2}$ hr. What is her rate in miles per hour?

**A.** $\frac{1}{3}$ mph      **B.** $1\frac{1}{2}$ mph

**C.** 3 mph      **D.** $4\frac{1}{2}$ mph

# Synthesis

Solve.

**26.** $\dfrac{x+3}{4} = \dfrac{5x+2}{8}$

**27.** $\dfrac{4-6x}{5} = \dfrac{x+3}{2}$

**28.** Nancy wants to win a gift card from the campus bookstore by guessing the number of marbles in an 8-gal jar. She knows that there are 128 oz in a gallon. She goes home and fills an 8-oz jar with 46 marbles. How many marbles should she guess are in the 8-gal jar?

# Cumulative Review

Calculate and simplify.

**1.**
$$\begin{array}{r} 2\ 7.6\ 8 \\ 3.0\ 1\ 9 \\ +\ 4\ 8\ 3.2\ 9\ 7 \\ \hline \end{array}$$

**2.**
$$\begin{array}{r} 2\frac{1}{3} \\ +\ 4\frac{5}{12} \\ \hline \end{array}$$

**3.** $\dfrac{6}{35} + \dfrac{5}{28}$

**4.**
$$\begin{array}{r} 4\ 0.2 \\ -\ \ \ 9.7\ 0\ 9 \\ \hline \end{array}$$

**5.** $-32 - (-15)$

**6.** $\dfrac{3}{20} - \dfrac{4}{15}$

**7.**
$$\begin{array}{r} 3\ 7.6\ 4 \\ \times\ \ \ \ \ 5.9 \\ \hline \end{array}$$

**8.** $-43(15)$

**9.** $2\dfrac{1}{3} \cdot 1\dfrac{2}{7}$

**10.** $2.3\overline{)\ 9\ 8.9}$

**11.** $-306 \div (-6)$

**12.** $\dfrac{7}{11} \div \dfrac{14}{33}$

**13.** Write expanded notation: 30,074.

**14.** Write a word name for 120.07.

Which number is larger?

**15.** 0.7, 0.698

**16.** $-0.799, -0.8$

**17.** Find the prime factorization of 144.

**18.** Find the LCM of 18 and 30.

**19.** What part is shaded?

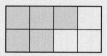

**20.** Simplify: $\dfrac{90}{144}$.

Calculate.

**21.** $\dfrac{3}{5} \times 9.53$

**22.** $7.2 \div 0.4(-1.5) + (1.2)^2$

**23.** Find the mean: 23, 49, 52, 71.

**24.** Graph on a plane: $y = -2x + 1$.

**25.** Evaluate $\dfrac{t - 7}{w}$ for $t = -3$ and $w = -2$.

**26.** Write fraction notation for the ratio 0.3 to 15. Do not simplify.

**27.** Determine whether the pairs 3, 9 and 25, 75 are proportional.

**28.** What is the rate in meters per second?

      660 meters, 12 seconds

**29.** *Unit Prices.* An 8-oz can of pineapple chunks costs $0.99. A 24.5-oz jar of pineapple chunks costs $3.29. Which has the lower unit price?

Solve.

**30.** $\dfrac{14}{25} = \dfrac{x}{54}$           **31.** $-423 = 16 \cdot t$

**32.** $\dfrac{2}{3} \cdot y = \dfrac{16}{27}$       **33.** $9x - 7 = -43$

**34.** $34.56 + n = 67.9$     **35.** $2(x - 3) + 9 = 5x - 6$

**36.** Ramona's recipe for fettuccini Alfredo has 520 calories in 1 cup. How many calories are there in $\frac{3}{4}$ cup?

**37.** A machine can stamp out 925 washers in 5 min. An order is placed for 1295 washers. How long will it take to stamp them out?

**38.** A 46-oz juice can contains $5\frac{3}{4}$ cups of juice. A recipe calls for $3\frac{1}{2}$ cups of juice. How many cups are left over?

**39.** It takes a carpenter $\frac{2}{3}$ hr to hang a door. How many doors can the carpenter hang in 8 hr?

**40.** *Car Travel.* A car travels 337.62 mi in 8 hr. How far does it travel in 1 hr?

**41.** *Shuttle Orbits.* A space shuttle made 16 orbits a day during an 8.25-day mission. How many orbits were made during the entire mission?

**42.** How many even prime numbers are there?

  **A.** 5           **B.** 3           **C.** 2

  **D.** 1          **E.** None

**43.** The gas mileage of a car is 28.16 mpg. How many gallons per mile is this?

  **A.** $\dfrac{704}{25}$      **B.** $\dfrac{25}{704}$      **C.** $\dfrac{2816}{100}$

  **D.** $\dfrac{250}{704}$     **E.** None

## Synthesis

**44.** A soccer goalie wishing to block an opponent's shot moves toward the shooter to reduce the shooter's view of the goal. If the goalie can only defend a region 10 ft wide, how far in front of the goal should the goalie be? (See the figure at right.)

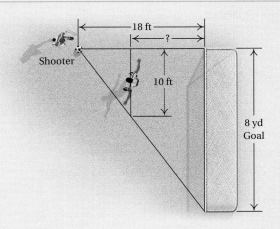

# Percent Notation

# Real-World Application

The Dow Jones Industrial Average (DJIA) plunged from 11,143 to 10,365 on
September 29, 2008. This was the largest one-day drop in its history. What was the
percent of decrease?

*Sources: Nightly Business Reports, September 29, 2008; DJIA*

***This problem appears as Example 1 in Section 8.5.***

# 8.1

# Percent Notation

**SKILL TO REVIEW**

Objective 5.3a: Multiply using decimal notation.

Multiply.

**1.** $68.3 \times 0.01$

**2.** $3013 \times 2.4$

## **a** Understanding Percent Notation

Of the total surface area of the earth, 70% is covered by water. What does this mean? It means that of every 100 square miles of the earth's surface area, 70 square miles are covered by water. Thus, 70% is a ratio of 70 to 100, or $\frac{70}{100}$.

**Source:** *The Handy Geography Answer Book*

70 of 100 squares are shaded.

70% or $\frac{70}{100}$ or 0.70 of the large square is shaded.

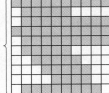

Percent notation is used extensively in our everyday lives. Here are some examples.

25.2% of the adult population in Washington, D.C., has a graduate degree.

Almost 22% of the glass that is produced is recycled.

A blood alcohol level of 0.08% is the standard used by most states as the legal limit for drunk driving.

71% of people in the United States use the Internet; 17% of people in China use the Internet.

In California, 42.5% of people age 5 and older speak a language other than English at home. In Kansas, the percentage is 10.3%.

Percent notation is often represented using a circle graph, or pie chart, to show how the parts of a quantity are related. For example, the circle graph at left illustrates the percentage of people in the United States with each of the four blood types.

> **PERCENT NOTATION**
>
> The notation **n%** means "n per hundred."

**Blood Types in the United States**

Type O 44%

Type AB 4%

Type B 10%

Type A 42%

SOURCE: bloodcenter.stanford.edu/about_blood/blood_types.html

*Answers*

*Skill to Review:*

**1.** 0.683    **2.** 7231.2

This definition leads us to the following equivalent ways of defining percent notation.

**NOTATION FOR *n*%**

**Percent notation, *n*%,** can be expressed using

ratio ⟶ $n\% = $ the ratio of $n$ to $100 = \dfrac{n}{100}$,

fraction notation ⟶ $n\% = n \times \dfrac{1}{100}$,   or

decimal notation ⟶ $n\% = n \times 0.01$.

At the airport in Oslo, Norway, 64% of all passengers arrive and leave using trains and buses. At the airport in San Francisco, the percentage is 23%.

SOURCE: Transportation Research Board

**EXAMPLE 1**   Write three kinds of notation for 35%.

| Using ratio: | $35\% = \dfrac{35}{100}$ | A ratio of 35 to 100 |
|---|---|---|
| Using fraction notation: | $35\% = 35 \times \dfrac{1}{100}$ | Replacing % with $\times \dfrac{1}{100}$ |
| Using decimal notation: | $35\% = 35 \times 0.01$ | Replacing % with $\times 0.01$ |

**EXAMPLE 2**   Write three kinds of notation for 67.8%.

| Using ratio: | $67.8\% = \dfrac{67.8}{100}$ | A ratio of 67.8 to 100 |
|---|---|---|
| Using fraction notation: | $67.8\% = 67.8 \times \dfrac{1}{100}$ | Replacing % with $\times \dfrac{1}{100}$ |
| Using decimal notation: | $67.8\% = 67.8 \times 0.01$ | Replacing % with $\times 0.01$ |

Do Exercises 1–4.

Write three kinds of notation as in Examples 1 and 2.

**1.** 70%            **2.** 23.4%

**3.** 100%           **4.** 0.6%

## b   Converting Between Percent Notation and Decimal Notation

To write decimal notation for a number like 78%, we can think of percent notation as a ratio and write

$$78\% = \dfrac{78}{100} \quad \text{Using the definition of percent as a ratio}$$
$$= 0.78. \quad \text{Dividing}$$

Similarly,

$$4.9\% = \dfrac{4.9}{100} \quad \text{Using the definition of percent as a ratio}$$
$$= 0.049. \quad \text{Dividing}$$

We could also convert 78% to decimal notation by replacing "%" with "× 0.01" and write

$$78\% = 78 \times 0.01 \quad \text{Replacing % with } \times 0.01$$
$$= 0.78. \quad \text{Multiplying}$$

*Answers*

1. $\dfrac{70}{100}$; $70 \times \dfrac{1}{100}$; $70 \times 0.01$
2. $\dfrac{23.4}{100}$; $23.4 \times \dfrac{1}{100}$; $23.4 \times 0.01$
3. $\dfrac{100}{100}$; $100 \times \dfrac{1}{100}$; $100 \times 0.01$
4. $\dfrac{0.6}{100}$; $0.6 \times \dfrac{1}{100}$; $0.6 \times 0.01$

Similarly,

$$4.9\% = 4.9 \times 0.01 \qquad \text{Replacing \% with } \times 0.01$$
$$= 0.049. \qquad \text{Multiplying}$$

Dividing by 100 amounts to moving the decimal point two places to the left, which is the same as multiplying by 0.01. This leads us to a quick way to convert from percent notation to decimal notation: We drop the percent symbol and move the decimal point two places to the left.

| To convert from percent notation to decimal notation, | 36.5% |
|---|---|
| **a)** replace the percent symbol % with $\times 0.01$, and | $36.5 \times 0.01$ |
| **b)** multiply by 0.01, which means move the decimal point two places to the left. | 0.36.5 — Move 2 places to the left.<br>36.5% = 0.365 |

**EXAMPLE 3**   Find decimal notation for 99.44%.

**a)** Replace the percent symbol with $\times 0.01$.     $99.44 \times 0.01$

**b)** Move the decimal point two places to the left.     0.99.44

Thus, 99.44% = 0.9944.

**EXAMPLE 4**   The interest rate on a $2\frac{1}{2}$-year certificate of deposit is $3\frac{3}{4}\%$. Find decimal notation for $3\frac{3}{4}\%$.

**a)** Convert $3\frac{3}{4}$ to decimal notation and replace the percent symbol with $\times 0.01$.     $3\frac{3}{4}\%$ <br> $3.75 \times 0.01$

**b)** Move the decimal point two places to the left.     0.03.75

Thus, $3\frac{3}{4}\% = 0.0375$.

> Do Exercises 5–9.

The procedure used in Examples 3 and 4 can be reversed to write a decimal, like 0.38, as an equivalent percent. To see why, note that

$$0.38 = 0.38 \times 100\% = (0.38 \times 100)\% = 38\%.$$

| To convert from decimal notation to percent notation, multiply by 100%. That is, | $0.675 = 0.675 \times 100\%$ |
|---|---|
| **a)** multiply by 100, which means move the decimal point two places to the right, and | 0.67.5 — Move 2 places to the right. |
| **b)** write a % symbol. | 67.5%<br>0.675 = 67.5% |

*Answers*

**5.** 0.34   **6.** 0.789   **7.** 0.035
**8.** 0.42   **9.** 0.0008

**EXAMPLE 5** Of the time that people declare as sick leave, 0.21 is actually used for family issues. Find percent notation for 0.21.

**Source:** CCH Inc.

a) Move the decimal point two places to the right.    0.21.

b) Write a % symbol.    21%

Thus, 0.21 = 21%.

**EXAMPLE 6** Find percent notation for 5.6.

a) Move the decimal point two places to the right, adding an extra zero.    5.60.

b) Write a % symbol.    560%

Thus, 5.6 = 560%.

**EXAMPLE 7** Find percent notation for 0.149.

a) Move the decimal point two places to the right.    0.14.9

b) Write a % symbol.    14.9%

Thus, 0.149 = 14.9%.

**EXAMPLE 8** Find percent notation for 0.00325.

a) Move the decimal point two places to the right.    0.00.325

b) Write a % symbol.    0.325%

Thus, 0.00325 = 0.325%.

Do Exercises 10–15.

It is thought that the Roman emperor Augustus began percent notation by taxing goods sold at a rate of $\frac{1}{100}$. In time, the symbol "%" evolved by interchanging the parts of the symbol "100" to "0/0" and then condensing them to "%."

Find percent notation.

**10.** 0.24    **11.** 3.47

**12.** 1    **13.** 0.005

Find percent notation for the decimal notation in each sentence.

**14. Women in Congress.** In 2010, 0.168 of the members of the United States Congress were women.

**Source:** Center for American Women and Politics at Rutgers University

**15. Soccer.** Of Americans in the 18–24 age group, 0.311 have played soccer; of the 12–17 age group, 0.396 have played.

**Source:** ESPN Sports Poll, a service of TNS Sport

## c Converting Between Fraction Notation and Percent Notation

| To convert from fraction notation to percent notation, | $\frac{3}{5}$ | Fraction notation |
|---|---|---|
| a) find decimal notation by division, and | $\begin{array}{r} 0.6 \\ 5\overline{)3.0} \\ 3\ 0 \\ \hline 0 \end{array}$ | |
| b) convert the decimal notation to percent notation. | $0.6 = 0.60 = 60\%$ <br> $\frac{3}{5} = 60\%$ | Percent notation |

**EXAMPLE 9** Find percent notation for $\frac{9}{16}$.

a) We first find decimal notation by division.

$$
\begin{array}{r}
0.5\ 6\ 2\ 5 \\
1\ 6\ )\overline{9.0\ 0\ 0\ 0} \\
\underline{8\ 0} \\
1\ 0\ 0 \\
\underline{9\ 6} \\
4\ 0 \\
\underline{3\ 2} \\
8\ 0 \\
\underline{8\ 0} \\
0
\end{array}
\qquad
\frac{9}{16} = 0.5625
$$

b) Next, we convert the decimal notation to percent notation. We move the decimal point two places to the right and write a % symbol.

$$0.56\underset{\curvearrowright}{.}25$$

$$\frac{9}{16} = 56.25\%, \text{ or } 56\tfrac{1}{4}\% \qquad 0.25 = \tfrac{1}{4}$$

Don't forget the % symbol.

Do Exercises 16 and 17.

Find percent notation.

**16.** $\dfrac{1}{4}$ **17.** $\dfrac{5}{8}$

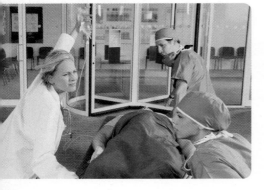

Fractions named by repeating decimals also can be converted to percent notation.

**EXAMPLE 10** *Without Health Insurance.* Approximately $\frac{1}{6}$ of all people in the United States are without health insurance. Find percent notation for $\frac{1}{6}$.
**Source:** U.S. Census Bureau, *Current Population Survey*

a) Find decimal notation by division.

$$
\begin{array}{r}
0.1\ 6\ 6 \\
6\ )\overline{1.0\ 0\ 0} \\
\underline{6} \\
4\ 0 \\
\underline{3\ 6} \\
4\ 0 \\
\underline{3\ 6} \\
4
\end{array}
$$

We get a repeating decimal: $0.16\overline{6}$.

b) Convert the answer to percent notation.

$$0.16\underset{\curvearrowright}{.}\overline{6}$$

$$\frac{1}{6} = 16.\overline{6}\%, \text{ or } 16\tfrac{2}{3}\% \qquad 0.\overline{6} = \tfrac{2}{3}$$

**18.** Water is the single most abundant chemical in the body. The human body is about $\frac{2}{3}$ water. Find percent notation for $\frac{2}{3}$.

**19.** Find percent notation: $\dfrac{5}{6}$.

Do Exercises 18 and 19.

*Answers*

**16.** 25% **17.** 62.5%, or $62\tfrac{1}{2}\%$

**18.** $66.\overline{6}\%$, or $66\tfrac{2}{3}\%$

**19.** $83.\overline{3}\%$, or $83\tfrac{1}{3}\%$

In some cases, division is not the fastest way to convert. The following are some optional ways in which conversion might be done.

**EXAMPLE 11**   Find percent notation for $\frac{69}{100}$.

We use the definition of percent as a ratio.

$$\frac{69}{100} = 69\%$$

**EXAMPLE 12**   Find percent notation for $\frac{17}{20}$.

We want to multiply by 1 to get 100 in the denominator. We think of what we must multiply 20 by in order to get 100. That number is 5, so we multiply by 1 using $\frac{5}{5}$.

$$\frac{17}{20} \cdot \frac{5}{5} = \frac{85}{100} = 85\%$$

Note that this shortcut works only when the denominator of a simplified fraction is a factor of 100.

**EXAMPLE 13**   Find percent notation for $\frac{18}{25}$.

$$\frac{18}{25} = \frac{18}{25} \cdot \frac{4}{4} = \frac{72}{100} = 72\%$$

> Do Exercises 20–23.

The method used in Example 11 is reversed when we convert from percent notation to fraction notation.

To convert from percent notation to fraction notation,

    30%   Percent notation

a)  use the definition of percent as a ratio, and

    $\dfrac{30}{100}$

b)  simplify, if possible.

    $\dfrac{3}{10}$   Fraction notation

    $30\% = \dfrac{3}{10}$

**EXAMPLE 14**   Write an equivalent fraction for 75% and simplify.

$$75\% = \frac{75}{100} \qquad \text{Using the definition of percent}$$

$$= \frac{3 \cdot 25}{4 \cdot 25}$$

$$= \frac{3}{4} \cdot \frac{25}{25} \qquad \text{Simplifying by removing a factor equal to 1: } \frac{25}{25} = 1$$

$$= \frac{3}{4}$$

Find percent notation.

**20.** $\dfrac{57}{100}$

**21.** $\dfrac{19}{25}$

**22.** $\dfrac{7}{10}$

**23.** $\dfrac{1}{4}$

---

**Calculator Corner**

**Converting from Fraction Notation to Percent Notation**   A calculator can be used to convert from fraction notation to percent notation. We simply perform the division on the calculator and then use the percent key. To convert $\frac{17}{40}$ to percent notation, for example, we press

  1  7  ÷  4  0  2nd  %  , or

  1  7  ÷  4  0  SHIFT  % .

The display reads   42.5  , so $\frac{17}{40} = 42.5\%$.

**Exercises:**   Use a calculator to find percent notation. Round to the nearest hundredth of a percent.

**1.** $\dfrac{13}{25}$      **2.** $\dfrac{5}{13}$

**3.** $\dfrac{43}{39}$      **4.** $\dfrac{12}{7}$

**5.** $\dfrac{217}{364}$      **6.** $\dfrac{2378}{8401}$

*Answers*

**20.** 57%   **21.** 76%   **22.** 70%
**23.** 25%

STUDY TIPS

## MEMORIZING

Memorizing is a very helpful tool in the study of mathematics. Don't underestimate its power as you memorize the table of decimal, fraction, and percent notation below and on the inside back cover.

**EXAMPLE 15** Write an equivalent fraction for 62.5% and simplify.

$$62.5\% = \frac{62.5}{100} \qquad \text{Using the definition of percent}$$

$$= \frac{62.5}{100} \times \frac{10}{10} \qquad \text{Multiplying by 1 to eliminate the decimal point in the numerator}$$

$$= \frac{625}{1000}$$

$$\left.\begin{array}{l} = \dfrac{5 \cdot 125}{8 \cdot 125} = \dfrac{5}{8} \cdot \dfrac{125}{125} \\[2mm] = \dfrac{5}{8} \end{array}\right\} \quad \text{Simplifying}$$

**EXAMPLE 16** Write an equivalent fraction for $16\frac{2}{3}\%$ and simplify.

$$16\frac{2}{3}\% = \frac{50}{3}\% \qquad \text{Converting from the mixed numeral to fraction notation}$$

$$= \frac{50}{3} \times \frac{1}{100} \qquad \text{Using the definition of percent}$$

$$\left.\begin{array}{l} = \dfrac{50 \cdot 1}{3 \cdot 50 \cdot 2} = \dfrac{1}{3 \cdot 2} \cdot \dfrac{50}{50} \\[2mm] = \dfrac{1}{6} \end{array}\right\} \quad \text{Simplifying}$$

Find fraction notation.

**24.** 60%  **25.** 3.25%

**26.** $66\frac{2}{3}\%$  **27.** $12\frac{1}{2}\%$

Do Exercises 24–27.

The table below lists fraction, decimal, and percent equivalents used so often that it would speed up your work if you memorized them. For example, $\frac{1}{3} = 0.\overline{3}$, so we say that the **decimal equivalent** of $\frac{1}{3}$ is $0.\overline{3}$ or that $0.\overline{3}$ has the **fraction equivalent** $\frac{1}{3}$. This table also appears on the inside back cover.

## FRACTION, DECIMAL, AND PERCENT EQUIVALENTS

| FRACTION NOTATION | $\frac{1}{10}$ | $\frac{1}{8}$ | $\frac{1}{6}$ | $\frac{1}{5}$ | $\frac{1}{4}$ | $\frac{3}{10}$ | $\frac{1}{3}$ | $\frac{3}{8}$ | $\frac{2}{5}$ | $\frac{1}{2}$ | $\frac{3}{5}$ | $\frac{5}{8}$ | $\frac{2}{3}$ | $\frac{7}{10}$ | $\frac{3}{4}$ | $\frac{4}{5}$ | $\frac{5}{6}$ | $\frac{7}{8}$ | $\frac{9}{10}$ | $\frac{1}{1}$ |
|---|---|---|---|---|---|---|---|---|---|---|---|---|---|---|---|---|---|---|---|---|
| DECIMAL NOTATION | 0.1 | 0.125 | $0.16\overline{6}$ | 0.2 | 0.25 | 0.3 | $0.33\overline{3}$ | 0.375 | 0.4 | 0.5 | 0.6 | 0.625 | $0.66\overline{6}$ | 0.7 | 0.75 | 0.8 | $0.83\overline{3}$ | 0.875 | 0.9 | 1 |
| PERCENT NOTATION | 10% | 12.5%, or $12\frac{1}{2}\%$ | $16.\overline{6}\%$, or $16\frac{2}{3}\%$ | 20% | 25% | 30% | $33.\overline{3}\%$, or $33\frac{1}{3}\%$ | 37.5%, or $37\frac{1}{2}\%$ | 40% | 50% | 60% | 62.5%, or $62\frac{1}{2}\%$ | $66.\overline{6}\%$, or $66\frac{2}{3}\%$ | 70% | 75% | 80% | $83.\overline{3}\%$, or $83\frac{1}{3}\%$ | 87.5%, or $87\frac{1}{2}\%$ | 90% | 100% |

**EXAMPLE 17** Find fraction notation for $16.\overline{6}\%$.

We can use the table above or recall that $16.\overline{6}\% = 16\frac{2}{3}\% = \frac{1}{6}$. We can also recall from our work with repeating decimals in Chapter 5 that $0.\overline{6} = \frac{2}{3}$. Then we have $16.\overline{6}\% = 16\frac{2}{3}\%$ and can proceed as in Example 16.

Do Exercises 28 and 29.

Find fraction notation.

**28.** $33.\overline{3}\%$  **29.** $83.\overline{3}\%$

*Answers*

24. $\frac{3}{5}$  25. $\frac{13}{400}$  26. $\frac{2}{3}$

27. $\frac{1}{8}$  28. $\frac{1}{3}$  29. $\frac{5}{6}$

**8.1** **Exercise Set**

For Extra Help

*MyMathLab*

Math  XL
PRACTICE

 WATCH

 DOWNLOAD

READ

 REVIEW

**a** Write three kinds of notation as in Examples 1 and 2 on p. 513.

**1.** 90%          **2.** 58.7%          **3.** 12.5%          **4.** 130%

**b** Write each percent as an equivalent decimal.

**5.** 67%          **6.** 17%          **7.** 45.6%          **8.** 76.3%

**9.** 59.01%          **10.** 30.02%          **11.** 10%          **12.** 80%

**13.** 1%          **14.** 100%          **15.** 200%          **16.** 300%

**17.** 0.1%          **18.** 0.4%          **19.** 0.09%          **20.** 0.12%

**21.** 0.18%          **22.** 5.5%          **23.** 23.19%          **24.** 87.99%

**25.** $56\frac{1}{2}$%          **26.** $61\frac{3}{4}$%          **27.** $14\frac{7}{8}$%          **28.** $93\frac{1}{8}$%

Find decimal notation for the percent notation(s) in each sentence.

**29.** *Video Games.*   According to a recent survey, 97% of the 12–17 age group play video games.
**Sources:** Pew Survey; *Time*, September 29, 2008

**30.** *Female Astronauts.*   Of the 466 astronauts who have flown in space, 10.52% are female.
**Source:** *Encyclopedia Astronautica*

**31.** *Fuel Efficiency.*   Speeding up by only 5 mph on the highway cuts fuel efficiency by approximately 7% to 8%.

**Source:** *Wall Street Journal*, "Pain Relief," by A. J. Miranda, September 15, 2008

**32.** *Foreign-Born Population.*   In 2008, the U.S. foreign-born population was 12.6%, the highest since 1920.

**Source:** U.S. Census Bureau

**33.** *High School Sports.*   During the 2007–2008 academic year, 54.8% of all high school students were involved in high school sports.

**Source:** National Federation of State High School Associations

**34.** *Eating Out.*   On a given day, 58% of all Americans eat meals and snacks away from home.

**Source:** U.S. Department of Agriculture

Write each decimal as an equivalent percent.

**35.** 0.47   **36.** 0.87   **37.** 0.03   **38.** 0.01   **39.** 8.7

**40.** 4   **41.** 0.334   **42.** 0.889   **43.** 12   **44.** 16.8

**45.** 0.4   **46.** 0.5   **47.** 0.006   **48.** 0.008   **49.** 0.017

**50.** 0.024   **51.** 0.2718   **52.** 0.8911   **53.** 0.0239   **54.** 0.00073

Find percent notation for the decimal notation(s) in each sentence.

**55.** *Wasting Food.*   Americans waste an estimated 0.27 of the food available for consumption. The waste occurs in restaurants, supermarkets, cafeterias, and household kitchens.

**Source:** *New York Times*, "One Country's Table Scraps, Another Country's Meal," by Andrew Martin, May 18, 2008

**56.** *Recycling Newspapers.*   Over 0.73 of all newspapers are recycled.

**Source:** Newspaper Association of America

**57.** *Age 65 and Older.*   In Alaska, 0.057 of the residents are age 65 and older. In Florida, 0.176 are age 65 and older.

**Source:** U.S. Census Bureau

**58.** *Dining Together.*   In 2008, 0.2 of families dined together every evening. This rate declined from 0.59 in 1987.

**Source:** Online polls at USATODAY.com

**59.** *Cancer Survival.* In 2005, the estimated 10-yr survival rate was 0.906 for children diagnosed with non-Hodgkin's lymphoma (NHL) and 0.88 for those diagnosed with acute lymphoblastic leukemia (ALL).

Source: *Journal of the National Cancer Institute*, news release, September 8, 2008

**60.** *Postsecondary Degrees.* In 2008, 0.308 of those 25 to 29 years old in the United States had attained a bachelor's degree or higher.

Source: National Center for Education Statistics, "The Condition of Education 2009"

**c** Write each fraction as an equivalent percent.

**61.** $\dfrac{41}{100}$

**62.** $\dfrac{36}{100}$

**63.** $\dfrac{5}{100}$

**64.** $\dfrac{1}{100}$

**65.** $\dfrac{2}{10}$

**66.** $\dfrac{7}{10}$

**67.** $\dfrac{1}{20}$

**68.** $\dfrac{3}{20}$

**69.** $\dfrac{1}{2}$

**70.** $\dfrac{3}{4}$

**71.** $\dfrac{7}{8}$

**72.** $\dfrac{1}{8}$

**73.** $\dfrac{4}{5}$

**74.** $\dfrac{2}{5}$

**75.** $\dfrac{2}{3}$

**76.** $\dfrac{1}{3}$

**77.** $\dfrac{1}{6}$

**78.** $\dfrac{5}{6}$

**79.** $\dfrac{3}{16}$

**80.** $\dfrac{11}{16}$

**81.** $\dfrac{31}{50}$

**82.** $\dfrac{17}{50}$

**83.** $\dfrac{4}{25}$

**84.** $\dfrac{17}{25}$

In Exercises 85–90, write percent notation for the fractions in the pie chart below. (The sum of the fractions is greater than 1 because of rounding.)

**How Food Dollars Are Spent**

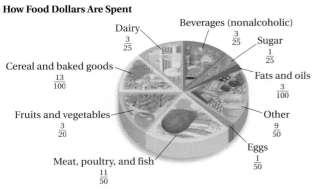

Dairy $\frac{3}{25}$

Beverages (nonalcoholic) $\frac{3}{25}$

Sugar $\frac{1}{25}$

Cereal and baked goods $\frac{13}{100}$

Fats and oils $\frac{3}{100}$

Fruits and vegetables $\frac{3}{20}$

Other $\frac{9}{50}$

Eggs $\frac{1}{50}$

Meat, poultry, and fish $\frac{11}{50}$

SOURCES: U.S. Bureau of Labor Statistics; Consumer Price Index; *The Hoosier Farmer*, Summer 2008

**85.** $\dfrac{11}{50}$

**86.** $\dfrac{9}{50}$

**87.** $\dfrac{3}{25}$

**88.** $\dfrac{1}{25}$

**89.** $\dfrac{3}{20}$

**90.** $\dfrac{13}{100}$

Find percent notation for the fraction notations in each sentence.

**91.** *Organ Transplants.* In the United States in 2008, $\frac{2}{25}$ of the organ transplants were heart transplants and $\frac{59}{100}$ were kidney transplants.

Source: National Organ Procurement and Transplantation Network

**92.** *Car Colors.* The four most popular colors for 2010 compact/sports cars were silver, black, red, and gray. Of all cars in this category, $\frac{19}{100}$ were silver, $\frac{4}{25}$ black, $\frac{4}{25}$ red, and $\frac{3}{20}$ gray.

Sources: Ward's Automotive Group; DuPont Automotive Products

Write an equivalent fraction for each percent and, if possible, simplify.

**93.** 85%

**94.** 55%

**95.** 62.5%

**96.** 12.5%

**97.** $33\frac{1}{3}\%$

**98.** $83\frac{1}{3}\%$

**99.** $16.\overline{6}\%$

**100.** $66.\overline{6}\%$

**101.** 7.25%

**102.** 4.85%

**103.** 0.8%

**104.** 0.2%

**105.** $25\frac{3}{8}\%$

**106.** $48\frac{7}{8}\%$

**107.** $78\frac{2}{9}\%$

**108.** $16\frac{5}{9}\%$

**109.** $64\frac{7}{11}\%$

**110.** $73\frac{3}{11}\%$

**111.** 150%

**112.** 110%

**113.** 0.0325%

**114.** 0.419%

**115.** $33.\overline{3}\%$

**116.** $83.\overline{3}\%$

In Exercises 117–122, find fraction notation for the percent notations in the table below.

**U.S. POPULATION BY SELECTED AGE CATEGORIES (Data have been rounded to the nearest percent.)**

| AGE CATEGORY | PERCENT OF POPULATION |
|---|---|
| 5–17 years | 18% |
| 18–24 years | 10 |
| 15–44 years | 42 |
| 18 years and older | 75 |
| 65 years and older | 12 |
| 75 years and older | 6 |

SOURCES: U.S. Census Bureau; 2006 American Community Survey

**117.** 6%

**118.** 18%

**119.** 12%

**120.** 42%

**121.** 75%

**122.** 10%

Find fraction notation for the percent notation in each sentence.

**123.** A $\frac{3}{4}$-cup serving of Post Selects Great Grains cereal with $\frac{1}{2}$ cup fat-free milk satisfies 15% of the minimum daily requirement for calcium.

**Source:** Kraft Foods Global, Inc.

**124.** A 1.8-oz serving of Kellogg's Frosted Mini-Wheats Blueberry Muffin with $\frac{1}{2}$ cup fat-free milk satisfies 35% of the minimum daily requirement for vitamin $B_{12}$.

**Source:** Kellogg, Inc.

**125.** In 2009, 19.8% of Americans age 18 and older smoked cigarettes.

**Source:** U.S. Centers for Disease Control and Prevention

**126.** In 2009, 14.3% of the adults age 18 and older in California smoked cigarettes.

**Source:** U.S. Centers for Disease Control and Prevention

Complete each table.

**127.**

| FRACTION NOTATION | DECIMAL NOTATION | PERCENT NOTATION |
|---|---|---|
| $\frac{1}{8}$ | | 12.5%, or $12\frac{1}{2}$% |
| $\frac{1}{6}$ | | |
| | | 20% |
| | 0.25 | |
| | | $33.\overline{3}$%, or $33\frac{1}{3}$% |
| | | 37.5%, or $37\frac{1}{2}$% |
| | | 40% |
| $\frac{1}{2}$ | | |

**128.**

| FRACTION NOTATION | DECIMAL NOTATION | PERCENT NOTATION |
|---|---|---|
| $\frac{3}{5}$ | | |
| | 0.625 | |
| $\frac{2}{3}$ | | |
| | 0.75 | 75% |
| $\frac{4}{5}$ | | |
| $\frac{5}{6}$ | | $83.\overline{3}$%, or $83\frac{1}{3}$% |
| $\frac{7}{8}$ | | 87.5%, or $87\frac{1}{2}$% |
| | | 100% |

**129.**

| FRACTION NOTATION | DECIMAL NOTATION | PERCENT NOTATION |
|---|---|---|
| | 0.5 | |
| $\frac{1}{3}$ | | |
| | | 25% |
| | | $16.\overline{6}$%, or $16\frac{2}{3}$% |
| | 0.125 | |
| $\frac{3}{4}$ | | |
| | $0.8\overline{3}$ | |
| $\frac{3}{8}$ | | |

**130.**

| FRACTION NOTATION | DECIMAL NOTATION | PERCENT NOTATION |
|---|---|---|
| | | 40% |
| | | 62.5%, or $62\frac{1}{2}$% |
| | 0.875 | |
| $\frac{1}{1}$ | | |
| | 0.6 | |
| | $0.\overline{6}$ | |
| $\frac{1}{5}$ | | |

## Skill Maintenance

Solve.

**131.** $13 \cdot x = 910$   [1.7b]     **132.** $15 \cdot y = 75$   [1.7b]     **133.** $0.05 \times b = 20$   [5.7a]     **134.** $3 = 0.16 \times b$   [5.7a]

**135.** $\dfrac{24}{37} = \dfrac{15}{x}$   [7.3b]     **136.** $\dfrac{17}{18} = \dfrac{x}{27}$   [7.3b]     **137.** $\dfrac{9}{10} = \dfrac{x}{5}$   [7.3b]     **138.** $\dfrac{7}{x} = \dfrac{4}{5}$   [7.3b]

## Synthesis

Find percent notation for each shaded area.

**139.**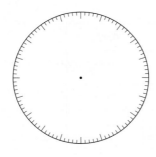

**140.**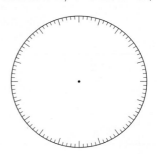

Write each number as an equivalent percent.

**141.** 🖩 $\dfrac{41}{369}$       **142.** $2.5\overline{74631}$

Write each percent as an equivalent decimal.

**143.** $\dfrac{14}{9}\%$       **144.** $\dfrac{19}{12}\%$

To draw a pie chart, or circle graph, think of a pie cut into 100 equal-sized pieces. Then to represent, say, 20%, draw a wedge equal in size to 20 of these pieces. Use the given information for Exercises 145 and 146 to complete a circle graph.

**145.** *How Teens Spend Their Money*

| | |
|---|---|
| Clothing | 34% |
| Entertainment | 22% |
| Food | 22% |
| Other | 22% |

**Source:** Rand Youth Poll, eMarketer

**146.** *Reasons for Drinking Coffee*

| | |
|---|---|
| To get going in the morning | 32% |
| Like the taste | 33% |
| Not sure | 2% |
| To relax | 4% |
| As a pick-me-up | 10% |
| It's a habit | 19% |

**Source:** LMK Associates survey for Au Bon Pain Co., Inc.

# 8.2

## Solving Percent Problems Using Percent Equations

### a) Translating to Equations

To solve a problem involving percents, it is helpful to translate first to an equation. To distinguish the method discussed in this section from that of Section 8.3, we will call these *percent equations*.

**EXAMPLES**   Translate each of the following.

**1.** 23%  of  5  is  what?

$0.23 \quad \cdot \quad 5 \quad = \quad a$     This is a *percent equation*.

**2.** What  is  11%  of  49?

$a \quad = \quad 0.11 \quad \cdot \quad 49$     Any letter can be used.

Do Margin Exercises 1 and 2.

**EXAMPLES**   Translate each of the following.

**3.** 3  is  10%  of  what?

$3 \quad = \quad 0.10 \quad \cdot \quad b$

**4.** 45%  of  what  is  23?

$0.45 \quad \times \quad b \quad = \quad 23$

-------- *Caution!* --------

Don't forget to translate percent notation to decimal notation!

Do Exercises 3 and 4.

**EXAMPLES**   Translate each of the following.

**5.** 10  is  what percent  of  20?

$10 \quad = \quad p \quad \times \quad 20$

**6.** What percent  of  50  is  7?

$p \quad \cdot \quad 50 \quad = \quad 7$

Do Exercises 5 and 6.

Each percent equation in Examples 1–6 can be written

**Amount  =  Percent number × Base**.

In each case, one of the three pieces of information is missing.

**SKILL TO REVIEW**
Objective 5.7a: Solve equations containing decimals and one variable term.

Solve.

**1.** $0.05 \cdot x = 830$

**2.** $8 \cdot y = 40.648$

Translate to an equation. Do not solve.

**1.** 12% of 50 is what?

**2.** What is 40% of 60?

Translate to an equation. Do not solve.

**3.** 45 is 20% of what?

**4.** 120% of what is 60?

Translate to an equation. Do not solve.

**5.** 16 is what percent of 40?

**6.** What percent of 84 is 10.5?

*Answers*

*Skill to Review:*
**1.** 16,600   **2.** 5.081

*Margin Exercises:*
**1.** $0.12 \cdot 50 = a$   **2.** $a = 0.4 \cdot 60$
**3.** $45 = 0.2 \cdot b$   **4.** $1.2 \cdot b = 60$
**5.** $16 = p \cdot 40$   **6.** $p \cdot 84 = 10.5$

Each year, Americans spend about $40 billion on pets; approximately 23.9% of that amount is spent on veterinary care. What is spent per year on veterinary care? (See Example 7.)

SOURCE: American Pet Products Manufacturers Association

# (b) Solving Percent Problems

In solving percent problems, we use the *Translate* and *Solve* steps in the problem-solving strategy used throughout this text.

Percent problems are actually of three different types. Although the method we present does *not* require that you be able to identify which type you are solving, it is helpful to know them. Each of the three types of percent problems depends on which of the three pieces of information is missing.

1. **Finding the *amount* (the result of taking the percent)**

   *Example:*   What   is   25%   of   60?

   *Translation:*   $a$   =   0.25   ·   60

2. **Finding the *base* (the number you are taking the percent of)**

   *Example:*   15   is   25%   of   what?

   *Translation:*   15   =   0.25   ·   $b$

3. **Finding the *percent number* (the percent itself)**

   *Example:*   15   is   what percent   of   60?

   *Translation:*   15   =   $p$   ·   60

## Finding the Amount

**EXAMPLE 7**   What is 23.9% of $40,000,000,000?

*Translate:* $a = 0.239 \cdot 40{,}000{,}000{,}000.$

*Solve:* The letter is by itself. To solve the equation, we multiply:

$$a = 0.239 \cdot 40{,}000{,}000{,}000 = 9{,}560{,}000{,}000.$$

Thus, $9,560,000,000 is 23.9% of $40,000,000,000. The answer is $9,560,000,000.

> **Do Exercise 7.**

**7.** Solve:

What is 12% of $50?

**EXAMPLE 8**   120% of 42 is what?

*Translate:* $1.20 \cdot 42 = a.$

*Solve:* The letter is by itself. To solve the equation, we carry out the calculation:

$$a = 1.20 \cdot 42 \qquad 120\% = 1.2$$
$$a = 50.4.$$

Thus, 120% of 42 is 50.4. The answer is 50.4.

> **Do Exercise 8.**

**8.** Solve:

64% of 55 is what?

*Answers*

**7.** $6    **8.** 35.20

## Finding the Base

**EXAMPLE 9**   5% of what is 20?

*Translate:* $0.05 \cdot b = 20$.

*Solve:* This time the letter is *not* by itself. To solve the equation, we divide both sides by 0.05:

$$\frac{0.05 \cdot b}{0.05} = \frac{20}{0.05} \qquad \text{Dividing both sides by } 0.05$$

$$b = \frac{20}{0.05}$$

$$b = 400.$$

Thus, 5% of 400 is 20. The answer is 400.

In a survey of a group of people, it was found that 5%, or 20 people, chose strawberry as their favorite ice cream flavor. How many people were surveyed? (See Example 9.)
SOURCE: International Ice Cream Association

**EXAMPLE 10**   $3 is 16% of what?

*Translate:*

| $3 | is | 16% | of | what? |
|-----|-----|-----|-----|-----|
| ↓ | ↓ | ↓ | ↓ | ↓ |
| 3 | = | 0.16 | · | $b$ |

*Solve:* Again, the variable is not by itself. To solve the equation, we divide both sides by 0.16:

$$\frac{3}{0.16} = \frac{0.16 \cdot b}{0.16} \qquad \text{Dividing both sides by } 0.16$$

$$\frac{3}{0.16} = b$$

$$18.75 = b.$$

Thus, $3 is 16% of $18.75. The answer is $18.75.

Do Exercises 9 and 10.

Solve.

**9.** 20% of what is 45?

**10.** $60 is 120% of what?

## Finding the Percent Number

In solving these problems, you *must* remember to convert to percent notation after you have solved the equation.

**EXAMPLE 11**   2000 is what percent of 15,000?

*Translate:*

| 2000 | is | what percent | of | 15,000? |
|------|-----|------|-----|------|
| ↓ | ↓ | ↓ | ↓ | ↓ |
| 2000 | = | $p$ | · | 15,000 |

*Solve:* To solve the equation, we divide both sides by 15,000 and convert the result to percent notation:

$$2000 = p \cdot 15{,}000$$

$$\frac{2000}{15{,}000} = \frac{p \cdot 15{,}000}{15{,}000} \qquad \text{Dividing both sides by } 15{,}000$$

$$0.13\overline{3} = p \qquad \text{Converting to decimal notation}$$

$$13\tfrac{1}{3}\% = p. \qquad \text{Converting to percent notation}$$

Thus, 2000 is $13\tfrac{1}{3}\%$ of 15,000. The answer is $13\tfrac{1}{3}\%$.

In 2009, there were about 15,000 earthquakes worldwide. Of this number, about 2000 earthquakes had magnitudes greater than 5.0. What percent of the 15,000 earthquakes had magnitudes greater than 5.0? (See Example 11.)
SOURCE: National Earthquake Information Center, U.S. Geological Survey

*Answers*
**9.** 225   **10.** $50

Make your study time efficient. Choose a location where you are unlikely to be interrupted. Avoid answering the phone and checking e-mail during your study sessions. If studying with a partner, keep the conversation focused on the material you are covering. Plan for time to socialize before or after the study time.

Solve.

**11.** 16 is what percent of 40?

**12.** What percent of $84 is $10.50?

**EXAMPLE 12**   What percent of $50 is $16?

*Translate:*   What percent   of   $50   is   $16?

$$p \quad\quad \cdot \quad 50 \quad = \quad 16$$

*Solve:*  To solve the equation, we divide both sides by 50 and convert the answer to percent notation:

$$\frac{p \cdot 50}{50} = \frac{16}{50} \qquad \text{Dividing both sides by 50}$$

$$p = \frac{16}{50}$$

$$p = \frac{16}{50} \cdot \frac{2}{2} \qquad \text{Multiplying by } \frac{2}{2}, \text{ or 1, to get a denominator of 100}$$

$$p = \frac{32}{100}$$

Had we preferred, we could have divided 16 by 50 to find decimal notation.

$$p = 32\%. \qquad \text{Converting to percent notation}$$

Thus, 32% of $50 is $16. The answer is 32%.

Do Exercises 11 and 12.

---

*Caution!*

When a question asks "what percent?", be sure to give the answer in percent notation.

---

### Calculator Corner

**Using Percents in Computations**   Many calculators have a $\boxed{\%}$ key that can be used in computations. (See the Calculator Corner on page 517.) For example, to find 11% of 49, we press $\boxed{1}\ \boxed{1}\ \boxed{\text{2nd}}\ \boxed{\%}\ \boxed{\times}\ \boxed{4}\ \boxed{9}\ \boxed{=}$, or $\boxed{4}\ \boxed{9}\ \boxed{\times}$ $\boxed{1}\ \boxed{1}\ \boxed{\text{SHIFT}}\ \boxed{\%}$. The display reads $\boxed{\phantom{xx}5.39\phantom{x}}$, so 11% of 49 is 5.39.

In Example 9, we perform the computation 20/0.05. To use the $\boxed{\%}$ key in this computation, we press $\boxed{2}\ \boxed{0}\ \boxed{\div}\ \boxed{5}\ \boxed{\text{2nd}}$ $\boxed{\%}\ \boxed{=}$, or $\boxed{2}\ \boxed{0}\ \boxed{\div}\ \boxed{5}\ \boxed{\text{SHIFT}}\ \boxed{\%}$. The result is 400.

We can also use the $\boxed{\%}$ key to find the percent number in a problem. In Example 11, for instance, we answer the question "2000 is what percent of 15,000?" On a calculator, we press $\boxed{2}\ \boxed{0}\ \boxed{0}\ \boxed{0}\ \boxed{\div}\ \boxed{1}\ \boxed{5}\ \boxed{0}\ \boxed{0}\ \boxed{0}\ \boxed{\text{2nd}}\ \boxed{\%}\ \boxed{=}$, or $\boxed{2}\ \boxed{0}\ \boxed{0}\ \boxed{0}$ $\boxed{\div}\ \boxed{1}\ \boxed{5}\ \boxed{0}\ \boxed{0}\ \boxed{0}\ \boxed{\text{SHIFT}}\ \boxed{\%}$. The result is $13.\overline{3}$, so 2000 is $13\frac{1}{3}$% of 15,000.

**Exercises:**   Use a calculator to find each of the following.

**1.** What is 12.6% of $40?

**2.** 0.04% of 28 is what?

**3.** 8% of what is 36?

**4.** $45 is 4.5% of what?

**5.** 23 is what percent of 920?

**6.** What percent of $442 is $53.04?

*Answers*

**11.** 40%   **12.** 12.5%

**a**   Translate to an equation. Do not solve.

**1.** What is 32% of 78?

**2.** 98% of 57 is what?

**3.** 89 is what percent of 99?

**4.** What percent of 25 is 8?

**5.** 13 is 25% of what?

**6.** 21.4% of what is 20?

**b**   Translate to an equation and solve.

**7.** What is 85% of 276?

**8.** What is 74% of 53?

**9.** 150% of 30 is what?

**10.** 100% of 13 is what?

**11.** What is 6% of $300?

**12.** What is 4% of $45?

**13.** 3.8% of 50 is what?

**14.** $33\frac{1}{3}$% of 480 is what?

$\left(Hint: 33\frac{1}{3}\% = \frac{1}{3}.\right)$

**15.** $39 is what percent of $50?

**16.** $16 is what percent of $90?

**17.** 20 is what percent of 10?

**18.** 60 is what percent of 20?

**19.** What percent of $300 is $150?

**20.** What percent of $50 is $40?

**21.** What percent of 80 is 100?

**22.** What percent of 60 is 15?

**23.** 20 is 50% of what?

**24.** 57 is 20% of what?

**25.** 40% of what is $16?

**26.** 100% of what is $74?

**27.** 56.32 is 64% of what?

**28.** 71.04 is 96% of what?

**29.** 70% of what is 14?

**30.** 70% of what is 35?

**31.** What is $62\frac{1}{2}$% of 10?

**32.** What is $35\frac{1}{4}$% of 1200?

**33.** What is 8.3% of $10,200?

**34.** What is 9.2% of $5600?

**35.** 2.5% of what is 30.4?

**36.** 8.2% of what is 328?

## Skill Maintenance

Write fraction notation.    [5.1b]

**37.** 0.09

**38.** 1.79

**39.** 0.875

**40.** 0.125

**41.** 0.9375

**42.** 0.6875

Write decimal notation.    [5.1b]

**43.** $\dfrac{89}{100}$

**44.** $\dfrac{7}{100}$

**45.** $\dfrac{3}{10}$

**46.** $\dfrac{17}{1000}$

## Synthesis

Solve.

**47.** What is 7.75% of $10,880?

Estimate _____

Calculate _____

**48.** 50,951.775 is what percent of 78,995?

Estimate _____

Calculate _____

**49.** *Recyclables.* It is estimated that 40% to 50% of all trash is recyclable. If a community produces 270 tons of trash, how much of their trash is recyclable?

**50.** *Batting.* An all-star baseball player gets a hit in 30% to 35% of all at-bats. If an all-star had 520 to 580 at-bats, how many hits would he have had?

**51.** 40% of $18\frac{3}{4}$% of $25,000 is what?

## a Translating to Proportions

A percent is a ratio of some number to 100. For example, 46% is the ratio $\frac{46}{100}$. The numbers 67,620,000 and 147,000,000 have the same ratio as 46 and 100.

$$\frac{46}{100} = \frac{67,620,000}{147,000,000}$$

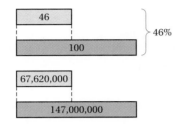

To solve a percent problem using a proportion, we translate as follows:

$$\text{Number} \rightarrow \frac{N}{100} = \frac{a}{b} \begin{array}{l} \leftarrow \text{Amount} \\ \leftarrow \text{Base} \end{array}$$
$$100 \rightarrow$$

You might find it helpful to read this as "part is to whole as part is to whole."

For example, 60% of 25 is 15 translates to

$$\frac{60}{100} = \frac{15}{25} \begin{array}{l} \leftarrow \text{Amount} \\ \leftarrow \text{Base} \end{array}$$

A clue in translating is that the base, $b$, corresponds to 100 and usually follows the wording "percent of." Also, $N\%$ always translates to $N/100$. Another aid in translating is to make a comparison drawing. To do this, we start with the percent side and list 0% at the top and 100% near the bottom. Then we estimate where the specified percent—in this case, 60%—is located. The corresponding quantities are then filled in. The base—in this case, 25—always corresponds to 100%, and the amount—in this case, 15—corresponds to the specified percent.

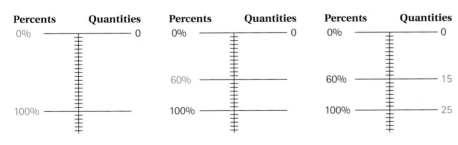

The proportion can then be read easily from the drawing: $\frac{60}{100} = \frac{15}{25}$.

## OBJECTIVES

**a** Translate percent problems to proportions.

**b** Solve basic percent problems using proportions.

### SKILL TO REVIEW

Objective 7.3b: Solve proportions.

Solve.

**1.** $\frac{3}{100} = \frac{27}{b}$

**2.** $\frac{4.3}{20} = \frac{N}{100}$

In the United States, 46% of the labor force is women. In 2007, there were approximately 147,000,000 people in the labor force. This means that about 67,620,000 were women.

SOURCES: U.S. Department of Labor; U.S. Bureau of Labor Statistics

*Answers*

*Skill to Review:*
**1.** 900    **2.** 21.5

In the examples that follow, we use comparison drawings as an aid for visualization. It is not necessary to always make a drawing to solve problems of this sort.

**EXAMPLE 1**   Translate to a proportion.

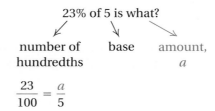

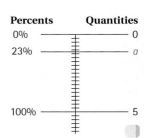

$$\frac{23}{100} = \frac{a}{5}$$

**EXAMPLE 2**   Translate to a proportion.

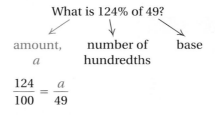

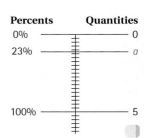

$$\frac{124}{100} = \frac{a}{49}$$

Do Exercises 1–3.

Translate to a proportion. Do not solve.

**1.** 12% of 50 is what?

**2.** What is 40% of 60?

**3.** 130% of 72 is what?

**EXAMPLE 3**   Translate to a proportion.

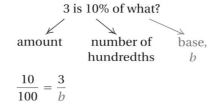

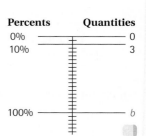

$$\frac{10}{100} = \frac{3}{b}$$

**EXAMPLE 4**   Translate to a proportion.

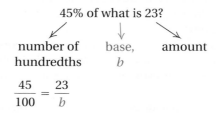

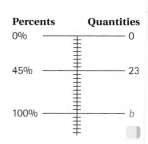

$$\frac{45}{100} = \frac{23}{b}$$

Do Exercises 4 and 5.

Translate to a proportion. Do not solve.

**4.** 45 is 20% of what?

**5.** 120% of what is 60?

**EXAMPLE 5**   Translate to a proportion.

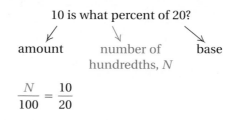

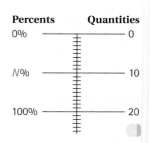

$$\frac{N}{100} = \frac{10}{20}$$

*Answers*

1. $\dfrac{12}{100} = \dfrac{a}{50}$   2. $\dfrac{40}{100} = \dfrac{a}{60}$   3. $\dfrac{130}{100} = \dfrac{a}{72}$

4. $\dfrac{20}{100} = \dfrac{45}{b}$   5. $\dfrac{120}{100} = \dfrac{60}{b}$

**EXAMPLE 6** Translate to a proportion.

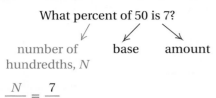

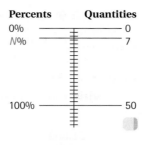

$$\frac{N}{100} = \frac{7}{50}$$

Do Exercises 6 and 7.

Translate to a proportion. Do not solve.

**6.** 16 is what percent of 40?

**7.** What percent of 84 is 10.5?

## b Solving Percent Problems

After a percent problem has been translated to a proportion, we solve as in Section 7.3.

**EXAMPLE 7** 5% of what is $20?

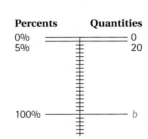

*Translate:* $\dfrac{5}{100} = \dfrac{20}{b}$

*Solve:* $5 \cdot b = 100 \cdot 20$    Equating cross products

$\dfrac{5 \cdot b}{5} = \dfrac{100 \cdot 20}{5}$    Dividing by 5

$b = \dfrac{2000}{5}$

$b = 400$    Simplifying

Thus, 5% of $400 is $20. The answer is $400.

Do Exercise 8.

**8.** Solve:

20% of what is $45?

**EXAMPLE 8** 120% of 42 is what?

*Translate:* $\dfrac{120}{100} = \dfrac{a}{42}$

*Solve:* $120 \cdot 42 = 100 \cdot a$    Equating cross products

$\dfrac{120 \cdot 42}{100} = \dfrac{100 \cdot a}{100}$    Dividing by 100

$\dfrac{5040}{100} = a$

$50.4 = a$    Simplifying

Thus, 120% of 42 is 50.4. The answer is 50.4.

Do Exercises 9 and 10.

Solve.

**9.** 64% of 55 is what?

**10.** What is 12% of 50?

*Answers*

**6.** $\dfrac{N}{100} = \dfrac{16}{40}$   **7.** $\dfrac{N}{100} = \dfrac{10.5}{84}$   **8.** $225

**9.** 35.2   **10.** 6

**EXAMPLE 9** 210 is $10\frac{1}{2}$% of what?

*Translate:* $\dfrac{210}{b} = \dfrac{10.5}{100}$      $10\frac{1}{2}\% = 10.5\%$

*Solve:* $210 \cdot 100 = b \cdot 10.5$      Equating cross products

$\dfrac{210 \cdot 100}{10.5} = \dfrac{b \cdot 10.5}{10.5}$      Dividing by 10.5

$\dfrac{21{,}000}{10.5} = b$      Multiplying and simplifying

$2000 = b$      Dividing

Thus, 210 is $10\frac{1}{2}$% of 2000. The answer is 2000.

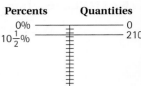

**11.** Solve:

60 is 120% of what?

Do Exercise 11.

**EXAMPLE 10** $10 is what percent of $20?

*Translate:* $\dfrac{10}{20} = \dfrac{N}{100}$

*Solve:* $10 \cdot 100 = 20 \cdot N$      Equating cross products

$\dfrac{10 \cdot 100}{20} = \dfrac{20 \cdot N}{20}$      Dividing by 20

$\dfrac{1000}{20} = N$      Multiplying and simplifying

$50 = N$      Dividing

Thus, $10 is 50% of $20. The answer is 50%.

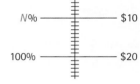

> Always "look before you leap." Many students can solve this problem mentally: $10 is half, or 50%, of $20.

**12.** Solve:

$12 is what percent of $40?

Do Exercise 12.

**EXAMPLE 11** What percent of 50 is 16?

*Translate:* $\dfrac{N}{100} = \dfrac{16}{50}$

*Solve:* $50 \cdot N = 100 \cdot 16$      Equating cross products

$\dfrac{50 \cdot N}{50} = \dfrac{100 \cdot 16}{50}$      Dividing by 50

$N = \dfrac{1600}{50}$      Multiplying and simplifying

$N = 32$      Dividing

Thus, 32% of 50 is 16. The answer is 32%.

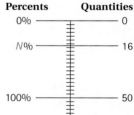

> Note when solving percent problems using proportions that $N$ is a percent and needs only a % sign.

**13.** Solve:

What percent of 84 is 10.5?

Do Exercise 13.

**Answers**

**11.** 50  **12.** 30%  **13.** 12.5%

*Caution!*

Don't forget to add the % sign when the problem asks "What percent . . . ?"

**a** Translate to a proportion. Do not solve.

**1.** What is 37% of 74?

**2.** 66% of 74 is what?

**3.** 4.3 is what percent of 5.9?

**4.** What percent of 6.8 is 5.3?

**5.** 14 is 25% of what?

**6.** 133% of what is 40?

**b** Translate to a proportion and solve.

**7.** What is 76% of 90?

**8.** What is 32% of 70?

**9.** 70% of 660 is what?

**10.** 80% of 920 is what?

**11.** What is 4% of 1000?

**12.** What is 6% of 2000?

**13.** 4.8% of 60 is what?

**14.** 63.1% of 80 is what?

**15.** $24 is what percent of $96?

**16.** $14 is what percent of $70?

**17.** 102 is what percent of 100?

**18.** 103 is what percent of 100?

**19.** What percent of $480 is $120?

**20.** What percent of $80 is $60?

**21.** What percent of 160 is 150?

**22.** What percent of 33 is 11?

**23.** $18 is 25% of what?

**24.** $75 is 20% of what?

**25.** 60% of what is 54?

**26.** 80% of what is 96?

**27.** 65.12 is 74% of what?

**28.** 63.7 is 65% of what?

**29.** 80% of what is 16?

**30.** 80% of what is 10?

**31.** What is $62\frac{1}{2}$% of 40?

**32.** What is $43\frac{1}{4}$% of 2600?

**33.** What is 9.4% of $8300?

**34.** What is 8.7% of $76,000?

**35.** 80.8 is $40\frac{2}{5}$% of what?

**36.** 66.3 is $10\frac{1}{5}$% of what?

## Skill Maintenance

Graph.   [6.4b]

**37.** $y = -\frac{1}{2}x$

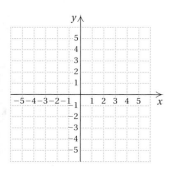

**38.** $y = 3x$

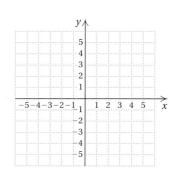

**39.** $y = 2x - 4$

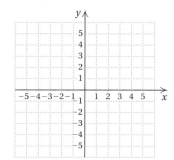

**40.** $y = \frac{1}{2}x - 3$

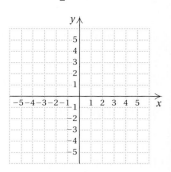

Solve.

**41.** A recipe for muffins calls for $\frac{1}{2}$ qt of buttermilk, $\frac{1}{3}$ qt of skim milk, and $\frac{1}{16}$ qt of oil. How many quarts of liquid ingredients does the recipe call for?   [4.2d]

**42.** The Ferristown School District purchased $\frac{3}{4}$ ton (T) of clay. If the clay is to be shared equally among the district's 6 art departments, how much will each art department receive?   [3.8b]

**43.** $\dfrac{5000}{t} = \dfrac{3000}{60}$   [7.3b]

**44.** $\dfrac{75}{100} = \dfrac{n}{20}$   [7.3b]

**45.** $\dfrac{x}{1.2} = \dfrac{36.2}{5.4}$   [7.3b]

**46.** $\dfrac{y}{1\frac{1}{2}} = \dfrac{2\frac{3}{4}}{22}$   [7.3b]

## Synthesis

Solve.

**47.**  What is 8.85% of $12,640?
Estimate _____
Calculate _____

**48.** ▦ 78.8% of what is 9809.024?
Estimate _____
Calculate _____

**49.** 30% of 80 is what percent of 120?

**50.** 40% of what is the same as 30% of 200?

**51.** ▦ What percent of 90 is the same as 26% of 135?

**52.** ▦ What percent of 80 is the same as 76% of 150?

# Mid-Chapter Review

## Concept Reinforcement

Determine whether each statement is true or false.

_____ **1.** When converting decimal notation to percent notation, move the decimal point two places to the right and write a percent symbol.   [8.1b]

_____ **2.** The symbol % is equivalent to $\times$ 0.10.   [8.1a]

_____ **3.** Of the numbers $\frac{1}{10}$, 1%, 0.1%, 10%, and $\frac{1}{100}$, the smallest number is 0.1%.   [8.1b, c]

## Guided Solutions

Fill in each blank with the number that creates a correct statement or solution.   [8.1b, c]

**4.** $\frac{1}{2}\% = \frac{1}{2} \cdot \frac{1}{\square} = \frac{1}{\square}$

**5.** $\frac{80}{1000} = \frac{\square}{100} = \square\%$

**6.** $5.5\% = \frac{\square}{100} = \frac{\square}{1000} = \frac{11}{\square}$

**7.** $0.375 = \frac{\square}{1000} = \frac{\square}{100} = \square\%$

**8.** Solve:  15 is what percent of 80?   [8.2b]

$15 = p \cdot \square$   Translating

$\dfrac{15}{\square} = \dfrac{p \cdot \square}{\square}$   Dividing both sides

$\dfrac{15}{\square} = p$   Simplifying

$\square = p$   Dividing

$\square\% = p$   Converting to percent notation

## Mixed Review

Find decimal notation.   [8.1b]

**9.** 28%

**10.** 0.15%

**11.** $5\frac{3}{8}\%$

**12.** 240%

Find percent notation.   [8.1b, c]

**13.** 0.71

**14.** $\frac{9}{100}$

**15.** 0.3891

**16.** $\frac{3}{16}$

**17.** 0.005

**18.** $\frac{37}{50}$

**19.** 6

**20.** $\frac{5}{6}$

Find fraction notation. Simplify.   [8.1c]

**21.** 85%

**22.** 0.048%

**23.** $22\frac{3}{4}\%$

**24.** $16.\overline{6}\%$

Write percent notation for the shaded area.   [8.1c]

**25.**

**26.**

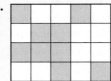

Solve.   [8.2b], [8.3b]

**27.** 25% of what is 14.5?

**28.** 220 is what percent of 1320?

**29.** What is 3.2% of 80,000?

**30.** $17.50 is 35% of what?

**31.** What percent of $800 is $160?

**32.** 130% of $350 is what?

**33.** Arrange the following numbers from smallest to largest.   [8.1b, c]

$$\frac{1}{2}\%, \ 5\%, \ 0.275, \ \frac{13}{100}, \ 1\%, \ 0.1\%, \ 0.05\%, \ \frac{3}{10}, \ \frac{7}{20}, \ 10\%$$

**34.** Solve: 8.5 is $2\frac{1}{2}\%$ of what?   [8.2b], [8.3b]

   **A.** 3.4             **B.** 21.25

   **C.** 0.2125         **D.** 340

**35.** Solve: $102,000 is what percent of $3.6 million?
   [8.2b], [8.3b]

   **A.** $2.8\overline{3}$ million       **B.** $2\frac{5}{6}\%$

   **C.** $0.028\overline{3}\%$         **D.** $28.\overline{3}$

# Understanding Through Discussion and Writing

**36.** Is it always best to convert from fraction notation to percent notation by first finding decimal notation? Why or why not?   [8.1c]

**37.** Suppose we know that 40% of 92 is 36.8. What is a quick way to find 4% of 92? 400% of 92? Explain.   [8.2b], [8.3b]

**38.** In solving Example 10 in Section 8.3, a student simplifies $\frac{10}{20}$ before solving. Is this a good idea? Why or why not? [8.3b]

**39.** What do the following have in common? Explain. [8.1b, c]

$$\frac{23}{16}, \ 1\frac{875}{2000}, \ 1.4375, \ \frac{207}{144}, \ 1\frac{7}{16}, \ 143.75\%, \ 1\frac{4375}{10,000}$$

# 8.4

## Applications of Percent

### (a) Applied Problems Involving Percent

Applied problems involving percent are not always stated in a manner easily translated to an equation. In such cases, it is helpful to rephrase the problem before translating. Sometimes it also helps to make a drawing.

**EXAMPLE 1** *Extinction of Mammals.* According to a study conducted for the International Union for the Conservation of Nature (IUCN), many of the world's mammals are in danger of extinction. Of the 5487 species of mammals on earth, 1141 are on the IUCN Red List of Threatened Species. Six of those mammals are shown below. What percent of all mammals are threatened with extinction?

**Sources:** Environment News Service, October 6, 2008; IUCN

**OBJECTIVE**

(a) Solve applied problems involving percent.

**SKILL TO REVIEW**
Objective 5.4a: Divide using decimal notation.

Divide.
1. $345 \div 57.5$
2. $111.87 \div 9.9$

*Top row, left to right:* Tasmanian devils, Père David's deer, African elephant. *Bottom row, left to right:* Iberian lynx, black-footed ferret, giant panda.

1. **Familiarize.** The question asks for the percent of the world's mammals that are in danger of extinction. We note that 5487 is approximately 5500 and 1141 is approximately 1100. Since 1100 is $\frac{1100}{5500}$, or $\frac{1}{5}$, or 20%, of 5500, our answer is close to 20%. We let $p =$ the percent of mammals that are in danger of extinction.

2. **Translate.** There are two ways in which we can translate this problem.

*Percent equation (see Section 8.2):*

$$1141 \text{ is what percent of } 5487?$$
$$1141 = p \cdot 5487$$

*Proportion (see Section 8.3):*

$$\frac{N}{100} = \frac{1141}{5487}$$

For proportions, $N\% = p$.

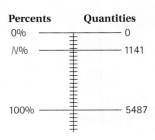

| Percents | Quantities |
|---|---|
| 0% | 0 |
| N% | 1141 |
| 100% | 5487 |

*Answers*
*Skill to Review:*
1. 6    2. 11.3

**3. Solve.**   We now have two ways in which to solve this problem.

*Percent equation (see Section 8.2):*

$$1141 = p \cdot 5487$$

$$\frac{1141}{5487} = \frac{p \cdot 5487}{5487} \qquad \text{Dividing both sides by 5487}$$

$$\frac{1141}{5487} = p$$

$$0.208 \approx p \qquad \text{Finding decimal notation and rounding to the nearest thousandth}$$

$$20.8\% \approx p \qquad \text{Remember to find percent notation.}$$

Note here that the solution, $p$, includes the % symbol.

*Proportion (see Section 8.3):*

$$\frac{N}{100} = \frac{1141}{5487}$$

$$N \cdot 5487 = 100 \cdot 1141 \qquad \text{Equating cross products}$$

$$\frac{N \cdot 5487}{5487} = \frac{114{,}100}{5487} \qquad \text{Dividing both sides by 5487}$$

$$N = \frac{114{,}100}{5487}$$

$$N \approx 20.8 \qquad \text{Dividing and rounding to the nearest tenth}$$

We use the solution of the proportion to express the answer to the problem as 20.8%. Note that in the proportion method, $N\% = p$.

**4. Check.**   To check, we note that the answer 20.8% is close to 20%, as estimated in the *Familiarize* step.

**5. State.**   About 20.8% of the world's mammals are threatened with extinction.

Do Exercise 1.

---

**EXAMPLE 2**   *Transportation to Work.*   In the United States, there are about 147,000,000 workers 16 years and older. Approximately 76.9% drive to work alone. How many workers drive to work alone?

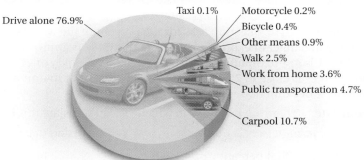

**Transportation to Work in the United States**

Drive alone 76.9%
Taxi 0.1%
Motorcycle 0.2%
Bicycle 0.4%
Other means 0.9%
Walk 2.5%
Work from home 3.6%
Public transportation 4.7%
Carpool 10.7%

SOURCES: U.S. Census Bureau; American Community Survey

---

**1. Presidential Assassinations in Office.**   Of the 43 different U.S. presidents, 4 have been assassinated in office. These were James A. Garfield, William McKinley, Abraham Lincoln, and John F. Kennedy. What percent have been assassinated in office?

*Answer*

**1.** About 9.3%

1. **Familiarize.** We can simplify the pie chart shown on the preceding page to help familiarize ourselves with the problem. We let $a =$ the total number of workers who drive to work alone.

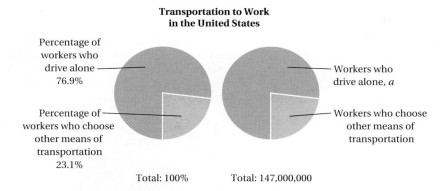

**Transportation to Work in the United States**

Percentage of workers who drive alone 76.9%

Percentage of workers who choose other means of transportation 23.1%

Workers who drive alone, $a$

Workers who choose other means of transportation

Total: 100%    Total: 147,000,000

2. **Translate.** There are two ways in which we can translate this problem.

*Percent equation:*

What number    is    76.9%    of    147,000,000?

$a$    $=$    0.769    $\cdot$    147,000,000

*Proportion:*

$$\frac{76.9}{100} = \frac{a}{147,000,000}$$

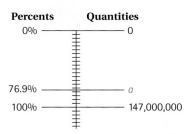

| Percents | Quantities |
|---|---|
| 0% | 0 |
| 76.9% | $a$ |
| 100% | 147,000,000 |

3. **Solve.** We now have two ways in which to solve this problem.

*Percent equation:*

$a = 0.769 \cdot 147,000,000$

We multiply:

$a = 0.769 \cdot 147,000,000 = 113,043,000.$

*Proportion:*

$$\frac{76.9}{100} = \frac{a}{147,000,000}$$

$76.9 \cdot 147,000,000 = 100 \cdot a$    Equating cross products

$\dfrac{76.9 \cdot 147,000,000}{100} = \dfrac{100 \cdot a}{100}$    Dividing by 100

$\dfrac{11,304,300,000}{100} = a$

$113,043,000 = a$    Simplifying

4. **Check.** To check, we can repeat the calculations. We can also do a partial check by estimating. Since 76.9% is about 75%, or $\frac{3}{4}$, and $\frac{3}{4}$ of 147,000,000 is 110,250,000 and 110,250,000 is close to 113,043,000, our answer is reasonable.

5. **State.** The number of workers who drive to work alone is about 113,043,000.

Do Exercise 2.

2. **Transportation to Work.** There are about 147,000,000 workers 16 years or older in the United States. Approximately 10.7% carpool to work. How many workers carpool to work?
**Sources:** U.S. Census Bureau; American Community Survey

*Answer*

2. About 15,729,000 workers

**8.4** Exercise Set

For Extra Help

*MyMathLab*

Math XL
PRACTICE

WATCH

DOWNLOAD

READ

REVIEW

**a** Solve.

**1.** *Foreign Students.* In the 2008–2009 school year, of the 15 million college students in the United States, 672,000 were from foreign countries. Approximately 11.18% of the foreign students were from South Korea and 4.36% were from Japan. How many foreign students were from South Korea? from Japan?

Source: Institute of International Education

**2.** *Mississippi River.* The Mississippi River, which extends from its source, at Lake Itasca in Minnesota, to the Gulf of Mexico, is 2348 mi long. Approximately 77% of the river is navigable. How many miles of the river are navigable?

Source: National Oceanic and Atmospheric Administration

Mississippi River

**3.** A person earns $43,200 one year and receives an 8% raise in salary. What is the new salary?

**4.** A person earns $28,600 one year and receives a 5% raise in salary. What is the new salary?

**5.** *Test Results.* On a test, Juan got 85%, or 119, of the items correct. How many items were on the test?

**6.** *Test Results.* On a test, Maj Ling got 86%, or 81.7, of the items correct. (There was partial credit on some items.) How many items were on the test?

**7.** *Farmland.* In Kansas, 47,000,000 acres are farmland. About 5% of all the farm acreage in the United States is in Kansas. What is the total number of acres of farmland in the United States?

Sources: U.S. Department of Agriculture; National Agricultural Statistics Service

**8.** *World Population.* World population is increasing by 1.2% each year. In 2010, it was 6.82 billion. What will the population be in 2015?

Sources: U.S. Census Bureau; International Data Base

**9.** *Car Depreciation.*   A car generally depreciates 25% of its original value in the first year. A car is worth $27,300 after the first year. What was its original cost?

**10.** *Car Depreciation.*   Given normal use, an American-made car will depreciate 25% of its original cost in the first year and 14% of its remaining value in the second year. What is the value of a car at the end of the second year if its original cost was $36,400? $28,400? $26,800?

**11.** *Test Results.*   On a test of 80 items, Pedro got 93% correct. (There was partial credit on some items.) How many items did he get correct? incorrect?

**12.** *Test Results.*   On a test of 40 items, Christina got 91% correct. (There was partial credit on some items.) How many items did she get correct? incorrect?

**13.** *Under 15 Years Old.*   In Egypt, 32.6% of the population is under 15 years old. In the United States, 20.4% of the population is under 15. The population of Egypt is 78,867,000, and the population of the United States is 309,126,000. How many are under 15 years old in Egypt? in the United States?
**Sources:** U.S. Census Bureau; International Data Base

**14.** *Age 65 and Older.*   In Egypt, 4.5% of the population is age 65 and older. In the United States, 12.5% of the population is age 65 and older. The population of Egypt is 78,867,000, and the population of the United States is 309,126,000. How many are age 65 and older in Egypt? in the United States?
**Sources:** U.S. Census Bureau; International Data Base

**15.** *Transplant Waiting List.*   The total number of patients waiting for transplants as of April 16, 2010, was 107,145. Of this number, 15,960 were waiting for a liver transplant. What percent were waiting for a liver transplant?
**Source:** United Network for Organ Sharing

**16.** *Doctors' Salaries.*   In 2010, the starting salary for a neurologist was approximately 59.7% of that of an anesthesiologist. The beginning salary for an anesthesiologist was $203,408. What was the beginning salary for a neurologist?
**Sources:** payscale.com; salarylist.com

**17.** *Tipping.*   For a party of 8 or more, some restaurants add an 18% tip to the bill. What is the total amount charged for a party of 10 if the cost of the meal, without tip, is $195?

**18.** *Tipping.*   Diners frequently add a 15% tip when charging a meal to a credit card. What is the total amount charged if the cost of the meal, without tip, is $18? $34? $49?

**19.** *Fast-Food Cooks.* The United States has 392,850 full-time farmers. This number is about 64.2% of the number of fast-food cooks. How many fast-food cooks are there in the United States?

Source: U.S. Department of Agriculture

**20.** *Spending in Restaurants.* Americans spent $364 billion in grocery stores in 2007. This amount is about $93\frac{1}{3}$% of the amount spent in restaurants. How much was spent in restaurants?

Source: U.S. Department of Agriculture

**21.** A lab technician has 540 mL of a solution of alcohol and water; 8% is alcohol. How many milliliters are alcohol? water?

**22.** A lab technician has 680 mL of a solution of water and acid; 3% is acid. How many milliliters are acid? water?

**23.** *U.S. Volunteerism.* There were 63,361,000 Americans who did volunteer work in 2009. The numbers who volunteered in several areas are listed in the table below. What percent of the total does each area represent? Round the answers to the nearest tenth of a percent.

### VOLUNTEERS IN AMERICA: 2009

| TOTAL | 63,361,000 |
|---|---|
| Civic | 3,486,000 |
| Educational | 16,537,000 |
| Environmental | 1,394,000 |
| Health Care | 5,397,000 |
| Religious | 21,543,000 |
| Community Service | 8,807,000 |
| Sports, Hobbies | 2,154,000 |

SOURCE: U.S. Bureau of Labor Statistics

**24.** *U.S. Volunteerism.* There were 63,361,000 Americans who did volunteer work in 2009. Numbers in various age groups are listed in the table below. What percent of the total does each age group represent? Round the answers to the nearest tenth of a percent.

### VOLUNTEERS IN AMERICA: 2009

| TOTAL | 63,361,000 |
|---|---|
| 16–25 years old | 8,290,000 |
| 25–34 years old | 9,511,000 |
| 35–44 years old | 12,835,000 |
| 45–54 years old | 13,703,000 |
| 55–64 years old | 9,894,000 |
| 65 years old and older | 9,129,000 |

SOURCE: U.S. Bureau of Labor Statistics

## Skill Maintenance

Convert to decimal notation.   [5.1b], [5.5a]

**25.** $\dfrac{25}{11}$

**26.** $\dfrac{11}{25}$

**27.** $\dfrac{27}{8}$

**28.** $\dfrac{43}{9}$

**29.** $\dfrac{23}{25}$

**30.** $\dfrac{20}{24}$

**31.** $\dfrac{14}{32}$

**32.** $\dfrac{2317}{1000}$

**33.** $\dfrac{34,809}{10,000}$

**34.** $\dfrac{27}{40}$

## Synthesis

**35.** A coupon allows a couple to have dinner and then have $10 subtracted from the bill. Before subtracting $10, however, the restaurant adds a tip of 20%. If the couple is presented with a bill for $40.40, how much would the dinner (without tip) have cost without the coupon?

**36.** If *p* is 120% of *q*, then *q* is what percent of *p*?

**37.** *Adult Height.* It has been determined that at the age of 10, a girl has reached 84.4% of her final adult height. Cynthia is 4 ft 8 in. at the age of 10. What will be her final adult height?

**38.** *Adult Height.* It has been determined that at the age of 15, a boy has reached 96.1% of his final adult height. Claude is 6 ft 4 in. at the age of 15. What will be his final adult height?

# Percent of Increase or Decrease

## (a) Applied Problems Involving Percent of Increase or Percent of Decrease

Percent is often used to state increase or decrease. Let's consider an example of each, using the price of a car as the original number.

### Percent of Increase

One year a car sold for $20,000. The manufacturer decides to raise the price of the following year's model by 5%. The increase is 0.05 × $20,000, or $1000. The new price is $20,000 + $1000, or $21,000. Note that the new price is 105% of the *former* price.

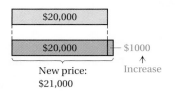

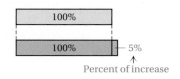

The increase, $1000, is 5% of the *former* price, $20,000. The *percent of increase* is 5%.

### Percent of Decrease

Abigail buys the car listed above for $20,000. After one year, the car depreciates in value by 25%. The decrease is 0.25 × $20,000, or $5000. This lowers the value of the car to $20,000 − $5000, or $15,000. Note that the new value is 75% of the original price. If Abigail decides to sell the car after one year, $15,000 might be the most she could expect to get for it.

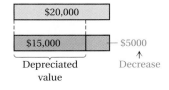

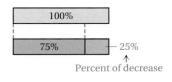

The decrease, $5000, is 25% of the *original* price, $20,000. The *percent of decrease* is 25%.

> Do Exercises 1 and 2.

When a quantity is decreased by a certain percent, we say that we have **percent of decrease**.

**EXAMPLE 1** *Dow Jones Industrial Average.* The Dow Jones Industrial Average (DJIA) plunged from 11,143 to 10,365 on September 29, 2008. This was the largest one-day drop in its history. What was the percent of decrease?
Sources: *Nightly Business Reports*, September 29, 2008; DJIA

New price: $21,000

Increase ↑

Former price: $20,000

Original price: $20,000

Decrease ↓

New value: $15,000

1. **Percent of Increase.** The price of a car is $36,875. The price is increased by 4%.
   a) How much is the increase?
   b) What is the new price?

2. **Percent of Decrease.** The value of a car is $36,875. The car depreciates in value by 25% after one year.
   a) How much is the decrease?
   b) What is the depreciated value of the car?

*Answers*
1. (a) $1475; (b) $38,350
2. (a) $9218.75; (b) $27,656.25

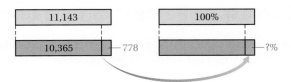

1. **Familiarize.** We first determine the amount of decrease and then make a drawing.

$$
\begin{array}{rl}
11{,}143 & \text{Opening average} \\
-10{,}365 & \text{Closing average} \\
\hline
778 & \text{Decrease}
\end{array}
$$

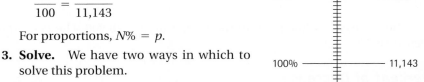

We are asking this question: The decrease is what percent of the opening average? We let $p =$ the percent of decrease.

2. **Translate.** There are two ways in which we can translate this problem.

*Percent equation:*

$$
\begin{array}{ccccc}
778 & \text{is} & \text{what percent} & \text{of} & 11{,}143? \\
\downarrow & \downarrow & \downarrow & \downarrow & \downarrow \\
778 & = & p & \cdot & 11{,}143
\end{array}
$$

*Proportion:*

$$
\frac{N}{100} = \frac{778}{11{,}143}
$$

For proportions, $N\% = p$.

3. **Solve.** We have two ways in which to solve this problem.

*Percent equation:*

$$
778 = p \cdot 11{,}143
$$

$$
\frac{778}{11{,}143} = \frac{p \cdot 11{,}143}{11{,}143} \qquad \text{Dividing both sides by } 11{,}143
$$

$$
\frac{778}{11{,}143} = p
$$

$$
0.07 \approx p
$$

$$
7\% \approx p \qquad \text{Converting to percent notation}
$$

*Proportion:*

$$
\frac{N}{100} = \frac{778}{11{,}143}
$$

$$
11{,}143 \times N = 100 \times 778 \qquad \text{Equating cross products}
$$

$$
\frac{11{,}143 \times N}{11{,}143} = \frac{100 \times 778}{11{,}143} \qquad \text{Dividing both sides by } 11{,}143
$$

$$
N = \frac{77{,}800}{11{,}143}
$$

$$
N \approx 7
$$

We use the solution of the proportion to express the answer to the problem as 7%.

**4. Check.** To check, we note that, with a 7% decrease, the closing Dow average should be 93% of the opening average. Since

$$93\% \cdot 11{,}143 = 0.93 \cdot 11{,}143 \approx 10{,}363$$

and 10,363 is close to 10,365, our answer checks. (Remember that we rounded to get 7%.)

**5. State.** The percent of decrease in the Dow average was approximately 7%.

> Do Exercise 3.

When a quantity is increased by a certain percent, we say we have **percent of increase**.

**EXAMPLE 2** *Costs for Moviegoers.* The average cost of movie tickets for a family of four was $16.56 in 1993. The cost rose to $30.00 in 2009. What was the percent of increase in the cost for a family of four to attend a movie?

**Source:** Motion Picture Association of America

**1. Familiarize.** We first determine the increase in the cost and then make a drawing.

$$
\begin{array}{rl}
\$30.00 & \text{Cost in 2009} \\
-\ 16.56 & \text{Cost in 1993} \\
\hline
\$13.44 & \text{Increase}
\end{array}
$$

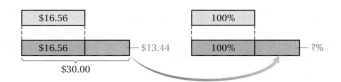

We are asking this question: The increase is what percent of the *original* cost? We let $p = $ the percent of increase.

**2. Translate.** There are two ways in which we can translate this problem.

*Percent equation:*

$$
\underbrace{13.44}_{\downarrow} \quad \underbrace{\text{is}}_{\downarrow} \quad \underbrace{\text{what percent}}_{\downarrow} \quad \underbrace{\text{of}}_{\downarrow} \quad \underbrace{16.56?}_{\downarrow}
$$
$$
13.44 \quad = \quad p \quad \cdot \quad 16.56
$$

*Proportion:*

$$\frac{N}{100} = \frac{13.44}{16.56}$$

For proportions, $N\% = p$.

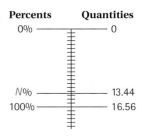

| Percents | Quantities |
|---|---|
| 0% | 0 |
| $N\%$ | 13.44 |
| 100% | 16.56 |

**3. Volume of Mail.** The volume of U.S. mail decreased from about 203 billion pieces of mail in 2008 to 176 billion pieces in 2009. What was the percent of decrease?

**Source:** U.S. Postal Service

*Answer*

**3.** About 13.3%

**4. Shipping Costs.** The total cost of shipping a 40-ft container from Shanghai to New York rose from $5100 in 2005 to $8350 in 2008. Find the percent of increase.

Source: CIBC World Markets, IMF

**3. Solve.** We have two ways in which to solve this problem.

*Percent equation:*

$$13.44 = p \cdot 16.56$$

$$\frac{13.44}{16.56} = \frac{p \cdot 16.56}{16.56} \qquad \text{Dividing both sides by 16.56}$$

$$\frac{13.44}{16.56} = p$$

$$0.81 \approx p$$

$$81\% \approx p \qquad \text{Converting to percent notation}$$

*Proportion:*

$$\frac{N}{100} = \frac{13.44}{16.56}$$

$$16.56 \times N = 100 \times 13.44 \qquad \text{Equating cross products}$$

$$\frac{16.56 \times N}{16.56} = \frac{100 \times 13.44}{16.56} \qquad \text{Dividing both sides by 16.56}$$

$$N = \frac{1344}{16.56}$$

$$N \approx 81$$

We use the solution of the proportion to express the answer to the problem as 81%.

**4. Check.** To check, we take 81% of 16.56:

$$81\% \times 16.56 = 0.81 \cdot 16.56 = 13.4136.$$

The approximation 13.4136 is close enough to 13.44 to be an excellent check. (Remember that we rounded to get 81%.)

**5. State.** The percent of increase in the cost for a family of four to attend a movie is 81%.

Do Exercise 4.

The percent of increase or decrease is *always* based on the original amount. To find a percent of increase or decrease, we need to know (1) the amount of increase or decrease and (2) the original amount. The "new" amount after the increase or decrease is not used in the calculation.

$$\text{Percent of increase/decrease} = \frac{\text{Amount of increase/decrease}}{\text{Original amount}}$$

*Answer*

4. About 64%

# Translating
## for Success

*Distance Walked.* After a knee replacement, Alex walked $\frac{1}{8}$ mi each morning and $\frac{1}{5}$ mi each afternoon. How much farther did he walk in the afternoon?

*Stock Prices.* A stock sold for $5 per share on Monday and only $2.125 per share on Friday. What was the percent of decrease from Monday to Friday?

*SAT Score.* After attending a class entitled *Improving Your SAT Scores*, Jacob raised his total score from 1450 to 1600. What was the percent of increase?

*Change in Population.* The population of a small farming community decreased from 1600 to 1450. What was the percent of decrease?

*Lawn Mowing.* During the summer, brothers Steve and Rob earned money for college by mowing lawns. The largest lawn that they mowed was $2\frac{1}{8}$ acres. Steve can mow $\frac{1}{5}$ acre per hour, and Rob can mow only $\frac{1}{8}$ acre per hour. Working together, how many acres did they mow per hour?

The goal of these matching questions is to practice step (2), *Translate*, of the five-step problem-solving process. Translate each word problem to an equation and select a correct translation from equations A–O.

**A.** $x + \frac{1}{5} = \frac{1}{8}$

**B.** $250 = x \cdot 1600$

**C.** $1450 = x \cdot 1600$

**D.** $\frac{250}{16.25} = \frac{1000}{x}$

**E.** $150 = x \cdot 1600$

**F.** $16.25 = 250 \cdot x$

**G.** $\frac{1}{5} + \frac{1}{8} = x$

**H.** $2\frac{1}{8} = x \cdot 5$

**I.** $5 = 2.875 \cdot x$

**J.** $\frac{1}{8} + x = \frac{1}{5}$

**K.** $1600 = x \cdot 1450$

**L.** $\frac{250}{16.25} = \frac{x}{1000}$

**M.** $2.875 = x \cdot 5$

**N.** $x \cdot 1450 = 150$

**O.** $x = 16.25 \cdot 250$

*Answers on page A-17*

**6.** *Land Sale.* Cole sold $2\frac{1}{8}$ acres of the 5 acres he inherited of his uncle. What percent did he sell?

**7.** *Travel Expenses.* A magazine photographer is reimbursed 16.25¢ per mile for business travel up to 1000 mi per week. In a recent week, he traveled 250 mi. What was the total reimbursement for travel?

**8.** *Trip Expenses.* The total expenses for Claire's recent business trip were $1600. She put $1450 on her credit card and paid the balance in cash. What percent did she place on her credit card?

**9.** *Cost of Copies.* During the first summer session at a community college, the campus copy center advertised 250 copies for $16.25. At this rate, what is the cost of 1000 copies?

**10.** *Cost of Insurance.* Following a raise in the cost of health insurance, 250 of a company's 1600 employees dropped their health coverage. What percent of the employees canceled their insurance?

**a** Solve.

**1.** *Mortgage Payment Increase.* A monthly mortgage payment increases from $840 to $882. What is the percent of increase?

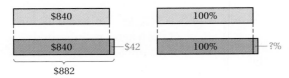

**2.** *Savings Increase.* The amount in a savings account increased from $200 to $216. What was the percent of increase?

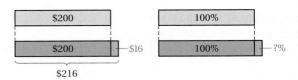

**3.** A person on a diet goes from a weight of 160 lb to a weight of 136 lb. What is the percent of decrease?

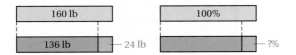

**4.** During a sale, a dress decreased in price from $90 to $72. What was the percent of decrease?

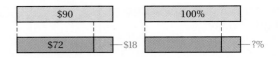

**5.** *Insulation.* A 15" roll of unfaced fiberglass insulation has a retail price of $23.43. For two weeks, it is on sale for $15.31. What is the percent of decrease?

**6.** *Set of Weights.* A 300-lb weight set retails for $199.95. For its grand opening, a sporting goods store reduced the price to $154.95. What is the percent of decrease?

**7.** *Mass Transit.* In 2008, people took 10.5 billion rides on public transit in the United States. The ridership was up from 8.8 billion rides in 1998. What is the percent of increase?

**Source:** American Public Transportation Association

**8.** *Pharmacists.* It is projected that there will be 296,000 people employed as pharmacists in 2016. In 2006, 243,000 members of the labor force were pharmacists. What is the percent of increase?

**Source:** EarnMyDegree.com

**9.** *Overdraft Fees.* Consumers are paying record amounts of fees for overdrawing their bank accounts. In 2009, banks collected $38.5 billion in overdraft fees, which is $18.6 billion more than in 2000. What is the percent of increase?

Source: CNNMoney.com

**10.** *Credit-Card Debt.* In 2009, the average credit-card debt per household with credit-card debt in the United States was $16,007. In 1990, the average credit-card debt was $2966. What is the percent of increase?

Sources: CardWeb.com; *USA TODAY*; creditcards.com

**11.** *Immigrant Applications.* In 2009, 47,537 immigrants applied for citizenship each month. During 2007, 114,469 immigrants applied each month. What is the percent of decrease?

Source: U.S. Citizenship and Immigration Services (USCIS)

**12.** *Miles of Railroad Track.* The greatest combined length of U.S.-owned operating railroad track was 254,037 mi in 1916, when industrial activity increased during World War I. By 2007, the number of miles of track had decreased to 140,695 mi. What is the percent of decrease from 1916 to 2007?

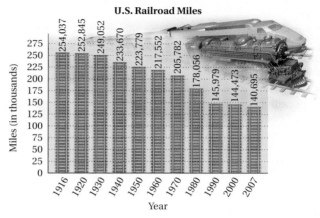

NOTE: The lengths exclude yard tracks, sidings, and parallel tracks.
SOURCE: Association of American Railroads

**13.** *Patents Issued.* The U.S. Patent and Trademark Office (USPTO) issued a total of 157,774 utility patents in 2008. This number of patents is down from 173,794 in 2006. What is the percent of decrease?

Source: IFI Patent Intelligence

**14.** *Highway Fatalities.* In 2009, there were 33,963 highway fatalities, which was 3298 fewer deaths than in 2008. What is the percent of decrease?

Source: National Highway Traffic Safety Administration

**15.** *Two-by-Four.* A cross-section of a standard or nominal "two-by-four" board actually measures $1\frac{1}{2}$ in. by $3\frac{1}{2}$ in. The rough board is 2 in. by 4 in. but is planed and dried to the finished size. What percent of the wood is removed in planing and drying?

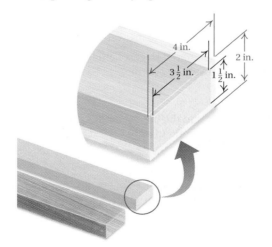

**16.** *Strike Zone.* In baseball, the *strike zone* is normally a 17-in. by 30-in. rectangle. Some batters give the pitcher an advantage by swinging at pitches thrown out of the strike zone. By what percent is the area of the strike zone increased if a 2-in. border is added to the outside?

Source: Major League Baseball

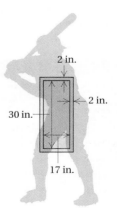

*Population Increase.* The table below provides data showing how the populations of various states increased from 2000 to 2009. Complete the table by filling in the missing numbers. Round percents to the nearest tenth of a percent.

| | STATE | POPULATION IN 2000 | POPULATION IN 2009 | CHANGE | PERCENT CHANGE |
|---|---|---|---|---|---|
| **17.** | Vermont | 608,827 | 621,760 | | |
| **18.** | Wisconsin | 5,363,675 | | 291,099 | |
| **19.** | Arizona | | 6,595,778 | 1,465,146 | |
| **20.** | Virginia | | 7,882,590 | 804,075 | |
| **21.** | Idaho | 1,293,953 | | 251,848 | |
| **22.** | Georgia | 8,186,453 | 9,829,211 | | |

SOURCE: U.S. Census Bureau

**23.** *Decrease in Population.* Between 2000 and 2008, the population of Youngstown, Ohio, decreased from 82,026 to 72,925. What was the percent of decrease?

Sources: U.S. Census Bureau; U.S. Department of Commerce

**24.** *Decrease in Population.* Between 2000 and 2008, the population of Baton Rouge, Louisiana, decreased from 227,818 to 223,689. What was the percent of decrease?

Sources: U.S. Census Bureau; U.S. Department of Commerce

## Skill Maintenance

Find the perimeter of each polygon. [2.7b]

**25.**
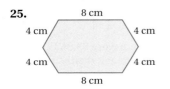
8 cm
4 cm    4 cm
4 cm    4 cm
8 cm

**26.**
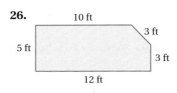
10 ft
3 ft
5 ft
3 ft
12 ft

**27.**
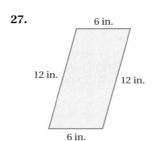
6 in.
12 in.    12 in.
6 in.

**28.**
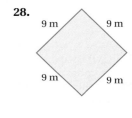
9 m    9 m
9 m    9 m

## Synthesis

The U.S. Census Bureau reported the following household incomes by state. The numbers are given as averages of median incomes over 2-yr periods and have been standardized to 2008 dollars. Use the information in the table to answer Exercises 29 and 30.

**Household Incomes 2005 to 2008**

| STATE | AVERAGE MEDIAN HOUSEHOLD INCOME 2007–2008 | CHANGE IN AVERAGE MEDIAN HOUSEHOLD INCOME FROM 2005 TO 2006 |
|---|---|---|
| Idaho | $49,247 | 0.4% |
| Maryland | 65,932 | −2.1% |
| Nebraska | 50,896 | −2.4% |
| Oregon | 51,947 | 5.0% |
| Virginia | 61,710 | 4.4% |
| Wyoming | 51,977 | 4.4% |

**29.** Which state had the largest actual amount of increase in median household income?

**30.** Which state had the largest actual amount of decrease in median household income?

**31.** 🖩 A worker receives raises of 3%, 6%, and then 9%. By what percent has the original salary increased?

# 8.6 Sales Tax, Commission, and Discount

## a Sales Tax

Sales tax computations represent a special type of percent of increase problem. For example, the sales tax rate in Pennsylvania is 6%. This means that the tax is 6% of the purchase price. Suppose the purchase price of a guitar is $839.95. The sales tax is then 6% of $839.95, or 0.06 × $839.95, or $50.397, or about $50.40.

**BILL:**

| | |
|---|---|
| Purchase price | = $839.95 |
| Sales tax (6% of $839.95) | = + 50.40 |
| Total price | $890.35 |

The total that you pay is the purchase price plus the sales tax:

$839.95 + $50.40,   or   $890.35.

---

**SALES TAX**

**Sales tax** = Sales tax rate × Purchase price

**Total price** = Purchase price + Sales tax

---

**EXAMPLE 1** *Wisconsin Sales Tax.* The sales tax rate in Wisconsin is 5%. How much tax is charged on the purchase of 3 gal of paint at $42.99 each? What is the total price?

**a)** We first find the cost of the paint. It is

3 × $42.99 = $128.97.

**b)** The sales tax on items costing $128.97 is

$$\underbrace{\text{Sales tax rate}}_{5\%} \times \underbrace{\text{Purchase price}}_{\$128.97},$$

or 0.05 × 128.97, or 6.4485. Thus, the tax is $6.45 (rounded to the nearest cent).

**c)** The total price is given by the purchase price plus the sales tax:

$128.97 + $6.45,   or   $135.42.

To check, note that the total price is the purchase price plus 5% of the purchase price. Thus, the total price is 105% of the purchase price. Since 1.05 × 128.97 ≈ 135.42, we have a check. The sales tax is $6.45, and the total price is $135.42.

Do Exercises 1 and 2.

1. **Texas Sales Tax.** The sales tax rate in Texas is 6.25%. In Texas, how much tax is charged on the purchase of an ultrasound toothbrush that sells for $139.95? What is the total price?

2. **Wyoming Sales Tax.** Samantha buys 7 copies of *The Last Lecture* by Randy Pausch with Jeffrey Zaslow for $14.95 each. The sales tax rate in Wyoming is 4%. In Wyoming, how much sales tax will be charged? What is the total price?

**EXAMPLE 2** The sales tax on the purchase of this GPS navigator, which costs $799, is $55.93. What is the sales tax rate?

GPS Navigator 4.3"-screen with speech recognition
**$799**
+ $55.93 sales tax

We rephrase and translate as follows:

*Rephrase:* Sales tax is what percent of purchase price?

*Translate:* 55.93 = $r$ · 799.

To solve the equation, we divide both sides by 799:

$$\frac{55.93}{799} = \frac{r \cdot 799}{799}$$

$$\frac{55.93}{799} = r$$

$$0.07 = r$$

$$7\% = r.$$

The sales tax rate is 7%.

Do Exercise 3.

3. The sales tax on the purchase of a set of holiday dishes that costs $449 is $26.94. What is the sales tax rate?

**EXAMPLE 3** The sales tax on the purchase of a stone-top firepit is $12.74 and the sales tax rate is 8%. Find the purchase price (the price before taxes are added).

We rephrase and translate as follows:

*Rephrase:* Sales tax is 8% of what?

*Translate:* 12.74 = 0.08 · $b$.

To solve, we divide both sides by 0.08:

$$\frac{12.74}{0.08} = \frac{0.08 \cdot b}{0.08}$$

$$\frac{12.74}{0.08} = b$$

$$159.25 = b.$$

The purchase price is $159.25.

**Price: ?**
$12.74 tax @ 8%

4. The sales tax on the purchase of a pair of designer jeans is $4.84 and the sales tax rate is 5.5%. Find the purchase price (the price before taxes are added).

Do Exercise 4.

**Answers**

1. $8.75; $148.70   2. $4.19; $108.84
3. 6%   4. $88

## b Commission

When you work for a **salary**, you receive the same amount of money each week or month. When you work for a **commission**, you are paid a percentage of the total sales for which you are responsible.

> **COMMISSION**
>
> **Commission** = Commission rate × Sales

**EXAMPLE 4** *Appliance Sales.* A salesperson's commission rate is 3%. What is the commission from the sale of $8300 worth of appliances?

$$\textit{Commission} = \textit{Commission rate} \times \textit{Sales}$$
$$C \quad = \quad 3\% \quad \times \quad 8300$$
$$C \quad = \quad 0.03 \quad \cdot \quad 8300$$
$$C = 249$$

The commission is $249.

Do Exercise 5.

**EXAMPLE 5** *Earth-Moving Equipment Sales.* Gavin earns a commission of $20,800 selling $320,000 worth of earth-moving equipment. What is the commission rate?

$$\textit{Commission} = \textit{Commission rate} \times \textit{Sales}$$
$$20,800 \quad = \quad r \quad \cdot \quad 320,000$$

**STUDY TIPS**

**FINISHING A CHAPTER**

Try to be aware of when you reach the end of a chapter. Sometimes the end of a chapter signals the arrival of a quiz or test. Almost always, the end of a chapter indicates the end of a particular area of study. Because future work will likely rely on your mastery of the chapter completed, make use of the chapter review and test to solidify your understanding before moving onward.

**5.** Hailey's commission rate is 15%. What is the commission from the sale of $9260 worth of exercise equipment?

*Answer*

5. $1389

To solve this equation, we divide both sides by 320,000:

$$\frac{20,800}{320,000} = \frac{r \cdot 320,000}{320,000}$$

$$0.065 = r$$

$$6.5\% = r.$$

The commission rate is 6.5%.

Do Exercise 6.

**6.** William earns a commission of $2040 selling $17,000 worth of concert tickets. What is the commission rate?

**EXAMPLE 6** *Cruise Vacations.* Mia's commission rate is 5.6%. She received a commission of $2457 on cruise vacation packages that she sold in November. How many dollars' worth of cruise vacations did she sell?

$$Commission = Commission\ rate \times Sales$$

$$2457 \quad = \quad 5.6\% \quad \times \quad S$$

$$2457 \quad = \quad 0.056 \quad \cdot \quad S$$

To solve this equation, we divide both sides by 0.056:

$$\frac{2457}{0.056} = \frac{0.056 \cdot S}{0.056}$$

$$\frac{2457}{0.056} = S$$

$$43,875 = S.$$

Mia sold $43,875 worth of cruise vacation packages.

Do Exercise 7.

**7.** Dylan's commission rate is 7.5%. He receives a commission of $2970 from the sale of winter ski passes. How many dollars' worth of ski passes did he sell?

## c Discount

Suppose that the regular price of a rug is $60, and the rug is on sale at 25% off. Since 25% of $60 is $15, the sale price is $60 − $15, or $45. We call $60 the **original**, or **marked**, **price**, 25% the **rate of discount**, $15 the **discount**, and $45 the **sale price**. Note that discount problems are a type of percent of decrease problem.

---

### DISCOUNT AND SALE PRICE

**Discount** = Rate of discount × Original price

**Sale price** = Original price − Discount

---

**EXAMPLE 7**   A leather sofa marked $2379 is on sale at $33\frac{1}{3}$% off. What is **(a)** the discount? **(b)** the sale price?

Leather sofa
$\$2379$ original price
Save $33\frac{1}{3}$%

**a)**  *Discount  =  Rate of discount  ×  Original price*

$$D \quad = \quad 33\frac{1}{3}\% \quad \times \quad 2379$$

$$D \quad = \quad \frac{1}{3} \quad \cdot \quad 2379$$

$$D = \frac{2379}{3} = 793$$

**b)**  *Sale price  =  Original price  −  Discount*

$$S \quad = \quad 2379 \quad - \quad 793$$

$$S = 1586$$

The discount is $793, and the sale price is $1586.

Do Exercise 8.

**8.** A computer marked $660 is on sale at $16\frac{2}{3}$% off. What is the discount? the sale price?

**EXAMPLE 8**   The price of a snowblower is marked down from $950 to $779. What is the rate of discount?

We first find the discount by subtracting the sale price from the original price:

$$950 - 779 = 171.$$

The discount is $171.

Next, we use the equation for discount:

*Discount  =  Rate of discount  ×  Original price*

$$171 \quad = \quad r \quad \cdot \quad 950.$$

To solve, we divide both sides by 950:

$$\frac{171}{950} = \frac{r \cdot 950}{950}$$

$$\frac{171}{950} = r$$

$$0.18 = r$$

$$18\% = r.$$

The discount rate is 18%.

> To check, note that an 18% discount rate means that 82% of the original price is paid:
>
> $$0.82 \cdot \$950 = \$779.$$

**9.** The price of a winter coat is reduced from $75 to $60. Find the rate of discount.

Do Exercise 9.

**a** Solve.

1. *Wyoming Sales Tax.* The sales tax rate in Wyoming is 4%. How much sales tax would be charged on a fireplace screen with doors that costs $239?

2. *Kansas Sales Tax.* The sales tax rate in Kansas is 5.3%. How much sales tax would be charged on a fireplace screen with doors that costs $239?

3. *Ohio Sales Tax.* The sales tax rate in Ohio is 5.5%. How much sales tax would be charged on a phone that sells for $89.99?

4. *New Mexico Sales Tax.* The sales tax rate in New Mexico is 5%. How much sales tax would be charged on a pair of sunglasses that sells for $29.50?

5. *California Sales Tax.* The sales tax rate in California is 7.25%. How much sales tax is charged on a purchase of 4 travel contour foam pillows at $39.95 each? What is the total price?

6. *Illinois Sales Tax.* The sales tax rate in Illinois is 6.25%. How much sales tax is charged on a purchase of 3 wet–dry vacs at $60.99 each? What is the total price?

7. The sales tax is $30 on the purchase of a diamond ring that sells for $750. What is the sales tax rate?

8. The sales tax is $48 on the purchase of a dining room set that sells for $960. What is the sales tax rate?

9. The sales tax is $9.12 on the purchase of a patio set that sells for $456. What is the sales tax rate?

10. The sales tax is $35.80 on the purchase of a refrigerator–freezer that sells for $895. What is the sales tax rate?

11. The sales tax on the purchase of a new fishing boat is $112 and the sales tax rate is 2%. What is the purchase price (the price before tax is added)?

12. The sales tax on the purchase of a used car is $100 and the sales tax rate is 5%. What is the purchase price?

**13.** The sales tax rate in New York City, New York, is 4.875% for the city and 4% for the state. Find the total amount paid for 6 boxes of chocolates at $17.95 each.

**14.** The sales tax rate in Nashville, Tennessee, is 2.25% for Davidson County and 7% for the state. Find the total amount paid for 2 ladders at $39 each.

**15.** The sales tax rate in Seattle, Washington, is 3% for King County and 6.5% for the state. Find the total amount paid for 3 ceiling fans at $84.49 each.

**16.** The sales tax rate in Miami, Florida, is 1% for Dade County and 6% for the state. Find the total amount paid for 2 tires at $49.95 each.

**17.** The sales tax rate in Atlanta, Georgia, is 1% for the city, 3% for Fulton County, and 4% for the state. Find the total amount paid for 6 basketballs at $29.95 each.

**18.** The sales tax rate in Dallas, Texas, is 1% for the city, 1% for Dallas County, and 6.25% for the state. Find the total amount paid for 5 shrubs at $19.95 each.

**b**    Solve.

**19.** Jose's commission rate is 21%. What is the commission from the sale of $12,500 worth of windows?

**20.** Jasmine's commission rate is 6%. What is the commission from the sale of $45,000 worth of lawn irrigation systems?

**21.** Olivia earns $408 selling $3400 worth of shoes. What is the commission rate?

**22.** Mitchell earns $120 selling $2400 worth of television sets. What is the commission rate?

**23.** *Real Estate Commission.*   A real estate agent's commission rate is 7%. She receives a commission of $12,950 from the sale of a home. How much did the home sell for?

**24.** *Clothing Consignment Commission.*   A clothing consignment shop's commission rate is 40%. The shop receives a commission of $552. How many dollars' worth of clothing was sold?

**25.** A real estate commission rate is 8%. What is the commission from the sale of a piece of land for $68,000?

**26.** A real estate commission rate is 6%. What is the commission from the sale of a $98,000 home?

**27.** David earns $1147.50 selling $7650 worth of car parts. What is the commission rate?

**28.** Isabella earns $280.80 selling $2340 worth of tee shirts. What is the commission rate?

**29.** Sabrina's commission is increased according to how much she sells. She receives a commission of 4% for the first $1000 of sales and 7% for the amount over $1000. What is the total commission on sales of $5500?

**30.** Miguel's commission is increased according to how much he sells. He receives a commission of 5% for the first $2000 of sales and 8% for the amount over $2000. What is the total commission on sales of $6200?

**c** Complete the table below by filling in the missing numbers.

| | MARKED PRICE | RATE OF DISCOUNT | DISCOUNT | SALE PRICE |
|---|---|---|---|---|
| **31.** | $300 | 10% | | |
| **32.** | $2000 | 40% | | |
| **33.** | $17 | 15% | | |
| **34.** | $20 | 25% | | |
| **35.** | | 10% | $12.50 | |
| **36.** | | 15% | $65.70 | |
| **37.** | $600 | | $240 | |
| **38.** | $12,800 | | $1920 | |

**39.** Find the marked price and the rate of discount for the steel log rack in this ad.

**40.** Find the marked price and the rate of discount for the digital photo frame in this ad.

**41.** Find the discount and the rate of discount for the pinball machine in this ad.

**42.** Find the discount and the rate of discount for the amaryllis in this ad.

## Skill Maintenance

Solve.  [7.3b]

**43.** $\dfrac{x}{12} = \dfrac{24}{16}$

**44.** $\dfrac{7}{2} = \dfrac{11}{x}$

Solve.  [5.7a]

**45.** $0.64 \cdot x = 170$

**46.** $29.44 = 25.6 \times y$

Find decimal notation.  [5.5a]

**47.** $\dfrac{5}{9}$

**48.** $\dfrac{23}{11}$

Graph.  [6.4b]

**49.** $y = \dfrac{4}{3}x$

**50.** $y = -\dfrac{4}{3}x + 1$

Convert to standard notation.  [5.3b]

**51.** 4.03 trillion

**52.** 5.8 million

Simplify.  [5.4b]

**53.** $80\left(1 + \dfrac{0.06}{2}\right)^2$

**54.** $70\left(1 + \dfrac{0.08}{2}\right)^2$

## Synthesis

**55.** *Magazine Subscriptions.*  In a recent subscription drive, *Car and Driver* offered a 1-year subscription of 12 issues for the price of $1.50 per issue. The company advertised that this was a savings of 69.94% off the cover price. What was the cover price?

Source: *Car and Driver*

**56.** Elijah collects baseball memorabilia. He bought two autographed plaques, but became short of funds and had to sell them quickly for $200 each. On one, he made a 20% profit, and on the other, he lost 20%. Did he make or lose money on the sale?

**57.** ▦ John receives a 10% commission on the first $5000 in sales and 15% on all sales beyond $5000. If John receives a commission of $2405, how much did he sell? Use a calculator and trial and error if you wish.

**58.** Tee shirts are being sold at the mall for $5 each or 3 for $10. If you buy three tee shirts, what is the rate of discount?

# 8.7

## Simple Interest and Compound Interest; Credit Cards

## OBJECTIVES

**a** Solve applied problems involving simple interest.

**b** Solve applied problems involving compound interest.

**c** Solve applied problems involving interest rates on credit cards.

---

**SKILL TO REVIEW**
Objective 8.1b: Convert between percent notation and decimal notation.

Find decimal notation.

**1.** $34\frac{5}{8}\%$          **2.** $5\frac{1}{4}\%$

---

**1.** What is the simple interest on $4300 invested at an interest rate of 4% for 1 year?

---

**2.** What is the simple interest on a principal of $4300 invested at an interest rate of 4% for 9 months?

---

**Answers**

*Skill to Review:*
**1.** 0.34625   **2.** 0.0525

*Margin Exercises:*
**1.** $172   **2.** $129

### a Simple Interest

Suppose you put $1000 into an investment for 1 year. The $1000 is called the **principal**. If the **interest rate** is 5%, in addition to the principal, you get back 5% of the principal, which is

5% of $1000,   or   0.05 · $1000,   or   $50.00.

The $50.00 is called **simple interest**. It is, in effect, the price that a financial institution pays for the use of the money over time.

---

**SIMPLE INTEREST FORMULA**

The **simple interest** $I$ on principal $P$, invested for $t$ years at interest rate $r$, is given by

$$I = P \cdot r \cdot t.$$

---

**EXAMPLE 1**   What is the simple interest on $2500 invested at an interest rate of 6% for 1 year?

We use the formula $I = P \cdot r \cdot t$:

$$I = P \cdot r \cdot t = \$2500 \cdot 6\% \cdot 1$$
$$= \$2500 \cdot 0.06$$
$$= \$150.$$

The simple interest for 1 year is $150.

Do Margin Exercise 1.

**EXAMPLE 2**   What is the simple interest on a principal of $2500 invested at an interest rate of 6% for 3 months?

We use the formula $I = P \cdot r \cdot t$ and express 3 months as a fraction of a year:

$$I = P \cdot r \cdot t = \$2500 \cdot 6\% \cdot \frac{3}{12} = \$2500 \cdot 0.06 \cdot \frac{1}{4}$$
$$= \frac{\$2500 \cdot 0.06}{4} = \$37.50.$$

The simple interest for 3 months is $37.50.

Do Margin Exercise 2.

When time is given in days, we generally divide it by 365 to express the time as a fractional part of a year.

**EXAMPLE 3**   To pay for a shipment of lawn furniture, Patio by Design borrows $8000 at $9\frac{3}{4}\%$ for 60 days. Find **(a)** the amount of simple interest that is due and **(b)** the total amount that must be paid after 60 days.

**a)** We express 60 days as a fractional part of a year:

$$I = P \cdot r \cdot t = \$8000 \cdot 9\frac{3}{4}\% \cdot \frac{60}{365}$$

$$= \$8000 \cdot 0.0975 \cdot \frac{60}{365}$$

$$\approx \$128.22.$$

The interest due for 60 days is $128.22.

**b)** The total amount to be paid after 60 days is the principal plus the interest:

$$\$8000 + \$128.22 = \$8128.22.$$

The total amount due is $8128.22.

Do Exercise 3.

**3.** The Glass Nook borrows $4800 at $8\frac{1}{2}\%$ for 30 days. Find **(a)** the amount of simple interest due and **(b)** the total amount that must be paid after 30 days.

## b  Compound Interest

When interest is paid *on interest*, we call it **compound interest**. This is the type of interest usually paid on investments. Suppose you have $5000 in a savings account at 6%. In 1 year, the account will contain the original $5000 plus 6% of $5000. Thus, the total in the account after 1 year will be

106% of $5000,   or   1.06 · $5000,   or   $5300.

Now suppose that the total of $5300 remains in the account for another year. At the end of this second year, the account will contain the $5300 plus 6% of $5300. The total in the account would thus be

106% of $5300,   or   1.06 · $5300,   or   $5618.

Note that in the second year, interest is also earned on the first year's interest. When this happens, we say that interest is **compounded annually**.

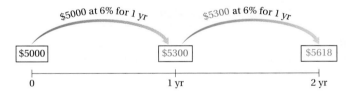

**EXAMPLE 4**   Find the amount in an account if $2000 is invested at 8%, compounded annually, for 2 years.

**a)** After 1 year, the account will contain 108% of $2000:

1.08 · $2000 = $2160.

**b)** At the end of the second year, the account will contain 108% of $2160:

1.08 · $2160 = $2332.80.

The amount in the account after 2 years is $2332.80.

Do Exercise 4.

**4.** Find the amount in an account if $2000 is invested at 9%, compounded annually, for 2 years.

*Answers*

**3. (a)** $33.53; **(b)** $4833.53    **4.** $2376.20

Suppose that the interest in Example 4 were **compounded semi-annually**—that is, every half year. Interest would then be calculated twice a year at a rate of 8% ÷ 2, or 4% each time. The approach used in Example 4 can then be adapted, as follows.

After the first $\frac{1}{2}$ year, the account will contain 104% of $2000:

$$1.04 \cdot \$2000 = \$2080.$$

After a second $\frac{1}{2}$ year (1 full year), the account will contain 104% of $2080:

$$1.04 \cdot \$2080 = \$2163.20.$$

After a third $\frac{1}{2}$ year $\left(1\frac{1}{2} \text{ full years}\right)$, the account will contain 104% of $2163.20:

$$1.04 \cdot \$2163.20 = \$2249.728$$
$$\approx \$2249.73. \qquad \text{Rounding to the nearest cent}$$

Finally, after a fourth $\frac{1}{2}$ year (2 full years), the account will contain 104% of $2249.73:

$$1.04 \cdot \$2249.73 = \$2339.7192$$
$$\approx \$2339.72. \qquad \text{Rounding to the nearest cent}$$

Let's summarize our results and look at them another way:

End of 1st $\frac{1}{2}$ year $\rightarrow 1.04 \cdot 2000 = 2000 \cdot (1.04)^1$;
End of 2nd $\frac{1}{2}$ year $\rightarrow 1.04 \cdot (1.04 \cdot 2000) = 2000 \cdot (1.04)^2$;
End of 3rd $\frac{1}{2}$ year $\rightarrow 1.04 \cdot (1.04 \cdot 1.04 \cdot 2000) = 2000 \cdot (1.04)^3$;
End of 4th $\frac{1}{2}$ year $\rightarrow 1.04 \cdot (1.04 \cdot 1.04 \cdot 1.04 \cdot 2000) = 2000 \cdot (1.04)^4$.

Note that each multiplication was by 1.04 and that

$$\$2000 \cdot 1.04^4 \approx \$2339.72. \qquad \text{Using a calculator and rounding to the nearest cent}$$

We have illustrated the following result.

---

**COMPOUND INTEREST FORMULA**

If a principal $P$ has been invested at interest rate $r$, compounded $n$ times a year, in $t$ years it will grow to an amount $A$ given by

$$A = P \cdot \left(1 + \frac{r}{n}\right)^{n \cdot t}.$$

---

In the compound interest formula, $n \cdot t$ is the total number of compounding periods, $r$ is written in decimal notation, and $\frac{r}{n}$ is the interest rate for each period.

Let's apply this formula to confirm our preceding discussion, where the amount invested is $P = \$2000$, the interest rate is $r = 8\%$, the number of years is $t = 2$, and the number of compounding periods each year is $n = 2$. Substituting into the compound interest formula, we have

$$A = P \cdot \left(1 + \frac{r}{n}\right)^{n \cdot t} = 2000 \cdot \left(1 + \frac{8\%}{2}\right)^{2 \cdot 2}$$
$$= \$2000 \cdot \left(1 + \frac{0.08}{2}\right)^4 = \$2000(1.04)^4$$
$$= \$2000 \cdot 1.16985856 \approx \$2339.72.$$

If you were using a calculator, you could perform this computation in one step.

**STUDY TIPS**

### WORKING WITH A CLASSMATE

If you are finding it difficult to master a particular topic or concept, try talking about it with a classmate. Verbalizing your questions about the material might help clarify it. If your classmate is also finding the material difficult, it is possible that others in your class are confused and you can ask your instructor to explain the concept again.

**EXAMPLE 5** The Ibsens invest $4000 in an account paying $5\frac{5}{8}\%$, compounded quarterly. Find the amount in the account after $2\frac{1}{2}$ years.

The compounding is quarterly, so $n$ is 4. We substitute $4000 for $P$, $5\frac{5}{8}\%$, or 0.05625, for $r$, 4 for $n$, and $2\frac{1}{2}$, or $\frac{5}{2}$, for $t$ and compute $A$:

$$A = P \cdot \left(1 + \frac{r}{n}\right)^{n \cdot t} = \$4000 \cdot \left(1 + \frac{5\frac{5}{8}\%}{4}\right)^{4 \cdot 5/2}$$

$$= \$4000 \cdot \left(1 + \frac{0.05625}{4}\right)^{10}$$

$$= \$4000(1.0140625)^{10}$$

$$\approx \$4599.46.$$

The amount in the account after $2\frac{1}{2}$ years is $4599.46.

 Do Exercise 5.

**5.** A couple invests $7000 in an account paying $6\frac{3}{8}\%$, compounded semiannually. Find the amount in the account after $1\frac{1}{2}$ years.

---

 **Calculator Corner**

**Compound Interest**   A calculator is useful in computing compound interest. Not only does it perform computations quickly but it also eliminates the need to round until the computation is completed. This minimizes "round-off errors" that occur when rounding is done at each stage of the computation. We must keep order of operations in mind when computing compound interest.

To find the amount due on a $20,000 loan made for 25 days at 11% interest, compounded daily, we would compute

$20,000\left(1 + \dfrac{0.11}{365}\right)^{25}$. To do this on a calculator, we press $\boxed{2}\boxed{0}\boxed{0}\boxed{0}\boxed{0}\boxed{\times}\boxed{(}\boxed{1}\boxed{+}\boxed{.}\boxed{1}\boxed{1}\boxed{\div}\boxed{3}\boxed{6}\boxed{5}\boxed{)}\boxed{y^x}$

(or $\boxed{\wedge}$) $\boxed{2}\boxed{5}\boxed{=}$. Using a calculator that does not have parentheses, we would first find $1 + \dfrac{0.11}{365}$, raise this result to the 25th power,

and then multiply by 20,000. To do this, we press $\boxed{1}\boxed{+}\boxed{.}\boxed{1}\boxed{1}\boxed{\div}\boxed{3}\boxed{6}\boxed{5}\boxed{=}\boxed{y^x}$ (or $\boxed{\wedge}$) $\boxed{2}\boxed{5}\boxed{=}\boxed{\times}\boxed{2}\boxed{0}\boxed{0}$ $\boxed{0}\boxed{0}\boxed{=}$. In either case, the result is 20,151.23, rounded to the nearest cent.

Some calculators have business keys that allow such computations to be done more quickly.

**Exercises:**

**1.** Find the amount due on a $16,000 loan made for 62 days at 13% interest, compounded daily.

**2.** An investment of $12,500 is made for 90 days at 8.5% interest, compounded daily. How much is the investment worth after 90 days?

---

## c  Credit Cards

According to creditcards.com, the average credit-card debt per household with credit-card debt in the United States was $16,007 in 2009. According to a 2007 survey by student lender Sallie Mae, the average college graduate with at least one credit card has a $4138 credit-card debt.

The money you obtain through the use of a credit card is not "free" money. There is a price (interest) to be paid for the privilege. A balance carried on a credit card is a type of loan. Comparing interest rates is essential if one is to become financially responsible. A small change in an interest rate can make a large difference in the cost of a loan. When you make a payment on a credit card, do you know how much of that payment is interest and how much is applied to reducing the principal?

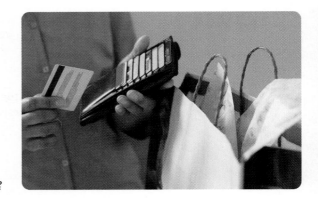

*Answer*
**5.** $7690.94

**EXAMPLE 6** *Credit Cards.* After the holidays, Sarah has a balance of $3216.28 on a credit card with an annual percentage rate (APR) of 19.7%. She decides not to make additional purchases with this card until she has paid off the balance.

**a)** Many credit cards require a minimum monthly payment of 2% of the balance. At this rate, what is Sarah's minimum payment on a balance of $3216.28? Round the answer to the nearest dollar.

**b)** Find the amount of interest and the amount applied to reduce the principal in the minimum payment found in part (a).

**c)** If Sarah had transferred her balance to a card with an APR of 12.5%, how much of her first payment would be interest and how much would be applied to reduce the principal?

**d)** Compare the amounts for 12.5% from part (c) with the amounts for 19.7% from part (b).

We solve as follows.

**a)** The minimum payment is 2% of $3216.28:

$$0.02 \cdot \$3216.28 = \$64.3256.$$ Sarah's minimum payment, rounded to the nearest dollar, is $64.

**b)** The amount of interest on $3216.28 at 19.7% for one month* is given by

$$I = P \cdot r \cdot t = \$3216.28 \cdot 0.197 \cdot \frac{1}{12} \approx \$52.80.$$

We subtract to find the amount applied to reduce the principal in the first payment:

$$\begin{array}{l} \text{Amount applied to} \\ \text{reduce the principal} \end{array} = \text{Minimum payment} - \begin{array}{l} \text{Interest for} \\ \text{the month} \end{array}$$

$$= \$64 - \$52.80$$

$$= \$11.20.$$

Thus, the principal of $3216.28 is decreased by only $11.20 with the first payment. (Sarah still owes $3205.08.)

**c)** The amount of interest on $3216.28 at 12.5% for one month is

$$I = P \cdot r \cdot t = \$3216.28 \cdot 0.125 \cdot \frac{1}{12} \approx \$33.50.$$

We subtract to find the amount applied to reduce the principal in the first payment:

$$\begin{array}{l} \text{Amount applied to} \\ \text{reduce the principal} \end{array} = \text{Minimum payment} - \begin{array}{l} \text{Interest for} \\ \text{the month} \end{array}$$

$$= \$64 - \$33.50$$

$$= \$30.50.$$

Thus, the principal of $3216.28 is decreased by $30.50 with the first payment. (Sarah still owes $3185.78.)

---

*Actually, the interest on a credit card is computed daily with a rate called a daily percentage rate (DPR). The DPR for Example 6 would be 19.7%/365 = 0.054%. When no payments or additional purchases are made during the month, the difference in total interest for the month is minimal and we will not deal with it here.

**d)** Let's organize the information for both rates in the following table.

| BALANCE BEFORE FIRST PAYMENT | FIRST MONTH'S PAYMENT | % APR | AMOUNT OF INTEREST | AMOUNT APPLIED TO PRINCIPAL | BALANCE AFTER FIRST PAYMENT |
|---|---|---|---|---|---|
| $3216.28 | $64 | 19.7% | $52.80 | $11.20 | $3205.08 |
| 3216.28 | 64 | 12.5 | 33.50 | 30.50 | 3185.78 |

Difference in balance after first payment → $19.30

At 19.7%, the interest is $52.80 and the principal is decreased by $11.20. At 12.5%, the interest is $33.50 and the principal is decreased by $30.50. Thus, the interest at 19.7% is $52.80 − $33.50, or $19.30, greater than the interest at 12.5%. The principal is decreased by $30.50 − $11.20, or $19.30, more with the 12.5% rate than with the 19.7% rate.

Do Exercise 6.

Even though the mathematics of the information in the table below is beyond the scope of this text, it is interesting to compare how long it takes to pay off the balance of Example 6 if Sarah continues to pay $64 for each payment with how long it takes if she pays double that amount, $128, for each payment. Financial consultants frequently tell clients that if they want to take control of their debt, they should pay double the minimum payment.

| RATE | PAYMENT PER MONTH | NUMBER OF PAYMENTS TO PAY OFF DEBT | TOTAL PAID BACK | ADDITIONAL COST OF PURCHASES |
|---|---|---|---|---|
| 19.7% | $64 | 107, or 8 yr 11 mo | $6848 | $3631.72 |
| 19.7 | 128 | 33, or 2 yr 9 mo | 4224 | 1007.72 |
| 12.5 | 64 | 72, or 6 yr | 4608 | 1391.72 |
| 12.5 | 128 | 29, or 2 yr 5 mo | 3712 | 495.72 |

As with most loans, if you pay an extra amount toward the principal with each payment, the length of the loan can be greatly reduced. Note that at the rate of 19.7%, it will take Sarah almost 9 yr to pay off her debt if she pays only $64 per month and does not make additional purchases. If she transfers her balance to a card with a 12.5% rate and pays $128 per month, she can eliminate her debt in approximately $2\frac{1}{2}$ yr. Debt can quickly get out of control if you continue to make purchases and pay only the minimum payment each month. The debt will never be eliminated.

**6. Credit Cards.** After the holidays, Jamal has a balance of $4867.59 on a credit card with an annual percentage rate (APR) of 21.3%. He decides not to make additional purchases with this card until he has paid off the balance.

**a)** Many credit cards require a minimum monthly payment of 2% of the balance. What is Jamal's minimum payment on a balance of $4867.59? Round the answer to the nearest dollar.

**b)** Find the amount of interest and the amount applied to reduce the principal in the minimum payment found in part (a).

**c)** If Jamal had transferred his balance to a card with an APR of 13.6%, how much of his first payment would be interest and how much would be applied to reduce the principal?

**d)** Compare the amounts for 13.6% from part (c) with the amounts for 21.3% from part (b).

**a** Find the simple interest.

|     | PRINCIPAL | RATE OF INTEREST | TIME | SIMPLE INTEREST |
|-----|-----------|------------------|------|-----------------|
| 1.  | $200      | 4%               | 1 year | |
| 2.  | $200      | 7.7%             | $\frac{1}{2}$ year | |
| 3.  | $4300     | 10.56%           | $\frac{1}{4}$ year | |
| 4.  | $80,000   | $6\frac{3}{4}\%$ | $\frac{1}{12}$ year | |
| 5.  | $20,000   | $4\frac{5}{8}\%$ | 1 year | |
| 6.  | $8000     | 9.42%            | 2 months | |
| 7.  | $50,000   | $5\frac{3}{8}\%$ | 3 months | |
| 8.  | $100,000  | $3\frac{1}{4}\%$ | 1 year | |

Solve. Assume that simple interest is being calculated in each case.

9. CopiPix, Inc. borrows $10,000 at 9% for 60 days. Find **(a)** the amount of interest due and **(b)** the total amount that must be paid after 60 days.

10. Sal's Laundry borrows $8000 at 10% for 90 days. Find **(a)** the amount of interest due and **(b)** the total amount that must be paid after 90 days.

11. Animal Instinct, a pet supply shop, borrows $6500 at $5\frac{1}{4}\%$ for 90 days. Find **(a)** the amount of interest due and **(b)** the total amount that must be paid after 90 days.

12. Andante's Cafe borrows $4500 at $12\frac{1}{2}\%$ for 60 days. Find **(a)** the amount of interest due and **(b)** the total amount that must be paid after 60 days.

13. Jean's Garage borrows $5600 at 10% for 30 days. Find **(a)** the amount of interest due and **(b)** the total amount that must be paid after 30 days.

14. Shear Delights Hair Salon borrows $3600 at 4% for 30 days. Find **(a)** the amount of interest due and **(b)** the total amount that must be paid after 30 days.

Interest is compounded annually. Find the amount in the account after the given length of time. Round to the nearest cent.

|      | PRINCIPAL | RATE OF INTEREST | TIME     | AMOUNT IN THE ACCOUNT |
|------|-----------|------------------|----------|-----------------------|
| 15.  | $400      | 5%               | 2 years  |                       |
| 16.  | $450      | 4%               | 2 years  |                       |
| 17.  | $2000     | 8.8%             | 4 years  |                       |
| 18.  | $4000     | 7.7%             | 4 years  |                       |
| 19.  | $4300     | 10.56%           | 6 years  |                       |
| 20.  | $8000     | 9.42%            | 6 years  |                       |
| 21.  | $20,000   | $6\frac{5}{8}\%$ | 25 years |                       |
| 22.  | $100,000  | $5\frac{7}{8}\%$ | 30 years |                       |

Interest is compounded semiannually. Find the amount in the account after the given length of time. Round to the nearest cent.

|      | PRINCIPAL | RATE OF INTEREST | TIME     | AMOUNT IN THE ACCOUNT |
|------|-----------|------------------|----------|-----------------------|
| 23.  | $4000     | 6%               | 1 year   |                       |
| 24.  | $1000     | 5%               | 1 year   |                       |
| 25.  | $20,000   | 8.8%             | 4 years  |                       |
| 26.  | $40,000   | 7.7%             | 4 years  |                       |
| 27.  | $5000     | 10.56%           | 6 years  |                       |
| 28.  | $8000     | 9.42%            | 8 years  |                       |
| 29.  | $20,000   | $7\frac{5}{8}\%$ | 25 years |                       |
| 30.  | $100,000  | $4\frac{7}{8}\%$ | 30 years |                       |

Solve.

**31.** The Martinez family invests $4000 in an account paying 6%, compounded monthly. How much is in the account after 5 months?

**32.** Emily and Blair invest $2500 in an account paying 3%, compounded monthly. How much is in the account after 6 months?

**33.** Bryn and Andy invest $1200 in an account paying 10%, compounded quarterly. How much is in the account after 1 year?

**34.** The O'Hares invest $6000 in an account paying 8%, compounded quarterly. How much is in the account after 18 months?

**35.** ▦ Emilio loans his niece's business $20,000 for 50 days at 6% interest, compounded daily. How much is Emilio owed after 50 days?

**36.** ▦ Elsa loans her nephew's business $25,000 for 40 days at 5% interest, compounded daily. How much is Elsa owed after 40 days?

**37.** *Credit Cards.* Amelia has a balance of $1278.56 on a credit card with an annual percentage rate (APR) of 19.6%. The minimum payment required in the current statement is $25.57. Find the amount of interest and the amount applied to reduce the principal in this payment and the balance after this payment.

**38.** *Credit Cards.* Lawson has a balance of $1834.90 on a credit card with an annual percentage rate (APR) of 22.4%. The minimum payment required in the current statement is $36.70. Find the amount of interest and the amount applied to reduce the principal in this payment and the balance after this payment.

**39.** *Credit Cards.* Antonio has a balance of $4876.54 on a credit card with an annual percentage rate (APR) of 21.3%.

   **a)** Many credit cards require a minimum monthly payment of 2% of the balance. What is Antonio's minimum payment on a balance of $4876.54? Round the answer to the nearest dollar.
   **b)** Find the amount of interest and the amount applied to reduce the principal in the minimum payment found in part (a).
   **c)** If Antonio had transferred his balance to a card with an APR of 12.6%, how much of his payment would be interest and how much would be applied to reduce the principal?
   **d)** Compare the amounts for 12.6% from part (c) with the amounts for 21.3% from part (b).

**40.** *Credit Cards.* Becky has a balance of $5328.88 on a credit card with an annual percentage rate (APR) of 18.7%.

   **a)** Many credit cards require a minimum monthly payment of 2% of the balance. What is Becky's minimum payment on a balance of $5328.88? Round the answer to the nearest dollar.
   **b)** Find the amount of interest and the amount applied to reduce the principal in the minimum payment found in part (a).
   **c)** If Becky had transferred her balance to a card with an APR of 13.2%, how much of her payment would be interest and how much would be applied to reduce the principal?
   **d)** Compare the amounts for 13.2% from part (c) with the amounts for 18.7% from part (b).

## Skill Maintenance

In each of Exercises 41–48, fill in the blank with the correct term from the given list. Some of the choices may not be used.

**41.** If the product of two numbers is 1, they are _____ of each other.   [3.7a]

**42.** A number is _____ if its ones digit is even and the sum of its digits is divisible by 3.   [3.1b]

**43.** The number 0 is the _____ identity.   [2.2a]

**44.** A(n) _____ is the ratio of price to the number of units. [7.2b]

**45.** The distance around an object is its _____.   [1.2b]

**46.** A number is _____ if the sum of its digits is divisible by 3.   [3.1b]

**47.** A natural number that has exactly two different factors, only itself and 1, is called a(n) _____ number.   [3.2b]

**48.** When two pairs of numbers have the same ratio, they are _____.   [7.3a]

divisible by 3

divisible by 5

divisible by 6

divisible by 9

perimeter

area

unit price

reciprocals

proportional

composite

prime

additive

multiplicative

## Synthesis

*Effective Yield.* The *effective yield* is the yearly rate of simple interest that corresponds to a rate for which interest is compounded two or more times a year. For example, if $P$ is invested at 12%, compounded quarterly, we would multiply $P$ by $(1 + 0.12/4)^4$, or $1.03^4$. Since $1.03^4 \approx 1.126$, the 12% compounded quarterly corresponds to an effective yield of approximately 12.6%. In Exercises 49 and 50, find the effective yield for the indicated account.

**49.** ▦ The account pays 9% compounded monthly.

**50.** ▦ The account pays 10% compounded daily.

## Key Terms and Formulas

percent notation, $n\%$, p. 512
percent of decrease, p. 545
percent of increase, p. 547
purchase price, p. 553
sales tax, p. 553
total price, p. 553

commission, p. 555
original price, p. 556
marked price, p. 556
rate of discount, p. 556
discount, p. 556
sale price, p. 556

principal, p. 562
interest rate, p. 562
simple interest, p. 562
compound interest, p. 563
compounded annually, p. 563
compounded semiannually, p. 564

Commission = Commission rate $\times$ Sales

Discount = Rate of discount $\times$ Original price

Sale price = Original price $-$ Discount

*Simple interest:*    $I = P \cdot r \cdot t$

*Compound interest:*    $A = P \cdot \left(1 + \dfrac{r}{n}\right)^{n \cdot t}$

## Concept Reinforcement

Determine whether each statement is true or false.

_____ **1.** A fixed principal invested for 4 years will earn more interest when interest is compounded quarterly than when interest is compounded semiannually.   [8.7b]

_____ **2.** Of the numbers 0.5%, $\dfrac{5}{1000}\%$, $\dfrac{1}{2}\%$, $\dfrac{1}{5}$, and $0.\overline{1}$, the largest number is $0.\overline{1}$.   [8.1b, c]

_____ **3.** If principal A equals principal B and principal A is invested for 2 years at 4%, compounded quarterly, while principal B is invested for 4 years at 2%, compounded semiannually, the interest earned from each investment is the same.   [8.7b]

## Important Concepts

**Objective 8.1b**   Convert between percent notation and decimal notation.

**Example**   Find decimal notation for $12\dfrac{3}{4}\%$.

$$12\dfrac{3}{4}\% = 12.75 \cdot 0.01 = 0.1275$$

**Practice Exercise**

**1.** Find decimal notation for $62\dfrac{5}{8}\%$.

**Objective 8.1c**   Convert between fraction notation and percent notation.

**Example**   Find percent notation for $\dfrac{5}{12}$.

$$
\begin{array}{r}
0.4\ 1\ 6 \\
1\ 2\ )\overline{\ 5.0\ 0\ 0} \\
\underline{4\ 8} \\
2\ 0 \\
\underline{1\ 2} \\
8\ 0 \\
\underline{7\ 2} \\
8
\end{array}
$$

$\dfrac{5}{12} = 0.41\overline{6} = 41.\overline{6}\%$, or $41\dfrac{2}{3}\%$

**Practice Exercise**

**2.** Find percent notation for $\dfrac{7}{11}$.

**Objective 8.2b** Solve basic percent problems using percent equations.

**Example** 165 is what percent of 3300?

We have

$$165 = p \cdot 3300 \qquad \text{Translating to a percent equation}$$

$$\frac{165}{3300} = \frac{p \cdot 3300}{3300}$$

$$\frac{165}{3300} = p$$

$$0.05 = p$$

$$5\% = p.$$

Thus, 165 is 5% of 3300.

**Practice Exercise**

**3.** 12 is what percent of 288?

---

**Objective 8.3b** Solve basic percent problems using proportions.

**Example** 18% of what is 1296?

$$\frac{18}{100} = \frac{1296}{b} \qquad \text{Translating to a proportion}$$

$$18 \cdot b = 100 \cdot 1296$$

$$\frac{18 \cdot b}{18} = \frac{129{,}600}{18}$$

$$b = 7200$$

Thus, 18% of 7200 is 1296.

**Practice Exercise**

**4.** 3% of what is 300?

---

**Objective 8.4a** Solve applied problems involving percent.

**Example** Some financial experts recommend spending no more than 30% of monthly income on mortgage payments. Elliott makes $2125 a month. What is the highest mortgage payment he should have?

*Rewording:* What number is 30% of 2125?

*Percent equation:*

$$a = 0.30 \cdot 2125$$

$$a = 637.5$$

*Proportion:*

$$\frac{30}{100} = \frac{a}{2125}$$

$$30 \cdot 2125 = 100 \cdot a$$

$$\frac{30 \cdot 2125}{100} = \frac{100 \cdot a}{100}$$

$$637.5 = a$$

The highest mortgage payment Elliott should have is $637.50.

**Practice Exercise**

**5.** Amanda makes $2650 a month and her monthly mortgage payment is $927.50. What percent of her monthly income is her mortgage payment?

---

**Objective 8.5a** Solve applied problems involving percent of increase or percent of decrease.

**Example** The total cost for 16 basic grocery items in the second quarter of 2008 averaged $46.67 nationally. The total cost of these 16 items in the second quarter of 2007 averaged $42.95. What was the percent of increase?

**Source:** American Farm Bureau Federation

**Practice Exercise**

**6.** In Indiana, the cost for 16 basic grocery items increased from $40.07 in the second quarter of 2007 to $46.20 in the second quarter of 2008. What was the percent of increase from 2007 to 2008?

## Objective 8.5a (continued)

We first determine the amount of increase: $46.67 − $42.95 = $3.72. Then we translate to a percent equation or a proportion and solve.

*Rewording:* $3.72 is what percent of $42.95?

*Percent equation:*

$$3.72 = p \cdot 42.95$$

$$\frac{3.72}{42.95} = \frac{p \cdot 42.95}{42.95}$$

$$\frac{3.72}{42.95} = p$$

$$0.087 \approx p$$

$$8.7\% \approx p$$

*Proportion:*

$$\frac{N}{100} = \frac{3.72}{42.95}$$

$$42.95N = 100 \cdot 3.72$$

$$\frac{42.95N}{42.95} = \frac{100 \cdot 3.72}{42.95}$$

$$N = \frac{372}{42.95}$$

$$N \approx 8.7$$

The percent of increase was 8.7%.

---

## Objective 8.6a  Solve applied problems involving sales tax and percent.

**Example**  The sales tax is $34.23 on the purchase of a flat-screen high-definition television that costs $489. What is the sales tax rate?

*Rephrase:*  Sales tax is what percent of purchase price?

*Translate:*  $34.23 = r \cdot 489$

*Solve:*

$$\frac{34.23}{489} = \frac{r \cdot 489}{489}$$

$$\frac{34.23}{489} = r$$

$$0.07 = r$$

$$7\% = r$$

The sales tax rate is 7%.

**Practice Exercise**

7. The sales tax is $1102.20 on the purchase of a new car that costs $18,370. What is the sales tax rate?

---

## Objective 8.6b  Solve applied problems involving commission and percent.

**Example**  A real estate agent's commission rate is $6\frac{1}{2}\%$. She received a commission of $17,160 on the sale of a home. For how much did the home sell?

*Rephrase:*  Commission is $6\frac{1}{2}\%$ of what selling price?

*Translate:*  $17{,}160 = 6\frac{1}{2}\% \times S$

*Solve:*

$$17{,}160 = 0.065 \cdot S$$

$$\frac{17{,}160}{0.065} = \frac{0.065 \cdot S}{0.065}$$

$$264{,}000 = S$$

The home sold for $264,000.

**Practice Exercise**

8. A real estate agent's commission rate is 7%. He received a commission of $12,950 on the sale of a home. For how much did the home sell?

**Objective 8.7a** Solve applied problems involving simple interest.

**Example** To meet its payroll, a business borrows $5200 at $4\frac{1}{4}\%$ for 90 days. Find the amount of simple interest that is due and the total amount that must be paid after 90 days.

$$I = P \cdot r \cdot t = \$5200 \cdot 4\tfrac{1}{4}\% \cdot \frac{90}{365}$$

$$= \$5200 \cdot 0.0425 \cdot \frac{90}{365}$$

$$\approx \$54.49$$

The interest due for 90 days = $54.49.

The total amount due = $5200 + $54.49 = $5254.49.

**Practice Exercise**

9. A student borrows $2500 for tuition at $5\frac{1}{2}\%$ for 60 days. Find the amount of simple interest that is due and the total amount that must be paid after 60 days.

---

**Objective 8.7b** Solve applied problems involving compound interest.

**Example** Find the amount in an account if $3200 is invested at 5%, compounded semiannually, for $1\frac{1}{2}$ years.

$$A = P \cdot \left(1 + \frac{r}{n}\right)^{n \cdot t}$$

$$= \$3200\left(1 + \frac{0.05}{2}\right)^{2 \cdot \frac{3}{2}}$$

$$= \$3200(1.025)^3$$

$$= \$3446.05$$

The amount in the account after $1\frac{1}{2}$ years is $3446.05.

**Practice Exercise**

10. Find the amount in an account if $6000 is invested at $4\frac{3}{4}\%$, compounded quarterly, for 2 years.

---

# Review Exercises

Find decimal notation for the percent notations in each sentence.   [8.1b]

1. In the 2008–2009 school year, about 3.7% of the 18 million college students in the United States were foreign students. Approximately 15.4% of the foreign students were from India.
   **Source:** Institute of International Education

2. Poland is 62.1% urban; Sweden is 84.2% urban.
   **Source:** The World Almanac, 2008

Find percent notation.   [8.1c]

3. $\dfrac{3}{8}$

4. $\dfrac{1}{3}$

Find percent notation.   [8.1b]

5. 1.7

6. 0.065

Find fraction notation.   [8.1c]

7. 24%

8. 6.3%

Translate to a percent equation. Then solve.   [8.2a, b]

9. 30.6 is what percent of 90?

10. 63 is 84% of what?

11. What is $38\frac{1}{2}\%$ of 168?

Translate to a proportion. Then solve. [8.3a, b]

**12.** 24 percent of what is 16.8?

**13.** 42 is what percent of 30?

**14.** What is 10.5% of 84?

Solve. [8.4a], [8.5a]

**15.** *Favorite Ice Creams.* According to a survey, 8.9% of those interviewed chose chocolate as their favorite ice cream flavor and 4.2% chose butter pecan. At this rate, of the 2000 students in a freshman class, how many would choose chocolate as their favorite ice cream? butter pecan?
Source: International Ice Cream Association

**16.** *Prescriptions.* Of the 305 million people in the United States, 140.3 million take at least one type of prescription drug per day. What percent take at least one type of prescription drug per day?
Source: William N. Kelly, *Pharmacy: What It Is and How It Works,* 2nd ed., CRC Press Pharmaceutical Education, 2006

**17.** *Water Output.* The average person expels 200 mL of water per day by sweating. This is 8% of the total output of water from the body. How much is the total output of water?
Source: Elaine N. Marieb, *Essentials of Human Anatomy and Physiology,* 6th ed. Boston: Addison Wesley Longman, Inc., 2000

**18.** *Test Scores.* After Sheila got a 75 on a math test, she was allowed to go to the math lab and take a retest. She increased her score to 84. What was the percent of increase?

**19.** *Test Scores.* James got an 80 on a math test. By taking a retest in the math lab, he increased his score by 15%. What was his new score?

Solve. [8.6a, b, c]

**20.** A state charges a meals tax of $7\frac{1}{2}\%$. What is the meals tax charged on a dinner party costing $320?

**21.** In a certain state, a sales tax of $453.60 is collected on the purchase of a used car for $7560. What is the sales tax rate?

**22.** Kim earns $753.50 selling $6850 worth of televisions. What is the commission rate?

**23.** An air conditioner has a marked price of $350. It is placed on sale at 12% off. What are the discount and the sale price?

**24.** The price of a smart phone is marked down from $305 to $262.30. What is the rate of discount?

**25.** An insurance salesperson receives a 7% commission. If $42,000 worth of life insurance is sold, what is the commission?

**26.** What is the rate of discount of this stepladder?

SPECIAL VALUE!
Now $67 Was $82
8' Aluminum Stepladder

Solve. [8.7a, b, c]

**27.** What is the simple interest on $1800 at 6% for $\frac{1}{3}$ year?

**28.** The Dress Shack borrows $24,000 at 10% simple interest for 60 days. Find **(a)** the amount of interest due and **(b)** the total amount that must be paid after 60 days.

**29.** What is the simple interest on $2200 principal at an interest rate of 5.5% for 1 year?

**30.** The Armstrongs invest $7500 in an investment account paying an annual interest rate of 4%, compounded monthly. How much is in the account after 3 months?

**31.** Find the amount in an investment account if $8000 is invested at 9%, compounded annually, for 2 years.

**32.** *Credit Cards.*   At the end of her junior year of college, Kasha has a balance of $6428.74 on a credit card with an annual percentage rate (APR) of 18.7%. She decides not to make additional purchases with this card until she has paid off the balance.

**a)** Many credit cards require a minimum payment of 2% of the balance. At this rate, what is Kasha's minimum payment on a balance of $6428.74? Round the answer to the nearest dollar.

**b)** Find the amount of interest and the amount applied to reduce the principal in the minimum payment found in part (a).

**c)** If Kasha had transferred her balance to a card with an APR of 13.2%, how much of her payment would be interest and how much would be applied to reduce the principal?

**d)** Compare the amounts for 13.2% from part (c) with the amounts for 18.7% from part (b).

**33.** A fishing boat listed at $16,500 is on sale at 15% off. What is the sale price?   [8.6c]

**A.** $14,025  **B.** $2475
**C.** 85%  **D.** $14,225

**34.** Find the amount in a money market account if $10,500 is invested at 6%, compounded semiannually, for $1\frac{1}{2}$ years.   [8.7b]

**A.** $11,139.45  **B.** $12,505.67
**C.** $11,473.63  **D.** $10,976.03

## Synthesis

**35.** Mike's Bike Shop reduces the price of a bicycle by 40% during a sale. By what percent must the store increase the sale price, after the sale, to get back to the original price?   [8.6c]

**36.** A $200 coat is marked up 20%. After 30 days, it is marked down 30% and sold. What is the final selling price of the coat?   [8.6c]

# Understanding Through Discussion and Writing

**1.** Which is the better deal for a consumer and why: a discount of 40% or a discount of 20% followed by another of 22%?   [8.6c]

**2.** Which is better for a wage earner and why: a 10% raise followed by a 5% raise a year later or a 5% raise followed by a 10% raise a year later?   [8.5a]

**3.** Ollie bought a microwave oven during a 10%-off sale. The sale price that Ollie paid was $162. To find the original price, Ollie calculates 10% of $162 and adds that to $162. Is this correct? Why or why not?   [8.6c]

**4.** You take 40% of 50% of a number. What percent of the number could you take to obtain the same result making only one multiplication? Explain your answer.   [8.4a]

**5.** A firm must choose between borrowing $5000 at 10% for 30 days and borrowing $10,000 at 8% for 60 days. Give arguments in favor of and against each option.   [8.7a]

**6.** On the basis of the mathematics presented in Section 8.7, discuss what you have learned about interest rates and credit cards.   [8.7c]

**CHAPTER**

**8**

Test    **For Extra Help**

Step-by-step test solutions are found on the Chapter Test Prep Videos available via the Video Resources on DVD, in **MyMathLab**, and on **You Tube** (search "BittingerPrealgebra" and click on "Channels").

1. *Television.*  In 2010, Americans spent about 45.8% of their media usage time watching television. Find decimal notation for 45.8%.

    **Source:** Veronis Suhler Stevenson, New York, NY

2. *Gravity.*  The gravity of Mars is 0.38 as strong as Earth's. Find percent notation for 0.38.

    **Source:** www.marsinstitute.info/epo/mermarsfacts.html

3. Find percent notation for $\dfrac{11}{8}$.

4. Find fraction notation for 65%.

5. Translate to a percent equation. Then solve.

    What is 40% of 55?

6. Translate to a proportion. Then solve.

    What percent of 80 is 65?

Solve.

7. *Organ Transplants.*  In 2008, there were 23,288 organ transplants in the United States. The pie chart below shows the percentages for the main transplants. How many kidney transplants were there in 2008? liver transplants? heart transplants?

    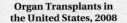

    **Organ Transplants in the United States, 2008**

    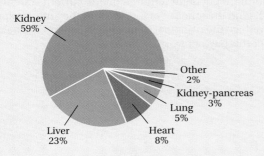

    SOURCES: Information Please® Database, National Organ Procurement and Transplantation Network

8. *Batting Average.*  Troy Tulowitzki, shortstop for the Colorado Rockies, got 161 hits during the 2009 baseball season. This was about 29.65% of his at-bats. How many at-bats did he have?

    **Source:** Major League Baseball

9. *Foreign Adoptions.*  The number of foreign children adopted by Americans declined from 17,438 in 2008 to 12,753 in 2009. Find the percent of decrease.

    **Sources:** U.S. State Department; www.1010wins.com

10. There are about 6,768,000,000 people living in the world today, and approximately 3,808,000,000 live in Asia. What percent of people live in Asia?

    **Source:** Population Division/International Programs Center, U.S. Census Bureau, U.S. Dept. of Commerce

11. *Oklahoma Sales Tax.*  The sales tax rate in Oklahoma is 4.5%. How much tax is charged on a purchase of $560? What is the total price?

12. Noah's commission rate is 15%. What is the commission from the sale of $4200 worth of merchandise?

13. The marked price of a DVD player is $200 and the item is on sale at 20% off. What are the discount and the sale price?

14. What is the simple interest on a principal of $120 at the interest rate of 7.1% for 1 year?

15. A city orchestra invests $5200 at 6% simple interest. How much is in the account after $\frac{1}{2}$ year?

16. Find the amount in an account if $1000 is invested at $5\frac{3}{8}\%$, compounded annually, for 2 years.

**17.** The Suarez family invests $10,000 at an annual interest rate of 4.9%, compounded monthly. How much is in the account after 3 years?

**18.** *Job Opportunities.* The table below lists job opportunities in 2006 and projected increases for 2016. Complete the table by filling in the missing numbers.

| OCCUPATION | TOTAL EMPLOYMENT IN 2006 | PROJECTED EMPLOYMENT IN 2016 | CHANGE | PERCENT OF INCREASE |
|---|---|---|---|---|
| Dental assistant | 280,000 | 362,000 | 82,000 | 29.3% |
| Plumber | 705,000 | | 52,000 | |
| Veterinary assistant | 71,000 | 100,000 | | |
| Motorcycle repair technician | | 24,000 | 3000 | |
| Fitness professional | | 298,000 | | 26.8% |

SOURCE: EarnMyDegree.com

**19.** Find the discount and the discount rate of the television in this ad.

**19" LCD HDTV**

**$299⁹⁹**

was $349⁹⁹

**20.** *Credit Cards.* Jayden has a balance of $2704.27 on a credit card with an annual percentage rate of 16.3%. The minimum payment required on the current statement is $54. Find the amount of interest and the amount applied to reduce the principal in this payment and the balance after this payment.

**21.** 0.75% of what number is 300?

   **A.** 2.25      **B.** 40,000      **C.** 400      **D.** 225

# Synthesis

**22.** By selling a home without using a realtor, Juan and Marie can avoid paying a 7.5% commission. They receive an offer of $180,000 from a potential buyer. In order to give a comparable offer, for what price would a realtor need to sell the house? Round to the nearest hundred.

**23.** Karen's commission rate is 16%. She invests her commission from the sale of $15,000 worth of merchandise at an interest rate of 12%, compounded quarterly. How much is Karen's investment worth after 6 months?

# Cumulative Review

1. *Passports.* In 2010, 57% of Canadians had passports while only 35% of Americans had passports. Find decimal notation for 57% and for 35%.

**Sources:** U.S. Department of State; www.sas.com

2. Find percent notation: 0.269.

3. Find percent notation: $\frac{9}{8}$.

4. Find decimal notation: $\frac{13}{6}$.

5. Write fraction notation for the ratio 5 to 0.5.

6. Find the rate in kilometers per hour: 350 km, 15 hr.

Use $<$, $>$, or $=$ for ☐ to write a true sentence.

7. $\frac{5}{7}$ ☐ $\frac{6}{8}$

8. $-3.78$ ☐ $-37.8$

Estimate the sum or the difference by first rounding to the nearest hundred.

9. $263{,}961 + 32{,}090 + 127.89$

10. $73{,}510 - 23{,}450$

11. Calculate: $46 - [4(6 + 4 \div 2) + 2 \times 3 - 5]$.

12. Combine like terms: $5x - 9 - 7x - 5$.

Perform the indicated operation and, if possible, simplify.

13. $\frac{6}{5} + 1\frac{5}{6}$

14. $-46.9 + 2.84$

15. $\begin{array}{r} 4\,8\,7{,}0\,9\,4 \\ 6{,}9\,3\,6 \\ +\ \ 2\,1{,}1\,2\,0 \\ \hline \end{array}$

16. $35 - 34.98$

17. $3\frac{1}{3} - 2\frac{2}{3}$

18. $-\frac{8}{9} - \frac{6}{7}$

19. $\frac{7}{9} \cdot \frac{3}{14}$

20. $(-32)(-4)(-3)$

21. $\begin{array}{r} 4\,6.0\,1\,2 \\ \times\ \ \ \ \ 0.0\,3 \\ \hline \end{array}$

22. $6\frac{3}{5} \div 4\frac{2}{5}$

23. $431.2 \div (-35.2)$

24. $15\,\overline{)\,1\,8\,5\,0}$

Solve.

25. $36 \cdot x = 3420$

26. $y + 142.87 = 151$

27. $\frac{3}{7}x - 5 = 16$

28. $\frac{3}{4} + x = \frac{5}{6}$

29. $3(x - 7) + 2 = 12x - 3$

30. $\frac{16}{n} = \frac{21}{11}$

**31. Museum Attendance.** Among leading art museums, the Indianapolis Museum of Art ranks 6th in the nation in the number of visitors as a percentage of the metropolitan population. In 2009, the number of visitors was 22.32% of the metropolitan population, which was 1,695,037. What was the total attendance?

Source: dashboard.imamuseum.org

**32. Salary of Medical Doctors.** In 2010, the average salary for a doctor in radiology was $81,368 higher than the average salary of a doctor in neurology. If the average salary for a radiologist was $234,065 what was the average salary of a neurologist?

Source: www.salarylist.com

**33.** At one point during the regular NBA season, the Cleveland Cavaliers had won 39 out of 49 games. At this rate, how many games would they win in the entire season of 82 games?

Source: National Basketball Association

**34. Unit Price.** A 200-oz bottle of liquid laundry detergent costs $14.99. What is the unit price?

**35.** Patty walked $\frac{7}{10}$ mi to school and then $\frac{8}{10}$ mi to the library. How far did she walk?

**36. Compound Interest.** The Bakers invest $8500 in an investment account paying 8%, compounded monthly. How much is in the account after 5 years?

**37. Ribbons.** How many pieces of ribbon $1\frac{4}{5}$ yd long can be cut from a length of ribbon 9 yd long?

**38.** In what quadrant does the point $(-3, -1)$ lie?

**39.** Graph on a plane: $y = -\frac{3}{5}x$.

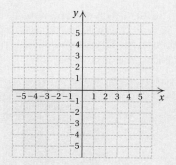

**40.** Find the mean: 19, 29, 34, 39, 45.

**41.** Find the median: 7, 7, 12, 15, 21.

**42.** Find the area of a 40-yd by 90-yd soccer field.

**43.** Subtract and simplify: $\frac{14}{25} - \frac{3}{20}$.

**A.** $\frac{11}{500}$    **B.** $\frac{11}{5}$

**C.** $\frac{41}{100}$    **D.** $\frac{205}{500}$

**44.** Find the perimeter of a 15-in. by 15-in. chessboard.

**A.** 60 in.    **B.** 225 in.
**C.** 30 in.    **D.** 90 in.

## Synthesis

**45.** How many successive 10% discounts are necessary to lower the price of an item to below 50% of its original price?

On a trip through the mountains, a Dodge Neon traveled 240 mi on $7\frac{1}{2}$ gal of gasoline. Going across the plains, the same car averaged 36 miles per gallon.

**46.** What was the percent of increase or decrease in miles per gallon when the car left the mountains for the plains?

**47.** How many miles per gallon did the Dodge average over the entire trip if it used 5 gal of gas to cross the plains?

# Geometry and Measurement

## Real-World Application

The Millau viaduct is part of the E11 expressway connecting Paris, France, and Barcelona, Spain. The viaduct has the highest bridge piers ever constructed. The tallest pier is 804 ft high and the overall height including the pylon is 1122 ft, making this the highest bridge in the world. Convert 804 feet and 1122 feet to meters.

*Source: www.abelard.org/france/viaduct-de-millau.php*

***This problem appears as Example 17 in Section 9.1.***

# 9.1

# Systems of Linear Measurement

## OBJECTIVES

**a** Convert from one American unit of length to another.

**b** Convert from one metric unit of length to another.

**c** Convert between American and metric units of length.

**SKILL TO REVIEW**
Objective 5.3a: Multiply using decimal notation.

Multiply.

**1.** $0.5603 \cdot 1000$

**2.** $1.87 \cdot 0.01$

Length, or distance, is one kind of measure. To find lengths, we start with some **unit segment** and assign to it a measure of 1. Suppose $\overline{AB}$ below is a unit segment.

Let's measure segment $\overline{CD}$ below, using $\overline{AB}$ as our unit segment.

Since 4 unit segments fit end to end along $\overline{CD}$, the measure of $\overline{CD}$ is 4.

Sometimes we have to use parts of units. For example, the measure of the segment $\overline{MN}$ below is $1\frac{1}{2}$. We place one unit segment and one half-unit segment end to end.

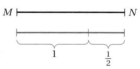

Do Margin Exercises 1–4.

## a American Measures

American units of length are related as follows.

| AMERICAN UNITS OF LENGTH | |
|---|---|
| 12 inches (in.) = 1 foot (ft) | 3 feet = 1 yard (yd) |
| 36 inches = 1 yard | 5280 feet = 1 mile (mi) |

(Actual size, in inches)

The symbols ′ and ″, as in 13 in. = 13″ and 27 ft = 27′, are also used for inches and feet. American units have also been called "English," or "British–American," because at one time they were used in both North America and Britain. Today, both Canada and England have officially converted to the metric system. However, if you travel in England, you will still see units such as miles on road signs.

Use the unit below to measure the length of each segment or object.

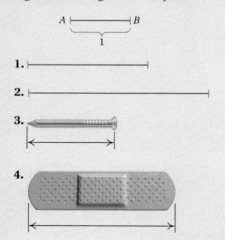

**1.**

**2.**

**3.**

**4.**

*Answers*

*Skill to Review:*
**1.** 560.3　**2.** 0.0187

*Margin Exercises:*
**1.** 2　**2.** 3　**3.** $1\frac{1}{2}$　**4.** $2\frac{1}{2}$

When we write a given length using a different unit, we say that we are "converting" from one unit to another. For example, we convert 1 yard to inches by saying that

$$1\text{ yd} = 36\text{ in.}$$

Conversions are made using either *substitution* or *multiplying by 1*. To use substitution, we replace one length with an equivalent length using another unit.

**EXAMPLE 1**   Complete:  5 yd = _____ in.

Since we are converting yards to inches, we use the fact that 1 yd = 36 in.

$$
\begin{aligned}
5\text{ yd} &= 5 \cdot 1\text{ yd} & &\text{We think of 5 yd as } 5 \cdot 1\text{ yd.} \\
&= 5 \cdot 36\text{ in.} & &\text{Substituting 36 in. for 1 yd} \\
&= 180\text{ in.} & &\text{Multiplying}
\end{aligned}
$$

**EXAMPLE 2**   Complete:  2 mi = _____ in.

We want to convert miles to inches, but we do not know from the box on the previous page how many inches are in a mile. Since we do know how many feet are in a mile and how many inches are in a foot, we substitute twice.

$$
\begin{aligned}
2\text{ mi} &= 2 \cdot 1\text{ mi} \\
&= 2 \cdot 5280\text{ ft} & &\text{Substituting 5280 ft for 1 mi} \\
&= 10{,}560 \cdot 1\text{ ft} & &\text{Multiplying} \\
&= 10{,}560 \cdot 12\text{ in.} & &\text{Substituting 12 in. for 1 ft} \\
&= 126{,}720\text{ in.} & &\text{The student should check the multiplication.}
\end{aligned}
$$

> Do Exercises 5–7.

In Examples 1 and 2, we converted from larger units to smaller ones and used substitution. Sometimes—especially when converting from smaller to larger units—it helps to multiply by 1. For example, 12 in. = 1 ft, so we might choose to write 1 as

$$\frac{12\text{ in.}}{1\text{ ft}} \quad\text{or}\quad \frac{1\text{ ft}}{12\text{ in.}}.$$

**EXAMPLE 3**   Complete:  48 in. = _____ ft.

Since we are converting "in." to "ft," we choose a symbol for 1 containing the equivalent lengths of "12 in." and "1 ft." We write "12 in." on the bottom to eliminate the inches unit.

$$
\begin{aligned}
48\text{ in.} &= \frac{48\text{ in.}}{1} \cdot \frac{1\text{ ft}}{12\text{ in.}} & &\text{Multiplying by 1 using } \frac{1\text{ ft}}{12\text{ in.}}\text{ to eliminate in.} \\[4pt]
&= \frac{48\text{ in.}}{12\text{ in.}} \cdot 1\text{ ft} & &\text{Note that in. appears in both the numerator and the denominator.} \\[4pt]
&= \frac{48}{12} \cdot \frac{\text{in.}}{\text{in.}} \cdot 1\text{ ft} & &\text{The } \frac{\text{in.}}{\text{in.}} \text{ acts like a factor equal to 1.} \\[4pt]
&= 4 \cdot 1\text{ ft} = 4\text{ ft} & &\text{Dividing}
\end{aligned}
$$

The conversion can also be regarded as "canceling" units:

$$48\text{ in.} = \frac{48\ \cancel{\text{in.}}}{1} \cdot \frac{1\text{ ft}}{12\ \cancel{\text{in.}}} = \frac{48}{12} \cdot 1\text{ ft} = 4\text{ ft.}$$

> Do Exercises 8 and 9.

Complete.

**5.** 8 yd = _____ in.

**6.** 14.5 yd = _____ ft

**7.** 3.8 mi = _____ in.

Complete.

**8.** 72 in. = _____ ft

**9.** 24 ft = _____ yd

***Answers***

**5.** 288   **6.** 43.5   **7.** 240,768
**8.** 6   **9.** 8

In Examples 4 and 5, we will use only the "canceling" method.

**EXAMPLE 4** Complete: 70 ft = _____ yd.

Since we are converting from "ft" to "yd," we choose a symbol for 1 with "yd" on the top and "ft" on the bottom:

$$70 \text{ ft} = 70 \text{ ft} \cdot \frac{1 \text{ yd}}{3 \text{ ft}} \qquad \text{If it helps, write 70 ft as } \frac{70 \text{ ft}}{1}.$$

$$= \frac{70}{3} \cdot 1 \text{ yd} = 23.\overline{3} \text{ yd, or } 23\frac{1}{3} \text{ yd}.$$

Do Exercises 10–13.

**EXAMPLE 5** *Rope Sculpture.* New York–based artist Orly Genger is known for transforming common nylon ropes into elaborate, monumental sculptures. For her work titled *Whole*, Genger hand-knotted and piled an estimated 113.6 mi of painted rope into nine separate sculptures, which were on display at the Indianapolis Museum of Art from November 21, 2008, to June 14, 2009. Convert 113.6 miles to yards.

Sources: www.art-directory.cn; *IMA Previews*, Winter 2008–2009

We have

$$113.6 \text{ mi} = 113.6 \text{ mi} \cdot \frac{5280 \text{ ft}}{1 \text{ mi}} \cdot \frac{1 \text{ yd}}{3 \text{ ft}}$$

$$= \frac{113.6 \cdot 5280}{1 \cdot 3} \cdot 1 \text{ yd}$$

$$= 199{,}936 \text{ yd}.$$

The nine sculptures of *Whole* contain 199,936 yd of rope.

Note that when we convert from smaller units to larger units, we divide. When converting from larger units to smaller units, we multiply.

Do Exercises 14–16.

---

Complete.

**10.** 24 in. = _____ ft

**11.** 35 ft = _____ yd

**12.** 26,400 ft = _____ mi

**13.** 2640 ft = _____ mi

Complete.

**14.** 6 mi = _____ ft

**15.** $2\frac{2}{3}$ yd = _____ in.

**16. Pedestrian Paths.** There are 23 mi of pedestrian paths in Central Park in New York City. Convert 23 miles to yards.

*Answers*

**10.** 2    **11.** $11\frac{2}{3}$, or $11.\overline{6}$    **12.** 5
**13.** 0.5, or $\frac{1}{2}$    **14.** 31,680    **15.** 96
**16.** 40,480 yd

# b The Metric System

The **metric system** is gradually replacing traditional systems of measurement. Because it is based on powers of 10, the metric system allows for easy conversion between units. The metric system does not use inches, feet, pounds, and so on, but the units for time and electricity are the same as those in the American system.

The basic unit of length is the **meter.** It is just over a yard. In fact, 1 meter ≈ 1.1 yd.

(Comparative sizes are shown.)

| 1 Meter |
| --- |

| 1 Yard |
| --- |

The other units of length are multiples of the length of a meter:

10 times a meter,   100 times a meter,   1000 times a meter,   and so on,

or fractions of a meter:

$\frac{1}{10}$ of a meter,   $\frac{1}{100}$ of a meter,   $\frac{1}{1000}$ of a meter,   and so on.

You should memorize these names and abbreviations. Think of *kilo-* for 1000, *hecto-* for 100, *deka-* for 10, *deci-* for $\frac{1}{10}$, *centi-* for $\frac{1}{100}$, and *milli-* for $\frac{1}{1000}$. (The units dekameter and decimeter are not used often.) We will also use these prefixes when considering units of area, capacity, and mass.

## Thinking Metric

To familiarize yourself with metric units, consider the following.

| | |
| --- | --- |
| 1 kilometer (1000 meters) | is a bit more than $\frac{1}{2}$ mile (≈0.6 mi). |
| 1 meter | is just over a yard (≈1.1 yd). |
| 1 centimeter (0.01 meter) | is a little more than the width of a jumbo paperclip (≈0.4 in.). |
| 1 millimeter (0.001 meter) | is about the diameter of a paperclip wire. |

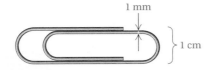

1 cm

1 mm

1 cm

**STUDY TIPS**

**BEING A TUTOR**

Try being a tutor for a fellow student. Explaining the material to someone else will help you maximize your own understanding and retention of concepts.

## METRIC UNITS OF LENGTH

1 *kilo*meter (km) = 1000 meters (m)

1 *hecto*meter (hm) = 100 meters (m)

1 *deka*meter (dam) = 10 meters (m)

1 meter (m)

1 *deci*meter (dm) = $\frac{1}{10}$ meter (m)

1 *centi*meter (cm) = $\frac{1}{100}$ meter (m)

1 *milli*meter (mm) = $\frac{1}{1000}$ meter (m)

1 inch is about 2.54 centimeters.

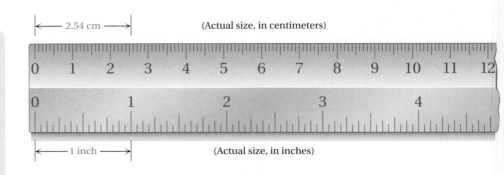

2.54 cm (Actual size, in centimeters)

1 inch (Actual size, in inches)

The millimeter (mm) is used to measure small distances, especially in industry.

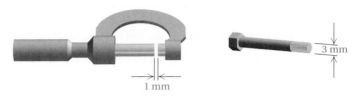

1 mm

3 mm

Centimeters (cm) are used for body dimensions and clothing sizes.

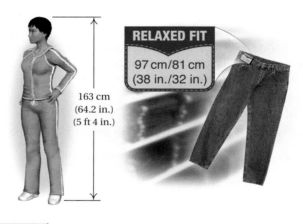

163 cm
(64.2 in.)
(5 ft 4 in.)

RELAXED FIT
97 cm/81 cm
(38 in./32 in.)

Do Exercises 17-19.

The meter (m) is used for expressing dimensions of larger objects—say, the height of a diving board—and for shorter distances, like the length of a rug.

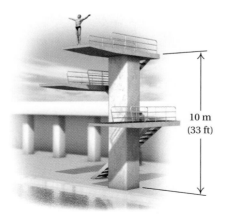

10 m
(33 ft)

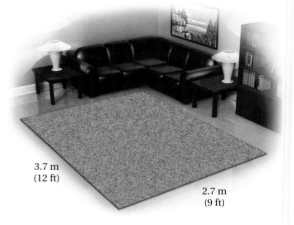

3.7 m
(12 ft)

2.7 m
(9 ft)

Using a centimeter ruler, measure each object.

**17.**

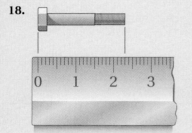

**18.**

**19.**

"Energy Plus"
1.5 Volts          AAA Size

*Answers*

**17.** 2 cm, or 20 mm
**18.** 2.3 cm, or 23 mm
**19.** 4.4 cm, or 44 mm

Kilometers (km) are used for longer distances, mostly in cases where miles are now being used. 1 mile is about 1.6 km.

| 1 km |
| 1 mi |

Do Exercises 20–25.

**Complete with mm, cm, m, or km.**

**20.** A stick of gum is 7 _____ long.

**21.** New Orleans is 2941 _____ from San Diego.

**22.** A penny is 1 _____ thick.

**23.** The halfback ran 7 _____ .

**24.** The book is 3 _____ thick.

**25.** The desk is 2 _____ long.

Metric conversions from larger units to smaller ones are often most easily made using substitution, much as we did in Examples 1 and 2.

**EXAMPLE 6**   Complete:  4 km = _____ m.

$$4 \text{ km} = 4 \cdot 1 \text{ km}$$
$$= 4 \cdot 1000 \text{ m} \qquad \text{Substituting 1000 m for 1 km}$$
$$= 4000 \text{ m} \qquad \text{Multiplying by 1000}$$

Do Exercises 26 and 27.

**Complete.**

**26.** 23 km = _____ m

**27.** 4 hm = _____ m

Since

$$\frac{1}{10} \text{ m} = 1 \text{ dm}, \qquad \frac{1}{100} \text{ m} = 1 \text{ cm}, \quad \text{and} \quad \frac{1}{1000} \text{ m} = 1 \text{ mm},$$

it follows that

**METRIC CONVERSIONS**

1 m = 10 dm,     1 m = 100 cm,   and   1 m = 1000 mm.

Remembering these equations will help you to write forms of 1 when canceling to make conversions. The procedure is the same as that used in Examples 3 and 4.

**EXAMPLE 7**   Complete:  93.4 m = _____ cm.

To convert from "m" to "cm," we multiply by 1 using a symbol for 1 with "m" on the bottom and "cm" on the top. This process introduces centimeters and at the same time eliminates meters.

$$93.4 \text{ m} = 93.4 \text{ m} \cdot \frac{100 \text{ cm}}{1 \text{ m}} \qquad \text{Multiplying by 1 using } \frac{100 \text{ cm}}{1 \text{ m}}$$

$$= 93.4 \text{ m} \cdot \frac{100 \text{ cm}}{1 \text{ m}} = 93.4 \cdot 100 \text{ cm} = 9340 \text{ cm}$$

*Answers*

**20.** cm   **21.** km   **22.** mm   **23.** m
**24.** cm   **25.** m   **26.** 23,000   **27.** 400

**EXAMPLE 8** Complete: 0.248 m = _____ mm.

We are converting from "m" to "mm," so we choose a symbol for 1 with "mm" on the top and "m" on the bottom:

$$0.248 \text{ m} = 0.248 \text{ m} \cdot \frac{1000 \text{ mm}}{1 \text{ m}} = 0.248 \cdot 1000 \text{ mm} = 248 \text{ mm}.$$

**Do Exercises 28 and 29.**

**EXAMPLE 9** Complete: 2347 m = _____ km.

We multiply by 1 using $\frac{1 \text{ km}}{1000 \text{ m}}$:

$$2347 \text{ m} = 2347 \text{ m} \cdot \frac{1 \text{ km}}{1000 \text{ m}} = \frac{2347}{1000} \cdot 1 \text{ km} = 2.347 \text{ km}.$$

It is helpful to remember that 1000 mm = 100 cm and, more simply, 10 mm = 1 cm.

**EXAMPLE 10** Complete: 8.42 mm = _____ cm.

We can multiply by 1 using either $\frac{1 \text{ cm}}{10 \text{ mm}}$ or $\frac{100 \text{ cm}}{1000 \text{ mm}}$. Both expressions for 1 will eliminate mm and leave cm:

$$8.42 \text{ mm} = 8.42 \text{ mm} \cdot \frac{1 \text{ cm}}{10 \text{ mm}} = \frac{8.42}{10} \cdot 1 \text{ cm} = 0.842 \text{ cm}.$$

**Do Exercises 30–33.**

## Mental Conversion

Changing from one unit to another in the metric system amounts only to the movement of a decimal point. That is because the metric system is based on 10. Let's find a faster way to convert. Look at the following table.

| 1000 m | 100 m | 10 m | 1 m | 0.1 m | 0.01 m | 0.001 m |
|--------|-------|------|-----|-------|--------|---------|
| 1 km | 1 hm | 1 dam | 1 m | 1 dm | 1 cm | 1 mm |

Each place in the table has a value $\frac{1}{10}$ that of the place to the left or 10 times that of the place to the right. Thus, moving one place in the table corresponds to moving one decimal place.

**EXAMPLE 11** Complete: 35.7 mm = _____ cm.

*Think*: Centimeters is the next larger unit after millimeters. Thus, we move the decimal point one place to the left.

35.7    3.5.7    35.7 mm = 3.57 cm    Converting to a larger unit shifts the decimal point to the left.

**EXAMPLE 12** Complete: 1.886 km = _____ cm.

*Think*: To go from km to cm in the table is a move of five places to the right. Thus, we move the decimal point five places to the right.

| 1000 m | 100 m | 10 m | 1 m | 0.1 m | 0.01 m | 0.001 m |
|--------|-------|------|-----|-------|--------|---------|
| 1 km | 1 hm | 1 dam | 1 m | 1 dm | 1 cm | 1 mm |

5 places to the right

1.886    1.88600.    1.886 km = 188,600 cm

**EXAMPLE 13** Complete: 3 m = _____ cm.

*Think*: To go from m to cm in the table is a move of two places to the right. Thus, we move the decimal point two places to the right.

| 1000 m | 100 m | 10 m | 1 m | 0.1 m | 0.01 m | 0.001 m |
|--------|-------|------|-----|-------|--------|---------|
| 1 km | 1 hm | 1 dam | 1 m | 1 dm | 1 cm | 1 mm |

2 places to the right

3    3.00.    3 m = 300 cm

You should try to make metric conversions mentally as much as possible. The fact that conversions can be done so easily is an important advantage of the metric system.

> The most commonly used metric units of length are km, m, cm, and mm. We have purposely used these more often than the others in the exercises and examples.

Do Exercises 34–37.

## c Converting Between American and Metric Units

We can make conversions between American units and metric units by using the following table. These listings are rounded approximations. Again, we either make a substitution or multiply by 1 appropriately.

| AMERICAN | METRIC |
|----------|--------|
| 1 in. | 2.540 cm |
| 1 ft | 0.305 m |
| 1 yd | 0.914 m |
| 1 mi | 1.609 km |
| 0.621 mi | 1 km |
| 1.094 yd | 1 m |
| 3.281 ft | 1 m |
| 39.370 in. | 1 m |

Complete. Try to do this mentally using the table.

**34.** 6780 m = _____ km

**35.** 9.74 cm = _____ mm

**36.** 1 mm = _____ cm

**37.** 845.1 mm = _____ dm

*Answers*

**34.** 6.78    **35.** 97.4    **36.** 0.1    **37.** 8.451

Complete.

**38.** 100 yd = _____ m
(The length of a football field excluding the end zones)

**39.** 2.5 mi = _____ km
(The length of the tri-oval track at Daytona International Speedway)

**40.** 2383 km = _____ mi
(The distance from St. Louis to Phoenix)

**41.** 3.175 mm = _____ in.
(The thickness of a quarter)

**42.** The distance from San Francisco to Las Vegas is 568 mi. Find the distance in kilometers.

**43.** The height of the Stratosphere Tower in Las Vegas, Nevada, is 1149 ft. Find the height in meters.

**EXAMPLE 14** Complete: 26.2 mi = _____ km (the length of the Olympic marathon).

Since we are given that 1 mi ≈ 1.609 km, we can convert using substitution.

$$26.2 \text{ mi} = 26.2 \cdot 1 \text{ mi}$$
$$\approx 26.2 \cdot 1.609 \text{ km} \qquad \text{Converting from mi to km}$$
$$\approx 42.16 \text{ km}$$

The table also indicates that 1 km ≈ 0.621 mi, so we can also convert by multiplying by 1 using $\dfrac{1 \text{ km}}{0.621 \text{ mi}}$.

$$26.2 \text{ mi} \approx \frac{26.2 \cancel{\text{mi}}}{1} \cdot \frac{1 \text{ km}}{0.621 \cancel{\text{mi}}}$$
$$\approx 42.19 \text{ km}$$

Note that, since the conversion factors given in the table are approximations, the answers found by the two methods differ slightly.

**EXAMPLE 15** Complete: 100 m = _____ yd (the length of a dash in track).

$$100 \text{ m} = 100 \cdot 1 \text{ m}$$
$$\approx 100 \cdot 1.094 \text{ yd} \qquad \text{Converting from m to yd}$$
$$\approx 109.4 \text{ yd}$$

**EXAMPLE 16** Complete: 170 cm = _____ in. (a common ski length).

We multiply by 1, using the fact that 1 in. ≈ 2.540 cm and writing 2.540 cm on the bottom to eliminate the centimeters unit.

$$170 \text{ cm} \approx 170 \cancel{\text{cm}} \cdot \frac{1 \text{ in.}}{2.540 \cancel{\text{cm}}} \qquad \text{Converting to inches from cm}$$
$$\approx 66.93 \text{ in.}$$

Do Exercises 38–41.

**EXAMPLE 17** *Millau Viaduct.* The Millau viaduct is part of the E11 expressway connecting Paris, France, and Barcelona, Spain. The viaduct has the highest bridge piers ever constructed. The tallest pier is 804 ft high and the overall height including the pylon is 1122 ft, making this the highest bridge in the world. Convert 804 feet and 1122 feet to meters.

Source: www.abelard.org/france/viaduct-de-millau.php

We let $P$ = the height of the pier and $H$ = the overall height of the bridge. To convert feet to meters, we substitute 0.305 m for 1 ft.

$$P = 804 \text{ ft} \qquad\qquad H = 1122 \text{ ft}$$
$$= 804 \cdot 1 \text{ ft} \qquad\qquad = 1122 \cdot 1 \text{ ft}$$
$$\approx 804 \cdot 0.305 \text{ m} \qquad \approx 1122 \cdot 0.305 \text{ m}$$
$$= 245.22 \text{ m} \qquad\qquad = 342.21 \text{ m}$$

Do Exercises 42 and 43.

*Answers*

**38.** 91.4   **39.** 4.0225   **40.** 1479.843
**41.** 0.125   **42.** 913.912 km   **43.** 350.445 m

For Extra Help

**MyMathLab**

Math XL
PRACTICE     WATCH     DOWNLOAD     READ     REVIEW

**a**   Complete.

**1.** 1 yd = _____ ft

**2.** 1 ft = _____ in.

**3.** 1 in. = _____ ft

**4.** 1 mi = _____ ft

**5.** 1 mi = _____ yd

**6.** 1 ft = _____ yd

**7.** 3 yd = _____ ft

**8.** 4 yd = _____ in.

**9.** 84 in. = _____ ft

**10.** 18 in. = _____ ft

**11.** 48 ft = _____ yd

**12.** 29 ft = _____ yd

**13.** 5 mi = _____ yd

**14.** 5 mi = _____ ft

**15.** 63 in. = _____ ft

**16.** 19 ft = _____ yd

**17.** 11,616 ft = _____ mi

**18.** 5.2 yd = _____ ft

**19.** 7.1 mi = _____ ft

**20.** 31,680 ft = _____ mi

**21.** $7\frac{1}{2}$ ft = _____ yd

**22.** 360 in. = _____ yd

**23.** 36 in. = _____ ft

**24.** 7.2 ft = _____ in.

**25.** 1760 yd = _____ mi

**26.** 3520 yd = _____ mi

**27.** 45 in. = _____ yd

**28.** $6\frac{1}{3}$ yd = _____ in.

**29.** 25 mi = _____ ft

**30.** 240 in. = _____ ft

**31.** 2 mi = _____ in.

**32.** 63,360 in. = _____ mi

Complete. Do as much as possible mentally.

**33. a)** 1 km = _____ m
   **b)** 1 m = _____ km

**34. a)** 1 hm = _____ m
   **b)** 1 m = _____ hm

**35. a)** 1 dam = _____ m
   **b)** 1 m = _____ dam

**36. a)** 1 dm = _____ m
   **b)** 1 m = _____ dm

**37. a)** 1 cm = _____ m
   **b)** 1 m = _____ cm

**38. a)** 1 mm = _____ m
   **b)** 1 m = _____ mm

**39.** 8.3 km = _____ m

**40.** 27 km = _____ m

**41.** 98 cm = _____ m

**42.** 0.789 cm = _____ m

**43.** 8921 m = _____ km

**44.** 8664 m = _____ km

**45.** 32.17 m = _____ km

**46.** 4.733 m = _____ km

**47.** 289 m = _____ cm

**48.** 869 m = _____ cm

**49.** 477 cm = _____ m

**50.** 6.27 mm = _____ m

**51.** 6.88 m = _____ cm

**52.** 6.88 m = _____ dm

**53.** 1 mm = _____ cm

**54.** 1 cm = _____ km

**55.** 1 km = _____ cm

**56.** 2 km = _____ cm

**57.** 14.2 cm = _____ mm

**58.** 25.3 cm = _____ mm

**59.** 8.2 mm = _____ cm

**60.** 9.7 mm = _____ cm

**61.** 4500 mm = _____ cm

**62.** 8,000,000 m = _____ km

**63.** 0.024 mm = _____ m

**64.** 60,000 mm = _____ dam

**65.** 6.88 m = _____ dam

**66.** 7.44 m = _____ hm

**67.** 2.3 dam = _____ dm

**68.** 9 km = _____ hm

Complete the following table.

| | OBJECT | MILLIMETERS (mm) | CENTIMETERS (cm) | METERS (m) |
|---|---|---|---|---|
| **69.** | Length of a calculator | | 18 | |
| **70.** | Width of a calculator | 85 | | |
| **71.** | Length of a piece of typing paper | | | 0.278 |
| **72.** | Length of a football field | | | 109.09 |
| **73.** | Width of a football field | | 4844 | |
| **74.** | Width of a credit card | 56 | | |

**C**     Complete. Answers may vary slightly, depending on the conversion used.

**75.** 10 km = _____ mi
(A common running distance)

**76.** 5 mi = _____ km
(A common running distance)

**77.** 14 in. = _____ cm
(A common paper length)

**78.** 400 m = _____ yd
(A common race distance)

**79.** 65 mph = _____ km/h
(A common speed limit in the United States)

**80.** 100 km/h = _____ mph
(A common speed limit in Canada)

**81.** 94 ft = _____ m
(The length of an NCAA basketball court)

**82.** 2.08 m = _____ in.
(The height of Amare Stoudemire of the Phoenix Suns)

**83.** 180 cm = _____ in.
(A common snowboard length)

**84.** 13 mm = _____ in.
(The thickness of a plastic case for a DVD)

**85.** 36 yd = _____ m
(A common length for a roll of tape)

**86.** 70 in. = _____ cm
(A common height for a man)

Complete the following table. Answers may vary, depending on the conversion factor used.

| | OBJECT | YARDS (yd) | CENTIMETERS (cm) | INCHES (in.) | METERS (m) | MILLIMETERS (mm) |
|---|---|---|---|---|---|---|
| **87.** | Thickness of an index card | | | | 0.00027 | |
| **88.** | Thickness of a piece of cardboard | | 0.23 | | | |
| **89.** | Height of the Willis Tower | | | | 442 | |
| **90.** | Height of Central Plaza, Hong Kong | 409 | | | | |

## Skill Maintenance

Solve.

**91.** $-7x - 9x = 24$   [5.7b]

**92.** $-2a + 9 = 5a + 23$   [5.7b]

**93.** If 3 calculators cost $43.50, how much would 7 calculators cost?   [7.4a]

**94.** A principal of $500 is invested at a rate of 8.9% for 1 year. Find the simple interest.   [8.7a]

Convert to percent notation.

**95.** 0.47   [8.1b]

**96.** $\dfrac{7}{20}$   [8.1c]

**97.** A living room is 12 ft by 16 ft. Find the perimeter and area of the room.   [1.2b], [1.4b]

**98.** A bedroom measures 10 ft by 12 ft. Find the perimeter and area of the room.   [1.2b], [1.4b]

## Synthesis

In Exercises 99–102, each sentence is incorrect. Insert or alter a decimal point to make the sentence correct.

**99.** When my right arm is extended, the distance from my left shoulder to the end of my right hand is 10 m.

**100.** The height of the Shanghai World Financial Center is 49.2 m.

**101.** A stack of ten quarters is 140 cm high.

**102.** The width of an adult's hand is 112 cm.

**103.** *Noah's Ark.* It is thought that the biblical measure cubit was approximately 18 in. The dimensions of Noah's ark are given as follows: "The length of the ark shall be three hundred cubits, the breadth of it fifty cubits, and the height of it thirty cubits." What were the dimensions of Noah's ark in inches? in feet?
Source: *Holy Bible, King James Version*, Gen. 6:15

**104.** *Goliath's Height.* In biblical measures, a span was considered to be half of a cubit (1 cubit ≈ 18 in.; see Exercise 103). The giant Goliath's height was "six cubits and a span." What was the height of Goliath in inches? in feet?
Source: *Holy Bible, King James Version*, 1 Sam. 17:4

Complete. Answers may vary, depending on the conversion used.

**105.** 🖩 2 mi = _____ cm

**106.** 🖩 10 km = _____ in.

**107.** 🖩 As of August 2010, the women's world record for the 100-m dash was 10.49 sec, set by Florence Griffith-Joyner in Indianapolis, Indiana, on July 16, 1988. How fast is this in miles per hour? Round to the nearest tenth of a mile per hour.
Source: International Association of Athletics Federations

**108.** 🖩 As of August 2010, the men's world record for the 100-m dash was 9.58 sec, set by Usain Bolt of Jamaica on August 16, 2009, in Berlin, Germany. How fast is this in miles per hour? Round to the nearest hundredth of a mile per hour.
Source: International Association of Athletics Federations

Use < or > to complete the following. Perform only approximate, mental calculations.

**109.** 59 in. ☐ 59 cm

**110.** 35 yd ☐ 35 m

**111.** 7 km ☐ 6 mi

**112.** 9 mi ☐ 18 km

**113.** 24 ft ☐ 6 m

**114.** 30 in. ☐ 90 cm

# 9.2 Converting Units of Area

## a) American Units

It is often necessary to convert units of area. First we will convert from one American unit of area to another.

**EXAMPLE 1**   Complete: $1 \text{ yd}^2 =$ _____ $\text{ft}^2$.

We recall that 1 yd = 3 ft and make a sketch. Note that $1 \text{ yd}^2 = 9 \text{ ft}^2$. The same result can be found as follows:

$$1 \text{ yd}^2 = 1 \cdot (3 \text{ ft})^2 \qquad \text{Substituting 3 ft for 1 yd}$$
$$= 3 \text{ ft} \cdot 3 \text{ ft}$$
$$= 9 \text{ ft}^2. \qquad \text{Note that ft} \cdot \text{ft} = \text{ft}^2.$$

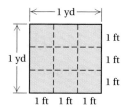

**EXAMPLE 2**   Complete: $2 \text{ ft}^2 =$ _____ $\text{in}^2$.

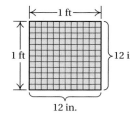

A sketch of $2 \text{ ft}^2$.
Note that $1 \text{ ft}^2 = 144 \text{ in}^2$.

$$2 \text{ ft}^2 = 2 \cdot (12 \text{ in.})^2 \qquad \text{Substituting 12 in. for 1 ft}$$
$$= 2 \cdot 12 \text{ in.} \cdot 12 \text{ in.} = 288 \text{ in}^2 \qquad \text{Note that in.} \cdot \text{in.} = \text{in}^2.$$

Do Margin Exercises 1–3.

American units of area are related as follows.

### AMERICAN UNITS OF AREA

| |
|---|
| 1 square yard ($\text{yd}^2$) = 9 square feet ($\text{ft}^2$) |
| 1 square foot ($\text{ft}^2$) = 144 square inches ($\text{in}^2$) |
| 1 square mile ($\text{mi}^2$) = 640 acres |
| 1 acre = 43,560 $\text{ft}^2$ |

**EXAMPLE 3**   Complete: $36 \text{ ft}^2 =$ _____ $\text{yd}^2$.

To convert from "$\text{ft}^2$" to "$\text{yd}^2$" we write 1 with $\text{yd}^2$ on top and $\text{ft}^2$ on the bottom.

$$36 \text{ ft}^2 = 36 \text{ ft}^2 \cdot \frac{1 \text{ yd}^2}{9 \text{ ft}^2} \qquad \text{Multiplying by 1 using } \frac{1 \text{ yd}^2}{9 \text{ ft}^2}$$
$$= \frac{36}{9} \cdot \text{yd}^2 = 4 \text{ yd}^2$$

Complete.
1. $1 \text{ ft}^2 =$ _____ $\text{in}^2$
2. $10 \text{ ft}^2 =$ _____ $\text{in}^2$
3. $7 \text{ yd}^2 =$ _____ $\text{ft}^2$

------- *Caution!* -------

Be aware of when you are converting units of length and when you are converting units of area.

$1 \text{ yd} = 3 \text{ ft} \Rightarrow 1 \text{ yd}^2 = 9 \text{ ft}^2$
$1 \text{ ft} = 12 \text{ in.} \Rightarrow 1 \text{ ft}^2 = 144 \text{ in}^2$

------------------------

***Answers***

*Skill to Review:*
1. 81   2. 1000

*Margin Exercises:*
1. 144   2. 1440   3. 63

When converting from larger units to smaller units, we can substitute directly from the table.

**EXAMPLE 4** Complete: $7 \text{ mi}^2 = $ _____ acres.

$$7 \text{ mi}^2 = 7 \cdot 640 \text{ acres} \qquad \text{Substituting 640 acres for 1 mi}^2$$
$$= 4480 \text{ acres}$$

Had we used canceling, we could have multiplied $7 \text{ mi}^2$ by $\dfrac{640 \text{ acres}}{1 \text{ mi}^2}$ to find the same answer.

Do Exercises 4 and 5.

## **b** Metric Units

We next convert from one metric unit of area to another.

**EXAMPLE 5** Complete: $1 \text{ cm}^2 = $ _____ $\text{mm}^2$.

Since $1 \text{ cm} = 10 \text{ mm}$, a square centimeter will have sides of length 10 mm.

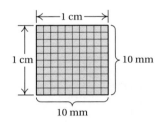

The area of the square can be written

$$(1 \text{ cm})(1 \text{ cm}) = 1 \text{ cm}^2$$

or

$$(10 \text{ mm})(10 \text{ mm}) = 100 \text{ mm}^2.$$

Thus, $1 \text{ cm}^2 = 100 \text{ mm}^2$.

**EXAMPLE 6** Complete: $1 \text{ km}^2 = $ _____ $\text{m}^2$.

$$1 \text{ km}^2 = (1 \text{ km})(1 \text{ km})$$
$$= (1000 \text{ m})(1000 \text{ m}) \qquad \text{Substituting 1000 m for 1 km}$$
$$= 1,000,000 \text{ m}^2 \qquad \text{Note that m} \cdot \text{m} = \text{m}^2.$$

**EXAMPLE 7** Complete: $1 \text{ cm}^2 = $ _____ $\text{m}^2$.

$$1 \text{ cm}^2 = (1 \text{ cm})(1 \text{ cm})$$
$$= \left(\frac{1}{100} \text{ m}\right)\left(\frac{1}{100} \text{ m}\right) \qquad \text{Substituting } \frac{1}{100} \text{ m for 1 cm}$$
$$= \frac{1}{10,000} \text{ m}^2 \qquad \text{Note that m} \cdot \text{m} = \text{m}^2.$$
$$= 0.0001 \text{ m}^2$$

Do Exercises 6 and 7.

Complete.

**4.** $360 \text{ in}^2 = $ _____ $\text{ft}^2$

**5.** $5 \text{ mi}^2 = $ _____ acres

Complete.

**6.** $1 \text{ m}^2 = $ _____ $\text{mm}^2$

**7.** $1 \text{ mm}^2 = $ _____ $\text{cm}^2$

*Answers*

**4.** 2.5   **5.** 3200
**6.** 1,000,000   **7.** 0.01

## Mental Conversion

Let's compare conversions of units of length to the conversions of units of area in Examples 6 and 7.

*Converting units of length*

**1 km $\rightarrow$ m**
1 km = 1000 m
Move the decimal point
3 places to the right.

**1 cm $\rightarrow$ m**
1 cm = 0.01 m
Move the decimal point
2 places to the left.

*Converting units of area*

**1 km$^2$ $\rightarrow$ m$^2$**
1 km$^2$ = 1,000,000 m$^2$
Move the decimal point
6 places to the right.

**1 cm$^2$ $\rightarrow$ m$^2$**
1 cm$^2$ = 0.0001 m$^2$
Move the decimal point
4 places to the left.

In general, a metric area conversion requires moving the decimal point twice as many places as the corresponding length conversion. We can use a table as before and multiply the number of places we move by 2 to determine the number of places to move the decimal point.

**EXAMPLE 8**   Complete: 3.48 cm$^2$ = _____ mm$^2$.

*Think:* To go from cm to mm in the table is a move of 1 place to the right.

| 1000 m | 100 m | 10 m | 1 m | 0.1 m | 0.01 m | 0.001 m |
|--------|-------|------|-----|-------|--------|---------|
| 1 km | 1 hm | 1 dam | 1 m | 1 dm | 1 cm | 1 mm |

1 move to the right

So we move the decimal point 2 · 1, or 2, places to the right.

$$3.48 \qquad 3.48. \qquad 3.48 \text{ cm}^2 = 348 \text{ mm}^2$$

2 places to the right

**EXAMPLE 9**   Complete: 586.78 mm$^2$ = _____ m$^2$.

*Think:* To go from mm to m in the table is a move of 3 places to the left.

| 1000 m | 100 m | 10 m | 1 m | 0.1 m | 0.01 m | 0.001 m |
|--------|-------|------|-----|-------|--------|---------|
| 1 km | 1 hm | 1 dam | 1 m | 1 dm | 1 cm | 1 mm |

3 moves to the left

So we move the decimal point 2 · 3, or 6, places to the left.

$$586.78 \qquad 0.000586.78 \qquad 586.78 \text{ mm}^2 = 0.00058678 \text{ m}^2$$

6 places to the left

Do Exercises 8–10.

Complete.

**8.** 2.88 m$^2$ = _____ cm$^2$

**9.** 4.3 mm$^2$ = _____ cm$^2$

**10.** 678,000 m$^2$ = _____ km$^2$

***Answers***
**8.** 28,800   **9.** 0.043   **10.** 0.678

9.2 **Exercise Set**

For Extra Help

*MathXL*
**MyMathLab**

Math XL
PRACTICE

WATCH

DOWNLOAD

READ

REVIEW

**a** Complete.

**1.** $1\,\text{ft}^2 = $ _____ $\text{in}^2$

**2.** $1\,\text{yd}^2 = $ _____ $\text{ft}^2$

**3.** $1\,\text{mi}^2 = $ _____ acres

**4.** $1\,\text{acre} = $ _____ $\text{ft}^2$

**5.** $1\,\text{in}^2 = $ _____ $\text{ft}^2$

**6.** $1\,\text{ft}^2 = $ _____ $\text{yd}^2$

**7.** $5\,\text{yd}^2 = $ _____ $\text{ft}^2$

**8.** $4\,\text{ft}^2 = $ _____ $\text{in}^2$

**9.** $7\,\text{ft}^2 = $ _____ $\text{in}^2$

**10.** $2\,\text{acres} = $ _____ $\text{ft}^2$

**11.** $432\,\text{in}^2 = $ _____ $\text{ft}^2$

**12.** $54\,\text{ft}^2 = $ _____ $\text{yd}^2$

**13.** $22\,\text{yd}^2 = $ _____ $\text{ft}^2$

**14.** $40\,\text{ft}^2 = $ _____ $\text{in}^2$

**15.** $15\,\text{ft}^2 = $ _____ $\text{in}^2$

**16.** $144\,\text{ft}^2 = $ _____ $\text{yd}^2$

**17.** $20\,\text{mi}^2 = $ _____ acres

**18.** $576\,\text{in}^2 = $ _____ $\text{ft}^2$

**19.** $1\,\text{mi}^2 = $ _____ $\text{ft}^2$

**20.** $1\,\text{mi}^2 = $ _____ $\text{yd}^2$

**21.** $720\,\text{in}^2 = $ _____ $\text{ft}^2$

**22.** $27\,\text{ft}^2 = $ _____ $\text{yd}^2$

**23.** $3\,\text{ft}^2 = $ _____ $\text{yd}^2$

**24.** $72\,\text{in}^2 = $ _____ $\text{ft}^2$

**25.** $1\,\text{acre} = $ _____ $\text{mi}^2$

**26.** $4\,\text{acres} = $ _____ $\text{ft}^2$

**27.** $40.3\,\text{mi}^2 = $ _____ acres

**28.** $1080\,\text{in}^2 = $ _____ $\text{ft}^2$

**29.** $216\,\text{in}^2 = $ _____ $\text{ft}^2$

**30.** $69\,\text{ft}^2 = $ _____ $\text{yd}^2$

**b** Complete.

**31.** $19 \, km^2 = $ _____ $m^2$

**32.** $39 \, km^2 = $ _____ $m^2$

**33.** $6.31 \, m^2 = $ _____ $cm^2$

**34.** $2.7 \, m^2 = $ _____ $mm^2$

**35.** $6.5432 \, mm^2 = $ _____ $cm^2$

**36.** $8.38 \, cm^2 = $ _____ $mm^2$

**37.** $349 \, cm^2 = $ _____ $m^2$

**38.** $125 \, mm^2 = $ _____ $m^2$

**39.** $250{,}000 \, mm^2 = $ _____ $cm^2$

**40.** $5900 \, mm^2 = $ _____ $cm^2$

**41.** $472{,}800 \, m^2 = $ _____ $km^2$

**42.** $1.37 \, cm^2 = $ _____ $mm^2$

## Skill Maintenance

In Exercises 43 and 44, find the simple interest.   [8.7a]

**43.** On $2000 at an interest rate of 8% for 1.5 years

**44.** On $2000 at an interest rate of 5.3% for 2 years

In each of Exercises 45–48, find **(a)** the amount of simple interest due and **(b)** the total amount that must be paid back.   [8.7a]

**45.** A firm borrows $15,500 at 9.5% for 120 days.

**46.** A firm borrows $8500 at 10% for 90 days.

**47.** A firm borrows $6400 at 8.4% for 150 days.

**48.** A firm borrows $4200 at 11% for 30 days.

## Synthesis

Complete.

**49.** ▦ $1 \, m^2 = $ _____ $ft^2$

**50.** ▦ $1 \, in^2 = $ _____ $cm^2$

**51.** ▦ $2 \, yd^2 = $ _____ $m^2$

**52.** ▦ $1 \, acre = $ _____ $m^2$

**53.** *Ronald Reagan Library.*   The largest of the presidential libraries is the Ronald Reagan Library in Simi Valley, California. It covers approximately $153{,}000 \, ft^2$. Convert $153{,}000 \, ft^2$ to square meters.

**54.**  A 30-ft by 60-ft ballroom is to be remodeled by placing an 18-ft by 42-ft dance floor in the middle and carpeting the rest of the room. The new dance floor is to be laid in tiles that are 8-in. by 8-in. squares. How many such tiles are needed? What percent of the area is the dance floor?

**55.** Janie's Rubik's Cube has 54 cm² of area. Each side of Norm's cube is twice as wide as Janie's. Find the area of Norm's cube.

**56.** In order to remodel an office, a carpenter needs to purchase carpeting, at $8.45 a square yard, and molding for the base of the walls, at $0.87 a foot. If the room is 9 ft by 12 ft, with a 3-ft-wide doorway, what will the materials cost?

Find the area of the shaded region of each figure. Give the answer in square feet. (Figures are not drawn to scale.)

**57.**

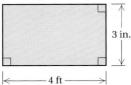

3 in.

4 ft

**58.**

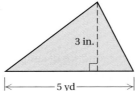

3 in.

5 yd

**59.**

13 ft

4 in.

**60.**

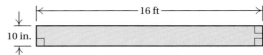

16 ft

10 in.

Find the area of the shaded region of each figure.

**61.**

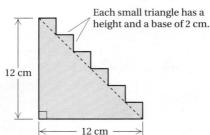

Each small triangle has a height and a base of 2 cm.

12 cm

12 cm

**62.**

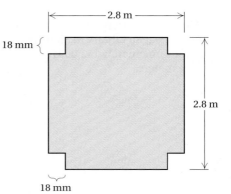

2.8 m

18 mm

2.8 m

18 mm

# More with Perimeter and Area

We have already discussed how to find the perimeter of polygons and the area of squares, rectangles, and triangles. In this section, we find the area of *parallelograms*, *trapezoids*, and *circles*. We also calculate the perimeter, or *circumference*, of a circle.

## a  Parallelograms and Trapezoids

A **parallelogram** is a four-sided figure with two pairs of parallel sides, as shown below.

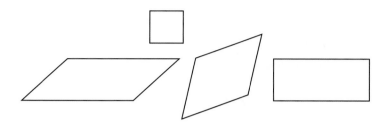

To understand how to find the area of a parallelogram, consider the one below.

If we cut off a piece and move it to the other end, we get a rectangle.

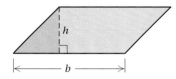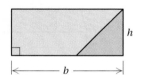

We can find the area by multiplying the length of a **base** $b$ by $h$, the **height**.

---

### AREA OF A PARALLELOGRAM

The **area of a parallelogram** is the product of the length of a base $b$ and the height $h$:

$$A = b \cdot h.$$

---

**EXAMPLE 1**  Find the area of this parallelogram.

$A = b \cdot h$
$= 7 \text{ km} \cdot 5 \text{ km}$
$= 35 \text{ km}^2$

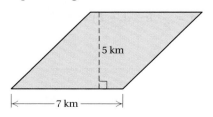

## OBJECTIVES

**a** Find the area of a parallelogram or trapezoid.

**b** Find the circumference, area, radius, or diameter of a circle, given the length of a radius or diameter.

---

**SKILL TO REVIEW**
Objective 2.7b: Determine the perimeter of a polygon.

1. Find the perimeter of a rectangular 12-ft by 14-ft bedroom.

2. Find the perimeter of a square frame that is 75 cm on each side.

---

## STUDY TIPS

### SKETCHES AND DRAWINGS

When a drawing appears in a mathematics text, study it carefully as you read the text, observing all the details marked. Be careful not to assume information that is not stated on the drawing or in the accompanying text. For example, even though one side of a figure may look twice as long as another side, do not assume that this is true unless it is stated. On the other hand, when you are making your own sketch for a problem, try to make sure that the sketch reflects, at least approximately, any information given in the problem.

***Answers***
*Skill to Review:*
**1.** 52 ft    **2.** 300 cm

Find the area.

**1.**

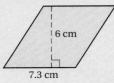

6 cm

7.3 cm

**2.**

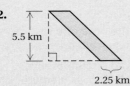

5.5 km

2.25 km

**EXAMPLE 2** Find the area of this parallelogram.

$$A = b \cdot h$$
$$= (1.2 \text{ m}) \cdot (6 \text{ m})$$
$$= 7.2 \text{ m}^2$$

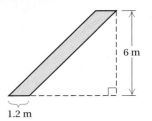

6 m

1.2 m

Do Exercises 1 and 2.

A **trapezoid** is a polygon with four sides, two of which, the **bases**, are parallel to each other.*

To understand how to find the area of a trapezoid, consider the one below.

$a$

$b$

Think of cutting out another just like it and placing the second one like this:

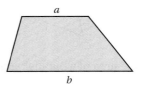

The resulting figure is a parallelogram with an area of

$$h \cdot (a + b).$$   The base of the parallelogram has length $a + b$.

The trapezoid we started with has half the area of the parallelogram, or

$$\frac{1}{2} \cdot h \cdot (a + b).$$

**AREA OF A TRAPEZOID**

The **area of a trapezoid** is half the product of the height and the sum of the lengths of the parallel sides, or the product of the height and the average length of the bases:

$$A = \frac{1}{2} \cdot h \cdot (a + b) = h \cdot \frac{a + b}{2}.$$

*Answers*

1. 43.8 cm² 2. 12.375 km²

*Some definitions of trapezoid specify *exactly* two parallel sides. We refrain from doing so. Thus, we consider a parallelogram a special type of trapezoid.

**EXAMPLE 3**   Find the area of this trapezoid.

$$A = \frac{1}{2} \cdot h \cdot (a + b)$$

$$= \frac{1}{2} \cdot 7 \text{ cm} \cdot (12 + 18) \text{ cm}$$

$$= \frac{7 \cdot 30}{2} \cdot \text{ cm}^2 = \frac{7 \cdot 15 \cdot 2}{1 \cdot 2} \text{ cm}^2$$

$$= 105 \text{ cm}^2 \quad \text{Removing a factor equal to 1} \left(\frac{2}{2} = 1\right) \text{ and multiplying}$$

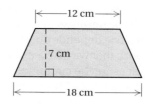
12 cm
7 cm
18 cm

Do Exercises 3 and 4.

## b  Circles

### Radius and Diameter

At right is a circle with center $O$. Segment $\overline{AC}$ is a *diameter*. A **diameter** is a segment that passes through the center of the circle and has endpoints on the circle. Segment $\overline{OB}$ is called a *radius*. A **radius** is a segment with one endpoint on the center and the other endpoint on the circle. The words *radius* and *diameter* are also used to represent the lengths of a circle's radius and diameter, respectively.

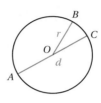

> **DIAMETER AND RADIUS**
>
> Suppose that $d$ is the diameter of a circle and $r$ is the radius. Then
>
> $$d = 2 \cdot r \quad \text{and} \quad r = \frac{d}{2}.$$

**EXAMPLE 4**   Find the length of a radius of this circle.

$$r = \frac{d}{2}$$

$$= \frac{12 \text{ m}}{2}$$

$$= 6 \text{ m}$$

12 m

The radius is 6 m.

**EXAMPLE 5**   Find the length of a diameter of this circle.

$$d = 2 \cdot r$$

$$= 2 \cdot \frac{1}{4} \text{ ft}$$

$$= \frac{1}{2} \text{ ft}$$

$\frac{1}{4}$ ft

The diameter is $\frac{1}{2}$ ft.

Do Exercises 5 and 6.

---

Find the area.

**3.**

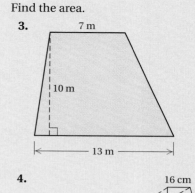
7 m
10 m
13 m

**4.**

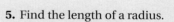
16 cm
35 cm
25 cm

**5.** Find the length of a radius.

18"

**6.** Find the length of a diameter.

$2\frac{1}{2}$ ft

***Answers***

**3.** 100 m$^2$   **4.** 717.5 cm$^2$   **5.** 9″   **6.** 5 ft

## Circumference

The perimeter of a circle is called its **circumference**. The circumference of a circle depends on its diameter. Suppose that we take a dinner plate and measure the circumference $C$ and the diameter $d$ with a tape measure. For the specific plate shown, we have

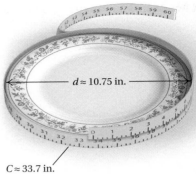

$d \approx 10.75$ in.

$C \approx 33.7$ in.

$$\frac{C}{d} = \frac{33.7 \text{ in.}}{10.75 \text{ in.}} \approx 3.1.$$

If we did this with plates and circles of several sizes, the result would always be a number close to 3.1. For any circle, if we divide the circumference $C$ by the diameter $d$, we get the same number. We call this number $\pi$ (pi). The number $\pi$ is a nonterminating, nonrepeating decimal. It is impossible to precisely express $\pi$ as a decimal or a fraction.

---

### CIRCUMFERENCE AND DIAMETER

The circumference $C$ of a circle of diameter $d$ is given by

$$C = \pi \cdot d.$$

The number $\pi$ is about 3.14, or about $\frac{22}{7}$.

---

**7.** Find the circumference of this circle. Use 3.14 for $\pi$.

20 m

**EXAMPLE 6** Find the circumference of this circle. Use 3.14 for $\pi$.

$$\begin{aligned}
C &= \pi \cdot d \\
&\approx 3.14 \cdot 6 \text{ cm} \\
&\approx 18.84 \text{ cm}
\end{aligned}$$

6 cm

The circumference is about 18.84 cm.

> Do Exercise 7.

Since $d = 2 \cdot r$, where $r$ is the length of a radius, it follows that

$$C = \pi \cdot d = \pi \cdot (2 \cdot r).$$

---

### CIRCUMFERENCE AND RADIUS

The circumference $C$ of a circle of radius $r$ is given by

$$C = 2 \cdot \pi \cdot r.$$

---

**EXAMPLE 7** Find the circumference of this circle. Use $\frac{22}{7}$ for $\pi$.

$$\begin{aligned}
C &= 2 \cdot \pi \cdot r \\
&= 2 \cdot \frac{22}{7} \cdot 70 \text{ in.} \\
&= 2 \cdot 22 \cdot \frac{70}{7} \text{ in.} \\
&= 44 \cdot 10 \text{ in.} \\
&= 440 \text{ in.}
\end{aligned}$$

70 in.

The circumference is about 440 in.

*Answer*

7. 62.8 m

**EXAMPLE 8** Find the perimeter of this figure. Use 3.14 for $\pi$.

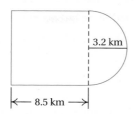

We let $P$ = the perimeter. Note that the figure has three straight edges (line segments) and one curved edge. The curved edge is half the circumference of a circle of radius 3.2 km. The top and bottom straight edges have lengths of 8.5 km. The left edge is the same length as twice the radius of the semicircle on the right, or 6.4 km.

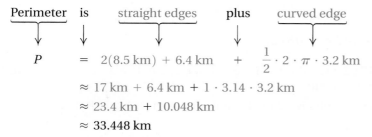

$$P = 2(8.5 \text{ km}) + 6.4 \text{ km} + \frac{1}{2} \cdot 2 \cdot \pi \cdot 3.2 \text{ km}$$
$$\approx 17 \text{ km} + 6.4 \text{ km} + 1 \cdot 3.14 \cdot 3.2 \text{ km}$$
$$\approx 23.4 \text{ km} + 10.048 \text{ km}$$
$$\approx 33.448 \text{ km}$$

The perimeter is about 33.448 km.

Do Exercises 8–10.

## Area

To understand how to find the area of a circle with radius $r$, think of cutting half a circular region into small slices and arranging them as shown below. Note that half of the circumference is half of $2\pi r$, or simply $\pi r$.

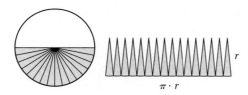

Then imagine slicing the other half of the circular region and arranging the pieces in with the others as shown below.

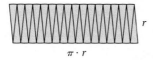

The thinner the slices, the closer this comes to being a rectangle. Its length is $\pi r$ and its width is $r$, so its area is

$$(\pi \cdot r) \cdot r.$$

This is the area of a circle.

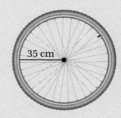
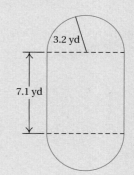

## AREA OF A CIRCLE

The **area of a circle** with radius of length $r$ is given by

$$A = \pi \cdot r \cdot r, \quad \text{or} \quad A = \pi \cdot r^2.$$

**EXAMPLE 9** Find the area of this circle. Use $\frac{22}{7}$ for $\pi$.

$$A = \pi \cdot r^2 = \pi \cdot r \cdot r$$

$$\approx \frac{22}{7} \cdot 14 \text{ cm} \cdot 14 \text{ cm}$$

$$\approx \frac{22}{7} \cdot \frac{7 \cdot 2}{1} \text{ cm} \cdot 14 \text{ cm}$$

$$\approx 616 \text{ cm}^2 \qquad \textit{Note: } r^2 \neq 2r.$$

The area is about 616 cm².

**EXAMPLE 10** Find the area of this circle. Use 3.14 for $\pi$. Round to the nearest hundredth.

The diameter is 4.2 m; the radius is 4.2 m ÷ 2, or 2.1 m.

$$A = \pi \cdot r \cdot r$$

$$\approx 3.14 \cdot 2.1 \text{ m} \cdot 2.1 \text{ m}$$

$$\approx 3.14 \cdot 4.41 \text{ m}^2$$

$$\approx 13.8474 \text{ m}^2 \approx 13.85 \text{ m}^2$$

The area is about 13.85 m².

Do Exercises 11 and 12.

We can often add or subtract areas to determine the area of a region.

**EXAMPLE 11** Find the area of the shaded region.

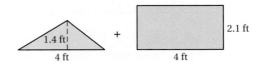

The region consists of a triangle and a rectangle, as illustrated below. Note that the base of the triangle is the same as the length of the rectangle. We calculate the areas separately and add to find the area of the shaded region.

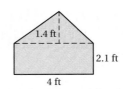

$$\text{Area} = \frac{1}{2} \cdot 4 \text{ ft} \cdot 1.4 \text{ ft} + 4 \text{ ft} \cdot 2.1 \text{ ft}$$

$$= 2.8 \text{ ft}^2 + 8.4 \text{ ft}^2$$

$$= 11.2 \text{ ft}^2$$

Do Exercise 13.

---

**11.** Find the area of this circle. Use $\frac{22}{7}$ for $\pi$.

**12.** Find the area of this circle. Use 3.14 for $\pi$.

**13.** Find the area of the shaded region. Use 3.14 for $\pi$. (*Hint*: The figure consists of a square and two semicircles, or one complete circle.)

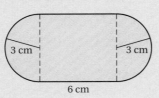

**Answers**

**11.** $78\frac{4}{7}$ km²   **12.** 339.6 cm²   **13.** 64.26 cm²

**EXAMPLE 12** *Area of Pizza Pans.* How much larger is a pizza made in a 16-in.-square pizza pan than a pizza made in a 16-in.-diameter circular pan?

First, we make a drawing of each.

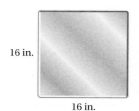  

16 in.

16 in.

16 in.

Then we compute areas.

The area of the square is

$$A = s \cdot s$$
$$= 16 \text{ in.} \cdot 16 \text{ in.} = 256 \text{ in}^2.$$

The diameter of the circle is 16 in., so the radius is 16 in./2, or 8 in. The area of the circle is

$$A = \pi \cdot r \cdot r$$
$$\approx 3.14 \cdot 8 \text{ in.} \cdot 8 \text{ in.} = 200.96 \text{ in}^2.$$

The square pizza is larger by about

$$256 \text{ in}^2 - 200.96 \text{ in}^2, \quad \text{or} \quad 55.04 \text{ in}^2.$$

Do Exercise 14.

---

### AREA AND CIRCUMFERENCE FORMULAS

Area of a parallelogram: $A = b \cdot h$

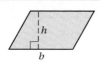

Area of a trapezoid: $A = \dfrac{1}{2} \cdot h \cdot (a + b)$

or $A = h \cdot \dfrac{a + b}{2}$

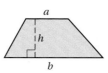

Circumference of a circle: $C = \pi \cdot d$
or $C = 2 \cdot \pi \cdot r$

Area of a circle: $A = \pi \cdot r^2$

*Note:* $d = 2r$ and $\pi \approx 3.14 \approx \frac{22}{7}$.

---

*Caution!*

Remember that circumference is always measured in linear units like ft, m, cm, yd, and so on. But area is measured in square units like ft², m², cm², yd², and so on.

---

**14.** Which is larger and by how much: a 10-ft-square flower bed or a 12-ft-diameter flower bed?

*Answer*

**14.** A 12-ft-diameter flower bed is 13.04 ft² larger.

**.9.3** **Exercise Set**

For Extra Help

MyMathLab

Math XL
PRACTICE

WATCH

DOWNLOAD

READ

REVIEW

**a** Find the area of each parallelogram or trapezoid.

**1.**
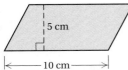
5 cm

10 cm

**2.**
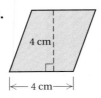
4 cm

4 cm

**3.**

6 ft

8 ft

20 ft

**4.**
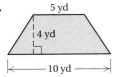
5 yd

4 yd

10 yd

**5.**
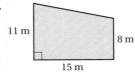
11 m

8 m

15 m

**6.**
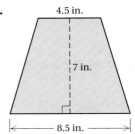
4.5 in.

7 in.

8.5 in.

**7.**
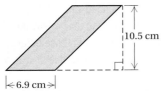
10.5 cm

6.9 cm

**8.**
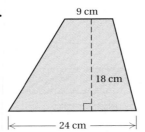
9 cm

18 cm

24 cm

**9.**
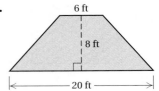
13 mi

9 mi

19 mi

**10.**
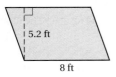
5.2 ft

8 ft

**11.**
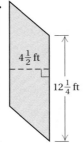
$4\frac{1}{2}$ ft

$12\frac{1}{4}$ ft

**12.**
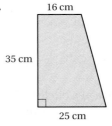
16 cm

35 cm

25 cm

**13.**
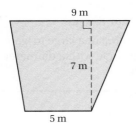
9 m

7 m

5 m

**14.**
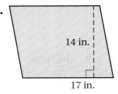
14 in.

17 in.

**15.**
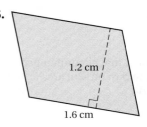
1.2 cm

1.6 cm

**16.**

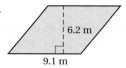
6.2 m
9.1 m

**17.**

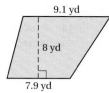
9.1 yd
8 yd
7.9 yd

**18.**

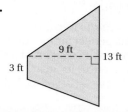
9 ft
13 ft
3 ft

**b** Find the length of a diameter of each circle.

**19.**

7 cm

**20.**

8 m

**21.**

$\frac{7}{8}$ in.

**22.**

$8\frac{2}{3}$ mi

Find the length of a radius of each circle.

**23.**

20 ft

**24.**

10 in.

**25.**

1.4 cm

**26.**

60.9 km

Find the circumference of each circle in Exercises 19–22. Use $\frac{22}{7}$ for $\pi$.

**27.** Exercise 19          **28.** Exercise 20          **29.** Exercise 21          **30.** Exercise 22

Find the circumference of each circle in Exercises 23–26. Use 3.14 for $\pi$.

**31.** Exercise 23          **32.** Exercise 24          **33.** Exercise 25          **34.** Exercise 26

Find the area of each circle in Exercises 19–22. Use $\frac{22}{7}$ for $\pi$.

**35.** Exercise 19          **36.** Exercise 20          **37.** Exercise 21          **38.** Exercise 22

Find the area of each circle in Exercises 23–26. Use 3.14 for $\pi$.

**39.** Exercise 23          **40.** Exercise 24          **41.** Exercise 25          **42.** Exercise 26

Solve. Use 3.14 for $\pi$.

**43.** *Treated Mosquito Net.* The Adventure II treated mosquito net is perfect for the rural or tropical traveler. This net is circular with a radius of 6.37 ft. What is its diameter? its circumference? its area?

**Source:** www.travelhealthhelp.com/nets1.html

**44.** *Penny.* A penny has a 1-cm radius. What is its diameter? its circumference? its area?

1 cm

**45.** *Earth.* The diameter of the earth at the equator is 7926.41 mi. What is the circumference of the earth at the equator?

**46.** *Circumference of a Baseball Bat.* In Major League Baseball, the diameter of the barrel of a bat cannot be more than $2\frac{3}{4}$ in., and the diameter of the bat handle can be no thinner than $\frac{16}{19}$ in. Find the maximum circumference of the barrel of a bat and the minimum circumference of the bat handle. Use $\frac{22}{7}$ for $\pi$.

**Source:** Major League Baseball

**47.** *Oil Spill.* The radius of an oil slick is 150 mi. How much area is covered by the slick?

**48.** *Radio.* A college radio station is allowed by the FCC to broadcast over an area with a radius of 7 mi. How much area is this?

**49.** *Masonry.* The Harris-Regency Hotel plans to install a 1-yd-wide walk around a circular swimming pool. The diameter of the pool is 20 yd. What is the area of the walk?

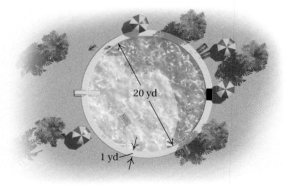

20 yd

1 yd

**50.** *Trampoline.* The standard backyard trampoline has a diameter of 14 ft. What is its area?

**Source:** International Trampoline Industry Association, Inc.

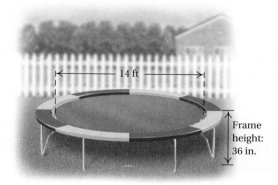

14 ft

Frame height: 36 in.

**51.** *Archaeology.* The columns in the Great Hypostyle Hall in the Temple of Amun in Luxor, Egypt, are 15 m in circumference. What is the diameter of the columns?

**52.** *Forestry.* To protect a tree, gypsy-moth caterpillar tape is wrapped around the trunk. It takes 47.1 in. of tape to go around the trunk once. What is the diameter of the tree?

**53.** *Pizza Slices.* Snappy Tomato Pizza sells a 12-in. square Chicago-style pizza and a 12-in. round New York–style pizza. The square pizza is cut into 9 slices and the round pizza into 8 slices. Which slice has the greater area?

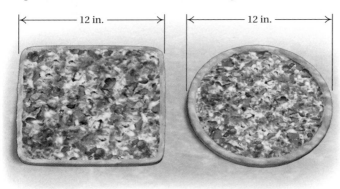

**54.** *Roller-Rink Floor.* A roller-rink floor is shown below. Each end is a semicircle. What is its area? If hardwood flooring costs $62.50 per square meter, how much will the flooring cost?

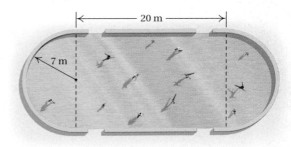

Find the perimeter of each figure. Use 3.14 for $\pi$.

**55.**

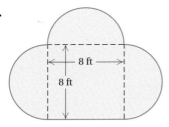

**56.**

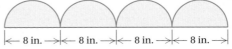

**57.**

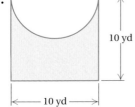

**58.**

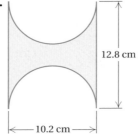

Find the area of the shaded region in each figure. Use 3.14 for $\pi$.

**59.**

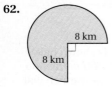

8 m

**60.**

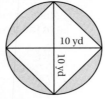

10 yd

10 yd

**61.**

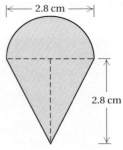

2.8 cm

2.8 cm

**62.**

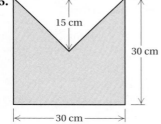

8 km

8 km

**63.**

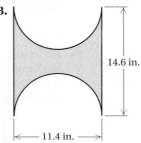

14.6 in.

11.4 in.

**64.**

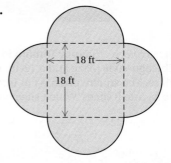

18 ft

18 ft

**65.**

15 cm

30 cm

30 cm

**66.**

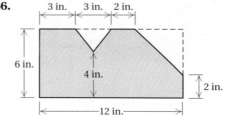

3 in.   3 in.   2 in.

6 in.

4 in.

2 in.

12 in.

**67.**

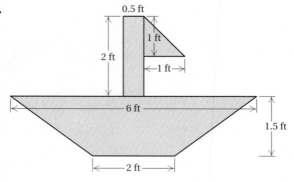

0.5 ft

1 ft

2 ft

1 ft

6 ft

1.5 ft

2 ft

**68.**

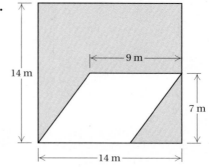

14 m

9 m

7 m

14 m

## Skill Maintenance

Convert to fraction notation.   [8.1c]

**69.** 9.25%                    **70.** $87\frac{1}{2}$%

Convert to percent notation.   [8.1c]

**71.** $\frac{5}{4}$            **72.** $\frac{2}{3}$

**73.** The weight of a human brain is 2.5% of total body weight. A person weighs 200 lb. What does his brain weigh?   [8.4a]

**74.** Jack's commission is increased according to how much he sells. He receives a commission of 6% for the first $3000 and 10% on the amount over $3000. What is the total commission on sales of $8500?   [8.6b]

## Synthesis

**75.** ▦ Calculate the surface area of an unopened steel can that has a height of 3.5 in. and a diameter of 2.5 in. (*Hint*: Make a sketch and "unroll" the sides of the can.) Use 3.14 for $\pi$.

**76.** ▦ The distance from Kansas City to Indianapolis is 500 mi. A car was driven this distance using tires with a radius of 14 in. How many revolutions of each tire occurred on the trip? Use $\frac{22}{7}$ for $\pi$.

**77.** *Sports Marketing.* Tennis balls are often packed vertically, three in a can, one on top of another. Without using a calculator, determine the larger measurement: the can's circumference or the can's height.

**78.** ▦ The sides of a cake box are trapezoidal, as shown in the figure. Determine the surface area of the box.

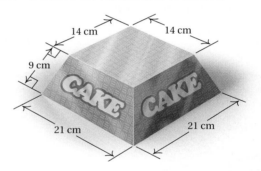

**79.** The radius of one circle is twice the size of another circle's radius. If the area of the smaller circle is $A$, what is the area of the larger circle?

**80.** The radius of one circle is twice the size of another circle's radius. If the circumference of the smaller circle is $C$, what is the circumference of the larger circle?

**81.** A cake recipe calls for a 9-inch round baking pan. If only square pans are available, what size pan should be used for the cake? Round to the nearest inch.

**82.** Peggy fenced a square garden with sides of 12 ft. If she decides to use the fence around a circular garden instead, what size garden can she fence? Which garden has the greater area?

# 9.4

## Volume and Capacity

### OBJECTIVES

**a** Find the volume of a rectangular solid, a cylinder, and a sphere.

**b** Convert from one unit of capacity to another.

**c** Solve applied problems involving volume and capacity.

---

**SKILL TO REVIEW**

Objective 9.3b: Find the circumference, area, radius, or diameter of a circle, given the length of a radius or a diameter.

1. Find the area of a circle whose radius is 21 in. Use $\frac{22}{7}$ for $\pi$.

2. Find the area of a circle whose diameter is $2\frac{4}{5}$ ft. Use $\frac{22}{7}$ for $\pi$.

---

1. Find the volume.

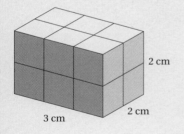

---

### a Volume

The **volume** of an object is the number of unit cubes needed to fill it.

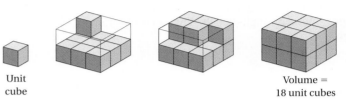

Unit cube

Volume = 18 unit cubes

Two common units are shown below (actual size).

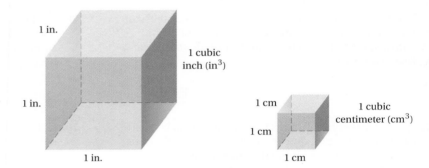

1 in.

1 in.

1 in.

1 cubic inch ($in^3$)

1 cm

1 cm

1 cm

1 cubic centimeter ($cm^3$)

**EXAMPLE 1** Find the volume.

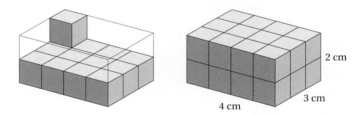

2 cm

4 cm

3 cm

The figure is made up of 2 layers of 12 cubes each, so its volume is 24 cubic centimeters ($cm^3$). Note that $24 = 4 \cdot 3 \cdot 2$.

Do Margin Exercise 1.

---

### VOLUME OF A RECTANGULAR SOLID

The **volume of a rectangular solid** is found by multiplying length by width by height:

$$V = \underbrace{l \cdot w} \cdot h.$$

Area    Height
of base

$h$

$w$

$l$

---

**Answers**

*Skill to Review:*

1. 1386 in²    2. $6\frac{4}{25}$ ft², or 6.16 ft²

*Margin Exercise:*

1. 12 cm³

---

**EXAMPLE 2** *Volume of a Safe.* For office security, William purchases a safe with dimensions 29 in. × 25 in. × 53 in. Find the volume of this rectangular solid.

$$V = l \cdot w \cdot h$$
$$= 29 \text{ in.} \cdot 25 \text{ in.} \cdot 53 \text{ in.}$$
$$= 38{,}425 \text{ in}^3$$

Do Exercises 2 and 3.

Do Exercises 2 and 3.

Volumes are described in units such as cubic centimeters (cm³) and cubic inches (in³). *Dimensional analysis* is an excellent way of determining the correct units for an answer.

- If measurements of length are added, use a one-dimensional unit of length: $3 \text{ ft} + 2 \text{ ft} + 7 \text{ ft} = 12 \text{ ft}$.

- If two measurements of length are multiplied, use a two-dimensional unit of area: $8 \text{ ft} \cdot 7 \text{ ft} = 56 \text{ ft}^2$.

- If three measurements of length are multiplied, use a three-dimensional unit of volume: $3 \text{ m} \cdot 2 \text{ m} \cdot 4 \text{ m} = 24 \text{ m}^3$.

A rectangular solid is shown below. Note that we can think of the volume as the product of the area of the base times the height:

$$V = l \cdot w \cdot h$$
$$= (l \cdot w) \cdot h$$
$$= (\text{Area of the base}) \cdot h$$
$$= B \cdot h,$$

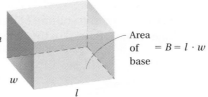

where *B* represents the area of the base.

Like rectangular solids, **circular cylinders** have bases of equal area that lie in parallel planes. The bases of circular cylinders are circular regions.

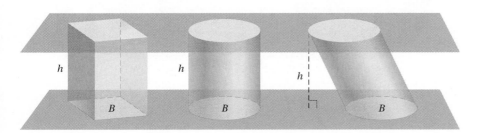

The volume of a circular cylinder is found in a manner similar to the way the volume of a rectangular solid is found. The volume is the product of the area of the base times the height. The height is always measured perpendicular to the base.

**2. Carry-on Luggage.** The largest piece of luggage that you can carry on an airplane measures 23 in. by 10 in. by 13 in. Find the volume of this solid.

**3. Cord of Wood.** A cord of wood measures 4 ft by 4 ft by 8 ft. What is the volume of a cord of wood?

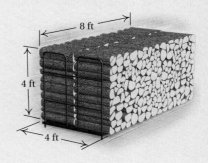

*Answers*

**2.** 2990 in³    **3.** 128 ft³

**4.** Find the volume of the cylinder. Use 3.14 for $\pi$.

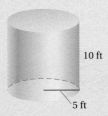

10 ft

5 ft

**5.** Find the volume of the cylinder. Use $\frac{22}{7}$ for $\pi$.

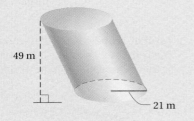

49 m

21 m

## VOLUME OF A CIRCULAR CYLINDER

The **volume of a circular cylinder** is the product of the area of the base $B$ and the height $h$:

$$V = B \cdot h, \quad \text{or} \quad V = \pi \cdot r^2 \cdot h.$$

**EXAMPLE 3** Find the volume of this circular cylinder. Use 3.14 for $\pi$.

12 cm

4 cm

$$V = B \cdot h = \pi \cdot r^2 \cdot h$$
$$\approx 3.14 \cdot 4 \text{ cm} \cdot 4 \text{ cm} \cdot 12 \text{ cm}$$
$$= 602.88 \text{ cm}^3$$

Do Exercises 4 and 5.

A **sphere** is the three-dimensional counterpart of a circle. It is the set of all points in space that are a given distance (the radius) from a given point (the center). The volume of a sphere depends on its radius.

## VOLUME OF A SPHERE

The **volume of a sphere** of radius $r$ is given by

$$V = \frac{4}{3} \cdot \pi \cdot r^3.$$

**EXAMPLE 4** *Bowling Ball.* The radius of a standard-sized bowling ball is 4.2915 in. Find the volume of a standard-sized bowling ball (disregarding the finger holes). Round to the nearest hundredth of a cubic inch. Use 3.14 for $\pi$.

**6.** Find the volume of the sphere. Use $\frac{22}{7}$ for $\pi$.

28 ft

**7.** The radius of a standard-sized golf ball is 2.1 cm. Find its volume. Use 3.14 for $\pi$.

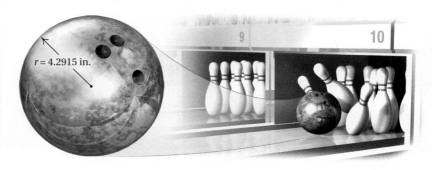

$r = 4.2915$ in.

We have
$$V = \frac{4}{3} \cdot \pi \cdot r^3 \approx \frac{4}{3} \cdot 3.14 \cdot (4.2915 \text{ in.})^3$$
$$\approx 330.90 \text{ in}^3. \qquad \text{Using a calculator}$$

Do Exercises 6 and 7.

*Answers*

**4.** 785 ft$^3$  **5.** 67,914 m$^3$

**6.** 91,989$\frac{1}{3}$ ft$^3$  **7.** 38.77272 cm$^3$

 **Capacity**

## American Units

To answer a question like "How much soda is in the bottle?" we need measures of **capacity**. American units of capacity are fluid ounces, cups, pints, quarts, and gallons. These units are related as follows.

---

**AMERICAN UNITS OF CAPACITY**

| | |
|---|---|
| 1 gallon (gal) = 4 quarts (qt) | 1 pt = 2 cups |
| | = 16 fluid ounces (fl oz) |
| 1 qt = 2 pints (pt) | 1 cup = 8 fluid oz |

---

Fluid ounces, abbreviated fl oz, are often referred to as ounces, or oz.

**EXAMPLE 5**  Complete: 24 qt = _____ gal.

Since we are converting from a *smaller* unit to a *larger* unit, we multiply by 1 using 1 gal in the numerator and 4 qt in the denominator:

$$24 \text{ qt} = 24 \text{ qt} \cdot \frac{1 \text{ gal}}{4 \text{ qt}} = \frac{24}{4} \cdot 1 \text{ gal} = 6 \text{ gal.}$$

**EXAMPLE 6**  Complete: 9 gal = _____ oz.

The box above does not list how many ounces are in 1 gal. We convert gallons to quarts, quarts to pints, and pints to ounces, using the relationships given in the box.

Since we are converting from a *larger* unit to a *smaller* unit, we use substitution:

| | |
|---|---|
| 9 gal = 9 · 1 gal = 9 · 4 qt = 36 qt | Substituting 4 qt for 1 gal |
| = 36 · 1 qt = 36 · 2 pt = 72 pt | Substituting 2 pt for 1 qt |
| = 72 · 1 pt = 72 · 16 oz = 1152 oz. | Substituting 16 oz for 1 pt |

Thus, 9 gal = 1152 oz.

> Do Exercises 8 and 9.

Complete.

**8.** 80 qt = _____ gal

**9.** 5 gal = _____ pt

## Metric Units

One unit of capacity in the metric system is a **liter**. A liter is just a bit more than a quart. It is defined as follows.

1 liter ≈ 1.06 quarts

1 liter     1 quart

*Answers*

**8.** 20    **9.** 40

## METRIC UNITS OF CAPACITY

> 1 liter (L) = 1000 cubic centimeters (1000 cm$^3$)
>
> The script letter $\ell$ is also used for "liter."

The metric prefixes are also used with liters. The most common is **milli-**. The milliliter (mL) is, then, $\frac{1}{1000}$ liter.

> 1 L = 1000 mL = 1000 cm$^3$
>
> 0.001 L = 1 mL = 1 cm$^3$

A common unit for drug dosage is the milliliter (mL) or the cubic centimeter (cm$^3$). The notation "cc" is also used for cubic centimeter, especially in medicine. The milliliter and the cubic centimeter represent the same measure of capacity. A milliliter is about $\frac{1}{5}$ of a teaspoon.

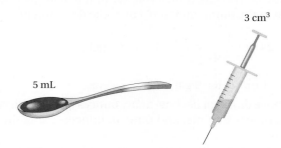

3 cm$^3$

5 mL

> 1 mL = 1 cm$^3$ = 1 cc

Volumes for which quarts and gallons are used are expressed in liters. Large volumes are expressed using measures of cubic meters (m$^3$).

Do Exercises 10–13.

**EXAMPLE 7**  Complete:  4.5 L = _____mL.

$$4.5 \text{ L} = 4.5 \cdot 1\text{L} = 4.5 \cdot 1000 \text{ mL} \quad \text{Substituting 1000 mL for 1 L}$$
$$= 4500 \text{ mL}$$

We can also convert units of capacity by multiplying by a form of 1.

**EXAMPLE 8**  Complete:  280 mL = _____ L.

$$280 \text{ mL} = 280 \text{ m\sout{L}} \cdot \frac{1\text{L}}{1000 \text{ m\sout{L}}}$$
$$= \frac{280}{1000} \text{ L}$$
$$= 0.28 \text{ L}$$

Do Exercises 14 and 15.

---

Complete with mL or L.

**10.** The patient received an injection of 2 _____ of penicillin.

**11.** There are 250 _____ in a coffee cup.

**12.** The gas tank holds 80 _____.

**13.** Bring home 8 _____ of milk.

Complete.

**14.** 0.97 L = _____ mL

**15.** 8990 mL = _____ L

*Answers*

**10.** mL  **11.** mL  **12.** L  **13.** L
**14.** 970  **15.** 8.99

# c Solving Applied Problems

**EXAMPLE 9**  At a self-service gas station, 89-octane gasoline sells for 102.6¢ a liter. Estimate the price of 1 gal in dollars.

Since 1 liter is about 1 quart and there are 4 quarts in a gallon, the price of a gallon is about 4 times the price of a liter.

$$4 \cdot 102.6¢ = 410.4¢ = \$4.104$$

Thus, 89-octane gasoline sells for about $4.10 a gallon.

Do Exercise 16.

**EXAMPLE 10**  *Propane Gas Tank.*  A propane gas tank is shaped like a circular cylinder with half of a sphere at each end. Find the volume of the tank if the cylindrical section is 5 ft long with a 4-ft diameter. Use 3.14 for $\pi$.

**1. Familiarize.**  We first make a drawing.

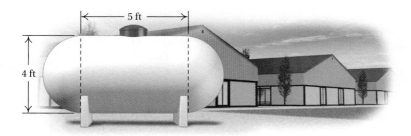

**2. Translate.**  This is a two-step problem. We first find the volume of the cylindrical portion. Then we find the volume of the two ends and add. Note that the radius is 2 ft and that together the two ends make a sphere. We let $V$ = the total volume.

$$
\underbrace{\text{Total volume}}_{V} \;\; \underbrace{\text{is}}_{=} \;\; \underbrace{\text{volume of the cylinder}}_{\substack{\pi \cdot r^2 \cdot h \\ 3.14 \cdot (2\text{ ft})^2 \cdot 5\text{ ft}}} \;\; \underbrace{\text{plus}}_{+} \;\; \underbrace{\text{volume of the two ends}}_{\substack{\frac{4}{3} \cdot \pi \cdot r^3 \\ \frac{4}{3} \cdot 3.14 \cdot (2\text{ ft})^3}}
$$

**3. Solve.**  The volume of the cylinder is approximately

$$3.14 \cdot (2\text{ ft})^2 \cdot 5\text{ ft} = 3.14 \cdot 2\text{ ft} \cdot 2\text{ ft} \cdot 5\text{ ft}$$
$$= 62.8\text{ ft}^3.$$

The volume of the two ends is approximately

$$\frac{4}{3} \cdot 3.14 \cdot (2\text{ ft})^3 = \frac{4}{3} \cdot 3.14 \cdot 2\text{ ft} \cdot 2\text{ ft} \cdot 2\text{ ft}$$
$$\approx 33.5\text{ ft}^3.$$

The total volume is approximately

$$62.8\text{ ft}^3 + 33.5\text{ ft}^3 = 96.3\text{ ft}^3.$$

**4. Check.**  We can repeat the calculations. The answer checks.

**5. State.**  The volume of the tank is about 96.3 ft$^3$.

Do Exercise 17.

---

**16.** At the same station, the price of 87-octane gasoline is 96.7 cents a liter. Estimate the price of 1 gal in dollars.

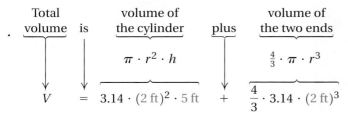

**Calculator Corner**

**Volumes Using Pi**  Many calculators have a $\boxed{\pi}$ key that can be used to give a more precise value of $\pi$ than 3.14.

To find the volume of the circular cylinder in Example 3, we press

$\boxed{\text{2nd}}\ \boxed{\pi}\ \boxed{\times}\ \boxed{4}\ \boxed{\times}\ \boxed{4}\ \boxed{\times}\ \boxed{1}\ \boxed{2}$
$\boxed{=}$, or $\boxed{\text{SHIFT}}\ \boxed{\pi}\ \boxed{\times}\ \boxed{4}\ \boxed{\times}\ \boxed{4}$
$\boxed{\times}\ \boxed{1}\ \boxed{2}\ \boxed{=}$. The result is approximately 603.19. Note that this is slightly different from the result found using 3.14 for $\pi$.

**Exercises:**

**1.** Use a calculator with a $\boxed{\pi}$ key to find the volume of a cylinder with radius 1.2 cm and height 0.25 cm.

**2.** Use a calculator with a $\boxed{\pi}$ key to find the volume of a sphere with radius $\frac{3}{8}$ in.

**17. Medicine Capsule.**  A cold capsule is 8 mm long and 4 mm in diameter. Find the volume of the capsule. Use 3.14 for $\pi$. (*Hint*: First find the length of the cylindrical section.)

*Answers*

**16.** $3.87     **17.** 83.7$\overline{3}$ mm$^3$

**a** Find each volume. Use 3.14 for π in Exercises 9–12.

**1.**
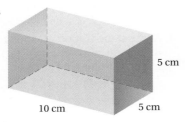
5 cm
10 cm   5 cm

**2.**
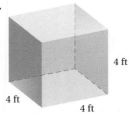
4 ft
4 ft   4 ft

**3.**
5 in.
9 in.
3 in.

**4.**
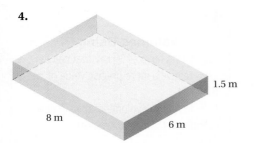
1.5 m
8 m   6 m

**5.**
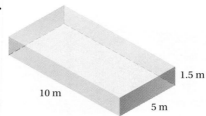
1.5 m
10 m   5 m

**6.**
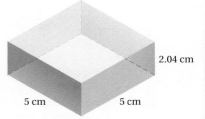
2.04 cm
5 cm   5 cm

**7.**
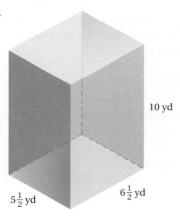
10 yd
$5\frac{1}{2}$ yd   $6\frac{1}{2}$ yd

**8.**
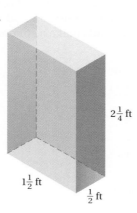
$2\frac{1}{4}$ ft
$1\frac{1}{2}$ ft   $\frac{1}{2}$ ft

**9.**
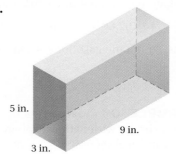
13 ft
10 ft

**10.**
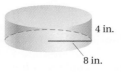
4 in.
8 in.

**11.**
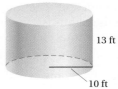
7.5 cm
4 cm

**12.**

15.1 m
3 m

Find each volume. Use $\frac{22}{7}$ for $\pi$ in Exercises 13, 14, 19, and 20. Use 3.14 for $\pi$ in Exercises 15–18.

**13.**

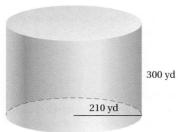

300 yd

210 yd

**14.**

28 km

4 km

**15.**

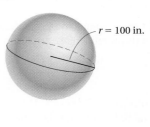

$r = 100$ in.

**16.**

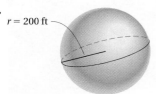

$r = 200$ ft

**17.**

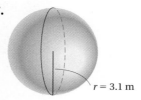

$r = 3.1$ m

**18.**

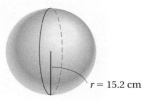

$r = 15.2$ cm

**19.**

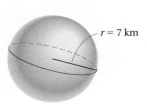

$r = 7$ km

**20.**

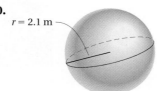

$r = 2.1$ m

**b** Complete.

**21.** 1 L = _____ mL = _____ cm$^3$

**22.** _____ L = 1 mL = _____ cm$^3$

**23.** 59 L = _____ mL

**24.** 714 L = _____ mL

**25.** 49 mL = _____ L

**26.** 43 mL = _____ L

**27.** 27.3 L = _____ cm$^3$

**28.** 49.2 L = _____ cm$^3$

**29.** 5 gal = _____ pt

**30.** 48 oz = _____ pt

**31.** 10 qt = _____ oz

**32.** 2 gal = _____ cups

**33.** 24 oz = _____ cups

**34.** 20 cups = _____ pt

**35.** 10 gal = _____ qt

**36.** 5 gal = _____ cups

**37.** 3 gal = _____ cups

**38.** 72 oz = _____ cups

**39.** 15 pt = _____ gal

**40.** 9 qt = _____ gal

**C** Solve.

**41.** *Oak Log.* An oak log has a diameter of 12 cm and a length (height) of 42 cm. Find the volume. Use 3.14 for $\pi$.

**42.** *Ladder Rung.* A rung of a ladder is 2 in. in diameter and 16 in. long. Find the volume. Use 3.14 for $\pi$.

**43.** *Times Square Crystal Ball.* In 2010, a Waterford Crystal ball featuring "Let There Be Courage" design triangles was used to usher in the new year in Times Square. The ball is 12 ft in diameter and weighs almost 6 tons. Find the volume of the crystal ball. Use 3.14 for $\pi$.

**44.** *Gas Pipeline.* The 638-mi Rockies Express–East gas pipeline from Colorado to Ohio was constructed with 80-ft sections of 42-in., or $3\frac{1}{2}$-ft, steel pipe. Find the volume of one section. Use $\frac{22}{7}$ for $\pi$.

**Source:** Rockies Express Pipeline

**45.** *Volume of a Trash Can.* The diameter of the base of a cylindrical trash can is 0.7 yd. The height is 1.1 yd. Find the volume. Use 3.14 for $\pi$ and round the answer to the nearest hundredth.

**46.** *Tennis Ball.* The diameter of a tennis ball is 6.5 cm. Find the volume. Use 3.14 for $\pi$ and round the answer to the nearest hundredth.

**47.** ▦ *Volume of Earth.* The radius of the earth is about 3980 mi. Find the volume of the earth. Use 3.14 for $\pi$. Round to the nearest ten thousand cubic miles.

**Source:** *The Cambridge Factfinder,* 4th ed.

**48.** ▦ *Astronomy.* The radius of the largest moon of Uranus is about 789 km. Find the volume of this satellite. Use $\frac{22}{7}$ for $\pi$. Round to the nearest ten thousand cubic kilometers.

**49.** ▦ A sphere with diameter 1 m is circumscribed by a cube. How much greater is the volume of the cube than the volume of the sphere? Use 3.14 for $\pi$.

**50.** ▦ *Golf-Ball Packaging.* The box shown is just big enough to hold 3 golf balls. If the radius of a golf ball is 2.1 cm, how much air surrounds the three balls? Use 3.14 for $\pi$.

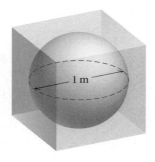

**51.** *Architecture.* The Torres de Hercúles in Cádiz, Spain, consists of two cylindrical towers. Each tower is 126 m high and 25 m in diameter. What is the volume of each tower? Use 3.14 for $\pi$, and round to the nearest cubic meter.

**52.** *Water Storage.* A water storage tank is a circular cylinder with a radius of 14 m and a height of 100 m. What is the tank's volume? Use $\frac{22}{7}$ for $\pi$.

**53.** *Oceanography.* The Deep-ESP (Environmental Sample Processor) is designed to study deep-sea environments. Its titanium pressure housing allows the ESP sphere to descend over 2 miles below the ocean surface. If the diameter of the sphere is 40 in., what is its volume? Use 3.14 for $\pi$ and round to the nearest cubic inch.

**Source:** "Extreme Life," by Henry Bortman, 06/15/09, on www.astrobio.net

The Deep-ESP (Environmental Sample Processor) was designed by the Monterey Bay Aquarium Research Institute to study the DNA of deep-sea life.

**54.** *Weather Forecasting.* Every day, the National Weather Service launches spherical weather balloons from 100 locations in the United States. Each balloon can rise up to 115,000 ft and travel up to 200 miles. Find the volume of a 6-ft-diameter weather balloon. Use 3.14 for $\pi$.

**Source:** Kaysam Worldwide, Inc.

**55.** *Metallurgy.* If all the gold in the world could be gathered together, it would form a cube 18 yd on a side. Find the volume of the world's gold.

**56.** ▦ The width of a dollar bill is 2.3125 in., the length is 6.0625 in., and the thickness is 0.0041 in. Find the volume occupied by 1 million one-dollar bills.

**57.** *Silo.* A silo is a circular cylinder with half a sphere on the top of the cylinder. If the diameter of the cylinder is 6 m and the height of the cylinder is 13 m, find the volume of the silo. Use 3.14 for $\pi$.

**58.** *Oceanography.* A research submarine is capsule-shaped. Find the volume of the submarine if it has a length of 10 m and a diameter of 8 m. Use 3.14 for $\pi$ and round the answer to the nearest hundredth. (*Hint*: First find the length of the cylindrical section.)

# Skill Maintenance

**59.** Find the simple interest on $600 at 8% for $\frac{1}{2}$ yr.  [8.7a]

**60.** Find the simple interest on $5000 at 7% for $\frac{1}{2}$ yr.  [8.7a]

**61.** If 9 pens cost $8.01, how much would 12 pens cost?  [7.4a]

**62.** Solve: $9(x - 1) = 3x + 5$.  [5.7b]

**63.** Solve: $-5y + 3 = -12y - 4$.  [5.7b]

**64.** A barge travels 320 km in 15 days. At this rate, how far will it travel in 21 days?  [7.4a]

**65.** Evaluate $\frac{9}{5}C + 32$ for $C = 15$.  [4.7c]

**66.** Evaluate $\frac{5}{9}(F - 32)$ for $F = 50$.  [4.7c]

# Synthesis

**67.** 🔲 *Truck Rental.*   The storage compartment of a moving truck is 9.83 ft by 5.67 ft by 5.83 ft, with an "attic" measuring 1.5 ft by 5.67 ft by 2.5 ft. Find the total volume of the compartment.

**68.** *Tennis-Ball Packaging.*   Tennis balls are generally packaged in circular cylinders that hold 3 balls each. The diameter of a tennis ball is 6.5 cm. If the tennis balls were "liquid" and could be poured into the cylinder, how many tennis balls would the cylinder hold?

**69.** The volume of a ball is $36\pi$ cm$^3$. Find the dimensions of a rectangular box that is just large enough to hold the ball.

**70.** 🔲 The volume of a basketball is $2304\pi$ cm$^3$. Find the volume of a cube-shaped box that is just large enough to hold the ball.

**71.** 🔲 A 2-cm-wide stream of water passes through a 30-m-long garden hose. At the instant that the water is turned off, how many liters of water are in the hose? Use 3.141593 for $\pi$.

**72.** *Remarkable Feat.*   In 1982, Larry Walters captured the world's imagination by riding a lawn chair attached to 42 helium-filled weather balloons to an altitude of 16,000 ft, before using a BB gun to pop a few balloons and safely descend. Walters used balloons measuring approximately 7 ft in diameter. Find the total volume of the balloons used. Use $\frac{22}{7}$ for $\pi$.
Source: MarkBarry.com

**73.** *Conservation.*   Many people leave the water running while brushing their teeth. Suppose that one person wastes 32 oz of water in such a manner each day. How much water, in gallons, would that person waste in a week? in 30 days? in a year? If each of 310 million Americans wastes water this way, estimate how much water is wasted in a year.

**74.** The volumes occupied by all the world's gold and by a million dollar bills are described in Exercises 55 and 56. How many million dollar bills would fit into the volume of all the world's gold?

# Mid-Chapter Review

## Concept Reinforcement

Determine whether each statement is true or false.

_____ **1.** Distances that are measured in miles in the American system would probably be measured in meters in the metric system.   [9.1c]

_____ **2.** One meter is slightly more than one yard.   [9.1c]

_____ **3.** One kilometer is longer than one mile.   [9.1c]

_____ **4.** The area of a parallelogram with base 8 cm and height 5 cm is the same as the area of a rectangle with length 8 cm and width 5 cm.   [9.3a]

_____ **5.** The exact value of the ratio $C/d$ is $\pi$.   [9.3b]

## Guided Solutions

Fill in each blank with the number that creates a correct solution.

**6.** Complete:  $16\frac{2}{3}$ yd = _____ ft.   [9.1a]

$$16\frac{2}{3} \text{ yd} = 16\frac{2}{3} \cdot 1\ \square = \frac{50}{3} \cdot 3\ \square = 50\ \square$$

**7.** Complete:  10,200 mm = _____ ft.   [9.1c]

$$10{,}200 \text{ mm} = 10{,}200 \text{ mm} \cdot \frac{1\ \square}{1000\ \square} \approx 10.2\ \square \cdot \frac{3.281\ \square}{1\ \square} = 33.4662\ \square$$

**8.** Find the circumference and the area. Use 3.14 for $\pi$.   [9.3b]

10.2 in.

$$
\begin{array}{ll}
C = \pi \cdot d & A = \pi \cdot r \cdot r \\
C \approx \square \cdot \square \text{ in.} & A \approx \square \cdot \square \text{ in.} \cdot \square \text{ in.} \\
C \approx \square \text{ in.} & A \approx \square \text{ in}^{\square}
\end{array}
$$

## Mixed Review

Complete.   [9.1a, b, c], [9.2a, b]

**9.** $5\frac{1}{2}$ mi = _____ yd

**10.** 840 in. = _____ ft

**11.** 24.05 cm = _____ dm

**12.** 0.15 m = _____ km

**13.** 3753 ft = _____ yd

**14.** 26,400 ft = _____ mi

**15.** 1800 m = _____ cm

**16.** 0.007 km = _____ cm

**17.** 2.5 yd = _____ m

**18.** 36 m = _____ ft

**19.** 6 in.= _____ cm

**20.** 100 mi = _____ km

**21.** 5 ft$^2$ = _____ in$^2$

**22.** 10 yd$^2$ = _____ ft$^2$

**23.** 60 ft$^2$ = _____ yd$^2$

**24.** 48 in$^2$ = _____ ft$^2$

**25.** 2 acres = _____ ft$^2$

**26.** 1.5 mi$^2$ = _____ acres

**27.** Arrange from smallest to largest:

100 in.,   430 ft,   $\frac{1}{100}$ mi,   3.5 ft,   6000 ft,   1000 in.,   2 yd.   [9.1a]

**28.** Arrange from largest to smallest:

3240 cm,   300 m,   250 dm,   150 hm,   33,000 mm,   310 dam,   13 km.   [9.1b]

Find the area.   [9.3a]

**29.**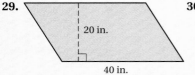
20 in.
40 in.

**30.**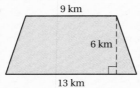
9 km
6 km
13 km

Find the circumference and the area. Use 3.14 for $\pi$.   [9.3b]

**31.**
7 in.

**32.**
8.6 cm

Find the volume. Use 3.14 for $\pi$ in Exercises 34–36.   [9.4a]

**33.**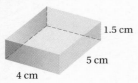
1.5 cm
5 cm
4 cm

**34.**
15 ft
6 ft

**35.**
3 in.
$\frac{1}{2}$ in.

**36.**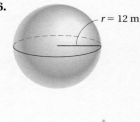
r = 12 m

Complete.   [9.4b]

**37.** 7.4 L = _____ mL

**38.** 62 cm$^3$ = _____ mL

**39.** 0.5 mL = _____ L

**40.** 3 gal = _____ qt

**41.** 18 cups = _____ pt

**42.** 32 oz = _____ cups

# Understanding Through Discussion and Writing

**43.** A student makes the following error:

23 in. = 23 · (12 ft) = 276 ft.

Explain the error.   [9.1a]

**44.** Explain why a 16-in.-diameter pizza that costs $16.25 is a better buy than a 10-in.-diameter pizza that costs $7.85.   [9.3b]

**45.** Explain how the area of a triangle can be found by considering the area of a parallelogram.   [9.3a]

**46.** The radius of one circle is twice the length of another circle's radius. Is the area of the first circle twice the area of the other circle? Why or why not?   [9.3b]

# 9.5 Angles and Triangles

## a Measuring Angles

We see a real-world application of *angles* of various types in the different back postures of the riders below.

### Style of Biking Determines Cycling Posture

**Road**
About 180° flat

Riders prefer a more aerodynamic flat-back position.

**Mountain**
About 45°

Riders prefer a semi-upright position to help lift the front wheel over obstacles.

**Comfort**
About 90°

Riders prefer an upright position that lessens stress on the lower back and neck.

SOURCE: USA TODAY research

An **angle** is a set of points consisting of two **rays**, or half-lines, with a common endpoint. The endpoint is called the **vertex**. The rays are called the **sides** of the angle.

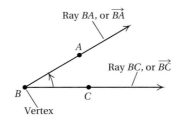

Ray *BA*, or $\vec{BA}$

Ray *BC*, or $\vec{BC}$

Vertex

The angle above can be named

angle *ABC*,     angle *CBA*,     angle *B*,     $\angle ABC$,     $\angle CBA$,     or     $\angle B$.

Note that the name of the vertex is either in the middle or, if no confusion results, listed by itself.

Do Exercises 1 and 2.

Name the angle in six different ways.

**1.**

*D*

*E*          *F*

**2.**

*P*

*Q*          *R*

***Answers***

**1.** Angle *DEF*, angle *FED*, angle *E*, $\angle DEF$, $\angle FED$, or $\angle E$     **2.** Angle *PQR*, angle *RQP*, angle *Q*, $\angle PQR$, $\angle RQP$, or $\angle Q$

**3.** Give another name for ∠1.

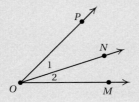

Angles may also be numbered. In the following figure, another name for ∠*ABC* is ∠1, and another name for ∠*CBD* is ∠2.

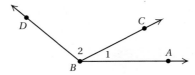

Do Exercise 3.

To measure angles, we start with some predetermined angle and assign to it a measure of 1. We call it a *unit angle*. Suppose that ∠*U* is a unit angle. Let's measure ∠*DEF*. If we made 3 copies of ∠*U*, they would "fill up" ∠*DEF*. Thus, the measure of ∠*DEF* would be 3 units.

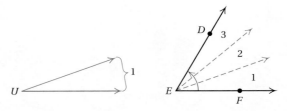

The unit most commonly used for angle measure is the degree. Below is such a unit. Its measure is 1 degree, or 1°.

A 1° angle:

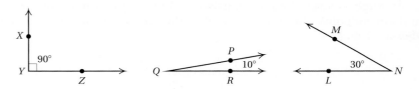

Here are some other angles with their degree measures.

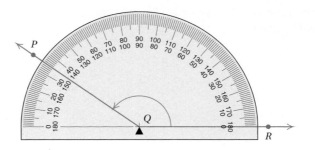

The symbol ⌐ is sometimes drawn on a figure to indicate a 90° angle.

A device called a **protractor** is used to measure angles. Protractors often have two scales. In the center of the protractor is a vertex indicator such as ▲ or a small hole. To measure an angle like ∠*Q* below, we place the protractor's ▲ at the vertex and line up one of the angle's sides at 0°. Then we check where the angle's other side crosses the scale. In the figure below, 0° is on the inside scale, so we check where the angle's other side crosses the inside scale. We see that *m*∠*Q* = 145°. The notation *m* ∠*Q* is read "the measure of angle *Q*."

**4.** Use a protractor to measure this angle.

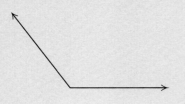

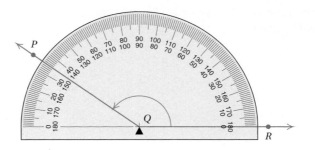

*Answers*

**3.** ∠*NOP* or ∠*PON*    **4.** 127°

Do Exercise 4.

Let's find the measure of ∠ABC. This time we will use the 0° on the outside scale. We see that $m \angle ABC = 68°$.

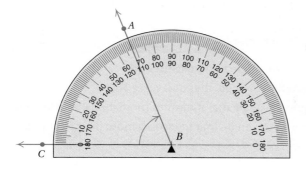

Do Exercise 5.

A protractor can be used to draw a circle graph.

**EXAMPLE 1** *Transportation.* According to a recent poll, 45% of adults believe that flying is the safest mode of transportation, 39% believe that cars are the safest, and 16% believe that trains are the safest. Draw a circle graph to represent these figures.

**Source:** *Marist Institute for Public Opinion*

Every circle graph contains a total of 360°. Thus,

    45% of the circle is a 0.45(360°), or 162°, angle;

    39% of the circle is a 0.39(360°), or 140.4°, angle;

    16% of the circle is a 0.16(360°), or 57.6°, angle.

We begin by drawing a 162° angle. Beginning at the center of the circle, we draw a horizontal segment to the circle. That segment is one side of the angle. We use a protractor to mark off a 162° angle. From that mark, we draw a segment to the center of the circle to complete the angle. This section of the circle graph we label with both the percent (45%) and the type of transportation (Airplanes).

From the second segment drawn, we repeat the above procedure to draw a 140.4° angle. Since protractors are marked in units of 1°, we must approximate this angle. This section we label with 39% and Cars.

The remainder of the circle represents Trains and should be a 57.6° angle; we measure to confirm this, and label the section with 16% and Trains.

Finally, we give a title to the graph: Safest Mode of Transportation.

**Safest Mode of Transportation**

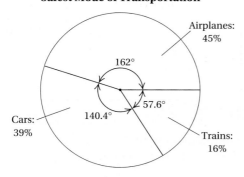

Do Exercise 6.

**5.** Use a protractor to measure this angle.

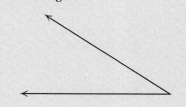

**6. Lengths of Engagement of Married Couples.** The data below list the percentages of married couples who were engaged for a certain time period before marriage. Use this information to draw a circle graph.

**Source:** Bruskin Goldring Research

| Less than 1 yr: | 24% |
|---|---|
| 1–2 yr: | 21% |
| More than 2 yr: | 35% |
| Never engaged: | 20% |

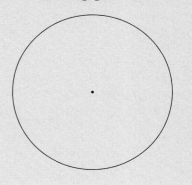

**Answers**

**5.** 33°
**6.** **Lengths of Engagement of Married Couples**

Arrangement of sections may vary.

Classify each angle as right, straight, acute, or obtuse. Use a protractor if necessary.

7.

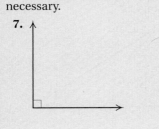

8.

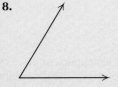

9.

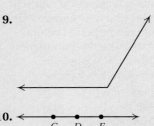

10.

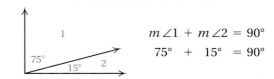

## b Classifying Angles

The following are ways in which we classify angles.

**TYPES OF ANGLES**

**Right angle:** An angle that measures 90°.

**Straight angle:** An angle that measures 180°.

**Acute angle:** An angle that measures more than 0° but less than 90°.

**Obtuse angle:** An angle that measures more than 90° but less than 180°.

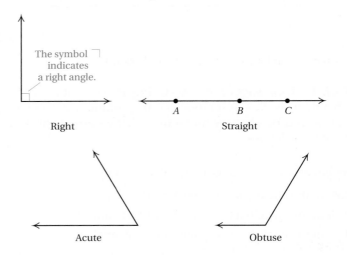

The symbol ⌐ indicates a right angle.

Right          Straight

Acute          Obtuse

Do Exercises 7-10.

## c Complementary, Supplementary, and Vertical Angles

Certain pairs of angles share special properties.

### Complementary Angles

When the sum of the measures of two angles is 90°, the angles are said to be **complementary**. For example, in the figure below, ∠1 and ∠2 are complementary. If two angles are complementary, each is an acute angle.

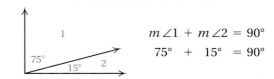

$$m \angle 1 + m \angle 2 = 90°$$
$$75° + 15° = 90°$$

**COMPLEMENTARY ANGLES**

Two angles are **complementary** if the sum of their measures is 90°. Each angle is called a **complement** of the other.

*Answers*

**7.** Right    **8.** Acute    **9.** Obtuse    **10.** Straight

When complementary angles are adjacent to each other (that is, they have a side in common), they form a right angle.

**EXAMPLE 2**   Identify each pair of complementary angles.

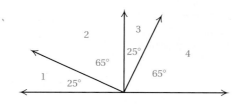

∠1  and  ∠2    25° + 65° = 90°        ∠2  and  ∠3
∠1  and  ∠4                           ∠3  and  ∠4

**EXAMPLE 3**   Find the measure of a complement of a 39° angle.

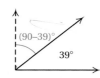

$$90° - 39° = 51°$$

The measure of a complement is 51°.

Do Exercises 11–14.

## Supplementary Angles

Next, consider ∠1 and ∠2 as shown below. Because the sum of their measures is 180°, ∠1 and ∠2 are said to be **supplementary**. Note that when supplementary angles are adjacent, they form a straight angle.

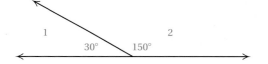

$$m∠1 + m∠2 = 180°$$
$$30° + 150° = 180°$$

---

**SUPPLEMENTARY ANGLES**

Two angles are **supplementary** if the sum of their measures is 180°. Each angle is called a **supplement** of the other.

---

**EXAMPLE 4**   Identify each pair of supplementary angles.

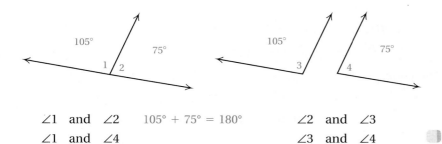

∠1  and  ∠2    105° + 75° = 180°        ∠2  and  ∠3
∠1  and  ∠4                             ∠3  and  ∠4

**11.** Identify each pair of complementary angles.

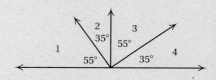

Find the measure of a complement of each angle.

**12.**

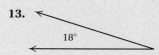

**13.**

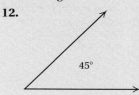

**14.**

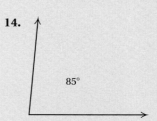

*Answers*
**11.** ∠1 and ∠2; ∠1 and ∠4; ∠2 and ∠3; ∠3 and ∠4   **12.** 45°   **13.** 72°   **14.** 5°

**15.** Identify each pair of supplementary angles.

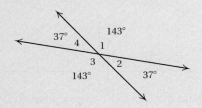

Find the measure of a supplement of an angle with each given measure.

**16.** 38°          **17.** 157°

**18.** 90°          **19.** 71°

**20.** Identify each pair of vertical angles.

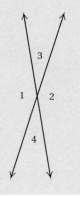

**EXAMPLE 5**   Find the measure of a supplement of an angle of 112°.

$$180° - 112° = 68°$$

The measure of a supplement is 68°.

Do Exercises 15–19.

## Vertical Angles

When two lines intersect, four angles are formed. The pairs of angles that do not share any side in common are said to be **vertical** (or *opposite*) angles. Thus, in the drawing below, ∠1 and ∠3 are vertical angles, as are ∠4 and ∠2. Note that $m \angle 1 = m \angle 3$ and $m \angle 4 = m \angle 2$.

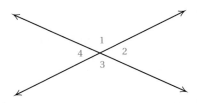

**EXAMPLE 6**   Identify each pair of vertical angles.

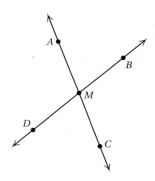

∠AMB and ∠CMD are vertical angles.

∠BMC and ∠DMA are vertical angles.

Do Exercise 20.

Note in the figure in Example 6 that ∠DMA and ∠AMB are supplementary; that is,

$$m \angle DMA + m \angle AMB = 180°.$$

Also, ∠CMD and ∠DMA are supplementary, so

$$m \angle CMD + m \angle DMA = 180°.$$

Therefore,

$$m \angle DMA + m \angle AMB = m \angle CMD + m \angle DMA$$

$$m \angle AMB = m \angle CMD. \quad \text{Subtracting } m \angle DMA \text{ from both sides}$$

This shows that vertical angles have the same measure.

*Answers*

**15.** ∠1 and ∠2; ∠1 and ∠4; ∠2 and ∠3; ∠3 and ∠4   **16.** 142°   **17.** 23°   **18.** 90°
**19.** 109°   **20.** ∠1 and ∠2; ∠3 and ∠4

## VERTICAL ANGLES

Two angles are **vertical** if they are formed by two intersecting lines and have no side in common. Vertical angles have the same measure.

> Do Exercise 21.

If two angles have the same measure, we say that they are **congruent**, denoted by the symbol ≅. In the figure below, the measures of angles *WVZ* and *XVY* are equal, and the angles are congruent:

$$m \angle WVZ = m \angle XVY$$
$$\angle WVZ \cong \angle XVY.$$

Note that we do not write that angles are equal: The *measures are equal* and the *angles are congruent*.

> Do Exercise 22.

## (d) Triangles

A **triangle** is a polygon made up of three segments, or sides. Consider these triangles. The triangle with vertices *A*, *B*, and *C* can be named △*ABC*.

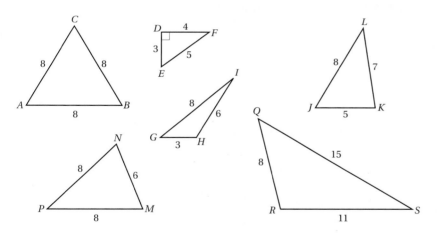

We can classify triangles according to sides and according to angles.

## TYPES OF TRIANGLES

**Equilateral triangle:** All sides are the same length.

**Isosceles triangle:** Two or more sides are the same length.

**Scalene triangle:** All sides are of different lengths.

**Right triangle:** One angle is a right angle.

**Obtuse triangle:** One angle is an obtuse angle.

**Acute triangle:** All three angles are acute.

> Do Exercises 23–26.

**21.** Complete:
$$m \angle ANC = \underline{\hspace{2cm}};$$
$$m \angle ANB = \underline{\hspace{2cm}}.$$

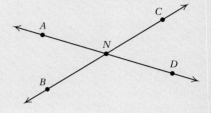

**22.** Complete:
$$\angle PMQ \cong \underline{\hspace{2cm}};$$
$$\angle SMP \cong \underline{\hspace{2cm}}.$$

**23.** Which triangles on this page are
  **a)** equilateral?
  **b)** isosceles?
  **c)** scalene?

**24.** Are all equilateral triangles isosceles?

**25.** Are all isosceles triangles equilateral?

**26.** Which triangles on this page are
  **a)** right triangles?
  **b)** obtuse triangles?
  **c)** acute triangles?

*Answers*

**21.** $m \angle BND$ or $m \angle DNB$; $m \angle CND$ or $m \angle DNC$
**22.** $\angle SMR$ or $\angle RMS$; $\angle QMR$ or $\angle RMQ$
**23.** **(a)** △*ABC*; **(b)** △*ABC*, △*MPN*;
**(c)** △*DEF*, △*GHI*, △*JKL*, △*QRS*
**24.** Yes  **25.** No  **26.** **(a)** △*DEF*;
**(b)** △*GHI*, △*QRS*; **(c)** △*ABC*, △*MPN*, △*JKL*

## e Sum of the Angle Measures of a Triangle

The sum of the angle measures of every triangle is 180°. To see this, note that we can think of cutting apart a triangle as shown on the left below. If we reassemble the pieces, we see that a straight angle is formed.

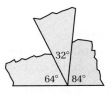

$$64° + 32° + 84° = 180°$$

**27.** Find $m \angle P + m \angle Q + m \angle R$.

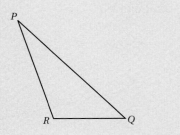

---

**SUM OF THE ANGLE MEASURES OF A TRIANGLE**

In any $\triangle ABC$, the sum of the measures of the angles is 180°:

$$m \angle A + m \angle B + m \angle C = 180°.$$

---

Do Exercise 27.

If we know the measures of two angles of a triangle, we can calculate the measure of the third angle.

**EXAMPLE 7** Find the missing angle measure.

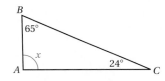

$$m \angle A + m \angle B + m \angle C = 180°$$
$$x + 65° + 24° = 180°$$
$$x + 89° = 180°$$
$$x = 180° - 89° \qquad \text{Subtracting 89° from both sides}$$
$$x = 91°$$

Thus, $m \angle A = 91°$.

Do Exercise 28.

**28.** Find the missing angle measure.

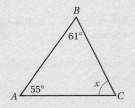

---

*Answers*

**27.** 180°     **28.** 64°

**a**   Name each angle in six different ways.

**1.**

**2.**

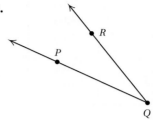

Give another name for ∠1 in each figure.

**3.**

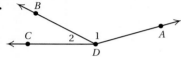

**4.**

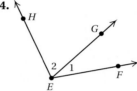

Use a protractor to measure each angle.

**5.**

**6.**

**7.**

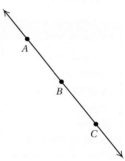

**8.**

**9.**

**10.**

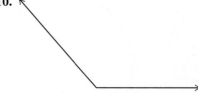

Use the given information and a protractor to draw a circle graph.

**11.** *Credit Score.* The FICO Credit Score is based on five types of information, as shown in the table below.

| PERSONAL INFORMATION | PERCENT |
|---|---|
| Payment history | 35% |
| Debt level | 30% |
| Length of credit history | 15% |
| Credit inquiries | 10% |
| Mix of credit | 10% |

SOURCE: credit.about.com

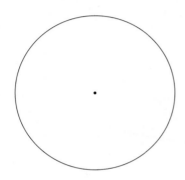

**12.** *Snacking Habits.* Some adults snack heavily and some not at all. The table below lists the frequency of snacking of American adults.

| FREQUENCY OF SNACKING | PERCENT |
|---|---|
| Never | 10% |
| Occasionally | 45% |
| Moderately | 35% |
| Heavily | 10% |

SOURCE: Market Facts for Hershey Foods

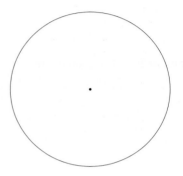

**13.** *Wind Energy.* The table below shows how wind-turbine capacity is distributed among countries.

| COUNTRY | PERCENT |
|---|---|
| United States | 22% |
| Germany | 16% |
| China | 16% |
| Spain | 12% |
| India | 7% |
| Rest of the world | 27% |

SOURCE: Zachary Shahan, "2009 Global Wind Power Report," 02/05/10, CleanTechnica.com

**14.** *Causes of Spinal Cord Injuries.* The table below lists the causes of spinal cord injury.

| CAUSES | PERCENT |
|---|---|
| Motor vehicle accidents | 44% |
| Acts of violence | 24% |
| Falls | 22% |
| Sports | 8% |
| Other | 2% |

SOURCE: National Spinal Cord Injury Association

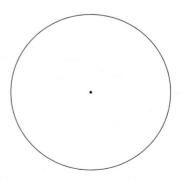

**15.–20.** Classify each of the angles in Exercises 5–10 as right, straight, acute, or obtuse.

**21.–24.** Classify each of the angles in Margin Exercises 1, 2, 4 and 5 as right, straight, acute, or obtuse.

 Identify two pairs of vertical angles for each figure.

**25.**

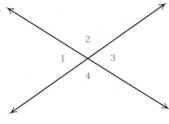

**26.**

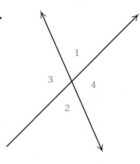

**27.**

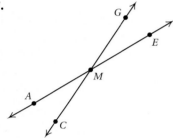

**28.**

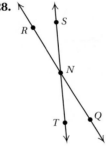

Complete.

**29.** Refer to Exercise 25.

      $m \angle 2 =$ _____

      $m \angle 3 =$ _____

**30.** Refer to Exercise 26.

      $m \angle 4 =$ _____

      $m \angle 2 =$ _____

**31.** Refer to Exercise 27.

      $\angle AMC \cong$ _____

      $\angle AMG \cong$ _____

**32.** Refer to Exercise 28.

      $\angle RNS \cong$ _____

      $\angle TNR \cong$ _____

Find the measure of a complement of an angle with the given measure.

**33.** 11°               **34.** 83°               **35.** 67°               **36.** 5°

**37.** 58°               **38.** 32°               **39.** 29°               **40.** 54°

Find the measure of a supplement of an angle with the given measure.

**41.** 3°               **42.** 54°               **43.** 139°               **44.** 13°

**45.** 75°               **46.** 128°               **47.** 104°               **48.** 49°

**d** Classify each triangle as equilateral, isosceles, or scalene. Then classify it as right, obtuse, or acute.

**49.**

**50.**

**51.**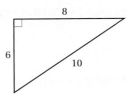

**52.**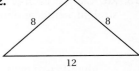

**53.**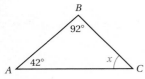

**54.**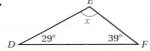

**55.**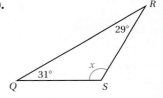

**56.**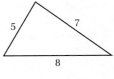

 **e** Find each missing angle measure.

**57.**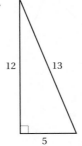

**58.**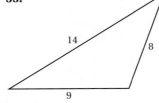

**59.**

**60.**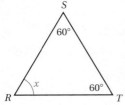

**61.**

**62.**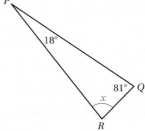

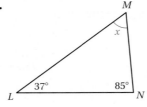 

## Skill Maintenance

Find the simple interest.   [8.7a]

|  | PRINCIPAL | RATE OF INTEREST | TIME | SIMPLE INTEREST |
|---|---|---|---|---|
| **63.** | $2000 | 8% | 1 year | |
| **64.** | $750 | 6% | $\frac{1}{2}$ year | |
| **65.** | $4000 | 7.4% | $\frac{1}{2}$ year | |
| **66.** | $200,000 | 6.7% | $\frac{1}{12}$ year | |

Interest is compounded semiannually. Find the amount in the account after the given length of time. Round to the nearest cent.   [8.7b]

|  | PRINCIPAL | RATE OF INTEREST | TIME | AMOUNT IN THE ACCOUNT |
|---|---|---|---|---|
| **67.** | $25,000 | 6% | 5 years | |
| **68.** | $150,000 | $6\frac{7}{8}$% | 15 years | |
| **69.** | $150,000 | 7.4% | 20 years | |
| **70.** | $160,000 | 7.4% | 20 years | |

## Synthesis

**71.** 🖩 In the figure, $m\angle 1 = 79.8°$ and $m\angle 3 = 33.07°$. Find $m\angle 2$, $m\angle 4$, $m\angle 5$, and $m\angle 6$.

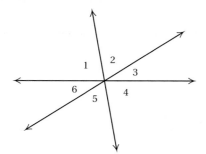

**72.** 🖩 In the figure, $m\angle 2 = 42.17°$ and $m\angle 6 = 81.9°$. Find $m\angle 1$, $m\angle 3$, $m\angle 4$, and $m\angle 5$.

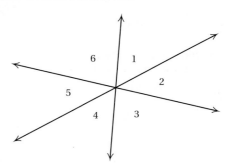

**73.** Find $m\angle ACB$, $m\angle CAB$, $m\angle EBC$, $m\angle EBA$, $m\angle AEB$, and $m\angle ADB$ in the rectangle shown below.

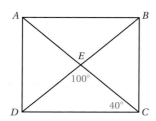

**74.** The angles in the figure are supplementary. Find the measure of each angle.

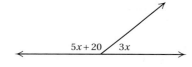

# 9.6

# Square Roots and the Pythagorean Theorem

## OBJECTIVES

**a** Simplify square roots of squares, such as $\sqrt{25}$.

**b** Approximate square roots.

**c** Given the lengths of any two sides of a right triangle, find the length of the third side.

**d** Solve applied problems involving right triangles.

---

Find each square.

1. $9^2$          2. $10^2$

3. $11^2$          4. $12^2$

5. $13^2$          6. $14^2$

7. $15^2$          8. $16^2$

Find all square roots. Use the results of Exercises 1–8 above, if necessary.

9. 100          10. 81

11. 49          12. 196

Simplify. Use the results of Exercises 1–8 above, if necessary.

13. $\sqrt{49}$          14. $\sqrt{16}$

15. $\sqrt{121}$          16. $\sqrt{100}$

17. $\sqrt{81}$          18. $\sqrt{64}$

19. $\sqrt{225}$          20. $\sqrt{169}$

21. $\sqrt{1}$          22. $\sqrt{0}$

Answers on p. 641

## a Square Roots

> **SQUARE ROOT**
>
> If a number is a product of a factor times itself, then that factor is a **square root** of the number. (If $c^2 = a$, then $c$ is a square root of $a$.)

The number 36 has two square roots, 6 and $-6$: $6 \cdot 6 = 36$ and $(-6) \cdot (-6) = 36$.

**EXAMPLE 1**  Find the square roots of 25.

The square roots of 25 are 5 and $-5$, because $5^2 = 25$ and $(-5)^2 = 25$.

---------------------------------- *Caution!* ----------------------------------

To find the *square* of a number, multiply the number by itself. To find a *square root* of a number, find a number that, when squared, gives the original number.

------------------------------------------------------------------------

Do Margin Exercises 1–12.

Every positive number has two square roots. However, the symbol $\sqrt{\phantom{n}}$ (called a **radical sign**) represents only the positive square root of the number underneath. Thus, $\sqrt{9}$ means 3, not $-3$.

> **RADICAL SIGN**
>
> If $n$ is a positive number, $\sqrt{n}$ means the positive square root of $n$.
>
> "The" square root of $n$ means $\sqrt{n}$.

**EXAMPLES**  Simplify.

**2.** $\sqrt{36} = 6$     The square root of 36 is 6 because $6^2 = 36$ and 6 is positive.

**3.** $\sqrt{144} = 12$     Note that $12^2 = 144$.

Do Exercises 13–22.

## b Approximating Square Roots

Many square roots can't be written as whole numbers or fractions. For example, $\sqrt{2}$ cannot be precisely represented in decimal notation. Each of the following gives a closer decimal approximation for $\sqrt{2}$, but none is exactly $\sqrt{2}$:

$$\sqrt{2} \approx 1.4 \qquad \text{because} \quad (1.4)^2 = 1.96;$$
$$\sqrt{2} \approx 1.41 \qquad \text{because} \quad (1.41)^2 = 1.9881;$$
$$\sqrt{2} \approx 1.414 \qquad \text{because} \quad (1.414)^2 = 1.999396.$$

Decimal approximations of square roots are commonly found by using a calculator.

**EXAMPLE 4**  Use a calculator to approximate $\sqrt{3}$, $\sqrt{27}$, and $\sqrt{180}$ to three decimal places.

We use a calculator to find each square root. Since the calculator displays more than three decimal places, we round back to three places.

$$\sqrt{3} \approx 1.732, \qquad \sqrt{27} \approx 5.196, \qquad \sqrt{180} \approx 13.416$$

As a check, note that $1 \cdot 1 = 1$ and $2 \cdot 2 = 4$, so we expect $\sqrt{3}$ to be between 1 and 2. Similarly, we expect $\sqrt{27}$ to be between 5 and 6 and $\sqrt{180}$ to be between 13 and 14.

> Do Exercises 23–26.

Use a calculator to approximate to three decimal places.

**23.** $\sqrt{5}$    **24.** $\sqrt{78}$

**25.** $\sqrt{168}$    **26.** $\sqrt{321}$

## (c) The Pythagorean Theorem

A **right triangle** is a triangle with a 90° angle, as shown here. In a right triangle, the longest side is called the **hypotenuse**. It is the side opposite the right angle. The other two sides are called **legs**. We generally use the letters $a$ and $b$ for the lengths of the legs and $c$ for the length of the hypotenuse. They are related as follows.

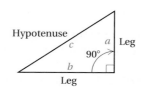

Hypotenuse — $c$    $a$  Leg
90°
$b$
Leg

---

### THE PYTHAGOREAN THEOREM

In any right triangle, if $a$ and $b$ are the lengths of the legs and $c$ is the length of the hypotenuse, then

$$a^2 + b^2 = c^2, \quad \text{or}$$

$$(\text{Leg})^2 + (\text{Other leg})^2 = (\text{Hypotenuse})^2.$$

The equation $a^2 + b^2 = c^2$ is called the **Pythagorean equation**.*

$c$    $a$
$b$

---

### 🖩 Calculator Corner

**Finding Square Roots**    Many calculators have a square root key, $\boxed{\sqrt{\ }}$. This is often the second function associated with the $\boxed{x^2}$ key, in which case it is accessed by pressing $\boxed{\text{2nd}}$ or $\boxed{\text{SHIFT}}$ followed by $\boxed{x^2}$. On some calculators, we enter the number under the $\sqrt{\ }$ sign first followed by the $\boxed{\sqrt{\ }}$ key. To find $\sqrt{30}$, for example, we press $\boxed{3}\ \boxed{0}\ \boxed{\sqrt{\ }}\ \boxed{=}$. On other calculators, we might press $\boxed{\sqrt{\ }}\ \boxed{3}\ \boxed{0}$ $\boxed{\text{ENTER}}$. In either case, the value 5.477225575 appears.

It is always best to wait until calculations are complete before rounding. For example, to find $9 \cdot \sqrt{30}$ rounded to the nearest tenth, we do not first determine that $\sqrt{30} \approx 5.5$ and then multiply by 9 to get 49.5. Rather, we press $\boxed{9}\ \boxed{\times}\ \boxed{3}\ \boxed{0}\ \boxed{\sqrt{\ }}\ \boxed{=}$, or $\boxed{9}\ \boxed{\times}\ \boxed{\sqrt{\ }}$ $\boxed{3}\ \boxed{0}\ \boxed{\text{ENTER}}$. The result is 49.29503018, so $9 \cdot \sqrt{30} \approx 49.3$.

**Exercises:**    Use a calculator to find each of the following. Round to the nearest tenth.

**1.** $\sqrt{43}$    **2.** $\sqrt{94}$    **3.** $7 \cdot \sqrt{8}$

**4.** $5 \cdot \sqrt{12}$    **5.** $\sqrt{35} + 19$    **6.** $17 + \sqrt{57}$

**7.** $13\sqrt{68} + 14$    **8.** $24 \cdot \sqrt{31} - 18$    **9.** $7 \cdot \sqrt{90} + 3 \cdot \sqrt{40}$

---

*The *converse* of the Pythagorean theorem is also true. That is, if $a^2 + b^2 = c^2$, then the triangle is a right triangle.

The Pythagorean theorem is named for the Greek mathematician Pythagoras (569?–500? B.C.). We can think of the relationship as adding areas.

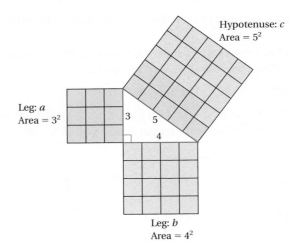

$$a^2 + b^2 = c^2$$
$$3^2 + 4^2 = 5^2$$
$$9 + 16 = 25$$

If we know the lengths of any two sides of a right triangle, we can use the Pythagorean equation to determine the length of the third side.

**EXAMPLE 5** Find the length of the hypotenuse of this right triangle.

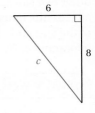

We substitute in the Pythagorean equation:

$$a^2 + b^2 = c^2$$
$$6^2 + 8^2 = c^2 \qquad \text{Substituting}$$
$$36 + 64 = c^2$$
$$100 = c^2.$$

The solution of this equation is the square root of 100, which is 10:

$$c = \sqrt{100} = 10.$$

Do Exercise 27.

**27.** Find the length of the hypotenuse of this right triangle.

----------------------------- *Caution!* -----------------------------

Before applying the Pythagorean theorem, determine which side of the triangle is the hypotenuse, or $c$. The hypotenuse is always the longest side of a right triangle and is always opposite the right angle.

----------------------------------------------------------------

**EXAMPLE 6** Find the length $b$ for the right triangle shown. Give an exact answer and an approximation to three decimal places.

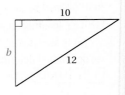

Recall that the leg opposite the right angle is the hypotenuse. Thus, for this triangle, $c = 12$. We substitute in the Pythagorean equation:

$$a^2 + b^2 = c^2$$
$$10^2 + b^2 = 12^2 \qquad \text{Substituting}$$
$$100 + b^2 = 144.$$

*Answer*

**27.** $c = 13$

Next, we solve for $b^2$ and then $b$:

$$100 + b^2 - 100 = 144 - 100 \qquad \text{Subtracting 100 from both sides}$$
$$b^2 = 144 - 100$$
$$b^2 = 44 \qquad \text{Solving for } b^2$$

*Exact answer:* $\quad b = \sqrt{44} \qquad$ Solving for $b$

*Approximation:* $\quad b \approx 6.633.$ $\qquad$ Using a calculator

> Do Exercises 28–30.

## (d) Applications

**EXAMPLE 7** *Height of Ladder.* A 12-ft ladder leans against a building. The bottom of the ladder is 7 ft from the building. How high is the top of the ladder? Give an exact answer and an approximation to the nearest tenth of a foot.

1. **Familiarize.** We first make a drawing. In it we see a right triangle. We let $h =$ the unknown height. We also note that the ladder forms the hypotenuse of the triangle, so $c = 12$.

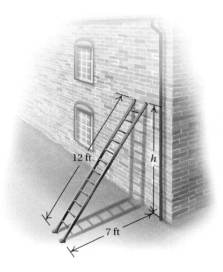

12 ft
h
7 ft

2. **Translate.** We substitute 7 for $a$, $h$ for $b$, and 12 for $c$ in the Pythagorean equation:

$$a^2 + b^2 = c^2 \qquad \text{Pythagorean equation}$$
$$7^2 + h^2 = 12^2.$$

3. **Solve.** We solve for $h^2$ and then $h$:

$$49 + h^2 = 144 \qquad 7^2 = 49 \text{ and } 12^2 = 144$$
$$49 + h^2 - 49 = 144 - 49 \qquad \text{Subtracting 49 from both sides}$$
$$h^2 = 144 - 49$$
$$h^2 = 95$$

*Exact answer:* $\quad h = \sqrt{95} \qquad$ Solving for $h$

*Approximation:* $\quad h \approx 9.7$ ft.

4. **Check.** $7^2 + (\sqrt{95})^2 = 49 + 95 = 144 = 12^2.$

5. **State.** The top of the ladder is $\sqrt{95}$ ft, or about 9.7 ft, from the ground.

> Do Exercise 31.

---

For each right triangle, find the length of the leg not given. Give an exact answer and an approximation to three decimal places.

**28.**

14
11
$a$

**29.**

1
11
$b$

**30.**

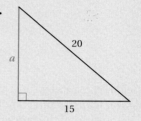

20
$a$
15

**31.** How long is a guy wire reaching from the top of an 18-ft pole to a point on the ground 10 ft from the pole? Give an exact answer and an approximation to the nearest tenth of a foot.

18 ft
$c = ?$
10 ft

*Answers*

**28.** $a = \sqrt{75}$; $a \approx 8.660$ **29.** $b = \sqrt{120}$;
$b \approx 10.954$ **30.** $a = \sqrt{175}$; $a \approx 13.229$
**31.** $\sqrt{424}$ ft $\approx 20.6$ ft

# Translating for Success

1. *Servings of Pork.* An 8-lb pork roast contains 37 servings of meat. How many pounds of pork would be needed for 55 servings?

2. *Height of a Ladder.* A 14.5-ft ladder leans against a house. The bottom of the ladder is 9.4 ft from the house. How high is the top of the ladder?

3. *Cruise Cost.* A group of 6 college students pays $4608 for a spring break cruise. What is each person's share?

4. *Sales Tax Rate.* The sales tax is $14.95 on the purchase of a new ladder that costs $299. What is the sales tax rate?

5. *Volume of a Sphere.* Find the volume of a sphere whose radius is 7.2 cm.

---

The goal of these matching questions is to practice step (2), *Translate*, of the five-step problem-solving process. Translate each word problem to an equation and select a correct translation from equations A–O.

**A.** $x = 50.5 \text{ km} \cdot \dfrac{1000 \text{ m}}{1 \text{ km}}$

**B.** $6 \cdot x = \$4608$

**C.** $x = \dfrac{4}{3} \cdot \pi \cdot 6^2 \cdot (7.2)$

**D.** $x = \pi \cdot \left(5\dfrac{1}{2} \div 2\right)^2 \cdot 7$

**E.** $x = 6\% \times 5 \times \$14.95$

**F.** $x = \pi \cdot 5\dfrac{1}{2} \cdot 7$

**G.** $(9.4)^2 + x^2 = (14.5)^2$

**H.** $\$14.95 = x \cdot \$299$

**I.** $x = 2(14.5 + 9.4)$

**J.** $(9.4 + 14.5)^2 = x$

**K.** $\dfrac{8}{37} = \dfrac{x}{55}$

**L.** $x = 50.5 \text{ km} \cdot \dfrac{1 \text{ km}}{1000 \text{ m}}$

**M.** $x = 6 \cdot \$4608$

**N.** $8 \cdot 37 = 55 \cdot x$

**O.** $x = \dfrac{4}{3}\pi(7.2)^3$

*Answers on page A-19*

---

6. *Inheritance.* Six children each inherit $4608 from their mother's estate. What is the total inheritance?

7. *Sales Tax.* Erica buys 5 pairs of earrings at $14.95 each. The sales tax rate is 6%. How much sales tax will be charged?

8. *Tunnel Length.* The Channel Tunnel linking England and France is 50.5 km long. Convert this distance to meters.

9. *Volume of a Storage Tank.* The diameter of a cylindrical grain-storage tank is $5\frac{1}{2}$ yd. Its height is 7 yd. Find its volume.

10. *Perimeter of a Photo.* A rectangular photo is 14.5 cm by 9.4 cm. What is the perimeter of the photo?

**a** Find both square roots for each number listed.

**1.** 16

**2.** 9

**3.** 121

**4.** 49

**5.** 169

**6.** 144

**7.** 2500

**8.** 3600

Simplify.

**9.** $\sqrt{64}$

**10.** $\sqrt{4}$

**11.** $\sqrt{81}$

**12.** $\sqrt{49}$

**13.** $\sqrt{225}$

**14.** $\sqrt{121}$

**15.** $\sqrt{625}$

**16.** $\sqrt{900}$

**17.** $\sqrt{400}$

**18.** $\sqrt{169}$

**19.** $\sqrt{10,000}$

**20.** $\sqrt{1,000,000}$

**b** Approximate each number to the nearest thousandth. Use a calculator.

**21.** $\sqrt{48}$

**22.** $\sqrt{17}$

**23.** $\sqrt{8}$

**24.** $\sqrt{7}$

**25.** $\sqrt{3}$

**26.** $\sqrt{6}$

**27.** $\sqrt{12}$

**28.** $\sqrt{18}$

**29.** $\sqrt{19}$

**30.** $\sqrt{75}$

**31.** $\sqrt{110}$

**32.** $\sqrt{10}$

**c** Find the length of the third side of each right triangle. Give an exact answer and, when appropriate, an approximation to the nearest thousandth.

**33.**

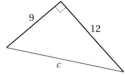

**34.**

**35.**

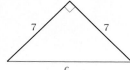

**36.**

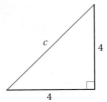

c, 4, 4

**37.**

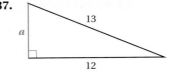

a, 13, 12

**38.**

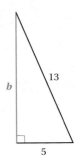

b, 13, 5

**39.**

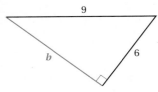

9, b, 6

**40.**

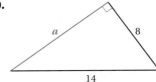

a, 8, 14

For each right triangle, find the length of the side not given. Assume that *c* represents the length of the hypotenuse. Give an exact answer and, when appropriate, an approximation to the nearest thousandth.

**41.** $a = 10, b = 24$

**42.** $a = 5, b = 12$

**43.** $a = 9, c = 15$

**44.** $a = 18, c = 30$

**45.** $a = 4, b = 5$

**46.** $a = 5, b = 6$

**47.** $a = 1, c = 32$

**48.** $b = 1, c = 20$

(**d**) In Exercises 49–56, give an exact answer and an approximation to the nearest tenth.

**49.** How long is a string of lights reaching from the top of a 12-ft pole to a point 8 ft from the base of the pole?

**50.** How long must a wire be in order to reach from the top of a 13-m telephone pole to a point on the ground 9 m from the base of the pole?

**51.** *Softball Diamond.* A slow-pitch softball diamond is actually a square 65 ft on a side. How far is it from home plate to second base?

**52.** *Baseball Diamond.* A baseball diamond is actually a square 90 ft on a side. How far is it from home plate to second base?

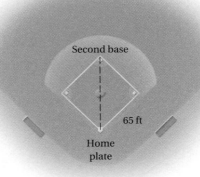

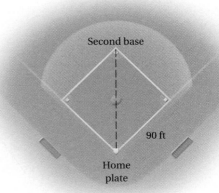

**53.** How tall is this tree?

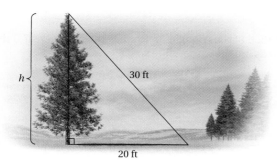

**54.** How far is the base of the fence post from point *A*?

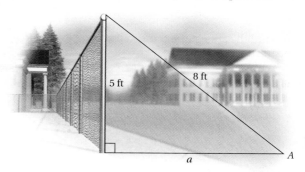

**55.** An airplane is flying at an altitude of 4100 ft. The slanted distance directly to the airport is 15,100 ft. How far is the airplane horizontally from the airport?

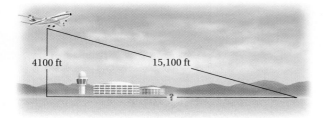

**56.** A surveyor had poles located at points *P*, *Q*, and *R* around a lake. The distances that the surveyor was able to measure are marked on the drawing. What is the distance from *P* to *R* across the lake?

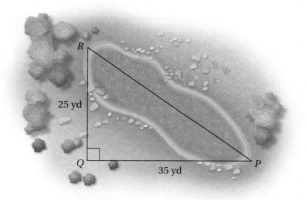

# Skill Maintenance

Solve.

**57.** Food expenses account for 26% of the average family's budget. A family makes $1800 one month. How much do they spend for food?   [8.4a]

**58.** Blakely County has a population that is increasing by 4% each year. This year the population is 180,000. What will it be next year?   [8.5a]

**59.** Dexter College has a student body of 1850 students. Of these, 17.5% are seniors. How many students are seniors?   [8.4a]

**60.** The price of a cell phone was reduced from $70 to $61.60. Find the percent of decrease in price.   [8.5a]

Simplify.   [1.9b]

**61.** $2^3$

**62.** $5^3$

**63.** $10^3$

**64.** $10^4$

# Synthesis

**65.** 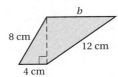 Find the area of the trapezoid shown. Round to the nearest hundredth.

8 cm

*b*

12 cm

4 cm

**66.** Which of the triangles below has the larger area? If the areas are the same, state so.

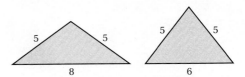

5    5

8

5    5

6

**67.** 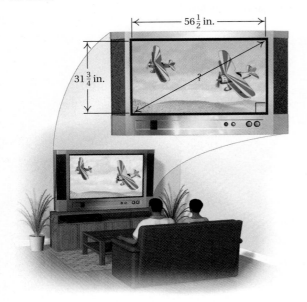 Caiden's new TV has a screen that measures $31\frac{3}{4}$ in. by $56\frac{1}{2}$ in. Determine the diagonal measurement of the screen. Round your answer to the nearest tenth of an inch.

$56\frac{1}{2}$ in.

$31\frac{3}{4}$ in.

?

**68.** A Philips 42-in. plasma television has a rectangular screen that measures 42 in. diagonally. The ratio of width to height is 16 to 9. Find the width and the height of the screen.

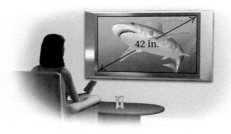

42 in.

**69.** A conventional 19-in. television set has a rectangular screen that measures 19 in. diagonally. The ratio of width to height in a conventional television set is 4 to 3. Find the width and the height of the screen.

19 in.

**70.** 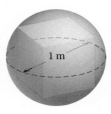 A cube is circumscribed by a sphere with a 1-m diameter. How much more volume is in the sphere?

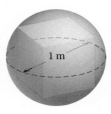

1 m

# 9.7

# Weight, Mass, and Temperature

## a) Weight: The American System

The American units of weight are as follows.

> **AMERICAN UNITS OF WEIGHT**
>
> $1\text{ lb} = 16\text{ ounces (oz)}$
> $1\text{ ton (T)} = 2000\text{ pounds (lb)}$

> **OBJECTIVES**
>
> **a** Convert from one American unit of weight to another.
>
> **b** Convert from one metric unit of mass to another.
>
> **c** Convert temperatures from Celsius to Fahrenheit and from Fahrenheit to Celsius.

The term "ounce" used here for weight is different from the "ounce" we used for capacity in Section 9.4. Often, however, 1 fluid ounce weighs approximately 1 ounce.

**EXAMPLE 1** A well-known hamburger is called a "quarter-pounder." Find its name in ounces: a "_____ ouncer."

$$\frac{1}{4}\text{ lb} = \frac{1}{4}\cdot 1\text{ lb}$$

$$= \frac{1}{4}\cdot 16\text{ oz} \qquad \text{Substituting 16 oz for 1 lb}$$

$$= 4\text{ oz}$$

A "quarter-pounder" could also be called a "four-ouncer."

**EXAMPLE 2** Complete: $15{,}360\text{ lb} = \text{_____ T}$.

$$15{,}360\text{ lb} = 15{,}360\,\cancel{\text{lb}}\cdot\frac{1\text{ T}}{2000\,\cancel{\text{lb}}} \qquad \text{Multiplying by 1}$$

$$= \frac{15{,}360}{2000}\text{ T}$$

$$= 7.68\text{ T} \qquad \text{Dividing by 2000}$$

> Do Margin Exercises 1–3.

## b) Mass: The Metric System

There is a difference between **mass** and **weight**, but the terms are often used interchangeably. People sometimes use the word "weight" when, technically, they are referring to "mass." Weight is related to the force of the earth's gravity. The farther you are from the center of the earth, the less you weigh. Your mass, on the other hand, stays the same no matter where you are.

The basic unit of mass is the **gram** (g), which is the mass of 1 cubic centimeter ($1\text{ cm}^3$ or $1\text{ mL}$) of water. Since a cubic centimeter is small, a gram is a small unit of mass.

$$1\text{ g} = 1\text{ gram} = \text{the mass of } 1\text{ cm}^3 \text{ of water}$$

> **SKILL TO REVIEW**
> Objective 3.4a: Multiply an integer and a fraction.
>
> Multiply.
>
> **1.** $\frac{3}{4}\cdot 16$
>
> **2.** $2480 \times \frac{1}{2000}$

Complete.

**1.** $5\text{ lb} = \text{_____ oz}$

**2.** $8640\text{ lb} = \text{_____ T}$

**3.** $1\text{ T} = \text{_____ oz}$

$1\text{ g} = 1\text{ cm}^3$ of water

***Answers***

*Skill to Review:*

**1.** 12  **2.** $\frac{31}{25}$, or 1.24

*Margin Exercises:*

**1.** 80  **2.** 4.32  **3.** 32,000

## METRIC UNITS OF MASS

1 metric ton (t) = 1000 kilograms (kg)

1 *kilo*gram (kg) = 1000 grams (g)

1 *hecto*gram (hg) = 100 grams (g)

1 *deka*gram (dag) = 10 grams (g)

1 gram (g)

1 *deci*gram (dg) = $\frac{1}{10}$ gram (g)

1 *centi*gram (cg) = $\frac{1}{100}$ gram (g)

1 *milli*gram (mg) = $\frac{1}{1000}$ gram (g)

The table at left lists the metric units of mass. The prefixes are the same as those for length.

### Thinking Metric

One gram is about the mass of 1 raisin or 1 package of artificial sweetener. Since 1 kg is about 2.2 lb, 1000 kg is about 2200 lb, or 1 metric ton (t), which is about 10% more than 1 American ton (T).

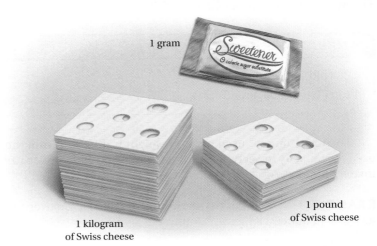

1 gram

1 kilogram of Swiss cheese

1 pound of Swiss cheese

Small masses, such as dosages of medicine and vitamins, may be measured in milligrams (mg). The gram (g) is used for objects ordinarily measured in ounces, such as the mass of a letter, a piece of candy, a coin, or a small package of food.

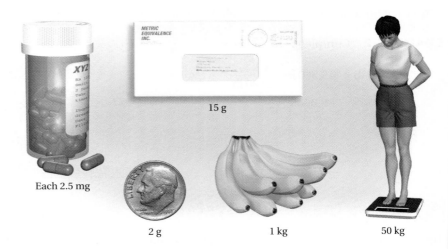

Each 2.5 mg

15 g

2 g

1 kg

50 kg

Complete with mg, g, kg, or t.

**4.** A laptop computer has a mass of 6 _____.

**5.** Eric has a body mass of 85.4 _____.

**6.** This is a 3-_____ vitamin.

**7.** A pen has a mass of 12 _____.

**8.** A sport utility vehicle has a mass of 3 _____.

The kilogram (kg) is used for larger food packages, such as fruit, or for human body mass. The metric ton (t) is used for very large masses, such as the mass of an automobile, a truckload of gravel, or an airplane.

Do Exercises 4–8.

*Answers*

**4.** kg  **5.** kg  **6.** mg  **7.** g  **8.** t

## Changing Units Mentally

As before, changing from one metric unit to another amounts to only the movement of a decimal point. We use this table.

| 1000 g | 100 g | 10 g | 1 g | 0.1 g | 0.01 g | 0.001 g |
|--------|-------|------|-----|-------|--------|---------|
| 1 kg | 1 hg | 1 dag | 1 g | 1 dg | 1 cg | 1 mg |

**EXAMPLE 3** Complete: 8 kg = _____ g.

*Think*: A kilogram is 1000 times the mass of a gram. To go from kg to g in the table is a move of three places to the right. Thus, we move the decimal point three places to the right.

| 1000 g | 100 g | 10 g | 1 g | 0.1 g | 0.01 g | 0.001 g |
|--------|-------|------|-----|-------|--------|---------|
| 1 kg | 1 hg | 1 dag | 1 g | 1 dg | 1 cg | 1 mg |

3 places to the right

8.0     8.000.     8 kg = 8000 g

**EXAMPLE 4** Complete: 4235 g = _____ kg.

*Think*: There are 1000 grams in 1 kilogram. To go from g to kg in the table is a move of three places to the left. Thus, we move the decimal point three places to the left.

| 1000 g | 100 g | 10 g | 1 g | 0.1 g | 0.01 g | 0.001 g |
|--------|-------|------|-----|-------|--------|---------|
| 1 kg | 1 hg | 1 dag | 1 g | 1 dg | 1 cg | 1 mg |

3 places to the left

4235.0     4.235.0     4235 g = 4.235 kg

**EXAMPLE 5** Complete: 6.98 cg = _____ mg.

*Think*: One centigram has the mass of 10 milligrams. To go from cg to mg is a move of one place to the right. Thus, we move the decimal point one place to the right.

| 1000 g | 100 g | 10 g | 1 g | 0.1 g | 0.01 g | 0.001 g |
|--------|-------|------|-----|-------|--------|---------|
| 1 kg | 1 hg | 1 dag | 1 g | 1 dg | 1 cg | 1 mg |

1 place to the right

6.98     6.9.8     6.98 cg = 69.8 mg

> The most commonly used metric units of mass are kg, g, and mg. We have intentionally used those more often than the others in the exercises.

Do Exercises 9–12.

Complete.

**9.** 6.2 kg = _____ g

**10.** 304.8 cg = _____ g

**11.** 7.7 cg = _____ mg

**12.** 2344 mg = _____ cg

*Answers*

**9.** 6200   **10.** 3.048   **11.** 77   **12.** 234.4

## c Temperature

### Estimated Conversions

Below are two temperature scales: **Fahrenheit** for American measure and **Celsius**, used internationally and in science.

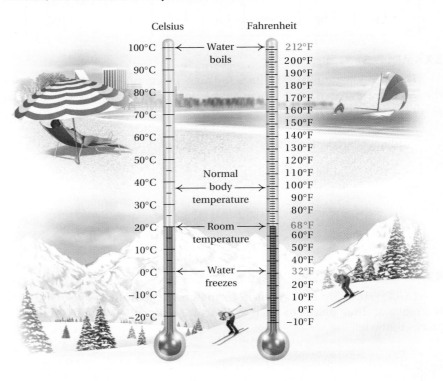

By laying a straightedge horizontally between the scales, we can make an approximate conversion from one measure of temperature to another and get an idea of how the temperature scales compare.

**EXAMPLES** Convert to Celsius using the scales shown above. Approximate to the nearest ten degrees.

| | | | |
|---|---|---|---|
| **6.** | 212°F (Boiling point of water) | 100°C | This is exact. |
| **7.** | 32°F (Freezing point of water) | 0°C | This is exact. |
| **8.** | 105°F | 40°C | This is approximate. |

Do Exercises 13–15.

**EXAMPLES** Make an approximate conversion to Fahrenheit.

| | | | |
|---|---|---|---|
| **9.** | 44°C (Hot bath) | 110°F | This is approximate. |
| **10.** | 20°C (Room temperature) | 68°F | This is exact. |
| **11.** | 83°C | 180°F | This is approximate. |

Do Exercises 16–18.

---

Convert to Celsius. Use a straightedge and the scales shown on this page. Approximate to the nearest ten degrees.

**13.** 180°F (Brewing coffee)

**14.** 25°F (Cold day)

**15.** −10°F (Miserably cold day)

Convert to Fahrenheit. Use a straightedge and the scales shown on this page. Approximate to the nearest ten degrees.

**16.** 25°C (Nice day at the park)

**17.** 40°C (Temperature of a patient with a high fever)

**18.** 10°C (Cold bath)

*Answers*

**13.** 80°C   **14.** 0°C   **15.** −20°C
**16.** 80°F   **17.** 100°F   **18.** 50°F

## Exact Conversions

A formula allows us to make exact conversions from Celsius to Fahrenheit.

---

### CELSIUS TO FAHRENHEIT

$$F = \frac{9}{5} \cdot C + 32, \quad \text{or} \quad F = 1.8 \cdot C + 32$$

$$\left( \text{Multiply the Celsius temperature by } \frac{9}{5}, \text{ or } 1.8, \text{ and add } 32. \right)$$

---

**EXAMPLES**  Convert to Fahrenheit.

**12.** 0°C (Freezing point of water)

$$F = \frac{9}{5} \cdot C + 32 = \frac{9}{5} \cdot 0 + 32 = 0 + 32 = 32$$

Thus, 0°C = 32°F.

**13.** 37°C (Normal body temperature)

$$F = 1.8 \cdot C + 32 = 1.8 \cdot 37 + 32 = 66.6 + 32 = 98.6$$

Thus, 37°C = 98.6°F.

Check the answers to Examples 12 and 13 using the scales on p. 652.

Do Exercises 19 and 20.

A second formula allows us to make exact conversions from Fahrenheit to Celsius.

---

### FAHRENHEIT TO CELSIUS

$$C = \frac{5}{9} \cdot (F - 32), \quad \text{or} \quad C = \frac{F - 32}{1.8}$$

$$\left( \text{Subtract } 32 \text{ from the Fahrenheit temperature and multiply by } \frac{5}{9} \text{ or divide by } 1.8. \right)$$

---

**EXAMPLES**  Convert to Celsius.

**14.** 212°F (Boiling point of water)

$$C = \frac{5}{9} \cdot (F - 32) = \frac{5}{9} \cdot (212 - 32) = \frac{5}{9} \cdot 180 = 100$$

Thus, 212°F = 100°C.

**15.** 77°F

$$C = \frac{F - 32}{1.8} = \frac{77 - 32}{1.8} = \frac{45}{1.8} = 25$$

Thus, 77°F = 25°C.

Check the answers to Examples 14 and 15 using the scales on p. 652.

Do Exercises 21 and 22.

---

Convert to Fahrenheit.
**19.** 80°C

**20.** 35°C

---

### Calculator Corner

**Temperature Conversions**  Temperature conversions can be done quickly using a calculator. To convert 37°C to Fahrenheit, for example, we press [1] [.] [8] [×] [3] [7] [+] [3] [2] [=]. The calculator displays [ 98.6 ], so 37°C = 98.6°F. We can convert 212°F to Celsius by pressing [(] [2] [1] [2] [−] [3] [2] [)] [÷] [1] [.] [8] [=]. The display reads [ 100 ], so 212°F = 100°C. Note that we must use parentheses when converting from Fahrenheit to Celsius in order to get the correct result.

**Exercises:**  Use a calculator to convert each temperature to Fahrenheit.

**1.** 5°C
**2.** 50°C

Use a calculator to convert each temperature to Celsius.

**3.** 68°F
**4.** 113°F

---

Convert to Celsius.
**21.** 95°F

**22.** 113°F

---

*Answers*
**19.** 176°F  **20.** 95°F  **21.** 35°C  **22.** 45°C

**9.7** **Exercise Set**

For Extra Help

**MyMathLab**

Math XL
PRACTICE

WATCH

DOWNLOAD

READ

REVIEW

**a** Complete.

**1.** 1 T = _____ lb

**2.** 1 lb = _____ oz

**3.** 6000 lb = _____ T

**4.** 8 T = _____ lb

**5.** 4 lb = _____ oz

**6.** 10 lb = _____ oz

**7.** 6.32 T = _____ lb

**8.** 8.07 T = _____ lb

**9.** 4800 lb = _____ T

**10.** 7500 lb = _____ T

**11.** 80 oz = _____ lb

**12.** 960 oz = _____ lb

**13.** *Excelsior.* Western Excelsior is a company that makes a packing material called excelsior. Excelsior is produced from the wood of aspen trees. The largest grove of aspen trees on record contained 13,000,000 tons of aspen. How many pounds of aspen were there?

**Source:** Western Excelsior Corporation, Mancos, Colorado

**14.** *Excelsior.* Western Excelsior buys 44,800 tons of aspen each year to make excelsior. How many pounds of aspen does it buy each year?

**Source:** Western Excelsior Corporation, Mancos, Colorado

**b** Complete.

**15.** 1 kg = _____ g

**16.** 1 hg = _____ g

**17.** 1 g = _____ kg

**18.** 1 dg = _____ g

**19.** 1 cg = _____ g

**20.** 1 mg = _____ g

**21.** 1 g = _____ mg

**22.** 1 g = _____ cg

**23.** 1 g = _____ dg

**24.** 25 kg = _____ g

**25.** 234 kg = _____ g

**26.** 9403 g = _____ kg

**27.** 5200 g = _____ kg

**28.** 1.506 kg = _____ g

**29.** 897 mg = _____ kg

**30.** 45 cg = _____ g

**31.** 7.32 kg = _____ g

**32.** 0.0025 cg = _____ mg

**33.** 8492 g = _____ kg

**34.** 9466 g = _____ kg

**35.** 585 mg = _____ cg

**36.** 96.1 mg = _____ cg

**37.** 8 kg = _____ cg

**38.** 0.06 kg = _____ mg

**39.** 1 t = _____ kg

**40.** 2 t = _____ kg

**41.** 3.4 cg = _____ dag

**42.** 115 mg = _____ g

**43.** 60.3 kg = _____ t

**44.** 15.68 kg = _____ t

**C** Convert to Celsius. Round the answer to the nearest ten degrees. Use the scales on p. 652.

**45.** 178°F

**46.** 195°F

**47.** 140°F

**48.** 107°F

**49.** 68°F

**50.** 45°F

**51.** 10°F

**52.** 120°F

Convert to Fahrenheit. Round the answer to the nearest ten degrees. Use the scales on p. 652.

**53.** 80°C

**54.** 93°C

**55.** 58°C

**56.** 33°C

**57.** −10°C

**58.** −5°C

**59.** 5°C

**60.** 15°C

Convert to Fahrenheit. Use the formula $F = \frac{9}{5} \cdot C + 32$.

**61.** 30°C           **62.** 85°C           **63.** 40°C           **64.** 90°C

**65.** 2°C           **66.** 8°C           **67.** −1°C           **68.** −15°C

**69.** 3000°C (The melting point of iron)           **70.** 1000°C (The melting point of gold)

Convert to Celsius. Use the formula $C = \frac{5}{9} \cdot (F - 32)$ or $C = \dfrac{F - 32}{1.8}$.

**71.** 77°F           **72.** 59°F           **73.** 131°F           **74.** 140°F

**75.** 178°F           **76.** 110°F           **77.** 5°F           **78.** −4°F

**79.** 98.6°F (Normal body temperature)           **80.** 104°F (High-fevered body temperature)

**81.** *Highest Temperatures.* The highest temperature ever recorded in the world is 136°F in the desert of Libya in 1922. The highest temperature ever recorded in the United States is $56\frac{2}{3}$°C in California's Death Valley in 1913.

Source: *The Handy Geography Answer Book*

  **a)** Convert each temperature to the other scale.
  **b)** How much higher in degrees Fahrenheit was the world record than the U.S. record?

**82.** *Boiling Point and Altitude.* The boiling point of water actually changes with altitude. The boiling point is 212°F at sea level, but lowers about 1°F for every 500 ft that the altitude increases above sea level.

Sources: *The Handy Geography Answer Book; The New York Times Almanac*

  **a)** What is the boiling point at an elevation of 1500 ft above sea level?
  **b)** The elevation of Tucson is 2564 ft above sea level and that of Phoenix is 1117 ft. What is the boiling point in each city?
  **c)** How much lower is the boiling point in Denver, whose elevation is 5280 ft, than in Tucson?
  **d)** What is the boiling point at the top of Mt. McKinley (Denali) in Alaska, the highest point in the United States, at 20,320 ft?

# Skill Maintenance

In each of Exercises 83–90, fill in the blank with the correct term from the given list. Some of the choices may not be used and some may be used more than once.

**83.** When interest is paid on interest, it is called _____ interest.   [8.7b]

**84.** A(n) _____ is the quotient of two quantities.   [7.1a]

**85.** The _____ of a set of data is the middle number if there is an odd number of data items.   [6.5b]

**86.** When you work for a(n) _____, you are paid a percentage of the total sales for which you are responsible.   [8.6b]

**87.** In _____ triangles, the lengths of their corresponding sides have the same ratio.   [7.5a]

**88.** To find the _____ of a set of data, add the numbers and then divide by the number of items of data.   [6.5a]

**89.** A natural number, other than 1, that is not prime is _____.   [3.2b]

**90.** The prefix _____ means 1000.   [9.1b]

mean
median
mode
complementary
composite
commission
salary
ratio
centi-
kilo-
simple
compound
isosceles
similar

# Synthesis

**91.** A box of gelatin-mix packages weighs $15\frac{3}{4}$ lb. Each package weighs $1\frac{3}{4}$ oz. How many packages are in the box?

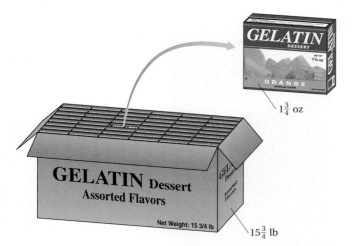

$1\frac{3}{4}$ oz

$15\frac{3}{4}$ lb

**92.** At $1.59 a dozen, the cost of eggs is $1.06 per pound. How much does an egg weigh?

**93.** Use the formula $F = \frac{9}{5} \cdot C + 32$ to find the temperature that is the same for both the Fahrenheit and Celsius scales.

**94.** Which represents a bigger change in temperature: a drop of 5°F or a drop of 5°C? Why?

**95.** *Chemistry.* Another temperature scale often used is the **Kelvin** scale. Conversions from Celsius to Kelvin can be carried out using the formula

$$K = C + 273.$$

A chemistry textbook describes an experiment in which a reaction takes place at a temperature of 400 Kelvin. A student wishes to perform the experiment, but has only a Fahrenheit thermometer. At what Fahrenheit temperature will the reaction take place?

**96.** A large egg is about $5\frac{1}{2}$ cm tall with a diameter of 4 cm. Estimate the mass of such an egg by averaging the volumes of two spheres. (*Hint*: 1 cc of water has a mass of 1 g.)

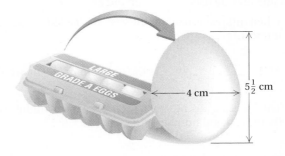

Complete. Use 453.6 g = 1 lb. Round to four decimal places.

**97.** 1 lb = _____ kg

**98.** 1 g = _____ lb

**99.** *Track and Field.* In shot put, a woman's shot weighs 8.8 lb and has a 4.5-in. diameter. Find its mass per cubic centimeter, given that 1 lb = 453.6 g.

**Source:** National Collegiate Athletic Association

**100.** *Track and Field.* In shot put, a man's shot weighs 16 lb and has a 5-in. diameter. Find its mass per cubic centimeter, given that 1 lb = 453.6 g.

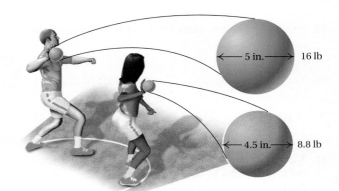

**101.** *Large Diamonds.* A **carat** (also spelled **karat**) is a unit of weight for precious stones; 1 carat = 200 mg. The Golden Jubilee Diamond weighs 545.67 carats and is the largest cut diamond in the world. The Hope Diamond, located at the Smithsonian Institution Museum of Natural History, weighs 45.52 carats.

**Source:** *National Geographic,* February 2001

**a)** How many grams does the Golden Jubilee Diamond weigh?

**b)** How many grams does the Hope Diamond weigh?

**c)** Given that 1 lb = 453.6 g, how many ounces does' each diamond weigh?

# Medical Applications

## (a) Measurements and Medicine

Measurements play a critical role in health care. Doctors, nurses, aides, technicians, and others all need to use the proper units and perform the proper calculations to assure the best possible care of patients.

Because of the ease with which conversions can be made and its extensive use in science—among other reasons—the metric system is the primary system of measurement in medicine.

**EXAMPLE 1** *Medical Dosage.* A physician ordered 3.5 L of 5% dextrose in water (abbrev. D5W) to be administered over a 24-hr period. How many milliliters were ordered?

We convert 3.5 L to milliliters:

$$3.5 \text{ L} = 3.5 \cdot 1 \text{ L}$$
$$= 3.5 \cdot 1000 \text{ mL} \qquad \text{Substituting}$$
$$= 3500 \text{ mL.}$$

The physician ordered 3500 mL of D5W.

Do Margin Exercise 1.

Liquids at a pharmacy are often labeled in liters or milliliters. Thus, if a physician's prescription is given in ounces, it must be converted. For conversion, a pharmacist knows that 1 oz ≈ 29.57 mL.*

**EXAMPLE 2** *Prescription Size.* A prescription calls for 3 oz of theophylline, a drug commonly used for children with asthma. For how many milliliters is the prescription?

We convert as follows:

$$3 \text{ oz} = 3 \cdot 1 \text{ oz}$$
$$\approx 3 \cdot 29.57 \text{ mL} \qquad \text{Substituting}$$
$$= 88.71 \text{ mL.}$$

The prescription calls for 88.71 mL of theophylline.

Do Exercise 2.

### STUDY TIPS

#### STUDYING THE EXAMPLES

The examples in each section prepare you for the exercise set. Study the step-by-step solutions. The time you spend studying the examples will save you valuable time when you do your homework.

---

## OBJECTIVE

(a) Make conversions and solve applied problems concerning medical dosages.

**SKILL TO REVIEW**
Objective 9.4b: Convert from one unit of capacity to another.

Complete.
1. 1.07 L = _____ mL
2. 350 mL = _____ L

1. **Medical Dosage.** A physician ordered 2400 mL of 0.9% saline solution to be administered intravenously over a 24-hr period. How many liters were ordered?

2. **Prescription Size.** A prescription calls for 2 oz of theophylline.
   a) For how many milliliters is the prescription?
   b) For how many liters is the prescription?

*Answers*

*Skill to Review:*
1. 1070  2. 0.35

*Margin Exercises:*
1. 2.4 L  2. (a) About 59.14 mL;
(b) about 0.059 L

**EXAMPLE 3** *Pill Splitting.* Chlorthalidone is a commonly prescribed drug used to treat hypertension. Tania's physician directs her to reduce her dosage from 25 mg to 12.5 mg. Tania's original prescription contained 30 tablets, each 25 mg.

**a)** How many milligrams of chlorthalidone, in total, were in the original prescription?

**b)** How many 12.5-mg doses can Tania obtain from the original prescription?

**a)** The original prescription contained 30 tablets, each containing 25 mg of chlorthalidone. To find the total amount in the prescription, we multiply:

$$30 \text{ tablets} \cdot 25 \text{ mg/tablet} = 750 \text{ mg}.$$

The original prescription contained 750 mg of chlorthalidone.

**b)** Since 12.5 mg is half of 25 mg ($25 \div 2 = 12.5$), Tania's original 30 doses can each be split in half, yielding $30 \cdot 2$, or 60, doses at 12.5 mg per dose.

> Do Exercise 3.

Another metric unit that is used in medicine is the microgram (mcg). It is defined as follows.

> **MICROGRAM**
>
> $$1 \text{ microgram} = 1 \text{ mcg} = \frac{1}{1,000,000} \text{ g} = 0.000001 \text{ g}$$
>
> $$1,000,000 \text{ mcg} = 1 \text{ g}$$

One microgram is one-millionth of a gram, so one million micrograms is one gram. A microgram is also one-thousandth of a milligram, so one thousand micrograms is one milligram.

**EXAMPLE 4** Complete: 1 mg = _____ mcg.

We convert to grams and then to micrograms:

$$\begin{aligned}
1 \text{ mg} &= 0.001 \text{ g} \\
&= 0.001 \cdot 1 \text{ g} \\
&= 0.001 \cdot 1,000,000 \text{ mcg} \qquad \text{Substituting 1,000,000 mcg for 1 g} \\
&= 1000 \text{ mcg}.
\end{aligned}$$

> Do Exercise 4.

**EXAMPLE 5** *Medical Dosage.* Sublingual nitroglycerin comes in 0.4-mg tablets. How many micrograms are in each tablet?
**Source:** Steven R. Smith, M.D.

We are to complete 0.4 mg = _____ mcg. Thus,

$$\begin{aligned}
0.4 \text{ mg} &= 0.4 \cdot 1 \text{ mg} \\
&= 0.4 \cdot 1000 \text{ mcg} \qquad \text{Substituting 1000 mcg for 1 mg} \\
& \qquad\qquad\qquad\qquad \text{(from Example 4)} \\
&= 400 \text{ mcg}.
\end{aligned}$$

We can also do this problem in a manner similar to that used in Example 4.

> Do Exercise 5.

**3. Pill Splitting.** If Tania's physician originally prescribed 14 tablets that were each 25 mg, how many milligrams were in the original prescription? How many 12.5-mg doses could be obtained from the original prescription?

**4.** Complete:
3 mg = _____ mcg.

**5. Medical Dosage.** A physician prescribes 500 mcg of alprazolam, an antianxiety medication. How many milligrams is this dosage?
**Source:** Steven R. Smith, M.D.

*Answers*

**3.** 350 mg; 28 doses   **4.** 3000 mcg   **5.** 0.5 mg

**9.8** | **Exercise Set**

For Extra Help

**MyMathLab**

   Math XL PRACTICE

 WATCH

DOWNLOAD

READ

 REVIEW

**a**　*Medical Dosage.*　Solve each of the following. (None of these medications should be taken without consulting your own physician.)

**1.** An emergency-room physician orders 2.0 L of Ringer's lactate to be administered over 2 hr for a patient in shock. How many milliliters is this?

**2.** Ingrid received 84 mL per hour of normal saline solution. How many liters did Ingrid receive in a 24-hr period?

**3.** To battle hypertension and prostate enlargement, Rick is directed to take 4 mg of doxazosin each day for 30 days. How many grams is this?

**4.** To battle high cholesterol, Kit is directed to take 40 mg of atorvastatin for 60 days. How many grams is this?

**5.** Cephalexin is an antibiotic that frequently is prescribed in 500-mg tablets. Dr. Bouvier prescribes 2 g of cephalexin per day for a patient with a skin abscess. How many 500-mg tablets would have to be taken in order to achieve this daily dosage?

**6.** Quinidine gluconate is a liquid mixture, part medicine and part water, which is administered intravenously. There are 80 mg of quinidine gluconate in each cubic centimeter (cc) of the liquid mixture. Dr. Nassat orders 500 mg of quinidine gluconate to be administered daily to a patient with malaria. How much of the solution would have to be administered in order to achieve the recommended daily dosage?

**7.** Albuterol is a medication used for the treatment of asthma. It comes in an inhaler that contains 17 mg of albuterol mixed with a liquid. One actuation (inhalation) from the mouthpiece delivers a 90-mcg dose of albuterol.

　**a)** Dr. Martinez orders 2 inhalations 4 times per day. How many micrograms of albuterol does the patient inhale per day?

　**b)** How many actuations/inhalations are contained in one inhaler?

　**c)** Daniel is going away for 4 months of college and wants to take enough albuterol to last for that time. His physician has prescribed 2 inhalations 4 times per day. Estimate how many inhalers Daniel will need to take with him for the 4-month period.

**8.** Amoxicillin is a common antibiotic prescribed for children. It is a liquid suspension composed of part amoxicillin and part water. In one formulation of amoxicillin suspension, there are 250 mg of amoxicillin in 5 cc of the liquid suspension. Dr. Scarlotti prescribes 400 mg per day for a 2-year-old child with an ear infection. How much of the amoxicillin liquid suspension would the child's parent need to administer in order to achieve the recommended daily dosage of amoxicillin?

9. Dr. Norris tells a patient to purchase 0.5 L of hydrogen peroxide. Commercially, hydrogen peroxide is found on the shelf in bottles that hold 4 oz, 8 oz, and 16 oz. Which bottle comes closest to filling the prescription?

10. Dr. Lopez wants a patient to receive 3 L of a normal glucose solution in a 24-hr period. How many milliliters per hour should the patient receive?

11. Remeron® is a commonly prescribed antianxiety drug. Joanne has a 14-tablet supply of 30-mg tablets and her physician now directs her to reduce her dosage to 15 mg.

    a) How many grams were originally prescribed?
    b) How many 15-mg doses can Joanne obtain from the original prescription?

12. Serzone® is a commonly prescribed antidepressant. Chad has a 90-tablet supply of 200-mg tablets when his physician directs him to cut his dosage to 100 mg.

    a) How many grams are in Chad's current supply?
    b) How many 100-mg doses can Chad obtain from his supply?

13. Amoxicillin is an antibiotic obtainable in a liquid suspension form, part medication and part water, and is frequently used to treat infections in infants. One formulation of the drug contains 125 mg of amoxicillin per 5 mL of liquid. A pediatrician orders 150 mg per day for a 4-month-old child with an ear infection. How much of the amoxicillin suspension would the parent need to administer to the infant in order to achieve the recommended daily dose?

14. Diphenhydramine HCL is an antihistamine available in liquid form, part medication and part water. One formulation contains 25 mg of medication in 5 mL of liquid. An allergist orders 40-mg doses for a high school student. How many milliliters should be in each dose?

Complete.

15. 1 mg = _____ mcg

16. 1 mcg = _____ mg

17. 325 mcg = _____ mg

18. 0.45 mg = _____ mcg

19. Dr. Djihn prescribes 0.25 mg of alprazolam, an antianxiety medication. How many micrograms are in this dose?

20. Dr. Kramer prescribes 0.4 mg of alprazolam, an antianxiety medication. How many micrograms are in this dose?

21. Digoxin is a medication used to treat heart problems. A cardiologist orders 0.125 mg of digoxin to be taken once daily. How many micrograms of digoxin are there in the daily dosage?

22. Digoxin is a medication used to treat heart problems. An internist orders 0.25 mg of digoxin to be taken once a day. How many micrograms of digoxin are there in the daily dosage?

**23.** Triazolam is a medication used for the short-term treatment of insomnia. A physician advises her patient to take one of the 0.125-mg tablets each night for 7 nights. How many milligrams of triazolam will the patient have ingested over that 7-day period? How many micrograms?

**24.** Clonidine is a medication used to treat high blood pressure. The usual starting dose of clonidine is one 0.1-mg tablet twice a day. If a patient is started on this dose by his physician, how many total milligrams of clonidine will the patient have taken before he returns to see his physician 14 days later? How many micrograms?

## Skill Maintenance

Subtract.  [1.3a]

**25.**  $\begin{array}{r} 5\ 7\ 8\ 9 \\ -\ 2\ 4\ 3\ 1 \\ \hline \end{array}$

**26.**  $\begin{array}{r} 8\ 4\ 2\ 9 \\ -\ 1\ 0\ 1\ 5 \\ \hline \end{array}$

**27.**  $\begin{array}{r} 4\ 0\ 9\ 7 \\ -\ 3\ 2\ 4\ 3 \\ \hline \end{array}$

**28.**  $\begin{array}{r} 8\ 3\ 9\ 0 \\ -\ 2\ 0\ 5\ 6 \\ \hline \end{array}$

Simplify.  [2.7a]

**29.** $7x + 9 - 2x - 1$

**30.** $8x + 12 - 2x - 7$

**31.** $8t - 5 - t - 4$

**32.** $9r - 6 - r - 4$

## Synthesis

**33.** 🖩 A patient is directed to take 200 mg of Serzone® three times a day for one week, then 200 mg twice a day for a week, and then 100 mg three times a day for a week.

    **a)** How many grams of medication are used altogether?
    **b)** What is the average dosage size?

**34.** 🖩 A patient is directed to take 200 mg of Wellbutrin® twice a day for a week, then 200 mg in the evening and 100 mg in the morning for a week, and then 100 mg twice a day for a week.

    **a)** How many grams of medication are used altogether?
    **b)** What is the average dosage size?

**35.** Naproxen sodium, sometimes sold under the brand name Aleve, is a painkiller that lasts approximately 12 hours. A typical dose is one 220-mg tablet. Ibuprofen is a similar painkiller that lasts about 6 hours and has a typical dosage of two 200-mg tablets. What would be a better purchase: a bottle containing 44 g of naproxen sodium costing $11 or a bottle containing 72 g of ibuprofen costing $11.24? Why?

# Summary and Review

## Key Terms and Formulas

parallelogram, p. 601
base (of parallelogram), p. 601
height, p. 601
trapezoid, p. 602
bases (of trapezoid), p. 602
diameter, p. 603
radius, p. 603
circumference, p. 604
pi ($\pi$), p. 604
volume, p. 614
rectangular solid, p. 614
circular cylinder, p. 615
sphere, p. 616
angle, p. 627

ray, p. 627
vertex, p. 627
sides, p. 627
protractor, p. 628
right angle, p. 630
straight angle, p. 630
acute angle, p. 630
obtuse angle, p. 630
complementary angles, p. 630
supplementary angles, p. 631
vertical angles, p. 632
congruent angles, p. 633
equilateral triangle, p. 633
isosceles triangle, p. 633

scalene triangle, p. 633
right triangle, p. 633
obtuse triangle, p. 633
acute triangle, p. 633
square root, p. 640
radical sign, p. 640
hypotenuse, p. 641
legs, p. 641
Pythagorean theorem, p. 641
Pythagorean equation, p. 641
mass, p. 649
weight, p. 649
Fahrenheit, p. 652
Celsius, p. 652

| | |
|---|---|
| *American Units of Length:* | 12 in. = 1 ft;  3 ft = 1 yd;  36 in. = 1 yd;  5280 ft = 1 mi |
| *Metric Units of Length:* | 1 km = 1000 m;  1 hm = 100 m;  1 dam = 10 m;  1 dm = 0.1 m;  1 cm = 0.01 m;  1 mm = 0.001 m |
| *American–Metric Conversion:* | 1 m = 39.370 in.;  1 m = 3.281 ft;  1 ft = 0.305 m;  1 in. = 2.540 cm;  1 km = 0.621 mi;  1 mi = 1.609 km;  1 m = 1.094 yd;  1 yd = 0.914 m |
| *American Units of Area:* | 1 yd$^2$ = 9 ft$^2$;  1 ft$^2$ = 144 in$^2$;  1 mi$^2$ = 640 acres;  1 acre = 43,560 ft$^2$ |
| *American Units of Capacity:* | 1 gal = 4 qt;  1 qt = 2 pt;  1 pt = 16 oz;  1 pt = 2 cups;  1 cup = 8 oz |
| *Metric Units of Capacity:* | 1 L = 1000 mL = 1000 cm$^3$ = 1000 cc |
| *American–Metric Conversion:* | 1 oz = 29.57 mL |
| *American Units of Weight:* | 1 T = 2000 lb;  1 lb = 16 oz |
| *Metric Units of Mass:* | 1 t = 1000 kg;  1 kg = 1000 g;  1 hg = 100 g;  1 dag = 10 g;  1 dg = 0.1 g;  1 cg = 0.01 g;  1 mg = 0.001 g;  1 mcg = 0.000001 g |

*Temperature Conversion:*
$$F = \frac{9}{5} \cdot C + 32, \text{ or } F = 1.8 \cdot C + 32;$$

$$C = \frac{5}{9} \cdot (F - 32), \text{ or } C = \frac{F - 32}{1.8}$$

| | | | |
|---|---|---|---|
| *Area of a Parallelogram:* | $A = b \cdot h$ | *Area of a Circle:* | $A = \pi \cdot r \cdot r$, or $A = \pi \cdot r^2$ |
| *Area of a Trapezoid:* | $A = \frac{1}{2} \cdot h \cdot (a + b)$ | *Volume of a Rectangular Solid:* | $V = l \cdot w \cdot h$ |
| *Radius and Diameter of a Circle:* | $d = 2 \cdot r$, or $r = \frac{d}{2}$ | *Volume of a Circular Cylinder:* | $V = \pi \cdot r^2 \cdot h$ |
| | | *Volume of a Sphere:* | $V = \frac{4}{3} \cdot \pi \cdot r^3$ |
| *Circumference of a Circle:* | $C = \pi \cdot d$, or $C = 2 \cdot \pi \cdot r$ | *Sum of Angle Measures of a Triangle:* | $m \angle A + m \angle B + m \angle C = 180°$ |
| | | *Pythagorean Equation:* | $a^2 + b^2 = c^2$ |

# Concept Reinforcement

Determine whether each statement is true or false.

_____ **1.** To convert mm$^2$ to cm$^2$, move the decimal point 2 places to the left.   [9.2b]

_____ **2.** Since 1 yd = 3 ft, we multiply by 3 to convert square yards to square feet.   [9.2a]

_____ **3.** The number $\pi$ is greater than 3.14 and $\frac{22}{7}$.   [9.3b]

_____ **4.** The acute angles of a right triangle are complementary.   [9.5b, c]

_____ **5.** The length of the hypotenuse of a right triangle is greater than the length of either of its legs.   [9.6c]

_____ **6.** You would probably use your furnace when the temperature was 40°C.   [9.7c]

# Important Concepts

**Objective 9.1a**   Convert from one American unit of length to another.

**Example**   Complete: 126 in. = _____ yd.

$$126 \text{ in.} = \frac{126 \text{ in.}}{1} \cdot \frac{1 \text{ yd}}{36 \text{ in.}} = 3.5 \text{ yd}$$

**Practice Exercise**

**1.** Complete: 7 ft = _____ yd.

**Objective 9.1b**   Convert from one metric unit of length to another.

**Example**   Complete: 38 km = _____ cm.

To go from km to cm, we move the decimal point 5 places to the right.

38      38.00000.      38 km = 3,800,000 cm

**Practice Exercise**

**2.** Complete: 4.6 cm = _____ km.

**Objective 9.1c**   Convert between American and metric units of length.

**Example**   Complete: 42 ft = _____ m.
(Note: 1 ft $\approx$ 0.305 m.)

$$42 \text{ ft} = 42 \cdot 1 \text{ ft} \approx 42 \cdot 0.305 \text{ m} = 12.81 \text{ m}$$

**Practice Exercise**

**3.** Complete: 10 m = _____ yd.
(Note: 1 m $\approx$ 1.094 yd.)

**Objective 9.2a**   Convert from one American unit of area to another.

**Example**   Complete: 14,400 in$^2$ = _____ ft$^2$.

$$14,400 \text{ in}^2 = \frac{14,400 \text{ in}^2}{1} \cdot \frac{1 \text{ ft}^2}{144 \text{ in}^2} = 100 \text{ ft}^2$$

**Practice Exercise**

**4.** Complete: 81 ft$^2$ = _____ yd$^2$.

**Objective 9.2b**   Convert from one metric unit of area to another.

**Example**   Complete: 9.6 m$^2$ = _____ cm$^2$.

To go from m$^2$ to cm$^2$, we move the decimal point 2 × 2, or 4, places to the right.

9.6      9.6000.      9.6 m$^2$ = 96,000 cm$^2$

**Practice Exercise**

**5.** Complete: 52.4 cm$^2$ = _____ mm$^2$.

**Objective 9.3a**   Find the area of a parallelogram or trapezoid.

**Examples**   Find the area of this parallelogram.

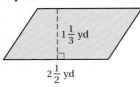

$$A = b \cdot h$$
$$= 2\frac{1}{2}\,\text{yd} \cdot 1\frac{1}{3}\,\text{yd}$$
$$= \frac{5}{2} \cdot \frac{4}{3} \cdot \text{yd} \cdot \text{yd}$$
$$= \frac{20}{6}\,\text{yd}^2 = \frac{10}{3}\,\text{yd}^2,$$
$$\text{or } 3\frac{1}{3}\,\text{yd}^2$$

Find the area of this trapezoid.

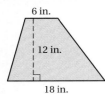

$$A = \frac{1}{2} \cdot h \cdot (a + b)$$
$$= \frac{1}{2} \cdot 12\,\text{in.} \cdot (6\,\text{in.} + 18\,\text{in.})$$
$$= \frac{12 \cdot 24}{2}\,\text{in}^2 = 144\,\text{in}^2$$

**Practice Exercises**

**6.** Find the area of this parallelogram.

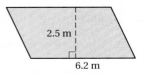

**7.** Find the area of this trapezoid.

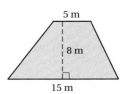

---

**Objective 9.3b**   Find the circumference, area, radius, or diameter of a circle, given the length of a radius or diameter.

**Examples**   Find the circumference of this circle. Use 3.14 for $\pi$.

$$C = \pi \cdot d, \quad \text{or} \quad 2 \cdot \pi \cdot r$$
$$\approx 2 \cdot 3.14 \cdot 4\,\text{ft}$$
$$= 25.12\,\text{ft}$$

Find the area of this circle. Use $\frac{22}{7}$ for $\pi$.

$$A = \pi \cdot r \cdot r, \quad \text{or} \quad \pi \cdot r^2$$
$$\approx \frac{22}{7} \cdot 21\,\text{mm} \cdot 21\,\text{mm}$$
$$= \frac{22 \cdot 21 \cdot 21}{7}\,\text{mm}^2 = 1386\,\text{mm}^2$$

**Practice Exercises**

**8.** Find the circumference of this circle. Use 3.14 for $\pi$.

**9.** Find the area of this circle. Use $\frac{22}{7}$ for $\pi$.

---

**Objective 9.4a**   Find the volume of a rectangular solid, a cylinder, and a sphere.

**Examples**   Find the volume of this rectangular solid.

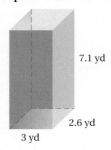

$$V = l \cdot w \cdot h$$
$$= 3\,\text{yd} \cdot 2.6\,\text{yd} \cdot 7.1\,\text{yd}$$
$$= 3 \cdot 2.6 \cdot 7.1\,\text{yd}^3$$
$$= 55.38\,\text{yd}^3$$

**Practice Exercises**

**10.** Find the volume of this rectangular solid.

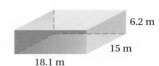

## Objective 9.4a (continued)

Find the volume of this circular cylinder. Use 3.14 for $\pi$.

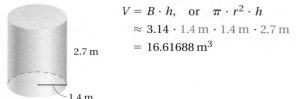

$$V = B \cdot h, \quad \text{or} \quad \pi \cdot r^2 \cdot h$$
$$\approx 3.14 \cdot 1.4\,\text{m} \cdot 1.4\,\text{m} \cdot 2.7\,\text{m}$$
$$= 16.61688\,\text{m}^3$$

2.7 m

1.4 m

**11.** Find the volume of this circular cylinder. Use $\frac{22}{7}$ for $\pi$.

$5\frac{2}{5}$ ft

$1\frac{1}{3}$ ft

Find the volume of this sphere. Use $\frac{22}{7}$ for $\pi$.

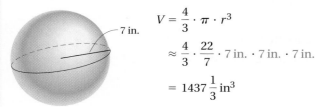

7 in.

$$V = \frac{4}{3} \cdot \pi \cdot r^3$$
$$\approx \frac{4}{3} \cdot \frac{22}{7} \cdot 7\,\text{in.} \cdot 7\,\text{in.} \cdot 7\,\text{in.}$$
$$= 1437\frac{1}{3}\,\text{in}^3$$

**12.** Find the volume of this sphere. Use 3.14 for $\pi$.

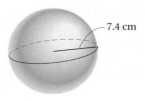

7.4 cm

---

## Objective 9.4b   Convert from one unit of capacity to another.

**Example**   Complete: 6 gal = _____ pt.

$$6\,\text{gal} = 6 \cdot 1\,\text{gal}$$
$$= 6 \cdot 4\,\text{qt}$$
$$= 24 \cdot 1\,\text{qt}$$
$$= 24 \cdot 2\,\text{pt} = 48\,\text{pt}$$

**Practice Exercise**

**13.** Complete: 16 qt = _____ cups.

---

## Objective 9.5c   Find the measure of a complement or a supplement of a given angle.

**Example**   Find the measure of a complement and a supplement of an angle that measures 65°.

The measure of the complement of an angle of 65° is 90° − 65°, or 25°.

The measure of the supplement of an angle of 65° is 180° − 65°, or 115°.

**Practice Exercise**

**14.** Find the measure of a complement and a supplement of an angle that measures 38°.

---

## Objective 9.5e   Given two of the angle measures of a triangle, find the third.

**Example**   Find the missing angle measure.

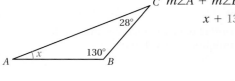

$$m\angle A + m\angle B + m\angle C = 180°$$
$$x + 130° + 28° = 180°$$
$$x + 158° = 180°$$
$$x = 180° - 158°$$
$$x = 22°$$

The measure of $\angle A$ is 22°.

**Practice Exercise**

**15.** Find the missing angle measure.

$x$

72°          21°

**Objective 9.6c** Given the lengths of any two sides of a right triangle, find the length of the third side.

**Example** Find the length of the third side of this triangle. Give an exact answer and an approximation to three decimal places.

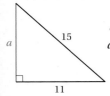

$$a^2 + b^2 = c^2 \quad \text{Pythagorean equation}$$
$$a^2 + 11^2 = 15^2$$
$$a^2 + 121 = 225$$
$$a^2 = 225 - 121$$
$$a^2 = 104$$
$$a = \sqrt{104} \approx 10.198$$

**Practice Exercise**

**16.** Find the length of the third side of this right triangle. Give an exact answer and an approximation to three decimal places.

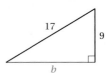

---

**Objective 9.7a** Convert from one American unit of weight to another.

**Example** Complete: 4020 oz = _____ lb.

$$4020 \, \text{oz} = 4020 \, \cancel{\text{oz}} \times \frac{1 \, \text{lb}}{16 \, \cancel{\text{oz}}} = 251.25 \, \text{lb}$$

**Practice Exercise**

**17.** Complete: 10,280 lb = _____ T.

---

**Objective 9.7b** Convert from one metric unit of mass to another.

**Example** Complete: 5.62 cg = _____ g.

To go from cg to g, we move the decimal point 2 places to the left.

5.62    0.05.62    5.62 cg = 0.0562 g

**Practice Exercise**

**18.** Complete: 9.78 mg = _____ g.

---

**Objective 9.7c** Convert temperatures from Celsius to Fahrenheit and from Fahrenheit to Celsius.

**Examples** Convert 18°C to Fahrenheit and 95°F to Celsius.

$$F = \frac{9}{5}C + 32 = 1.8 \cdot 18 + 32$$
$$= 32.4 + 32 = 64.4$$

Thus, 18°C = 64.4°F.

$$C = \frac{5}{9} \cdot (F - 32) = \frac{5}{9} \cdot (95 - 32)$$
$$= \frac{5}{9} \cdot 63 = 35$$

Thus, 95°F = 35°C.

**Practice Exercises**

**19.** Convert 68°C to Fahrenheit.

**20.** Convert 104°F to Celsius.

---

**Objective 9.8a** Make conversions and solve applied problems concerning medical dosages.

**Example** A physician ordered 320 mL of 5% dextrose in water (D5W) to be administered intravenously over 4 hr. How many liters of D5W is this?

We convert 320 mL to liters:

$$320 \, \text{mL} = 320 \cdot 1 \, \text{mL}$$
$$= 320 \cdot 0.001 \, \text{L}$$
$$= 0.32 \, \text{L}.$$

The physician ordered 0.32 L of D5W.

**Practice Exercise**

**21.** A physician orders 0.5 L of normal saline solution. How many milliliters are ordered?

# Review Exercises

Complete.

**1.** 10 ft = _____ yd
[9.1a]

**2.** $\frac{5}{6}$ yd = _____ in.
[9.1a]

**3.** 1.7 mm = _____ cm
[9.1b]

**4.** 2 yd = _____ in.
[9.1a]

**5.** 4 km = _____ cm
[9.1b]

**6.** 14 in. = _____ ft
[9.1a]

**7.** 200 m = _____ yd
[9.1c]

**8.** 20 mi = _____ km
[9.1c]

**9.** 5 lb = _____ oz
[9.7a]

**10.** 3 g = _____ kg
[9.7b]

**11.** 50 qt = _____ gal
[9.4b]

**12.** 28 gal = _____ pt
[9.4b]

**13.** 60 mL = _____ L
[9.4b]

**14.** 0.4 L = _____ mL
[9.4b]

**15.** 0.7 T = _____ lb
[9.7a]

**16.** 0.2 g = _____ mg
[9.7b]

**17.** 4.7 kg = _____ g
[9.7b]

**18.** 4 cg = _____ g
[9.7b]

**19.** 4 yd$^2$ = _____ ft$^2$
[9.2a]

**20.** 0.7 km$^2$ = _____ m$^2$
[9.2b]

**21.** 1008 in$^2$ = _____ ft$^2$
[9.2a]

**22.** 570 cm$^2$ = _____ m$^2$
[9.2b]

**23.** Find the circumference of a circle of radius 5 m. Use 3.14 for $\pi$.　[9.3b]

**24.** Find the length of a radius of the circle.　[9.3b]

$\frac{28}{11}$ in.

**25.** Find the length of a diameter of the circle.　[9.3b]

12 m

**26.** *Track and Field.*　Track meets are held on a track similar to the one shown below. Find the shortest distance around the track. Use 3.14 for $\pi$.　[9.3b]

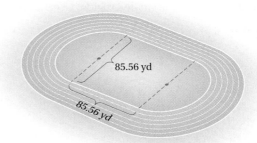

85.56 yd

85.56 yd

Find the area of each figure in Exercises 27–32.　[9.3a, b]

**27.**

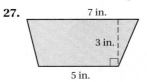

7 in.

3 in.

5 in.

**28.**

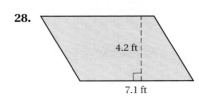

4.2 ft

7.1 ft

**29.**

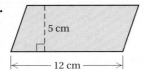

5 cm

12 cm

**30.**

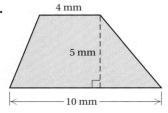

4 mm

5 mm

10 mm

**31.** Use $\frac{22}{7}$ for $\pi$.

7 ft

**32.** Use 3.14 for $\pi$.

20 cm

**33.** Find the area of the shaded region. Use 3.14 for $\pi$. [9.3b]

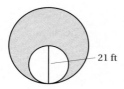

21 ft

**34.** A "Norman" window is designed with dimensions as shown. Find its area. Use 3.14 for $\pi$. [9.3b]

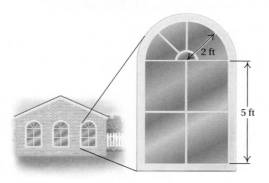

2 ft

5 ft

Use a protractor to measure each angle. [9.5a]

**35.**

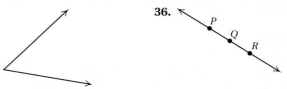

**36.**

P   Q   R

**37.**

**38.**

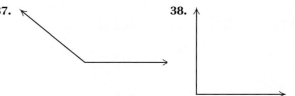

**39–42.** Classify each of the angles in Exercises 35–38 as right, straight, acute, or obtuse. [9.5b]

**43.** Find the measure of a complement of $\angle BAC$. [9.5c]

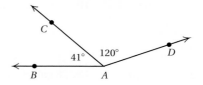

C

41°   120°   D

B   A

**44.** Find the measure of a supplement of a 44° angle. [9.5c]

Use the following triangle for Exercises 45–47.

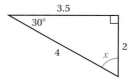

3.5

30°

4

$x$   2

**45.** Find the missing angle measure. [9.5e]

**46.** Classify the triangle as equilateral, isosceles, or scalene. [9.5d]

**47.** Classify the triangle as right, obtuse, or acute. [9.5d]

Find the volume of each figure. Use 3.14 for $\pi$.   [9.4a]

**48.**

2.6 m

12 m

3 m

**49.**

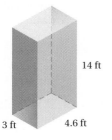

14 ft

3 ft

4.6 ft

**50.**

90 cm

10 cm

**51.**

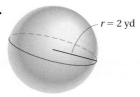

$r = 2$ yd

**52.**

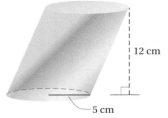

12 cm

5 cm

**53.** Simplify: $\sqrt{64}$.   [9.6a]

**54.** Use a calculator to approximate $\sqrt{83}$ to three decimal places.   [9.6b]

For each right triangle, find the length of the side not given. Find an exact answer and an approximation to three decimal places. Assume that $c$ represents the length of the hypotenuse.   [9.6c]

**55.** $a = 15, b = 25$          **56.** $a = 4, c = 10$

**57.**

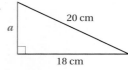

$c$

5 ft

8 ft

**58.**

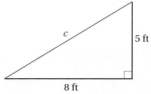

20 cm

$a$

18 cm

Solve.   [9.6d]

**59.** How long is a wire reaching from the top of a 24-ft pole to a point on the ground 16 ft from the base of the pole? Round to the nearest tenth of a foot.

**60.** *Muir Woods.*   On a tour of Muir Woods in Marin County, California, Eliza photographed the tallest red-wood in the park. Standing 24 m from the base of the tree, she estimates that it is 83 m to the top of the tree. How tall is this redwood? Round to the nearest meter.
**Source:** National Register of Historic Places

83 m

$h$

24 m

**61.** Convert 45°C to Fahrenheit.   [9.7c]

**62.** Convert 68°F to Celsius.   [9.7c]

*Medical Dosage.* Solve. [9.8a]

**63.** Amoxicillin is an antibiotic obtainable in a liquid suspension form, part medication and part water, and is frequently used to treat infections in infants. One formulation of the drug contains 125 mg of amoxicillin per 5 mL of liquid. A pediatrician orders 200 mg per day for a child with an ear infection. How much of the amoxicillin suspension would the parent need to administer to the child in order to achieve the recommended daily dose?

**64.** An emergency-room physician orders 3 L of Ringer's lactate to be administered over 4 hr for a patient suffering from shock and severe low blood pressure. How many milliliters is this?

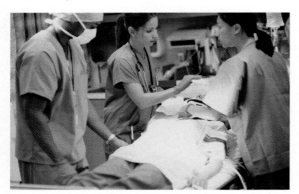

**65.** A physician prescribes 0.25 mg of alprazolam, an antianxiety medication. How many micrograms are in this dose?

**66.** Find the measure of a supplement of a $20\frac{3}{4}^\circ$ angle. [9.5c]

**A.** $339\frac{1}{4}^\circ$       **B.** $159\frac{1}{4}^\circ$

**C.** $69\frac{1}{4}^\circ$       **D.** $70\frac{1}{4}^\circ$

**67.** Complete: 0.16 gal = _____ cups. [9.4b]

**A.** 1.28      **B.** 2.56

**C.** 0.64      **D.** 160

## Synthesis

**68.** *Running Record.* The world record for running the 200-m dash is 19.19 sec, set by Usain Bolt of Jamaica in the 2009 World Championships in Berlin. How would the record be changed if the run were a 200-yd dash? [9.1c]
**Source:** wikipedia.com

**69.** It is known that 1 gal of water weighs 8.3453 lb. Which weighs more: an ounce of pennies or an ounce (as capacity) of water? Explain. [9.4b], [9.7a]

**70.** A community center has a rectangular swimming pool that is 50 ft wide, 100 ft long, and 10 ft deep. The pool is filled with water to a line that is 1 ft from the top. Water costs $2.25 per 1000 ft³. How much does it cost to fill the pool? [9.4c]

**71.** 🖩 One lap around a standard running track is 440 yd. A marathon is 26 mi, 385 yd long. How many laps around a track does a marathon require? [9.1a]

**72.** 🖩 Find the area of the largest round pizza that can be baked in a 35-cm by 50-cm pan. Use 3.14 for $\pi$. [9.3b]

# Understanding Through Discussion and Writing

**1.** Give at least two reasons why someone might prefer the use of grams to the use of ounces. [9.7a, b]

**2.** Is it possible for a triangle to contain two 90° angles? Why or why not? [9.5e]

**3.** Which is larger and why: one square meter or one square yard? [9.1c], [9.2a, b]

**4.** Explain a procedure that could be used to determine the measure of an angle's supplement from the measure of the angle's complement. [9.5c]

**5.** Explain how the Pythagorean theorem can be used to prove that a triangle is a *right* triangle. [9.6c]

**6.** Which occupies more volume: two spheres, each with radius *r*, or one sphere with radius 2*r*? Explain why. [9.4c]

Test

For Extra Help

 CHAPTER
Test Prep
VIDEOS

Step-by-step test solutions are found on the Chapter Test Prep Videos available via the Video Resources on DVD, in **MyMathLab** , and on **YouTube** (search "BittingerPrealgebra" and click on "Channels").

**Complete.**

**1.** 8 ft = _____ in.

**2.** 280 cm = _____ m

**3.** 2 yd² = _____ ft²

**4.** 5 km = _____ m

**5.** 9.1 mm = _____ cm

**6.** 4520 m² = _____ km²

**7.** 2983 mL = _____ L

**8.** 3.8 kg = _____ g

**9.** 10 gal = _____ oz

**10.** 0.69 L = _____ mL

**11.** 9 lb = _____ oz

**12.** 4.11 T = _____ lb

**13.** Find the length of a radius of this circle.

16 cm

**14.** Find the area of circle of radius 4 m. Use 3.14 for $\pi$.

**15.** Find the circumference of a circle of radius 14 ft. Use $\frac{22}{7}$ for $\pi$.

**Find each area. Use 3.14 for $\pi$.**

**16.**

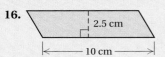

2.5 cm

10 cm

**17.**

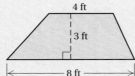

4 ft

3 ft

8 ft

**18.**

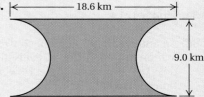

18.6 km

9.0 km

**19.** Find the measure of a complement and a supplement of $\angle CAD$.

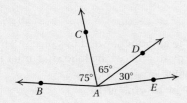

C

D

65°

75°   30°

B    A    E

Use a protractor to measure each angle.

**20.**

**21.**

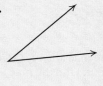

**22.** Classify the angle in Exercise 20 as right, straight, acute, or obtuse.

**23.** Classify the angle in Exercise 21 as right, straight, acute, or obtuse.

Use the following triangle for Exercises 24–26.

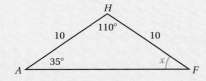

H

110°

10       10

35°              x

A                    F

**24.** Find the missing angle measure.

**25.** Classify the triangle as equilateral, isosceles, or scalene.

**26.** Classify the triangle as right, obtuse, or acute.

**27.** A twelve-box rectangular carton of 12-oz juice boxes measures $10\frac{1}{2}$ in. by 8 in. by 5 in. What is the volume of the carton?

In Exercises 28–30, find the volume in each figure. Use 3.14 for $\pi$.

**28.**

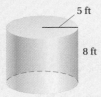
5 ft
8 ft

**29.**

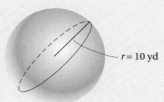
$r = 10$ yd

**30.**

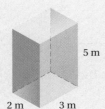
5 m
2 m   3 m

**31.** How long must a wire be in order to reach from the top of a 13-m antenna to a point on the ground 9 m from the base of the antenna? Round to the nearest tenth of a meter.

For each right triangle, find the length of the side not given. Find an exact answer and an approximation to three decimal places.

**32.**

$c$
1
1

**33.**

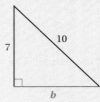
7
10
$b$

**34.** Find $\sqrt{121}$.

**35.** Convert 32°F to Celsius.

**36.** An antihistamine solution contains 25 mg of medication in 5 mL of liquid. Pat is directed to take a 30-mg dose of the antihistamine. How many mL's of the solution should Pat take?

  **A.** 6 mL                 **B.** 55 mL
  **C.** 125 mL              **D.** 150 mL

## Synthesis

**37.** The measure of $\angle SMC$ is three times that of its complement. Find the measure of the supplement of $\angle SMC$.

**38.** ⌨ A *board foot* is the amount of wood in a piece 12 in. by 12 in. by 1 in. A carpenter places the following order for a certain kind of lumber:

    25 pieces: 2 in. by 4 in. by 8 ft;

    32 pieces: 2 in. by 6 in. by 10 ft;

    24 pieces: 2 in. by 8 in. by 12 ft.

The price of this type of lumber is $225 per thousand board feet. What is the total cost of the carpenter's order?

Find the volume of the solid. (Note that the solids are not drawn in perfect proportion.) Give the answer in cubic feet. Use 3.14 for $\pi$.

**39.**

12 ft
2.6 in.
3 in.

**40.**

18 ft
$\frac{3}{4}$ in.

# Cumulative Review

Solve.

**1.** *Neon Tubing.* The highest concentration of neon tubing in the world is found in Las Vegas. There are approximately 79.2 million feet of neon tubing in the city. Find standard notation for 79.2 million.

**Source:** *Las Vegas: An Unconventional History,* by Michelle Ferrari and Stephen Ives, Bulfinch Press

**2.** *Firefighting.* During a fire, the firefighters get a 1-ft layer of water on the 25-ft by 60-ft first floor of a 5-floor building. Water weighs $62\frac{1}{2}$ lb per cubic foot. What is the total weight of the water on the floor?

Calculate.

**3.** $1\frac{1}{2} + 2\frac{2}{3}$

**4.** $120.5 - 32.98$

**5.** $-27{,}148 \div 22$

**6.** $8^3 + 45 \cdot 24 - 9^2 \div 3$

**7.** $\left(\frac{1}{4}\right)^2 \div \left(\frac{1}{2}\right)^3 \times 2^4 + (10.3)(4)$

**8.** $14 \div [33 \div 11 + 8 \times 2 - (15 - 3)]$

Find fraction notation.

**9.** $1.209$

**10.** $17\%$

Use $<$, $>$, or $=$ for $\square$ to write a true sentence.

**11.** $\frac{5}{6} \square \frac{7}{8}$

**12.** $\frac{15}{18} \square \frac{10}{12}$

Complete.

**13.** $2.5 \text{ yd} = \underline{\hspace{1cm}} \text{ in.}$

**14.** $6 \text{ oz} = \underline{\hspace{1cm}} \text{ lb}$

**15.** $15°\text{C} = \underline{\hspace{1cm}} °\text{F}$

**16.** $0.087 \text{ L} = \underline{\hspace{1cm}} \text{ mL}$

**17.** $3 \text{ yd}^2 = \underline{\hspace{1cm}} \text{ ft}^2$

**18.** $17 \text{ cm} = \underline{\hspace{1cm}} \text{ m}$

Solve.

**19.** $x + \frac{3}{4} = \frac{7}{8}$

**20.** $\frac{3}{x} = \frac{7}{10}$

**21.** $1 - 7x = 4 - (x + 9)$

**22.** $-15x = 240$

**23.** Find the perimeter and the area.

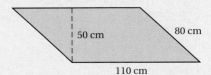

50 cm

80 cm

110 cm

**24.** Find the diameter, the circumference, and the area of this circle. Use $\frac{22}{7}$ for $\pi$.

35 in.

**25.** Find the volume of this sphere. Use $\frac{22}{7}$ for $\pi$.

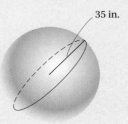

35 in.

**26.** Find the mean: 49, 53, 60, 62, 69.

Solve.

**27.** What is the simple interest on $8000 at 4.2% for $\frac{1}{4}$ year?

**28.** What is the amount in an account after 25 years if $8000 is invested at 4.2%, compounded annually?

**29.** How long must a rope be in order to reach from the top of an 8-m tree to a point on the ground 15 m from the bottom of the tree?

**30.** The sales tax on an office supply purchase of $5.50 is $0.33. What is the sales tax rate?

**31.** A bolt of fabric in a fabric store has $10\frac{3}{4}$ yd on it. A customer purchases $8\frac{5}{8}$ yd. How many yards remain on the bolt?

**32.** What is the cost, in dollars, of 15.6 gal of gasoline at 329.9¢ per gallon? Round to the nearest cent.

**33.** A box of powdered milk that makes 20 qt costs $4.99. A box that makes 8 qt costs $1.99. Which size has the lower unit price?

**34.** It is $\frac{7}{10}$ km from Maria's dormitory to the library. Maria started to walk from the dorm to the library, changed her mind after going $\frac{1}{4}$ of the distance, and returned to the dorm. How far did she walk?

**35.** Find the missing angle measure.

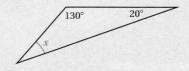

130°    20°

$x$

**36.** Combine like terms: $12a - 7 - 3a - 9$.

**37.** Graph: $y = -\dfrac{1}{3}x + 2$.

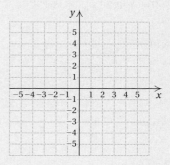

## Synthesis

Find the volume in cubic feet. Use 3.14 for $\pi$.

**38.**

100 yd

10 ft

**39.**

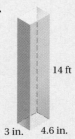

14 ft

3 in.    4.6 in.

# Exponents and Polynomials

# Real-World Application

There are 10 million bacteria per square centimeter of coral in a coral reef. The coral reefs near the Hawaiian Islands cover 14,000 km². How many bacteria are there in Hawaii's coral reef?

*Sources:* livescience.com; U.S. Geological Survey

***This problem appears as Exercise 35 in Section 10.3.***

# 10.1

# Integers as Exponents

## OBJECTIVES

**a** Evaluate algebraic expressions containing whole-number exponents.

**b** Express exponential expressions involving negative exponents as equivalent expressions containing positive exponents.

**SKILL TO REVIEW**
Objective 1.9b: Evaluate exponential notation.

Evaluate.

**1.** $4^3$          **2.** $12^2$

---------- *Caution!* ----------

Don't confuse the powers 1 and 0. Be careful: $8^0 = 1$, but $8^1 = 8$.

---

Evaluate.

**1.** $16^1$          **2.** $-16^1$

**3.** $(-35)^1$          **4.** $7^0$

**5.** $(-9)^0$          **6.** $-8^0$

**7.** Evaluate $(2x - 9)^0$ for $x = 3$.

**8.** Evaluate $t^0 - 4$ for $t = 7$.

*Answers*

*Skill to Review:*
1. 64     2. 144

*Margin Exercises:*
1. 16     2. −16     3. −35     4. 1     5. 1
6. −1     7. 1     8. −3

---

We have already used the numbers 2, 3, . . . , as exponents. In this section, we consider 1 and 0, as well as negative integers, as exponents.

## **a** One and Zero as Exponents

Look for a pattern in the following:

$$8 \cdot 8 \cdot 8 \cdot 8 = 8^4$$
$$8 \cdot 8 \cdot 8 = 8^3$$
$$8 \cdot 8 = 8^2$$
$$8 = 8^?$$
$$1 = 8^?$$

We divide by 8 each time.

The exponents decrease by 1 each time. Continuing the pattern, we have

$$8 = 8^1 \quad \text{and} \quad 1 = 8^0.$$

We make the following definitions.

> **THE EXPONENTS 1 AND 0**
>
> $b^1 = b$, for any number $b$.
> $b^0 = 1$, for any nonzero number $b$.

**EXAMPLE 1**  Evaluate $23^1$, $-23^1$, and $(-23)^1$.

$$23^1 = 23;$$
$$-23^1 = -23; \quad \text{We read } -23^1 \text{ as "the opposite of } 23^1\text{."}$$
$$(-23)^1 = -23 \quad \text{We read } (-23)^1 \text{ as "negative 23 to the first."}$$

**EXAMPLE 2**  Evaluate $3^0$, $(-3)^0$, and $-3^0$.

$$3^0 = 1; \qquad\qquad 3 \text{ is the base; 0 is the exponent.}$$
$$(-3)^0 = 1; \qquad -3 \text{ is the base; 0 is the exponent.}$$
$$-3^0 = -1 \cdot 3^0 = -1 \cdot 1 = -1 \quad \text{Note that } -3^0 \neq (-3)^0.$$

Do Margin Exercises 1–6.

**EXAMPLE 3**  Evaluate $m^0 + 5$ for $m = 9$.

$$m^0 + 5 = 9^0 + 5 = 1 + 5 = 6$$

**EXAMPLE 4**  Evaluate $(3x + 2)^0$ for $x = -5$.

We substitute $-5$ for $x$ and follow the rules for order of operations:

$$(3x + 2)^0 = (3(-5) + 2)^0 \quad \text{Substituting}$$
$$= (-15 + 2)^0 \quad \text{Multiplying}$$
$$= (-13)^0 = 1.$$

Do Exercises 7 and 8.

## b Negative Integers as Exponents

The pattern used to help define the exponents 1 and 0 can also be used to define negative-integer exponents:

$$8 \cdot 8 \cdot 8 = 8^3$$
$$8 \cdot 8 = 8^2$$
$$8 = 8^1$$
$$1 = 8^0$$
$$\frac{1}{8} = 8^?$$
$$\frac{1}{8 \cdot 8} = 8^?.$$

We divide by 8 each time.

The exponents decrease by 1 each time. To continue the pattern, we would say that

$$\frac{1}{8} = 8^{-1} \quad \text{and} \quad \frac{1}{8 \cdot 8} = 8^{-2}.$$

Thus, if we are to preserve the above pattern, we must have

$$\frac{1}{8^1} = 8^{-1} \quad \text{and} \quad \frac{1}{8^2} = 8^{-2}.$$

This leads to our definition of negative exponents.

---

### NEGATIVE EXPONENTS

For any nonzero numbers $a$ and $b$, and any integer $n$,

$$a^{-n} = \frac{1}{a^n} \quad \text{and} \quad \left(\frac{a}{b}\right)^{-n} = \left(\frac{b}{a}\right)^{n}.$$

(A base raised to a negative exponent is equal to the reciprocal of the base raised to a positive exponent.)

---

**EXAMPLES** Write an equivalent expression using positive exponents. Then simplify.

**5.** $4^{-2} = \dfrac{1}{4^2} = \dfrac{1}{16}$  Note that $4^{-2}$ represents a *positive* number.

**6.** $(-3)^{-2} = \dfrac{1}{(-3)^2} = \dfrac{1}{(-3)(-3)} = \dfrac{1}{9}$

**7.** $m^{-3} = \dfrac{1}{m^3}$

**8.** $ab^{-1} = a \cdot b^{-1} = a\left(\dfrac{1}{b^1}\right) = a\left(\dfrac{1}{b}\right) = \dfrac{a}{b}$  Think of $a$ as $\dfrac{a}{1}$ if you wish.

**9.** $\left(\dfrac{5}{6}\right)^{-2} = \left(\dfrac{6}{5}\right)^{2}$  $\quad\quad \left(\dfrac{a}{b}\right)^{-n} = \left(\dfrac{b}{a}\right)^{n}$

$\quad\quad = \dfrac{6}{5} \cdot \dfrac{6}{5} = \dfrac{36}{25}$

### STUDY TIPS

**DON'T GIVE UP NOW!**

Often students are tempted to ease up once the end of a course is in sight. Don't let this happen to you! You have invested a great deal of time and energy thus far. Don't tarnish that hard effort by doing anything less than your best work as the course winds down.

### Calculator Corner

**Expressions with Negative Exponents** Most calculators can evaluate expressions like

$$\left(-\frac{2}{5}\right)^{-6}$$

with just a few keystrokes. Although the keys may be labeled differently, on most calculators with a $\boxed{+/-}$ key, the following keystrokes will work.

$\boxed{2}\ \boxed{\div}\ \boxed{5}\ \boxed{=}\ \boxed{+/-}$
$\boxed{x^y}\ \boxed{6}\ \boxed{+/-}\ \boxed{=}$

On most calculators with a $\boxed{(-)}$ key, the following keystrokes will work.

$\boxed{(}\ \boxed{(-)}\ \boxed{2}\ \boxed{\div}\ \boxed{5}\ \boxed{)}$
$\boxed{\wedge}\ \boxed{(-)}\ \boxed{6}\ \boxed{\text{ENTER}}$

The same result, 244.140625, is found when $\left(-\frac{5}{2}\right)^6$ is evaluated.

**Exercises:** Calculate the value of each pair of expressions.

**1.** $\left(\dfrac{3}{4}\right)^{-5}; \left(\dfrac{4}{3}\right)^{5}$

**2.** $\left(\dfrac{4}{5}\right)^{-7}; \left(\dfrac{5}{4}\right)^{7}$

**3.** $\left(-\dfrac{5}{4}\right)^{-7}; \left(-\dfrac{4}{5}\right)^{7}$

**4.** $\left(-\dfrac{7}{5}\right)^{-9}; \left(-\dfrac{5}{7}\right)^{9}$

**5.** $\left(-\dfrac{5}{6}\right)^{-4}; \left(-\dfrac{6}{5}\right)^{4}$

**6.** $\left(-\dfrac{9}{8}\right)^{-6}; \left(-\dfrac{8}{9}\right)^{6}$

Note in Example 5 that

$$4^{-2} \neq 4(-2) \quad \text{and} \quad 4^{-2} \neq \frac{1}{-4^2}.$$

In general, $a^{-n} \neq a(-n)$. The negative exponent also does not indicate that a negative number is in the result. That is,

$$4^{-2} = \frac{1}{16}, \quad not \quad \frac{1}{-16}.$$

Do Exercises 9–14.

**EXAMPLES** Write an equivalent expression using negative exponents.

**10.** $\dfrac{1}{7^2} = 7^{-2}$    Reading $a^{-n} = \dfrac{1}{a^n}$ from right to left: $\dfrac{1}{a^n} = a^{-n}$

**11.** $\dfrac{5}{x^8} = 5 \cdot \dfrac{1}{x^8} = 5x^{-8}$

Do Exercises 15 and 16.

Consider an expression like

$$\frac{a^2}{b^{-3}},$$

in which the denominator is a negative power. We can simplify as follows:

$$\frac{a^2}{b^{-3}} = \frac{a^2}{\dfrac{1}{b^3}} \qquad \text{Rewriting } b^{-3} \text{ as } \dfrac{1}{b^3}$$

$$= a^2 \cdot \frac{b^3}{1} \qquad \text{To divide by a fraction, we multiply by its reciprocal.}$$

$$= a^2 b^3.$$

Do Exercises 17 and 18.

Our work above indicates that to divide by a base raised to a negative power, we can instead multiply by the opposite power of the same base. This will shorten our work.

**EXAMPLES** Write an equivalent expression using positive exponents.

**12.** $\dfrac{x^3}{y^{-2}} = x^3 y^2$    Instead of dividing by $y^{-2}$, multiply by $y^2$.

**13.** $\dfrac{a^2 b^5}{c^{-6}} = a^2 b^5 c^6$    Instead of dividing by $c^{-6}$, multiply by $c^6$.

**14.** $\dfrac{x^{-2} y}{z^{-3}} = x^{-2} y z^3 = \dfrac{y z^3}{x^2}$

Do Exercises 19–21.

Write an equivalent expression with positive exponents. Then simplify.

**9.** $4^{-3}$      **10.** $5^{-2}$

**11.** $2^{-4}$      **12.** $(-2)^{-3}$

**13.** $\left(\dfrac{5}{3}\right)^{-2}$      **14.** $xy^{-2}$

Write an equivalent expression with negative exponents.

**15.** $\dfrac{1}{9^2}$      **16.** $\dfrac{7}{x^4}$

Write an equivalent expression with positive exponents.

**17.** $\dfrac{m^3}{n^{-5}}$      **18.** $\dfrac{ab}{c^{-1}}$

Write an equivalent expression with positive exponents.

**19.** $\dfrac{a^4}{b^{-6}}$      **20.** $\dfrac{x^7 y}{z^{-4}}$

**21.** $\dfrac{a^4 b^{-7}}{c^{-3}}$

*Answers*

**9.** $\dfrac{1}{4^3}; \dfrac{1}{64}$   **10.** $\dfrac{1}{5^2}; \dfrac{1}{25}$   **11.** $\dfrac{1}{2^4}; \dfrac{1}{16}$
**12.** $\dfrac{1}{(-2)^3}; -\dfrac{1}{8}$   **13.** $\left(\dfrac{3}{5}\right)^2; \dfrac{9}{25}$
**14.** $x \cdot \dfrac{1}{y^2}; \dfrac{x}{y^2}$   **15.** $9^{-2}$   **16.** $7x^{-4}$
**17.** $m^3 n^5$   **18.** $abc$   **19.** $a^4 b^6$
**20.** $x^7 y z^4$   **21.** $\dfrac{a^4 c^3}{b^7}$

**a** Evaluate.

**1.** $4^0$

**2.** $17^0$

**3.** $3.14^1$

**4.** $2.67^1$

**5.** $(-19.57)^1$

**6.** $(-34.6)^0$

**7.** $(-98.6)^0$

**8.** $(-98.6)^1$

**9.** $x^0, x \neq 0$

**10.** $a^0, a \neq 0$

**11.** $x^1$, for $x = 97$

**12.** $a^1$, for $a = -10$

**13.** $(3x - 17)^0$, for $x = 10$

**14.** $(7x - 45)^0$, for $x = 8$

**15.** $(5x - 3)^1$, for $x = 4$

**16.** $(35 - 4x)^1$, for $x = 8$

**17.** $(4m - 19)^0$, for $m = 3$

**18.** $(9 - 2x)^0$, for $x = 5$

**19.** $3x^0 + 4$, for $x = -2$

**20.** $7x^0 + 6$, for $x = -3$

**21.** $(3x)^0 + 4$, for $x = -2$

**22.** $(7x)^0 + 6$, for $x = -3$

**23.** $(5 - 3x^0)^1$, for $x = 19$

**24.** $(5x^1 - 29)^0$, for $x = 4$

**b** Write an equivalent expression with positive exponents. Then simplify, if possible.

**25.** $3^{-2}$

**26.** $2^{-3}$

**27.** $10^{-4}$

**28.** $5^{-6}$

**29.** $t^{-4}$

**30.** $x^{-2}$

**31.** $7^{-1}$

**32.** $10^{-1}$

**33.** $(-5)^{-2}$

**34.** $(-4)^{-3}$

**35.** $(-x)^{-2}$

**36.** $(-n)^{-2}$

Simplify.

**37.** $3x^{-7}$

**38.** $-6y^{-2}$

**39.** $mn^{-5}$

**40.** $cd^{-1}$

**41.** $xy^{-1}$

**42.** $pw^{-3}$

**43.** $-7a^{-9}$

**44.** $9p^{-4}$

**45.** $\dfrac{1}{x^{-3}}$

**46.** $\dfrac{1}{x^{-1}}$

**47.** $\dfrac{x}{y^{-4}}$

**48.** $\dfrac{r}{t^{-7}}$

**49.** $\dfrac{r^5}{t^{-3}}$

**50.** $\dfrac{x^7}{y^{-5}}$

**51.** $\dfrac{xy}{z^{-1}}$

**52.** $\dfrac{ab}{c^{-6}}$

**53.** $\dfrac{a^8b}{c^{-2}}$

**54.** $\dfrac{x^7y}{t^{-6}}$

**55.** $\dfrac{x^2y^3}{z^{-8}}$

**56.** $\dfrac{x^8y^{10}}{z^{-1}}$

**57.** $\dfrac{x^{-4}y^3}{t^{-2}}$

**58.** $\dfrac{ab^{-6}}{c^{-2}}$

**59.** $\dfrac{cd^{-4}}{a^{-1}}$

**60.** $\dfrac{x^3y^{-8}}{z^{-4}}$

**61.** $\left(\dfrac{2}{5}\right)^{-2}$

**62.** $\left(\dfrac{3}{7}\right)^{-2}$

**63.** $\left(\dfrac{5}{a}\right)^{-3}$

**64.** $\left(\dfrac{x}{3}\right)^{-4}$

Write an equivalent expression using negative exponents.

**65.** $\dfrac{1}{7^3}$

**66.** $\dfrac{1}{5^2}$

**67.** $\dfrac{9}{x^3}$

**68.** $\dfrac{4}{y^2}$

## Skill Maintenance

**69.** When used for a singles match, a regulation tennis court is 27 ft by 78 ft. Find its perimeter.   [2.7b]

**70.** The Floral Doctor's delivery van traveled 147 mi on 10.5 gal of gas. How many miles per gallon did the van get?   [7.2a]

**71.** Ramon's new truck gets 21 mpg. This is 20% more than the mileage his old truck got. What mileage did the old truck get?   [8.5a]

**72.** A 5% sales tax is added to the price of a two-speed washing machine. If the machine is priced at $399, find the total amount paid.   [8.6a]

**73.** Of the 8 fish Mac caught, 3 were trout. What percentage were not trout?   [8.4a]

**74.** The diameter of a compact disc is 12 cm. What is its circumference? (Use 3.14 for $\pi$.)   [9.3b]

**75.** Multiply:  $-57 \cdot 48$.   [2.4a]

**76.** Multiply:  $(-72)(-46)$.   [2.4a]

## Synthesis

Simplify.

**77.** $3 + 4^{-1} \cdot 10^2$

**78.** $(2 + 3)^{-2} \cdot 10^2 \div 2^{-1}$

**79.** $4^2 \div 2^{-1} \cdot 5^1 - (12 \cdot 10)^0$

**80.** $(42 - 6 \cdot 7)^8 \cdot (13 - 5 \cdot 4)^{-3}$

**81.** ▦ Evaluate $\dfrac{3^x}{3^{x-1}}$ for $x = -4$ and then for $x = -40$.

**82.** ▦ Evaluate $\dfrac{5^x}{5^{x+1}}$ for $x = -3$ and then for $x = -30$.

# 10.2

# Working with Exponents

There are several rules that we can use to simplify exponential expressions. We first consider multiplying powers with like bases. Throughout this section, we assume that a variable in a denominator is not zero.

## a Multiplying Powers with Like Bases

When we multiply expressions like $a^3$ and $a^2$, we can use the associative law to combine the factors:

$$a^3 \cdot a^2 = \underbrace{(a \cdot a \cdot a)}_{3 \text{ factors}}\underbrace{(a \cdot a)}_{2 \text{ factors}} = \underbrace{a \cdot a \cdot a \cdot a \cdot a}_{5 \text{ factors}} = a^5.$$

Note that the exponent in $a^5$ is the sum of the exponents in $a^3$ and $a^2$. Similarly,

$$b^4 \cdot b^3 = (b \cdot b \cdot b \cdot b)(b \cdot b \cdot b) = b^7, \quad \text{where} \quad 4 + 3 = 7.$$

Adding the exponents gives the correct result.

---

**THE PRODUCT RULE**

For any nonzero number $a$ and any integers $m$ and $n$,

$$a^m \cdot a^n = a^{m+n}.$$

(To multiply powers with the same base, keep the base and add the exponents.)

---

**EXAMPLES** Multiply and simplify. Here "simplify" means to express the product as one base with a positive exponent whenever possible.

**1.** $3^6 \cdot 3^2 = 3^{6+2}$    Adding exponents: $a^m \cdot a^n = a^{m+n}$
     $= 3^8$

**2.** $x^4 \cdot x^5 = x^{4+5} = x^9$

**3.** $5 \cdot 5^6 \cdot 5^2 = 5^1 \cdot 5^6 \cdot 5^2$    Writing 5 as $5^1$
     $= 5^{1+6+2}$
     $= 5^9$

**4.** $x^3 \cdot x^{-8} = x^{3+(-8)}$    The product rule works with negative exponents.

     $= x^{-5}$

     $= \dfrac{1}{x^5}$    Writing with a positive exponent

**5.** $(a^3b^2)(a^3b^5) = (a^3a^3)(b^2b^5)$    Using the associative law and the commutative law

     $= a^6b^7$    Adding exponents

> Do Margin Exercises 1–5.

## OBJECTIVES

**a** Use the product rule to multiply exponential expressions with like bases.

**b** Use the quotient rule to divide exponential expressions with like bases.

**c** Use the power rule to raise powers to powers.

**d** Raise a product to a power and a quotient to a power.

**SKILL TO REVIEW**
Objective 2.3a: Subtract integers and simplify combinations of additions and subtractions.

Simplify.

**1.** $2 - (-4)$      **2.** $-10 - (-6)$

------ *Caution!* ------

Use the Product Rule correctly.

$2^4 \cdot 2^{10} = 2^{14}$

$2^4 \cdot 2^{10} \neq 4^{14}$    Incorrect!

$2^4 \cdot 2^{10} \neq 2^{40}$    Incorrect!

Simplify.

**1.** $x^4 \cdot x^9$      **2.** $y^{10} \cdot y$

**3.** $2^4 \cdot 2^2 \cdot 2^3$      **4.** $7^{-1} \cdot 7^4$

**5.** $(xy^3)(x^4y^5)$

***Answers***

*Skill to Review:*
1. 6    2. $-4$

*Margin Exercises:*
1. $x^{13}$    2. $y^{11}$    3. $2^9$    4. $7^3$    5. $x^5y^8$

## b Dividing Powers with Like Bases

Recall that any expression that is divided or multiplied by 1 is unchanged. This, together with the fact that anything (besides 0) divided by itself is 1, can lead to a rule for division:

$$\frac{a^5}{a^2} = \frac{a \cdot a \cdot a \cdot a \cdot a}{a \cdot a} = \frac{a \cdot a \cdot a}{1} \cdot \frac{a \cdot a}{a \cdot a} = \frac{a \cdot a \cdot a}{1} \cdot 1 = a^3.$$

Note that the exponent in $a^3$ is the difference of the exponents in $\frac{a^5}{a^2}$; that is, $5 - 2 = 3$. Similarly,

$$\frac{x^4}{x^3} = \frac{x \cdot x \cdot x \cdot x}{x \cdot x \cdot x} = \frac{x}{1} \cdot \frac{x \cdot x \cdot x}{x \cdot x \cdot x} = \frac{x}{1} \cdot 1 = x^1, \text{ or } x.$$

Subtracting the exponents gives the correct result: $4 - 3 = 1$.

------- *Caution!* -------

Use the Quotient Rule correctly.

$$\frac{7^{10}}{7^2} = 7^8$$

$$\frac{7^{10}}{7^2} \ne 1^8 \qquad \text{Incorrect!}$$

$$\frac{7^{10}}{7^2} \ne 7^5 \qquad \text{Incorrect!}$$

### THE QUOTIENT RULE

For any nonzero number $a$ and any integers $m$ and $n$,

$$\frac{a^m}{a^n} = a^{m-n}.$$

(To divide powers with the same base, subtract the exponent of the denominator from the exponent of the numerator.)

**EXAMPLES** Divide and simplify. Here "simplify" means to write the expression as one base with a positive exponent whenever possible.

**6.** $\dfrac{4^7}{4^2} = 4^{7-2}$  Subtracting exponents

$\qquad = 4^5$

**7.** $\dfrac{x^9}{x^{-3}} = x^{9-(-3)} = x^{12}$  Use parentheses when subtracting negative exponents.

**8.** $\dfrac{p}{p^5} = \dfrac{p^1}{p^5}$  Writing $p$ as $p^1$

$\qquad = p^{1-5}$

$\qquad = p^{-4}$

$\qquad = \dfrac{1}{p^4}$  Writing with a positive exponent

**9.** $\dfrac{3^{10}}{3^{10}} = 3^{10-10} = 3^0 = 1$   $a^0 = 1$

**10.** $\dfrac{c^8 d^{12}}{c^3 d^{10}} = \dfrac{c^8}{c^3} \cdot \dfrac{d^{12}}{d^{10}}$

$\qquad = c^{8-3} d^{12-10}$

$\qquad = c^5 d^2$

Do Exercises 6–10.

Simplify.

**6.** $\dfrac{x^5}{x^3}$

**7.** $\dfrac{a^8}{a}$

**8.** $\dfrac{4^{-3}}{4^{10}}$

**9.** $\dfrac{x^{11}}{x^{11}}$

**10.** $\dfrac{x^2 y^5}{x y^4}$

**Answers**

**6.** $x^2$  **7.** $a^7$  **8.** $\dfrac{1}{4^{13}}$

**9.** 1  **10.** $xy$

## c Raising a Power to a Power

Consider an expression like $(7^2)^4$:

$$(7^2)^4 = (7^2)(7^2)(7^2)(7^2) \qquad \text{There are four factors of } 7^2.$$
$$= (7 \cdot 7)(7 \cdot 7)(7 \cdot 7)(7 \cdot 7)$$
$$= 7 \cdot 7 \cdot 7 \cdot 7 \cdot 7 \cdot 7 \cdot 7 \cdot 7 \qquad \text{Using an associative law}$$
$$= 7^8.$$

Note that the exponent in $7^8$ is the product of the exponents in $(7^2)^4$. Similarly,

$$(y^5)^3 = y^5 \cdot y^5 \cdot y^5 \qquad \text{There are three factors of } y^5.$$
$$= (y \cdot y \cdot y \cdot y \cdot y)(y \cdot y \cdot y \cdot y \cdot y)(y \cdot y \cdot y \cdot y \cdot y)$$
$$= y^{15}.$$

Once again, we get the same result if we multiply exponents:

$$(y^5)^3 = y^{5 \cdot 3} = y^{15}.$$

> ### THE POWER RULE
>
> For any nonzero number $a$ and any integers $m$ and $n$,
>
> $$(a^m)^n = a^{mn}.$$
>
> (To raise a power to a power, multiply the exponents and leave the base unchanged.)

**EXAMPLES** Simplify. Express the answer using positive exponents.

**11.** $(2^4)^3 = 2^{4 \cdot 3} \qquad$ Multiplying exponents
$\phantom{(2^4)^3} = 2^{12}$

**12.** $(y^{-3})^8 = y^{-3 \cdot 8} = y^{-24} = \dfrac{1}{y^{24}}$

**13.** $(x^{-2})^{-10} = x^{(-2)(-10)} = x^{20}$

> Do Exercises 11–13.

Simplify.

**11.** $(3^4)^5$

**12.** $(5^{-2})^8$

**13.** $(t^{-8})^{-5}$

## d Raising a Product or a Quotient to a Power

When an expression inside parentheses is raised to a power, the inside expression is the base. Let's compare $2a^3$ and $(2a)^3$.

$2a^3 = 2 \cdot a \cdot a \cdot a \qquad$ The base is $a$. $\qquad (2a)^3 = (2a)(2a)(2a) \qquad$ The base is $2a$.
$$= (2 \cdot 2 \cdot 2)(a \cdot a \cdot a)$$
$$= 2^3 a^3$$
$$= 8a^3$$

We see that $2a^3$ and $(2a)^3$ are *not* equivalent. Note too that $(2a)^3$ can be simplified by cubing each factor in $2a$. This leads to the following rule for raising a product to a power.

> ### RAISING A PRODUCT TO A POWER
>
> For any nonzero numbers $a$ and $b$ and any integer $n$,
>
> $$(ab)^n = a^n b^n.$$
>
> (To raise a product to a power, raise each factor to that power.)

*Answers*

**11.** $3^{20}$    **12.** $\dfrac{1}{5^{16}}$    **13.** $t^{40}$

**EXAMPLES** Simplify.

**14.** $(3a^4)^2 = (3^1a^4)^2$       Writing 3 as $3^1$

           $= (3^1)^2 \cdot (a^4)^2$       Raising each factor to the second power

           $= 3^2 \cdot a^8 = 9a^8$       Multiplying exponents and simplifying

**15.** $(-5x^5y^8)^3 = (-5)^3(x^5)^3(y^8)^3$       Raising each factor to the third power

           $= -125x^{15}y^{24}$

Do Exercises 14 and 15.

Simplify.

**14.** $(x^3y^8)^5$

**15.** $(-2ab^4)^2$

There is a similar rule for raising a quotient to a power.

---

### RAISING A QUOTIENT TO A POWER

For any nonzero numbers $a$ and $b$ and any integer $n$,

$$\left(\frac{a}{b}\right)^n = \frac{a^n}{b^n}.$$

(To raise a quotient to a power, raise the numerator to the power and divide by the denominator to the power.)

---

**EXAMPLES**

**16.** $\left(\frac{x}{7}\right)^2 = \frac{x^2}{7^2} = \frac{x^2}{49}$       Squaring the numerator and the denominator

**17.** $\left(\frac{5}{a^4}\right)^3 = \frac{5^3}{(a^4)^3}$       Raising a quotient to a power

           $= \frac{5^3}{a^{12}} = \frac{125}{a^{12}}$       Using the power rule (multiplying exponents) and simplifying

Do Exercises 16 and 17.

Simplify.

**16.** $\left(\frac{a}{b}\right)^8$          **17.** $\left(\frac{x^4}{y^5}\right)^2$

In the following summary of definitions and rules, we assume that no denominators are 0 and $0^0$ is not considered.

---

### DEFINITIONS AND PROPERTIES OF EXPONENTS

For any integers $m$ and $n$,

| | |
|---|---|
| 1 as an exponent: | $a^1 = a$ |
| 0 as an exponent: | $a^0 = 1$ |
| Negative exponents: | $a^{-n} = \dfrac{1}{a^n}$ |
| The Product Rule: | $a^m \cdot a^n = a^{m+n}$ |
| The Quotient Rule: | $\dfrac{a^m}{a^n} = a^{m-n}$ |
| The Power Rule: | $(a^m)^n = a^{mn}$ |
| Raising a product to a power: | $(ab)^n = a^nb^n$ |
| Raising a quotient to a power: | $\left(\dfrac{a}{b}\right)^n = \dfrac{a^n}{b^n}$ |

---

*Answers*

**14.** $x^{15}y^{40}$    **15.** $4a^2b^8$    **16.** $\dfrac{a^8}{b^8}$    **17.** $\dfrac{x^8}{y^{10}}$

**a**  Simplify. Assume that no denominator is 0.

**1.** $3^4 \cdot 3^2$

**2.** $4^5 \cdot 4^7$

**3.** $x^2 \cdot x^{12}$

**4.** $y^3 \cdot y^9$

**5.** $a^6 \cdot a$

**6.** $x \cdot x^4$

**7.** $8 \cdot 8^5$

**8.** $2^{10} \cdot 2$

**9.** $5^{17} \cdot 5^{21}$

**10.** $4^{32} \cdot 4^{16}$

**11.** $x^2 \cdot x \cdot x^5$

**12.** $y \cdot y^4 \cdot y^8$

**13.** $a^{-5} \cdot a^8$

**14.** $c^{12} \cdot c^{-9}$

**15.** $p^6 \cdot p^{-5}$

**16.** $t^{-4} \cdot t^5$

**17.** $x^{-5} \cdot x^{-7}$

**18.** $y^{-3} \cdot y^{-4}$

**19.** $5^4 \cdot 5^{-10}$

**20.** $6^{-12} \cdot 6^2$

**21.** $(a^3b^5)(ab^9)$

**22.** $(x^2y^{10})(x^5y)$

**23.** $a^{-4} \cdot a^9 \cdot a^{-6}$

**24.** $x^{-11} \cdot x^4 \cdot x^6$

**b**  Simplify. Assume that no denominator is 0.

**25.** $\dfrac{8^9}{8^2}$

**26.** $\dfrac{3^{11}}{3^3}$

**27.** $\dfrac{x^4}{x^3}$

**28.** $\dfrac{x^{12}}{x^{11}}$

**29.** $\dfrac{a^{10}}{a}$

**30.** $\dfrac{c^{12}}{c}$

**31.** $\dfrac{x^2}{x^5}$

**32.** $\dfrac{y^4}{y^{12}}$

**33.** $\dfrac{7^8}{7^8}$

**34.** $\dfrac{10^6}{10^6}$

**35.** $\dfrac{t^2}{t^{-1}}$

**36.** $\dfrac{n^{-5}}{n^8}$

**37.** $\dfrac{c^{-2}}{c^{-6}}$

**38.** $\dfrac{t^{-2}}{t}$

**39.** $\dfrac{x^4y^{12}}{xy^7}$

**40.** $\dfrac{u^9v^{15}}{u^8v}$

**c** Simplify. Assume that no denominator is 0.

**41.** $(x^2)^5$

**42.** $(y^5)^7$

**43.** $(2^{-3})^6$

**44.** $(3^{-2})^4$

**45.** $(7^{-1})^{-6}$

**46.** $(9^{-4})^{-3}$

**47.** $(y^{12})^{-1}$

**48.** $(x^{-18})^{-1}$

**d** Simplify. Assume that no denominator is 0.

**49.** $(x^2y^5)^2$

**50.** $(a^3b^4)^2$

**51.** $(2x^2y)^3$

**52.** $(3xy^3)^2$

**53.** $(-2a^2b^5)^3$

**54.** $(-5a^4b^2)^3$

**55.** $(-9x^2y^2)^2$

**56.** $(-10x^5y)^2$

**57.** $\left(\dfrac{a}{b}\right)^7$

**58.** $\left(\dfrac{x}{y}\right)^{10}$

**59.** $\left(\dfrac{a^2}{2}\right)^3$

**60.** $\left(\dfrac{x^4}{5}\right)^2$

**61.** $\left(\dfrac{x^4}{y^5}\right)^3$

**62.** $\left(\dfrac{a^6}{b^{10}}\right)^4$

**63.** $\left(\dfrac{2^4}{3^5}\right)^2$

**64.** $\left(\dfrac{5^9}{3^5}\right)^4$

## Skill Maintenance

**65.** A sidewalk of uniform width is built around three sides of a store, as shown in the figure. What is the area of the sidewalk? [1.8a]

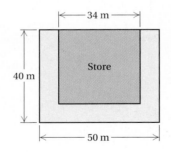

**66.** A real estate agent's commission rate is 6%. A commission of $7380 is received on the sale of a home. For how much did the home sell? [8.6b]

Add. [2.2a]

**67.** $-5 + (-12)$

**68.** $-19 + 12$

**69.** $17 + (-24)$

**70.** $-15 + (-2)$

## Synthesis

Simplify. Assume that no denominator is 0.

**71.** $(3y)^{10}(3y)^6$

**72.** $(x^2y^3z^8)^0$

**73.** $\left(\dfrac{x}{y}\right)^{-1}$

**74.** $a^x \cdot a^{2x}$

**75.** $\dfrac{x^{3y}}{x^{2y}}$

**76.** $(a^{-3}b^2)^{-4}$

**77.** $\dfrac{a^{33}}{a^{3(11)}}$

**78.** $\left(\dfrac{1}{a}\right)^{-n}$

# 10.3 Scientific Notation

## a  Writing Scientific Notation

There are many kinds of symbols, or notation, for numbers. You are already familiar with fraction notation, decimal notation, and percent notation. Now we study another, **scientific notation,** which makes use of exponential notation. Scientific notation is especially useful when calculations involve very large or very small numbers. The following are examples of scientific notation:

① *The diameter of the Milky Way:*

$$9.46 \times 10^{17} \text{ km} = 946{,}000{,}000{,}000{,}000{,}000 \text{ km}$$

② *The mass of the water in the earth's oceans:*

$$1.39 \times 10^{21} \text{ kg} = 1{,}390{,}000{,}000{,}000{,}000{,}000{,}000 \text{ kg}$$

③ *The length of a hepatitis B virus:*

$$4.5 \times 10^{-8} \text{ m} = 0.000000045 \text{ m}$$

**OBJECTIVES**

**a** Convert between scientific notation and decimal notation.

**b** Multiply and divide using scientific notation.

**c** Solve applied problems using scientific notation.

①

②

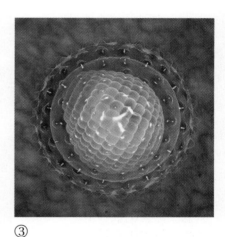

③

---

### SCIENTIFIC NOTATION

**Scientific notation** for a number is an expression of the type

$$M \times 10^n,$$

where $n$ is an integer, $M$ is greater than or equal to 1 and less than 10 ($1 \leq M < 10$), and $M$ is expressed in decimal notation.

---

You should try to make conversions to scientific notation mentally as much as possible. To do so, note the following.

In scientific notation, a positive power of 10 indicates a large number (greater than or equal to 10) and a negative power of 10 indicates a small number (between 0 and 1).

Converting from scientific notation to decimal notation involves multiplying by a power of 10. To convert $M \times 10^n$ to decimal notation, we move the decimal point.

- When $n$ is positive, we move the decimal point right $n$ places.
- When $n$ is negative, we move the decimal point left $|n|$ places.

**EXAMPLES** Convert each number to decimal notation.

**1.** $7.893 \times 10^5 = 789{,}300$     The decimal point moves right 5 places.

*Think*: Positive exponent indicates large number.

*Check*: $7.893 \times 10^5 = 7.893 \times 100{,}000 = 789{,}300$

**2.** $4.7 \times 10^{-8} = 0.000000047$     The decimal point moves left 8 places.

*Think*: Negative exponent indicates small number.

*Check*: $4.7 \times 10^{-8} = 4.7 \times 0.00000001 = 0.000000047$

Do Exercises 1 and 2.

We reverse the process when converting from decimal notation to scientific notation. To convert a number to scientific notation, we write it in the form $M \times 10^n$, where $M$ looks like the original number but with exactly one nonzero digit to the immediate left of the decimal point. Compare the following.

| *Scientific Notation* | *NOT Scientific Notation* | |
|---|---|---|
| $1.98 \times 10^{12}$ | $19.8 \times 10^{11}$ | 19.8 is greater than 10. |
| $2.007 \times 10^{-4}$ | $0.2007 \times 10^{-3}$ | 0.2007 is less than 1. |

Do Exercises 3–6.

**EXAMPLES** Convert each number to scientific notation.

**3.** $78{,}000 = 78{,}000. = 7.8000 \times 10^4 = 7.8 \times 10^4$

            4 places

*Check*: $7.8 \times 10^4 = 7.8 \times 10{,}000 = 78{,}000$

**4.** $0.0000923 = 000009.23 \times 10^{-5} = 9.23 \times 10^{-5}$

            5 places

*Check*: $9.23 \times 10^{-5} = 9.23 \times 0.00001 = 0.0000923$

Do Exercises 7 and 8.

## (b) Multiplying and Dividing Using Scientific Notation

### Multiplying

Consider the product $400 \cdot 2000 = 800{,}000$. In scientific notation, this is

$$(4 \times 10^2) \cdot (2 \times 10^3) = (4 \cdot 2)(10^2 \cdot 10^3) = 8 \times 10^5.$$

By applying the commutative and associative laws, we found this product by multiplying $4 \cdot 2$, to get 8, and $10^2 \cdot 10^3$, to get $10^5$ (adding exponents).

---

Convert each number to decimal notation.

**1.** $6.893 \times 10^{11}$

**2.** $5.67 \times 10^{-5}$

Determine whether each number is written in scientific notation.

**3.** $8.06 \times 10^3$

**4.** $0.49 \times 10^{12}$

**5.** $10.608 \times 10^{-3}$

**6.** $1.008 \times 10^{-6}$

Convert each number to scientific notation.

**7.** $0.000517$

**8.** $523{,}000{,}000$

**Answers**

**1.** 689,300,000,000    **2.** 0.0000567
**3.** Yes   **4.** No   **5.** No   **6.** Yes
**7.** $5.17 \times 10^{-4}$    **8.** $5.23 \times 10^8$

**EXAMPLE 5**   Multiply: $(1.8 \times 10^6) \cdot (2.3 \times 10^{-4})$.

We apply the commutative and associative laws to get

$$(1.8 \times 10^6) \cdot (2.3 \times 10^{-4}) = (1.8 \cdot 2.3) \times (10^6 \cdot 10^{-4})$$
$$= 4.14 \times 10^{6+(-4)}$$
$$= 4.14 \times 10^2.$$

**EXAMPLE 6**   Multiply: $(3.1 \times 10^5) \cdot (4.5 \times 10^{-3})$.

$(3.1 \times 10^5) \cdot (4.5 \times 10^{-3}) = (3.1 \times 4.5)(10^5 \cdot 10^{-3})$

$\qquad = 13.95 \times 10^2$   Not scientific notation; 13.95 is greater than 10.

$\qquad = (1.395 \times 10^1) \times 10^2$   Substituting $1.395 \times 10^1$ for 13.95

$\qquad = 1.395 \times (10^1 \times 10^2)$   Using the associative law

$\qquad = 1.395 \times 10^3$   Adding exponents; the answer is now in scientific notation.

Do Exercises 9 and 10.

Do Exercises 9 and 10.

Multiply and write scientific notation for the result.

**9.** $(1.12 \times 10^{-8})(5 \times 10^{-7})$

**10.** $(9.1 \times 10^{-17})(8.2 \times 10^3)$

## Dividing

Consider the quotient $800{,}000 \div 400 = 2000$. In scientific notation, this is

$$(8 \times 10^5) \div (4 \times 10^2) = \frac{8 \times 10^5}{4 \times 10^2} = \frac{8}{4} \times 10^{5-2} = 2 \times 10^3.$$

**EXAMPLE 7**   Divide: $(3.41 \times 10^5) \div (1.1 \times 10^{-3})$.

$$(3.41 \times 10^5) \div (1.1 \times 10^{-3}) = \frac{3.41 \times 10^5}{1.1 \times 10^{-3}}$$
$$= \frac{3.41}{1.1} \times \frac{10^5}{10^{-3}}$$
$$= 3.1 \times 10^{5-(-3)}$$
$$= 3.1 \times 10^8$$

**EXAMPLE 8**   Divide: $(3.2 \times 10^{-7}) \div (8.0 \times 10^6)$.

$(3.2 \times 10^{-7}) \div (8.0 \times 10^6) = \dfrac{3.2 \times 10^{-7}}{8.0 \times 10^6}$

$\qquad = \dfrac{3.2}{8.0} \times \dfrac{10^{-7}}{10^6}$

$\qquad = 0.4 \times 10^{-13}$   Not scientific notation; 0.4 is less than 1.

$\qquad = (4.0 \times 10^{-1}) \times 10^{-13}$   Substituting $4.0 \times 10^{-1}$ for 0.4

$\qquad = 4.0 \times (10^{-1} \times 10^{-13})$   Using the associative law

$\qquad = 4.0 \times 10^{-14}$   Adding exponents

Do Exercises 11 and 12.

Do Exercises 11 and 12.

### STUDY TIPS

#### GET SOME REST

The final exam is probably your most important math test of the semester. Do yourself a favor and see to it that you get a good night's sleep the night before. Being well rested will help guarantee that you put forth your best work.

Divide and write scientific notation for each result.

**11.** $\dfrac{4.2 \times 10^5}{2.1 \times 10^2}$

**12.** $\dfrac{1.1 \times 10^{-4}}{2.0 \times 10^{-7}}$

***Answers***

**9.** $5.6 \times 10^{-15}$   **10.** $7.462 \times 10^{-13}$
**11.** $2.0 \times 10^3$   **12.** $5.5 \times 10^2$

# (c) Applications with Scientific Notation

**EXAMPLE 9** *Distance from the Sun to Earth.* Light from the sun traveling at a rate of 300,000 km/s (kilometers per second) reaches Earth in 499 sec. Find the distance, expressed in scientific notation, from the sun to Earth.

The time it takes light to reach Earth from the sun is $4.99 \times 10^2$ sec. The speed is $3.0 \times 10^5$ km/s. Recall that distance can be expressed in terms of speed and time as

$$\text{Distance} = \text{Speed} \cdot \text{Time}.$$

We substitute $3.0 \times 10^5$ for the speed and $4.99 \times 10^2$ for the time:

$$
\begin{aligned}
\text{Distance} &= (3.0 \times 10^5)(4.99 \times 10^2) && \text{Substituting}\\
&= 14.97 \times 10^7 && \text{Note that 14.97 is greater than 10.}\\
&= (1.497 \times 10^1) \times 10^7 \\
&= 1.497 \times (10^1 \times 10^7) && \text{Converting to scientific notation}\\
&= 1.497 \times 10^8 \text{ km.}
\end{aligned}
$$

Thus, the distance from the sun to Earth is $1.497 \times 10^8$ km.

**EXAMPLE 10** *Information Storage.* As of April 2010, Facebook users were uploading 3 billion photographs each month to the social networking site. If one DVD can store about 1500 photographs, how many DVDs would it take to store a month's worth of Facebook photograph uploads?
Source: Facebook

We first write each number in scientific notation:

$$3 \text{ billion} = 3{,}000{,}000{,}000 = 3 \times 10^9;$$
$$1500 = 1.5 \times 10^3.$$

We then divide to find the number of DVDs needed to store the photographs:

$$
\begin{aligned}
\frac{3 \times 10^9}{1.5 \times 10^3} &= \frac{3}{1.5} \times \frac{10^9}{10^3} \\
&= 2 \times 10^6.
\end{aligned}
$$

It would take about $2 \times 10^6$, or 2 million, DVDs to store a month's worth of Facebook photograph uploads.

Do Exercises 13 and 14.

## Calculator Corner

### Scientific Notation

Most scientific calculators have a key labeled EE which is used for scientific notation. We simply enter the decimal portion of the number, press EE , and then enter the exponent. Addition, subtraction, multiplication, and division are performed as usual. Many calculators abbreviate scientific notation. A calculator display like 5ᴇ16 means $5 \times 10^{16}$.

**Exercises:** Multiply or divide and express each answer in scientific notation.

**1.** $(3.15 \times 10^7)(4.3 \times 10^{-12})$

**2.** $(4 \times 10^4)(9 \times 10^7)$

**3.** $\dfrac{4 \times 10^{-9}}{5 \times 10^{16}}$

**4.** $\dfrac{9 \times 10^{11}}{3 \times 10^{-2}}$

**13. Niagara Falls Water Flow.**
During the summer, the amount of water that spills over Niagara Falls in 1 min is about $1.68 \times 10^8$ L. How many liters of water spill over the falls in one day? Express the answer in scientific notation.
Source: www.niagaraparks.com

**14. Earth vs. Saturn.** The mass of Earth is about $6 \times 10^{21}$ metric tons. The mass of Saturn is about $5.7 \times 10^{23}$ metric tons. About how many times the mass of Earth is the mass of Saturn? Express the answer in scientific notation.

*Answers*

**13.** $2.4192 \times 10^{11}$ L    **14.** The mass of Saturn is $9.5 \times 10$ times the mass of Earth.

# Translating
# for Success

Test Items.  On a test of 90 items, Sally got 80% correct. How many items did she get correct?

Suspension Bridge.  The San Francisco/Oakland Bay suspension bridge is 0.4375 mi long. Convert this distance to yards.

Population Growth.  Brookdale's growth rate per year is 0.9%. If the population was 1,500,000 in 2009, what is the population in 2010?

Roller Coaster Drop.  The Manhattan Express Roller Coaster at the New York–New York Hotel and Casino, Las Vegas, Nevada, has a 144-ft drop. The California Screamin' Roller Coaster at Disney's California Adventure, Anaheim, California, has a 32.635-m drop. How much larger in meters is the drop of the Manhattan Express than the drop of the California Screamin'?

Driving Distance.  Nate drives the company car 675 mi in 15 days. At this rate, how far will he drive in 20 days?

---

The goal of these matching questions is to practice step (2), *Translate*, of the five-step problem-solving process. Translate each word problem to an equation and select a correct translation from equations A–O.

**A.**  $32.635 \text{ m} + x = 144 \text{ ft} \cdot \dfrac{0.305 \text{ m}}{1 \text{ ft}}$

**B.**  $x = 0.4375 \text{ mi} \times \dfrac{5280 \text{ ft}}{1 \text{ mi}} \times \dfrac{12 \text{ in.}}{1 \text{ ft}}$

**C.**  $80\% \cdot x = 90$

**D.**  $x = 0.89 \text{ km} \cdot \dfrac{1000 \text{ m}}{1 \text{ km}} \cdot \dfrac{1 \text{ m}}{3.281 \text{ ft}}$

**E.**  $x = 80\% \cdot 90$

**F.**  $x = 420 \text{ m} + 75 \text{ ft} \cdot \dfrac{0.305 \text{ m}}{1 \text{ ft}}$

**G.**  $\dfrac{x}{20} = \dfrac{675}{15}$

**H.**  $x = 0.4375 \text{ mi} \times \dfrac{5280 \text{ ft}}{1 \text{ mi}} \times \dfrac{1 \text{ yd}}{3 \text{ ft}}$

**I.**  $x = 0.89 \text{ km} \cdot \dfrac{0.621 \text{ mi}}{1 \text{ km}} \cdot \dfrac{5280 \text{ ft}}{1 \text{ mi}}$

**J.**  $\dfrac{x}{15} = \dfrac{675}{20}$

**K.**  $x = 420 \text{ m} \cdot \dfrac{3.281 \text{ ft}}{1 \text{ m}} + 75 \text{ ft}$

**L.**  $144 \text{ ft} + x = 32.635 \text{ m} \cdot \dfrac{1 \text{ ft}}{0.305 \text{ m}}$

**M.**  $x = 1{,}500{,}000 - 0.9\%(1{,}500{,}000)$

**N.**  $20 \cdot x = 675$

**O.**  $x = 1{,}500{,}000 + 0.9\%(1{,}500{,}000)$

*Answers on page A-21*

---

**6.** Test Items.  Jason answered correctly 90 items on a recent test. These items represented 80% of the total number of questions. How many items were on the test?

**7.** Population Decline.  Flintville's growth rate per year is −0.9%. If the population was 1,500,000 in 2009, what is the population in 2010?

**8.** Bridge Length.  The Tatara Bridge in Onomichi-Imabari, Japan, is 0.89 km long. Convert this distance to feet.

**9.** Height of Tower.  The Willis Tower in Chicago is 75 ft taller than the Jin Mao Building in Shanghai. The height of the Jin Mao Building is 420 m. What is the height of the Willis Tower in feet?

**10.** Gasoline Usage.  Nate's company car gets 20 miles to the gallon in city driving. How many gallons will it use in 675 mi of city driving?

**a**    Convert each number to scientific notation.

**1.** 28,000,000,000

**2.** 4,900,000,000,000

**3.** 907,000,000,000,000,000

**4.** 168,000,000,000,000

**5.** 0.00000304

**6.** 0.000000000865

**7.** 0.000000018

**8.** 0.00000000002

**9.** 100,000,000,000

**10.** 0.0000001

Convert the number in each sentence to scientific notation.

**11.** *Population of the United States.*   In 2010, the population of the United States was about 309 million. $(1 \text{ million} = 10^6)$.
   **Source:** U.S. Bureau of the Census

**12.** *Racing.*   Total revenue of Speedway Motorsports Inc. was $551 million in 2009.
   **Source:** finance.yahoo.com

**13.** *Hair.*   A human hair is about 4 hundred-thousandths of a meter in diameter.

**14.** *DNA.*   A strand of DNA is about 2 ten-billionths of a centimeter in diameter.

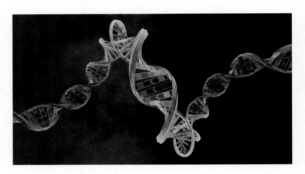

Convert each number to decimal notation.

**15.** $8.74 \times 10^7$

**16.** $1.85 \times 10^8$

**17.** $5.704 \times 10^{-8}$

**18.** $8.043 \times 10^{-4}$

**19.** $10^7$

**20.** $10^6$

**21.** $10^{-5}$

**22.** $10^{-8}$

**b**    Multiply or divide and write scientific notation for each result.

**23.** $(3 \times 10^4)(2 \times 10^5)$

**24.** $(3.9 \times 10^8)(8.4 \times 10^{-3})$

**25.** $(5.2 \times 10^5)(6.5 \times 10^{-2})$

**26.** $(7.1 \times 10^{-7})(8.6 \times 10^{-5})$

**27.** $(9.9 \times 10^{-6})(8.23 \times 10^{-8})$

**28.** $(1.123 \times 10^4) \times 10^{-9}$

**29.** $\dfrac{8.5 \times 10^8}{3.4 \times 10^{-5}}$

**30.** $\dfrac{5.6 \times 10^{-2}}{2.5 \times 10^5}$

**31.** $(3.0 \times 10^6) \div (6.0 \times 10^9)$

**32.** $(1.5 \times 10^{-3}) \div (1.6 \times 10^{-6})$

**33.** $\dfrac{7.5 \times 10^{-9}}{2.5 \times 10^{12}}$

**34.** $\dfrac{4.0 \times 10^{-3}}{8.0 \times 10^{20}}$

 Solve.

**35.** *Coral Reefs.* There are 10 million bacteria per square centimeter of coral in a coral reef. The coral reefs near the Hawaiian Islands cover 14,000 km². How many bacteria are there in Hawaii's coral reef?

Source: livescience.com; U.S. Geological Survey

**36.** *River Discharge.* The average discharge at the mouths of the Amazon River is 4,200,000 cubic feet per second. How much water is discharged from the Amazon River in 1 yr? Express the answer in scientific notation.

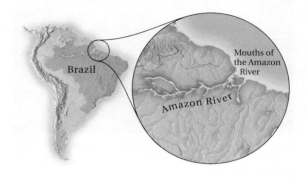

**37.** *Earth vs. Jupiter.* The mass of Earth is about $6 \times 10^{21}$ metric tons. The mass of Jupiter is about $1.908 \times 10^{24}$ metric tons. About how many times the mass of Earth is the mass of Jupiter? Express the answer in scientific notation.

**38.** *Water Contamination.* In the United States, 200 million gal of used motor oil are improperly disposed of each year. One gallon of used oil can contaminate 1 million gal of drinking water. How many gallons of drinking water can 200 million gal of oil contaminate? Express the answer in scientific notation.

Source: *The Macmillan Visual Almanac*

*Space Travel.* Use the following information for Exercises 39 and 40.

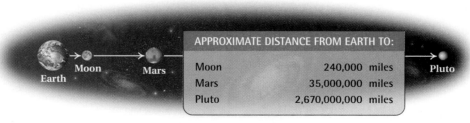

| APPROXIMATE DISTANCE FROM EARTH TO: | |
| --- | --- |
| Moon | 240,000 miles |
| Mars | 35,000,000 miles |
| Pluto | 2,670,000,000 miles |

**39.** *Time to Reach Mars.* Suppose that it takes about 3 days for a spacecraft to travel from Earth to the moon. About how long would it take the same spacecraft traveling at the same speed to reach Mars? Express the answer in scientific notation.

**40.** *Time to Reach Pluto.* Suppose that it takes about 3 days for a spacecraft to travel from Earth to the moon. About how long would it take the same spacecraft traveling at the same speed to reach Pluto? Express the answer in scientific notation.

**41.** *Stars.* It is estimated that there are 10 billion trillion stars in the universe. Express the number of stars in scientific notation (1 billion = $10^9$; 1 trillion = $10^{12}$).

**42.** *Closest Star.* Excluding the sun, the closest star to Earth is Proxima Centauri, which is 4.3 light-years away (one light-year = $5.88 \times 10^{12}$ mi). How far, in miles, is Proxima Centauri from Earth? Express the answer in scientific notation.

**43.** *Red Light.* The wavelength of light is given by the velocity divided by the frequency. The velocity of red light is 300,000,000 m/sec, and its frequency is 400,000,000,000,000 cycles per second. What is the wavelength of red light? Express the answer in scientific notation.

**44.** *Earth vs. Sun.* The mass of Earth is about $6 \times 10^{21}$ metric tons. The mass of the sun is about $1.998 \times 10^{27}$ metric tons. About how many times the mass of Earth is the mass of the sun? Express the answer in scientific notation.

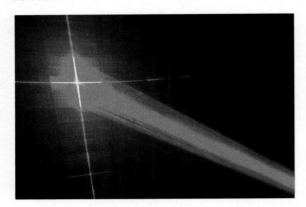

## Skill Maintenance

Solve. [5.7b]

**45.** $2x - 4 - 5x + 8 = x - 3$

**46.** $4(x - 3) + 5 = 6(x + 2) - 8$

Graph. [6.4b]

**47.** $y = x - 5$

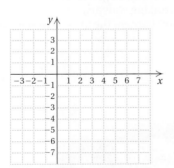

**48.** $y = -2x + 8$

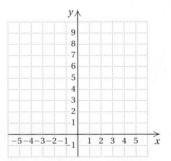

## Synthesis

**49.** 🖩 Carry out the indicated operations. Express the result in scientific notation.
$$\frac{(5.2 \times 10^6)(6.1 \times 10^{-11})}{1.28 \times 10^{-3}}$$

**50.** Find the reciprocal and express it in scientific notation.
$$6.25 \times 10^{-3}$$

**51.** Find the LCM for $6.4 \times 10^8$ and $1.28 \times 10^4$.

# Mid-Chapter Review

## Concept Reinforcement

Determine whether each statement is true or false.

_____ **1.** Raising a number to a negative power may not result in a negative number.   [10.1b]

_____ **2.** Any number raised to the zero power is zero.   [10.1a]

_____ **3.** In scientific notation, if the exponent is negative, the number is very large.   [10.3a]

## Guided Solutions

Fill in each blank with the number that creates a correct statement or solution.

**4.** $y^2 \cdot y^{10} = y^{\square+10} = y^{\square}$   [10.2a]

**5.** $\dfrac{y^2}{y^{10}} = y^{\square-10} = y^{\square} = \dfrac{1}{y^{\square}}$   [10.2b]

**6.** Convert to scientific notation.
$$4{,}032{,}000{,}000{,}000 = \square \times 10^{\square}$$   [10.3a]

**7.** Convert to scientific notation.
$$0.000038 = \square \times 10^{\square}$$   [10.3a]

## Mixed Review

Evaluate.   [10.1a]

**8.** $(-12)^0$

**9.** $(-12)^1$

**10.** $-12^0$

**11.** $(12 - 6t)^1$ for $t = 5$

Write an equivalent expression with positive exponents.   [10.1b]

**12.** $x^{-7}$

**13.** $3^{-1}$

**14.** $9x^{-6}$

**15.** $\dfrac{1}{x^{-4}}$

Simplify. Assume that no denominator is 0.   [10.2a, b, c, d]

**16.** $3^4 \cdot 3^{11}$

**17.** $6 \cdot 6^9$

**18.** $x^{-6} \cdot x^{17}$

**19.** $y^{-2} \cdot y \cdot y^{-3}$

**20.** $(x^2y^8)(x^4y^3)$

**21.** $\dfrac{8^{11}}{8^2}$

**22.** $\dfrac{x^8}{x^9}$

**23.** $\dfrac{d^5}{d^5}$

**24.** $\dfrac{n^4}{n^{-1}}$

**25.** $\dfrac{a^{12}b^{11}}{a^4b^7}$

**26.** $(x^4)^8$

**27.** $(y^{-2})^{-6}$

**28.** $(x^{-3})^1$

**29.** $(5^4)^{-12}$

**30.** $(x^3y^4)^5$

**31.** $(3x^2y^4)^2$

**32.** $(-2ab^5)^3$

**33.** $\left(\dfrac{c}{d}\right)^{10}$

**34.** $\left(\dfrac{2}{x^3}\right)^2$

**35.** $\left(\dfrac{a^4}{b^8}\right)^7$

Convert each number to scientific notation.   [10.3a]

**36.** 0.000907

**37.** 431,080,000,000

Convert each number to decimal notation.   [10.3a]

**38.** $2.08 \times 10^8$

**39.** $4.1 \times 10^{-6}$

**40.** The wavelength of a radio wave is given by the velocity divided by the frequency. The velocity of radio waves is approximately 300,000,000 m/sec, and the frequency of Rick's ham radio repeater is 1,200,000,000 cycles per sec. What is the wavelength of Rick's repeater frequency? Express the answer in scientific notation.   [10.3c]

**41.** A good length for a radio antenna is one fourth of the wavelength of the frequency it is designed to receive. What should the length of such a *quarterwave* antenna be for the frequency described in Exercise 40?   [10.3c]

**42.** Multiply. Write scientific notation for the result: $(4.5 \times 10^{12})(5.2 \times 10^{20})$.   [10.3b]
  **A.** $23.4 \times 10^{240}$    **B.** $23.4 \times 10^{32}$
  **C.** $2.34 \times 10^{33}$    **D.** $2.34 \times 10^{31}$

**43.** Divide. Write scientific notation for the result:
$\dfrac{2.4 \times 10^{-6}}{4.8 \times 10^{-20}}$.   [10.3b]
  **A.** $5.0 \times 10^{13}$    **B.** $0.5 \times 10^{14}$
  **C.** $0.5 \times 10^{-26}$    **D.** $5.0 \times 10^{-15}$

# Understanding Through Discussion and Writing

**44.** Under what circumstances is $a^0 > a^1$?   [10.1a]

**45.** Consider the expresssion $x^{-3}$. When evaluated, will the expression ever be negative? Explain.   [10.1b]

**46.** Which number is larger and why: $5^{-8}$ or $6^{-8}$? Do not use a calculator.   [10.1b]

**47.** Emma can give an answer using the unit km or mm. Which measurement would require the larger exponent if scientific notation were used? Why?   [10.3c]

# 10.4

## Addition and Subtraction of Polynomials

In Section 2.7, we defined a *term* as a number, a variable, a product of numbers and/or variables, or a quotient of numbers and/or variables. Thus, expressions like

$$5x^2, \quad -34, \quad \frac{3}{4}ab^2, \quad xy^3z^5, \quad \text{and} \quad \frac{7n}{m}$$

are terms. A term is called a **monomial** if there is no division by a variable expression. All of the terms above, except for $\frac{7n}{m}$, are monomials. Monomials are used to form **polynomials** like the following:

$$a^2b + c^3, \quad 5y + 3, \quad 3x^2 + 2x - 5, \quad -7a^3 + \tfrac{1}{2}a, \quad 37p^4, \quad x, \quad 0.$$

> ### POLYNOMIAL
>
> A **polynomial** is a monomial or a combination of sums and/or differences of monomials.

In a monomial, the number multiplied by the variable or variables is called the **coefficient**. The coefficient of $-6a^2b$ is $-6$. The coefficient of $\frac{1}{3}x^4$ is $\frac{1}{3}$. Since $x^2 = 1 \cdot x^2$, the coefficient of $x^2$ is 1.

Do Margin Exercises 1 and 2.

## a Adding Polynomials

Addition is commutative and associative; that is, we can reorder and regroup addends without changing the sum. We use these properties when adding polynomials in order to pair "like" terms. Recall that when two terms have the same variable(s) raised to the same power(s), they are like terms and can be combined.

**EXAMPLE 1** Add: $(5x^3 + 4x^2 + 3x) + (2x^3 + 5x^2 - x)$.

$(5x^3 + 4x^2 + 3x) + (2x^3 + 5x^2 - x)$      $5x^3$ and $2x^3$ are like terms; $4x^2$ and $5x^2$ are like terms; $3x$ and $-x$ are like terms

$= (5x^3 + 2x^3) + (4x^2 + 5x^2) + (3x - x)$      Using the commutative and associative laws to pair like terms

$= 7x^3 + 9x^2 + 2x$      Combining like terms. Remember that $x = 1x$.

## OBJECTIVES

**a** Add polynomials.

**b** Find the opposite of a polynomial.

**c** Subtract polynomials.

**d** Evaluate a polynomial.

### SKILL TO REVIEW
Objective 2.7a: Combine like terms.

Combine like terms.
 1. $8x^2 - 3x + 4x - x^2$
 2. $2x^3 - 6 - 10 - 3x^3 + 9x$

1. What is the coefficient of $19x^3$?

2. What is the coefficient of $-a^2$?

*Answers*

*Skill to Review:*
**1.** $7x^2 + x$   **2.** $-x^3 + 9x - 16$

*Margin Exercises:*
**1.** 19   **2.** $-1$

**EXAMPLE 2** Add: $(3a^2 + 7a^2b) + (5a^2 - 6ab^2)$.

$$(3a^2 + 7a^2b) + (5a^2 - 6ab^2) = (3a^2 + 5a^2) + 7a^2b - 6ab^2$$
$$= 8a^2 + 7a^2b - 6ab^2 \qquad \text{Combining like terms}$$

**EXAMPLE 3** Add: $(7x^2 + 5) + (5x^3 + 4x)$.

$$(7x^2 + 5) + (5x^3 + 4x) = 7x^2 + 5 + 5x^3 + 4x \qquad \text{There are no like terms here.}$$
$$= 5x^3 + 7x^2 + 4x + 5 \qquad \text{Rearranging the order}$$

Note that in Example 3 we wrote the answer so that the powers of $x$ decrease as we read from left to right. This **descending order** is the traditional way of expressing an answer.

Do Exercises 3–5.

**b)  Opposites of Polynomials**

To subtract a number, we can add its opposite. We can similarly subtract a polynomial by adding its opposite. To check if two polynomials are opposites, recall that 5 and $-5$ are opposites because $5 + (-5) = 0$.

> **THE OPPOSITE OF A POLYNOMIAL**
>
> Two polynomials are **opposites**, or **additive inverses**, of each other if their sum is zero.

The opposite of $-9x^2 + x - 7$ is $9x^2 - x + 7$ because

$$(-9x^2 + x - 7) + (9x^2 - x + 7) = -9x^2 + 9x^2 + x + (-x) + (-7) + 7$$
$$= 0.$$

This can be said using algebraic symbolism:

The opposite of $(-9x^2 + x - 7)$ is $9x^2 - x + 7$.

$$-(-9x^2 + x - 7) = 9x^2 - x + 7$$

Note that in the opposite polynomials, the terms in each pair are opposites.

$-9x^2$ and $9x^2$ are opposites.

$x$ and $-x$ are opposites.

$-7$ and $7$ are opposites.

> **TO FIND THE OPPOSITE OF A POLYNOMIAL**
>
> We can find an equivalent polynomial for the opposite, or additive inverse, of a polynomial by replacing each term with its opposite—that is, *changing the sign of every term*.

---

Add.

**3.** $(7a^2 + 2a + 8) + (2a^2 + a - 9)$

**4.** $(5x^2y + 3x^2 + 4) + (2x^2y + 4x)$

**5.** $(2a^3 + 17) + (2a^2 - 9a)$

---

**STUDY TIPS**

**ASK TO SEE YOUR FINAL**

Once the course is over, many students neglect to find out how they fared on the final exam. Please don't overlook this valuable opportunity to extend your learning. It is important for you to find out what mistakes you may have made and to be certain no grading errors occurred.

*Answers*

**3.** $9a^2 + 3a - 1$
**4.** $7x^2y + 3x^2 + 4x + 4$
**5.** $2a^3 + 2a^2 - 9a + 17$

**EXAMPLE 4** Find two equivalent expressions for the opposite of

$$4x^5 - 7x^3 - 8x + \tfrac{5}{6}.$$

**a)** $-\left(4x^5 - 7x^3 - 8x + \tfrac{5}{6}\right)$     This is one expression for the opposite of $4x^5 - 7x^3 - 8x + \tfrac{5}{6}$.

**b)** $-4x^5 + 7x^3 + 8x - \tfrac{5}{6}$     Changing the sign of every term

Thus, $-\left(4x^5 - 7x^3 - 8x + \tfrac{5}{6}\right)$ is equivalent to $-4x^5 + 7x^3 + 8x - \tfrac{5}{6}$, and each is the opposite of the original polynomial $4x^5 - 7x^3 - 8x + \tfrac{5}{6}$.

> Do Exercises 6–9.

Find two equivalent expressions for the opposite of each polynomial.

**6.** $12x^4 - 3x^2 + 4x$

**7.** $-4x^4 + 3x^2 - 4x$

**8.** $-13x^6 + 2x^4 - 3x^2 + x - \tfrac{5}{13}$

**9.** $-8a^3b + 5ab^2 - 2ab$

**EXAMPLE 5** Simplify: $-\left(-7x^4 - \tfrac{5}{9}x^3 + 8x^2 - x + 67\right)$.

$$-\left(-7x^4 - \tfrac{5}{9}x^3 + 8x^2 - x + 67\right) = 7x^4 + \tfrac{5}{9}x^3 - 8x^2 + x - 67$$

> Do Exercises 10–12.

### (c) Subtracting Polynomials

We can now subtract a polynomial by adding the opposite of that polynomial. That is, for any polynomials $P$ and $Q$, $P - Q = P + (-Q)$.

Simplify.

**10.** $-(4x^3 - 6x + 3)$

**11.** $-(5x^3y + 3x^2y^2 - 7xy^3)$

**12.** $-\left(14x^{10} - \tfrac{1}{2}x^5 + 5x^3 - x^2 + 3x\right)$

**EXAMPLE 6** Subtract:

$$(9x^5 + x^3 - 2x^2 + 4) - (2x^5 + x^4 - 4x^3 - 3x^2).$$

We have

$(9x^5 + x^3 - 2x^2 + 4) - (2x^5 + x^4 - 4x^3 - 3x^2)$

$= (9x^5 + x^3 - 2x^2 + 4) + [-(2x^5 + x^4 - 4x^3 - 3x^2)]$    Adding the opposite

$= (9x^5 + x^3 - 2x^2 + 4) + [-2x^5 - x^4 + 4x^3 + 3x^2]$    Changing the sign of every term

$= 9x^5 + x^3 - 2x^2 + 4 - 2x^5 - x^4 + 4x^3 + 3x^2$

$= 7x^5 - x^4 + 5x^3 + x^2 + 4.$    Combining like terms

> Do Exercises 13 and 14.

To shorten our work, we often begin by changing the sign of each term in the polynomial being subtracted.

**EXAMPLE 7** Subtract: $(5a^4 - 7a^3 + 5a^2b) - (-3a^4 + 4a^2b + 6)$.

We have

$(5a^4 - 7a^3 + 5a^2b) - (-3a^4 + 4a^2b + 6)$

$= 5a^4 - 7a^3 + 5a^2b + 3a^4 - 4a^2b - 6$    Removing all parentheses; changing the sign of every term in the second polynomial

$= 8a^4 - 7a^3 + a^2b - 6.$    Combining like terms

> Do Exercise 15.

Subtract.

**13.** $(7x^3 + 2x + 4) - (5x^3 - 4)$

**14.** $(-3x^2 + 5x - 4) - (-4x^2 + 11x - 2)$

**15.** Subtract:

$(7x^3 + 3x^2 - xy) - (5x^3 + 3xy + 2).$

**Answers**

**6.** $-(12x^4 - 3x^2 + 4x);\ -12x^4 + 3x^2 - 4x$
**7.** $-(-4x^4 + 3x^2 - 4x);\ 4x^4 - 3x^2 + 4x$
**8.** $-\left(-13x^6 + 2x^4 - 3x^2 + x - \dfrac{5}{13}\right);$
$\quad 13x^6 - 2x^4 + 3x^2 - x + \dfrac{5}{13}$
**9.** $-(-8a^3b + 5ab^2 - 2ab);$
$\quad 8a^3b - 5ab^2 + 2ab$
**10.** $-4x^3 + 6x - 3$
**11.** $-5x^3y - 3x^2y^2 + 7xy^3$
**12.** $-14x^{10} + \dfrac{1}{2}x^5 - 5x^3 + x^2 - 3x$
**13.** $2x^3 + 2x + 8$    **14.** $x^2 - 6x - 2$
**15.** $2x^3 + 3x^2 - 4xy - 2$

## d Evaluating Polynomials and Applications

It is important to keep in mind that when we add or subtract polynomials, we are *not* solving an equation. Rather, we are finding an equivalent expression that is usually more concise. One reason we do this is to make it easier to evaluate.

**EXAMPLE 8** Evaluate both $(5x^3 + 4x^2 + 3x) + (2x^3 + 5x^2 - x)$ and $7x^3 + 9x^2 + 2x$ for $x = 2$ (see Example 1).

a) When $x$ is replaced by 2 in $(5x^3 + 4x^2 + 3x) + (2x^3 + 5x^2 - x)$, we have

$$5 \cdot 2^3 + 4 \cdot 2^2 + 3 \cdot 2 + 2 \cdot 2^3 + 5 \cdot 2^2 - 2,$$

or $5 \cdot 8 + 4 \cdot 4 + 6 + 2 \cdot 8 + 5 \cdot 4 - 2,$

or $40 + 16 + 6 + 16 + 20 - 2,$ which is 96.

b) Similarly, when $x$ is replaced by 2 in $7x^3 + 9x^2 + 2x$, we have

$$7 \cdot 2^3 + 9 \cdot 2^2 + 2 \cdot 2,$$

or $7 \cdot 8 + 9 \cdot 4 + 4,$

or $56 + 36 + 4.$ As expected, this is also 96.

Note how much easier it is to evaluate the simplified sum in part (b) rather than the original expression.

> Do Exercise 16.

Polynomials are frequently evaluated in real-world situations.

**EXAMPLE 9** *Athletics.* In a sports league of $n$ teams in which all teams play each other twice, the total number of games played is given by the polynomial

$$n^2 - n.$$

A women's softball league has 10 teams. If each team plays every other team twice, what is the total number of games played?

We evaluate the polynomial for $n = 10$:

$$n^2 - n = 10^2 - 10 = 100 - 10 = 90.$$

The league plays 90 games.

> Do Exercises 17–19.

---

**16.** Evaluate each expression for $a = 2$. (See Margin Exercise 3.)

a) $(7a^2 + 2a + 8) + (2a^2 + a - 9)$

b) $9a^2 + 3a - 1$

**17.** In the situation of Example 9, how many games are played in a league with 12 teams?

Wendy is pedaling down a hill. Her distance from the top of the hill, in meters, can be approximated by

$$\frac{1}{2}t^2 + 3t,$$

where $t$ is the number of seconds she has been pedaling and $t < 30$.

**18.** How far has Wendy traveled in 4 sec?

**19.** How far has Wendy traveled in 10 sec?

---

*Answers*

**16.** (a) 41; (b) 41    **17.** 132 games
**18.** 20 m    **19.** 80 m

**a**  Add.

**1.** $(3x + 7) + (-7x + 3)$

**2.** $(6x + 1) + (-7x + 2)$

**3.** $(-9x + 7) + (x^2 + x - 2)$

**4.** $(x^2 - 5x + 4) + (5x - 9)$

**5.** $(x^2 - 7) + (x^2 + 7)$

**6.** $(x^3 + x^2) + (2x^3 - 5x^2)$

**7.** $(6t^4 + 4t^3 - 1) + (5t^2 - t + 1)$

**8.** $(5t^2 - 3t + 12) + (2t^2 + 8t - 30)$

**9.** $(2 + 4x + 6x^2 + 7x^3) + (5 - 4x + 6x^2 - 7x^3)$

**10.** $(3x^4 - 6x - 5x^2 + 5) + (6x^2 - 4x^3 - 1 + 7x)$

**11.** $(9x^8 - 7x^4 + 2x^2 + 5) + (8x^7 + 4x^4 - 2x)$

**12.** $(4x^5 - 6x^3 - 9x + 1) + (6x^3 + 9x^2 + 9x)$

**13.** $\left(\frac{1}{2}t^3 - \frac{3}{4}t^2 + \frac{1}{3}\right) + \left(\frac{3}{2}t^3 - \frac{1}{4}t^2 + \frac{1}{6}t\right)$

**14.** $\left(\frac{2}{3}t^5 - t^4 + \frac{1}{2}t^2 + 4\right) + \left(\frac{7}{3}t^5 + \frac{1}{2}t^4 - \frac{1}{2}t^2 + \frac{1}{3}t\right)$

**15.** $(9x^3 - 3x + x^2 - 10) + (3x - x^2 - 9x^3 + 10)$

**16.** $(5 - x^4 + 3x^2 + 2x) + (x^4 - 3x^2 - 2x - 5)$

**17.** $(8a^3b^2 + 5a^2b^2 + 6ab^2) + (5a^3b^2 - a^2b^2 - 4a^2b)$

**18.** $(6x^3y^3 - 4x^2y^2 + 3xy^2) + (x^3y^3 + 7x^3y^2 - 2xy^2)$

**19.** $(17.5abc^3 + 4.3a^2bc) + (-4.9a^2bc - 5.2abc)$

**20.** $(23.9x^3yz - 19.7x^2y^2z) + (-14.6x^3yz - 8x^2yz)$

**b**　Find two equivalent expressions for the opposite of each polynomial.

**21.** $-5x$

**22.** $x^2 - 3x$

**23.** $-x^2 + 13x - 7$

**24.** $-7x^3 - x^2 - x$

**25.** $12x^4 - 3x^3 + 3$

**26.** $4x^3 - 6x^2 - 8x + 1$

Simplify.

**27.** $-(3x - 5)$

**28.** $-(-2x + 4)$

**29.** $-(4x^2 - 3x + 2)$

**30.** $-(-6a^3 + 2a^2 - 9a + 1)$

**31.** $-\left(-4x^4 + 6x^2 + \frac{3}{4}x - 8\right)$

**32.** $-(-5x^4 + 4x^3 - x^2 + 0.9)$

**c**　Subtract.

**33.** $(3x + 2) - (-4x + 3)$

**34.** $(6x + 1) - (-7x + 2)$

**35.** $(9t^2 + 7t + 5) - (5t^2 + 7t - 1)$

**36.** $(8t^2 - 5t + 7) - (3t^2 - 5t - 8)$

**37.** $(-8x + 2) - (x^2 + x - 3)$

**38.** $(x^2 - 5x + 4) - (8x - 9)$

**39.** $(7a^2 + 5a - 9) - (8a^2 + 7)$

**40.** $(8a^2 - 6a + 5) - (2a^2 - 19a)$

**41.** $(8x^4 + 3x^3 - 1) - (4x^2 - 3x + 5)$

**42.** $(-4x^2 + 2x) - (3x^3 - 5x^2 + 3)$

**43.** $(1.2x^3 + 4.5x^2 - 3.8x) - (-3.4x^3 - 4.7x^2 + 23)$

**44.** $(0.5x^4 - 0.6x^2 + 0.7) - (2.3x^4 + 1.8x - 3.9)$

**45.** $\left(\frac{5}{8}x^3 - \frac{1}{4}x - \frac{1}{3}\right) - \left(-\frac{1}{8}x^3 + \frac{1}{4}x - \frac{1}{3}\right)$

**46.** $\left(\frac{1}{5}x^3 + 2x^2 - 0.1\right) - \left(-\frac{2}{5}x^3 + 2x^2 + 0.01\right)$

**47.** $(9x^3y^3 + 8x^2y^2 + 7xy) - (3x^3y^3 - 2x^2y + 6xy)$

**48.** $(3x^4y + 2x^3y - 7x^2y) - (5x^4y + 2x^2y^2 - 8x^2y)$

**d**    Evaluate each polynomial for $x = 4$.

**49.** $-7x + 5$

**50.** $-3x + 1$

**51.** $2x^2 - 5x + 7$

**52.** $3x^2 + x + 7$

**53.** $x^3 - 5x^2 + x$

**54.** $7 - x + 3x^2$

Evaluate each polynomial for $x = -1$.

**55.** $2x + 9$

**56.** $6 - 2x$

**57.** $x^2 - 2x + 1$

**58.** $5x - 6 + x^2$

**59.** $-3x^3 + 7x^2 - 3x - 2$

**60.** $-2x^3 - 5x^2 + 4x + 3$

*Falling Distance.*    The distance, in feet, traveled by a body falling freely from rest in $t$ seconds is approximated by the polynomial $16t^2$.

$16t^2$

**61.** A stone is dropped from a cliff and takes 8 sec to hit the ground. How high is the cliff?

**62.** A brick falls from the top of a building and takes 3 sec to hit the ground. How high is the building?

*Minutes of Daylight.*    The number of minutes of daylight in Chicago, on a date $n$ days after December 21, can be approximated by

$$-0.01096n^2 + 4n + 548.$$

**63.** ▦ Determine the number of minutes of daylight in Chicago 92 days after December 21.

**64.** ▦ Determine the number of minutes of daylight in Chicago 123 days after December 21.

*Medicine.*    Ibuprofen is a medication used to relieve pain. The polynomial

$$0.5t^4 + 3.45t^3 - 96.65t^2 + 347.7t$$

can be used to estimate the number of milligrams of ibuprofen in the bloodstream $t$ hours after 400 mg of the medication has been swallowed, where $0 \leq t \leq 6$.

**Source:** Based on data from Dr. P. Carey, Burlington, VT

**65.** ▦ Determine the number of milligrams of ibuprofen in the bloodstream 1 hour after the medication has been swallowed.

**66.** ▦ Determine the number of milligrams of ibuprofen in the bloodstream 5 hours after the medication has been swallowed.

*Daily Accidents.* The average number of accidents per day involving drivers who are $a$ years old is approximated by the polynomial

$$0.4a^2 - 40a + 1039.$$

**67.** Evaluate the polynomial for $a = 18$ to find the daily number of accidents involving 18-year-old drivers.

**68.** Evaluate the polynomial for $a = 20$ to find the daily number of accidents involving 20-year-old drivers.

*Total Revenue.* Cutting Edge Electronics is marketing a new kind of phone. *Total revenue* is the total amount of money taken in. The firm determines that when it sells $x$ phones, it takes in

$$280x - 0.4x^2 \text{ dollars.}$$

**69.** What is the total revenue from the sale of 75 phones?

**70.** What is the total revenue from the sale of 100 phones?

*Total Cost.* Cutting Edge Electronics determines that the total cost of producing $x$ phones is given by

$$5000 + 0.6x^2 \text{ dollars.}$$

**71.** What is the total cost of producing 500 phones?

**72.** What is the total cost of producing 650 phones?

# Skill Maintenance

**73.** A 10-lb fish serves 7 people. What is the ratio of servings to pounds?   [7.1a]

**74.** A bicycle salesperson's commission rate is 22%. A commission of \$783.20 is received. How many dollars' worth of bicycles were sold?   [8.6b]

**75.** The sales tax rate in Pennsylvania is 6%. How much tax would be paid in Pennsylvania for a laptop computer that sold for \$1350?   [8.6a]

**76.** Find the area of a rectangle that is 6.5 m by 4 m.   [5.8b]

**77.** Find the area of a circle with radius 20 cm. Use 3.14 for $\pi$.   [9.3b]

**78.** Melanie earned \$4740 for working 12 weeks. What was the rate of pay?   [7.2a]

Write the prime factorization for each number.   [3.2c]

**79.** 168

**80.** 192

**81.** 735

**82.** 117

# Synthesis

*Minutes of Daylight.* The number of minutes of daylight in Los Angeles, on a date $n$ days after December 21, can be approximated by

$$-0.0085n^2 + 3.1014n + 593.$$

**83.** ▦ Determine the number of minutes of daylight in Los Angeles on "Ground Hog Day" (February 2).

**84.** ▦ How much more daylight is available in Chicago than in Los Angeles on July 4? (See Exercises 63 and 64.)

**85.** ▦ *Medicine.* The polynomial

$$0.5t^4 + 3.45t^3 - 96.65t^2 + 347.7t$$

used in Exercises 65 and 66 describes the number of milligrams of ibuprofen in the bloodstream $t$ hours after 400 mg of the medication has been swallowed. To visualize this, make a line graph, with the number of hours on the horizontal axis and the amount of ibuprofen in the bloodstream on the vertical axis. Use values for $t = 0, 1, 2, 3, 4, 5,$ and 6.

**86.** ▦ *World Wide Web.* The polynomial

$$4.03t^2 + 6.78t + 42.86$$

can be used to estimate the number of web sites, in millions, $t$ years after 2003. To visualize this, make a vertical bar graph, with the number of years after 2003 on the horizontal axis and the number of web sites on the vertical axis. Show values for $t = 1, 2, 3, 4, 5, 6, 7,$ and 8.

**Source:** Based on information from Web Server Surveys, www.news.netcraft.com

**87.** *Total Profit.* Total profit is defined as total revenue minus total cost. Find a polynomial giving the total profit for Cutting Edge Electronics, described in Exercises 69–72, when $x$ phones are produced and sold.

**88.** *Total Profit.* Use the polynomial found in Exercise 87 to find **(a)** the total profit when 200 phones are produced and sold and **(b)** the total profit when 500 phones are produced and sold.

Perform the indicated operations and simplify.

**89.** $(7y^2 - 5y + 6) - (3y^2 + 8y - 12) + (8y^2 - 10y + 3)$

**90.** $(3x^2 - 4x + 6) - (-2x^2 + 4) + (-5x - 3)$

**91.** $(-y^4 - 7y^3 + y^2) + (-2y^4 + 5y - 2) - (-6y^3 + y^2)$

**92.** $(-4 + x^2 + 2x^3) - (-6 - x + 3x^3) - (-x^2 - 5x^3)$

**93.** Complete: $9x^4 + \underline{\hspace{1cm}} + 5x^2 - 7x^3 + \underline{\hspace{1cm}} - 9 + \underline{\hspace{1cm}} = 12x^4 - 5x^3 + 5x^2 - 16.$

**94.** Complete: $8t^4 + \underline{\hspace{1cm}} - 2t^3 + \underline{\hspace{1cm}} - 2t^2 + t - \underline{\hspace{1cm}} - 3 + \underline{\hspace{1cm}} = 8t^4 + 7t^3 - 3t + 4.$

# 10.5

# Introduction to Multiplying and Factoring Polynomials

## OBJECTIVES

**a** Multiply monomials.

**b** Multiply a monomial and any polynomial.

**c** Use the distributive law to factor.

**SKILL TO REVIEW**

Objective 10.2a: Use the product rule to multiply exponential expressions with like bases.

Multiply.

**1.** $x^2 \cdot x^3$　　　**2.** $x \cdot x^4$

## a Multiplying Monomials

Recall that the area of a square with sides of length $x$ is $x^2$.

Area $= x^2$

If a rectangle is 3 times as long as it is wide, we can represent its width by $x$ and its length by $3x$.

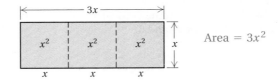

Area $= 3x^2$

The area, $3x^2$, is the product of $3x$ and $x$. This product can be found using an associative law:

$$(3x)x = 3(xx) = 3x^2.$$

To find other products of monomials, we may need to use a commutative law as well.

**EXAMPLE 1**　Multiply: $(4x)(5x)$.

$$
\begin{aligned}
(4x)(5x) &= 4 \cdot x \cdot 5 \cdot x && \text{Using an associative law} \\
&= 4 \cdot 5 \cdot x \cdot x && \text{Using a commutative law} \\
&= (4 \cdot 5)(x \cdot x) && \text{Using an associative law} \\
&= 20x^2
\end{aligned}
$$

The multiplication in Example 1 can be regarded as finding the area of a rectangle of width $4x$ and length $5x$. Note that the area consists of 20 squares, each of which has area $x^2$.

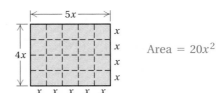

Area $= 20x^2$

Do Margin Exercises 1 and 2.

Usually the steps in Example 1 are combined: We multiply coefficients and we multiply variables.

Multiply.

**1.** $(6a)(3a)$　　　**2.** $(-7x)(2x)$

*Answers*

*Skill to Review:*
1. $x^5$　　2. $x^5$

*Margin Exercises:*
1. $18a^2$　　2. $-14x^2$

**EXAMPLES** Multiply.

**2.** $(5x)(6x) = (5 \cdot 6)(x \cdot x)$     Multiplying the coefficients

$= 30x^2$     Simplifying

**3.** $(3x)(-y) = (3x)(-1y)$     Rewriting $-y$ as $-1y$

$= (3)(-1)(x \cdot y)$

$= -3xy$

Do Exercises 3–5.

We can use the product rule when multiplying monomials.

**EXAMPLES** Multiply and simplify.

**4.** $x^2 \cdot x^5 = x^{2+5}$     Adding exponents

$= x^7$

**5.** $(3a^4)(5a^2) = (3 \cdot 5)(a^4 \cdot a^2)$   ⎫    Multiplying coefficients;

$= 15a^6$   ⎭    adding exponents

**6.** $(-4x^2y^3)(3x^6y^7) = (-4 \cdot 3)(x^2 \cdot x^6)(y^3 \cdot y^7)$

$= -12x^8y^{10}$

Do Exercises 6–9.

Multiply.

**3.** $(4a)(12a)$

**4.** $(-m)(5m)$

**5.** $(-6a)(-7b)$

Multiply.

**6.** $a^5 \cdot a^4$

**7.** $(2x^8)(4x^5)$

**8.** $(-7m^4)(-5m^7)$

**9.** $(3a^5b^4)(5a^2b^8)$

## b   Multiplying a Monomial and Any Polynomial

The product of $x$ and $x + 2$ can be visualized as the area of a rectangle with width $x$ and length $x + 2$, as illustrated in the figure at right.

The distributive law is used to find products of polynomials algebraically.

**EXAMPLE 7** Multiply: $2x$ and $5x + 3$.

$2x(5x + 3) = 2x \cdot 5x + 2x \cdot 3$     Using the distributive law

$= 10x^2 + 6x$     Multiplying each pair of monomials

Area $= x(x + 2)$

$= x \cdot x + x \cdot 2$

$= x^2 + 2x$

**EXAMPLE 8** Multiply: $5x(2x^2 - 3x + 4)$.

$5x(2x^2 - 3x + 4) = 5x \cdot 2x^2 - 5x \cdot 3x + 5x \cdot 4$

$= 10x^3 - 15x^2 + 20x$     Note that $x \cdot x^2 = x^1 \cdot x^2 = x^3$.

**EXAMPLE 9** Multiply: $-3r^2s(2r^3s^2 - 5rs)$.

$-3r^2s(2r^3s^2 - 5rs) = -3r^2s \cdot 2r^3s^2 - (-3r^2s)5rs$

$= -6r^5s^3 + 15r^3s^2$

Do Exercises 10–12.

Multiply.

**10.** $4x$ and $3x + 5$

**11.** $3a(2a^2 - 5a + 7)$

**12.** $4a^3b^2(2a^2 + 5b^4)$

*Answers*

**3.** $48a^2$    **4.** $-5m^2$    **5.** $42ab$    **6.** $a^9$
**7.** $8x^{13}$    **8.** $35m^{11}$    **9.** $15a^7b^{12}$
**10.** $12x^2 + 20x$    **11.** $6a^3 - 15a^2 + 21a$
**12.** $8a^5b^2 + 20a^3b^6$

## (c) Factoring

**Factoring** is the reverse of multiplying. We use the distributive law, beginning with a sum or a difference of terms that contain a common factor:

$$ab + ac = a(b + c) \quad \text{and} \quad rs - rt = r(s - t).$$

---

**FACTORING**

To **factor** an expression is to find an equivalent expression that is a product.

---

To *factor* an expression like $10y + 15$, we find an equivalent expression that is a product. To do this, we look to see if both terms have a factor in common. If there *is* a common factor, we can "factor it out" using the distributive law. Note the following:

The prime factorization of $10y$ is $2 \cdot 5 \cdot y$.

The prime factorization of $15$ is $3 \cdot 5$.

We factor out the common factor, 5:

$$10y + 15 = 5 \cdot 2y + 5 \cdot 3 \qquad \text{Try to do this step mentally.}$$
$$= 5(2y + 3). \qquad \text{Using the distributive law}$$

We generally factor out the *largest* common factor. This is the product of all factors common to all terms.

**EXAMPLE 10** Factor $12a - 30$.

The prime factorization of $12a$ is $2 \cdot 2 \cdot 3 \cdot a$.

The prime factorization of $30$ is $2 \cdot 3 \cdot 5$.

Both factorizations include a factor of 2 and a factor of 3. Thus, 2 is a common factor, 3 is a common factor, and $2 \cdot 3$ is a common factor. The largest common factor is $2 \cdot 3$, or 6:

$$12a - 30 = 6(2a - 5). \qquad \text{Try to go directly to this step.}$$

**EXAMPLE 11** Factor $9x + 27y - 9$.

The prime factorization of $9x$ is $3 \cdot 3 \cdot x$.

The prime factorization of $27y$ is $3 \cdot 3 \cdot 3 \cdot y$.

The prime factorization of $9$ is $3 \cdot 3$.

$$9x + 27y - 9 = 9 \cdot x + 9 \cdot 3y - 9 \cdot 1 \qquad \text{The largest common factor is } 3 \cdot 3, \text{ or } 9.$$
$$= 9(x + 3y - 1)$$

In Example 11, the 1 in the factorization is necessary. To see this, reverse the factorization process by multiplying. This provides a check for the answer.

$$9(x + 3y - 1) = 9 \cdot x + 9 \cdot 3y - 9 \cdot 1 = 9x + 27y - 9$$

---

Factorizations can always be checked by multiplying.

---

Note in Example 11 that although $3(3x + 9y - 3)$ is also equivalent to $9x + 27y - 9$, it is not factored "completely." However, we can complete the process by factoring out another factor of 3:

$$9x + 27y - 9 = 3(3x + 9y - 3) = 3 \cdot 3(x + 3y - 1) = 9(x + 3y - 1).$$

Remember to factor out the *largest common factor*.

---

## EXAMPLES Factor. Try to write just the answer.

**12.** $-3x + 6y - 9z = -3(x - 2y + 3z)$

We generally factor out a negative factor when the first coefficient is negative. We might also factor as $-3x + 6y - 9z = 3(-x + 2y - 3z)$.

**13.** $18z - 12x - 24 = 6(3z - 2x - 4)$

The largest common factor is $2 \cdot 3$. ←
$$\begin{cases} 18z = 2 \cdot 3 \cdot 3 \cdot z; \\ 12x = 2 \cdot 2 \cdot 3 \cdot x; \\ 24 = 2 \cdot 2 \cdot 2 \cdot 3 \end{cases}$$

Check: $6(3z - 2x - 4) = 6 \cdot 3z - 6 \cdot 2x - 6 \cdot 4 = 18z - 12x - 24$

---

*Remember*: An expression is factored when it is written as a product.

---

Do Exercises 13-16.

## EXAMPLE 14 Factor each of the following:

**a)** $10x^6 + 15x^2$

**b)** $8xy^3 - 6xy^2 + 4xy$

**a)** The prime factorization of $10x^6$ is $2 \cdot 5 \cdot x \cdot x \cdot x \cdot x \cdot x \cdot x$.
The prime factorization of $15x^2$ is $3 \cdot 5 \cdot x \cdot x$.

$$10x^6 + 15x^2 = 5x^2 \cdot 2x^4 + 5x^2 \cdot 3 \quad \text{The largest common factor is } 5x^2.$$
$$= 5x^2(2x^4 + 3)$$

**b)** The prime factorization of $8xy^3$ is $2 \cdot 2 \cdot 2 \cdot x \cdot y \cdot y \cdot y$.
The prime factorization of $6xy^2$ is $2 \cdot 3 \cdot x \cdot y \cdot y$.
The prime factorization of $4xy$ is $2 \cdot 2 \cdot x \cdot y$.

$$8xy^3 - 6xy^2 + 4xy = 2xy \cdot 4y^2 - 2xy \cdot 3y + 2xy \cdot 2 \quad \text{The largest common factor is } 2xy.$$

$$= 2xy(4y^2 - 3y + 2)$$

The checks are left for the student.

The largest common factor can be determined by considering the coefficients and the variables separately. The largest common factor of the coefficients is found using prime factorizations. The largest common variable factors can be found by examining the exponents.

---

When a variable appears in every term of a polynomial, the *largest* common factor of that variable is the *smallest* of the powers of that variable in the polynomial.

---

Do Exercises 17-19.

Factor.

**13.** $6z - 12$

**14.** $3x - 6y + 12$

**15.** $16a - 36b + 42$

**16.** $-12x + 32y - 16z$

Factor.

**17.** $5a^3 + 10a$

**18.** $14x^3 - 7x^2 + 21x$

**19.** $9a^2b - 6ab^2$

**Answers**

**13.** $6(z - 2)$ **14.** $3(x - 2y + 4)$
**15.** $2(8a - 18b + 21)$ **16.** $-4(3x - 8y + 4z)$
**17.** $5a(a^2 + 2)$ **18.** $7x(2x^2 - x + 3)$
**19.** $3ab(3a - 2b)$

**10.5** | **Exercise Set**

For Extra Help

**MyMathLab**

Math XL
PRACTICE

WATCH

DOWNLOAD

READ

REVIEW

**a** Multiply.

**1.** $(4a)(7a)$

**2.** $(7x)(6x)$

**3.** $(-4x)(15x)$

**4.** $(-9a)(10a)$

**5.** $(7x^5)(4x^3)$

**6.** $(10a^2)(3a^2)$

**7.** $(-0.1x^6)(0.7x^3)$

**8.** $(0.3x^3)(-0.4x^6)$

**9.** $(5x^2y^3)(7x^4y^9)$

**10.** $(9a^5b^4)(2a^4b^7)$

**11.** $(4a^3b^4c^2)(3a^5b^4)$

**12.** $(7x^3y^5z^2)(8x^3z^4)$

**13.** $(3x^2)(-4x^3)(2x^6)$

**14.** $(-2y^5)(10y^4)(-3y^3)$

**b** Multiply.

**15.** $3x(-x + 7)$

**16.** $2x(4x - 6)$

**17.** $-3x(x - 2)$

**18.** $-9x(-x - 1)$

**19.** $x^2(x^3 + 1)$

**20.** $-2x^3(x^2 - 1)$

**21.** $5x(2x^2 - 6x + 1)$

**22.** $-4x(2x^3 - 6x^2 - 5x + 1)$

**23.** $4xy(3x^2 + 2y)$

**24.** $7xy(3x^2 - 6y^2)$

**25.** $3a^2b(4a^5b^2 - 3a^2b^2)$

**26.** $4a^2b^2(2a^3b - 5ab^2)$

**c** Factor. Check by multiplying.

**27.** $2x + 8$

**28.** $3x + 12$

**29.** $7a - 35$

**30.** $9a - 18$

**31.** $28x + 21y$      **32.** $8x - 10y$      **33.** $9a - 27b + 81$      **34.** $5x + 10 + 15y$

**35.** $18 - 6m$      **36.** $28 - 4y$      **37.** $-16 - 8x + 40y$      **38.** $-35 + 14x - 21y$

**39.** $9x^5 + 9x$      **40.** $5x^6 + 5x$      **41.** $a^3 - 8a^2$      **42.** $a^5 - 9a^2$

**43.** $8x^3 - 6x^2 + 2x$      **44.** $9x^4 - 12x^3 + 3x$      **45.** $12a^4b^3 + 18a^5b^2$      **46.** $15a^5b^2 + 20a^2b^3$

## Skill Maintenance

In each of Exercises 47–54, fill in the blank with the correct term from the given list. Some of the choices may not be used and some may be used more than once.

**47.** A(n) _____ is a polynomial with one term. [10.4a]

**48.** A parallelogram is a four-sided figure with two pairs of _____ sides. [9.3a]

**49.** In the metric system, the _____ is the basic unit of mass. [9.7b]

**50.** A natural number, other than 1, that is not _____ is composite. [3.2b]

**51.** A(n) _____ is a set of points consisting of two rays with a common endpoint. [9.5a]

**52.** To convert from _____ to _____, move the decimal point two places to the left and change the ¢ sign at the end to the $ sign in front. [5.3b]

**53.** The _____ of a polygon is the sum of the lengths of its sides. [2.7b]

**54.** The number 1 is known as the _____ identity, and the number 0 is known as the _____ identity. [1.4a], [2.2a]

prime
composite
monomial
coefficient
dollars
cents
perimeter
parallel
perpendicular
additive
multiplicative
meter
gram
vertex
angle
area

## Synthesis

Factor.

**55.** 🖩 $391x^{391} + 299x^{299}$      **56.** 🖩 $703a^{437} + 437a^{703}$

**57.** $84a^7b^9c^{11} - 42a^8b^6c^{10} + 49a^9b^7c^8$

**58.** Draw a figure similar to those preceding Examples 1 and 7 to show that $2x \cdot 3x = 6x^2$.

# Summary and Review

## Key Terms and Properties

scientific notation, p. 689
monomial, p. 699

polynomial, p. 699
opposite of a polynomial, p. 700

factor, p. 710

*Exponents:*      $b^1 = b;$
                 $b^0 = 1$ for $b \neq 0$

*Negative exponents:*   $a^{-n} = \dfrac{1}{a^n}$ and $\left(\dfrac{a}{b}\right)^{-n} = \left(\dfrac{b}{a}\right)^n$

*The Product Rule:*    $a^m \cdot a^n = a^{m+n}$

*The Quotient Rule:*    $\dfrac{a^m}{a^n} = a^{m-n}$

*The Power Rule:*          $(a^m)^n = a^{mn}$

*Raising a product to a power:*   $(ab)^n = a^n b^n$

*Raising a quotient to a power:*   $\left(\dfrac{a}{b}\right)^n = \dfrac{a^n}{b^n}$

## Concept Reinforcement

————————   **1.** If $x^7$ is divided by $x^7$, the result is the same as $x^0$.   [10.1a]

————————   **2.** We subtract a polynomial by adding its opposite.   [10.4c]

————————   **3.** If a polynomial is written as a product, it is factored.   [10.5c]

## Important Concepts

**Objective 10.1a**   Evaluate algebraic expressions containing whole-number exponents.

**Examples**   Evaluate $-6^1$.
          Since $6^1 = 6$, $-6^1 = -6$.
          Evaluate $(-5)^0$.
          Since $b^0 = 1$ for all $b$, $(-5)^0 = 1$.

**Practice Exercises**

Evaluate.
 **1.** $(-3)^1$

 **2.** $12^0$

**Objective 10.1b**   Express exponential expressions involving negative exponents as equivalent expressions containing positive exponents.

**Example**   Write an expression equivalent to $x^{-10}$ using a positive exponent.

$$x^{-10} = \frac{1}{x^{10}}$$

**Practice Exercise**

 **3.** Write an expression equivalent to $3^{-12}$ using a positive exponent.

**Objective 10.2a**   Use the product rule to multiply exponential expressions with like bases.

**Example**   Simplify $x^2 \cdot x^8$.
     $x^2 \cdot x^8 = x^{2+8} = x^{10}$

**Practice Exercise**

 **4.** Simplify $y \cdot y^{11}$.

**Objective 10.2b**    Use the quotient rule to divide exponential expressions with like bases.

**Example**    Simplify $\dfrac{7^{12}}{7^{10}}$.

$$\frac{7^{12}}{7^{10}} = 7^{12-10} = 7^2 = 49$$

**Practice Exercise**

**5.** Simplify $\dfrac{x^9}{x^2}$.

---

**Objective 10.2c**    Use the power rule to raise powers to powers.

**Example**    Simplify $(x^3)^{-5}$.

$$(x^3)^{-5} = x^{3(-5)} = x^{-15} = \frac{1}{x^{15}}$$

**Practice Exercise**

**6.** Simplify $(3^{-2})^{-6}$.

---

**Objective 10.2d**    Raise a product to a power and a quotient to a power.

**Example**    Simplify $(a^2b)^9$.
$$(a^2b)^9 = (a^2)^9(b)^9 = a^{18}b^9$$

**Practice Exercise**

**7.** Simplify $(3x^7)^2$.

---

**Objective 10.3a**    Convert between scientific notation and decimal notation.

**Examples**    Convert 96,000,000,000 to scientific notation.
$$96{,}000{,}000{,}000 = 9.6 \times 10^{10}$$
Convert $6.02 \times 10^{-5}$ to decimal notation.
$$6.02 \times 10^{-5} = 0.0000602$$

**Practice Exercises**

**8.** Convert 0.000803 to scientific notation.

**9.** Convert $3.48 \times 10^3$ to decimal notation.

---

**Objective 10.4a**    Add polynomials.

**Example**    Add: $(3x^3 - 7x^2 - 9) + (2x^3 + x^2 - 1)$.
$$(3x^3 - 7x^2 - 9) + (2x^3 + x^2 - 1)$$
$$= 3x^3 + 2x^3 - 7x^2 + x^2 - 9 - 1$$
$$= 5x^3 - 6x^2 - 10$$

**Practice Exercise**

**10.** Add: $(2a^2 - 5a + 3) + (a^3 + a - 5)$.

---

**Objective 10.4c**    Subtract polynomials.

**Example**    Subtract:
$(9y^3 - 8y^2 + 11y) - (4y^3 - 2y^2 - 12y)$.
$$(9y^3 - 8y^2 + 11y) - (4y^3 - 2y^2 - 12y)$$
$$= 9y^3 - 8y^2 + 11y - 4y^3 + 2y^2 + 12y$$
$$= 5y^3 - 6y^2 + 23y$$

**Practice Exercise**

**11.** Subtract: $(x^2 - x - 1) - (7x^2 - 6x + 2)$.

---

**Objective 10.5a**    Multiply monomials.

**Example**    Multiply: $(-3m^2)(9m^5)$.
$$(-3m^2)(9m^5) = (-3 \cdot 9)(m^2 \cdot m^5)$$
$$= -27m^7$$

**Practice Exercise**

**12.** Multiply: $(-2x^3)(-7x)$.

**Objective 10.5b** Multiply a monomial and any polynomial.

| | |
|---|---|
| **Example** Multiply: $2x^3(3x^4 - 5)$. <br> $\quad 2x^3(3x^4 - 5) = 2x^3 \cdot 3x^4 - 2x^3 \cdot 5$ <br> $\quad\quad\quad\quad\quad = 6x^7 - 10x^3$ | **Practice Exercise** <br> **13.** Multiply: $5x^2(x^3 + 2x^2 - 3x + 9)$. |

**Objective 10.5c** Use the distributive law to factor.

| | |
|---|---|
| **Example** Factor $6x^3 - 20x^2 + 2x$. <br> $\quad 6x^3 - 20x^2 + 2x = 2x(3x^2 - 10x + 1)$ | **Practice Exercise** <br> **14.** Factor $20a^4 - 30a$. |

## Review Exercises

Evaluate.  [10.1a]

**1.** $(-53)^0$

**2.** $46^1$

**3.** $(5x + 7)^1$ for $x = -2$

**4.** $(3x - 2)^0$ for $x = 5$

Write an equivalent expression using positive exponents. Then simplify, if possible.  [10.1b]

**5.** $12^{-2}$

**6.** $8a^{-7}$

**7.** $\dfrac{x^{-3}}{y^5 z^{-6}}$

**8.** $\left(\dfrac{4}{5}\right)^{-2}$

**9.** Write an expression equivalent to $\dfrac{1}{x^7}$ using a negative exponent.  [10.1b]

Simplify.  [10.2a, b, c, d]

**10.** $x^4 \cdot x^{11}$

**11.** $\dfrac{x^{16}}{x^4}$

**12.** $(3^4)^{11}$

**13.** $(x^2 y^4)^3$

**14.** $\dfrac{x^3}{x^{11}}$

**15.** $\dfrac{u^2 v^8}{uv^7}$

**16.** $x^{-5} \cdot x^{-6}$

**17.** $(3x^4)^2$

**18.** $a^{-2} \cdot a \cdot a^7$

**19.** $\left(\dfrac{y^2}{2}\right)^5$

**20.** Write scientific notation for 42,700,000.  [10.3a]

**21.** Write scientific notation for 0.0001924.  [10.3a]

Simplify. Write the answer in scientific notation.  [10.3b]

**22.** $(5.1 \times 10^6)(2.3 \times 10^4)$

**23.** $\dfrac{1.6 \times 10^2}{6.4 \times 10^{18}}$

**24.** Every day about 12.4 billion spam e-mails are sent. If each spam e-mail wastes 4 sec, how many hours are wasted each day because of spam? Write the answer in scientific notation.

Source: spam-filter-review.toptenreviews.com    [10.3c]

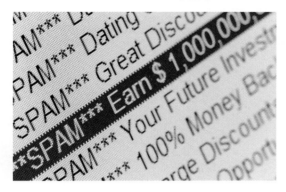

Perform the indicated operation.    [10.4a, c]

**25.** $(-4x + 9) + (7x - 15)$

**26.** $(7x^4 - 5x^3 + 3x - 5) + (x^3 - 4x + 2)$

**27.** $(9a^5 + 8a^3 + 4a + 7) - (a^5 - 4a^3 + a^2 - 2)$

**28.** $(8a^3b^3 + 9a^2b^3) - (3a^3b^3 - 2a^2b^3 + 7)$

**29.** Find two equivalent expressions for the opposite of
$$12x^3 - 4x^2 + 9x - 3.    [10.4b]$$

**30.** Evaluate $5t^3 + t$ for $t = -2$.    [10.4d]

**31.** The altitude, in feet, of a falling golf ball $t$ seconds after it reaches the peak of its flight can be estimated by $-16t^2 + 200$. Find the ball's altitude 3 sec after it has reached its peak.    [10.4d]

Multiply.

**32.** $(5x^3)(6x^4)$    [10.5a]

**33.** $(-2xy)(9x^2y^4)$    [10.5a]

**34.** $3x(6x^3 - 4x - 1)$    [10.5b]

**35.** $2a^4b(7a^3b^3 + 5a^2b^3)$    [10.5b]

Factor.    [10.5c]

**36.** $45x^3 - 10x$

**37.** $7a - 35b - 49ac$

**38.** Evaluate $6 - 2x - x^2$ for $x = -1$.    [10.4d]
   **A.** 3                   **B.** 5
   **C.** 7                   **D.** 9

**39.** Factor out the largest common factor: $6x^3y - 9x^2y^5$.    [10.5c]
   **A.** $3x^2y(2x - 3y^4)$       **B.** $3(2x^2y - 3x^2y^5)$
   **C.** $3x^2(2xy - 3y^5)$       **D.** $y(6x^3 - 9x^2y^4)$

## Synthesis

**40.** Simplify: $-3x^5 \cdot 3x^3 - x^6(2x)^2 + (3x^4)^2 + (2x^4)^2 - 40x^2(x^3)^2$.    [10.4a], [10.4c]

Factor.    [10.5c]

**41.** $39a^3b^7c^6 - 130a^2b^5c^8 + 52a^4b^6c^5$

**42.** $w^5x^6y^4z^5 - w^7x^3y^7z^3 + w^6x^2y^5z^6 - w^6x^7y^3z^4$

**43.** $10a^4b^{-5} + 12a^7b^{-3}$

# Understanding Through Discussion and Writing

**1.** Can $x^{-2}$ represent a negative number? Why or why not?    [10.1b]

**2.** Is it true that if $a > b$, then $a^{-1} < b^{-1}$? Explain your answer.    [10.1b]

**3.** Adi claims that $(3x^{-5})(-4x^{-2}) = -x^{10}$. What mistake(s) is she probably making?    [10.1b], [10.5a]

**4.** Using the quotient rule, explain why $7^0$ is 1.    [10.1a], [10.2b]

**5.** Is every term a monomial? Why or why not?    [10.4a]

**6.** If all of a polynomial's coefficients are prime, is it still possible to factor the polynomial? Why or why not?    [10.5c]

**For Extra Help**

CHAPTER
Test Prep
VIDEOS

Step-by-step test solutions are found on the Chapter Test Prep Videos available via the Video Resources on DVD, in *MyMathLab*, and on YouTube (search "BittingerPrealgebra" and click on "Channels").

Evaluate.

**1.** $193^1$

**2.** $-9^0$

**3.** $(3x - 7)^0$ for $x = 2$

Write an equivalent expression with positive exponents. Then simplify, if possible.

**4.** $5^{-3}$

**5.** $\dfrac{5a^{-3}}{b^{-2}}$

**6.** $\left(\dfrac{3}{5}\right)^{-3}$

Simplify. Express the answer using positive exponents.

**7.** $a^{12} \cdot a^{13}$

**8.** $\dfrac{x}{x^2}$

**9.** $(5x^3y^4)^2$

**10.** $\left(\dfrac{3}{x^9}\right)^3$

**11.** $x^{-1} \cdot x^{-7} \cdot x$

**12.** $\dfrac{a^{10}b^{12}}{a^4b^{10}}$

**13.** Write scientific notation for 0.00047.

**14.** Write scientific notation for 8,250,000.

**15.** Find the product and write the answer using scientific notation:
$(3.2 \times 10^{-8})(5.7 \times 10^{-9})$.

**16.** Add: $(12a^3 - 9a^2 + 8) + (6a^3 + 4a^2 - a)$.

**17.** Find two equivalent expressions for the opposite of $-9a^4 + 7b^2 - ab + 3$.

**18.** Subtract: $(12x^4 + 7x^2 - 6) - (9x^4 + 8x^2 + 5)$.

**19.** The height, in meters, of a ball $t$ sec after it has been thrown is approximated by $-4.9t^2 + 15t + 2$. How high is the ball 2 sec after it has been thrown?

Multiply.

**20.** $(-5x^4y^3)(2x^2y^5)$

**21.** $2a(5a^2 - 4a + 3)$

Factor.

**22.** $35x^6 - 25x^3 + 15x^2$

**23.** $6ab - 9bc + 12ac$

**24.** Americans throw away 2.5 million plastic beverage bottles every hour. How many bottles do they throw away in a week? Give your answer in scientific notation.
**Source:** www.grabstats.com

**A.** $6.0 \times 10^7$ bottles    **B.** $1.75 \times 10^7$ bottles
**C.** $2.52 \times 10^9$ bottles    **D.** $4.2 \times 10^8$ bottles

## Synthesis

**25.** The polynomial

$$0.041h - 0.018A - 2.69$$

can be used to estimate the lung capacity, in liters, of a female of height $h$, in centimeters, and age $A$, in years. Find the lung capacity of a 30-yr-old woman who is 150 cm tall.

**26.** Simplify: $x^{8y} \cdot x^{2y} \cdot x^{-y}$.

This exam reviews the entire textbook. Many answers to exercises (and to real-world problems) can be written in several kinds of notation. For this exam, here is the guideline we follow: Use the notation given in the problem. That is, if the problem is given using mixed numerals, give the answer as a mixed numeral. If the problem is given in decimal notation, give the answer in decimal notation.

1. *Area.* There are 5.1 million km² of forest in the Amazon jungle. Find standard notation for 5.1 million.

   **Source:** www.outbackbrazil.com.

Add and, if possible, simplify.

2.  $\begin{array}{r} 4\ 9\ 0\ 3 \\ 5\ 2\ 7\ 8 \\ 6\ 3\ 9\ 1 \\ +\ 4\ 5\ 1\ 3 \\ \hline \end{array}$

3.  $\begin{array}{r} 5\dfrac{4}{9} \\ +\ 3\dfrac{1}{3} \\ \hline \end{array}$

4. $-29 + 53$

5. $-543 + (-219)$

6. $-34.56 + 2.783 + 0.433 + (-13.02)$

7. $(4x^5 + 7x^4 - 3x^2 + 9) + (6x^5 - 8x^4 + 2x^3 - 7)$

Subtract and, if possible, simplify.

8.  $\begin{array}{r} 6\ 7\ 4 \\ -\ 4\ 3\ 1 \\ \hline \end{array}$

9. $-4x - 13x$

10. $\dfrac{2}{5} - \dfrac{7}{8}$

11. $\begin{array}{r} 4\dfrac{1}{3} \\ -\ 1\dfrac{5}{8} \\ \hline \end{array}$

12.  $\begin{array}{r} 2\ 0.0 \\ -\ \ \ 0.0\ 0\ 2\ 7 \\ \hline \end{array}$

13. $(7x^3 + 2x^2 - x) - (5x^3 - 3x^2 - 8x)$

Multiply and, if possible, simplify.

14.  $\begin{array}{r} 2\ 9\ 7 \\ \times\ \ \ 1\ 6 \\ \hline \end{array}$

15. $349 \cdot (-213)$

16. $2\dfrac{3}{4} \cdot 1\dfrac{2}{3}$

17. $-\dfrac{9}{7} \cdot \dfrac{14}{15}$

18. $12 \cdot \dfrac{5}{6}$

19.  $\begin{array}{r} 3\ 4.0\ 9 \\ \times\ \ \ \ \ 7.6 \\ \hline \end{array}$

20. $3(8x - 5)$

21. $(9a^3b^2)(3a^5b)$

22. $7x^2(3x^3 - 2x + 8)$

Divide and simplify. State the answer using a remainder when appropriate.

23. $6\overline{)3\ 4\ 3\ 8}$

24. $34\overline{)1\ 9\ 1\ 4}$

Divide and, if possible, simplify.

**25.** $\dfrac{4}{5} \div \left(-\dfrac{8}{15}\right)$        **26.** $-2\dfrac{1}{3} \div (-30)$

**27.** $2.7\ \overline{)\ 1\ 0\ 5.3}$

Simplify.

**28.** $10 \div 2 \times 20 - 5^2$      **29.** $\dfrac{|3^2 - 5^2|}{2 - 2 \cdot 5}$

**30.** Write exponential notation: $14 \cdot 14 \cdot 14$.

**31.** Round 68,489 to the nearest thousand.

**32.** Round $21.\overline{83}$ to the nearest hundredth.

**33.** Determine whether 1368 is divisible by 3.

**34.** Find all the factors of 15.

**35.** Find the LCM of 15 and 35.

**36.** Simplify $\dfrac{24}{33}$.

**37.** Convert to a mixed numeral: $-\dfrac{18}{5}$.

**38.** Use < or > for $\square$ to write a true sentence:
$-17 \ \square \ -29$.

**39.** Use < or > for $\square$ to write a true sentence:
$\dfrac{4}{7} \ \square \ \dfrac{3}{5}$.

**40.** Which number is greater: $-1.001$ or $-0.9976$?

**41.** Evaluate $\dfrac{a^2 - b}{3}$ for $a = -9$ and $b = -6$.

Factor.

**42.** $40 - 5t$

**43.** $18a^3 - 15a^2 + 6a$

**44.** What part is shaded?

Write decimal notation for each number.

**45.** $\dfrac{429}{10,000}$        **46.** $-\dfrac{13}{25}$

**47.** $\dfrac{8}{9}$        **48.** $7\%$

Write each number in fraction notation.

**49.** $6.71$        **50.** $-7\dfrac{1}{4}$

**51.** $40\%$

Write each number in percent notation.

**52.** $\dfrac{17}{20}$        **53.** $1.5$

**54.** Estimate the sum $9.389 + 4.2105$ to the nearest tenth.

Solve.

**55.** $234 + y = 789$

**56.** $3.9a = 249.6$

**57.** $\frac{2}{3} \cdot t = \frac{5}{6}$

**58.** $\frac{8}{17} = \frac{36}{x}$

**59.** $7x - 9 = 26$

**60.** $-2(x - 5) = 3x + 12$

*Egg Consumption.* The line graph below shows egg consumption per person in the United States for recent years. Use the graph to solve Exercises 61–64.

**Egg Consumption**

SOURCE: United Egg Producers

**61.** Find the lowest egg consumption and the year(s) in which it occurred.

**62.** Find the highest egg consumption and the year(s) in which it occurred.

**63.** Find the mean, the median, and the mode(s) of the egg consumptions.

**64.** What was the percent of decrease in egg consumption from 2006 to 2010?

**65.** *Dead Sea.* The lowest point in the world is the Dead Sea on the border of Israel and Jordan. It is 1312 ft below sea level. Convert 1312 ft to yards; to meters.
**Source:** *The Handy Geography Answer Book*

**66.** In Sam's writing lab, 3 of the 20 students are left-handed. If a student is randomly selected, what is the probability that he or she is left-handed?

**67.** Find the missing angle measure.

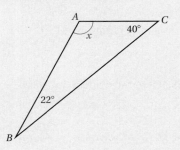

**68.** Margie donated $20 to the Humane Society, $30 to the Red Cross, $25 to the Salvation Army, and $20 to Amnesty International. What was the average size of the donations?

**69.** On Monday morning, a bolt of fabric contained $8\frac{1}{4}$ yd. Madison sold $3\frac{5}{8}$ yd from the bolt. How much fabric remains?

**70.** From Indira's income of $32,000, amounts of $6400 and $1600 are paid for federal and state taxes. How much remains after these taxes have been paid?

**71.** A toddler walks $\frac{3}{5}$ km per hour. At this rate, how far would the child walk in $\frac{1}{2}$ hr?

**72.** Eight gallons of paint covers 2000 ft$^2$. How much paint is needed to cover 3250 ft$^2$?

**73.** What is the simple interest on $4000 principal at 8% for $\frac{3}{4}$ yr?

**74.** The population of Bridgeton is 29,000 this year and is increasing at a rate of 4% per year. What will the population be next year?

**75.** *Firefighting.* During a fire, firefighters get a 1-ft layer of water on the 25-ft by 60-ft first floor of a 5-floor building. Water weighs $62\frac{1}{2}$ lb per cubic foot. What is the total weight of the water on the floor?

**76.** *Medical Dosage.* A doctor suggests that a child who weighs 24 kg be given 42 mg Phenytoin. If the dosage is proportional to the child's weight, how much Phenytoin is recommended for a child who weighs 32 kg?

**77.** A machine wraps 134 candy bars per minute. How long does it take this machine to wrap 8710 bars?

**78.** At the start of a trip, the odometer on the Oquendos' minivan read 27,428.6 mi, and at the end of the trip, the reading was 27,914.5 mi. How long was the trip?

**79.** Shannon is paid $85 a day for 7 days' work as a lifeguard. How much will she be paid?

**80.** Eight identical dresses cost a total of $679.68. What is the cost of each dress?

**81.** Eighteen ounces of a fruit "smoothie" cost $3.06. Find the unit price in cents per ounce.

**82.** Baldacci Real Estate received $5880 commission on the sale of an $84,000 home. What was the rate of commission?

**83.** Luis paid $35 a day plus 15¢ a mile for a van rental. If his one-day van rental cost $68, how many miles did he drive?

Evaluate.

**84.** $18^2$

**85.** $37^0$

**86.** $42^1$

**87.** $\sqrt{121}$

Write an equivalent expression with positive exponents. Then simplify, if possible.

**88.** $4^{-3}$

**89.** $\left(\dfrac{5}{4}\right)^{-2}$

Express each of the following in scientific notation.

**90.** 4,357,000

**91.** $(6.2 \times 10^7)(4.3 \times 10^{-23})$

Complete.

**92.** $\frac{1}{3}$ yd = _____ in.

**93.** 5.8 km = _____ m

**94.** 10 lb = _____ oz

**95.** 8190 mL = _____ L

**96.** 3917 mm = _____ cm

**97.** 60,000 g = _____ kg

**98.** 2.3 g = _____ mg

**99.** 28 qt = _____ gal

**100.** 10 yd$^2$ = _____ ft$^2$

**101.** 200 cm$^2$ = _____ m$^2$

The data in the following table show the percent of people who eat salad a certain number of times per week.

| NUMBER OF SALADS PER WEEK | PERCENT |
|---|---|
| None | 3% |
| 2 or fewer | 37% |
| 3–6 | 47% |
| At least one a day | 13% |

SOURCE: Market Facts for the Association of Dressings and Sauces

**102.** Make a bar graph of the data

**103.** Make a circle graph of the data.

**104.** Plot the following points:

$(-5, 2), (4, 0), (3, -4), (0, 2)$

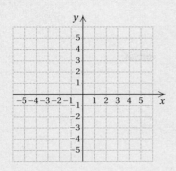

**105.** Graph: $y = -\dfrac{1}{3}x$.

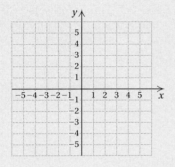

**106.** Graph: $y = 3$.

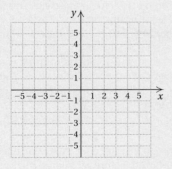

**107.** These triangles are similar. Find the missing lengths.

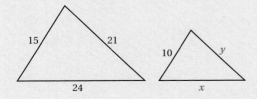

**108.** A rectangular mirror measures 20 in. by 24 in. Find its perimeter.

Find the area of each figure. Use 3.14 for $\pi$.

**109.**

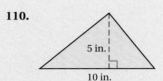

15.4 cm

**110.**

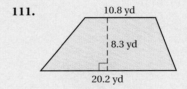

5 in.

10 in.

**111.**

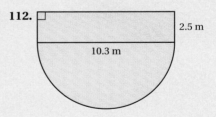

10.8 yd

8.3 yd

20.2 yd

**112.**

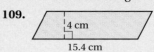

2.5 m

10.3 m

**113.** Find the diameter, the circumference, and the area of this circle. Use 3.14 for $\pi$.

10.4 in.

Find the volume of each shape. Use 3.14 for $\pi$.

**114.**

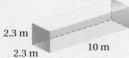

2.3 m
2.3 m
10 m

**115.**

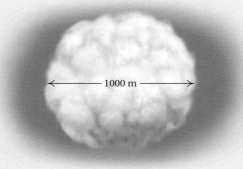

16 ft

4 ft

**116.** *Clouds.* Find the volume of a spherical cloud with a 1000-m diameter. Use 3.14 for $\pi$.

1000 m

**117.** Find the length of the third side of this right triangle. Give an exact answer and an approximation to three decimal places.

11 ft
$a$
6 ft

**118.** Find the measure of a complement of $\angle CBA$, if $m\angle CBA = 40°$.

  **A.** 140°          **B.** 40°
  **C.** 50°           **D.** 90°

**119.** Convert 100°C to Fahrenheit.

  **A.** $37\frac{7}{9}°F$        **B.** 237.6°F
  **C.** 180°F           **D.** 212°F

**120.** A month of the year is randomly selected. What is the probability that the name of the month contains an "$r$" in its spelling?

  **A.** $\frac{1}{2}$           **B.** $\frac{1}{3}$
  **C.** $\frac{2}{3}$           **D.** 8

## Synthesis

**121.** A housing development is constructed on a dead-end road that runs along a river and ends in a cul-de-sac, as shown in the figure. The property owners agree to share the cost of maintaining the road in the following manner. The cost of the first fifth of the road in front of lot 1 is to be shared equally among all five lot owners. The cost of the second fifth in front of lot 2 is to be shared equally among the owners of lots 2–5, and so on. Assume that all five sections of the road cost the same to maintain.

**a)** What fractional part of the cost is paid by each owner?

**b)** What percent of the cost is paid by each owner?

**c)** If lots 3, 4, and 5 were all owned by the same person, what percent of the cost of maintenance would this person pay?

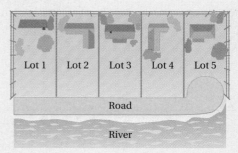

# Developmental Units

**A**    Addition

**S**    Subtraction

**M**    Multiplication

**D**    Division

# A Addition

## OBJECTIVES

**a** Add any two of the numbers 0, 1, 2, 3, 4, 5, 6, 7, 8, 9.

**b** Find certain sums of three numbers such as 1 + 7 + 9.

**c** Add two whole numbers when carrying is not necessary.

**d** Add two whole numbers when carrying is necessary.

## **a** Basic Addition

Basic addition can be explained by counting. The sum

$$3 + 4$$

can be found by counting out a set of 3 objects and a separate set of 4 objects, putting them together, and counting all the objects.

| A set of | + | A set of | = | A set of |
| 3 | | 4 | | 7 |

The numbers to be added are called **addends**. The result is the **sum**.

$$\underset{\text{Addend}}{3} \; + \; \underset{\text{Addend}}{4} \; = \; \underset{\text{Sum}}{7}$$

**EXAMPLES** Add. Think of putting sets of objects together.

**1.** 5 + 6 = 11

$$\begin{array}{r} 5 \\ + \, 6 \\ \hline 11 \end{array}$$

**2.** 8 + 5 = 13

$$\begin{array}{r} 8 \\ + \, 5 \\ \hline 13 \end{array}$$

We can also do these problems by counting up from one of the numbers. For example, in Example 2, we start at 8 and count up 5 times: 9, 10, 11, 12, 13.

Do Exercises 1–6.

What happens when we add 0? Think of a set of 5 objects. If we add 0 objects to it, we still have 5 objects. Similarly, if we have a set with 0 objects in it and add 5 objects to it, we have a set with 5 objects. Thus,

$$5 + 0 = 5 \quad \text{and} \quad 0 + 5 = 5.$$

---

**ADDITION OF 0**

Adding 0 to a number does not change the number:

$$a + 0 = 0 + a = a.$$

We say that 0 is the **additive identity**.

---

Add; think of joining sets of objects.

**1.** 4 + 5

**2.** 5 + 2

**3.** $\begin{array}{r} 9 \\ + \, 5 \\ \hline \end{array}$

**4.** $\begin{array}{r} 8 \\ + \, 8 \\ \hline \end{array}$

**5.** $\begin{array}{r} 9 \\ + \, 7 \\ \hline \end{array}$

**6.** $\begin{array}{r} 7 \\ + \, 9 \\ \hline \end{array}$

The first printed use of the + symbol was in a book by a German, Johann Widmann, in 1498.

*Answers*

**1.** 9 **2.** 7 **3.** 14 **4.** 16
**5.** 16 **6.** 16

**EXAMPLES** Add.

**3.** $0 + 9 = 9$

$$\begin{array}{r} 0 \\ + 9 \\ \hline 9 \end{array}$$

**4.** $0 + 0 = 0$

$$\begin{array}{r} 0 \\ + 0 \\ \hline 0 \end{array}$$

**5.** $97 + 0 = 97$

$$\begin{array}{r} 97 \\ + 0 \\ \hline 97 \end{array}$$

Do Exercises 7–12.

Add.

**7.** $8 + 0$      **8.** $0 + 8$

**9.** $\begin{array}{r} 7 \\ + 0 \\ \hline \end{array}$      **10.** $\begin{array}{r} 46 \\ + 0 \\ \hline \end{array}$

**11.** $0 + 13$      **12.** $58 + 0$

Your objective for this part of the section is to be able to add any two of the numbers 0, 1, 2, 3, 4, 5, 6, 7, 8, 9. Adding 0 is easy. The rest of the sums are listed in this table. Memorize the table by saying it to yourself over and over or by using flash cards.

| + | 1 | 2 | 3 | 4 | 5 | 6 | 7 | 8 | 9 |
|---|---|---|---|---|---|---|---|---|---|
| 1 | 2 | 3 | 4 | 5 | 6 | 7 | 8 | 9 | 10 |
| 2 | 3 | 4 | 5 | 6 | 7 | 8 | 9 | 10 | 11 |
| 3 | 4 | 5 | 6 | 7 | 8 | 9 | 10 | 11 | 12 |
| 4 | 5 | 6 | 7 | 8 | 9 | 10 | 11 | 12 | 13 |
| 5 | 6 | 7 | 8 | 9 | 10 | 11 | 12 | 13 | 14 |
| 6 | 7 | 8 | 9 | 10 | 11 | 12 | 13 | 14 | 15 |
| 7 | 8 | 9 | 10 | 11 | 12 | 13 | 14 | 15 | 16 |
| 8 | 9 | 10 | 11 | 12 | 13 | 14 | 15 | 16 | 17 |
| 9 | 10 | 11 | 12 | 13 | 14 | 15 | 16 | 17 | 18 |

$6 + 7 = 13$
Find 6 at the left, and 7 at the top.

$7 + 6 = 13$
Find 7 at the left, and 6 at the top.

It is very important that you *memorize* the basic addition facts! If you do not, you will always have trouble with addition.

Note the following.

$3 + 4 = 7$    $7 + 6 = 13$    $7 + 2 = 9$

$4 + 3 = 7$    $6 + 7 = 13$    $2 + 7 = 9$

We can add whole numbers in any order. This is the *commutative law of addition*. Because of this law, you need to learn only about half the table above, as shown by the shading.

Do Exercises 13 and 14.

Complete the table.

**13.**

| + | 1 | 2 | 3 | 4 | 5 |
|---|---|---|---|---|---|
| 1 |   |   | 4 |   |   |
| 2 |   |   |   |   |   |
| 3 |   |   | 7 |   |   |
| 4 |   |   |   |   |   |
| 5 |   |   |   |   |   |

**14.**

| + | 6 | 5 | 7 | 4 | 9 |
|---|---|---|---|---|---|
| 7 |   |   | 14 |   |   |
| 9 |   |   |   |   |   |
| 5 |   |   |   | 9 |   |
| 8 |   |   |   |   |   |
| 4 |   |   |   |   |   |

## b   Certain Sums of Three Numbers

To add $3 + 5 + 4$, we can add 3 and 5, then 4:

$$3 + 5 + 4$$
$$8 + 4$$
$$12.$$

We can also add 5 and 4, then 3:

$$3 + 5 + 4$$
$$3 + 9$$
$$12.$$

Either way we get 12.

*Answers*

**7.** 8   **8.** 8   **9.** 7   **10.** 46   **11.** 13   **12.** 58

**13.**

| + | 1 | 2 | 3 | 4 | 5 |
|---|---|---|---|---|---|
| 1 | 2 | 3 | 4 | 5 | 6 |
| 2 | 3 | 4 | 5 | 6 | 7 |
| 3 | 4 | 5 | 6 | 7 | 8 |
| 4 | 5 | 6 | 7 | 8 | 9 |
| 5 | 6 | 7 | 8 | 9 | 10 |

**14.**

| + | 6 | 5 | 7 | 4 | 9 |
|---|---|---|---|---|---|
| 7 | 13 | 12 | 14 | 11 | 16 |
| 9 | 15 | 14 | 16 | 13 | 18 |
| 5 | 11 | 10 | 12 | 9 | 14 |
| 8 | 14 | 13 | 15 | 12 | 17 |
| 4 | 10 | 9 | 11 | 8 | 13 |

A Addition   **729**

**EXAMPLE 6**  Add from the top mentally.

$$\begin{array}{r} 1 \\ 7 \\ + \ 9 \\ \hline \end{array}$$

We first add 1 and 7, getting 8. Then we add 8 and 9, getting 17.

$$\begin{array}{r} 1 \\ 7 \to 8 \\ + \ 9 \quad 9 \to 17 \\ \hline 17 \longleftarrow \end{array}$$

**EXAMPLE 7**  Add from the top mentally.

$$\begin{array}{r} 2 \\ 4 \to 6 \\ + \ 8 \quad 8 \to 14 \\ \hline 14 \longleftarrow \end{array}$$

Do Exercises 15–18.

## (c) Addition (No Carrying)

We now move to a gradual, conceptual development of addition of whole numbers. It is intended to provide you with a greater understanding so that your skill level will increase.

To add larger numbers, we can add the ones first, then the tens, then the hundreds, and so on.

**EXAMPLE 8**  Add: 5722 + 3234.

$$\begin{array}{r} 5\ 7\ 2\ \underline{2} \\ + \ 3\ 2\ 3\ \underline{4} \\ \hline 6 \end{array}$$  Add ones.

$$\begin{array}{r} 5\ 7\ \underline{2}\ 2 \\ + \ 3\ 2\ \underline{3}\ 4 \\ \hline 5\ 6 \end{array}$$  Add tens.

$$\begin{array}{r} 5\ \underline{7}\ 2\ 2 \\ + \ 3\ \underline{2}\ 3\ 4 \\ \hline 9\ 5\ 6 \end{array}$$  Add hundreds.

This is for explanation.

$$\begin{array}{r} \underline{5}\ 7\ 2\ 2 \\ + \ \underline{3}\ 2\ 3\ 4 \\ \hline 8\ 9\ 5\ 6 \end{array}$$  Add thousands.

$$\begin{array}{r} 5\ 7\ 2\ 2 \\ + \ 3\ 2\ 3\ 4 \\ \hline 8\ 9\ 5\ 6 \end{array}$$  You should write only this.

Do Exercises 19–22.

### Add from the top mentally.

**15.**
$$\begin{array}{r} 1 \\ 6 \\ + \ 9 \\ \hline \end{array}$$

**16.**
$$\begin{array}{r} 2 \\ 3 \\ + \ 4 \\ \hline \end{array}$$

**17.**
$$\begin{array}{r} 6 \\ 1 \\ + \ 4 \\ \hline \end{array}$$

**18.**
$$\begin{array}{r} 5 \\ 2 \\ + \ 8 \\ \hline \end{array}$$

### Add.

**19.**
$$\begin{array}{r} 2\ 4 \\ + \ 3\ 5 \\ \hline \end{array}$$

**20.**
$$\begin{array}{r} 3\ 4\ 6 \\ + \ 2\ 0\ 3 \\ \hline \end{array}$$

**21.**
$$\begin{array}{r} 8\ 3\ 2\ 7 \\ + \ 1\ 6\ 5\ 2 \\ \hline \end{array}$$

**22.**
$$\begin{array}{r} 3\ 4\ 6\ 1 \\ + \ 2\ 0\ 3\ 5 \\ \hline \end{array}$$

*Answers*

**15.** 16   **16.** 9   **17.** 11   **18.** 15
**19.** 59   **20.** 549   **21.** 9979   **22.** 5496

## (d) Addition (with Carrying)

### Carrying Tens

**EXAMPLE 9** Add: $18 + 27$.

$$\begin{array}{r} 1\;8 \\ +\;2\;7 \\ \hline ? \end{array}$$

Add ones. *Think:*

$$\begin{array}{r} 8 \\ +\;7 \\ \hline 1\;5 \end{array}$$

15 ones = 10 ones + 5 ones
= 1 ten + 5 ones

$$\begin{array}{r} {}^{1} \\ 1\;8 \\ +\;2\;7 \\ \hline 5 \end{array}$$

Write 5 in the ones column.
Write 1 as a reminder above the tens.
This is called *carrying*.

$$\begin{array}{r} {}^{1} \\ 1\;8 \\ +\;2\;7 \\ \hline 4\;5 \end{array}$$

Add tens.

We can use money to help explain Example 9.

$$\begin{array}{r} 1\;8¢ \\ +\;2\;7¢ \end{array}$$ $\longrightarrow$ 1 dime and 8 pennies
$\longrightarrow$ 2 dimes and 7 pennies

We first add the pennies: $8¢ + 7¢ = 15¢$.

1 dime
$$\begin{array}{r} 1\;8 \\ +\;2\;7 \\ \hline 5\text{ pennies} \end{array}$$

We regard ten pennies as one dime and write $15¢$ as 1 dime and 5 pennies.

$$\begin{array}{r} {}^{1} \\ 1\;8 \\ +\;2\;7 \\ \hline 4\;5 \end{array}$$

We now add the dimes. The result is 4 dimes and 5 pennies.

Do Exercises 23 and 24.

**Add.**

**23.**
$$\begin{array}{r} 1\;9 \\ +\;3\;7 \end{array}$$

**24.**
$$\begin{array}{r} 4\;6 \\ +\;3\;9 \end{array}$$

### Carrying Hundreds

**EXAMPLE 10** Add: $256 + 391$.

$$\begin{array}{r} 2\;5\;6 \\ +\;3\;9\;1 \\ \hline 7 \end{array}$$

Add ones.

$$\begin{array}{r} {}^{1} \\ 2\;5\;6 \\ +\;3\;9\;1 \\ \hline 4\;7 \end{array}$$

Add tens. We get 14 tens.
Now 14 tens = 10 tens + 4 tens = 1 hundred + 4 tens.
Write 4 in the tens column and a 1 above the hundreds.

> The carrying here is like exchanging 14 dimes for 1 dollar bill and 4 dimes.

$$\begin{array}{r} {}^{1} \\ 2\;5\;6 \\ +\;3\;9\;1 \\ \hline 6\;4\;7 \end{array}$$

Add hundreds.

Do Exercises 25 and 26.

**Add.**

**25.**
$$\begin{array}{r} 3\;4\;1 \\ +\;4\;8\;8 \end{array}$$

**26.**
$$\begin{array}{r} 7\;3\;0 \\ +\;2\;9\;6 \end{array}$$

*Answers*
**23.** 56   **24.** 85   **25.** 829   **26.** 1026

### Carrying Thousands

**EXAMPLE 11**   Add: 4803 + 3792.

$$\begin{array}{r} 4\;8\;0\;\boxed{3} \\ +\;3\;7\;9\;\boxed{2} \\ \hline 5 \end{array}$$   Add ones.

$$\begin{array}{r} 4\;8\;0\;3 \\ +\;3\;7\;9\;2 \\ \hline 9\;5 \end{array}$$   Add tens.

$$\begin{array}{r} {}^{1}\;\;\;\;\;\; \\ 4\;8\;0\;3 \\ +\;3\;7\;9\;2 \\ \hline 5\;9\;5 \end{array}$$   Add hundreds. We get 15 hundreds. Now 15 hundreds = 10 hundreds + 5 hundreds = 1 thousand + 5 hundreds. Write 5 in the hundreds column and 1 above the thousands.

$$\begin{array}{r} {}^{1}\;\;\;\;\;\; \\ 4\;8\;0\;3 \\ +\;3\;7\;9\;2 \\ \hline 8\;5\;9\;5 \end{array}$$   Add thousands.

> Do Exercise 27.

### Carrying More Than Once

Sometimes we must carry more than once.

**EXAMPLE 12**   Add: 5767 + 4993.

$$\begin{array}{r} 5\;7\;6\;{}^{1}7 \\ +\;4\;9\;9\;3 \\ \hline 0 \end{array}$$   Add ones. We get 10 ones. Now 10 ones = 1 ten + 0 ones. Write 0 in the ones column and 1 above the tens.

$$\begin{array}{r} 5\;7\;{}^{1}6\;{}^{1}7 \\ +\;4\;9\;9\;3 \\ \hline 6\;0 \end{array}$$   Add tens. We get 16 tens. Now 16 tens = 1 hundred + 6 tens. Write 6 in the tens column and 1 above the hundreds.

$$\begin{array}{r} 5\;{}^{1}7\;{}^{1}6\;{}^{1}7 \\ +\;4\;9\;9\;3 \\ \hline 7\;6\;0 \end{array}$$   Add hundreds. We get 17 hundreds. Now 17 hundreds = 1 thousand + 7 hundreds. Write 7 in the hundreds column and 1 above the thousands.

$$\begin{array}{r} {}^{1}5\;{}^{1}7\;{}^{1}6\;7 \\ +\;4\;9\;9\;3 \\ \hline 1\;0\;7\;6\;0 \end{array}$$   Add thousands. We get 10 thousands.

> Do Exercises 28 and 29.

---

**27.** Add.

$$\begin{array}{r} 7\;8\;5\;0 \\ +\;4\;8\;4\;8 \\ \hline \end{array}$$

**TO THE STUDENT**

If you had trouble with Section 1.2 and have studied Developmental Unit A, you should go back and work through Section 1.2 after completing Exercise Set A.

Add.

**28.**
$$\begin{array}{r} 7\;9\;8\;9 \\ +\;5\;6\;7\;2 \\ \hline \end{array}$$

**29.**
$$\begin{array}{r} 5\;6,7\;8\;9 \\ +\;1\;4,5\;3\;9 \\ \hline \end{array}$$

---

*Answers*

**27.** 12,698   **28.** 13,661   **29.** 71,328

**732**   DEVELOPMENTAL UNITS

**a** Add. Try to do these mentally. If you have trouble, think of putting sets of objects together.

| | | | | | |
|---|---|---|---|---|---|
| **1.** 8 <br> + 9 | **2.** 8 <br> + 7 | **3.** 6 <br> + 7 | **4.** 9 <br> + 5 | **5.** 5 <br> + 7 | **6.** 5 <br> + 6 |
| **7.** 9 <br> + 8 | **8.** 9 <br> + 7 | **9.** 8 <br> + 4 | **10.** 9 <br> + 1 | **11.** 8 <br> + 2 | **12.** 3 <br> + 8 |
| **13.** 0 <br> + 7 | **14.** 4 <br> + 3 | **15.** 2 <br> + 9 | **16.** 0 <br> + 0 | **17.** 3 <br> + 0 | **18.** 9 <br> + 9 |
| **19.** 8 <br> + 6 | **20.** 3 <br> + 7 | **21.** 2 <br> + 2 | **22.** 7 <br> + 7 | **23.** 6 <br> + 5 | **24.** 7 <br> + 8 |
| **25.** 8 <br> + 8 | **26.** 8 <br> + 1 | **27.** 5 <br> + 8 | **28.** 5 <br> + 9 | **29.** 4 <br> + 7 | **30.** 6 <br> + 1 |

**31.** $6 + 7$   **32.** $7 + 7$   **33.** $3 + 9$   **34.** $6 + 0$   **35.** $6 + 4$

**36.** $9 + 3$   **37.** $5 + 5$   **38.** $5 + 3$   **39.** $1 + 1$   **40.** $4 + 5$

**41.** $9 + 4$   **42.** $0 + 8$   **43.** $4 + 6$   **44.** $2 + 7$   **45.** $3 + 7$

**46.** $3 + 3$   **47.** $5 + 8$   **48.** $3 + 6$   **49.** $4 + 4$   **50.** $4 + 7$

**b** Add from the top mentally.

| | | | | |
|---|---|---|---|---|
| **51.** 1 <br> 8 <br> + 3 | **52.** 1 <br> 7 <br> + 5 | **53.** 3 <br> 2 <br> + 5 | **54.** 4 <br> 3 <br> + 5 | **55.** 1 <br> 7 <br> + 9 |
| **56.** 5 <br> 2 <br> + 6 | **57.** 4 <br> 5 <br> + 1 | **58.** 1 <br> 9 <br> + 6 | **59.** 1 <br> 8 <br> + 7 | **60.** 1 <br> 6 <br> + 8 |

**c** Add.

61.  23
    + 16

62.  54
    + 35

63.  67
    + 20

64.  496
    + 503

65.  700
    + 200

66.  801
    +  67

67.  666
    + 333

68.  523
    + 325

69.  747
    + 130

70.  8250
    + 9430

71.  6552
    + 4321

72.  3406
    + 1293

73.  7340
    + 3527

74.  4825
    + 5070

75.  2073
    + 1925

76.  9111
    + 9111

77.  7889
    + 9000

78.  52,433
    + 12,056

79.  43,723
    + 56,276

80.  51,670
    + 26,107

**d** Add.

81.  38
    +  8

82.  17
    +  9

83.  17
    + 38

84.  95
    +  6

85.  862
    + 781

86.  613
    + 799

87.  355
    + 491

88.  280
    + 348

89.  814
    + 390

90.  274
    + 333

91.  9990
    +   10

92.  999
    +  11

93.  999
    + 111

94.  839
    + 388

95.  909
    + 202

96.  808
    + 909

97.  8718
    + 1420

98.  3854
    + 2700

99.  4828
    + 1283

100.  6995
     + 1432

101.  9889
     +    1

102.  6889
     + 4723

103.  9128
     + 1997

104.  8898
     + 6645

105.  9989
     + 6785

106.  46,889
     + 21,786

107.  23,448
     + 10,989

108.  67,658
     + 98,786

109.  77,548
     + 23,767

110.  44,684
     +  4,765

# S Subtraction

## (a) Basic Subtraction

Subtraction can be explained by taking away part of a set.

**EXAMPLE 1**   Subtract: $7 - 3$.

We can do this by counting out 7 objects and then taking away 3 of them. Then we count the number that remain: $7 - 3 = 4$.

We could also do this mentally by starting at 7 and counting down 3 times: 6, 5, 4.

**EXAMPLES**   Subtract. Think of "take away."

**2.** $11 - 6 = 5$   *Take away:* "11 take away 6 is 5."

$$\begin{array}{r} 11 \\ -\ 6 \\ \hline 5 \end{array}$$

**3.** $17 - 9 = 8$

$$\begin{array}{r} 17 \\ -\ 9 \\ \hline 8 \end{array}$$

> Do Exercises 1–4.

The addition table in Developmental Unit A will enable you to subtract also. First, recall how addition and subtraction are related.

*An addition*:

*Two related subtractions*:

**A.**

**B.**

---

### OBJECTIVES

**a** Find basic differences such as $5 - 3$, $13 - 8$, and so on.

**b** Subtract one whole number from another when borrowing is not necessary.

**c** Subtract one whole number from another when borrowing is necessary.

Subtract.

**1.** $10 - 6$

**2.** $11 - 4$

**3.** $\begin{array}{r} 16 \\ -\ 8 \\ \hline \end{array}$

**4.** $\begin{array}{r} 10 \\ -\ 7 \\ \hline \end{array}$

*Answers*

1. 4   2. 7   3. 8   4. 3

Since we know that

$$4 + 3 = 7, \quad \text{A basic addition fact}$$

we also know the two subtraction facts

$$7 - 3 = 4 \quad \text{and} \quad 7 - 4 = 3.$$

**EXAMPLE 4** From $8 + 9 = 17$, write two subtraction facts.

**a)** The addend 8 is subtracted from the sum 17.

$8 + 9 = 17.$   The related sentence is   $17 - 8 = 9.$

**b)** The addend 9 is subtracted from the sum 17.

$8 + 9 = 17.$   The related sentence is   $17 - 9 = 8.$

Do Exercises 5 and 6.

We can use the idea that subtraction is defined in terms of addition to think of subtraction as "how much more."

**EXAMPLE 5** Find: $13 - 6$.

To find $13 - 6$, we ask, "6 plus what number is 13?"

$$6 + \square = 13$$

| + | 1 | 2 | 3 | 4 | 5 | 6 | 7 | 8 | 9 |
|---|---|---|---|---|---|---|---|---|---|
| 1 | 2 | 3 | 4 | 5 | 6 | 7 | 8 | 9 | 10 |
| 2 | 3 | 4 | 5 | 6 | 7 | 8 | 9 | 10 | 11 |
| 3 | 4 | 5 | 6 | 7 | 8 | 9 | 10 | 11 | 12 |
| 4 | 5 | 6 | 7 | 8 | 9 | 10 | 11 | 12 | 13 |
| 5 | 6 | 7 | 8 | 9 | 10 | 11 | 12 | 13 | 14 |
| 6 | 7 | 8 | 9 | 10 | 11 | 12 | 13 | 14 | 15 |
| 7 | 8 | 9 | 10 | 11 | 12 | 13 | 14 | 15 | 16 |
| 8 | 9 | 10 | 11 | 12 | 13 | 14 | 15 | 16 | 17 |
| 9 | 10 | 11 | 12 | 13 | 14 | 15 | 16 | 17 | 18 |

$13 - 6 = 7$

Using the addition table above, we find 13 inside the table and 6 at the left. Then we read the answer, 7, from the top. Thus, we have $13 - 6 = 7$. Strive to do this kind of thinking mentally as fast as you can, without having to use the table.

Do Exercises 7–10.

For each addition fact, write two subtraction facts.

**5.** $8 + 4 = 12$

**6.** $6 + 7 = 13$

Subtract. Try to do these mentally.

**7.** $14 - 6$      **8.** $12 - 5$

**9.** $\begin{array}{r} 13 \\ -\ 4 \\ \hline \end{array}$      **10.** $\begin{array}{r} 11 \\ -\ 7 \\ \hline \end{array}$

*Answers*

**5.** $12 - 8 = 4$; $12 - 4 = 8$   **6.** $13 - 6 = 7$; $13 - 7 = 6$   **7.** 8   **8.** 7   **9.** 9   **10.** 4

## b Subtraction (No Borrowing)

We now move to a gradual, conceptual development of subtraction of whole numbers. It is intended to provide you with a greater understanding so that your skill level will increase.

To subtract larger numbers, we can subtract the ones first, then the tens, then the hundreds, and so on.

**EXAMPLE 6**  Subtract: $5787 - 3214$.

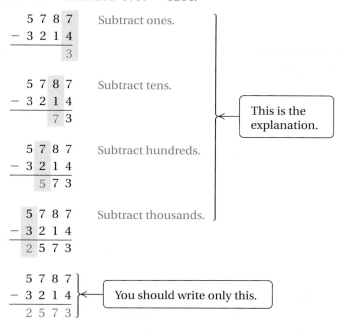

Do Exercises 11–14.

Do Exercises 11–14.

Subtract.

**11.**  $\begin{array}{r} 7\,8 \\ -\ 6\,4 \\ \hline \end{array}$  **12.**  $\begin{array}{r} 2\,9 \\ -\ \ \ 9 \\ \hline \end{array}$

**13.**  $\begin{array}{r} 5\,4\,2 \\ -\ 3\,0\,1 \\ \hline \end{array}$  **14.**  $\begin{array}{r} 6\,8\,9\,6 \\ -\ 4\,8\,7\,1 \\ \hline \end{array}$

## c Subtraction (with Borrowing)

We now consider subtraction when borrowing, or regrouping, is necessary.

### Borrowing from the Tens Place

**EXAMPLE 7**  Subtract: $37 - 18$.

$\begin{array}{r} 3\,7 \\ -\ 1\,8 \\ \hline ? \end{array}$   Try to subtract ones: $7 - 8$ is not a whole number.

$\begin{array}{r} {}^{2}\,{}^{17} \\ 3\,7 \\ -\ 1\,8 \\ \hline \end{array}$   Borrow a ten. That is, 1 ten = 10 ones, and 10 ones + 7 ones = 17 ones. Write 2 above the tens column and 17 above the ones. We regard 37 as $20 + 17$.

$\begin{array}{r} {}^{2}\,{}^{17} \\ 3\,7 \\ -\ 1\,8 \\ \hline 9 \end{array}$   Subtract ones.

> The borrowing here is like exchanging 3 dimes and 7 pennies for 2 dimes and 17 pennies.

$$\begin{array}{r} \overset{2}{\cancel{3}}\ \overset{17}{\cancel{7}} \\ -\ 1\ 8 \\ \hline 1\ 9 \end{array}$$ Subtract tens.

$$\begin{array}{r} \overset{2}{\cancel{3}}\ \overset{17}{\cancel{7}} \\ -\ 1\ 8 \\ \hline 1\ 9 \end{array}$$ You should write only this.

Do Exercises 15 and 16.

Subtract.

**15.**
$$\begin{array}{r} 4\ 6 \\ -\ 2\ 9 \end{array}$$

**16.**
$$\begin{array}{r} 7\ 4 \\ -\ 3\ 8 \end{array}$$

### Borrowing Hundreds

**EXAMPLE 8**   Subtract: $538 - 275$.

$$\begin{array}{r} 5\ 3\ 8 \\ -\ 2\ 7\ 5 \\ \hline 3 \end{array}$$ Subtract ones.

$$\begin{array}{r} 5\ 3\ 8 \\ -\ 2\ 7\ 5 \\ \hline ?\ 3 \end{array}$$ Try to subtract tens: 3 tens − 7 tens is not a whole number.

$$\begin{array}{r} \overset{4}{\cancel{5}}\ \overset{13}{\cancel{3}}\ 8 \\ -\ 2\ 7\ 5 \\ \hline 3 \end{array}$$ Borrow a hundred. That is, 1 hundred = 10 tens, and 10 tens + 3 tens = 13 tens. Write 4 above the hundreds column and 13 above the tens.

> The borrowing is like exchanging 5 dollars and 3 dimes for 4 dollars and 13 dimes.

$$\begin{array}{r} \overset{4}{\cancel{5}}\ \overset{13}{\cancel{3}}\ 8 \\ -\ 2\ 7\ 5 \\ \hline 6\ 3 \end{array}$$ Subtract tens.

$$\begin{array}{r} \overset{4}{\cancel{5}}\ \overset{13}{\cancel{3}}\ 8 \\ -\ 2\ 7\ 5 \\ \hline 2\ 6\ 3 \end{array}$$ Subtract hundreds.

$$\begin{array}{r} \overset{4}{\cancel{5}}\ \overset{13}{\cancel{3}}\ 8 \\ -\ 2\ 7\ 5 \\ \hline 2\ 6\ 3 \end{array}$$ You should write only this.

Do Exercises 17 and 18.

Subtract.

**17.**
$$\begin{array}{r} 6\ 4\ 6 \\ -\ 1\ 9\ 2 \end{array}$$

**18.**
$$\begin{array}{r} 7\ 3\ 3 \\ -\ 4\ 8\ 3 \end{array}$$

*Answers*

**15.** 17   **16.** 36   **17.** 454   **18.** 250

## Borrowing More Than Once

Sometimes we must borrow more than once.

**EXAMPLE 9**  Subtract: $672 - 394$.

$$
\begin{array}{r}
6\overset{6}{7}\overset{12}{2} \\
-\ 3\ 9\ 4 \\
\hline
8
\end{array}
$$
Borrowing a ten to subtract ones

$$
\begin{array}{r}
\overset{16}{} \\
\overset{5}{6}\overset{}{7}\overset{12}{2} \\
-\ 3\ 9\ 4 \\
\hline
2\ 7\ 8
\end{array}
$$
Borrowing a hundred to subtract tens

*Do Exercises 19 and 20.*

Do Exercises 19 and 20.

Subtract.

**19.**  $\begin{array}{r} 5\ 6\ 3 \\ -\ 1\ 8\ 7 \end{array}$   **20.**  $\begin{array}{r} 7\ 3\ 3 \\ -\ 4\ 8\ 8 \end{array}$

**EXAMPLE 10**  Subtract: $6357 - 1769$.

$$
\begin{array}{r}
6\ 3\overset{4}{5}\overset{17}{7} \\
-\ 1\ 7\ 6\ 9 \\
\hline
8
\end{array}
$$
$7 - 9$ is not a whole number. We borrow a ten.

$$
\begin{array}{r}
\overset{14}{} \\
6\overset{2}{3}\overset{4}{5}\overset{17}{7} \\
-\ 1\ 7\ 6\ 9 \\
\hline
8\ 8
\end{array}
$$
4 tens minus 6 tens is not a whole number. We borrow a hundred.

$$
\begin{array}{r}
\overset{12}{}\overset{14}{} \\
\overset{5}{6}\overset{2}{3}\overset{4}{5}\overset{17}{7} \\
-\ 1\ 7\ 6\ 9 \\
\hline
4\ 5\ 8\ 8
\end{array}
$$
2 hundreds minus 7 hundreds is not a whole number. We borrow a thousand.

We can always check by adding the answer to the number being subtracted.

**EXAMPLE 11**  Subtract: $8341 - 2673$. Check by adding.

We check by adding 5668 and 2673.

$$
\begin{array}{r}
\overset{7}{8}\overset{12}{3}\overset{13}{4}\overset{11}{1} \\
-\ 2\ 6\ 7\ 3 \\
\hline
5\ 6\ 6\ 8
\end{array}
$$
*Check:*
$$
\begin{array}{r}
\overset{1}{5}\overset{1}{6}\overset{1}{6}\ 8 \\
+\ 2\ 6\ 7\ 3 \\
\hline
8\ 3\ 4\ 1
\end{array}
$$

Do Exercises 21 and 22.

Subtract. Check by adding.

**21.**  $\begin{array}{r} 4\ 2\ 3\ 6 \\ -\ 1\ 6\ 7\ 9 \end{array}$   **22.**  $\begin{array}{r} 7\ 5\ 4\ 1 \\ -\ 3\ 8\ 6\ 7 \end{array}$

## Zeros in Subtraction

Before subtracting, note the following:

50 is 5 tens;

70 is 7 tens.

Then

100 is 10 tens;

200 is 20 tens.

Complete.

**23.** $80 = $ _____ tens

**24.** $60 = $ _____ tens

**25.** $300 = $ _____ tens

**26.** $900 = $ _____ tens

Do Exercises 23–26.

*Answers*

**19.** 376  **20.** 245  **21.** 2557  **22.** 3674
**23.** 8  **24.** 6  **25.** 30  **26.** 90

**Complete.**

**27.** $5000 = \underline{\hspace{1.5cm}}$ tens

**28.** $9000 = \underline{\hspace{1.5cm}}$ tens

**29.** $5380 = \underline{\hspace{1.5cm}}$ tens

**30.** $6770 = \underline{\hspace{1.5cm}}$ tens

**Subtract.**

**31.**
$$\begin{array}{r} 6\,0 \\ -\,1\,8 \\ \hline \end{array}$$

**32.**
$$\begin{array}{r} 4\,8\,0 \\ -\,2\,5\,6 \\ \hline \end{array}$$

**Subtract.**

**33.**
$$\begin{array}{r} 6\,0\,2 \\ -\,4\,6\,4 \\ \hline \end{array}$$

**34.**
$$\begin{array}{r} 4\,0\,8 \\ -\,3\,6\,4 \\ \hline \end{array}$$

**Subtract.**

**35.**
$$\begin{array}{r} 4\,0\,0\,6 \\ -\,1\,2\,3\,8 \\ \hline \end{array}$$

**36.**
$$\begin{array}{r} 9\,0\,0\,1 \\ -\,7\,8\,0\,4 \\ \hline \end{array}$$

**Subtract.**

**37.**
$$\begin{array}{r} 3\,0\,0\,0 \\ -\,1\,7\,5\,4 \\ \hline \end{array}$$

**38.**
$$\begin{array}{r} 8\,0\,1\,7 \\ -\,3\,2\,8\,9 \\ \hline \end{array}$$

---

**TO THE STUDENT**

If you had trouble with Section 1.3 and have studied Developmental Unit S, you should go back and work through Section 1.3 after completing Exercise Set S.

---

Also,

230 is 2 hundreds + 3 tens

or 20 tens + 3 tens

or 23 tens.

Similarly,

1000 is 100 tens;

2000 is 200 tens;

4670 is 467 tens.

> Do Exercises 27–30.

**EXAMPLE 12**   Subtract: $50 - 37$.

$$\begin{array}{r} \overset{4}{\cancel{5}}\;\overset{10}{\cancel{0}} \\ -\,3\;7 \\ \hline 1\;3 \end{array}$$

We have 5 tens. We keep 4 of them in the tens column and put 1 ten, or 10 ones, with the ones.

> Do Exercises 31 and 32.

**EXAMPLE 13**   Subtract: $803 - 547$.

$$\begin{array}{r} \overset{7}{\cancel{8}}\;\overset{9}{\cancel{0}}\;\overset{13}{\cancel{3}} \\ -\,5\;4\;7 \\ \hline 2\;5\;6 \end{array}$$

We have 8 hundreds, or 80 tens. We keep 79 tens and put 1 ten, or 10 ones, with the ones.

> Do Exercises 33 and 34.

**EXAMPLE 14**   Subtract: $9003 - 2789$.

$$\begin{array}{r} \overset{8}{\cancel{9}}\;\overset{9}{\cancel{0}}\;\overset{9}{\cancel{0}}\;\overset{13}{\cancel{3}} \\ -\,2\;7\;8\;9 \\ \hline 6\;2\;1\;4 \end{array}$$

We have 9 thousands, or 900 tens. We keep 899 tens and put 1 ten, or 10 ones, with the ones.

> Do Exercises 35 and 36.

**EXAMPLES**   Subtract.

**15.**
$$\begin{array}{r} \overset{4}{\cancel{5}}\,\overset{9}{\cancel{0}}\,\overset{9}{\cancel{0}}\,\overset{10}{\cancel{0}} \\ -\,2\,8\,6\,1 \\ \hline 2\,1\,3\,9 \end{array}$$

**16.**
$$\begin{array}{r} \overset{4}{\cancel{5}}\,\overset{9}{\cancel{0}}\,\overset{10}{\underset{\cancel{1}}{\cancel{0}}}\,\overset{13}{\cancel{3}} \\ -\,1\,8\,5\,7 \\ \hline 3\,1\,5\,6 \end{array}$$

We have 5 thousands, or 49 hundreds and 10 tens.

> Do Exercises 37 and 38.

---

*Answers*

**27.** 500   **28.** 900   **29.** 538   **30.** 677
**31.** 42   **32.** 224   **33.** 138   **34.** 44
**35.** 2768   **36.** 1197   **37.** 1246   **38.** 4728

## a   Subtract. Try to do these mentally.

**1.**  $\begin{array}{r} 7 \\ -\,0 \\ \hline \end{array}$

**2.**  $\begin{array}{r} 8 \\ -\,8 \\ \hline \end{array}$

**3.**  $\begin{array}{r} 7 \\ -\,7 \\ \hline \end{array}$

**4.**  $\begin{array}{r} 8 \\ -\,3 \\ \hline \end{array}$

**5.**  $\begin{array}{r} 5 \\ -\,2 \\ \hline \end{array}$

**6.**  $\begin{array}{r} 1\,6 \\ -\quad 8 \\ \hline \end{array}$

**7.**  $\begin{array}{r} 1\,7 \\ -\quad 9 \\ \hline \end{array}$

**8.**  $\begin{array}{r} 1\,2 \\ -\quad 6 \\ \hline \end{array}$

**9.**  $\begin{array}{r} 1\,1 \\ -\quad 4 \\ \hline \end{array}$

**10.**  $\begin{array}{r} 1\,2 \\ -\quad 9 \\ \hline \end{array}$

**11.**  $\begin{array}{r} 1\,4 \\ -\quad 7 \\ \hline \end{array}$

**12.**  $\begin{array}{r} 1\,8 \\ -\quad 9 \\ \hline \end{array}$

**13.**  $\begin{array}{r} 1\,3 \\ -\quad 7 \\ \hline \end{array}$

**14.**  $\begin{array}{r} 1\,5 \\ -\quad 9 \\ \hline \end{array}$

**15.**  $\begin{array}{r} 9 \\ -\,7 \\ \hline \end{array}$

**16.** $7 - 3$

**17.** $4 - 1$

**18.** $2 - 0$

**19.** $3 - 3$

**20.** $6 - 3$

**21.** $7 - 6$

**22.** $9 - 8$

**23.** $10 - 3$

**24.** $6 - 6$

**25.** $11 - 7$

**26.** $12 - 8$

**27.** $5 - 0$

**28.** $4 - 0$

**29.** $13 - 9$

**30.** $14 - 9$

**31.** $11 - 2$

**32.** $12 - 3$

**33.** $16 - 9$

**34.** $18 - 9$

**35.** $11 - 5$

**36.** $10 - 4$

**37.** $10 - 8$

**38.** $14 - 8$

**39.** $15 - 8$

**40.** $10 - 2$

## b   Subtract.

**41.**  $\begin{array}{r} 6\,4 \\ -\,3\,1 \\ \hline \end{array}$

**42.**  $\begin{array}{r} 5\,5 \\ -\,3\,4 \\ \hline \end{array}$

**43.**  $\begin{array}{r} 5\,4\,8 \\ -\,3\,0\,1 \\ \hline \end{array}$

**44.**  $\begin{array}{r} 5\,9\,6 \\ -\,4\,0\,3 \\ \hline \end{array}$

**45.**  $\begin{array}{r} 7\,0\,0 \\ -\,2\,0\,0 \\ \hline \end{array}$

46.
$$\begin{array}{r} 765 \\ -\ 111 \\ \hline \end{array}$$

47.
$$\begin{array}{r} 525 \\ -\ 323 \\ \hline \end{array}$$

48.
$$\begin{array}{r} 747 \\ -\ 130 \\ \hline \end{array}$$

49.
$$\begin{array}{r} 988 \\ -\ 700 \\ \hline \end{array}$$

50.
$$\begin{array}{r} 9450 \\ -\ 8230 \\ \hline \end{array}$$

51.
$$\begin{array}{r} 6552 \\ -\ 4321 \\ \hline \end{array}$$

52.
$$\begin{array}{r} 7547 \\ -\ 3421 \\ \hline \end{array}$$

53.
$$\begin{array}{r} 5875 \\ -\ 2111 \\ \hline \end{array}$$

54.
$$\begin{array}{r} 38{,}695 \\ -\ 37{,}004 \\ \hline \end{array}$$

55.
$$\begin{array}{r} 67{,}899 \\ -\ 66{,}673 \\ \hline \end{array}$$

56.
$$\begin{array}{r} 99{,}999 \\ -\ \ \ \ \ \ 1 \\ \hline \end{array}$$

57.
$$\begin{array}{r} 56{,}780 \\ -\ 56{,}770 \\ \hline \end{array}$$

58.
$$\begin{array}{r} 42{,}111 \\ -\ 32{,}010 \\ \hline \end{array}$$

59.
$$\begin{array}{r} 77{,}654 \\ -\ 66{,}611 \\ \hline \end{array}$$

60.
$$\begin{array}{r} 23{,}456 \\ -\ 12{,}345 \\ \hline \end{array}$$

C  Subtract.

61.
$$\begin{array}{r} 93 \\ -\ 28 \\ \hline \end{array}$$

62.
$$\begin{array}{r} 42 \\ -\ 13 \\ \hline \end{array}$$

63.
$$\begin{array}{r} 86 \\ -\ 78 \\ \hline \end{array}$$

64.
$$\begin{array}{r} 98 \\ -\ 89 \\ \hline \end{array}$$

65.
$$\begin{array}{r} 625 \\ -\ 317 \\ \hline \end{array}$$

66.
$$\begin{array}{r} 735 \\ -\ 609 \\ \hline \end{array}$$

67.
$$\begin{array}{r} 853 \\ -\ 236 \\ \hline \end{array}$$

68.
$$\begin{array}{r} 961 \\ -\ 747 \\ \hline \end{array}$$

69.
$$\begin{array}{r} 787 \\ -\ 698 \\ \hline \end{array}$$

70.
$$\begin{array}{r} 6769 \\ -\ 2367 \\ \hline \end{array}$$

71.
$$\begin{array}{r} 6431 \\ -\ 2876 \\ \hline \end{array}$$

72.
$$\begin{array}{r} 7654 \\ -\ 1765 \\ \hline \end{array}$$

73.
$$\begin{array}{r} 5246 \\ -\ 2859 \\ \hline \end{array}$$

74.
$$\begin{array}{r} 6328 \\ -\ 2679 \\ \hline \end{array}$$

75.
$$\begin{array}{r} 7641 \\ -\ 3809 \\ \hline \end{array}$$

76.
$$\begin{array}{r} 8743 \\ -\ \ 599 \\ \hline \end{array}$$

77.
$$\begin{array}{r} 12{,}647 \\ -\ \ 4{,}897 \\ \hline \end{array}$$

78.
$$\begin{array}{r} 16{,}222 \\ -\ \ 5{,}777 \\ \hline \end{array}$$

79.
$$\begin{array}{r} 46{,}781 \\ -\ 12{,}988 \\ \hline \end{array}$$

80.
$$\begin{array}{r} 470 \\ -\ 189 \\ \hline \end{array}$$

81.
$$\begin{array}{r} 690 \\ -\ 235 \\ \hline \end{array}$$

82.
$$\begin{array}{r} 703 \\ -\ 132 \\ \hline \end{array}$$

83.
$$\begin{array}{r} 6406 \\ -\ \ 258 \\ \hline \end{array}$$

84.
$$\begin{array}{r} 2309 \\ -\ \ 109 \\ \hline \end{array}$$

85.
$$\begin{array}{r} 3406 \\ -\ 1293 \\ \hline \end{array}$$

86.
$$\begin{array}{r} 6807 \\ -\ 3059 \\ \hline \end{array}$$

87.
$$\begin{array}{r} 8000 \\ -\ 2794 \\ \hline \end{array}$$

88.
$$\begin{array}{r} 8002 \\ -\ 6543 \\ \hline \end{array}$$

89.
$$\begin{array}{r} 38{,}000 \\ -\ 37{,}695 \\ \hline \end{array}$$

90.
$$\begin{array}{r} 16{,}043 \\ -\ 11{,}588 \\ \hline \end{array}$$

# M  Multiplication

## a  Basic Multiplication

To multiply, we begin with two numbers, called **factors**, and get a third number, called a **product**. Multiplication of whole numbers can be explained by counting. The product $3 \times 5$ can be found by counting out 3 sets of 5 objects each, joining them (in a rectangular array if desired), and counting all the objects.

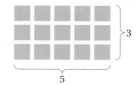

$$3 \times 5 = 15$$
Factor    Factor    Product

We can also think of multiplication as repeated addition.

$$3 \times 5 = \underbrace{5 + 5 + 5}_{\text{3 addends of 5}} = 15$$

**EXAMPLES**  Multiply. Think either of putting sets of objects together in a rectangular array or of repeated addition.

**1.** $5 \times 6 = 30$

$$\begin{array}{r} 6 \\ \times\ 5 \\ \hline 30 \end{array}$$

**2.** $8 \times 4 = 32$

$$\begin{array}{r} 4 \\ \times\ 8 \\ \hline 32 \end{array}$$

Do Exercises 1–4.

### Multiplying by 0

How do we multiply by 0? Consider $4 \cdot 0$. Using repeated addition, we see that

$$4 \cdot 0 = \underbrace{0 + 0 + 0 + 0}_{\text{4 addends of 0}} = 0.$$

We can also think of $4 \cdot 0$ as 4 sets with 0 objects in each set, so the total is 0.

Consider $0 \cdot 4$. Using repeated addition, this is 0 addends of 4, which is 0. Using sets, this is 0 sets with 4 objects in each set, which is 0. Thus, we have the following.

> **MULTIPLICATION BY 0**
>
> Multiplying by 0 gives 0.

**EXAMPLES**  Multiply.

**3.** $13 \times 0 = 0$

$$\begin{array}{r} 0 \\ \times\ 13 \\ \hline 0 \end{array}$$

**4.** $0 \cdot 11 = 0$

$$\begin{array}{r} 11 \\ \times\ 0 \\ \hline 0 \end{array}$$

**5.** $0 \cdot 0 = 0$

$$\begin{array}{r} 0 \\ \times\ 0 \\ \hline 0 \end{array}$$

Do Exercises 5 and 6.

---

Multiply. Think of joining sets in a rectangular array or of repeated addition.

**1.** $7 \cdot 8$ (The dot "·" means the same as "×".)

**2.** $\begin{array}{r} 9 \\ \times\ 4 \\ \hline \end{array}$

**3.** $4 \cdot 7$

**4.** $\begin{array}{r} 7 \\ \times\ 6 \\ \hline \end{array}$

Multiply.

**5.** $8 \cdot 0$

**6.** $\begin{array}{r} 17 \\ \times\ 0 \\ \hline \end{array}$

*Answers*

1. 56    2. 36    3. 28    4. 42
5. 0    6. 0

## Multiplying by 1

How do we multiply by 1? Consider $5 \cdot 1$. Using repeated addition, we see that

$$5 \cdot 1 = \underbrace{1 + 1 + 1 + 1 + 1}_{5 \text{ addends of } 1} = 5.$$

We can also think of $5 \cdot 1$ as 5 sets with 1 object in each set, for a total of 5 objects.

Consider $1 \cdot 5$. Using repeated addition, this is 1 addend of 5, which is 5. Using sets, this is 1 set of 5 objects, which is again 5 objects. Thus, we have the following.

---

**MULTIPLICATION BY 1**

Multiplying a number by 1 does not change the number:

$$a \cdot 1 = 1 \cdot a = a.$$

We say that 1 is the **multiplicative identity**.

---

Multiply.

**7.** $8 \cdot 1$

**8.**
$$\begin{array}{r} 2\,3 \\ \times \quad 1 \\ \hline \end{array}$$

**EXAMPLES**  Multiply.

**6.** $13 \cdot 1 = 13$
$$\begin{array}{r} 1 \\ \times\ 13 \\ \hline 13 \end{array}$$

**7.** $1 \cdot 7 = 7$
$$\begin{array}{r} 7 \\ \times\ 1 \\ \hline 7 \end{array}$$

**8.** $1 \cdot 1 = 1$
$$\begin{array}{r} 1 \\ \times\ 1 \\ \hline 1 \end{array}$$

Do Exercises 7 and 8.

You should be able to multiply any two of the numbers 0, 1, 2, 3, 4, 5, 6, 7, 8, 9. Multiplying by 0 and 1 is easy. The rest of the products are listed in the following table.

Complete the table.

**9.**

| × | 2 | 3 | 4 | 5 |
|---|---|---|---|---|
| 2 | | | | |
| 3 | | 12 | | |
| 4 | | | | |
| 5 | | 15 | | |
| 6 | | | | |

**10.**

| × | 6 | 7 | 8 | 9 |
|---|---|---|---|---|
| 5 | | | | |
| 6 | | | 48 | |
| 7 | | | | |
| 8 | | 56 | | |
| 9 | | | | |

| × | 2 | 3 | 4 | 5 | 6 | 7 | 8 | 9 |
|---|---|---|---|---|---|---|---|---|
| 2 | 4 | 6 | 8 | 10 | 12 | 14 | 16 | 18 |
| 3 | 6 | 9 | 12 | 15 | 18 | 21 | 24 | 27 |
| 4 | 8 | 12 | 16 | 20 | 24 | 28 | 32 | 36 |
| 5 | 10 | 15 | 20 | 25 | 30 | 35 | 40 | 45 |
| 6 | 12 | 18 | 24 | 30 | 36 | 42 | 48 | 54 |
| 7 | 14 | 21 | 28 | 35 | 42 | 49 | 56 | 63 |
| 8 | 16 | 24 | 32 | 40 | 48 | 56 | 64 | 72 |
| 9 | 18 | 27 | 36 | 45 | 54 | 63 | 72 | 81 |

$5 \times 7 = 35$
Find 5 at the left, and 7 at the top.

$8 \cdot 4 = 32$
Find 8 at the left, and 4 at the top.

It is *very* important that you have the basic multiplication facts *memorized*. If you do not, you will always have trouble with multiplication.

The *commutative law of multiplication* says that we can multiply numbers in any order. Thus, you need to learn only about half the table, as shown by the shading.

Do Exercises 9 and 10.

**Answers**

**7.** 8   **8.** 23

**9.**

| × | 2 | 3 | 4 | 5 |
|---|---|---|---|---|
| 2 | 4 | 6 | 8 | 10 |
| 3 | 6 | 9 | 12 | 15 |
| 4 | 8 | 12 | 16 | 20 |
| 5 | 10 | 15 | 20 | 25 |
| 6 | 12 | 18 | 24 | 30 |

**10.**

| × | 6 | 7 | 8 | 9 |
|---|---|---|---|---|
| 5 | 30 | 35 | 40 | 45 |
| 6 | 36 | 42 | 48 | 54 |
| 7 | 42 | 49 | 56 | 63 |
| 8 | 48 | 56 | 64 | 72 |
| 9 | 54 | 63 | 72 | 81 |

# b Multiplying by 10, 100, and 1000

We now move to a gradual, conceptual development of multiplication of whole numbers. It is intended to provide you with a greater understanding so that your skill level will increase.

We begin by considering multiplication by 10, 100, and 1000.

## Multiplying by 10

We know that

$$50 = 5 \text{ tens} \qquad 340 = 34 \text{ tens} \quad \text{and} \quad 2340 = 234 \text{ tens}$$
$$= 5 \cdot 10, \qquad = 34 \cdot 10, \qquad \qquad = 234 \cdot 10.$$

Turning this around, we see that to multiply any number by 10, all we need to do is write a 0 on the end of the number.

> ### MULTIPLICATION BY 10
>
> To multiply a number by 10, write 0 on the end of the number.

**EXAMPLES** Multiply.

**9.** $10 \cdot 6 = 60$

**10.** $10 \cdot 47 = 470$

**11.** $10 \cdot 583 = 5830$

Do Exercises 11–15.

> Multiply.
>
> **11.** $10 \cdot 7$   **12.** $10 \cdot 45$
>
> **13.** $10 \cdot 273$   **14.** $10 \cdot 10$
>
> **15.** $10 \cdot 100$

Let's find $4 \cdot 90$. This is $4 \cdot (9 \text{ tens})$, or 36 tens. The procedure is the same as multiplying 4 and 9 and writing a 0 on the end. Thus, $4 \cdot 90 = 360$.

**EXAMPLES** Multiply.

**12.** $5 \cdot 70 = 350$
   — $5 \cdot 7$, then write a 0

**13.** $8 \cdot 80 = 640$

**14.** $5 \cdot 60 = 300$

Do Exercises 16 and 17.

> Multiply.
>
> **16.** $\begin{array}{r} 70 \\ \times \quad 8 \\ \hline \end{array}$   **17.** $\begin{array}{r} 60 \\ \times \quad 6 \\ \hline \end{array}$

## Multiplying by 100

Note the following:

$$300 = 3 \text{ hundreds} \qquad 4700 = 47 \text{ hundreds} \quad \text{and} \quad 56{,}800 = 568 \text{ hundreds}$$
$$= 3 \cdot 100, \qquad \qquad = 47 \cdot 100, \qquad \qquad = 568 \cdot 100.$$

Turning this around, we see that to multiply any number by 100, all we need to do is write two 0's on the end of the number.

> ### MULTIPLICATION BY 100
>
> To multiply a number by 100, write two 0's on the end of the number.

*Answers*

**11.** 70   **12.** 450   **13.** 2730   **14.** 100
**15.** 1000   **16.** 560   **17.** 360

**EXAMPLES** Multiply.

**15.** $100 \cdot 6 = 600$

**16.** $100 \cdot 39 = 3900$

**17.** $100 \cdot 448 = 44{,}800$

Do Exercises 18–22.

Let's find $4 \cdot 900$. This is $4 \cdot (9$ hundreds$)$, or 36 hundreds. The procedure is the same as multiplying 4 and 9 and writing two 0's on the end. Thus, $4 \cdot 900 = 3600$.

**EXAMPLES** Multiply.

**18.** $6 \cdot 800 = 4800$

        $6 \cdot 8$, then write 00

**19.** $9 \cdot 700 = 6300$

**20.** $5 \cdot 500 = 2500$

Do Exercises 23 and 24.

## Multiplying by 1000

Note the following:

$$6000 = 6 \text{ thousands} \quad \text{and} \quad 19{,}000 = 19 \text{ thousands}$$
$$= 6 \cdot 1000 \quad\quad\quad\quad\quad = 19 \cdot 1000.$$

Turning this around, we see that to multiply any number by 1000, all we need to do is write three 0's on the end of the number.

> **MULTIPLYING BY 1000**
>
> To multiply a number by 1000, write three 0's on the end of the number.

**EXAMPLES** Multiply.

**21.** $1000 \cdot 8 = 8000$

**22.** $2000 \cdot 13 = 26{,}000$     $2 \cdot 13$, then write 000

**23.** $1000 \cdot 567 = 567{,}000$

Do Exercises 25–29.

## Multiplying Multiples by Multiples

Let's multiply 50 and 30. This is $50 \cdot (3$ tens$)$, or 150 tens, or 1500. The procedure is the same as multiplying 5 and 3 and writing two 0's on the end.

To multiply multiples of tens, hundreds, thousands, and so on:

a) Multiply the one-digit numbers.

b) Count the number of zeros.

c) Write that many 0's on the end.

**EXAMPLES** Multiply.

**24.**
```
      80    1 zero at end
×     60    1 zero at end
    4800
```
↑——— 6 · 8, then write 00

**25.**
```
     800    2 zeros at end
×     60    1 zero at end
  48,000
```
↑——— 6 · 8, then write 000

**26.**
```
     800    2 zeros at end
×    600    2 zeros at end
 480,000
```
↑——— 6 · 8, then write 0,000

**27.**
```
     800    2 zeros at end
×     50    1 zero at end
  40,000
```
↑——— 5 · 8, then write 000

Do Exercises 30–33.

Multiply.

**30.**
```
  9 0 0 0
×       6
```

**31.**
```
    8 0
×   7 0
```

**32.**
```
  8 0 0
×   7 0
```

**33.**
```
  6 0 0
×   3 0
```

## (c) Multiplying Larger Numbers

The product $3 \times 24$ can be represented as

$$3 \times (2 \text{ tens} + 4) = (2 \text{ tens} + 4) + (2 \text{ tens} + 4) + (2 \text{ tens} + 4)$$
$$= 6 \text{ tens} + 12$$
$$= 6 \text{ tens} + 1 \text{ ten} + 2$$
$$= 7 \text{ tens} + 2$$
$$= 72.$$

We multiply the 4 ones by 3, getting        12
We multiply the 2 tens by 3, getting     + 60
        Then we add:          72

**EXAMPLE 28** Multiply: $3 \times 24$.

```
      2 4    We use the approach described above.
×       3
      1 2 ←—Multiply the 4 ones by 3.
      6 0 ←—Multiply the 2 tens by 3.
      7 2 ←—Add.
```

Do Exercises 34–36.

Multiply.

**34.**
```
    1 4
×     2
```

**35.**
```
    5 8
×     2
```

**36.**
```
    3 7
×     4
```

**EXAMPLE 29** Multiply: $5 \times 734$.

```
      7 3 4
×         5
        2 0 ←—Multiply the 4 ones by 5.
      1 5 0 ←—Multiply the 3 tens by 5.
    3 5 0 0 ←—Multiply the 7 hundreds by 5.
    3 6 7 0 ←—Add.
```

Do Exercises 37 and 38.

Multiply.

**37.**
```
    8 2 3
×       6
```

**38.**
```
  1 3 4 8
×       5
```

*Answers*

**30.** 54,000   **31.** 5600   **32.** 56,000
**33.** 18,000   **34.** 28   **35.** 116
**36.** 148   **37.** 4938   **38.** 6740

Let's look at Example 29 again. Instead of writing each product on a separate line, we can use a shorter form.

**EXAMPLE 30**  Multiply: $5 \times 734$.

$$
\begin{array}{r}
7\ \overset{2}{3}\ 4 \\
\times \quad\quad 5 \\
\hline
0
\end{array}
$$

Multiply the ones by 5: $5 \cdot (4 \text{ ones}) = 20 \text{ ones} = 2 \text{ tens} + 0 \text{ ones}$. Write 0 in the ones column and 2 above the tens.

$$
\begin{array}{r}
7\ \overset{1}{\phantom{}}\overset{2}{3}\ 4 \\
\times \quad\quad 5 \\
\hline
7\ 0
\end{array}
$$

Multiply 3 tens by 5 and add 2 tens: $5 \cdot (3 \text{ tens}) = 15 \text{ tens}$; $15 \text{ tens} + 2 \text{ tens} = 17 \text{ tens} = 1 \text{ hundred} + 7 \text{ tens}$. Write 7 in the tens column and 1 above the hundreds.

$$
\begin{array}{r}
\overset{1}{7}\ \overset{2}{3}\ 4 \\
\times \quad\quad 5 \\
\hline
3\ 6\ 7\ 0
\end{array}
$$

Multiply the 7 hundreds by 5 and add 1 hundred: $5 \cdot (7 \text{ hundreds}) = 35 \text{ hundreds}$, $35 \text{ hundreds} + 1 \text{ hundred} = 36 \text{ hundreds}$.

$$
\left.\begin{array}{r}
\overset{1}{7}\ \overset{2}{3}\ 4 \\
\times \quad\quad 5 \\
\hline
3\ 6\ 7\ 0
\end{array}\right\}
$$
You should write only this.

Do Exercises 39–42.

Multiply using the short form.

**39.**
$$
\begin{array}{r}
5\ 8 \\
\times\quad 2 \\
\hline
\end{array}
$$

**40.**
$$
\begin{array}{r}
3\ 7 \\
\times\quad 4 \\
\hline
\end{array}
$$

**41.**
$$
\begin{array}{r}
8\ 2\ 3 \\
\times\quad\quad 6 \\
\hline
\end{array}
$$

**42.**
$$
\begin{array}{r}
1\ 3\ 4\ 8 \\
\times\quad\quad\quad 5 \\
\hline
\end{array}
$$

## (d) Multiplying by Multiples of 10, 100, and 1000

To multiply 327 by 50, we multiply by 10 (write a 0) and then multiply 327 by 5.

$$
\begin{array}{r}
3\ 2\ 7 \\
\times\quad 5\,\boxed{0} \\
\hline
1\ 6{,}3\ 5\ 0
\end{array}
$$
← Write a 0.
Multiply $5 \cdot 327$.

**EXAMPLE 31**  Multiply: $400 \times 289$.

$$
\begin{array}{r}
2\ 8\ 9 \\
\times\ 4\,\boxed{0\ 0} \\
\hline
0\ 0
\end{array}
$$
← Write two 0's.

$$
\begin{array}{r}
2\ 8\ 9 \\
\times\quad 4\ 0\ 0 \\
\hline
1\ 1\ 5{,}6\ 0\ 0
\end{array}
$$
Multiply 4 and 289:
$$
\begin{array}{r}
\overset{3}{2}\ \overset{3}{8}\ 9 \\
\times\quad\quad 4 \\
\hline
1\ 1\ 5\ 6
\end{array}
$$

$$
\left.\begin{array}{r}
\overset{3}{2}\ \overset{3}{8}\ 9 \\
\times\quad 4\ 0\ 0 \\
\hline
1\ 1\ 5{,}6\ 0\ 0
\end{array}\right\}
$$
Try to write only this.

Do Exercises 43–45.

Multiply.

**43.**
$$
\begin{array}{r}
7\ 4\ 6 \\
\times\quad\quad 8 \\
\hline
\end{array}
$$

**44.**
$$
\begin{array}{r}
7\ 4\ 6 \\
\times\quad 8\ 0 \\
\hline
\end{array}
$$

**45.**
$$
\begin{array}{r}
7\ 4\ 6 \\
\times\ 8\ 0\ 0 \\
\hline
\end{array}
$$

**TO THE STUDENT**

If you had trouble with Section 1.4 and have studied Developmental Unit M, you should go back and work through Section 1.4 after completing Exercise Set M.

*Answers*

**39.** 116  **40.** 148  **41.** 4938  **42.** 6740
**43.** 5968  **44.** 59,680  **45.** 596,800

**a**    Multiply. Try to do these mentally.

1.    3
  $\times\ 4$

2.    6
  $\times\ 0$

3.    7
  $\times\ 1$

4.    0
  $\times\ 2$

5.    10
  $\times\ 1$

6.    6
  $\times\ 5$

7.    5
  $\times\ 2$

8.    9
  $\times\ 7$

9.    9
  $\times\ 6$

10.    2
  $\times\ 6$

11.    7
  $\times\ 0$

12.    8
  $\times\ 9$

13.    1
  $\times\ 8$

14.    8
  $\times\ 0$

15.    4
  $\times\ 7$

16.    3
  $\times\ 8$

17.    5
  $\times\ 9$

18.    2
  $\times\ 9$

19.    0
  $\times\ 7$

20.    5
  $\times\ 7$

21.    9
  $\times\ 5$

22.    5
  $\times\ 8$

23.    0
  $\times\ 0$

24.    2
  $\times\ 8$

25. $5 \cdot 5$

26. $9 \cdot 9$

27. $1 \cdot 1$

28. $0 \cdot 0$

29. $2 \cdot 2$

30. $6 \cdot 6$

31. $1 \cdot 8$

32. $0 \cdot 1$

33. $3 \cdot 9$

34. $2 \cdot 9$

35. $6 \cdot 0$

36. $10 \cdot 1$

37. $6 \cdot 8$

38. $9 \cdot 6$

39. $8 \cdot 0$

40. $9 \cdot 8$

41. $3 \cdot 5$

42. $1 \cdot 8$

43. $1 \cdot 9$

44. $2 \cdot 1$

45. $8 \cdot 4$

46. $3 \cdot 2$

47. $5 \cdot 3$

48. $1 \cdot 6$

49. $4 \cdot 2$

50. $4 \cdot 5$

51. $5 \cdot 4$

52. $4 \cdot 4$

53. $5 \cdot 2$

54. $8 \cdot 0$

**b** Multiply.

55. 
```
  1 0
×    8
```

56. 
```
    7
× 1 0
```

57. 
```
  2 0
×    8
```

58. 
```
  3 0
×    7
```

59. 
```
  4 5
× 1 0
```

60. 
```
  7 8
× 1 0
```

61. 
```
  8 0
×    7
```

62. 
```
  9 0
×    4
```

63. 
```
  1 0 0
×      8
```

64. 
```
  1 0 0
×      3
```

65. 
```
  1 0 0
×      9
```

66. 
```
  1 0 0
×    1 0
```

67. 
```
  3 4 5 7
×    1 0 0
```

68. 
```
  4 0 0
×      3
```

69. 
```
  7 0 0
×      7
```

70. 
```
  5 0 0
×      8
```

71. 
```
  1 0 0
×  1 0 0
```

72. 
```
  1 0 0 0
×        7
```

73. 
```
  1 0 0 0
×        9
```

74. 
```
  1 0 0 0
×        2
```

75. 
```
      4 5 7
×  1 0 0 0
```

76. 
```
      6 7 6 9
×    1 0 0 0
```

77. 
```
  2 0 0 0
×        9
```

78. 
```
  5 0 0 0
×        4
```

79. 
```
  6 0 0 0
×        8
```

80. 
```
  8 0 0 0
×        2
```

81. 
```
  3 0 0 0
×        2
```

82. 
```
  1 0 0 0
×  1 0 0 0
```

83. 
```
  4 0
× 3 0
```

84. 
```
  2 0
× 1 0
```

85. 
```
  8 0
× 5 0
```

86. 
```
  5 0
× 5 0
```

87. 
```
  4 0 0
×    3 0
```

88. 
```
  2 0 0
×    3 0
```

89. 
```
  7 0 0
×    9 0
```

90. 
```
  4 0 0
× 3 0 0
```

91. 
```
  4 0 0 0
×    2 0 0
```

92. 
```
  6 0 0 0
×      2 0
```

93. 
```
  4 0 0 0
× 4 0 0 0
```

94. 
```
  8 0 0 0
×      1 0
```

**c** Multiply.

95. 
```
  4 9
×    3
```

96. 
```
  7 4
×    6
```

97. 
```
  5 9 3
×      5
```

98. 
```
  6 0 9
×      8
```

99. 
```
  8 9 9
×      7
```

100. 
```
  8 6 5
×      4
```

101. 
```
  8 1 1 8
×        2
```

102. 
```
  6 7 5 4
×        2
```

103. 
```
  4 3,7 7 7
×          2
```

104. 
```
  3 2,5 6 4
×          6
```

**d** Multiply.

105. 
```
  5 8
× 6 0
```

106. 
```
  9 3
× 3 0
```

107. 
```
  4 2
× 8 0
```

108. 
```
  7 8
× 9 0
```

109. 
```
  3 4 6
×    6 0
```

110. 
```
  2 6 7
×    4 0
```

111. 
```
  8 9 7
×  4 0 0
```

112. 
```
  3 6 6
×  3 0 0
```

113. 
```
  8 3 4
×  7 0 0
```

114. 
```
  3 3 3
×  9 0 0
```

115. 
```
  5 6 7 3
×  2 0 0 0
```

116. 
```
  4 6 7 8
×  5 0 0 0
```

117. 
```
  6 7 8 8
×  9 0 0 0
```

118. 
```
  9 1 2 9
×  8 0 0 0
```

# D Division

## OBJECTIVES

**a)** Find basic quotients such as $20 \div 5$, $56 \div 7$, and so on.

**b)** Divide by estimating multiples of thousands, hundreds, tens, and ones.

## a Basic Division

Division can be explained by arranging a set of objects in a rectangular array. This can be done in two ways.

**EXAMPLE 1**  Divide: $18 \div 6$.

**Method 1**  We can do this division by taking 18 objects and determining how many rows, each with 6 objects, we can form.

3 rows of 6 objects

Since there are 3 rows of 6 objects, we have

$$18 \div 6 = 3.$$

**Method 2**  We can also arrange the objects into 6 rows and determine how many objects are in each row.

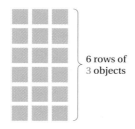

6 rows of 3 objects

Since there are 3 objects in each of the 6 rows, we have

$$18 \div 6 = 3.$$

We can also use fraction notation for division. That is,

$$18 \div 6 = 18/6 = \frac{18}{6}.$$

**EXAMPLES**  Divide.

**2.**  $9 \overline{\smash{)}\ 3\ 6}$  with quotient $4$
    *Think*: 36 objects: How many rows, each with 9 objects?
    *or* 36 objects: How many objects in each of 9 rows?

**3.**  $42 \div 7 = 6$

**4.**  $\dfrac{24}{3} = 8$

Do Exercises 1–4.

Divide.

**1.** $24 \div 6$        **2.** $64 \div 8$

**3.** $\dfrac{63}{7}$        **4.** $\dfrac{27}{9}$

*Answers*

**1.** 4  **2.** 8  **3.** 9  **4.** 3

The multiplication table in Developmental Unit M will enable you to divide also. First, recall how multiplication and division are related.

*A multiplication*:  $5 \cdot 4 = 20$.

*Two related divisions*:

**A.**   4 rows of 5 objects     $20 \div 5 = 4.$

**B.** 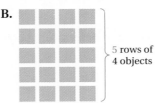  5 rows of 4 objects     $20 \div 4 = 5.$

Since we know that

$$5 \cdot 4 = 20, \quad \text{A basic multiplication fact}$$

we also know the two division facts

$$20 \div 5 = 4 \quad \text{and} \quad 20 \div 4 = 5.$$

**EXAMPLE 5**   From $7 \cdot 8 = 56$, write two division facts.

**a)** We have

$$7 \cdot 8 = 56 \qquad \text{Multiplication sentence}$$

$$7 = 56 \div 8. \qquad \text{Related division sentence}$$

**b)** We also have

$$7 \cdot 8 = 56 \qquad \text{Multiplication sentence}$$

$$8 = 56 \div 7. \qquad \text{Related division sentence}$$

Do Exercises 5 and 6.

For each multiplication fact, write two division facts.

**5.** $6 \cdot 2 = 12$

**6.** $7 \times 6 = 42$

We can use the idea that division is defined in terms of multiplication to do basic divisions.

**EXAMPLE 6**  Find: $35 \div 5$.

To find $35 \div 5$, we ask, "5 times what number is 35?"

$$5 \cdot \square = 35$$

| × | 2 | 3 | 4 | 5 | 6 | 7 | 8 | 9 |
|---|---|---|---|---|---|---|---|---|
| 2 | 4 | 6 | 8 | 10 | 12 | 14 | 16 | 18 |
| 3 | 6 | 9 | 12 | 15 | 18 | 21 | 24 | 27 |
| 4 | 8 | 12 | 16 | 20 | 24 | 28 | 32 | 36 |
| 5 | 10 | 15 | 20 | 25 | 30 | 35 | 40 | 45 |
| 6 | 12 | 18 | 24 | 30 | 36 | 42 | 48 | 54 |
| 7 | 14 | 21 | 28 | 35 | 42 | 49 | 56 | 63 |
| 8 | 16 | 24 | 32 | 40 | 48 | 56 | 64 | 72 |
| 9 | 18 | 27 | 36 | 45 | 54 | 63 | 72 | 81 |

$35 \div 5 = 7$

Using the multiplication table above, we find 35 inside the table and 5 at the left. Then we read the answer, 7, from the top. Thus, we have $35 \div 5 = 7$. Strive to do this kind of thinking mentally as fast as you can, without having to use the table.

Do Exercises 7–10.

Divide.

**7.** $28 \div 4$   **8.** $81 \div 9$

**9.** $\dfrac{16}{2}$   **10.** $\dfrac{54}{6}$

## Division by 1

Note that

$$3 \div 1 = 3 \quad \text{because} \quad 3 \cdot 1 = 3; \qquad \frac{14}{1} = 14 \quad \text{because} \quad 14 \cdot 1 = 14.$$

**DIVISION BY 1**

Any number divided by 1 is that same number:

$$a \div 1 = \frac{a}{1} = a.$$

**EXAMPLES**  Divide.

**1.** $\dfrac{8}{1} = 8$   **2.** $6 \div 1 = 6$   **3.** $34 \div 1 = 34$

Do Exercises 11–13.

Divide.

**11.** $6 \div 1$   **12.** $\dfrac{13}{1}$

**13.** $1 \div 1$

## Division by 0

Why can't we divide by 0? Suppose the number 4 *could* be divided by 0. Then if $\square$ were the answer,

$$4 \div 0 = \square,$$

we would have

$$4 = \square \cdot 0 = 0, \qquad \text{False!}$$

since 0 times any number is 0.

Similarly, suppose 12 could be divided by 0. If $\square$ were the answer,

$$12 \div 0 = \square,$$

we would have

$$12 = \square \cdot 0 = 0, \qquad \text{False!}$$

since 0 times any number is 0. Thus, $a \div 0$ would have to be some number $\square$ such that $a = \square \cdot 0 = 0$. So the only number that could possibly be divided by 0 would be 0 itself.

But such a division would give us any number we wish, for

$$0 \div 0 = 8 \quad \text{because} \quad 8 \cdot 0 = 0;$$
$$0 \div 0 = 3 \quad \text{because} \quad 3 \cdot 0 = 0; \qquad \text{All true!}$$
$$0 \div 0 = 7 \quad \text{because} \quad 7 \cdot 0 = 0.$$

We avoid these difficulties by agreeing to not divide *any* number by 0.

---

**DIVISION BY 0**

Division by 0 is not defined. (We agree not to divide by 0.)

---

### Dividing 0 by Other Numbers

Note that

$$0 \div 3 = 0 \quad \text{because} \quad 0 \cdot 3 = 0; \qquad \frac{0}{12} = 0 \quad \text{because} \quad 0 \cdot 12 = 0.$$

---

**DIVISION INTO 0**

Zero divided by any number other than 0 is 0:

$$\frac{0}{a} = 0, \quad a \neq 0.$$

---

Do Exercises 14–21.

**EXAMPLES** Divide.

**10.** $0 \div 8 = 0$

**11.** $0 \div 22 = 0$

**12.** $\dfrac{0}{9} = 0$

**13.** $12 \div 0$ is undefined.

**14.** $\dfrac{137}{0}$ is undefined.

### Dividing a Number by Itself

Note that

$$3 \div 3 = 1 \quad \text{because} \quad 1 \cdot 3 = 3; \qquad \frac{34}{34} = 1 \quad \text{because} \quad 1 \cdot 34 = 34.$$

Divide, if possible. If not possible, write "undefined."

**14.** $\dfrac{8}{4}$

**15.** $\dfrac{5}{0}$

**16.** $\dfrac{0}{5}$

**17.** $\dfrac{0}{0}$

**18.** $12 \div 0$

**19.** $100 \div 10$

**20.** $\dfrac{5}{3 - 3}$

**21.** $\dfrac{8 - 8}{4}$

**Answers**

**14.** 2    **15.** Undefined    **16.** 0
**17.** Undefined    **18.** Undefined
**19.** 10    **20.** Undefined    **21.** 0

## DIVISION OF A NUMBER BY ITSELF

Any number other than 0 divided by itself is 1:

$$\frac{a}{a} = 1, \quad a \neq 0.$$

**EXAMPLES** Divide.

**15.** $8 \div 8 = 1$     *Check:* $1 \cdot 8 = 8$

**16.** $27 \div 27 = 1$     *Check:* $1 \cdot 27 = 27$

**17.** $\dfrac{32}{32} = 1$     *Check:* $1 \cdot 32 = 32$

Do Exercises 22–27.

Divide.

**22.** $23 \div 23$     **23.** $\dfrac{67}{67}$

**24.** $\dfrac{41}{41}$     **25.** $17 \div 17$

**26.** $17 \div 1$     **27.** $\dfrac{54}{54}$

## b Dividing by Estimating Multiples

We divide using a process of estimating multiples of 10, 100, 1000, and so on.

**EXAMPLE 18** Divide: $7643 \div 3$.

**a)** Are there any thousands in the quotient? Yes, $3 \cdot 1000 = 3000$, which is less than 7643. To find how many thousands, we find products of 3 and multiples of 1000.

$$3 \cdot 1000 = 3000$$
$$3 \cdot 2000 = 6000 \quad \longleftarrow 7643$$
$$3 \cdot 3000 = 9000$$

```
        2 0 0 0
   3 )7 6 4 3
        6 0 0 0  ← Multiply 3 × 2000.
        1 6 4 3  ← Subtract.
```

**b)** Now look at 1643 and go to the hundreds place. Are there any hundreds in the quotient?

$$3 \cdot 100 = 300$$
$$3 \cdot 200 = 600$$
$$3 \cdot 300 = 900$$
$$3 \cdot 400 = 1200$$
$$3 \cdot 500 = 1500 \quad \longleftarrow 1643$$
$$3 \cdot 600 = 1800$$

```
          5 0 0
        2 0 0 0
   3 ) 7 6 4 3
        6 0 0 0
        1 6 4 3
        1 5 0 0  ← Multiply 3 × 500.
          1 4 3  ← Subtract.
```

**c)** Now look at 143 and go to the tens place. Are there any tens in the quotient?

$$3 \cdot 10 = 30$$
$$3 \cdot 20 = 60$$
$$3 \cdot 30 = 90$$
$$3 \cdot 40 = 120$$
$$3 \cdot 50 = 150 \quad \longleftarrow 143$$

```
            4 0
          5 0 0
        2 0 0 0
   3 ) 7 6 4 3
        6 0 0 0
        1 6 4 3
        1 5 0 0
          1 4 3
          1 2 0  ← Multiply 3 × 40.
            2 3  ← Subtract.
```

*Answers*

**22.** 1   **23.** 1   **24.** 1
**25.** 1   **26.** 17   **27.** 1

**d)** Now look at 23 and go to the ones place. Are there any ones in the quotient?

$$3 \cdot 1 = 3$$
$$3 \cdot 2 = 6$$
$$3 \cdot 3 = 9$$
$$3 \cdot 4 = 12$$
$$3 \cdot 5 = 15$$
$$3 \cdot 6 = 18$$
$$3 \cdot 7 = 21$$
$$3 \cdot 8 = 24$$

```
        2 5 4 7
              7
            4 0
          5 0 0
        2 0 0 0
    3 ) 7 6 4 3
        6 0 0 0
        1 6 4 3
        1 5 0 0
          1 4 3
          1 2 0
            2 3
            2 1  ←── Multiply 3 × 7.
             2  ←── Subtract. This is the
                    remainder, since
                    2 is less than 3.
```

← 23

The answer is 2547 R 2.

Do Exercises 28 and 29.

Do Exercises 28 and 29.

Divide.

**28.** 4 ) 3 8 5      **29.** 7 ) 8 8 4 6

## A Short Form

Here is a shorter way to write Example 18.

*Short form*

```
    2 5 4 7
3 ) 7 6 4 3
    6 0 0 0
    1 6 4 3
    1 5 0 0
      1 4 3
      1 2 0
        2 3
        2 1
         2
```

We write a 2 above the thousands digit in the dividend to record 2000.
We write a 5 to record 500.
We write a 4 to record 40.
We write a 7 to record 7.

Divide using the short form.

**30.** 2 ) 6 4 8      **31.** 9 ) 3 7 5 8

Do Exercises 30 and 31.

Do Exercises 30 and 31.

**EXAMPLE 19** Divide: $2637 \div 41$. Use the short form.

```
           6
  4 1 ) 2 6 3 7
        2 4 6 0
          1 7 7
```

60 times 41 is 2460. Since the remainder, 177, is greater than 41, we continue.

```
          6 4
  4 1 ) 2 6 3 7
        2 4 6 0
          1 7 7
          1 6 4
            1 3
```

Try to just write this.

The answer is 64 R 13.

Divide.

**32.** 1 1 ) 4 1 5

**33.** 4 6 ) 1 0 7 5

Do Exercises 32 and 33.

Do Exercises 32 and 33.

## TO THE STUDENT

If you had trouble with Section 1.5 and have studied Developmental Unit D, you should go back and work through Section 1.5 after completing Exercise Set D.

*Answers*

**28.** 96 R 1     **29.** 1263 R 5     **30.** 324
**31.** 417 R 5     **32.** 37 R 8     **33.** 23 R 17

**a** Divide, if possible.

**1.** $24 \div 8$

**2.** $72 \div 9$

**3.** $28 \div 7$

**4.** $22 \div 22$

**5.** $32 \div 1$

**6.** $45 \div 5$

**7.** $14 \div 2$

**8.** $40 \div 8$

**9.** $37 \div 1$

**10.** $10 \div 2$

**11.** $36 \div 4$

**12.** $12 \div 3$

**13.** $54 \div 9$

**14.** $18 \div 2$

**15.** $20 \div 4$

**16.** $16 \div 2$

**17.** $72 \div 8$

**18.** $42 \div 7$

**19.** $12 \div 4$

**20.** $8 \div 4$

**21.** $54 \div 6$

**22.** $18 \div 9$

**23.** $9 \div 3$

**24.** $28 \div 4$

**25.** $56 \div 7$

**26.** $24 \div 6$

**27.** $14 \div 2$

**28.** $14 \div 7$

**29.** $21 \div 7$

**30.** $36 \div 6$

**31.** $8 \div 8$

**32.** $32 \div 8$

**33.** $30 \div 5$

**34.** $18 \div 6$

**35.** $49 \div 7$

**36.** $81 \div 9$

**37.** $0 \div 7$

**38.** $9 \div 0$

**39.** $16 \div 0$

**40.** $42 \div 6$

**41.** $\dfrac{48}{6}$

**42.** $\dfrac{35}{5}$

**43.** $\dfrac{9}{9}$

**44.** $\dfrac{45}{9}$

**45.** $\dfrac{0}{5}$

**46.** $\dfrac{0}{8}$

**47.** $\dfrac{6}{2}$

**48.** $\dfrac{3}{3}$

**49.** $\dfrac{8}{2}$

**50.** $\dfrac{7}{1}$

**51.** $\dfrac{5}{5}$

**52.** $\dfrac{6}{1}$

**53.** $\dfrac{2}{2}$

**54.** $\dfrac{25}{5}$

**55.** $\dfrac{4}{2}$

**56.** $\dfrac{24}{3}$

**57.** $\dfrac{0}{9}$

**58.** $\dfrac{0}{4}$

**59.** $\dfrac{40}{5}$

**60.** $\dfrac{3}{1}$

**61.** $\dfrac{16}{4}$

**62.** $\dfrac{9}{0}$

**63.** $\dfrac{32}{8}$

**64.** $\dfrac{9}{9}$

**b**    Divide.

**65.** $4\overline{)277}$

**66.** $2\overline{)399}$

**67.** $8\overline{)737}$

**68.** $6\overline{)831}$

**69.** $5\overline{)105}$

**70.** $6\overline{)708}$

**71.** $9\overline{)820}$

**72.** $3\overline{)965}$

**73.** $5\overline{)8619}$

**74.** $3\overline{)8775}$

**75.** $9\overline{)7777}$

**76.** $8\overline{)4179}$

**77.** $5\overline{)4823}$

**78.** $8\overline{)5437}$

**79.** $2\overline{)5794}$

**80.** $7\overline{)9298}$

**81.** $20\overline{)875}$

**82.** $30\overline{)987}$

**83.** $50\overline{)6893}$

**84.** $40\overline{)9946}$

**85.** $21\overline{)999}$

**86.** $23\overline{)975}$

**87.** $85\overline{)7757}$

**88.** $54\overline{)2821}$

**89.** $46\overline{)1058}$

**90.** $24\overline{)7722}$

**91.** $38\overline{)8522}$

**92.** $81\overline{)2247}$

**93.** $111\overline{)3219}$

**94.** $102\overline{)5610}$

**95.** $346\overline{)78,910}$

**96.** $781\overline{)15,999}$

**97.** $94\overline{)2153}$

**98.** $82\overline{)4064}$

**99.** $117\overline{)44,902}$

**100.** $740\overline{)55,200}$

# Answers

## CHAPTER 1

### Exercise Set 1.1, p. 6

**1.** 5 thousands **3.** 5 hundreds **5.** 9 **7.** 7
**9.** 5 thousands + 7 hundreds + 0 tens + 2 ones, or
5 thousands + 7 hundreds + 2 ones
**11.** 9 ten thousands + 3 thousands + 9 hundreds + 8 tens +
6 ones **13.** 2 thousands + 0 hundreds + 5 tens + 8 ones, or
2 thousands + 5 tens + 8 ones **15.** 1 thousand +
5 hundreds + 7 tens + 6 ones **17.** 1 billion +
4 hundred millions + 2 ten millions + 4 millions +
1 hundred thousand + 6 ten thousands + 1 thousand +
9 hundreds + 4 tens + 8 ones **19.** 9 ten millions +
9 millions + 8 hundred thousands + 8 ten thousands +
6 thousands + 5 hundreds + 6 tens + 8 ones
**21.** 6 hundred thousands + 1 ten thousand + 7 thousands +
2 hundreds + 4 tens + 9 ones **23.** Eighty-five
**25.** Eighty-eight thousand **27.** One hundred twenty-three
thousand, seven hundred sixty-five **29.** Seven billion, seven
hundred fifty-four million, two hundred eleven thousand, five
hundred seventy-seven **31.** Seven hundred thousand, six
hundred thirty-four **33.** Three million, forty-eight thousand,
five **35.** 2,233,812 **37.** 8,000,000,000
**39.** 50,324 **41.** 632,896 **43.** 1,600,000,000
**45.** 64,186,000 **47.** 138

### Calculator Corner, p. 10

**1.** 121 **2.** 1602 **3.** 1932 **4.** 864

### Exercise Set 1.2, p. 12

**1.** 387 **3.** 164 **5.** 5198 **7.** 100 **9.** 8503 **11.** 5266
**13.** 4466 **15.** 6608 **17.** 34,432 **19.** 101,310 **21.** 230
**23.** 18,424 **25.** 31,685 **27.** 132 yd **29.** 1661 ft
**31.** 570 ft **33.** 8 ten thousands **34.** Nine billion, three
hundred forty-six million, three hundred ninety-nine
thousand, four hundred sixty-eight
**35.** $1 + 99 = 100, 2 + 98 = 100, \ldots, 49 + 51 = 100.$ Then
49 100's = 4900 and 4900 + 50 + 100 = 5050.

### Calculator Corner, p. 15

**1.** 28 **2.** 47 **3.** 67 **4.** 119 **5.** 2128 **6.** 2593

### Exercise Set 1.3, p. 17

**1.** 44 **3.** 533 **5.** 39 **7.** 14 **9.** 369 **11.** 26 **13.** 234
**15.** 417 **17.** 5382 **19.** 2778 **21.** 3069 **23.** 1089
**25.** 7748 **27.** 4144 **29.** 56 **31.** 454 **33.** 3831
**35.** 3749 **37.** 2191 **39.** 4418 **41.** 43,028 **43.** 95,974
**45.** 4206 **47.** 1305 **49.** 9989 **51.** 48,017 **53.** 1024

**54.** 12,732 **55.** 90,283 **56.** 29,364 **57.** 1345 **58.** 924
**59.** 22,692 **60.** 10,920 **61.** Six million, three hundred
seventy-five thousand, six hundred two **62.** 7 ten thousands
**63.** 3; 4

### Calculator Corner, p. 23

**1.** 448 **2.** 21,970 **3.** 6380 **4.** 39,564 **5.** 180,480
**6.** 2,363,754

### Exercise Set 1.4, p. 24

**1.** 520 **3.** 564 **5.** 1527 **7.** 64,603 **9.** 4770 **11.** 3995
**13.** 870 **15.** 1920 **17.** 46,296 **19.** 14,652 **21.** 258,312
**23.** 798,408 **25.** 20,723,872 **27.** 362,128 **29.** 302,220
**31.** 49,101,136 **33.** 25,236,000 **35.** 20,064,048
**37.** 529,984 sq mi **39.** 8100 sq ft **41.** 12,685 **42.** 10,834
**43.** 427,477 **44.** 111,110 **45.** 1241 **46.** 8889
**47.** 254,119 **48.** 66,444 **49.** 247,464 sq ft

### Calculator Corner, p. 30

**1.** 28 **2.** 123 **3.** 323 **4.** 36

### Exercise Set 1.5, p. 32

**1.** 12 **3.** 1 **5.** 22 **7.** 0 **9.** Not defined **11.** 6
**13.** 55 R 2 **15.** 108 **17.** 307 **19.** 753 R 3 **21.** 74 R 1
**23.** 92 R 2 **25.** 1703 **27.** 987 R 5 **29.** 12,700 **31.** 127
**33.** 52 R 52 **35.** 29 R 5 **37.** 40 R 12 **39.** 90 R 22
**41.** 29 **43.** 105 R 3 **45.** 1609 R 2 **47.** 1007 R 1 **49.** 23
**51.** 107 R 1 **53.** 370 **55.** 609 R 15 **57.** 304
**59.** 3508 R 219 **61.** 8070 **63.** Perimeter **64.** Minuend
**65.** Digits; periods **66.** Dividend **67.** Factors; product
**68.** Additive **69.** Associative **70.** Divisor; remainder;
dividend **71.** 54, 122; 33, 2772; 4, 8 **73.** 30 buses

### Mid-Chapter Review: Chapter 1, p. 35

**1.** False **2.** True **3.** True **4.** False **5.** True **6.** False
**7.**                                             95,406,237

Ninety-five million,

four hundred six thousand,

two hundred thirty-seven
**8.** $\begin{array}{r} {\scriptstyle 5\ 9\ 14} \\ 6\ 0\ 4 \\ -\ 4\ 9\ 7 \\ \hline 1\ 0\ 7 \end{array}$ **9.** 6 hundreds **10.** 6 ten thousands
**11.** 6 thousands **12.** 6 ones **13.** 2 **14.** 6 **15.** 5 **16.** 1
**17.** 5 thousands + 6 hundreds + 0 tens + 2 ones, or
5 thousands + 6 hundreds + 2 ones

**18.** 6 ten thousands + 9 thousands + 3 hundreds + 4 tens + 5 ones   **19.** One hundred thirty-six thousand, three hundred twenty-five   **20.** Sixty-four   **21.** 308,716   **22.** 4,567,216   **23.** 798   **24.** 1030   **25.** 7922   **26.** 7534   **27.** 465   **28.** 339   **29.** 1854   **30.** 4328   **31.** 216   **32.** 15,876   **33.** 132,275   **34.** 5,679,870   **35.** 253   **36.** 112 R 5   **37.** 23 R 19   **38.** 144 R 31   **39.** 25 m   **40.** 8 sq in.   **41.** When numbers are being added, it does not matter how they are grouped.   **42.** Subtraction is not commutative. For example, $5 - 2 = 3$, but $2 - 5 \neq 3$.   **43.** Answers will vary. Suppose one coat costs $150. Then the multiplication $4 \cdot \$150$ gives the cost of four coats. Or, suppose one ream of copy paper costs $4. Then the multiplication $\$4 \cdot 150$ gives the cost of 150 reams.   **44.** If we use the definition of division, $0 \div 0 = a$ such that $a \cdot 0 = 0$. We see that $a$ could be *any* number since $a \cdot 0 = 0$ for any number $a$. Thus we cannot say that $0 \div 0 = 0$. This is why we agree not to allow division by 0.

### Exercise Set 1.6, p. 43

**1.** 50   **3.** 460   **5.** 730   **7.** 900   **9.** 100   **11.** 1000   **13.** 9100   **15.** 32,800   **17.** 6000   **19.** 8000   **21.** 45,000   **23.** 373,000   **25.** $80 + 90 = 170$   **27.** $8070 - 2350 = 5720$   **29.** 220; incorrect   **31.** 890; incorrect   **33.** $7300 + 9200 = 16,500$   **35.** $6900 - 1700 = 5200$   **37.** 1600; correct   **39.** 1500; correct   **41.** $10,000 + 5000 + 9000 + 7000 = 31,000$   **43.** $92,000 - 23,000 = 69,000$   **45.** $50 \cdot 70 = 3500$   **47.** $30 \cdot 30 = 900$   **49.** $900 \cdot 300 = 270,000$   **51.** $400 \cdot 200 = 80,000$   **53.** $350 \div 70 = 5$   **55.** $8450 \div 50 = 169$   **57.** $1200 \div 200 = 6$   **59.** $8400 \div 300 = 28$   **61.** $11,200   **63.** $18,900; no   **65.** Answers will vary depending on the options chosen.   **67.** (a) $2,190,000; (b) $2,170,000   **69.** 90 people   **71.** $<$   **73.** $>$   **75.** $<$   **77.** $>$   **79.** $>$   **81.** $>$   **83.** $172,000 < 284,370$, or $284,370 > 172,000$   **85.** $1663 < 9453$, or $9453 > 1663$   **87.** 86,754   **88.** 13,589   **89.** 48,824   **90.** 4415   **91.** 1702   **92.** 17,748   **93.** 54 R 4   **94.** 208   **95.** Left to the student   **97.** Left to the student

### Exercise Set 1.7, p. 52

**1.** 14   **3.** 0   **5.** 90,900   **7.** 450   **9.** 352   **11.** 25   **13.** 29   **15.** 0   **17.** 79   **19.** 45   **21.** 8   **23.** 14   **25.** 32   **27.** 143   **29.** 17,603   **31.** 37   **33.** 1035   **35.** 66   **37.** 324   **39.** 743   **41.** 175   **43.** 335   **45.** 18,252   **47.** 104   **49.** 45   **51.** 4056   **53.** 2847   **55.** 15   **57.** 205   **59.** 457   **61.** 142 R 5   **62.** 142   **63.** 334   **64.** 334 R 11   **65.** $<$   **66.** $>$   **67.** $>$   **68.** $<$   **69.** 6,376,000   **70.** 6,375,600   **71.** 347

### Translating for Success, p. 63

**1.** E   **2.** M   **3.** D   **4.** G   **5.** A   **6.** O   **7.** F   **8.** K   **9.** J   **10.** H

### Exercise Set 1.8, p. 64

**1.** 318 ft   **3.** 7450 ft   **5.** 95 milligrams   **7.** 18 rows   **9.** 1502 hr   **11.** 2054 mi   **13.** 2,073,600 pixels   **15.** 275,000 men   **17.** 168 hr   **19.** $23 per month   **21.** $233   **23.** $9276   **25.** 151,500   **27.** 1,190,000 motorcycles   **29.** $78   **31.** $40 per month   **33.** $24,456   **35.** 35 weeks; 2 episodes   **37.** 250 gal   **39.** 21 columns   **41.** (a) 4200 sq ft; (b) 268 ft   **43.** $247   **45.** 645 mi; 5 in.   **47.** 56 cartons   **49.** 32 $10 bills   **51.** $400   **53.** 525 min, or 8 hr 45 min   **55.** 700 min, or 11 hr 40 min   **57.** 104 seats   **59.** 106 bones   **61.** 234,600   **62.** 234,560   **63.** 235,000   **64.** 22,000   **65.** 16,000   **66.** 4000   **67.** 8000   **68.** 320,000   **69.** 720,000   **70.** 46,800,000   **71.** 792,000 mi; 1,386,000 mi

### Calculator Corner, p. 72

**1.** 243   **2.** 15,625   **3.** 20,736   **4.** 2048

### Calculator Corner, p. 74

**1.** 49   **2.** 85   **3.** 36   **4.** 0   **5.** 73   **6.** 49

### Exercise Set 1.9, p. 77

**1.** $3^4$   **3.** $5^2$   **5.** $7^5$   **7.** $10^3$   **9.** 49   **11.** 729   **13.** 20,736   **15.** 243   **17.** 22   **19.** 20   **21.** 100   **23.** 1   **25.** 49   **27.** 434   **29.** 41   **31.** 88   **33.** 4   **35.** 303   **37.** 1   **39.** 5   **41.** 5   **43.** 32   **45.** 906   **47.** 62   **49.** 102   **51.** 295   **53.** 32   **55.** 57   **57.** 54   **59.** $94   **61.** 401   **63.** 106   **65.** 30   **67.** 155   **69.** 19   **71.** 8   **73.** 12   **75.** 110   **77.** 7   **79.** 544   **81.** 7   **83.** 696   **85.** 452   **86.** 835   **87.** 13   **88.** 37   **89.** 2342   **90.** 4898   **91.** 25   **92.** 100   **93.** $24; 1 + 5 \cdot (4 + 3) = 36$   **95.** $7; 12 \div (4 + 2) \cdot 3 - 2 = 4$

### Summary and Review: Chapter 1, p. 80

#### Concept Reinforcement

**1.** True   **2.** True   **3.** False   **4.** False   **5.** True   **6.** False

#### Important Concepts

**1.** 2 thousands   **2.** 65,302   **3.** 3237   **4.** 225,036   **5.** 315 R 14   **6.** 36,500   **7.** 36,000   **8.** $<$   **9.** 36   **10.** 216

#### Review Exercises

**1.** 8 thousands   **2.** 3   **3.** 2 thousands + 7 hundreds + 9 tens + 3 ones   **4.** 5 ten thousands + 6 thousands + 0 hundreds + 7 tens + 8 ones, or 5 ten thousands + 6 thousands + 7 tens + 8 ones   **5.** 4 millions + 0 hundred thousands + 0 ten thousands + 7 thousands + 1 hundred + 0 tens + 1 one, or 4 millions + 7 thousands + 1 hundred + 1 one   **6.** Sixty-seven thousand, eight hundred nineteen   **7.** Two million, seven hundred eighty-one thousand, four hundred twenty-seven   **8.** Four million, eight hundred seventeen thousand, nine hundred forty-one   **9.** 476,588   **10.** 1,620,000,000   **11.** 14,272   **12.** 66,024   **13.** 21,788   **14.** 98,921   **15.** 5148   **16.** 1689   **17.** 2274   **18.** 17,757   **19.** 5,100,000   **20.** 6,276,800   **21.** 506,748   **22.** 27,589   **23.** 5,331,810   **24.** 12 R 3   **25.** 5   **26.** 913 R 3   **27.** 384 R 1   **28.** 4 R 46   **29.** 54   **30.** 452   **31.** 5008   **32.** 4389   **33.** 320   **34.** 345,800   **35.** 345,760   **36.** 346,000   **37.** $>$   **38.** $<$   **39.** $41,300 + 19,700 = 61,000$   **40.** $38,700 - 24,500 = 14,200$   **41.** $400 \cdot 700 = 280,000$   **42.** 8   **43.** 45   **44.** 58   **45.** 0   **46.** $4^3$   **47.** 10,000   **48.** 36   **49.** 65   **50.** 233   **51.** 260   **52.** 165   **53.** $502   **54.** $484   **55.** 1982   **56.** 19 cartons   **57.** $13,585   **58.** 14 beehives   **59.** 98 sq ft; 42 ft   **60.** 137 beakers filled; 13 mL left over   **61.** $27,598   **62.** B   **63.** A   **64.** D   **65.** 8   **66.** $a = 8, b = 4$   **67.** 6 days

#### Understanding Through Discussion and Writing

**1.** No; if subtraction were associative, then $a - (b - c) = (a - b) - c$ for any $a$, $b$, and $c$. But, for example,

$$12 - (8 - 4) = 12 - 4 = 8,$$

whereas

$$(12 - 8) - 4 = 4 - 4 = 0.$$

Since $8 \neq 0$, this example shows that subtraction is not associative.   **2.** By rounding prices and estimating their sum, a shopper can estimate the total grocery bill while shopping. This is particularly useful if the shopper wants to spend no more than a certain amount.   **3.** Answers will vary. Anthony is driving from Kansas City to Minneapolis, a distance of 512 mi. He stops for gas after driving 183 mi. How much farther must he drive?

**4.** The parentheses are not necessary in the expression $9 - (4 \cdot 2)$. Using the rules for order of operations, the multiplication would be performed before the subtraction even if the parentheses were not present. The parentheses are necessary in the expression $(3 \cdot 4)^2$; $(3 \cdot 4)^2 = 12^2 = 144$, but $3 \cdot 4^2 = 3 \cdot 16 = 48$.

## Test: Chapter 1, p. 85

**1.** [1.1a] 5　**2.** [1.1b] 8 thousands + 8 hundreds + 4 tens + 3 ones　**3.** [1.1c] Thirty-eight million, four hundred three thousand, two hundred seventy-seven
**4.** [1.2a] 9989　**5.** [1.2a] 63,791　**6.** [1.2a] 3165
**7.** [1.2a] 10,515　**8.** [1.3a] 3630　**9.** [1.3a] 1039
**10.** [1.3a] 6848　**11.** [1.3a] 5175　**12.** [1.4a] 41,112
**13.** [1.4a] 5,325,600　**14.** [1.4a] 2405　**15.** [1.4a] 534,264
**16.** [1.5a] 3 R 3　**17.** [1.5a] 70　**18.** [1.5a] 97
**19.** [1.5a] 805 R 8　**20.** [1.6a] 35,000　**21.** [1.6a] 34,530
**22.** [1.6a] 34,500　**23.** [1.6b] 23,600 + 54,700 = 78,300
**24.** [1.6b] 54,800 − 23,600 = 31,200
**25.** [1.6b] 800 · 500 = 400,000　**26.** [1.6c] >
**27.** [1.6c] <　**28.** [1.7b] 46　**29.** [1.7b] 13　**30.** [1.7b] 14
**31.** [1.7b] 381　**32.** [1.8a] 83 calories　**33.** [1.8a] 20 staplers
**34.** [1.8a] 1,256,615 sq mi　**35. (a)** [1.2b], [1.4b] 300 in., 5000 sq in., 3872 sq in.; 228 in., 2888 sq in.;
**(b)** [1.8a] 2112 sq in.　**36.** [1.8a] 1852 12-packs; 7 cakes left over　**37.** [1.8a] $95　**38.** [1.9a] $12^4$　**39.** [1.9b] 343
**40.** [1.9b] 100,000　**41.** [1.9c] 31　**42.** [1.9c] 98
**43.** [1.9c] 2　**44.** [1.9c] 18　**45.** [1.9d] 216　**46.** [1.9c] A
**47.** [1.4b], [1.8a] 336 sq in.　**48.** [1.9c] 9　**49.** [1.8a] 80 payments

# CHAPTER 2

## Exercise Set 2.1, p. 92

**1.** −282　**3.** 820; −541　**5.** 950,000,000; −460　**7.** 40; −15
**9.** <　**11.** >　**13.** >　**15.** <　**17.** <　**19.** <　**21.** >
**23.** 57　**25.** 0　**27.** 24　**29.** 53　**31.** 8　**33.** 7　**35.** −7
**37.** 0　**39.** 21　**41.** −53　**43.** 1　**45.** 7　**47.** −9
**49.** −17　**51.** 23　**53.** −1　**55.** 85　**57.** −345　**59.** 0
**61.** −8　**63.** 825　**64.** 125　**65.** 7106　**66.** 4　**67.** 42
**68.** 69　**69.** >　**71.** =　**73.** −1, 0, 1
**75.** −100, −5, 0, |3|, 4, |−6|, $7^2$, $10^2$, $2^7$, $2^{10}$

## Calculator Corner, p. 96

**1.** 13　**2.** −8

## Exercise Set 2.2, p. 97

**1.** −5　**3.** −4　**5.** 6　**7.** 0　**9.** −4　**11.** −12　**13.** −11
**15.** −15　**17.** 42　**19.** 5　**21.** −4　**23.** 0　**25.** 0
**27.** −9　**29.** 7　**31.** 0　**33.** 45　**35.** −3　**37.** 0
**39.** −10　**41.** −24　**43.** −5　**45.** −21　**47.** −30　**49.** 6
**51.** −21　**53.** 25　**55.** −17　**57.** 6　**59.** −65
**61.** −160　**63.** −62　**65.** −23　**67.** 6681　**68.** 73
**69.** 3 ten thousands + 9 thousands + 4 hundreds + 1 ten + 7 ones　**70.** 2352　**71.** 32　**72.** 3500　**73.** −40
**75.** −6483　**77.** All negative　**79.** Negative　**81.** Negative

## Exercise Set 2.3, p. 102

**1.** −4　**3.** −7　**5.** −6　**7.** 0　**9.** −4　**11.** −7　**13.** −6
**15.** 0　**17.** 0　**19.** 14　**21.** 11　**23.** −14　**25.** 5　**27.** −7
**29.** −1　**31.** 18　**33.** −10　**35.** −3　**37.** −21　**39.** 5
**41.** −8　**43.** 12　**45.** −19　**47.** −68　**49.** −81　**51.** 116
**53.** 0　**55.** 55　**57.** 19　**59.** −62　**61.** −139　**63.** 6
**65.** 107　**67.** 219　**69.** 25 pages　**71.** 17 lb　**73.** −3°
**75.** $13,000　**77.** 100°F　**79.** 5676 ft　**81.** −$85
**83.** −10,011 ft　**85.** −83, or $83 in debt　**87.** 64　**88.** 4896
**89.** 1　**90.** 4147　**91.** 8 cans　**92.** 288 oz　**93.** 35
**94.** 3　**95.** 32　**96.** 165　**97.** −309,882

**99.** False; $3 - 0 \neq 0 - 3$　**101.** True　**103.** True　**105.** 17
**107.** Up 15 points

## Calculator Corner, p. 109

**1.** 148,035,889　**2.** −1,419,857　**3.** −1,124,864　**4.** 1,048,576
**5.** −531,441　**6.** −117,649　**7.** −7776　**8.** −19,683

## Exercise Set 2.4, p. 110

**1.** −16　**3.** −60　**5.** −48　**7.** −30　**9.** 15　**11.** 18
**13.** 42　**15.** 20　**17.** −120　**19.** 0　**21.** 72　**23.** −340
**25.** 400　**27.** 0　**29.** 24　**31.** 420　**33.** −70　**35.** 30
**37.** 0　**39.** −294　**41.** 36　**43.** −125　**45.** 10,000
**47.** −16　**49.** −243　**51.** 3　**53.** −121　**55.** −64
**57.** The opposite of eight to the fourth power
**59.** Negative nine to the tenth power　**61.** 532,500
**62.** 60,000,000　**63.** 80　**64.** 2550　**65.** 5　**66.** 48
**67.** 40 sq ft　**68.** 240 cartons　**69.** 5 trips　**70.** 4 trips
**71.** 243　**73.** 0　**75.** 7　**77.** −2209　**79.** 130,321
**81.** −2197　**83.** 116,875　**85.** −$23　**87. (a)** Both $m$ and $n$ must be odd. **(b)** At least one of $m$ and $n$ must be even.

## Calculator Corner, p. 114

**1.** −4　**2.** −2　**3.** 787

## Exercise Set 2.5, p. 115

**1.** −6　**3.** −13　**5.** −2　**7.** 4　**9.** −9　**11.** 2　**13.** −12
**15.** −8　**17.** Undefined　**19.** −8　**21.** −23　**23.** 0
**25.** −19　**27.** −41　**29.** −7　**31.** −7　**33.** −334　**35.** 23
**37.** 8　**39.** 12　**41.** −1　**43.** 0　**45.** −10　**47.** −86
**49.** −9　**51.** 18　**53.** −10　**55.** −67　**57.** 10　**59.** −25
**61.** −7988　**63.** −3000　**65.** 60　**67.** 1　**69.** −37
**71.** −22　**73.** 2　**75.** 7　**77.** Undefined　**79.** 3　**81.** 2
**83.** 0　**85.** 28 sq in.　**86.** 248 rooms　**87.** 12 gal　**88.** 27 gal
**89.** 150 cal　**90.** 672 g　**91.** 4 pieces; 2 pieces
**92.** 4 lozenges; 4 lozenges　**93.** 0　**95.** 0　**97.** −2　**99.** 992
**101.** $\boxed{(}\ \boxed{1}\ \boxed{5}\ \boxed{x^2}\ \boxed{-}\ \boxed{5}\ \boxed{y^x}\ \boxed{3}\ \boxed{)}\ \boxed{\div}$
$\boxed{(}\ \boxed{3}\ \boxed{x^2}\ \boxed{+}\ \boxed{4}\ \boxed{x^2}\ \boxed{)}\ \boxed{=}$　**103.** 5　**105.** Positive
**107.** Negative　**109.** Positive

## Mid-Chapter Review: Chapter 2, p. 118

**1.** False　**2.** False　**3.** True
**4.** $-x = -(-4) = 4$; $-(-x) = -(-(-4)) = -(4) = -4$
**5.** $5 - 13 = 5 + (-13) = -8$　**6.** $-6 - (-7) = -6 + 7 = 1$
**7.** 450; −79　**8.** −9　**9.** <　**10.** <　**11.** <　**12.** >
**13.** 38　**14.** 18　**15.** 0　**16.** 12　**17.** 56　**18.** −3
**19.** 0　**20.** 49　**21.** 19　**22.** 23　**23.** −2　**24.** −16
**25.** 0　**26.** −17　**27.** 1　**28.** 2　**29.** −26　**30.** −4
**31.** −13　**32.** 16　**33.** 6　**34.** −12　**35.** −36　**36.** −54
**37.** 26　**38.** 82　**39.** 81　**40.** −81　**41.** 25　**42.** −5
**43.** 75　**44.** 14　**45.** 42　**46.** −38　**47.** −13　**48.** 2
**49.** 33°C　**50.** $54.80
**51.** Answers may vary. The student may be confusing distance from 0 on the number line with position on the number line. Although −45 is farther from 0 than −21, it is less than −21 because it is to the left of −21 on the number line.
**52.** Answers may vary. Subtraction of integers is not associative, as can be illustrated by an example: Compare $3 - (9 - 10) = 4$ with $(3 - 9) - 10 = -16$.　**53.** Answers may vary. If we think of the addition on the number line, we start at a negative number and move to the left. This always brings us to a point on the negative portion of the number line.　**54.** Yes: consider $m - (-n)$, where both $m$ and $n$ are positive. Then $m - (-n) = m + n$. Now $m + n$, the sum of two positive numbers, is positive.

## Calculator Corner, p. 122

**1.** 243　**2.** 1024　**3.** −32　**4.** −3125

## Exercise Set 2.6, p. 124

**1.** 20¢  **3.** $-2$  **5.** 1  **7.** 18 yr  **9.** $-7$  **11.** 14 ft
**13.** 14 ft  **15.** $-21$  **17.** $-21$  **19.** 400 ft  **21.** 22
**23.** 0  **25.** 36  **27.** $-3$  **29.** 11  **31.** $-1100$
**33.** $\dfrac{-5}{t}; \dfrac{5}{-t}$  **35.** $\dfrac{n}{-b}; -\dfrac{n}{b}$  **37.** $\dfrac{-9}{p}; -\dfrac{9}{p}$  **39.** $\dfrac{14}{-w}; -\dfrac{14}{w}$
**41.** $-5; -5; -5$  **43.** $-27; -27; -27$  **45.** 36; $-12$
**47.** 45; 45  **49.** 216; $-216$  **51.** 1; 1  **53.** 32; $-32$
**55.** $5a + 5b$  **57.** $4x + 4$  **59.** $2b + 10$  **61.** $7 - 7t$
**63.** $30x - 12$  **65.** $8x + 56 + 48y$  **67.** $-7y + 14$
**69.** $3x + 6$  **71.** $-4x + 12y + 8z$  **73.** $8a - 24b + 8c$
**75.** $4x - 12y - 28z$  **77.** $20a - 25b + 5c - 10d$
**79.** $-3m - 2n$  **81.** $-2a + 3b - 4$  **83.** $-x + y + z$
**85.** Twenty-three million, forty-three thousand, nine hundred
twenty-one  **86.** 901  **87.** $5280 - 2480 = 2800$  **88.** 994
**89.** 5 in.  **90.** \$63  **91.** 698°F  **93.** 4438  **95.** 279
**97.** 2  **99.** 2,560,000  **101.** $-32$ ☒ $(88$ ☐ $29) = -1888$
**103.** True  **105.** True

## Translating for Success, p. 131

**1.** J  **2.** G  **3.** A  **4.** K  **5.** H  **6.** N  **7.** F  **8.** M
**9.** D  **10.** C

## Exercise Set 2.7, p. 132

**1.** $2a, 5b, -7c$  **3.** $mn, -6n, 8$  **5.** $3x^2y, -4y^2, -2z^3$
**7.** $14x$  **9.** $-5a$  **11.** $3x + 6y$  **13.** $-13a + 62$
**15.** $-4 + 4t + 6y$  **17.** $6a + 4b - 2$  **19.** $-1 + a - 12b$
**21.** $7x^2 + 3y$  **23.** $7x^4 + y^3$  **25.** $6a^2 - 3a$
**27.** $3x^3 - 8x^2 + 4$  **29.** $9x^3y - xy^3 + 3xy$
**31.** $-4a^6 - 3b^4 + 2a^6b^4$  **33.** 10 ft  **35.** 42 km  **37.** 8 m
**39.** 210 ft  **41.** 138 ft  **43.** 36 ft  **45.** 56 in.  **47.** 260 cm
**49.** 64 ft  **51.** 17 servings  **52.** 210  **53.** 29  **54.** 7
**55.** 8  **56.** 27  **57.** 16  **58.** 26  **59.** 16  **60.** 13  **61.** 15
**62.** 25  **63.** $7x + 1$  **65.** $-29 - 3a$  **67.** $-10 - x - 27y$
**69.** \$29.75  **71.** 912 mm

## Exercise Set 2.8, p. 141

**1.** Equivalent equations  **3.** Equivalent expressions
**5.** Equivalent expressions  **7.** Equivalent equations
**9.** Equivalent expressions  **11.** Equivalent equations
**13.** $-3$  **15.** $-8$  **17.** 18  **19.** $-15$  **21.** 32  **23.** $-17$
**25.** 17  **27.** 0  **29.** 10  **31.** $-14$  **33.** 5  **35.** 0
**37.** $-11$  **39.** $-56$  **41.** 12  **43.** 390  **45.** 4  **47.** $-9$
**49.** $-15$  **51.** $-26$  **53.** $-5$  **55.** $-4$  **57.** $-172$
**59.** $-4$  **61.** 7  **63.** 3  **65.** $-3$  **67.** $-7$  **69.** $-8$
**71.** 8  **73.** 24  **75.** 6  **77.** 5  **79.** $-8$  **81.** Polygon
**82.** Similar  **83.** Factors  **84.** Equivalent  **85.** Sum
**86.** Variable  **87.** Absolute value  **88.** Substitute  **89.** 8
**91.** 0  **93.** $-5$  **95.** 29  **97.** $-20$  **99.** 1027  **101.** $-343$
**103.** 17

## Summary and Review: Chapter 2, p. 144

### Concept Reinforcement

**1.** True  **2.** True  **3.** False

### Important Concepts

**1.** **(a)** 17; **(b)** 300  **2.** 21  **3.** 14  **4.** $-90$  **5.** $-11$
**6.** $-12$  **7.** $30x - 40y - 5z$  **8.** $17a - 7b$  **9.** $-6$

### Review Exercises

**1.** $-45; 72$  **2.** $>$  **3.** $<$  **4.** $>$  **5.** 39  **6.** 23  **7.** 0
**8.** 72  **9.** 59  **10.** $-9$  **11.** $-11$  **12.** 6  **13.** $-24$
**14.** $-12$  **15.** 23  **16.** $-1$  **17.** 0  **18.** $-4$  **19.** 12
**20.** 92  **21.** $-84$  **22.** $-40$  **23.** $-3$  **24.** $-5$  **25.** 0
**26.** $-5$  **27.** 2  **28.** $-180$  **29.** $-20$  **30.** $-62$  **31.** 7
**32.** $-6, -6, -6$  **33.** $20x + 36$  **34.** $6a - 12b + 15$
**35.** $-20x - 10y$  **36.** $17a$  **37.** $6x$  **38.** $-3m + 6$

**39.** 36 in.  **40.** 100 cm  **41.** $-8$  **42.** $-9$  **43.** $-13$
**44.** 11  **45.** 15  **46.** $-7$  **47.** 8-yd gain  **48.** $-\$130$
**49.** A  **50.** B  **51.** 403 and 397
**52.** **(a)** $-7 + (-6) + (-5) + (-4) + (-3) + (-2) +$
$(-1) + 0 + 1 + 2 + 3 + 4 + 5 + 6 + 7 + 8 = 8$; **(b)** 0
**53.** 662,582  **54.** $-88,174$  **55.** $-240$  **56.** $x < -2$
**57.** $x < 0$

### Understanding Through Discussion and Writing

**1.** We know that the product of an even number of negative
numbers is positive, and the product of an odd number of
negative numbers is negative. Since $(-7)^8$ is equivalent to the
product of eight negative numbers, it will be a positive number.
Similarly, since $(-7)^{11}$ is equivalent to the product of eleven
negative numbers, it will be a negative number.  **2.** The
expression $-x$ does not always represent a negative number.
When $x = 0, -x = 0$, and when $x$ is negative, $-x$ is positive.
**3.** Jake is expecting the multiplication to be performed before
the division.  **4.** The expression $-x^2$ represents a negative
number, except when $x = 0$. For all other values of $x, x^2$ is
positive, and the opposite of $x^2$ is negative.

## Test: Chapter 2, p. 149

**1.** [2.1a] $-542; 307$  **2.** [2.1b] $>$  **3.** [2.1c] 739  **4.** [2.1d] $-19$
**5.** [2.2a] $-11$  **6.** [2.2a] $-21$  **7.** [2.2a] 9  **8.** [2.3a] $-12$
**9.** [2.3a] $-15$  **10.** [2.3a] $-24$  **11.** [2.3a] 19  **12.** [2.3a] 38
**13.** [2.4b] $-64$  **14.** [2.4a] $-270$  **15.** [2.4a] 0  **16.** [2.5a] 8
**17.** [2.5a] $-8$  **18.** [2.5b] $-1$  **19.** [2.5b] 25
**20.** [2.3b] 14°F higher  **21.** [2.6a] $-3$
**22.** [2.6b] $14x + 21y - 7$  **23.** [2.7a] $4x - 17$  **24.** [2.7b] 20 ft
**25.** [2.8b] 5  **26.** [2.8a] $-12$  **27.** [2.8b] $-95$  **28.** [2.8d] 4
**29.** [2.6b] C  **30.** [2.7b] 66 ft  **31.** [2.5b] $35x - 7$
**32.** [2.5b] $-24x - 57$  **33.** [2.5b] 103,097  **34.** [2.5b] 1086

## Cumulative Review: Chapters 1–2, p. 151

**1.** [1.1c] 2,294,392  **2.** [1.1c] Five billion, three hundred eighty
million, one thousand, four hundred thirty-seven
**3.** [1.2a] 18,827  **4.** [1.2a] 8857  **5.** [2.2a] 56  **6.** [2.2a] $-102$
**7.** [1.3a] 7846  **8.** [1.3a] 2428  **9.** [2.3a] 14  **10.** [2.3a] $-43$
**11.** [1.4a] 16,767  **12.** [1.4a] 8,266,500  **13.** [2.4a] $-344$
**14.** [2.4a] 72  **15.** [1.5a] 104  **16.** [1.5a] 62  **17.** [2.5a] 0
**18.** [2.5a] $-5$  **19.** [1.6a] 428,000  **20.** [1.6a] 5300
**21.** [1.6b] $749,600 + 301,400 = 1,051,000$
**22.** [1.6b] $700 \times 500 = 350,000$  **23.** [2.1b] $<$  **24.** [2.1c] 279
**25.** [1.9c], [2.5b] 36  **26.** [1.9d], [2.5b] 2  **27.** [1.9c], [2.5b] $-86$
**28.** [1.9b] 125  **29.** [2.6a] 3  **30.** [2.6a] 28
**31.** [2.6b] $-2x - 10$  **32.** [2.6b] $18x - 12y + 24$
**33.** [2.2a] $-2$  **34.** [2.4b] $-8$  **35.** [2.3a] $-15$  **36.** [2.5a] 32
**37.** [2.3a] 13  **38.** [2.4b] $-30$  **39.** [2.3a] 16  **40.** [2.5b] $-57$
**41.** [1.7b], [2.8a] 27  **42.** [2.8b] $-3$  **43.** [2.8d] 15
**44.** [2.8d] $-8$  **45.** [1.8a] 104 yr  **46.** [1.8a] 16,677 rooms
**47.** [2.3b] 155°C  **48.** [1.8a] Westside Appliance
**49.** [2.7a] $10x - 14$  **50.** [1.8a] Cases: 6; six-packs: 3; loose
cans: 4  **51.** [2.5b], [2.7a] $4a$  **52.** [2.5b] $-1071$
**53.** [2.1c], [2.8d] $\pm 3$

# CHAPTER 3

## Calculator Corner, p. 155

**1.** Yes  **2.** No  **3.** No  **4.** Yes

## Exercise Set 3.1, p. 158

**1.** 7, 14, 21, 28, 35, 42, 49, 56, 63, 70  **3.** 20, 40, 60, 80, 100,
120, 140, 160, 180, 200  **5.** 3, 6, 9, 12, 15, 18, 21, 24, 27, 30
**7.** 12, 24, 36, 48, 60, 72, 84, 96, 108, 120  **9.** 10, 20, 30, 40,
50, 60, 70, 80, 90, 100  **11.** 25, 50, 75, 100, 125, 150, 175, 200,
225, 250  **13.** No  **15.** Yes  **17.** Yes  **19.** Yes; the sum of
the digits is 12, which is divisible by 3.

**21.** No; the ones digit is not 0 or 5.   **23.** Yes; the ones digit is 0.
**25.** Yes; the sum of the digits is 18, which is divisible by 9.
**27.** No; the ones digit is not even.   **29.** No; the ones digit is
not even.   **31.** 6825 is divisible by 3 and 5.   **33.** 119,117 is
divisible by none of these numbers.   **35.** 127,575 is divisible
by 3, 5, and 9.   **37.** 9360 is divisible by 2, 3, 5, 6, 9, and 10.
**39.** 555; 300; 36; 45,270; 711; 13,251; 8064   **41.** 300; 45,270
**43.** 300; 36; 45,270; 8064   **45.** 56; 324; 784; 200; 42; 812; 402
**47.** 55,555; 200; 75; 2345; 35; 1005   **49.** 324   **51.** 53
**52.** 5   **53.** −8   **54.** −24   **55.** 45 gal   **56.** 4320 min
**57.** 125   **58.** 16   **59.** $9^3$   **60.** $3^6$   **61.** 99,969   **63.** 30
**65.** 210   **67.** 840   **69.** (a) $999a + 99b + 9c = 9(111a +$
$11b + c) = 3(333a + 33b + 3c)$; therefore, $999a + 99b + 9c$ is
divisible by both 9 and 3; (b) if $a + b + c + d$ is divisible by 9,
then $a + b + c + d = 9n$ for some number $n$. Then
$abcd = 999a + 99b + 9c + 9n = 9(111a + 11b + c + n)$, so
$abcd$ is divisible by 9. A similar argument holds for 3.
**71.** 332,986,412 is divisible by 4 and 11.

### Exercise Set 3.2, p. 165

**1.** 1, 2, 3, 6, 9, 18   **3.** 1, 2, 3, 6, 9, 18, 27, 54   **5.** 1, 3, 9
**7.** 1, 13   **9.** 1, 2, 7, 14, 49, 98   **11.** 1, 3, 5, 15, 17, 51, 85, 255
**13.** Prime   **15.** Composite   **17.** Composite   **19.** Prime
**21.** Neither   **23.** Composite   **25.** Prime   **27.** Prime
**29.** $3 \cdot 3 \cdot 3$   **31.** $2 \cdot 7$   **33.** $2 \cdot 2 \cdot 2 \cdot 2 \cdot 5$   **35.** $5 \cdot 5$
**37.** $2 \cdot 31$   **39.** $2 \cdot 2 \cdot 5 \cdot 5$   **41.** $11 \cdot 13$   **43.** $11 \cdot 11$
**45.** $3 \cdot 7 \cdot 13$   **47.** $5 \cdot 5 \cdot 7$   **49.** $11 \cdot 19$
**51.** $2 \cdot 2 \cdot 2 \cdot 2 \cdot 3 \cdot 5 \cdot 5$   **53.** $3 \cdot 7 \cdot 13$   **55.** $2 \cdot 2 \cdot 7 \cdot 103$
**57.** $2 \cdot 3 \cdot 11 \cdot 17$   **59.** −26   **60.** 256   **61.** 8   **62.** −23
**63.** 1   **64.** −98   **65.** 0   **66.** 0   **67.** $946   **68.** 201 min, or
3 hr 21 min   **69.** $13 \cdot 19 \cdot 19 \cdot 29$   **71.** $23 \cdot 31 \cdot 61 \cdot 73 \cdot 149$
**73.** Answers may vary. One arrangement is a three-dimensional
rectangular array consisting of 2 tiers of 12 objects each, where
each tier consists of a rectangular array of 4 rows with 3 objects
each.
**75.**

| Product | 56 | 63 | 36 | 72 | 140 | 96 |
|---|---|---|---|---|---|---|
| Factor | 7 | 7 | 2 | 2 | 10 | 8 |
| Factor | 8 | 9 | 18 | 36 | 14 | 12 |
| Sum | 15 | 16 | 20 | 38 | 24 | 20 |

| Product | 48 | 168 | 110 | 90 | 432 | 63 |
|---|---|---|---|---|---|---|
| Factor | 6 | 21 | 10 | 9 | 24 | 3 |
| Factor | 8 | 8 | 11 | 10 | 18 | 21 |
| Sum | 14 | 29 | 21 | 19 | 42 | 24 |

### Exercise Set 3.3, p. 172

**1.** Numerator: 3; denominator: 4   **3.** Numerator: 7;
denominator: −9   **5.** Numerator: $2x$; denominator: $3y$
**7.** $\frac{2}{4}$   **9.** $\frac{1}{8}$   **11.** $\frac{4}{9}$   **13.** $\frac{3}{4}$   **15.** $\frac{4}{8}$   **17.** $\frac{12}{12}$   **19.** $\frac{4}{3}$
**21.** $\frac{7}{5}$   **23.** $\frac{5}{8}$   **25.** $\frac{4}{7}$   **27.** $\frac{12}{16}$   **29.** $\frac{36}{16}$   **31.** (a) $\frac{2}{8}$; (b) $\frac{6}{8}$
**33.** (a) $\frac{3}{8}$; (b) $\frac{5}{8}$   **35.** (a) $\frac{5}{8}$; (b) $\frac{5}{3}$; (c) $\frac{3}{8}$; (d) $\frac{3}{5}$   **37.** (a) $\dfrac{1068}{100,000}$;
(b) $\dfrac{680}{100,000}$; (c) $\dfrac{797}{100,000}$; (d) $\dfrac{962}{100,000}$; (e) $\dfrac{866}{100,000}$; (f) $\dfrac{1561}{100,000}$
**39.** (a) $\frac{4}{15}$; (b) $\frac{4}{11}$; (c) $\frac{11}{15}$   **41.** $\frac{16}{100}, \frac{18.5}{100}$   **43.** 0   **45.** 7   **47.** 1
**49.** 1   **51.** 0   **53.** $19x$   **55.** 1   **57.** −87   **59.** 0
**61.** Undefined   **63.** $7n$   **65.** Undefined   **67.** −210
**68.** −322   **69.** 0   **70.** 0   **71.** 24,465 members   **72.** 71 gal
**73.** $\frac{1}{6}$   **75.** $\frac{2}{16}$, or $\frac{1}{8}$   **77.** ⬡⬡⬡⬡⬡

**79.** 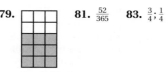   **81.** $\frac{52}{365}$   **83.** $\frac{3}{4}; \frac{1}{4}$

### Exercise Set 3.4, p. 181

**1.** $\frac{3}{8}$   **3.** $\frac{-5}{6}$, or $-\frac{5}{6}$   **5.** $\frac{14}{3}$   **7.** $\frac{-7}{9}$, or $-\frac{7}{9}$   **9.** $\frac{5x}{6}$
**11.** $\frac{-6}{5}$, or $-\frac{6}{5}$   **13.** $\frac{2a}{7}$   **15.** $\frac{-17m}{6}$, or $-\frac{17m}{6}$   **17.** $\frac{6}{5}$
**19.** $\frac{2x}{7}$   **21.** $\frac{4}{15}$   **23.** $\frac{-1}{40}$, or $-\frac{1}{40}$   **25.** $\frac{2}{15}$   **27.** $\frac{2x}{9y}$
**29.** $\frac{9}{16}$   **31.** $\frac{14}{39}$   **33.** $\frac{-3}{50}$, or $-\frac{3}{50}$   **35.** $\frac{7a}{64}$   **37.** $\frac{100}{y}$
**39.** $\frac{-147}{20}$, or $-\frac{147}{20}$   **41.** $\frac{1}{24}$   **43.** $\frac{1}{2625}$   **45.** $\frac{40}{3}$ yd
**47.** $\frac{9}{20}$   **49.** 204   **50.** 700   **51.** 13   **52.** −90   **53.** 50
**54.** 6399   **55.** $\frac{71,269}{180,433}$   **57.** $\frac{-56}{1125}$, or $-\frac{56}{1125}$   **59.** $\frac{1}{80}$ gal

### Calculator Corner, p. 186

**1.** $\frac{14}{15}$   **2.** $\frac{7}{8}$   **3.** $\frac{138}{167}$   **4.** $\frac{7}{25}$

### Exercise Set 3.5, p. 188

**1.** $\frac{5}{10}$   **3.** $\frac{-36}{-48}$   **5.** $\frac{35}{50}$   **7.** $\frac{11t}{5t}$   **9.** $\frac{20}{4}$   **11.** $\frac{-51}{54}$
**13.** $\frac{15}{-40}$   **15.** $\frac{-42}{132}$   **17.** $\frac{x}{8x}$   **19.** $\frac{-10a}{7a}$   **21.** $\frac{4ab}{9ab}$
**23.** $\frac{12b}{27b}$   **25.** $\frac{1}{2}$   **27.** $-\frac{2}{3}$   **29.** $\frac{2}{5}$   **31.** 3   **33.** $\frac{3}{4}$   **35.** $-\frac{12}{7}$
**37.** $\frac{3}{4}$   **39.** $\frac{-1}{3}$   **41.** −5   **43.** $\frac{7}{8}$   **45.** $\frac{-2}{3}$   **47.** $\frac{2}{5}$   **49.** $\frac{9}{8}$
**51.** $\frac{12}{13}$   **53.** $\frac{17}{19}$   **55.** $\frac{3}{8}$   **57.** $\frac{3y}{2}$   **59.** $\frac{-9}{10b}$   **61.** =
**63.** ≠   **65.** =   **67.** ≠   **69.** ≠   **71.** =   **73.** ≠
**75.** =   **77.** 526,761 yd²; 3020 yd   **78.** $928   **79.** 60   **80.** 65
**81.** −63   **82.** −64   **83.** 5   **84.** 89   **85.** 3520   **86.** 9001
**87.** $\frac{17}{29}$   **89.** $-\frac{29x}{15y}$   **91.** $\frac{137}{149}$   **93.** (a) $\frac{4}{10} = \frac{2}{5}$; (b) $\frac{6}{10} = \frac{3}{5}$
**95.** No. $\frac{197}{576} \neq \frac{191}{523}$ because $197 \cdot 523 \neq 576 \cdot 191$.

### Mid-Chapter Review: Chapter 3, p. 190

**1.** True   **2.** False   **3.** False   **4.** True   **5.** $\frac{25}{25} = 1$
**6.** $\frac{0}{9} = 0$   **7.** $\frac{8}{1} = 8$   **8.** $\frac{6}{13} = \frac{18}{39}$
**9.** $\frac{70}{225} = \frac{2 \cdot 5 \cdot 7}{3 \cdot 3 \cdot 5 \cdot 5} = \frac{5}{5} \cdot \frac{2 \cdot 7}{3 \cdot 3 \cdot 5} = 1 \cdot \frac{14}{45} = \frac{14}{45}$
**10.** 84; 17,576; 224; 132; 594; 504; 1632   **11.** 84; 300; 132;
500; 180   **12.** 17,576; 224; 500   **13.** 84; 300; 132; 120; 1632
**14.** 300; 180; 120   **15.** Prime   **16.** Prime   **17.** Composite
**18.** Neither   **19.** 1, 2, 4, 5, 8, 10, 16, 20, 32, 40, 80, 160;
$2 \cdot 2 \cdot 2 \cdot 2 \cdot 2 \cdot 5$   **20.** 1, 2, 3, 6, 37, 74, 111, 222; $2 \cdot 3 \cdot 37$
**21.** 1, 2, 7, 14, 49, 98; $2 \cdot 7 \cdot 7$   **22.** 1, 3, 5, 7, 9, 15, 21, 35, 45,
63, 105, 315; $3 \cdot 3 \cdot 5 \cdot 7$   **23.** $\frac{8}{24}$, or $\frac{1}{3}$   **24.** $\frac{5}{4}$   **25.** $\frac{7}{9}$
**26.** $\frac{8}{45}$   **27.** $\frac{-40}{11}$, or $-\frac{40}{11}$   **28.** $\frac{2}{5}$   **29.** $\frac{11}{3}$   **30.** 1   **31.** 0
**32.** $\frac{9}{31}$   **33.** $\frac{-9}{7}$   **34.** $\frac{5}{42}$   **35.** $\frac{21}{29}$   **36.** Undefined   **37.** =
**38.** ≠   **39.** $\frac{60}{500}$, or $\frac{3}{25}$   **40.** $\frac{21}{10,000}$ mi²   **41.** Find the product
of two prime numbers.   **42.** If we use the divisibility tests, it is
quickly clear that none of the even-numbered years is a prime
number. In addition, the divisibility tests for 5 and 3 show that
2001, 2005, 2007, 2013, 2015, and 2019 are not prime numbers.
Then the years 2003, 2009, 2011, and 2017 can be divided by
prime numbers to determine whether they are prime. When we
do this, we find that 2003, 2011, and 2017 are prime numbers. If
the divisibility tests are not used, each of the numbers from
2000 to 2020 can be divided by prime numbers to determine if it

is prime. **43.** It is possible to cancel only when identical *factors* appear in the numerator and the denominator of a fraction. Situations in which it is not possible to cancel include the occurrence of identical *addends* or *digits* in the numerator and the denominator. **44.** No; since the only factors of a prime number are the number itself and 1, two different prime numbers cannot contain a common factor (other than 1).

## Exercise Set 3.6, p. 196

**1.** $\frac{1}{3}$  **3.** $-\frac{1}{8}$  **5.** $\frac{4}{7}$  **7.** $\frac{1}{15}$  **9.** $-\frac{27}{10}$  **11.** $\frac{4x}{9}$
**13.** 1  **15.** 1  **17.** 3  **19.** 7  **21.** 12  **23.** $3a$  **25.** 1
**27.** 1  **29.** $\frac{1}{5}$  **31.** $\frac{9}{25}$  **33.** $60n$  **35.** $-\frac{10}{3}$  **37.** 1  **39.** 3
**41.** $\frac{119}{750}$  **43.** $-\frac{20}{187}$  **45.** $-\frac{42}{275}$  **47.** $-\frac{16}{5x}$  **49.** $-\frac{11}{40}$  **51.** $\frac{5a}{28b}$
**53.** $\frac{5}{8}$ in.  **55.** \$48,280  **57.** 625 addresses  **59.** $\frac{1}{3}$ cup
**61.** \$115,500  **63.** 160 mi  **65.** Food: \$8400; housing: \$10,500; clothing: \$4200; savings: \$3000; taxes: \$8400; other expenses: \$7500  **67.** $60 \, \text{in}^2$  **69.** $\frac{35}{4}\,\text{mm}^2$  **71.** $\frac{63}{8}\,\text{m}^2$
**73.** $92\,\text{mi}^2$  **75.** $56\,\text{mm}^2$  **77.** 35  **78.** 85  **79.** 125
**80.** 120  **81.** 4989  **82.** 8546  **83.** 6498  **84.** 6407
**85.** $\frac{129}{485}$  **87.** $\frac{1}{12}$  **89.** $13,380\,\text{mm}^2$

## Calculator Corner, p. 202

**1.** $\frac{1}{6}$  **2.** $\frac{20}{9}$  **3.** $-\frac{9}{7}$  **4.** $\frac{3}{2}$

## Exercise Set 3.7, p. 204

**1.** $\frac{3}{7}$  **3.** $\frac{1}{9}$  **5.** 7  **7.** $-\frac{9}{8}$  **9.** $\frac{c}{a}$  **11.** $\frac{m}{-3n}$  **13.** $\frac{-15}{8}$
**15.** $\frac{1}{7m}$  **17.** $4a$  **19.** $-3z$  **21.** $\frac{4}{7}$  **23.** $\frac{4}{15}$  **25.** 4
**27.** $-2$  **29.** $\frac{25}{7}$  **31.** $\frac{1}{8}$  **33.** $\frac{3}{28}$  **35.** $-8$  **37.** $\frac{x}{2}$
**39.** $\frac{1}{9x}$  **41.** $35a$  **43.** 1  **45.** $-\frac{2}{3}$  **47.** $\frac{99}{224}$  **49.** $\frac{112a}{3}$
**51.** $\frac{14}{15}$  **53.** $\frac{7}{32}$  **55.** $-\frac{25}{12}$  **57.** Associative  **58.** Factors
**59.** Prime  **60.** Denominator  **61.** Additive
**62.** Reciprocals  **63.** Opposites  **64.** Equation  **65.** $\frac{100}{9}$
**67.** 36  **69.** $\frac{121}{900}$  **71.** $\frac{9}{19}$  **73.** $\frac{220}{51}$

## Translating for Success, p. 209

**1.** C  **2.** H  **3.** A  **4.** N  **5.** O  **6.** F  **7.** I  **8.** L
**9.** D  **10.** M

## Exercise Set 3.8, p. 210

**1.** 15  **3.** 9  **5.** $-45$  **7.** $\frac{2}{17}$  **9.** $\frac{12}{5}$  **11.** $-\frac{16}{21}$  **13.** $-\frac{2}{25}$
**15.** $\frac{1}{6}$  **17.** $-80$  **19.** $-\frac{1}{6}$  **21.** $-\frac{7}{13}$  **23.** $\frac{27}{31}$  **25.** $\frac{6}{7}$  **27.** $\frac{12}{5}$
**29.** $-\frac{7}{15}$  **31.** $\frac{9}{5}$  **33.** 6  **35.** $\frac{10}{7}$  **37.** 960 extension cords
**39.** 1800 gal  **41.** 9 bees  **43.** 20 packages  **45.** $\frac{1}{8}$ T
**47.** 288 km; 108 km  **49.** 45 customers  **51.** 32 pairs
**53.** $\frac{1}{16}$ in.  **55.** 26  **56.** $-42$  **57.** $-67$  **58.** $-65$  **59.** 20
**60.** 6  **61.** $17x$  **62.** $4a$  **63.** $7a+3$  **64.** $4x-7$  **65.** $\frac{2}{9}$
**67.** $\frac{7}{8}$ lb  **69.** 103 slices  **71.** \$850

## Summary and Review: Chapter 3, p. 214

### Concept Reinforcement

**1.** True  **2.** False  **3.** True  **4.** True

### Important Concepts

**1.** 1, 2, 4, 8, 13, 26, 52, 104  **2.** $2 \cdot 2 \cdot 2 \cdot 13$  **3.** 0, 1, 18
**4.** $\frac{56}{96}$  **5.** $\frac{5}{14}$  **6.** $\ne$  **7.** $\frac{70}{9}$  **8.** $\frac{7}{10}$  **9.** $\frac{7}{3}$ cups

### Review Exercises

**1.** 8, 16, 24, 32, 40, 48, 56, 64, 72, 80  **2.** No  **3.** No  **4.** No
**5.** Yes  **6.** Yes  **7.** 1, 2, 3, 4, 5, 6, 10, 12, 15, 20, 30, 60  **8.** 1, 2, 4, 8, 11, 16, 22, 44, 88, 176  **9.** Prime  **10.** Neither  **11.** Composite

**12.** $2 \cdot 5 \cdot 7$  **13.** $2 \cdot 2 \cdot 2 \cdot 3 \cdot 3$  **14.** $3 \cdot 3 \cdot 5$
**15.** $2 \cdot 3 \cdot 5 \cdot 5$  **16.** $2 \cdot 2 \cdot 2 \cdot 3 \cdot 3 \cdot 3 \cdot 3$
**17.** $2 \cdot 2 \cdot 2 \cdot 2 \cdot 3 \cdot 5 \cdot 5$  **18.** Numerator: 9; denominator: 7
**19.** $\frac{3}{5}$  **20.** $\frac{7}{6}$  **21.** (a) $\frac{3}{5}$; (b) $\frac{5}{3}$; (c) $\frac{3}{8}$  **22.** 0  **23.** 1
**24.** 48  **25.** 1  **26.** $-\frac{2}{3}$  **27.** $\frac{1}{4}$  **28.** $-1$  **29.** $\frac{3}{4}$  **30.** $\frac{2}{5}$
**31.** Undefined  **32.** $\frac{2}{3}$  **33.** $\frac{32}{225}$  **34.** $\frac{15}{225}$  **35.** $\frac{-30}{55}$
**36.** $\frac{15}{100}=\frac{3}{20}; \frac{38}{100}=\frac{19}{50}; \frac{23}{100}=\frac{23}{100}; \frac{24}{100}=\frac{6}{25}$  **37.** $\ne$  **38.** $=$
**39.** $\ne$  **40.** $=$  **41.** $\frac{13}{2}$  **42.** $-\frac{1}{7}$  **43.** 8  **44.** $\frac{5y}{3x}$
**45.** $\frac{14}{45}$  **46.** $\frac{3y}{7x}$  **47.** $\frac{2}{3}$  **48.** $-14$  **49.** $\frac{10}{7}$  **50.** 1
**51.** $24x$  **52.** $\frac{1}{4}$  **53.** $\frac{80}{3}$  **54.** $\frac{1}{15}$  **55.** $6a$  **56.** $-1$
**57.** $\frac{3}{4}$  **58.** $\frac{4}{9}$  **59.** 240  **60.** $\frac{-3}{8}$  **61.** 28  **62.** $\frac{2}{3}$
**63.** $42\,\text{m}^2$  **64.** $\frac{35}{2}\,\text{ft}^2$  **65.** 9 days  **66.** About 13,560,000 bales  **67.** 1000 km  **68.** $\frac{1}{3}$ cup  **69.** $\frac{1}{6}$ mi  **70.** 60 bags
**71.** D  **72.** B  **73.** $\frac{17}{6}$  **74.** 2, 8  **75.** $a = 11,176; b = 9887$
**76.** 13, 11, 101, 37

### Understanding Through Discussion and Writing

**1.** The student is probably multiplying the divisor by the reciprocal of the dividend rather than multiplying the dividend by the reciprocal of the divisor.
**2.** $9432 = 9 \cdot 1000 + 4 \cdot 100 + 3 \cdot 10 + 2 \cdot 1 = 9(999 + 1) + 4(99 + 1) + 3(9 + 1) + 2 \cdot 1 = 9 \cdot 999 + 9 \cdot 1 + 4 \cdot 99 + 4 \cdot 1 + 3 \cdot 9 + 3 \cdot 1 + 2 \cdot 1$. Since 999, 99, and 9 are each a multiple of 9, $9 \cdot 999$, $4 \cdot 99$, and $3 \cdot 9$ are multiples of 9. This leaves $9 \cdot 1 + 4 \cdot 1 + 3 \cdot 1 + 2 \cdot 1$, or $9 + 4 + 3 + 2$. If $9 + 4 + 3 + 2$, the sum of the digits, is divisible by 9, then 9432 is divisible by 9.  **3.** Taking $\frac{1}{2}$ of a number is equivalent to multiplying the number by $\frac{1}{2}$. Dividing by $\frac{1}{2}$ is equivalent to multiplying by the reciprocal of $\frac{1}{2}$, or 2. Thus taking $\frac{1}{2}$ of a number is not the same as dividing by $\frac{1}{2}$.
**4.** We first consider an object and take $\frac{4}{7}$ of it. We divide the object into 7 parts and take 4 them, as shown by the shading below in the left figure.

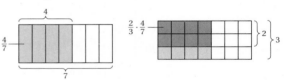

Next, we take $\frac{2}{3}$ of the shaded area in the left figure. We divide it into 3 parts and take two of them, as shown in the right figure. The entire object has been divided into 21 parts, 8 of which have been shaded twice. Thus, $\frac{2}{3} \cdot \frac{4}{7} = \frac{8}{21}$.  **5.** Since $\frac{1}{7}$ is a smaller number than $\frac{2}{3}$, there are more $\frac{1}{7}$'s in 5 than $\frac{2}{3}$'s. Thus, $5 \div \frac{1}{7}$ is a greater number than $5 \div \frac{2}{3}$.  **6.** No; in order to simplify a fraction, we must be able to remove a factor of the type $\frac{n}{n}, n \ne 0$, where $n$ is a factor that the numerator and the denominator have in common.

### Test: Chapter 3, p. 219

**1.** [3.1b] Yes  **2.** [3.1b] No  **3.** [3.2a] 1, 2, 3, 5, 6, 9, 10, 15, 18, 30, 45, 90  **4.** [3.2b] Composite  **5.** [3.2c] $2 \cdot 2 \cdot 3 \cdot 3$
**6.** [3.2c] $2 \cdot 2 \cdot 3 \cdot 5$  **7.** [3.3a] Numerator: 4; denominator: 9
**8.** [3.3a] $\frac{3}{4}$  **9.** [3.3a] $\frac{3}{7}$  **10.** [3.3a] (a) $\frac{180}{47}$; (b) $\frac{47}{93}$
**11.** [3.3b] 32  **12.** [3.3b] 1  **13.** [3.3b] 0  **14.** [3.5b] $\frac{-1}{3}$
**15.** [3.5b] 6  **16.** [3.5b] $\frac{1}{5}$  **17.** [3.3b] Undefined
**18.** [3.5b] $\frac{2}{3}$  **19.** [3.5c] $=$  **20.** [3.5c] $\ne$  **21.** [3.5a] $\frac{15}{40}$
**22.** [3.7a] $\frac{42}{a}$  **23.** [3.7a] $\frac{-1}{9}$  **24.** [3.6a] $\frac{5}{2}$  **25.** [3.7b] $\frac{8}{33}$
**26.** [3.4a] $\frac{3x}{8}$  **27.** [3.7b] $\frac{-3}{14}$  **28.** [3.7b] 18  **29.** [3.6a] $\frac{2}{9}$
**30.** [3.8b] $\frac{3}{20}$ lb  **31.** [3.6b] 125 lb  **32.** [3.8a] 64

**33.** [3.8a] $\frac{-7}{4}$   **34.** [3.6b] $\frac{91}{2}$ m²   **35.** [3.3a] C
**36.** [3.6b] $\frac{7}{48}$ acre   **37.** [3.6a], [3.7b] $\frac{-7}{960}$

## Cumulative Review: Chapters 1–3, p. 221

**1.** [1.1c] Two million, fifty-six thousand, seven hundred eighty-three   **2.** [1.2a] 10,982   **3.** [2.2a] −43   **4.** [2.2a] −33   **5.** [1.3a] 2129   **6.** [2.3a] −23   **7.** [2.3a] −8   **8.** [1.4a] 16,905   **9.** [2.4a] −312   **10.** [3.6a] −30$x$   **11.** [3.6a] $\frac{7}{15}$   **12.** [1.5a] 235 R 3   **13.** [2.5a] −17   **14.** [3.7b] −28   **15.** [3.7b] $\frac{2}{3}$   **16.** [1.6a] 4510   **17.** [1.6b] 900 × 500 = 450,000   **18.** [2.1c] 479   **19.** [2.5b] 8   **20.** [3.2b] Composite   **21.** [2.6a] −21   **22.** [1.7b], [2.8a] 25   **23.** [1.7b], [2.8b] 7   **24.** [3.8a] −45   **25.** [2.8b] −12   **26.** [2.8d] −5   **27.** [2.8b] 10   **28.** [2.7a] 5$x$ − 5   **29.** [2.7a] 3$x$ + 7$y$   **30.** [3.3b] 1   **31.** [3.3b] 0   **32.** [3.3b] 63$x$   **33.** [3.5b] $-\frac{5}{27}$   **34.** [3.7a] $\frac{5}{2}$   **35.** [3.7a] $\frac{1}{57}$   **36.** [3.5a] $\frac{21}{70}$   **37.** [1.8a] 8 oz   **38.** [1.8a] 8 mpg   **39.** [3.6b] 4375 students   **40.** [3.8b] 5 qt   **41.** [3.6b] $\frac{3}{5}$ mi   **42.** [2.6a], [3.6a], [3.7b] $\frac{-54}{169}$   **43.** [2.1c], [2.6a], [3.6a] $\frac{-9}{100}$   **44.** [3.6b] $468

# CHAPTER 4

## Exercise Set 4.1, p. 229

**1.** 4   **3.** 50   **5.** 40   **7.** 54   **9.** 150   **11.** 120   **13.** 72   **15.** 420   **17.** 144   **19.** 180   **21.** 42   **23.** 30   **25.** 72   **27.** 60   **29.** 36   **31.** 900   **33.** 48   **35.** 210   **37.** 300   **39.** 60   **41.** $abc$   **43.** $9x^2$   **45.** $4x^3y$   **47.** $24r^3s^2t^4$   **49.** $a^3b^2c^2$   **51.** Every 60 yr   **53.** Every 420 yr   **55.** 1690 tornadoes   **56.** 975,000,000   **57.** 14   **58.** −27   **59.** 7935   **60.** $\frac{2}{3}$   **61.** $-\frac{8}{7}$   **62.** −167   **63.** 2592   **65.** 54,033   **67.** 24 ft   **69.** 24 strands   **71.** **(a)** Not the LCM because $a^2b^5$ is not a factor of $a^3b^3$; **(b)** Not the LCM because $a^3b^2$ is not a factor of $a^2b^5$; **(c)** The LCM because both $a^3b^2$ and $a^2b^5$ are factors of $a^3b^5$ and it is the smallest such expression   **73.** 2520   **75.** 27 and 2; 27 and 6; 27 and 18

## Exercise Set 4.2, p. 237

**1.** $\frac{5}{9}$   **3.** 1   **5.** $\frac{2}{5}$   **7.** $\frac{13}{a}$   **9.** $-\frac{1}{2}$   **11.** $\frac{7}{9}x$   **13.** $\frac{1}{2}t$   **15.** $-\frac{9}{x}$   **17.** $\frac{7}{24}$   **19.** $-\frac{1}{10}$   **21.** $\frac{23}{24}$   **23.** $\frac{83}{20}$   **25.** $\frac{5}{24}$   **27.** $\frac{37}{100}x$   **29.** $\frac{19}{20}$   **31.** $\frac{7}{8}$   **33.** $-\frac{99}{100}$   **35.** $-\frac{1}{30}x$   **37.** $-\frac{33}{7}t$   **39.** $-\frac{17}{24}$   **41.** $\frac{3}{4}$   **43.** $\frac{437}{1000}$   **45.** $\frac{5}{4}$   **47.** $\frac{239}{78}$   **49.** $\frac{59}{90}$   **51.** $-\frac{5}{4}$   **53.** >   **55.** <   **57.** >   **59.** <   **61.** >   **63.** >   **65.** $\frac{4}{15}, \frac{3}{10}, \frac{5}{12}$   **67.** $\frac{37}{40}$ mi   **69.** $\frac{77}{40}$ mi   **71.** $\frac{7}{8}$ in.   **73.** $\frac{33}{20}$ mi   **75.** $\frac{13}{12}$ lb   **77.** $\frac{27}{32}''$   **79.** $\frac{4}{5}$ qt; $\frac{8}{5}$ qt; $\frac{2}{5}$ qt   **81.** −13   **82.** 4   **83.** −8   **84.** −31   **85.** $\frac{10}{3}$   **86.** 42; 42   **87.** 544,683 votes   **88.** 5 votes   **89.** 84 votes   **90.** 510,314 votes   **91.** 21,484,549 votes   **92.** 33,128,217 votes   **93.** $\frac{13}{30}t + \frac{31}{35}$   **95.** $7t^2 + \frac{9}{a}t$   **97.** >   **99.** $\frac{4}{15}$; $320   **101.** $a = 2, b = 8$

## Calculator Corner, p. 244

**1.** $\frac{5}{8}$   **2.** $\frac{43}{60}$   **3.** $\frac{17}{21}$   **4.** $\frac{13}{28}$   **5.** $-\frac{17}{50}$   **6.** $\frac{410}{667}$

## Translating for Success, p. 245

**1.** J   **2.** E   **3.** D   **4.** B   **5.** I   **6.** N   **7.** A   **8.** C   **9.** L   **10.** F

## Exercise Set 4.3, p. 246

**1.** $\frac{2}{3}$   **3.** $-\frac{1}{4}$   **5.** $\frac{2}{a}$   **7.** $-\frac{1}{2}$   **9.** $-\frac{4}{5a}$   **11.** $\frac{2}{t}$   **13.** $\frac{13}{16}$   **15.** $-\frac{1}{3}$   **17.** $\frac{7}{10}$   **19.** $-\frac{17}{60}$   **21.** $\frac{47}{100}$   **23.** $\frac{26}{75}$   **25.** $-\frac{71}{100}$   **27.** $\frac{13}{24}$   **29.** $-\frac{29}{50}$   **31.** $-\frac{1}{24}$   **33.** $-\frac{41}{72}$   **35.** $\frac{1}{360}$   **37.** $\frac{2}{9}x$   **39.** $-\frac{7}{20}a$   **41.** $\frac{7}{9}$   **43.** $\frac{4}{11}$   **45.** $\frac{1}{15}$   **47.** $\frac{9}{8}$   **49.** $\frac{2}{15}$   **51.** $-\frac{7}{24}$   **53.** $\frac{2}{15}$   **55.** $-\frac{5}{4}$   **57.** $\frac{5}{12}$ hr   **59.** $\frac{1}{32}$ in.   **61.** $\frac{11}{60}$ lb   **63.** $\frac{7}{20}$ hr

**65.** $\frac{3}{16}$ in.   **67.** $\frac{7}{24}$ cup   **69.** $\frac{4}{21}$   **70.** $\frac{3}{2}$   **71.** 21   **72.** $\frac{1}{32}$   **73.** About 12.8 billion, or 12,800,000,000, crayons   **74.** 9 cups   **75.** 11   **76.** 3   **77.** $\frac{1}{16}$   **79.** $-\frac{11}{16}$   **81.** $-\frac{64}{35}$   **83.** $-\frac{37}{1000}$   **85.** $\frac{43}{12}$   **87.** $\frac{1}{8}$ of the dealership   **89.** $\frac{21}{40}$ km   **91.** $\frac{14}{3553}$

**93.** *Day 1*: Cut off $\frac{1}{7}$ of bar and pay the contractor.
*Day 2*: Cut off $\frac{2}{7}$ of the bar's original length and trade it for the $\frac{1}{7}$.
*Day 3*: Give the $\frac{1}{7}$ back to the contractor.
*Day 4*: Trade the $\frac{4}{7}$ remaining for the contractor's $\frac{3}{7}$.
*Day 5*: Give the contractor the $\frac{1}{7}$ again.
*Day 6*: Trade the $\frac{2}{7}$ for the $\frac{1}{7}$.
*Day 7*: Give the contractor the $\frac{1}{7}$ again. This assumes that the contractor does not spend parts of the gold bar immediately.

## Exercise Set 4.4, p. 254

**1.** 3   **3.** $\frac{3}{5}$   **5.** −1   **7.** $\frac{27}{2}$   **9.** $\frac{1}{2}$   **11.** $\frac{3}{2}$   **13.** $-\frac{4}{3}$   **15.** $\frac{1}{2}$   **17.** $\frac{8}{7}$   **19.** −6   **21.** $\frac{21}{2}$   **23.** 6   **25.** $-\frac{17}{4}$   **27.** $-\frac{3}{4}$   **29.** $\frac{3}{2}$   **31.** $\frac{9}{2}$   **33.** $-\frac{10}{3}$   **35.** $-\frac{1}{5}$   **37.** $-\frac{1}{6}$   **39.** $\frac{35}{12}$   **41.** −13   **42.** −8   **43.** 18   **44.** 27   **45.** The balance has decreased $150.   **46.** $1180 profit   **47.** $\frac{5}{7m}$   **48.** 20$n$   **49.** 4   **51.** $-\frac{290}{697}$   **53.** $\frac{436}{35}$   **55.** 2 cm

## Calculator Corner, p. 258

**1.** $5\frac{4}{7}$   **2.** $8\frac{2}{5}$   **3.** $1476\frac{1}{6}$   **4.** $676\frac{4}{9}$   **5.** $134\frac{1}{15}$   **6.** $12\frac{169}{454}$

## Exercise Set 4.5, p. 260

**1.** $\frac{23}{3}$   **3.** $\frac{25}{4}$   **5.** $-\frac{161}{8}$   **7.** $\frac{51}{10}$   **9.** $\frac{103}{5}$   **11.** $-\frac{100}{3}$   **13.** $\frac{13}{8}$   **15.** $-\frac{51}{4}$   **17.** $\frac{57}{10}$   **19.** $-\frac{507}{100}$   **21.** $5\frac{1}{3}$   **23.** $7\frac{1}{2}$   **25.** $5\frac{7}{10}$   **27.** $7\frac{2}{9}$   **29.** $-5\frac{1}{2}$   **31.** $11\frac{1}{2}$   **33.** $-1\frac{1}{2}$   **35.** $61\frac{2}{5}$   **37.** $-8\frac{13}{50}$   **39.** $108\frac{5}{8}$   **41.** $906\frac{3}{7}$   **43.** $40\frac{4}{7}$   **45.** $-20\frac{2}{15}$   **47.** $-22\frac{3}{7}$   **49.** $6\frac{2}{5}$   **51.** $3\frac{1}{5}$   **53.** $\frac{8}{9}$   **54.** $\frac{3}{8}$   **55.** $\frac{1}{4}$   **56.** $\frac{5}{4}$   **57.** $-\frac{3}{10}$   **58.** $-\frac{9}{28}$   **59.** $237\frac{19}{541}$   **61.** $8\frac{2}{3}$   **63.** $3\frac{2}{3}$   **65.** $52\frac{1}{7}$

## Mid-Chapter Review: Chapter 4, p. 262

**1.** True   **2.** True   **3.** False   **4.** False

**5.**
$$\frac{11}{42} - \frac{3}{35} = \frac{11}{2 \cdot 3 \cdot 7} - \frac{3}{5 \cdot 7}$$
$$= \frac{11}{2 \cdot 3 \cdot 7} \cdot \left(\frac{5}{5}\right) - \frac{3}{5 \cdot 7} \cdot \left(\frac{2 \cdot 3}{2 \cdot 3}\right)$$
$$= \frac{11 \cdot 5}{2 \cdot 3 \cdot 7 \cdot 5} - \frac{3 \cdot 2 \cdot 3}{5 \cdot 7 \cdot 2 \cdot 3}$$
$$= \frac{55}{2 \cdot 3 \cdot 5 \cdot 7} - \frac{18}{2 \cdot 3 \cdot 5 \cdot 7}$$
$$= \frac{55 - 18}{2 \cdot 3 \cdot 5 \cdot 7} = \frac{37}{210}$$

**6.**
$$x + \frac{1}{8} = \frac{2}{3}$$
$$x + \frac{1}{8} - \frac{1}{8} = \frac{2}{3} - \frac{1}{8}$$
$$x + 0 = \frac{2}{3} \cdot \frac{8}{8} - \frac{1}{8} \cdot \frac{3}{3}$$
$$x = \frac{16}{24} - \frac{3}{24}$$
$$x = \frac{13}{24}$$
The solution is $\frac{13}{24}$.

**7.**
| 45 and 50 | 120 |
| 50 and 80 | 720 |
| 30 and 24 | 400 |
| 18, 24, and 80 | 450 |
| 30, 45, and 50 | |

**8.** $\frac{16}{45}$   **9.** $\frac{25}{12}$   **10.** $\frac{1}{18}$   **11.** $-\frac{19}{90}$   **12.** $\frac{7}{240}$   **13.** $-\frac{7}{24}x$   **14.** $-\frac{17}{60}$   **15.** $\frac{6}{91}$   **16.** $\frac{22}{15}$ mi   **17.** $\frac{101}{20}$ hr   **18.** $\frac{1}{5}, \frac{2}{7}, \frac{3}{10}, \frac{4}{9}$   **19.** $\frac{13}{80}$   **20.** $-\frac{8}{9}$   **21.** $17\frac{8}{15}$   **22.** C   **23.** C   **24.** No; if one number is a multiple of the other, for example, the LCM is the larger of the numbers.   **25.** We multiply by 1, using the notation $n/n$, to express each fraction in terms of the least common denominator.   **26.** Write $\frac{8}{5}$ as $\frac{16}{10}$ and $\frac{8}{2}$ as $\frac{40}{10}$ and since taking 40 tenths away from 16 tenths would give a result less than 0, it cannot possibly be $\frac{8}{3}$. You could also find the sum $\frac{8}{3} + \frac{8}{2}$ and show that it is not $\frac{8}{5}$.   **27.** No; $2\frac{1}{3} = \frac{7}{3}$ but $2 \cdot \frac{1}{3} = \frac{2}{3}$.

## Exercise Set 4.6, p. 270

**1.** $11\frac{2}{5}$ **3.** $9\frac{1}{2}$ **5.** $5\frac{1}{3}$ **7.** $13\frac{7}{12}$ **9.** $12\frac{1}{10}$ **11.** $17\frac{5}{24}$
**13.** $21\frac{1}{2}$ **15.** $27\frac{7}{8}$ **17.** $7\frac{1}{5}$ **19.** $6\frac{1}{10}$ **21.** $1\frac{3}{5}$ **23.** $13\frac{1}{4}$
**25.** $15\frac{3}{8}$ **27.** $7\frac{5}{12}$ **29.** $11\frac{5}{18}$ **31.** $8\frac{3}{4}t$ **33.** $2x$ **35.** $8\frac{3}{10}y$
**37.** $11\frac{8}{9}t$ **39.** $10\frac{5}{24}x$ **41.** $2\frac{1}{3}a$ **43.** $19\frac{1}{16}$ in. **45.** $7\frac{5}{12}$ lb
**47.** $4\frac{5}{6}$ ft **49.** $6\frac{5}{12}$ in. **51.** $3\frac{4}{5}$ hr **53.** $134\frac{1}{4}$ in.
**55.** $14\frac{13}{24}$ flats **57.** $28\frac{3}{4}$ yd **59.** $4\frac{5}{6}$ ft **61.** $7\frac{3}{8}$ ft **63.** $5\frac{3}{8}$ yd
**65.** $4\frac{5}{6}$ ft $\times$ $10\frac{7}{12}$ ft **67.** $-\frac{2}{5}$ **69.** $-3\frac{1}{4}$ **71.** $-3\frac{13}{15}$
**73.** $-7\frac{3}{8}$ **75.** $2\frac{7}{12}$ **77.** $-1\frac{8}{9}$ **79.** 16 packages
**80.** 286 cartons; 2 oz left over **81.** Yes **82.** No **83.** No
**84.** Yes **85.** No **86.** Yes **87.** Yes **88.** Yes **89.** $\frac{10}{13}$
**90.** $\frac{1}{10}$ **91.** $8568\frac{786}{1189}$ **93.** $10\frac{7}{12}$ **95.** $-28\frac{3}{8}$ **97.** $55\frac{3}{4}$ in.

## Calculator Corner, p. 280

**1.** $10\frac{2}{15}$ **2.** $1\frac{1}{28}$ **3.** $2\frac{5}{63}$ **4.** $-\frac{60}{209}$ **5.** $10\frac{11}{15}$ **6.** $2\frac{91}{115}$
**7.** $-1\frac{136}{189}$ **8.** $-27\frac{1}{112}$

## Translating for Success, p. 281

**1.** O **2.** K **3.** F **4.** D **5.** H **6.** G **7.** L **8.** E
**9.** M **10.** J

## Exercise Set 4.7, p. 282

**1.** $22\frac{2}{3}$ **3.** $1\frac{2}{3}$ **5.** $-56\frac{2}{3}$ **7.** $16\frac{1}{3}$ **9.** $-10\frac{3}{5}$ **11.** $35\frac{91}{100}$
**13.** $6\frac{1}{4}$ **15.** $1\frac{1}{5}$ **17.** $1\frac{1}{3}$ **19.** $-1\frac{1}{8}$ **21.** $1\frac{8}{43}$ **23.** $\frac{9}{40}$
**25.** $23\frac{2}{5}$ **27.** $15\frac{5}{7}$ **29.** $-27\frac{2}{9}$ **31.** $-1\frac{1}{3}$ **33.** $12\frac{1}{4}$ **35.** $8\frac{3}{20}$
**37.** 45,000 beagles **39.** 40 tsp **41.** $12\frac{4}{5}$ tiles
**43.** About 690,000 **45.** 68°F **47.** About 2,128,000
**49.** $62\frac{1}{2}$ ft$^2$ **51.** About 306,400,000 **53.** $5\frac{1}{2}$ cups of flour,
$2\frac{2}{3}$ cups of sugar **55.** 400 cu ft **57.** $16\frac{1}{2}$ servings **59.** 15 mpg
**61.** Yes; $\frac{7}{8}$ in. **63.** $441\frac{1}{4}$ ft$^2$ **65.** $76\frac{1}{4}$ ft$^2$ **67.** $27\frac{5}{16}$ cm$^2$
**69.** Integers **70.** Common **71.** Composite
**72.** Divisible; divisible **73.** Least common multiple
**74.** Addends **75.** Numerator **76.** Reciprocal **77.** $16\frac{25}{64}$
**79.** $\frac{4}{9}$ **81.** $r = \frac{240}{13}$, or $18\frac{6}{13}$ **83.** 88 gal

## Exercise Set 4.8, p. 292

**1.** $\frac{7}{24}$ **3.** $\frac{3}{2}$, or $1\frac{1}{2}$ **5.** $\frac{7}{8}$ **7.** $\frac{59}{30}$, or $1\frac{29}{30}$ **9.** $\frac{7}{16}$ **11.** $\frac{3}{20}$
**13.** $-6$ **15.** $\frac{1}{36}$ **17.** $-\frac{3}{100}$ **19.** $-1$ **21.** $\frac{19}{4}$, or $4\frac{3}{4}$ **23.** $\frac{3}{8}$
**25.** $\frac{2}{9}$ **27.** $\frac{3}{11}$ **29.** $-\frac{14}{3}$, or $-4\frac{2}{3}$ **31.** $-\frac{5}{4}$, or $-1\frac{1}{4}$ **33.** $\frac{1}{100}$
**35.** $-\frac{1}{6}$ **37.** $\frac{2x}{35}$ **39.** $-\frac{3n}{28}$ **41.** $\frac{6}{7x}$ **43.** $-\frac{5}{18}$ **45.** $-\frac{7}{12}$
**47.** $\frac{1}{3}$ **49.** $\frac{37}{48}$ **51.** $\frac{25}{72}$ **53.** $\frac{103}{16}$, or $6\frac{16}{16}$ **55.** $16\frac{11}{96}$ mi
**57.** $9\frac{19}{40}$ lb **59.** $-\frac{8}{3}$ **60.** $\frac{3}{8}$ **61.** Prime: 5, 7, 23, 43;
composite: 9, 14; neither: 1 **62.** 59 R 77 **63.** 16 people
**64.** 43 mg **65.** $3\frac{11}{14}$ **67.** $\frac{8x}{147}$ **69.** 0 **71.** $\frac{1}{2}$ **73.** $\frac{1}{2}$

## Summary and Review: Chapter 4, p. 295

### Concept Reinforcement

**1.** True **2.** True **3.** True **4.** False

### Important Concepts

**1.** 156 **2.** $\frac{112}{180}$, or $\frac{28}{45}$ **3.** $<$ **4.** $\frac{4}{35}$ **5.** $-\frac{1}{12}$ **6.** $\frac{26}{3}$ **7.** $7\frac{5}{6}$
**8.** $7\frac{27}{28}$ **9.** $14\frac{14}{25}$ **10.** About 4,500,000 **11.** $\frac{9}{2}$, or $4\frac{1}{2}$

### Review Exercises

**1.** 36 **2.** 90 **3.** 30 **4.** 1404 **5.** $\frac{7}{9}$ **6.** $\frac{9}{x}$ **7.** $\frac{-7}{15}$
**8.** $\frac{7}{16}$ **9.** $\frac{2}{9}$ **10.** $-\frac{1}{8}$ **11.** $\frac{4}{27}$ **12.** $\frac{1}{18}$ **13.** $>$ **14.** $<$
**15.** $\frac{19}{40}$ **16.** 4 **17.** $-\frac{5}{6}$ **18.** $\frac{12}{25}$ **19.** $\frac{2}{5}$ **20.** $\frac{15}{2}$ **21.** $\frac{67}{8}$
**22.** $\frac{13}{3}$ **23.** $-\frac{12}{7}$ **24.** $2\frac{1}{3}$ **25.** $-6\frac{3}{4}$ **26.** $12\frac{3}{5}$ **27.** $3\frac{1}{2}$
**28.** $-877\frac{1}{3}$ **29.** $82\frac{1}{3}$ **30.** $10\frac{2}{5}$ **31.** $11\frac{11}{15}$ **32.** $-9$

**33.** $1\frac{3}{4}$ **34.** $7\frac{7}{9}$ **35.** $4\frac{11}{15}$ **36.** $-5\frac{1}{8}$ **37.** $-14\frac{1}{4}$ **38.** $\frac{7}{9}x$
**39.** $3\frac{7}{40}a$ **40.** 16 **41.** $-3\frac{1}{2}$ **42.** $2\frac{21}{50}$ **43.** 6 **44.** $-24$
**45.** $-1\frac{7}{17}$ **46.** $\frac{1}{8}$ **47.** $\frac{9}{10}$ **48.** $13\frac{5}{7}$ **49.** $2\frac{8}{11}$ **50.** $4\frac{1}{4}$ yd
**51.** $3\frac{1}{8}$ pizzas **52.** 24 lb **53.** $177\frac{3}{4}$ in$^2$ **54.** $50\frac{1}{4}$ in$^2$
**55.** $850 **56.** $8\frac{3}{8}$ cups **57.** $63\frac{2}{3}$ pies; $19\frac{1}{3}$ pies
**58.** $13\frac{1}{4}$ in. $\times$ $13\frac{1}{4}$ in.: perimeter $= 53$ in., area $= 175\frac{9}{16}$ sq in.;
$13\frac{1}{4}$ in. $\times$ $3\frac{1}{4}$ in.: perimeter $= 33$ in., area $= 43\frac{1}{16}$ sq in.
**59.** 1 **60.** $\frac{39}{40}$ **61.** 3 **62.** $\frac{77}{240}$ **63.** $-\frac{8}{9}$ **64.** $-\frac{2x}{3}$
**65.** A **66.** D **67.** $\frac{600}{13}$, or $46\frac{2}{13}$ **68.** $\frac{6}{3} + \frac{5}{4} = 3\frac{1}{4}$
**69. (a)** 6; **(b)** 10; **(c)** $-28$; **(d)** $-1$

### Understanding Through Discussion and Writing

**1.** No; if the sum of the fractional parts of the mixed numerals is $n/n$, then the sum of the mixed numerals is an integer. For example, $1\frac{1}{5} + 6\frac{4}{5} = 7\frac{5}{5} = 8$. **2.** A wheel makes $33\frac{1}{3}$ revolutions per minute. It rotates for $4\frac{1}{2}$ min. How many revolutions does it make? Answers may vary. **3.** The student is multiplying the whole numbers to get the whole-number portion of the answer and multiplying fractions to get the fraction part of the answer. The student should have converted each mixed numeral to fraction notation, multiplied, simplified, and then converted back to a mixed numeral. The correct answer is $4\frac{6}{7}$. **4.** It might be necessary to find the least common denominator before adding or subtracting. The least common denominator is the least common multiple of the denominators. **5.** Suppose that a room has dimensions $15\frac{3}{4}$ ft by $28\frac{5}{8}$ ft. The equation $2 \cdot 15\frac{3}{4} + 2 \cdot 28\frac{5}{8} = 88\frac{3}{4}$ gives the perimeter of the room, in feet. Answers may vary. **6.** Note that $5 \cdot 3\frac{2}{7} = 5\left(3 + \frac{2}{7}\right) = 5 \cdot 3 + 5 \cdot \frac{2}{7}$. The products $5 \cdot 3$ and $5 \cdot \frac{2}{7}$ should be added rather than multiplied together. The student could also have converted $3\frac{2}{7}$ to fraction notation, multiplied, simplified, and converted back to a mixed numeral. The correct answer is $16\frac{3}{7}$.

## Test: Chapter 4, p. 301

**1.** [4.1a] 48 **2.** [4.2a] 3 **3.** [4.2b] $\frac{-5}{24}$ **4.** [4.3a] $\frac{2}{t}$
**5.** [4.3a] $\frac{1}{12}$ **6.** [4.3a] $-\frac{1}{12}$ **7.** [4.3b] $\frac{1}{4}$ **8.** [4.4a] $\frac{-12}{5}$
**9.** [4.4a] $\frac{3}{20}$ **10.** [4.2c] $>$ **11.** [4.5a] $\frac{7}{2}$ **12.** [4.5a] $-\frac{75}{8}$
**13.** [4.5a] $-8\frac{2}{9}$ **14.** [4.5b] $162\frac{7}{11}$ **15.** [4.6a] $14\frac{1}{5}$
**16.** [4.6a] $12\frac{5}{24}$ **17.** [4.6b] $4\frac{7}{24}$ **18.** [4.6d] $8\frac{4}{7}$ **19.** [4.6d] $-5\frac{7}{16}$
**20.** [4.3a] $-\frac{1}{8}x$ **21.** [4.6b] $1\frac{54}{55}a$ **22.** [4.7a] 39 **23.** [4.7a] $-18$
**24.** [4.7b] 6 **25.** [4.7b] 2 **26.** [4.7c] $19\frac{3}{5}$ **27.** [4.7c] $28\frac{1}{20}$
**28.** [4.7d] About 105 kg **29.** [4.7d] 80 books
**30.** [4.6c] **(a)** 3 in.; **(b)** $4\frac{1}{2}$ in. **31.** [4.3c] $1\frac{1}{16}$ in. **32.** [4.7d] $6\frac{11}{36}$ ft
**33.** [4.8a] $3\frac{1}{2}$ **34.** [4.8a] $-\frac{9}{4}$, or $-2\frac{1}{4}$ **35.** [4.8b] $-\frac{2}{3}$
**36.** [4.1a] D **37.** [4.1a] **(a)** 24, 48, 72; **(b)** 24
**38.** [4.3c] Rebecca walks $\frac{17}{56}$ mi farther.

## Cumulative Review: Chapters 1–4, p. 303

**1. (a)** [4.6c] $14\frac{13}{24}$ mi; **(b)** [4.7d] $4\frac{61}{72}$ mi **2.** [1.8a] 31 people
**3.** [3.6b] $\frac{2}{5}$ tsp; 4 tsp **4.** [4.7d] 16 pieces **5.** [1.8a] $108
**6.** [4.2d] $\frac{33}{20}$ mi **7.** [3.3a] $\frac{5}{16}$ **8.** [3.3a] $\frac{4}{3}$ **9.** [1.2a] 8982
**10.** [1.3a] 4518 **11.** [1.4a] 5004 **12.** [2.4a] $-145$
**13.** [4.2b] $\frac{5}{12}$ **14.** [4.6a] $8\frac{1}{4}$ **15.** [4.3a] $\frac{-5}{t}$ **16.** [4.6b] $1\frac{1}{6}$
**17.** [3.6a] $\frac{3}{2}$ **18.** [3.6a] $-15$ **19.** [2.3a] 16 **20.** [4.7b] $7\frac{1}{3}$
**21.** [1.5a] 715 **22.** [1.5a] 56 R 11 **23.** [4.5b] $56\frac{11}{45}$
**24.** [1.1a] 5 **25. (a)** [4.7d] $142\frac{1}{4}$ ft$^2$; **(b)** [4.6c] 54 ft
**26.** [1.6a] 38,500 **27.** [4.1a] 72 **28.** [4.8a] $\frac{1377}{100}$, or $13\frac{77}{100}$
**29.** [4.2c] $>$ **30.** [3.5c] $=$ **31.** [4.2c] $>$ **32.** [2.6a] 4
**33.** [3.5b] $\frac{4}{5}$ **34.** [3.3b] 0 **35.** [3.5b] $-32$ **36.** [4.5a] $\frac{37}{8}$
**37.** [4.5a] $-5\frac{2}{3}$ **38.** [1.7b] 93 **39.** [4.3b] $\frac{5}{9}$ **40.** [3.8a] $-\frac{12}{7}$

**41.** $[4.4a] \frac{2}{21}$    **42.** [3.1a, b], [3.2a, b, c]
Factors of 68: 1, 2, 4, 17, 34, 68
Factorization of 68: $2 \cdot 2 \cdot 17$, or $2 \cdot 34$
Prime factorization of 68: $2 \cdot 2 \cdot 17$
Numbers divisible by 6: 12, 54, 72, 300
Numbers divisible by 8: 8, 16, 24, 32, 40, 48, 64, 864
Numbers divisible by 5: 70, 95, 215
Prime numbers: 2, 3, 17, 19, 23, 31, 47, 101
**43.** [3.2b] 2003    **44.** [4.4a], [4.6b] $\frac{3}{7}$

## CHAPTER 5

### Exercise Set 5.1, p. 313

**1.** Four hundred eighty-six and thirty-four hundredths
**3.** One hundred forty-six thousandths    **5.** Two hundred
forty-nine and eighty-nine hundredths    **7.** Three and seven
thousand, eight hundred fifty-four ten-thousandths
**9.** Five and four tenths    **11.** Five hundred twenty-four and
$\frac{95}{100}$ dollars    **13.** Thirty-six and $\frac{72}{100}$ dollars    **15.** $\frac{73}{10}$; $7\frac{3}{10}$
**17.** $\frac{2167}{100}$; $21\frac{67}{100}$    **19.** $-\frac{2703}{1000}$; $-2\frac{703}{1000}$    **21.** $\frac{109}{10,000}$
**23.** $-\frac{40,003}{10,000}$; $-4\frac{3}{10,000}$    **25.** $-\frac{207}{10,000}$    **27.** $\frac{7,000,105}{100,000}$; $70\frac{105}{100,000}$
**29.** 0.3    **31.** $-0.59$    **33.** 3.798    **35.** 0.0078    **37.** $-0.00018$
**39.** 0.486197    **41.** 7.013    **43.** $-8.431$    **45.** 2.1739
**47.** 8.953073    **49.** 0.58    **51.** 0.410    **53.** $-5.043$
**55.** 235.07    **57.** $\frac{7}{100}$    **59.** $-0.872$    **61.** 0.2    **63.** $-0.4$
**65.** 3.0    **67.** $-327.2$    **69.** 0.89    **71.** $-0.67$    **73.** 1.00
**75.** $-0.03$    **77.** 0.572    **79.** 17.002    **81.** $-20.202$
**83.** 9.985    **85.** 809.5    **87.** 809.47    **89.** 830    **90.** $\frac{830}{1000}$, or $\frac{83}{100}$
**91.** 182    **92.** $\frac{182}{100}$, or $\frac{91}{50}$    **93.** $-\frac{12}{55}$    **94.** $-\frac{15}{34}$    **95.** 32,958
**96.** 10,726    **97.** $-1.09, -1.009, -0.989, -0.898, -0.098$
**99.** 6.78346    **101.** 99.99999    **103.** 1998, 2006    **105.** 1996

### Calculator Corner, p. 319

**1.** 317.645    **2.** 49.08    **3.** 33.83    **4.** 0.99    **5.** 242.93
**6.** $-11.692$

### Exercise Set 5.2, p. 320

**1.** 464.37    **3.** 1576.015    **5.** 132.56, or 132.560    **7.** 7.823
**9.** 50.7124    **11.** 10.06    **13.** 771.967    **15.** 20.8649
**17.** 227.468, or 227.4680    **19.** 41.381    **21.** 49.02    **23.** 3.564
**25.** 85.921    **27.** 1.6666    **29.** 4.0622    **31.** 29.999    **33.** 3.37
**35.** 1.045    **37.** 3.703    **39.** 0.9092    **41.** 605.21    **43.** 53.203
**45.** 161.62    **47.** 44.001    **49.** $-3.29$    **51.** $-2.5$    **53.** $-7.2$
**55.** 3.379    **57.** $-16.6$    **59.** 2.5    **61.** $-3.519$    **63.** 9.601
**65.** 75.5    **67.** 9.7    **69.** $-10.292$    **71.** $-0.3$    **73.** $5.7x$
**75.** $4.86a$    **77.** $21.1t + 7.9$    **79.** $-2.917x$    **81.** $8.106y - 7.1$
**83.** $-0.9x + 3.1y$    **85.** $-1.1 - 8.4t$    **87.** $\frac{12}{35}$    **88.** $\frac{14}{45}$    **89.** $\frac{63}{1000}$
**90.** $-10$    **91.** $-7$    **92.** 31    **93.** $-12.001 - 12.2698a +$
$10.366b$    **95.** $4.593a - 10.996b - 59.491$    **97.** $-138.5$
**99.** 2

### Calculator Corner, p. 326

**1.** 142.803    **2.** $-0.5076$    **3.** 7916.4    **4.** 20.4153

### Exercise Set 5.3, p. 330

**1.** 47.6    **3.** 6.72    **5.** 0.252    **7.** 0.522    **9.** 426.3
**11.** $-783,686.852$    **13.** $-780$    **15.** 7.918    **17.** 0.09768
**19.** $-0.287$    **21.** 43.68    **23.** 0.030504    **25.** 89.76
**27.** $-322.07$    **29.** 55.68    **31.** 3487.5    **33.** 0.1155
**35.** $-9420$    **37.** 0.00953    **39.** 5706¢    **41.** 95¢    **43.** 1¢
**45.** \$0.72    **47.** \$0.02    **49.** \$63.99    **51.** 11,100,000
**53.** 152,700,000,000    **55.** 3,156,000,000    **57.** 11,000
**59.** 26.025    **61. (a)** 44 ft; **(b)** 118.75 sq ft
**63. (a)** 37.8 m; **(b)** 88.2 m²    **65.** 2.7625 million nurses, or
2,762,500 nurses    **67.** $-27$    **68.** 36    **69.** 69    **70.** $-141$
**71.** $-21$    **72.** $-27$    **73.** 1257    **74.** 1176 R 14
**75.** $10^{21} = 1$ sextillion    **77.** $10^{24} = 1$ septillion

**79.** 6,600,000,000,000    **81.** 366.5488175    **83.** 72.996 cm²
**85.** \$61.45

### Calculator Corner, p. 334

**1.** 14.3    **2.** 2.56    **3.** $-0.064$    **4.** 75.8

### Calculator Corner, p. 335

**1.** 28 R 2    **2.** 116 R 3    **3.** 74 R 10    **4.** 415 R 3

### Exercise Set 5.4, p. 339

**1.** 2.99    **3.** 23.78    **5.** 7.08    **7.** 1.2    **9.** $-0.9$    **11.** $-6000$
**13.** 140    **15.** 40    **17.** $-0.15$    **19.** 3.2    **21.** $-3.9$
**23.** 0.625    **25.** 0.26    **27.** 2.34    **29.** $-0.3045$
**31.** 2.134567    **33.** $-2.359$    **35.** 1023.7    **37.** $-9236$
**39.** 0.08172    **41.** 9.7    **43.** $-0.0527$    **45.** $-75,300$
**47.** $-0.0753$    **49.** 2107    **51.** $-302.997$    **53.** $-178.1$
**55.** 206.0176    **57.** 0.5    **59.** $-400.0108$    **61.** 0.6725
**63.** 5.383    **65.** 10.5    **67.** 5.14 million stays    **69.** 30.425 mpg
**71.** 31.24 mi    **73.** $\frac{3}{4}$    **74.** $\frac{7}{8}$    **75.** $-\frac{3}{2}$    **76.** $-\frac{3}{10}$
**77.** $\frac{a}{3}$    **78.** $\frac{2x}{5}$    **79.** $\frac{1}{5}$    **80.** $\frac{2}{3}$    **81.** $-56.6916$
**83.** 6.254194585    **85.** 1000    **87.** 100    **89.** 68 points
**91.** 450 kWh

### Mid-Chapter Review: Chapter 5, p. 343

**1.** False    **2.** True    **3.** True
**4.** $P(1 + r) = 5000(1 + 0.045)$
$\qquad\qquad = 5000(1.045)$
$\qquad\qquad = 5225$
**5.** $5.6 + 4.3 \times (6.5 - 0.25)^2 = 5.6 + 4.3 \times (6.25)^2$
$\qquad\qquad\qquad\qquad\qquad = 5.6 + 4.3 \times 39.0625$
$\qquad\qquad\qquad\qquad\qquad = 5.6 + 167.96875$
$\qquad\qquad\qquad\qquad\qquad = 173.56875$
**6.** Nine and sixty-nine hundredths    **7.** 1,050,000
**8.** $\frac{453}{100}$; $4\frac{53}{100}$    **9.** $\frac{287}{1000}$    **10.** 0.13    **11.** $-5.09$    **12.** 0.7
**13.** 6.39    **14.** $-35.67$    **15.** 8.002    **16.** 28.462    **17.** 28.46
**18.** 28.5    **19.** 28    **20.** 50.095    **21.** 1214.862    **22.** $-10.23$
**23.** 18.24    **24.** 272.19    **25.** 5.593    **26.** 15.55    **27.** $-19.9$
**28.** 4.14    **29.** 92.871    **30.** 8123.6    **31.** $-0.0483$    **32.** 5.06
**33.** 3.2    **34.** 763    **35.** 0.914036    **36.** 2045¢    **37.** \$1.47
**38.** $-1.22x - 7.1$    **39.** 4.2    **40.** 59.774    **41.** 33.33
**42.** The student probably rounded over successively from the
thousandths place as follows: $236.448 \approx 236.45 \approx 236.5 \approx 237$.
The student should have considered only the tenths place and
rounded down.    **43.** The decimal points were not lined up
before the subtraction was carried out.    **44.** $10 \div 0.2 = \frac{10}{0.2} =$
$\frac{10}{0.2} \cdot \frac{10}{10} = \frac{100}{2} = 100 \div 2$.    **45.** $0.247 \div 0.1 = \frac{247}{1000} \div \frac{1}{10} =$
$\frac{247}{1000} \cdot \frac{10}{1} = \frac{247 \cdot 10}{10 \cdot 100} = \frac{247}{100} = 2.47 \neq 0.0247$;
$0.247 \div 10 = \frac{247}{1000} \div 10 = \frac{247}{1000} \cdot \frac{1}{10} = \frac{247}{10,000} = 0.0247 \neq 2.47$

### Calculator Corner, p. 347

**1.** $-0.1\overline{6}$    **2.** $0.\overline{63}$    **3.** $6.\overline{3}$    **4.** $-57.\overline{1}$

### Calculator Corner, p. 348

**1.** 123.150432    **2.** 52.59026102

### Exercise Set 5.5, p. 350

**1.** 0.375    **3.** $-0.5$    **5.** 0.12    **7.** 0.225    **9.** 0.52    **11.** $-0.85$
**13.** $-0.5625$    **15.** 1.4    **17.** 1.12    **19.** $-1.375$    **21.** $-0.975$
**23.** 0.605    **25.** $0.5\overline{3}$    **27.** $0.\overline{3}$    **29.** $-1.\overline{3}$    **31.** $1.1\overline{6}$
**33.** $-1.\overline{27}$    **35.** $-0.41\overline{6}$    **37.** 0.254    **39.** $0.\overline{12}$    **41.** $-0.2\overline{18}$
**43.** $0.\overline{571428}$    **45.** 0.4; 0.36; 0.364    **47.** $-1.7$; $-1.67$; $-1.667$
**49.** $-0.5$; $-0.47$; $-0.471$    **51.** 0.6; 0.58; 0.583
**53.** $-0.2$; $-0.19$; $-0.193$    **55.** $-0.8$; $-0.78$; $-0.778$
**57.** 0.2; 0.18; 0.182    **59.** 0.3; 0.28; 0.278    **61. (a)** 0.429;
**(b)** 0.75; **(c)** 0.571; **(d)** 1.333    **63.** 15.8 mpg    **65.** 17.8 mpg

**67.** 11.06   **69.** 2.736   **71.** $-417.51\overline{6}$   **73.** 0   **75.** 0.09705   **77.** $-1.5275$   **79.** 24.375   **81.** 1.08 m$^2$   **83.** 5.78 cm$^2$   **85.** 790.92 in$^2$   **87.** 3570   **88.** 4000   **89.** 79,000   **90.** 19,830,000   **91.** $-95$   **92.** $-10$   **93.** $-7$   **94.** 1   **95.** $0.1\overline{42857}$   **97.** $0.\overline{428571}$   **99.** $0.\overline{714285}$   **101.** $0.\overline{1}$   **103.** $0.\overline{001}$   **105.** 13.86 cm$^2$   **107.** 1.76625 ft$^2$ or 1.767145868 ft$^2$

### Exercise Set 5.6, p. 357

**1.** (d)   **3.** (c)   **5.** (a)   **7.** (c)   **9.** 1.6   **11.** 6   **13.** 60   **15.** 2.3   **17.** 180   **19.** (a)   **21.** (c)   **23.** (b)   **25.** (b)   **27.** $1800 \div 9 = 200$ posts; answers may vary   **29.** $\$2 \cdot 12 = \$24$; answers may vary   **31.** Repeating   **32.** Multiple   **33.** Distributive   **34.** Solution   **35.** Multiplicative   **36.** Commutative   **37.** Denominator; multiple   **38.** Divisible   **39.** Yes   **41.** No   **43.** (a) $+, \times$; (b) $+, \times, -$

### Exercise Set 5.7, p. 363

**1.** 5.4   **3.** $-12.6$   **5.** 6   **7.** 1.8   **9.** $-3.7$   **11.** $-4.7$   **13.** 1.7   **15.** 2.94   **17.** 4.5   **19.** 9   **21.** 3.2   **23.** $-1.75$   **25.** 30   **27.** 3.2   **29.** 9   **31.** 2.1   **33.** 4.5   **35.** 4.12   **37.** 5.6   **39.** $-1.9$   **41.** 13   **43.** 0   **45.** $-1.5$   **47.** 14 m$^2$   **48.** 27 cm$^2$   **49.** $\frac{25}{2}$ in$^2$   **50.** 24 ft$^2$   **51.** 5 ft$^2$   **52.** 12 m$^2$   **53.** $-\frac{29}{50}$   **54.** 0   **55.** $-2$   **56.** 8   **57.** 3.1   **59.** 36   **61.** 1.1212963

### Translating for Success, p. 374

**1.** I   **2.** C   **3.** N   **4.** A   **5.** G   **6.** B   **7.** D   **8.** O   **9.** F   **10.** M

### Exercise Set 5.8, p. 375

**1.** Let $a =$ Ron's age: $a + 5$, or $5 + a$   **3.** $b + 6$, or $6 + b$   **5.** $c - 9$   **7.** Let $n =$ the number; $n - 16$   **9.** Let $s =$ Nate's speed; $8s$   **11.** $\frac{x}{17}$   **13.** Let $x =$ the number; $\frac{1}{2}x + 20$, or $20 + \frac{1}{2}x$   **15.** Let $x =$ the number; $4x - 20$   **17.** Let $l =$ the length and $w =$ the width; $l + w$, or $w + l$   **19.** Let $r =$ the rate and $t =$ the time; $rt + 10$, or $10 + rt$   **21.** Let $n =$ the number; $10n + n$   **23.** Let $x$ and $y =$ the numbers: $5(x - y)$   **25.** $\$1.85$ billion   **27.** $\$43.1$ billion   **29.** $\$45.88$   **31.** $\$0.51$   **33.** $\$64,333,333.33$   **35.** Area: 8.125 cm$^2$; perimeter: 11.5 cm   **37.** 22,691.5 mi   **39.** 20.2 mpg   **41.** $\$24.33$   **43.** 2.66 cc   **45.** 10.8¢   **47.** 180.26 in$^2$   **49.** 193.04 cm$^2$   **51.** 2.31 cm   **53.** 331.74 ft$^2$   **55.** 125 gigabytes   **57.** 17 bottles   **59.** 16.5 hr as waiter, 10.5 hr doing lawn maintenance   **61.** 2.5 million protected acres, 2.9 million unprotected acres   **63.** 175 billion spam messages, 35 billion non-spam messages   **65.** $\$65.30$ for food, $\$195.90$ for lodging   **67.** 0   **68.** Undefined   **69.** 1   **70.** $-\frac{1}{10}$   **71.** $-\frac{20}{33}$   **72.** $6\frac{5}{6}$   **73.** $-4$   **74.** $-54$   **75.** $0.\overline{6}$ million viewers per year, or $\frac{2}{3}$ million viewers per year   **77.** $\$0.19$   **79.** 25 cm$^2$. We assume that the figures are nested squares formed by connecting the midpoints of consecutive sides of the next larger square.

### Summary and Review: Chapter 5, p. 381

#### Concept Reinforcement

**1.** True   **2.** False   **3.** True   **4.** False   **5.** True

#### Important Concepts

**1.** $\frac{3}{100}$   **2.** 81.7   **3.** 42.159   **4.** 153.35   **5.** 38.611   **6.** 207.848   **7.** 19.11   **8.** 0.176   **9.** 60,437   **10.** 7.4   **11.** 0.047   **12.** 15,690

#### Review Exercises

**1.** 6,590,000   **2.** 3,100,000,000   **3.** Three and forty-seven hundredths   **4.** Thirty-one thousandths   **5.** Twenty-seven and one ten-thousandth   **6.** Nine tenths

**7.** $\frac{9}{100}$   **8.** $-\frac{4561}{1000}$; $-4\frac{561}{1000}$   **9.** $-\frac{89}{1000}$   **10.** $\frac{30,227}{10,000}$; $3\frac{227}{10,000}$   **11.** 0.034   **12.** 4.2603   **13.** 27.91   **14.** $-867.006$   **15.** 0.034   **16.** $-0.19$   **17.** 0.741   **18.** 1.041   **19.** 17.4   **20.** 17.43   **21.** 17.429   **22.** 17   **23.** 499.829   **24.** 29.148   **25.** 229.1   **26.** 685.0519   **27.** $-57.3$   **28.** 2.37   **29.** 12.96   **30.** $-1.073$   **31.** 24,680   **32.** 3.2   **33.** $-1.6$   **34.** 0.2763   **35.** $2.2x - 9.1y$   **36.** $-2.84a + 12.57$   **37.** 925   **38.** 40.84   **39.** $\$15.49$   **40.** 248.27   **41.** 2.6   **42.** 1.28   **43.** 3.25   **44.** $-1.1\overline{6}$   **45.** 21.08   **46.** $-3.2$   **47.** $-3$   **48.** $-7.5$   **49.** 6.5   **50.** 11.16 poles   **51.** $\$15.52$   **52.** 249.76 ft$^2$   **53.** $\$5788.56$   **54.** 95 transactions   **55.** 195.7 lb of newspaper, 65.7 lb of glass   **56.** 14.5 mpg   **57.** (a) 98.1 lb; (b) 16.35 lb   **58.** 8.4 mi   **59.** $\$1.33$   **60.** 61.5 ft; 235.625 sq ft   **61.** B   **62.** A   **63.** (a) $+$; (b) $-$   **64.** $\frac{-13}{15}, \frac{-17}{20}, -\frac{11}{13}, -\frac{15}{19}, \frac{-5}{7}, -\frac{2}{3}$   **65.** 26,260 mi   **66.** The rectangular pizza, at $\frac{4.4¢}{\text{in}^2}$, is a better buy than the round pizza, which costs $\frac{5.5¢}{\text{in}^2}$.

#### Understanding Through Discussion and Writing

**1.** Count the number of decimal places. Move the decimal point that many places to the right and write the result over a denominator of 1 followed by that many zeros.   **2.** $346.708 \times 0.1 = \frac{346,708}{1000} \times \frac{1}{10} = \frac{346,708}{10,000} = 34.6708 \neq 3467.08$   **3.** When the denominator of a fraction is a multiple of 10, long division is not the fastest way to convert the fraction to decimal notation. Many times this is also the case when the denominator has only 2's or 5's or both as factors.   **4.** Multiply by 1 to get a denominator that is a power of 10:

$$\frac{44}{125} = \frac{44}{125} \cdot \frac{8}{8} = \frac{352}{1000} = 0.352.$$

We can also divide to find that $\frac{44}{125} = 0.352$.

### Test: Chapter 5, p. 386

**1.** [5.3b] 18,400,000   **2.** [5.3b] 13,100,000,000   **3.** [5.1a] Two and thirty-four hundredths   **4.** [5.1a] One hundred five and five ten-thousandths   **5.** [5.1b] $-\frac{91}{100}$   **6.** [5.1b] $\frac{2769}{1000}$; $2\frac{769}{1000}$   **7.** [5.1b] 0.074   **8.** [5.1b] $-3.7047$   **9.** [5.1b] 756.09   **10.** [5.1b] 91.703   **11.** [5.1c] 0.162   **12.** [5.1c] 0.078   **13.** [5.1c] $-0.09$   **14.** [5.1d] 6   **15.** [5.1d] 5.68   **16.** [5.1d] 5.678   **17.** [5.1d] 5.7   **18.** [5.2a] 405.219   **19.** [5.3a] 0.03   **20.** [5.3a] 0.21345   **21.** [5.2b] 44.746   **22.** [5.2a] 356.37   **23.** [5.2c] $-2.2$   **24.** [5.2b] 1.9946   **25.** [5.3a] 73,962   **26.** [5.4a] 4.75   **27.** [5.4a] 30.4   **28.** [5.4a] $-0.34682$   **29.** [5.4a] 34,682   **30.** [5.3b] 17,982¢   **31.** [5.2d] $9.8x - 3.9y - 4.6$   **32.** [5.3c] 11.6   **33.** [5.4b] 7.6   **34.** [5.5c] 1.6   **35.** [5.5c] 5.25   **36.** [5.5a] $-0.4375$   **37.** [5.5a] $1.\overline{5}$   **38.** [5.5b] 1.56   **39.** [5.4b] $-22.25$   **40.** [5.4b] 1.8045   **41.** [5.5d] 9.72   **42.** [5.7a] $-3.24$   **43.** [5.7b] 10   **44.** [5.7b] 1.4   **45.** [5.8b] $\$133.99$   **46.** [5.8b] 28.3 mpg   **47.** [5.8b] $\$592.45$   **48.** [5.8b] $\$293.93$   **49.** [5.8b] 2.864 million passengers   **50.** [5.6a] C   **51.** [5.3a] (a) Always; (b) never; (c) sometimes; (d) sometimes   **52.** [5.8b] $\$35$   **53.** [5.8b] (a) Fly; (b) fly; (c) drive

### Cumulative Review: Chapters 1–5, p. 389

**1.** [4.5a] $\frac{20}{9}$   **2.** [5.1b] $\frac{3051}{1000}$   **3.** [5.5a] $-1.4$   **4.** [5.5a] $0.\overline{54}$   **5.** [3.2b] Prime   **6.** [3.1b] No   **7.** [1.9c] 1754   **8.** [5.4b] 4.364   **9.** [5.1d] 584.97   **10.** [5.5b] 218.56   **11.** [5.6a] 160   **12.** [5.6a] 4   **13.** [2.6a] 12   **14.** [2.7a] $7p - 8$   **15.** [4.6a] $6\frac{1}{20}$   **16.** [1.2a] 139,116   **17.** [4.2b] $\frac{31}{18}$   **18.** [5.2a] 145.953   **19.** [1.3a] 710,137   **20.** [5.2b] $-13.097$   **21.** [4.6b] $\frac{5}{7}$   **22.** [4.3a] $-\frac{1}{110}$   **23.** [3.6a] $-\frac{1}{6}$   **24.** [1.4a] 5,317,200   **25.** [5.3a] 4.78   **26.** [5.3a] 0.0279431

**27.** [5.4a] 2.122    **28.** [1.5a] 1843    **29.** [5.4a] 13,862.1
**30.** [3.7b] $\frac{5}{6}$    **31.** [5.7b] $-11.5$    **32.** [2.8b] $-28$
**33.** [5.7a] 3.8125    **34.** [1.7b] 367,251    **35.** [4.3b] $\frac{1}{18}$
**36.** [3.8a] $-\frac{1}{2}$    **37.** [1.8a] 24,929 transplants
**38.** [1.8a] $177 billion    **39.** [3.8b] $1500    **40.** [3.6b] $2400
**41.** [5.8b] $258.77    **42.** [4.6c] $6\frac{1}{2}$ lb    **43.** [3.6b] 88 ft$^2$
**44.** [5.8b] 43.585 in$^2$    **45.** [4.8a] $\frac{9}{32}$    **46.** [5.4b] 527.04
**47.** [5.8b] $2.39    **48.** [4.7d] 144 packages

# CHAPTER 6

## Exercise Set 6.1, p. 395

**1.** 100 calories    **3.** Boca All American Flame Grilled Meatless;
Franklin Farms Portabella Fresh; Gardenburger Portabella
**5.** 3.5 g    **7.** Most expensive: Lightlife Meatless Light; least
expensive: Boca All American Flame Grilled Meatless
**9.** Greatest fat: Veggie Patch Garlic Portabella; least fat: Franklin
Farms Portabella Fresh and Lightlife Meatless Light
**11.** 92°    **13.** 108°    **15.** 85°, 60%; 90°, 40%; 100°, 10%
**17.** 90° and higher    **19.** 30% and higher    **21.** 15°
**23.** 483,612,200 mi    **25.** Neptune    **27.** All    **29.** 11 earth
diameters    **31.** 31,191.75 mi    **33.** White rhino
**35.** About 1350 rhinos    **37.** About 4100 rhinos    **39.** $-\frac{1}{16}$
**40.** $-\frac{31}{35}$    **41.** $-\frac{4}{3}$    **42.** $-2$    **43.** $-3$    **44.** 1    **45.** 80 min

## Exercise Set 6.2, p. 405

**1.** Miniature tall bearded    **3.** About 23.2 in.    **5.** 16 in. to
26 in.    **7.** Tall bearded    **9.** 25 in.    **11.** 190 calories
**13.** 1 slice of chocolate cake with fudge frosting    **15.** 1 cup of
premium chocolate ice cream    **17.** About 125 calories
**19.** 1950 and 1970    **21.** About 175,000 bachelor's degrees
**23.**

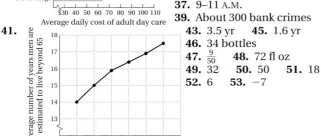

Average daily cost of adult day care

**25.** Chicago, Denver,
Pittsburgh, and San
Antonio    **27.** $43 higher
**29.** New York City
**31.** 17.4 min    **33.** 3.2 million
tourists; 4.3 million tourists
**35.** 2007 to 2008
**37.** 9–11 A.M.
**39.** About 300 bank crimes
**43.** 3.5 yr    **45.** 1.6 yr
**46.** 34 bottles
**47.** $\frac{9}{50}$    **48.** 72 fl oz
**49.** 32    **50.** 50    **51.** 18
**52.** 6    **53.** $-7$

**41.**

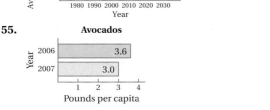

**55.**

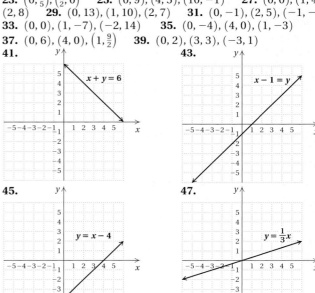

Avocados

## Calculator Corner, p. 413

**1.** Yes    **2.** No    **3.** No    **4.** Yes    **5.** No    **6.** Yes
**7.** Yes    **8.** No

## Exercise Set 6.3, p. 414

**1.**

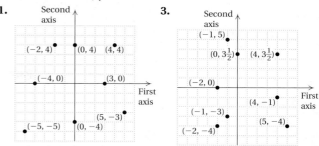

**3.**

**5.** $A$: (3, 3); $B$: (0, $-4$); $C$: ($-5$, 0); $D$: ($-1$, $-1$); $E$: (2, 0); $F$: ($-3$, 5)
**7.** $A$: (5, 0); $B$: (0, 5); $C$: ($-3$, 4); $D$: (2, $-4$); $E$: (2, 3); $F$: ($-4$, $-2$)
**9.** II    **11.** IV    **13.** III    **15.** I    **17.** Positive; negative
**19.** III    **21.** IV; positive    **23.** Yes    **25.** No    **27.** Yes
**29.** No    **31.** No    **33.** Yes    **35.** 7    **36.** 3    **37.** 0
**38.** $\frac{30}{17}$    **39.** $1\frac{28}{33}a$    **40.** $7x - 24$    **41.** Yes    **43.** I, IV
**45.** I, III    **47.** ($-1$, $-5$)    **49.** 26 units

## Calculator Corner, p. 418

**1.** $y = \frac{2}{3}x + 1$    **2.** $y = x + 1$

**3.** $y = -2x + 1$    **4.** $y = \frac{3}{5}x$

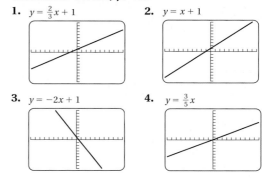

## Exercise Set 6.4, p. 423

**1.** (5, 3)    **3.** (3, 1)    **5.** (5, 14)    **7.** (10, $-3$)    **9.** (0, 3)
**11.** (2, $-1$)    **13.** (1, 3); ($-1$, 5)    **15.** (7, 3); (10, 6)
**17.** (0, 10); (15, 0)    **19.** (3, 1); ($-2$, 4)    **21.** (1, 4); ($-2$, $-8$)
**23.** $\left(0, \frac{3}{5}\right)$; $\left(\frac{3}{2}, 0\right)$    **25.** (0, 9), (4, 5), (10, $-1$)    **27.** (0, 0), (1, 4),
(2, 8)    **29.** (0, 13), (1, 10), (2, 7)    **31.** (0, $-1$), (2, 5), ($-1$, $-4$)
**33.** (0, 0), (1, $-7$), ($-2$, 14)    **35.** (0, $-4$), (4, 0), (1, $-3$)
**37.** (0, 6), (4, 0), $\left(1, \frac{9}{2}\right)$    **39.** (0, 2), (3, 3), ($-3$, 1)
**41.**    **43.**
**45.**    **47.**

**49.**

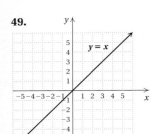

**51.**

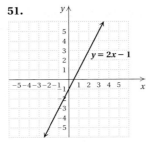

**53.**

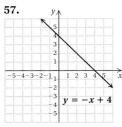

**55.**

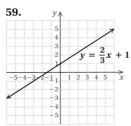

**57.**

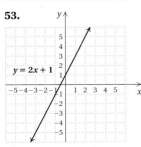

**59.**

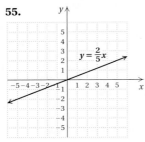

**61.**

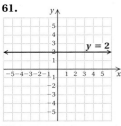

**63.**

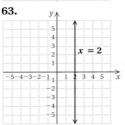

**65.**

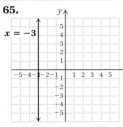

**67.**

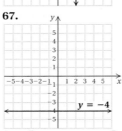

**69.** 9 min    **70.** 319.75 pages    **71.** $1\frac{7}{8}$ cups    **72.** $-\frac{7}{11}$
**73.** 42    **74.** $-\frac{5}{8}$    **75.** $(0, 0.2)$, $(-4, -1)$, $(1, 0.5)$;
answers may vary

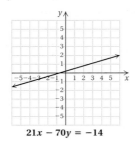

$21x - 70y = -14$

**77.** $(-3, 4.4)$, $(-3.9, 5)$, $(3, 0.4)$; answers may vary

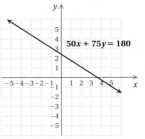

**79.** $(0, 6)$, $(1, 5)$, $(2, 4)$, $(3, 3)$, $(4, 2)$, $(5, 1)$, $(6, 0)$
**81.** **(a)** $y_1 = -0.63x + 2.8$          **(b)** $y_1 = 2.3x - 4.1$

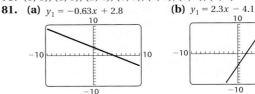

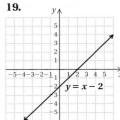

### Mid-Chapter Review: Chapter 6, p. 428
**1.** True    **2.** True    **3.** False
**4.** $3 \cdot 2 + (-1)$; $6 + (-1)$; 5    **5.** $1 - y = 6$; $-y = 5$; $y = -5$;
Thus, $(1, -5)$ is a solution of $x - y = 6$.    **6.** 8 oz    **7.** 6%
**8.** Hershey's Special Dark chocolate bar    **9.** 7 oz
**10.** Nabisco Chips Ahoy cookies    **11.** III    **12.** II
**13.** IV    **14.** No    **15.** $(-10, -7)$    **16.** $(-1, -10)$
**17.** $(1, 3)$, $(0, 5)$, $(-1, 7)$
**18.**

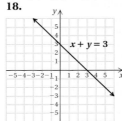

**19.**

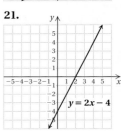

**20.**

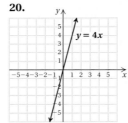

**21.**

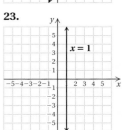

**22.**

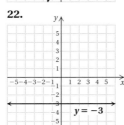

**23.**

**24.** The points $(a, b)$ and $(b, a)$ will be in the same quadrant when both $a$ and $b$ are negative or when both are positive.
**25.** The point $(5, 0)$ is not in any quadrant. It lies on the $x$-axis, and the axes are not considered part of any quadrant.
**26.** The student should check each point carefully in the original equation. At least one of the points will not be a solution of the equation. After identifying any points that are not on the graph of the line, the student should calculate and plot new points to replace them. If these points do line up with each other, the student can then draw the line.

**27.** A line can pass through at most three quadrants. For example, the line $y = x - 2$ passes through quadrants I, III, and IV. A straight line cannot pass through four quadrants.

## Calculator Corner, p. 433

**1.** $203.\overline{3}$   **2.** The answers are the same.

## Translating for Success, p. 435

**1.** F   **2.** N   **3.** A   **4.** O   **5.** G   **6.** D   **7.** C   **8.** L
**9.** H   **10.** E

## Exercise Set 6.5, p. 436

**1.** Mean: 21; median: 18.5; mode: 29
**3.** Mean: 21; median: 20; modes: 5, 20
**5.** Mean: 5.38; median: 5.7; no mode exists
**7.** Mean: 239.5; median: 234; mode: 234
**9.** Mean: $39.\overline{3}$; median: 22; no mode exists   **11.** 23 mpg
**13.** 90   **15.** 263 days   **17.** 2.7
**19.** Mean: $4.19; median: $3.99; mode: $3.99
**21.** Bulb A: mean time = 1171.25 hr;
bulb B: mean time $\approx$ 1251.58 hr; bulb B is better
**23.** 196   **24.** $\frac{4}{9}$   **25.** 1.96   **26.** 1.999396   **27.** 225.05
**28.** 126.0516   **29.** $\frac{3}{35}$   **30.** $\frac{14}{15}$   **31.** 2.5 hr   **32.** 6 hr
**33.** 181   **35.** 10 home runs   **37.** $3475   **39.** 480.375 mi

## Exercise Set 6.6, p. 443

**1.** 83   **3.** About 26,000 kiosks   **5.** About $3.1 million
**7.** About $42 billion   **9.** $\frac{1}{6}$, or $0.1\overline{6}$   **11.** $\frac{1}{2}$, or 0.5
**13.** $\frac{1}{52}$   **15.** $\frac{2}{13}$   **17.** $\frac{3}{26}$   **19.** $\frac{4}{39}$   **21.** $\frac{34}{39}$   **23.** Natural
**24.** Fraction; decimal   **25.** Mean   **26.** Repeating
**27.** Interpolation   **28.** Distributive   **29.** Commutative
**30.** Axes   **31.** $\frac{1}{4}$, or 0.25   **33.** $\frac{1}{36}$   **35.** 0

## Summary and Review: Chapter 6, p. 446

### Concept Reinforcement

**1.** False   **2.** True   **3.** True

### Important Concepts

**1.** Quaker Organic Maple & Brown Sugar; $0.54 per serving
**2.** 12 g   **3.** Arrowhead Stadium   **4.** About $110 million more
**5.** No   **6.**

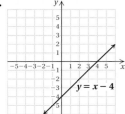

**7.** Mean: 8; median: 8; mode: 8   **8.** $\frac{1}{6}$

### Review Exercises

**1.** $30.37   **2.** $9.32   **3.** $3.76   **4.** 18 champions
**5.** 30–34 years   **6.** 19 more champions   **7.** 2007, 2008, 2009, 2010   **8.** 60%   **9.** About 107% − 55% = 52%   **10.** 2006
**11.** 2005   **12.** 2007   **13.** 2007, 2008, 2009   **14.** 2006
**15.** About 4200 more visitors
**16.**

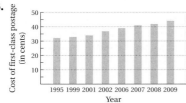

**17.**

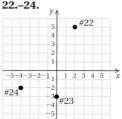

**18.** $(-5, -1)$   **19.** $(-2, 5)$   **20.** $(3, 0)$   **21.** $(4, -2)$
**22.–24.**   **25.** IV   **26.** III   **27.** I
**28.** $(1, 2); (9, -2)$

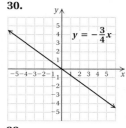

**29.**   **30.**

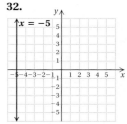

**31.**   **32.**

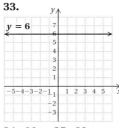

**33.**

**34.** 96   **35.** 28 mpg   **36.** 38.5   **37.** 13.4   **38.** 1.55
**39.** 1840   **40.** $16.\overline{6}$   **41.** $321.\overline{6}$   **42.** 38.5   **43.** 14
**44.** 1.8   **45.** 1900   **46.** $17   **47.** 375   **48.** 26
**49.** 11 and 17   **50.** 0.2   **51.** 700 and 800   **52.** $17
**53.** 20   **54.** Mean: $260; median: $228   **55.** 3.1
**56.** Battery A: mean $\approx$ 43.04 hr; battery B: mean = 41.55 hr; battery A is better   **57.** About 2500 visitors   **58.** $\frac{1}{52}$
**59.** $\frac{1}{2}$   **60.** D   **61.** A
**62.** $\left(0, \frac{100}{47}\right), \left(\frac{100}{34}, 0\right), \left(2, \frac{32}{47}\right)$ (Ordered pairs may vary.)

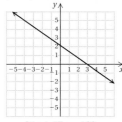

$34x + 47y = 100$

**63.** $11.52/hr   **64.** $a = 316, b = 349$

**65.** 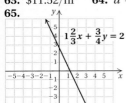 $1\frac{2}{3}x + \frac{3}{4}y = 2$

**66.** 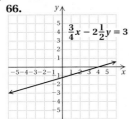 $\frac{3}{4}x - 2\frac{1}{2}y = 3$

## Understanding Through Discussion and Writing

**1.** The equation could represent a person's average income during a 4-yr period. Answers may vary.   **2.** Bar graphs that show change over time can be successfully converted to line graphs. Other bar graphs cannot be successfully converted to line graphs.   **3.** It is possible for the mean of a set of numbers to be larger than all but one number in the set. To see this, note that the mean of the set $\{6, 8\}$ is 7, which is larger than all of the numbers in the set but one.   **4.** The median of a set of four numbers *can* be in the set. For example, the median of the set $\{11, 15, 15, 17\}$ is 15, which is in the set.   **5.** The median income is often used instead of the mean income because it is not artificially raised by the small number of people with very high incomes. The mean of the medians over a 3-yr period would lessen the impact of a sudden drop or rise in income for one year.   **6.** A company considering expansion would probably be more interested in extrapolation than in interpolation because it will be future sales or other activity that will influence profit after expansion.

## Test: Chapter 6, p. 453

**1.** [6.1a] Hiking at 3 mph with 20-lb load   **2.** [6.1a] Hiking at 3 mph with 10-lb load, hiking at 3 mph with 20-lb load
**3.** [6.1b] Spain   **4.** [6.1b] Norway and the United States
**5.** [6.1b] 900 lb   **6.** [6.1b] 1000 lb
**7.** [6.2b]

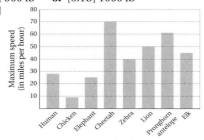

Animal

**8.** [6.1a] 61 mph   **9.** [6.1a] $2\frac{1}{2}$ times faster
**10.** [6.1a], [6.5a] 41 mph   **11.** [6.1a], [6.5b] 42.5 mph
**12.** [6.2c] 53%   **13.** [6.2c] 41%   **14.** [6.2c] 1967
**15.** [6.6a] About 58%   **16.** [6.3b] II   **17.** [6.3b] III
**18.** [6.3a] $(3, 4)$   **19.** [6.3a] $(0, -4)$   **20.** [6.3a] $(-4, 2)$
**21.** [6.4a] $(4, 2)$   **22.** [6.4b]

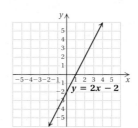 $y = 2x - 2$

**23.** [6.4b]

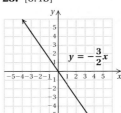 $y = -\frac{3}{2}x$

**24.** [6.4c]

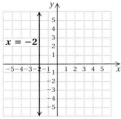 $x = -2$

**25.** [6.5a] 50   **26.** [6.5a] 3   **27.** [6.5a] 15.5
**28.** [6.5b, c] Median: 50.5; mode: 54   **29.** [6.5b, c] Median: 3; no mode exists   **30.** [6.5b, c] Median: 17.5; modes: 17, 18
**31.** [6.5a] 33 mpg   **32.** [6.5a] 76   **33.** [6.5d] Bar A: mean $\approx$ 8.417; bar B: mean $\approx$ 8.417; equal quality
**34.** [6.5a] 2.9   **35.** [6.6b] $\frac{2}{5}$
**36.** [6.4b]   **37.** [6.4b]

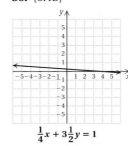 $\frac{1}{4}x + 3\frac{1}{2}y = 1$

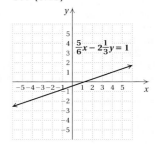 $\frac{5}{6}x - 2\frac{1}{3}y = 1$

**38.** [6.3a] 56 sq units

## Cumulative Review: Chapters 1–6, p. 457

**1.** [5.3b] 62,100,000,000   **2.** [6.5a] 24 mpg
**3.** [1.1a] 5 hundreds   **4.** [1.9c] 128
**5.** [3.2a] 1, 2, 3, 4, 5, 6, 10, 12, 15, 20, 30, 60   **6.** [5.1d] 52.0
**7.** [4.5a] $\frac{33}{10}$   **8.** [5.3b] $2.10   **9.** [2.1d] 9   **10.** [2.6a] $-2$
**11.** [4.2b] $\frac{1}{6}$   **12.** [2.2a] $-39$   **13.** [4.3a] $\frac{1}{3}$   **14.** [5.2b] 325.43
**15.** [4.7a] 15   **16.** [5.3a] $-42.282$   **17.** [3.7b] $\frac{9}{10}$
**18.** [5.4a] 62.345   **19.** [4.3b] $-\frac{13}{15}$   **20.** [3.8a] $\frac{3}{4}$   **21.** [4.4a] 24
**22.** [5.7b] $-\frac{17}{4}$   **23.** [1.8a] 2572 billion kWh   **24.** [4.7d] $\frac{1}{4}$ yd
**25.** [3.6b] $\frac{3}{8}$ cup   **26.** [5.8b] 6.2 lb   **27.** [5.8b] $9.55
**28.** [1.4b], [2.7b] 22 cm; 28 cm$^2$   **29.** [6.3b] II
**30.** [6.4b]

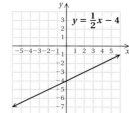 $y = \frac{1}{2}x - 4$

**31.** [2.7a] $-2x + y$   **32.** [2.6b] $10a - 15b + 5$
**33.** [6.5a, b] $56.52; $56.76
**34.** [6.2b]   **35.** [6.2d]

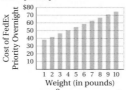

Weight (in pounds)

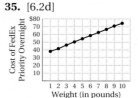

Weight (in pounds)

**36.** [4.8a] $\frac{9}{32}$   **37.** [4.6a] $7\frac{47}{1000}$
**38.** [6.3a] $(-2, -1), (-2, 7), (6, 7), (6, -1)$

# CHAPTER 7

## Exercise Set 7.1, p. 464

**1.** $\frac{4}{5}$ **3.** $\frac{178}{572}$ **5.** $\frac{0.4}{12}$ **7.** $\frac{3.8}{7.4}$ **9.** $\frac{56.78}{98.35}$ **11.** $\frac{8\frac{3}{4}}{9\frac{5}{6}}$ **13.** $\frac{21}{4}$; $\frac{4}{21}$

**15.** $\frac{100}{84.7}, \frac{84.7}{100}$ **17.** $\frac{113}{366}$ **19.** $\frac{1000}{126}, \frac{126}{1000}$ **21.** $\frac{60}{100}, \frac{100}{60}$ **23.** $\frac{2}{3}$

**25.** $\frac{3}{4}$ **27.** $\frac{12}{25}$ **29.** $\frac{7}{9}$ **31.** $\frac{2}{3}$ **33.** $\frac{14}{25}$ **35.** $\frac{1}{2}$ **37.** $\frac{3}{4}$

**39.** $\frac{32}{41}$ **41.** $\frac{72}{23}$ **43.** $\frac{478}{213}, \frac{213}{478}$ **45.** = **46.** ≠ **47.** ≠

**48.** = **49.** 50 **50.** 9.5 **51.** 14.5 **52.** 152 **53.** $6\frac{7}{20}$ cm

**54.** $17\frac{11}{20}$ cm **55.** $\frac{7107}{6629}$ **57.** 1:2:3

## Calculator Corner, p. 469

**1.** $0.\overline{3}$, or $\frac{1}{3}$ home run per strikeout

## Exercise Set 7.2, p. 470

**1.** 40 km/h **3.** 7.48 mi/sec **5.** 25 mpg **7.** 33 mpg
**9.** 43,728 people/sq mi **11.** 25 mph; 0.04 hr/mi
**13.** About 26.6 points/game **15.** 0.623 gal/ft² **17.** 186,000 mi/sec **19.** 124 km/h **21.** 25 beats/min
**23.** 26.188¢/oz; 26.450¢/oz; 16 oz **25.** 8.867¢/oz; 7.320¢/oz;
75 oz **27.** 13.889¢/oz; 17.464¢/oz; 18 oz **29.** 12.991¢/oz;
10.346¢/oz; 26 oz **31.** B: 18.719¢/oz; E: 14.563¢/oz; Brand E
**33.** A: 10.375¢/oz; B: 9.139¢/oz; H: 8.022¢/oz; Brand H **35.** <
**36.** < **37.** > **38.** > **39.** < **40.** > **41.** 1.7 million
people **42.** $25\frac{1}{2}$ servings **43.** 6-oz: 10.833¢/oz;
5.5-oz: 10.909¢/oz **45.** 2 min; 0.0000022 min **47.** The unit
price increases by 0.76¢/oz.

## Calculator Corner, p. 477

**1.** 27.5625 **2.** 25.6 **3.** 15.140625 **4.** 40.03952941
**5.** 39.74857143 **6.** 119

## Exercise Set 7.3, p. 478

**1.** No **3.** Yes **5.** Yes **7.** No **9.** 45 **11.** 12 **13.** 10
**15.** 20 **17.** 5 **19.** 18 **21.** 22 **23.** 28 **25.** $\frac{28}{3}$, or $9\frac{1}{3}$
**27.** $\frac{26}{9}$, or $2\frac{8}{9}$ **29.** 5 **31.** 5 **33.** 0 **35.** 14 **37.** 2.7
**39.** 1.8 **41.** 0.06 **43.** 0.7 **45.** 12.5725 **47.** 1
**49.** $\frac{1}{20}$ **51.** $\frac{3}{8}$ **53.** $\frac{16}{75}$ **55.** $\frac{51}{16}$, or $3\frac{3}{16}$ **57.** $\frac{546}{185}$, or $2\frac{176}{185}$
**59.** Quotient **60.** Sum **61.** Mean **62.** Dollars; cents
**63.** Opposites **64.** Terminating **65.** Commutative
**66.** Cross products **67.** Approximately 2731.4 **69.** −2

**71.** $\frac{a}{b} = \frac{c}{d} \Rightarrow \frac{dba}{b} = \frac{dbc}{d} \Rightarrow da = bc \Rightarrow \frac{da}{ba} = \frac{bc}{ba} \Rightarrow \frac{d}{b} = \frac{c}{a}$

## Mid-Chapter Review: Chapter 7, p. 481

**1.** True **2.** True **3.** False **4.** False
**5.** $\frac{120 \text{ mi}}{2 \text{ hr}} = \frac{120}{2} \frac{\text{mi}}{\text{hr}} = 60 \text{ mi/hr}$
**6.** $\frac{x}{4} = \frac{3}{6}$
$x \cdot 6 = 4 \cdot 3$
$\frac{x \cdot 6}{6} = \frac{4 \cdot 3}{6}$
$x = 2$

**7.** $\frac{4}{7}$ **8.** $\frac{313}{199}$ **9.** $\frac{35}{17}$ **10.** $\frac{59}{101}$ **11.** $\frac{2}{3}$ **12.** $\frac{1}{3}$ **13.** $\frac{8}{7}$
**14.** $\frac{25}{19}$ **15.** $\frac{2}{7}$ **16.** $\frac{5}{1}$ **17.** $\frac{2}{7}$ **18.** $\frac{3}{5}$ **19.** 60.75 mi/hr, or
60.75 mph **20.** 48.67 km/h **21.** 13 m/sec **22.** 16.17 ft/sec
**23.** 27 in./day **24.** About 0.897 free throw made/attempt
**25.** 11.611¢/oz **26.** 49.917¢/oz **27.** Yes **28.** No
**29.** No **30.** Yes **31.** 12 **32.** 40 **33.** 9 **34.** 35
**35.** 2.2 **36.** 4.32 **37.** $\frac{1}{2}$ **38.** $\frac{65}{4}$, or $16\frac{1}{4}$ **39.** Yes; every
ratio $\frac{a}{b}$ can be written as $\frac{\frac{a}{b}}{1}$. **40.** By making some sketches, we
see that the rectangle's length must be twice the width.
**41.** The student's approach will work. However, when we use
the approach of equating cross products, we eliminate the need
to find the least common denominator.

**42.** The instructor thinks that the longer a student studies, the
higher his or her grade will be. An example is the situation in
which one student gets a test grade of 96 after studying for 8 hr
while another student gets a score of 78 after studying for $6\frac{1}{2}$ hr.
This is represented by the proportion $\frac{96}{8} = \frac{78}{6\frac{1}{2}}$.

## Translating for Success, p. 488

**1.** N **2.** I **3.** A **4.** K **5.** J **6.** F **7.** M **8.** B
**9.** G **10.** E

## Exercise Set 7.4, p. 489

**1.** 11.04 hr **3.** 60 students **5.** (a) About 122 gal; (b) 3080 mi
**7.** 212.52 million, or 212,520,000 **9.** 175 bulbs **11.** 2975 ft²
**13.** 450 pages **15.** (a) 32.3687 British pounds; (b) $13,346.23
**17.** (a) 11,698.5 Japanese yen; (b) $54.71 **19.** 13,500 mi
**21.** 880 calories **23.** 120 lb **25.** 64 gal **27.** 100 oz
**29.** 954 deer **31.** 58.1 mi **33.** 9.75 gal **35.** (a) 134 games;
(b) about 61 home runs **37.** Neither **38.** Prime
**39.** Composite **40.** Composite **41.** Prime
**42.** $2 \cdot 2 \cdot 2 \cdot 101$, or $2^3 \cdot 101$ **43.** $2 \cdot 2 \cdot 2 \cdot 7$, or $2^3 \cdot 7$
**44.** $2 \cdot 433$ **45.** $3 \cdot 31$ **46.** $2 \cdot 2 \cdot 5 \cdot 101$, or $2^2 \cdot 5 \cdot 101$
**47.** 17 positions **49.** 2150 earned runs **51.** CD player:
$150; receiver: $450; speakers: $300

## Exercise Set 7.5, p. 498

**1.** 25 **3.** $\frac{4}{3}$, or $1\frac{1}{3}$ **5.** $x = \frac{27}{4}$, or $6\frac{3}{4}$; $y = 9$ **7.** $x = 7.5$;
$y = 7.2$ **9.** 1.25 m **11.** 36 ft **13.** 7 ft **15.** 100 ft
**17.** 4 **19.** $10\frac{1}{2}$ **21.** $x = 6$; $y = 5.25$; $z = 3$
**23.** $x = 5\frac{1}{3}$, or $5.\overline{3}$; $y = 4\frac{2}{3}$, or $4.\overline{6}$; $z = 5\frac{1}{3}$, or $5.\overline{3}$ **25.** 20 ft
**27.** 152 ft **29.** $59.81 **30.** 9.63 **31.** −679.4928
**32.** 2.74568 **33.** 27,456.8 **34.** 0.549136 **35.** 0.85
**36.** −1.825 **37.** −0.909 **38.** 0.843 **39.** 13.75 ft
**41.** 1.25 cm **43.** 3681.437 **45.** $x = 0.4$; $y \approx 0.35$

## Summary and Review: Chapter 7, p. 502

### Concept Reinforcement

**1.** True **2.** True **3.** False **4.** True

### Important Concepts

**1.** $\frac{17}{3}$ **2.** $\frac{8}{7}$ **3.** $7.50/hr **4.** A: 9.964¢/oz; B: 10.281¢/oz;
Brand A **5.** Yes **6.** $\frac{27}{8}$ **7.** 175 mi **8.** 21

### Review Exercises

**1.** $\frac{47}{84}$ **2.** $\frac{46}{1.27}$ **3.** $\frac{83}{100}$ **4.** $\frac{0.72}{197}$ **5.** (a) $\frac{12,480}{16,640}$, or $\frac{3}{4}$;
(b) $\frac{16,640}{29,120}$, or $\frac{4}{7}$ **6.** $\frac{3}{4}$ **7.** $\frac{9}{16}$ **8.** 26 mpg
**9.** 6300 revolutions/min **10.** 0.638 gal/ft²
**11.** 6.33¢/tablet **12.** 11.208¢/oz **13.** 14.969¢/oz;
12.479¢/oz; 15.609¢/oz; 48 oz **14.** Yes **15.** No
**16.** 32 **17.** 7 **18.** $\frac{1}{40}$ **19.** 24 **20.** 27 circuits
**21.** (a) 249.2 Canadian dollars; (b) 50.16 U.S. dollars
**22.** 832 mi **23.** 27 acres **24.** About 3,418,140 lb **25.** 6 in.
**26.** About 13,644 lawyers **27.** $x = \frac{14}{3}$, or $4\frac{2}{3}$ **28.** $x = \frac{56}{5}$, or
$11\frac{1}{5}$; $y = \frac{63}{5}$, or $12\frac{3}{5}$ **29.** 40 ft **30.** $x = 3$; $y = \frac{21}{2}$, or
$10\frac{1}{2}$; $z = \frac{15}{2}$, or $7\frac{1}{2}$ **31.** B **32.** C **33.** 1.329¢/sheet;
1.554¢/sheet; 1.110¢/sheet; 6 big rolls **34.** 105 min, or 1 hr
45 min **35.** 4 bracelets; 100 lavender beads **36.** 240 min
**37.** Finishing paint: 11 gal; primer: 16.5 gal

### Understanding Through Discussion and Writing

**1.** In terms of cost, a low faculty-to-student ratio is less
expensive than a high faculty-to-student ratio. In terms of
quality of education and student satisfaction, a high faculty-
to-student ratio is more desirable. A college president must
balance the cost and quality issues.
**2.** Yes; unit prices can be used to solve proportions involving
money. In Example 3 of Section 7.4, for instance, we could have

divided $90 by the unit price, or the price per ticket, to find the number of tickets that could be purchased for $90.
**3.** Leslie used 4 gal of gasoline to drive 92 mi. At the same rate, how many gallons would be needed to travel 368 mi?
**4.** Yes; consider the following pair of triangles.

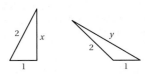

Two pairs of sides are proportional, but we can see that $x$ is shorter than $y$, so the ratio of $x$ to $y$ is clearly not the same as the ratio of 1 to 1 (or 2 to 2).

### Test: Chapter 7, p. 507
**1.** [7.1a] $\frac{85}{97}$   **2.** [7.1a] $\frac{0.34}{124}$   **3.** [7.1b] $\frac{9}{10}$   **4.** [7.1b] $\frac{25}{32}$
**5.** [7.2a] 0.625 ft/sec   **6.** [7.2a] $1\frac{1}{3}$ servings/lb
**7.** [7.2a] 22 mpg   **8.** [7.2b] About 15.563¢/oz
**9.** [7.2b] 16.475¢/oz; 13.980¢/oz; 11.490¢/oz; 16.660¢/oz; 100 oz
**10.** [7.1b] $\frac{32}{15}$   **11.** [7.3a] Yes   **12.** [7.3a] No   **13.** [7.3b] 12
**14.** [7.3b] 360   **15.** [7.3b] 42.1875   **16.** [7.3b] 100
**17.** [7.4a] 1512 km   **18.** [7.4a] 4.8 min   **19.** [7.4a] 525 mi
**20.** [7.5a] 66 m   **21.** [7.4a] **(a)** 3492.45 Hong Kong dollars;
**(b)** $102.44   **22.** [7.4a] About $59.17   **23.** [7.5a] $x = 8$;
$y = 8.8$   **24.** [7.5b] $x = \frac{24}{5}$, or 4.8; $y = \frac{32}{5}$, or 6.4; $z = 12$
**25.** [7.2a] C   **26.** [7.3b] $\frac{4}{3}$, or $1.\overline{3}$   **27.** [7.3b] $-\frac{7}{17}$
**28.** [7.4a] 5888 marbles

### Cumulative Review: Chapters 1–7, p. 509
**1.** [5.2a] 513.996   **2.** [4.6a] $6\frac{3}{4}$   **3.** [4.2b] $\frac{7}{20}$
**4.** [5.2b] 30.491   **5.** [2.3a] $-17$   **6.** [4.3a] $-\frac{7}{60}$
**7.** [5.3a] 222.076   **8.** [2.4a] $-645$   **9.** [4.7a] 3
**10.** [5.4a] 43   **11.** [2.5a] 51   **12.** [3.7b] $\frac{3}{2}$
**13.** [1.1b] 3 ten thousands + 0 thousands + 0 hundreds + 7 tens + 4 ones, or 3 ten thousands + 7 tens + 4 ones
**14.** [5.1a] One hundred twenty and seven hundredths
**15.** [5.1c] 0.7   **16.** [5.1c] $-0.799$
**17.** [3.2c] $2 \cdot 2 \cdot 2 \cdot 2 \cdot 3 \cdot 3$, or $2^4 \cdot 3^2$   **18.** [4.1a] 90
**19.** [3.3a] $\frac{5}{8}$   **20.** [3.5b] $\frac{5}{8}$   **21.** [5.5d] 5.718   **22.** [5.4b] $-25.56$
**23.** [6.5a] 48.75
**24.** [6.4b]

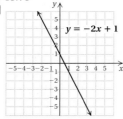

**25.** [2.6a] 5   **26.** [7.1a] $\frac{0.3}{15}$   **27.** [7.3a] Yes   **28.** [7.2a] 55 m/sec
**29.** [7.2b] 8-oz can   **30.** [7.3b] 30.24   **31.** [5.7a] $-26.4375$
**32.** [3.8a] $\frac{8}{9}$   **33.** [2.8d] $-4$   **34.** [5.7a] 33.34   **35.** [5.7b] 3
**36.** [3.6b] 390 cal   **37.** [7.4a] 7 min   **38.** [4.6c] $2\frac{1}{4}$ cups
**39.** [3.8b] 12 doors   **40.** [5.8b], [7.2a] 42.2025 mi   **41.** [5.8b]
132 orbits   **42.** [3.2b] D   **43.** [7.2a] B   **44.** [7.5a] $10\frac{1}{2}$ ft

## CHAPTER 8

### Calculator Corner, p. 517
**1.** 52%   **2.** 38.46%   **3.** 110.26%   **4.** 171.43%
**5.** 59.62%   **6.** 28.31%

### Exercise Set 8.1, p. 519
**1.** $\frac{90}{100}$; $90 \times \frac{1}{100}$; $90 \times 0.01$   **3.** $\frac{12.5}{100}$; $12.5 \times \frac{1}{100}$; $12.5 \times 0.01$
**5.** 0.67   **7.** 0.456   **9.** 0.5901   **11.** 0.1   **13.** 0.01   **15.** 2
**17.** 0.001   **19.** 0.0009   **21.** 0.0018   **23.** 0.2319   **25.** 0.565

**27.** 0.14875   **29.** 0.97   **31.** 0.07; 0.08   **33.** 0.548   **35.** 47%
**37.** 3%   **39.** 870%   **41.** 33.4%   **43.** 1200%   **45.** 40%
**47.** 0.6%   **49.** 1.7%   **51.** 27.18%   **53.** 2.39%   **55.** 27%
**57.** 5.7%; 17.6%   **59.** 90.6%; 88%   **61.** 41%   **63.** 5%
**65.** 20%   **67.** 5%   **69.** 50%   **71.** 87.5%, or $87\frac{1}{2}$%   **73.** 80%
**75.** $66.\overline{6}$%, or $66\frac{2}{3}$%   **77.** $16.\overline{6}$%, or $16\frac{2}{3}$%   **79.** 18.75%, or
$18\frac{3}{4}$%   **81.** 62%   **83.** 16%   **85.** 22%   **87.** 12%   **89.** 15%
**91.** 8%; 59%   **93.** $\frac{17}{20}$   **95.** $\frac{5}{8}$   **97.** $\frac{1}{3}$   **99.** $\frac{1}{6}$   **101.** $\frac{29}{400}$
**103.** $\frac{1}{125}$   **105.** $\frac{203}{800}$   **107.** $\frac{176}{225}$   **109.** $\frac{711}{1100}$   **111.** $\frac{3}{2}$
**113.** $\frac{13}{40,000}$   **115.** $\frac{1}{3}$   **117.** $\frac{3}{50}$   **119.** $\frac{3}{25}$   **121.** $\frac{3}{4}$   **123.** $\frac{3}{20}$
**125.** $\frac{99}{500}$
**127.**

| Fraction Notation | Decimal Notation | Percent Notation |
|---|---|---|
| $\frac{1}{8}$ | 0.125 | 12.5%, or $12\frac{1}{2}$% |
| $\frac{1}{6}$ | $0.1\overline{6}$ | $16.\overline{6}$%, or $16\frac{2}{3}$% |
| $\frac{1}{5}$ | 0.2 | 20% |
| $\frac{1}{4}$ | 0.25 | 25% |
| $\frac{1}{3}$ | $0.\overline{3}$ | $33.\overline{3}$%, or $33\frac{1}{3}$% |
| $\frac{3}{8}$ | 0.375 | 37.5%, or $37\frac{1}{2}$% |
| $\frac{2}{5}$ | 0.4 | 40% |
| $\frac{1}{2}$ | 0.5 | 50% |

**129.**

| Fraction Notation | Decimal Notation | Percent Notation |
|---|---|---|
| $\frac{1}{2}$ | 0.5 | 50% |
| $\frac{1}{3}$ | $0.\overline{3}$ | $33.\overline{3}$%, or $33\frac{1}{3}$% |
| $\frac{1}{4}$ | 0.25 | 25% |
| $\frac{1}{6}$ | $0.1\overline{6}$ | $16.\overline{6}$%, or $16\frac{2}{3}$% |
| $\frac{1}{8}$ | 0.125 | 12.5%, or $12\frac{1}{2}$% |
| $\frac{3}{4}$ | 0.75 | 75% |
| $\frac{5}{6}$ | $0.8\overline{3}$ | $83.\overline{3}$%, or $83\frac{1}{3}$% |
| $\frac{3}{8}$ | 0.375 | 37.5%, or $37\frac{1}{2}$% |

**131.** 70    **132.** 5    **133.** 400    **134.** 18.75    **135.** 23.125
**136.** 25.5    **137.** 4.5    **138.** 8.75    **139.** 20%    **141.** 11.$\overline{1}$%
**143.** 0.01$\overline{5}$    **145.**

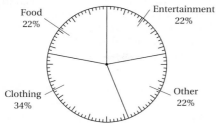

Food 22%
Entertainment 22%
Clothing 34%
Other 22%

### Calculator Corner, p. 528

**1.** $5.04    **2.** 0.0112    **3.** 450    **4.** $1000    **5.** 2.5%    **6.** 12%

### Exercise Set 8.2, p. 529

**1.** $a = 0.32 \cdot 78$    **3.** $89 = p \cdot 99$    **5.** $13 = 0.25 \cdot b$
**7.** 234.6    **9.** 45    **11.** $18    **13.** 1.9    **15.** 78%    **17.** 200%
**19.** 50%    **21.** 125%    **23.** 40    **25.** $40    **27.** 88    **29.** 20
**31.** 6.25    **33.** $846.60    **35.** 1216    **37.** $\frac{9}{100}$    **38.** $\frac{179}{100}$
**39.** $\frac{875}{1000}$, or $\frac{7}{8}$    **40.** $\frac{125}{1000}$, or $\frac{1}{8}$    **41.** $\frac{9375}{10,000}$, or $\frac{15}{16}$    **42.** $\frac{6875}{10,000}$, or $\frac{11}{16}$
**43.** 0.89    **44.** 0.07    **45.** 0.3    **46.** 0.017    **47.** $800 (can vary); $843.20    **49.** 108 to 135 tons    **51.** $1875

### Exercise Set 8.3, p. 535

**1.** $\frac{37}{100} = \frac{a}{74}$    **3.** $\frac{N}{100} = \frac{4.3}{5.9}$    **5.** $\frac{25}{100} = \frac{14}{b}$    **7.** 68.4    **9.** 462
**11.** 40    **13.** 2.88    **15.** 25%    **17.** 102%    **19.** 25%
**21.** 93.75%, or 93$\frac{3}{4}$%    **23.** $72    **25.** 90    **27.** 88    **29.** 20
**31.** 25    **33.** $780.20    **35.** 200
**37.** 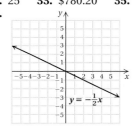    **38.**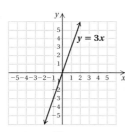

$y = -\frac{1}{2}x$    $y = 3x$

**39.** 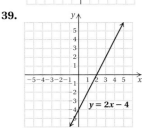    **40.**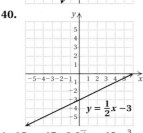

$y = 2x - 4$    $y = \frac{1}{2}x - 3$

**41.** $\frac{43}{48}$ qt    **42.** $\frac{1}{8}$ T    **43.** 100    **44.** 15    **45.** 8.0$\overline{4}$    **46.** $\frac{3}{16}$, or 0.1875    **47.** $1170 (can vary); $1118.64    **49.** 20%    **51.** 39%

### Mid-Chapter Review: Chapter 8, p. 537

**1.** True    **2.** False    **3.** True    **4.** $\frac{1}{2}$% = $\frac{1}{2} \cdot \frac{1}{100}$ = $\frac{1}{200}$
**5.** $\frac{80}{1000} = \frac{8}{100} = 8\%$    **6.** 5.5% = $\frac{5.5}{100} = \frac{55}{1000} = \frac{11}{200}$
**7.** 0.375 = $\frac{375}{1000} = \frac{37.5}{100}$ = 37.5%
**8.**    15 = $p \cdot 80$    **9.** 0.28    **10.** 0.0015
   $\frac{15}{80} = \frac{p \cdot 80}{80}$
   $\frac{15}{80} = p$
   0.1875 = $p$
   18.75% = $p$
**11.** 0.05375    **12.** 2.4    **13.** 71%    **14.** 9%    **15.** 38.91%
**16.** 18.75%, or 18$\frac{3}{4}$%    **17.** 0.5%    **18.** 74%    **19.** 600%
**20.** 83.$\overline{3}$%, or 83$\frac{1}{3}$%    **21.** $\frac{17}{20}$    **22.** $\frac{3}{6250}$    **23.** $\frac{91}{400}$    **24.** $\frac{1}{6}$
**25.** 62.5%, or 62$\frac{1}{2}$%    **26.** 45%    **27.** 58    **28.** 16.$\overline{6}$%, or 16$\frac{2}{3}$%

**29.** 2560    **30.** $50    **31.** 20%    **32.** $455
**33.** 0.05%, 0.1%, $\frac{1}{2}$%, 1%, 5%, 10%, $\frac{13}{100}$, 0.275, $\frac{3}{10}$, $\frac{7}{20}$    **34.** D
**35.** B    **36.** Some will say that the conversion will be done most accurately by first finding decimal notation. Others will say that it is more efficient to become familiar with some or all of the fraction and percent equivalents that appear inside the back cover and to make the conversion by going directly from fraction notation to percent notation.
**37.** Since 40% ÷ 10 = 4%, we can divide 36.8 by 10, obtaining 3.68. Since 400% = 40% × 10, we can multiply 36.8 by 10, obtaining 368.    **38.** Answers may vary. Some will say this is a good idea since it makes the computations in the solution easier. Others will say it is a poor idea since it adds an extra step to the solution.    **39.** They all represent the same number.

### Exercise Set 8.4, p. 542

**1.** South Korea: 75,130 students; Japan: 29,299 students
**3.** $46,656    **5.** 140 items    **7.** 940,000,000 acres    **9.** $36,400
**11.** 74.4 items correct; 5.6 items incorrect    **13.** Egypt: 25,710,642; United States: 63,061,704    **15.** About 14.9%
**17.** $230.10    **19.** About 612,000 fast-food cooks
**21.** Alcohol: 43.2 mL; water: 496.8 mL    **23.** Civic: 5.5%; educational: 26.1%; environmental: 2.2%; health care: 8.5%; religious: 34.0%; community service: 13.9%; sports, hobbies: 3.4%
**25.** 2.$\overline{27}$    **26.** 0.44    **27.** 3.375    **28.** 4.$\overline{7}$    **29.** 0.92
**30.** 0.8$\overline{3}$    **31.** 0.4375    **32.** 2.317    **33.** 3.4809    **34.** 0.675
**35.** $42    **37.** About 5 ft 6 in.

### Calculator Corner, p. 546

**1.** Left to the student    **2.** $80,040

### Translating for Success, p. 549

**1.** J    **2.** M    **3.** N    **4.** E    **5.** G    **6.** H    **7.** O    **8.** C
**9.** D    **10.** B

### Exercise Set 8.5, p. 550

**1.** 5%    **3.** 15%    **5.** About 34.7%    **7.** About 19.3%
**9.** About 93.5%    **11.** About 58.5%    **13.** About 9.2%
**15.** 34.375%, or 34$\frac{3}{8}$%    **17.** 12,933; 2.1%    **19.** 5,130,632; 28.6%    **21.** 1,545,801; 19.5%    **23.** 11.1%    **25.** 32 cm
**26.** 33 ft    **27.** 36 in.    **28.** 36 m    **29.** Virginia
**31.** About 19%

### Exercise Set 8.6, p. 558

**1.** $9.56    **3.** $4.95    **5.** $11.59; $171.39    **7.** 4%    **9.** 2%
**11.** $5600    **13.** $117.26    **15.** $277.55    **17.** $194.08
**19.** $2625    **21.** 12%    **23.** $185,000    **25.** $5440    **27.** 15%
**29.** $355    **31.** $30; $270    **33.** $2.55; $14.45    **35.** $125; $112.50
**37.** 40%; $360    **39.** $50; 30%    **41.** $849; 21.2%    **43.** 18
**44.** $\frac{22}{7}$    **45.** 265.625    **46.** 1.15    **47.** 0.$\overline{5}$    **48.** 2.0$\overline{9}$
**49.** 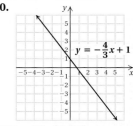    **50.**

$y = \frac{4}{3}x$    $y = -\frac{4}{3}x + 1$

**51.** 4,030,000,000,000    **52.** 5,800,000    **53.** 84.872
**54.** 75.712    **55.** $4.99    **57.** $17,700

### Calculator Corner, p. 565

**1.** $16,357.18    **2.** $12,764.72

### Exercise Set 8.7, p. 568

**1.** $8    **3.** $113.52    **5.** $925    **7.** $671.88    **9.** (a) $147.95; (b) $10,147.95    **11.** (a) $84.14; (b) $6584.14    **13.** (a) $46.03;

**(b)** $5646.03    **15.** $441    **17.** $2802.50    **19.** $7853.38
**21.** $99,427.40    **23.** $4243.60    **25.** $28,225.00
**27.** $9270.87    **29.** $129,871.09    **31.** $4101.01    **33.** $1324.58
**35.** $20,165.05    **37.** Interest: $20.88; amount applied to principal: $4.69; balance after the payment: $1273.87
**39. (a)** $98; **(b)** interest: $86.56; amount applied to principal: $11.44; **(c)** interest: $51.20; amount applied to principal: $46.80; **(d)** At 12.6%, the principal is reduced by $35.36 more than at the 21.3% rate. The interest at 12.6% is $35.36 less than at 21.3%.
**41.** Reciprocals    **42.** Divisible by 6    **43.** Additive
**44.** Unit price    **45.** Perimeter    **46.** Divisible by 3
**47.** Prime    **48.** Proportional    **49.** 9.38%

### Summary and Review: Chapter 8, p. 571

#### Concept Reinforcement

**1.** True    **2.** False    **3.** True

#### Important Concepts

**1.** 0.62625    **2.** $63.\overline{63}$%, or $63\frac{7}{11}$%    **3.** $4.1\overline{6}$%, or $4\frac{1}{6}$%
**4.** 10,000    **5.** 35%    **6.** About 15.3%    **7.** 6%
**8.** $185,000    **9.** Simple interest: $22.60; total amount due: $2522.60    **10.** $6594.26

#### Review Exercises

**1.** 0.037; 0.154    **2.** 0.621; 0.842    **3.** 37.5%, or $37\frac{1}{2}$%
**4.** $33.\overline{3}$%, or $33\frac{1}{3}$%    **5.** 170%    **6.** 6.5%    **7.** $\frac{6}{25}$    **8.** $\frac{63}{1000}$
**9.** $30.6 = p \cdot 90$; 34%    **10.** $63 = 0.84 \cdot b$; 75
**11.** $a = 0.385 \cdot 168$; 64.68    **12.** $\frac{24}{100} = \frac{16.8}{b}$; 70
**13.** $\frac{42}{30} = \frac{N}{100}$; 140%    **14.** $\frac{10.5}{100} = \frac{a}{84}$; 8.82    **15.** 178 students; 84 students    **16.** 46%    **17.** 2500 mL    **18.** 12%    **19.** 92
**20.** $24    **21.** 6%    **22.** 11%    **23.** $42; $308    **24.** 14%
**25.** $2940    **26.** About 18.3%    **27.** $36    **28. (a)** $394.52; **(b)** $24,394.52    **29.** $121    **30.** $7575.25    **31.** $9504.80
**32. (a)** $129; **(b)** interest: $100.18; amount applied to principal: $28.82; **(c)** interest: $70.72; amount applied to principal: $58.28; **(d)** At 13.2%, the principal is decreased by $29.46 more than at the 18.7% rate. The interest at 13.2% is $29.46 less than at 18.7%.    **33.** A    **34.** C    **35.** $66.\overline{6}$%, or $66\frac{2}{3}$%    **36.** $168

#### Understanding Through Discussion and Writing

**1.** A 40% discount is better. When successive discounts are taken, each is based on the previous discounted price rather than on the original price. A 20% discount followed by a 22% discount is the same as a 37.6% discount off the original price.    **2.** Let $S$ = the original salary. After both raises have been given, the two situations yield the same salary: $1.05 \cdot 1.1S = 1.1 \cdot 1.05S$. However, the first situation is better for the wage earner, because $1.1S$ is earned the first year when a 10% raise is given while in the second situation $1.05S$ is earned that year.
**3.** No; the 10% discount was based on the original price rather than on the sale price.    **4.** For a number $n$, 40% of 50% of $n$ is $0.4\ (0.5n)$, or $0.2n$, or 20% of $n$. Thus taking 40% of 50% of a number is the same as taking 20% of the number.
**5.** The interest due on the 30-day loan will be $41.10 while that due on the 60-day loan will be $131.51. This could be an argument in favor of the 30-day loan. On the other hand, the 60-day loan puts twice as much cash at the firm's disposal for twice as long as the 30-day loan does. This could be an argument in favor of the 60-day loan.    **6.** Answers will vary.

#### Test: Chapter 8, p. 577

**1.** [8.1b] 0.458    **2.** [8.1b] 38%    **3.** [8.1c] 137.5%    **4.** [8.1c] $\frac{13}{20}$
**5.** [8.2 a, b] $a = 0.40 \cdot 55$; 22    **6.** [8.3a, b] $\frac{N}{100} = \frac{65}{80}$; 81.25%
**7.** [8.4a] 13,740 kidney transplants; 5356 liver transplants; 1863 heart transplants    **8.** [8.4a] About 543 at-bats    **9.** [8.5a] 26.9%
**10.** [8.4a] 56.3%    **11.** [8.6a] $25.20; $585.20    **12.** [8.6b] $630

**13.** [8.6c] $40; $160    **14.** [8.7a] $8.52    **15.** [8.7a] $5356
**16.** [8.7b] $1110.39    **17.** [8.7b] $11,580.07    **18.** [8.5a] Plumber: 757,000, 7.4%; veterinary assistant: 29,000, 40.8%; motorcycle repair technician: 21,000, 14.3%; fitness professional: 235,000, 63,000    **19.** [8.6c] $50; about 14.3%    **20.** [8.7c] Interest: $36.73; amount applied to the principal: $17.27; balance after payment: $2687    **21.** [8.2a, b], [8.3a, b] B    **22.** [8.6b] $194,600
**23.** [8.6b], [8.7b] $2546.16

### Cumulative Review: Chapters 1–8, p. 579

**1.** [8.1b] 0.57; 0.35    **2.** [8.1b] 26.9%    **3.** [8.1c] 112.5%
**4.** [5.5a] $2.1\overline{6}$    **5.** [7.1a] $\frac{5}{0.5}$, or $\frac{10}{1}$    **6.** [7.2a] $\frac{70\text{ km}}{3\text{ hr}}$, or $23.\overline{3}$ km/hr, or $23\frac{1}{3}$ km/hr    **7.** [4.2c] <    **8.** [5.1c] >
**9.** [5.6a] 296,200    **10.** [1.6b] 50,000    **11.** [1.9d] 13
**12.** [2.7a] $-2x - 14$    **13.** [4.6a] $3\frac{1}{30}$    **14.** [5.2c] $-44.06$
**15.** [1.2a] 515,150    **16.** [5.2b] 0.02    **17.** [4.6b] $\frac{2}{3}$
**18.** [4.3a] $-\frac{110}{63}$    **19.** [3.6a] $\frac{1}{6}$    **20.** [2.4b] $-384$
**21.** [5.3a] 1.38036    **22.** [4.7b] $\frac{3}{2}$, or $1\frac{1}{2}$    **23.** [5.4a] $-12.25$
**24.** [1.5a] 123 R 5    **25.** [1.7b] 95    **26.** [5.7a] 8.13
**27.** [4.4a] 49    **28.** [4.3b] $\frac{1}{12}$    **29.** [5.7b] $-1\frac{7}{9}$    **30.** [7.3b] $\frac{176}{21}$
**31.** [8.4a] 378,332 visitors    **32.** [1.8a] $152,697
**33.** [7.4a] About 65 games    **34.** [7.2b] 7.495 cents/oz
**35.** [4.2d] $\frac{3}{2}$ mi, or $1\frac{1}{2}$ mi    **36.** [8.7b] $12,663.69
**37.** [4.7d] 5 pieces    **38.** [6.3b] III
**39.** [6.4b]

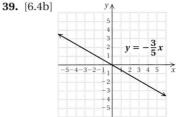

**40.** [6.5a] 33.2    **41.** [6.5b] 12    **42.** [1.4b] 3600 yd$^2$
**43.** [4.3a] C    **44.** [2.7b] A    **45.** [8.6c] 7 discounts
**46.** [8.5a] 12.5% increase    **47.** [4.7d], [6.5a] 33.6 mpg

## CHAPTER 9

### Exercise Set 9.1, p. 591

**1.** 3    **3.** $\frac{1}{12}$    **5.** 1760    **7.** 9    **9.** 7    **11.** 16    **13.** 8800
**15.** $5\frac{1}{4}$, or 5.25    **17.** $2\frac{1}{5}$, or 2.2    **19.** 37,488    **21.** $2\frac{1}{2}$, or 2.5
**23.** 3    **25.** 1    **27.** $1\frac{1}{4}$, or 1.25    **29.** 132,000    **31.** 126,720
**33. (a)** 1000; **(b)** 0.001    **35. (a)** 10; **(b)** 0.1    **37. (a)** 0.01; **(b)** 100    **39.** 8300    **41.** 0.98    **43.** 8.921    **45.** 0.03217
**47.** 28,900    **49.** 4.77    **51.** 688    **53.** 0.1    **55.** 100,000
**57.** 142    **59.** 0.82    **61.** 450    **63.** 0.000024    **65.** 0.688
**67.** 230    **69.** 180; 0.18    **71.** 278; 27.8    **73.** 48,440; 48.44
**75.** 6.21    **77.** 35.56    **79.** 104.585    **81.** 28.67
**83.** 70.866    **85.** 32.904

|    | yd | cm | in. | m | mm |
|----|-----|-----|-----|-----|-----|
| **87.** | 0.000295 | 0.027 | 0.0106299 | 0.00027 | 0.27 |
| **89.** | 483.548 | 44,200 | 17,401.54 | 442 | 442,000 |

**91.** $-\frac{3}{2}$    **92.** $-2$    **93.** $101.50    **94.** $44.50    **95.** 47%
**96.** 35%    **97.** 56 ft; 192 ft$^2$    **98.** 44 ft; 120 ft$^2$    **99.** 1.0 m
**101.** 1.4 cm    **103.** Length: 5400 in., or 450 ft; breadth: 900 in., or 75 ft; height: 540 in., or 45 ft    **105.** 321,800    **107.** 21.3 mph
**109.** >    **111.** <    **113.** >

## Section 9.2, p. 598

**1.** 144 **3.** 640 **5.** $\frac{1}{144}$ **7.** 45 **9.** 1008 **11.** 3
**13.** 198 **15.** 2160 **17.** 12,800 **19.** 27,878,400 **21.** 5
**23.** $\frac{1}{3}$ **25.** $\frac{1}{640}$ **27.** 25,792 **29.** $1\frac{1}{2}$ **31.** 19,000,000
**33.** 63,100 **35.** 0.065432 **37.** 0.0349 **39.** 2500
**41.** 0.4728 **43.** $240 **44.** $212 **45.** (a) $484.11;
**(b)** $15,984.11 **46.** (a) $209.59; (b) $8709.59
**47.** (a) $220.93; (b) $6620.93 **48.** (a) $37.97; (b) $4237.97
**49.** 10.76 **51.** 1.67 **53.** About 14,233 m² **55.** 216 cm²
**57.** 1 ft² **59.** $4\frac{1}{3}$ ft² **61.** 84 cm²

## Calculator Corner, p. 607

**1.** Answers will vary. **2.** 1417.99 in.; 160,005.91 in²
**3.** 1729.27 in² **4.** 125,663.71 ft²

## Exercise Set 9.3, p. 608

**1.** 50 cm² **3.** 104 ft² **5.** 142.5 m² **7.** 72.45 cm²
**9.** 144 mi² **11.** $55\frac{1}{8}$ ft² **13.** 49 m² **15.** 1.92 cm²
**17.** 68 yd² **19.** 14 cm **21.** $1\frac{3}{4}$ in. **23.** 10 ft **25.** 0.7 cm
**27.** 44 cm **29.** $5\frac{1}{2}$ in. **31.** 62.8 ft **33.** 4.396 cm
**35.** 154 cm² **37.** $2\frac{13}{32}$ in² **39.** 314 ft² **41.** 1.5386 cm²
**43.** 12.74 ft; about 40 ft; about 127.41 ft² **45.** About 24,889 mi
**47.** 70,650 mi² **49.** 65.94 yd² **51.** About 4.8 m
**53.** A slice of the square pizza; the areas are 16 in² and 14.13 in².
**55.** 45.68 ft **57.** 45.7 yd **59.** 100.48 m² **61.** 6.9972 cm²
**63.** 64.4214 in² **65.** 675 cm² **67.** 7.5 ft² **69.** $\frac{37}{400}$
**70.** $\frac{7}{8}$ **71.** 125% **72.** 66.$\overline{6}$%, or $66\frac{2}{3}$% **73.** 5 lb
**74.** $730 **75.** 37.2875 in² **77.** Circumference **79.** $4A$
**81.** An 8-inch square pan

## Calculator Corner, p. 619

**1.** 1.13 cm³ **2.** 0.22 in³

## Exercise Set 9.4, p. 620

**1.** 250 cm³ **3.** 135 in³ **5.** 75 m³ **7.** $357\frac{1}{2}$ yd³
**9.** 4082 ft³ **11.** 376.8 cm³ **13.** 41,580,000 yd³
**15.** 4,186,666.$\overline{6}$ in³ **17.** Approximately 124.725 m³
**19.** $1437\frac{1}{3}$ km³ **21.** 1000; 1000 **23.** 59,000 **25.** 0.049
**27.** 27,300 **29.** 40 **31.** 320 **33.** 3 **35.** 40 **37.** 48
**39.** $1\frac{7}{8}$ **41.** 4747.68 cm³ **43.** About 904 ft³ **45.** 0.42 yd³
**47.** 263,947,530,000 mi³ **49.** 0.477 m³ **51.** 61,819 m³
**53.** 33,493 in³ **55.** 5832 yd³ **57.** 423.9 m³ **59.** $24
**60.** $175 **61.** $10.68 **62.** $\frac{7}{3}$, or 2.$\overline{3}$ **63.** −1 **64.** 448 km
**65.** 59 **66.** 10 **67.** About 346.2 ft³ **69.** 6 cm by 6 cm
by 6 cm **71.** 9.424779 L **73.** $1\frac{3}{4}$ gal; 7.5 gal; $91\frac{1}{4}$ gal; about
28,000,000,000 gal

## Mid-Chapter Review: Chapter 9, p. 625

**1.** False **2.** True **3.** False **4.** True **5.** True
**6.** $16\frac{2}{3}$ yd = $16\frac{2}{3}$ · 1 yd = $\frac{50}{3}$ · 3 ft = 50 ft
**7.** 10,200 mm = 10,200 mm · $\frac{1\,m}{1000\,mm}$ ≈
10.2 m · $\frac{3.281\,ft}{1\,m}$ = 33.4662 ft
**8.** C ≈ 3.14 · 10.2 in.
  C ≈ 32.028 in.;
  A ≈ 3.14 · 5.1 in. · 5.1 in.
  A ≈ 81.6714 in²
**9.** 9680 **10.** 70 **11.** 2.405 **12.** 0.00015 **13.** 1251
**14.** 5 **15.** 180,000 **16.** 700 **17.** 2.285 **18.** 118.116
**19.** 15.24 **20.** 160.9 **21.** 720 **22.** 90 **23.** $6\frac{2}{3}$ **24.** $\frac{1}{3}$
**25.** 87,120 **26.** 960 **27.** 3.5 ft, 2 yd, 100 in., $\frac{1}{100}$ mi, 1000 in.,
430 ft, 6000 ft **28.** 150 hm, 13 km, 310 dam, 300 m, 33,000 mm,
3240 cm, 250 dm **29.** 800 in² **30.** 66 km²
**31.** C = 43.96 in.; A = 153.86 in² **32.** C = 27.004 cm;
A = 58.0586 cm² **33.** 30 cm³ **34.** 1695.6 ft³
**35.** 2.355 in³ **36.** 7234.56 m³ **37.** 7400 **38.** 62
**39.** 0.0005 **40.** 12 **41.** 9 **42.** 4

**43.** The student should have multiplied by $\frac{1}{12}$ (or divided by 12)
to convert inches to feet. The correct procedure is as follows:
$$23\text{ in.} = 23\text{ in.} \cdot \frac{1\text{ ft}}{12\text{ in.}} = \frac{23\text{ in.}}{12\text{ in.}} \cdot 1\text{ ft}$$
$$= \frac{23}{12} \cdot \frac{\text{in.}}{\text{in.}} \cdot 1\text{ ft} = \frac{23}{12} \cdot 1\text{ ft} = \frac{23}{12}\text{ ft.}$$
**44.** The area of a 16-in-diameter pizza is approximately
3.14 · 8 in. · 8 in., or 200.96 in². At $16.25, its unit price is
$\frac{\$16.25}{200.96\text{ in}^2}$, or about $0.08/in². The area of 10-in.-diameter pizza
is approximately 3.14 · 5 in. · 5 in., or 78.5 in². At $7.85, its unit
price is $\frac{\$7.85}{78.5\text{ in}^2}$, or $0.10/in². Since the 16-in.-diameter pizza has
the lower unit price, it is a better buy. **45.** A parallelogram
with base $b$ and height $h$ can be divided into two triangles, each
with base $b$ and height $h$. Since the area of a parallelogram is
given by $A = bh$, the area of one of the triangles is given by
$A = \frac{1}{2}bh$. **46.** No; let $r$ = radius of the smaller circle. Then
its area is $\pi \cdot r \cdot r$, or $\pi r^2$. The radius of the larger circle is $2r$,
and its area is $\pi \cdot 2r \cdot 2r$, or $4\pi r^2$, or $4 \cdot \pi r^2$. Thus, the area of
the larger circle is 4 times the area of the smaller circle.

## Exercise Set 9.5, p. 635

**1.** Angle GHI, angle IHG, angle H, ∠GHI, ∠IHG, or ∠H
**3.** ∠ADB or ∠BDA **5.** 10° **7.** 180° **9.** 90°
**11.**

Credit Score
Arrangement of sections may vary.

**13.**

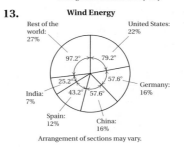

Wind Energy
Arrangement of sections may vary.

**15.** Acute **17.** Straight **19.** Right **21.** Acute
**23.** Obtuse **25.** ∠1 and ∠3; ∠2 and ∠4
**27.** ∠GME (or ∠EMG) and ∠AMC (or ∠CMA);
∠AMG (or ∠GMA) and ∠EMC (or ∠CME) **29.** m∠4; m∠1
**31.** ∠GME or ∠EMG; ∠CME or ∠EMC **33.** 79° **35.** 23°
**37.** 32° **39.** 61° **41.** 177° **43.** 41° **45.** 105° **47.** 76°
**49.** Scalene; obtuse **51.** Scalene; right **53.** Equilateral;
acute **55.** Scalene; obtuse **57.** 46° **59.** 120°
**61.** 58° **63.** $160 **64.** $22.50 **65.** $148 **66.** $1116.67
**67.** $33,597.91 **68.** $413,458.31 **69.** $641,566.26
**70.** $684,337.34 **71.** m∠2 = 67.13°; m∠4 = 79.8°;
m∠5 = 67.13°; m∠6 = 33.07° **73.** m∠ACB = 50°;
m∠CAB = 40°; m∠EBC = 50°; m∠EBA = 40°; m∠AEB = 100°;
m∠ADB = 50°

## Calculator Corner, p. 641

**1.** 6.6 **2.** 9.7 **3.** 19.8 **4.** 17.3 **5.** 24.9 **6.** 24.5
**7.** 121.2 **8.** 115.6 **9.** 85.4

## Translating for Success, p. 644

**1.** K **2.** G **3.** B **4.** H **5.** O **6.** M **7.** E **8.** A
**9.** D **10.** I

## Exercise Set 9.6, p. 645

**1.** $-4, 4$ **3.** $-11, 11$ **5.** $-13, 13$ **7.** $-50, 50$ **9.** 8
**11.** 9 **13.** 15 **15.** 25 **17.** 20 **19.** 100 **21.** 6.928
**23.** 2.828 **25.** 1.732 **27.** 3.464 **29.** 4.359 **31.** 10.488
**33.** $c = 15$ **35.** $c = \sqrt{98}; c \approx 9.899$ **37.** $a = 5$
**39.** $b = \sqrt{45}; b \approx 6.708$ **41.** $c = 26$ **43.** $b = 12$
**45.** $c = \sqrt{41}; c \approx 6.403$ **47.** $b = \sqrt{1023}; b \approx 31.984$
**49.** $\sqrt{208}$ ft $\approx 14.4$ ft **51.** $\sqrt{8450}$ ft $\approx 91.9$ ft
**53.** $h = \sqrt{500}$ ft $\approx 22.4$ ft **55.** $\sqrt{211,200,000}$ ft $\approx 14,532.7$ ft
**57.** \$468 **58.** 187,200 **59.** About 324 students **60.** 12%
**61.** 8 **62.** 125 **63.** 1000 **64.** 10,000 **65.** 47.80 cm$^2$
**67.** 64.8 in. **69.** Width: 15.2 in.; height: 11.4 in.

## Calculator Corner, p. 653

**1.** 41°F **2.** 122°F **3.** 20°C **4.** 45°C

## Exercise Set 9.7, p. 654

**1.** 2000 **3.** 3 **5.** 64 **7.** 12,640 **9.** 2.4 **11.** 5
**13.** 26,000,000,000 lb **15.** 1000 **17.** $\frac{1}{1000}$, or 0.001
**19.** $\frac{1}{100}$, or 0.01 **21.** 1000 **23.** 10 **25.** 234,000 **27.** 5.2
**29.** 0.000897 **31.** 7320 **33.** 8.492 **35.** 58.5
**37.** 800,000 **39.** 1000 **41.** 0.0034 **43.** 0.0603 **45.** 80°C
**47.** 60°C **49.** 20°C **51.** $-10$°C **53.** 180°F **55.** 140°F
**57.** 10°F **59.** 40°F **61.** 86°F **63.** 104°F **65.** 35.6°F
**67.** 30.2°F **69.** 5432°F **71.** 25°C **73.** 55°C **75.** 81.$\overline{1}$°C
**77.** $-15$°C **79.** 37°C **81.** (a) 136°F $= 57.\overline{7}$°C,
$56\frac{2}{3}$°C $= 134$°F; (b) 2°F **83.** Compound **84.** Ratio
**85.** Median **86.** Commission **87.** Similar **88.** Mean
**89.** Composite **90.** Kilo- **91.** 144 packages **93.** $-40$°
**95.** 260.6°F **97.** 0.4536 kg **99.** About 5.1 g/cm$^3$
**101.** (a) 109.134 g; (b) 9.104 g; (c) Golden Jubilee: 3.85 oz;
Hope: 0.321 oz

## Exercise Set 9.8, p. 661

**1.** 2000 mL **3.** 0.12 g **5.** 4 tablets **7.** (a) 720 mcg;
(b) about 189 actuations; (c) 6 inhalers **9.** 16 oz
**11.** (a) 0.42 g; (b) 28 doses **13.** 6 mL **15.** 1000 **17.** 0.325
**19.** 250 mcg **21.** 125 mcg **23.** 0.875 mg; 875 mcg
**25.** 3358 **26.** 7414 **27.** 854 **28.** 6334 **29.** $5x + 8$
**30.** $6x + 5$ **31.** $7t - 9$ **32.** $8r - 10$ **33.** (a) 9.1 g;
(b) 162.5 mg **35.** Naproxen. The Naproxen costs 11 cents/day
and the ibuprofen costs 25 cents/day.

## Summary and Review: Chapter 9, p. 664

### Concept Reinforcement

**1.** True **2.** False **3.** False **4.** True **5.** True **6.** False

### Important Concepts

**1.** $\frac{7}{3}$, or $2\frac{1}{3}$, or $2.\overline{3}$ **2.** 0.000046 **3.** 10.94 **4.** 9
**5.** 5240 **6.** 15.5 m$^2$ **7.** 80 m$^2$ **8.** 37.68 in. **9.** 616 cm$^2$
**10.** 1683.3 m$^3$ **11.** $30\frac{6}{35}$ ft$^3$ **12.** 1696.537813 cm$^3$
**13.** 64 **14.** Complement: 52°; supplement: 142° **15.** 87°
**16.** $\sqrt{208} \approx 14.422$ **17.** 5.14 **18.** 0.00978 **19.** 154.4°F
**20.** 40°C **21.** 500 mL

### Review Exercises

**1.** $3.\overline{3}$ **2.** 30 **3.** 0.17 **4.** 72 **5.** 400,000 **6.** $1\frac{1}{6}$
**7.** 218.8 **8.** 32.18 **9.** 80 **10.** 0.003 **11.** 12.5 **12.** 224
**13.** 0.06 **14.** 400 **15.** 1400 **16.** 200 **17.** 4700
**18.** 0.04 **19.** 36 **20.** 700,000 **21.** 7 **22.** 0.057
**23.** 31.4 m **24.** $\frac{14}{11}$ in. **25.** 24 m **26.** 439.7784 yd
**27.** 18 in$^2$ **28.** 29.82 ft$^2$ **29.** 60 cm$^2$ **30.** 35 mm$^2$
**31.** 154 ft$^2$ **32.** 314 cm$^2$ **33.** 1038.555 ft$^2$ **34.** 26.28 ft$^2$
**35.** 54° **36.** 180° **37.** 140° **38.** 90° **39.** Acute
**40.** Straight **41.** Obtuse **42.** Right **43.** 49° **44.** 136°
**45.** 60° **46.** Scalene **47.** Right **48.** 93.6 m$^3$
**49.** 193.2 ft$^3$ **50.** 28,260 cm$^3$ **51.** $33\frac{37}{75}$ yd$^3$ **52.** 942 cm$^3$

**53.** 8 **54.** 9.110 **55.** $c = \sqrt{850}; c \approx 29.155$
**56.** $b = \sqrt{84}; b \approx 9.165$ **57.** $c = \sqrt{89}$ ft; $c \approx 9.434$ ft
**58.** $a = \sqrt{76}$ cm; $a \approx 8.718$ cm **59.** About 28.8 ft
**60.** About 79 m **61.** 113°F **62.** 20°C **63.** 8 mL
**64.** 3000 mL **65.** 250 mcg **66.** B **67.** B **68.** 17.54 sec
**69.** 1 gal $= 128$ oz, so 1 oz of water (as capacity) weighs $\frac{8.3453}{128}$ lb,
or about 0.0652 lb. An ounce of pennies weighs $\frac{1}{16}$ lb, or
0.0625 lb. Thus, an ounce of water (as capacity) weighs more
than an ounce of pennies. **70.** \$101.25
**71.** 104.875 laps **72.** 961.625 cm$^2$

### Understanding Through Discussion and Writing

**1.** Grams are more easily converted to other units of mass than
ounces. Since 1 gram is much smaller than 1 ounce, masses that
might be expressed using fractional or decimal parts of ounces
can often be expressed by whole numbers when grams are
used. **2.** No, the sum of the three angles of a triangle is 180°.
If you have two 90° angles, totaling 180°, the third angle would
be 0°. A triangle can't have an angle of 0°. **3.** Since 1 m is
slightly longer than 1 yd, it follows that 1 m$^2$ is larger than 1 yd$^2$.
**4.** Add 90° to the measure of the angle's complement.
**5.** Show that the sum of the squares of the lengths of the legs is
the same as the square of the length of the hypotenuse.
**6.** Volume of two spheres, each with radius $r$: $2\left(\frac{4}{3}\pi r^3\right) = \frac{8}{3}\pi r^3$;
volume of one sphere with radius $2r$: $\frac{4}{3}\pi(2r)^3 = \frac{32}{3}\pi r^3$. The
volume of the sphere with radius $2r$ is four times the volume of
the two spheres, each with radius $r$: $\frac{32}{3}\pi r^3 = 4 \cdot \frac{8}{3}\pi r^3$.

### Test: Chapter 9, p. 673

**1.** [9.1a] 96 **2.** [9.1b] 2.8 **3.** [9.2a] 18 **4.** [9.1b] 5000
**5.** [9.1b] 0.91 **6.** [9.2b] 0.00452 **7.** [9.4b] 2.983
**8.** [9.7b] 3800 **9.** [9.4b] 1280 **10.** [9.4b] 690 **11.** [9.7a] 144
**12.** [9.7a] 8220 **13.** [9.3b] 8 cm **14.** [9.3b] 50.24 m$^2$
**15.** [9.3b] 88 ft **16.** [9.3a] 25 cm$^2$ **17.** [9.3a] 18 ft$^2$
**18.** [9.3b] 103.815 km$^2$ **19.** [9.5c] Complement: 25°;
supplement: 115° **20.** [9.5a] 90° **21.** [9.5a] 35°
**22.** [9.5b] Right **23.** [9.5b] Acute **24.** [9.5e] 35°
**25.** [9.5d] Isosceles **26.** [9.5d] Obtuse **27.** [9.4c] 420 in$^3$
**28.** [9.4a] 628 ft$^3$ **29.** [9.4a] 4186.$\overline{6}$ yd$^3$ **30.** [9.4a] 30 m$^3$
**31.** [9.6d] About 15.8 m **32.** [9.6c] $c = \sqrt{2}; c \approx 1.414$
**33.** [9.6c] $b = \sqrt{51}; b \approx 7.141$ **34.** [9.6a] 11 **35.** [9.7c] 0°C
**36.** [9.8a] A **37.** [9.5c] 112.5° **38.** [9.4c] \$188.40
**39.** [9.4a] 0.65 ft$^3$ **40.** [9.4a] 0.055 ft$^3$

### Cumulative Review: Chapters 1–9, p. 675

**1.** [5.3b] 79,200,000 **2.** [9.4c] 93,750 lb **3.** [4.6a] $4\frac{1}{6}$
**4.** [5.2b] 87.52 **5.** [2.5a] $-1234$ **6.** [1.9c] 1565
**7.** [5.5d] 49.2 **8.** [1.9d] 2 **9.** [5.1b] $\frac{1209}{1000}$ **10.** [8.1c] $\frac{17}{100}$
**11.** [4.2c] $<$ **12.** [3.5c] $=$ **13.** [9.1a] 90 **14.** [9.7a] $\frac{3}{8}$
**15.** [9.7c] 59 **16.** [9.4b] 87 **17.** [9.2a] 27 **18.** [9.1b] 0.17
**19.** [4.3b] $\frac{1}{8}$ **20.** [7.3b] $4\frac{2}{7}$ **21.** [5.7b] 1
**22.** [2.8b], [3.8a] $-16$ **23.** [2.7b], [9.3a] 380 cm; 5500 cm$^2$
**24.** [9.3b] 70 in.; 220 in.; 3850 in$^2$ **25.** [9.4a] $179,666\frac{2}{3}$ in$^3$
**26.** [6.5a] 58.6 **27.** [8.7a] \$84 **28.** [8.7b] \$22,376.03
**29.** [9.6d] 17 m **30.** [8.6a] 6% **31.** [4.6c] $2\frac{1}{8}$ yd
**32.** [5.8b] \$51.46 **33.** [7.2b] The 8-qt box
**34.** [3.6b] $\frac{7}{20}$ km **35.** [9.5e] 30° **36.** [2.7a] $9a - 16$
**37.** [6.4b]

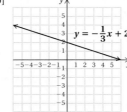

**38.** [9.1a], [9.4a] 94,200 ft$^3$ **39.** [9.1a], [9.4a] 1.342 ft$^3$

# CHAPTER 10

### Calculator Corner, p. 679
**1.** 4.21399177; 4.21399177  **2.** 4.768371582; 4.768371582
**3.** −0.2097152; −0.2097152  **4.** −0.0484002582;
−0.0484002582  **5.** 2.0736; 2.0736  **6.** 0.4932701843;
0.4932701843

### Exercise Set 10.1, p. 681
**1.** 1  **3.** 3.14  **5.** −19.57  **7.** 1  **9.** 1  **11.** 97  **13.** 1
**15.** 17  **17.** 1  **19.** 7  **21.** 5  **23.** 2  **25.** $\frac{1}{3^2}$; $\frac{1}{9}$

**27.** $\frac{1}{10^4}$; $\frac{1}{10,000}$  **29.** $\frac{1}{t^4}$  **31.** $\frac{1}{7^1}$; $\frac{1}{7}$  **33.** $\frac{1}{(-5)^2}$; $\frac{1}{25}$

**35.** $\frac{1}{(-x)^2}$; $\frac{1}{x^2}$  **37.** $\frac{3}{x^7}$  **39.** $\frac{m}{n^5}$  **41.** $\frac{x}{y}$  **43.** $\frac{-7}{a^9}$
**45.** $x^3$  **47.** $xy^4$  **49.** $r^5t^3$  **51.** $xyz$  **53.** $a^8bc^2$
**55.** $x^2y^3z^8$  **57.** $\frac{t^2y^3}{x^4}$  **59.** $\frac{ac}{d^4}$  **61.** $\frac{25}{4}$  **63.** $\frac{a^3}{125}$
**65.** $7^{-3}$  **67.** $9x^{-3}$  **69.** 210 ft  **70.** 14 mpg  **71.** 17.5 mpg
**72.** \$418.95  **73.** 62.5%  **74.** 37.68 cm  **75.** −2736
**76.** 3312  **77.** 28  **79.** 159  **81.** 3; 3

### Exercise Set 10.2, p. 687
**1.** $3^6$  **3.** $x^{14}$  **5.** $a^7$  **7.** $8^6$  **9.** $5^{38}$  **11.** $x^8$  **13.** $a^3$
**15.** $p$  **17.** $\frac{1}{x^{12}}$  **19.** $\frac{1}{5^6}$  **21.** $a^4b^{14}$  **23.** $\frac{1}{a}$  **25.** $8^7$
**27.** $x$  **29.** $a^9$  **31.** $\frac{1}{x^3}$  **33.** 1  **35.** $t^3$  **37.** $c^4$  **39.** $x^3y^5$
**41.** $x^{10}$  **43.** $\frac{1}{2^{18}}$  **45.** $7^6$  **47.** $\frac{1}{y^{12}}$  **49.** $x^4y^{10}$  **51.** $8x^6y^3$
**53.** $-8a^6b^{15}$  **55.** $81x^4y^4$  **57.** $\frac{a^7}{b^7}$  **59.** $\frac{a^6}{8}$  **61.** $\frac{x^{12}}{y^{15}}$
**63.** $\frac{2^8}{3^{10}}$  **65.** 912 m²  **66.** \$123,000  **67.** −17  **68.** −7
**69.** −7  **70.** −17  **71.** $(3y)^{16}$  **73.** $\frac{y}{x}$  **75.** $x^y$  **77.** 1

### Calculator Corner, p. 692
**1.** $1.3545 \times 10^{-4}$  **2.** $3.6 \times 10^{12}$  **3.** $8 \times 10^{-26}$  **4.** $3 \times 10^{13}$

### Translating for Success, p. 693
**1.** E  **2.** H  **3.** O  **4.** A  **5.** G  **6.** C  **7.** M  **8.** I
**9.** K  **10.** N

### Exercise Set 10.3, p. 694
**1.** $2.8 \times 10^{10}$  **3.** $9.07 \times 10^{17}$  **5.** $3.04 \times 10^{-6}$
**7.** $1.8 \times 10^{-8}$  **9.** $10^{11}$  **11.** $3.09 \times 10^8$  **13.** $4 \times 10^{-5}$
**15.** 87,400,000  **17.** 0.00000005704  **19.** 10,000,000
**21.** 0.00001  **23.** $6 \times 10^9$  **25.** $3.38 \times 10^4$
**27.** $8.1477 \times 10^{-13}$  **29.** $2.5 \times 10^{13}$  **31.** $5.0 \times 10^{-4}$
**33.** $3.0 \times 10^{-21}$  **35.** $1.4 \times 10^{21}$ bacteria  **37.** The mass of
Jupiter is $3.18 \times 10^2$ times the mass of Earth.
**39.** $4.375 \times 10^2$ days  **41.** $1 \times 10^{22}$ stars  **43.** $7.5 \times 10^{-7}$ m
**45.** $\frac{7}{4}$, or 1.75  **46.** $-\frac{11}{2}$, or −5.5

**47.**
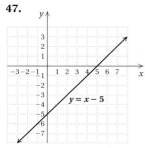
$y = x - 5$

**48.**
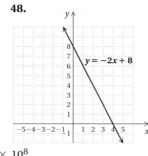
$y = -2x + 8$

**49.** $2.478125 \times 10^{-1}$  **51.** $6.4 \times 10^8$

### Mid-Chapter Review: Chapter 10, p. 697
**1.** True  **2.** False  **3.** False  **4.** $y^2 \cdot y^{10} = y^{2+10} = y^{12}$
**5.** $\frac{y^2}{y^{10}} = y^{2-10} = y^{-8} = \frac{1}{y^8}$
**6.** $4,032,000,000,000 = 4.032 \times 10^{12}$
**7.** $0.000038 = 3.8 \times 10^{-5}$  **8.** 1  **9.** −12  **10.** −1
**11.** −18  **12.** $\frac{1}{x^7}$  **13.** $\frac{1}{3}$  **14.** $\frac{9}{x^6}$  **15.** $x^4$  **16.** $3^{15}$
**17.** $6^{10}$  **18.** $x^{11}$  **19.** $\frac{1}{y^4}$  **20.** $x^6y^{11}$  **21.** $8^9$  **22.** $\frac{1}{x}$
**23.** 1  **24.** $n^5$  **25.** $a^8b^4$  **26.** $x^{32}$  **27.** $y^{12}$  **28.** $\frac{1}{x^3}$
**29.** $\frac{1}{5^{48}}$  **30.** $x^{15}y^{20}$  **31.** $9x^4y^8$  **32.** $-8a^3b^{15}$  **33.** $\frac{c^{10}}{d^{10}}$
**34.** $\frac{4}{x^6}$  **35.** $\frac{a^{28}}{b^{56}}$  **36.** $9.07 \times 10^{-4}$  **37.** $4.3108 \times 10^{11}$
**38.** 208,000,000  **39.** 0.0000041  **40.** $2.5 \times 10^{-1}$ m
**41.** $6.25 \times 10^{-2}$ m, or 6.25 cm  **42.** C  **43.** A
**44.** When $a$ is not zero, $a^0 = 1$. Thus, $a^0 > a^1$ is equivalent to
$1 > a$. So $a^0 > a^1$ for all numbers less than 1.

**45.** The expression $x^{-3}$ is equivalent to $\frac{1}{x^3}$. This expression will
be negative whenever $x$ is negative.  **46.** Since $5^{-8} = \frac{1}{5^8}$ and
$6^{-8} = \frac{1}{6^8}$, we know that $5^{-8} > 6^{-8}$ because $5^8 < 6^8$.
**47.** An answer given in mm would require a larger exponent in
scientific notation because mm is a smaller unit than km.

### Exercise Set 10.4, p. 703
**1.** $-4x + 10$  **3.** $x^2 - 8x + 5$  **5.** $2x^2$
**7.** $6t^4 + 4t^3 + 5t^2 - t$  **9.** $7 + 12x^2$
**11.** $9x^8 + 8x^7 - 3x^4 + 2x^2 - 2x + 5$
**13.** $2t^3 - t^2 + \frac{1}{6}t + \frac{1}{3}$  **15.** 0
**17.** $13a^3b^2 + 4a^2b^2 - 4a^2b + 6ab^2$
**19.** $-0.6a^2bc + 17.5abc^3 - 5.2abc$  **21.** $-(-5x)$; $5x$
**23.** $-(-x^2 + 13x - 7)$; $x^2 - 13x + 7$
**25.** $-(12x^4 - 3x^3 + 3)$; $-12x^4 + 3x^3 - 3$  **27.** $-3x + 5$
**29.** $-4x^2 + 3x - 2$  **31.** $4x^4 - 6x^2 - \frac{3}{4}x + 8$
**33.** $7x - 1$  **35.** $4t^2 + 6$  **37.** $-x^2 - 9x + 5$
**39.** $-a^2 + 5a - 16$  **41.** $8x^4 + 3x^3 - 4x^2 + 3x - 6$
**43.** $4.6x^3 + 9.2x^2 - 3.8x - 23$  **45.** $\frac{3}{4}x^3 - \frac{1}{2}x$
**47.** $6x^3y^3 + 8x^2y^2 + 2x^2y + xy$  **49.** −23  **51.** 19
**53.** −12  **55.** 7  **57.** 4  **59.** 11  **61.** 1024 ft
**63.** About 823 min  **65.** 255 mg  **67.** About 449 accidents
**69.** \$18,750  **71.** \$155,000  **73.** $\frac{7}{10}$ serving per pound
**74.** \$3560  **75.** \$81  **76.** 26 m²  **77.** 1256 cm²
**78.** \$395 per week  **79.** $2 \cdot 2 \cdot 2 \cdot 3 \cdot 7$
**80.** $2 \cdot 2 \cdot 2 \cdot 2 \cdot 2 \cdot 2 \cdot 3$  **81.** $3 \cdot 5 \cdot 7 \cdot 7$  **82.** $3 \cdot 3 \cdot 13$
**83.** About 711 min
**85.**

(graph: Milligrams of ibuprofen vs. Time (in hours), y-axis 0–400 in increments of 40, x-axis 0–6)

**87.** $-x^2 + 280x - 5000$  **89.** $12y^2 - 23y + 21$
**91.** $-3y^4 - y^3 + 5y - 2$  **93.** $3x^4, 2x^3, (-7)$; order of answers
may vary.

## Exercise Set 10.5, p. 712

**1.** $28a^2$ **3.** $-60x^2$ **5.** $28x^8$ **7.** $-0.07x^9$ **9.** $35x^6y^{12}$
**11.** $12a^8b^8c^2$ **13.** $-24x^{11}$ **15.** $-3x^2 + 21x$
**17.** $-3x^2 + 6x$ **19.** $x^5 + x^2$ **21.** $10x^3 - 30x^2 + 5x$
**23.** $12x^3y + 8xy^2$ **25.** $12a^7b^3 - 9a^4b^3$ **27.** $2(x + 4)$
**29.** $7(a - 5)$ **31.** $7(4x + 3y)$ **33.** $9(a - 3b + 9)$
**35.** $6(3 - m)$ **37.** $-8(2 + x - 5y)$ **39.** $9x(x^4 + 1)$
**41.** $a^2(a - 8)$ **43.** $2x(4x^2 - 3x + 1)$ **45.** $6a^4b^2(2b + 3a)$
**47.** Monomial **48.** Parallel **49.** Gram **50.** Prime
**51.** Angle **52.** Cents; dollars **53.** Perimeter
**54.** Multiplicative; additive **55.** $23x^{299}(17x^{92} + 13)$
**57.** $7a^7b^6c^8(12b^3c^3 - 6ac^2 + 7a^2b)$

## Summary and Review: Chapter 10, p. 714

### Concept Reinforcement

**1.** True **2.** True **3.** True

### Important Concepts

**1.** $-3$ **2.** $1$ **3.** $\dfrac{1}{3^{12}}$ **4.** $y^{12}$ **5.** $x^7$ **6.** $3^{12}$ **7.** $9x^{14}$
**8.** $8.03 \times 10^{-4}$ **9.** $3480$ **10.** $a^3 + 2a^2 - 4a - 2$
**11.** $-6x^2 + 5x - 3$ **12.** $14x^4$
**13.** $5x^5 + 10x^4 - 15x^3 + 45x^2$ **14.** $10a(2a^3 - 3)$

### Review Exercises

**1.** $1$ **2.** $46$ **3.** $-3$ **4.** $1$ **5.** $\dfrac{1}{12^2}; \dfrac{1}{144}$ **6.** $\dfrac{8}{a^7}$
**7.** $\dfrac{z^6}{x^3y^5}$ **8.** $\left(\dfrac{5}{4}\right)^2; \dfrac{25}{16}$ **9.** $x^{-7}$ **10.** $x^{15}$ **11.** $x^{12}$
**12.** $3^{44}$ **13.** $x^6y^{12}$ **14.** $\dfrac{1}{x^8}$ **15.** $uv$ **16.** $\dfrac{1}{x^{11}}$ **17.** $9x^8$
**18.** $a^6$ **19.** $\dfrac{y^{10}}{32}$ **20.** $4.27 \times 10^7$ **21.** $1.924 \times 10^{-4}$
**22.** $1.173 \times 10^{11}$ **23.** $2.5 \times 10^{-17}$ **24.** About $1.4 \times 10^7$ hr
**25.** $3x - 6$ **26.** $7x^4 - 4x^3 - x - 3$
**27.** $8a^5 + 12a^3 - a^2 + 4a + 9$ **28.** $5a^3b^3 + 11a^2b^3 - 7$
**29.** $-(12x^3 - 4x^2 + 9x - 3); -12x^3 + 4x^2 - 9x + 3$
**30.** $-42$ **31.** $56$ ft **32.** $30x^7$ **33.** $-18x^3y^5$
**34.** $18x^4 - 12x^2 - 3x$ **35.** $14a^7b^4 + 10a^6b^4$
**36.** $5x(9x^2 - 2)$ **37.** $7(a - 5b - 7ac)$ **38.** C **39.** A
**40.** $-40x^8$ **41.** $13a^2b^5c^5(3ab^2c - 10c^3 + 4a^2b)$
**42.** $w^5x^2y^3z^3(x^4yz^2 - w^2xy^4 + wy^2z^3 - wx^5z)$
**43.** $2a^4b^{-5}(5 + 6a^3b^2)$

### Understanding Through Discussion and Writing

**1.** Because $x^{-2} = \dfrac{1}{x^2}$ and because $x^2$ is never negative, it follows that $x^{-2}$ is never negative. **2.** It is not true that if $a > b$, then $a^{-1} < b^{-1}$. For example, $3 > -4$, but $\frac{1}{3} \not< -\frac{1}{4}$. **3.** Adi is probably adding coefficients and multiplying exponents instead of the other way around. **4.** Consider the expression $\dfrac{7^2}{7^2}$. Since this is equivalent to $\dfrac{49}{49}$, $\dfrac{7^2}{7^2} = 1$. And, using the quotient rule, we have $\dfrac{7^2}{7^2} = 7^{2-2} = 7^0$. Therefore, $7^0 = 1$.

**5.** Not every term is a monomial. For example, $\dfrac{3}{x^2}$ is a term but not a monomial. **6.** A polynomial whose coefficients are all prime may still be factored. For example, $3x + 3y = 3(x + y)$, and $5x^3 + 7x^2 + 13x = x(5x^2 + 7x + 13)$.

## Test: Chapter 10, p. 718

**1.** [10.1a] $193$ **2.** [10.1a] $-1$ **3.** [10.1a] $1$
**4.** [10.1b] $\dfrac{1}{5^3}; \dfrac{1}{125}$ **5.** [10.1b] $\dfrac{5b^2}{a^3}$ **6.** [10.1b] $\left(\dfrac{5}{3}\right)^3; \dfrac{125}{27}$
**7.** [10.2a] $a^{25}$ **8.** [10.2b] $\dfrac{1}{x}$ **9.** [10.2d] $25x^6y^8$
**10.** [10.2d] $\dfrac{27}{x^{27}}$ **11.** [10.2a] $\dfrac{1}{x^7}$ **12.** [10.2b] $a^6b^2$
**13.** [10.3a] $4.7 \times 10^{-4}$ **14.** [10.3a] $8.25 \times 10^6$
**15.** [10.3b] $1.824 \times 10^{-16}$ **16.** [10.4a] $18a^3 - 5a^2 - a + 8$
**17.** [10.4b] $-(-9a^4 + 7b^2 - ab + 3); 9a^4 - 7b^2 + ab - 3$
**18.** [10.4c] $3x^4 - x^2 - 11$ **19.** [10.4d] $12.4$ m
**20.** [10.5a] $-10x^6y^8$ **21.** [10.5b] $10a^3 - 8a^2 + 6a$
**22.** [10.5c] $5x^2(7x^4 - 5x + 3)$
**23.** [10.5c] $3(2ab - 3bc + 4ac)$ **24.** [10.3c] D
**25.** [10.4d] $2.92$ L **26.** [10.2a] $x^{9y}$

## Cumulative Review/Final Examination: Chapters 1-10, p. 720

**1.** [5.3b] $5,100,000$ **2.** [1.2a] $21,085$ **3.** [4.6a] $8\dfrac{7}{9}$
**4.** [2.2a] $24$ **5.** [2.2a] $-762$ **6.** [5.2c] $-44.364$
**7.** [10.4a] $10x^5 - x^4 + 2x^3 - 3x^2 + 2$ **8.** [1.3a] $243$
**9.** [2.7a] $-17x$ **10.** [4.3a] $-\dfrac{19}{40}$ **11.** [4.6b] $2\dfrac{17}{24}$
**12.** [5.2b] $19.9973$ **13.** [10.4c] $2x^3 + 5x^2 + 7x$
**14.** [1.4a] $4752$ **15.** [2.4a] $-74,337$ **16.** [4.7a] $4\dfrac{7}{12}$
**17.** [3.6a] $-\dfrac{6}{5}$ **18.** [3.6a] $10$ **19.** [5.3a] $259.084$
**20.** [2.6b] $24x - 15$ **21.** [10.5a] $27a^8b^3$
**22.** [10.5b] $21x^5 - 14x^3 + 56x^2$ **23.** [1.5a] $573$
**24.** [1.5a] $56$ R $10$ **25.** [3.7b] $-\dfrac{3}{2}$ **26.** [4.7b] $\dfrac{7}{90}$
**27.** [5.4a] $39$ **28.** [1.9c] $75$ **29.** [2.1c], [2.5b] $-2$
**30.** [1.9a] $14^3$ **31.** [1.6a] $68,000$ **32.** [5.5b] $21.84$
**33.** [3.1b] Yes **34.** [3.2a] $1, 3, 5, 15$ **35.** [4.1a] $105$
**36.** [3.5b] $\dfrac{8}{11}$ **37.** [4.5a] $-3\dfrac{3}{5}$ **38.** [2.1b] $>$ **39.** [4.2c] $<$
**40.** [5.1c] $-0.9976$ **41.** [2.6a] $29$ **42.** [10.5c] $5(8 - t)$
**43.** [10.5c] $3a(6a^2 - 5a + 2)$ **44.** [3.3a] $\dfrac{3}{5}$ **45.** [5.1b] $0.0429$
**46.** [5.5a] $-0.52$ **47.** [5.5a] $0.\overline{8}$ **48.** [8.1b] $0.07$
**49.** [5.1b] $\dfrac{671}{100}$ **50.** [4.5a] $-\dfrac{29}{4}$ **51.** [8.1c] $\dfrac{2}{5}$ **52.** [8.1c] $85\%$
**53.** [8.1b] $150\%$ **54.** [5.6a] $13.6$ **55.** [1.7b] $555$
**56.** [5.7a] $64$ **57.** [3.8a] $\dfrac{5}{4}$ **58.** [7.3b] $76.5$ **59.** [2.8d] $5$
**60.** [5.7b] $-\dfrac{2}{5}$ **61.** [6.2c] $246; 2010$ **62.** [6.2c] $258; 2006$
**63.** [6.5a, b, c] Mean: $252.375$; median: $253.5$; modes: $248, 255$
**64.** [8.5a] About $4.7\%$ **65.** [9.1a], [9.1c] $437\dfrac{1}{3}$ yd; about $400$ m
**66.** [6.6b] $\dfrac{3}{20}$, or $0.15$ **67.** [9.5e] $118°$ **68.** [6.5a] $\$23.75$
**69.** [4.6c] $4\dfrac{5}{8}$ yd **70.** [1.8a] $\$24,000$ **71.** [3.4c] $\dfrac{3}{10}$ km
**72.** [7.4a] $13$ gal **73.** [8.7a] $\$240$ **74.** [8.5a] $30,160$
**75.** [9.4c] $93,750$ lb **76.** [9.8a] $56$ mg **77.** [1.8a] $65$ min
**78.** [5.8b] $485.9$ mi **79.** [1.8a] $\$595$ **80.** [5.8b] $\$84.96$
**81.** [7.2b] $17¢/$oz **82.** [8.6b] $7\%$ **83.** [5.8b] $220$ mi
**84.** [1.9b] $324$ **85.** [10.1a] $1$ **86.** [10.1a] $42$ **87.** [9.6a] $11$
**88.** [10.1b] $\dfrac{1}{4^3}; \dfrac{1}{64}$ **89.** [10.1b] $\left(\dfrac{4}{5}\right)^2; \dfrac{16}{25}$
**90.** [10.3a] $4.357 \times 10^6$ **91.** [10.3b] $2.666 \times 10^{-15}$
**92.** [9.1a] $12$ **93.** [9.1b] $5800$ **94.** [9.7a] $160$

**95.** [9.4b] 8.19  **96.** [9.1b] 391.7  **97.** [9.7b] 60
**98.** [9.7b] 2300  **99.** [9.4b] 7  **100.** [9.2a] 90
**101.** [9.2b] 0.02
**102.** [6.2b]

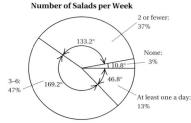

Number of Salads per Week

**103.** [8.1a], [9.5a]

**Number of Salads per Week**

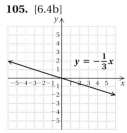

Arrangement of sections may vary.

**104.** [6.3a]

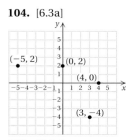

**105.** [6.4b]

$y = -\frac{1}{3}x$

**106.** [6.4c]

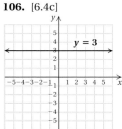

$y = 3$

**107.** [7.5a] $x = 16$, $y = 14$  **108.** [2.7b] 88 in.
**109.** [9.3a] 61.6 cm$^2$  **110.** [3.6b] 25 in$^2$  **111.** [9.3a] 128.65 yd$^2$
**112.** [9.3b] 67.390325 m$^2$  **113.** [9.3b] Diameter: 20.8 in.;
circumference: 65.312 in.; area: 339.6224 in$^2$
**114.** [9.4a] 52.9 m$^3$  **115.** [9.4a] 803.84 ft$^3$
**116.** [9.4c] 523,333,333.$\overline{3}$ m$^3$  **117.** [9.6c] $\sqrt{85}$ ft; 9.220 ft
**118.** [9.5c] C  **119.** [9.7c] D  **120.** [6.6b] C

**121.** **(a)** [3.6a], [4.2d] #1 pays $\frac{1}{25}$, #2 pays $\frac{9}{100}$, #3 pays $\frac{47}{300}$,

#4 pays $\frac{77}{300}$, #5 pays $\frac{137}{300}$; **(b)** [8.1c], [8.4a] 4%, 9%, 15$\frac{2}{3}$%,

25$\frac{2}{3}$%, 45$\frac{2}{3}$%; **(c)** [8.1c], [8.4a] 87%

## APPENDIX

### Exercise Set A, p. 773

**1.** 17  **2.** 15  **3.** 13  **4.** 14  **5.** 12  **6.** 11  **7.** 17
**8.** 16  **9.** 12  **10.** 10  **11.** 10  **12.** 11  **13.** 7  **14.** 7
**15.** 11  **16.** 0  **17.** 3  **18.** 18  **19.** 14  **20.** 10  **21.** 4

**22.** 14  **23.** 11  **24.** 15  **25.** 16  **26.** 9  **27.** 13
**28.** 14  **29.** 11  **30.** 7  **31.** 13  **32.** 14  **33.** 12  **34.** 6
**35.** 10  **36.** 12  **37.** 10  **38.** 8  **39.** 2  **40.** 9  **41.** 13
**42.** 8  **43.** 10  **44.** 9  **45.** 10  **46.** 6  **47.** 13  **48.** 9
**49.** 8  **50.** 11  **51.** 12  **52.** 13  **53.** 10  **54.** 12
**55.** 17  **56.** 13  **57.** 10  **58.** 16  **59.** 16  **60.** 15
**61.** 39  **62.** 89  **63.** 87  **64.** 999  **65.** 900  **66.** 868
**67.** 999  **68.** 848  **69.** 877  **70.** 17,680  **71.** 10,873
**72.** 4699  **73.** 10,867  **74.** 9895  **75.** 3998  **76.** 18,222
**77.** 16,889  **78.** 64,489  **79.** 99,999  **80.** 77,777  **81.** 46
**82.** 26  **83.** 55  **84.** 101  **85.** 1643  **86.** 1412  **87.** 846
**88.** 628  **89.** 1204  **90.** 607  **91.** 10,000  **92.** 1010
**93.** 1110  **94.** 1227  **95.** 1111  **96.** 1717  **97.** 10,138
**98.** 6554  **99.** 6111  **100.** 8427  **101.** 9890  **102.** 11,612
**103.** 11,125  **104.** 15,543  **105.** 16,774  **106.** 68,675
**107.** 34,437  **108.** 166,444  **109.** 101,315  **110.** 49,449

### Exercise Set S, p. 741

**1.** 7  **2.** 0  **3.** 0  **4.** 5  **5.** 3  **6.** 8  **7.** 8  **8.** 6
**9.** 7  **10.** 3  **11.** 7  **12.** 9  **13.** 6  **14.** 6  **15.** 2
**16.** 4  **17.** 3  **18.** 2  **19.** 0  **20.** 3  **21.** 1  **22.** 1
**23.** 7  **24.** 0  **25.** 4  **26.** 4  **27.** 5  **28.** 4  **29.** 4
**30.** 5  **31.** 9  **32.** 9  **33.** 7  **34.** 9  **35.** 6  **36.** 6
**37.** 2  **38.** 6  **39.** 7  **40.** 8  **41.** 33  **42.** 21  **43.** 247
**44.** 193  **45.** 500  **46.** 654  **47.** 202  **48.** 617  **49.** 288
**50.** 1220  **51.** 2231  **52.** 4126  **53.** 3764  **54.** 1691
**55.** 1226  **56.** 99,998  **57.** 10  **58.** 10,101  **59.** 11,043
**60.** 11,111  **61.** 65  **62.** 29  **63.** 8  **64.** 9  **65.** 308
**66.** 126  **67.** 617  **68.** 214  **69.** 89  **70.** 4402  **71.** 3555
**72.** 5889  **73.** 2387  **74.** 3649  **75.** 3832  **76.** 8144
**77.** 7750  **78.** 10,445  **79.** 33,793  **80.** 281  **81.** 455
**82.** 571  **83.** 6148  **84.** 2200  **85.** 2113  **86.** 3748
**87.** 5206  **88.** 1459  **89.** 305  **90.** 4455

### Exercise Set M, p. 749

**1.** 12  **2.** 0  **3.** 7  **4.** 0  **5.** 10  **6.** 30  **7.** 10  **8.** 63
**9.** 54  **10.** 12  **11.** 0  **12.** 72  **13.** 8  **14.** 0  **15.** 28
**16.** 24  **17.** 45  **18.** 18  **19.** 0  **20.** 35  **21.** 45  **22.** 40
**23.** 0  **24.** 16  **25.** 25  **26.** 81  **27.** 1  **28.** 0  **29.** 4
**30.** 36  **31.** 8  **32.** 0  **33.** 27  **34.** 18  **35.** 0  **36.** 10
**37.** 48  **38.** 54  **39.** 0  **40.** 72  **41.** 15  **42.** 8  **43.** 9
**44.** 2  **45.** 32  **46.** 6  **47.** 15  **48.** 6  **49.** 8  **50.** 20
**51.** 20  **52.** 16  **53.** 10  **54.** 0  **55.** 80  **56.** 70
**57.** 160  **58.** 210  **59.** 450  **60.** 780  **61.** 560  **62.** 360
**63.** 800  **64.** 300  **65.** 900  **66.** 1000  **67.** 345,700
**68.** 1200  **69.** 4900  **70.** 4000  **71.** 10,000  **72.** 7000
**73.** 9000  **74.** 2000  **75.** 457,000  **76.** 6,769,000
**77.** 18,000  **78.** 20,000  **79.** 48,000  **80.** 16,000
**81.** 6000  **82.** 1,000,000  **83.** 1200  **84.** 200  **85.** 4000
**86.** 2500  **87.** 12,000  **88.** 6000  **89.** 63,000  **90.** 120,000
**91.** 800,000  **92.** 120,000  **93.** 16,000,000  **94.** 80,000
**95.** 147  **96.** 444  **97.** 2965  **98.** 4872  **99.** 6293
**100.** 3460  **101.** 16,236  **102.** 13,508  **103.** 87,554
**104.** 195,384  **105.** 3480  **106.** 2790  **107.** 3360
**108.** 7020  **109.** 20,760  **110.** 10,680  **111.** 358,800
**112.** 109,800  **113.** 583,800  **114.** 299,700  **115.** 11,346,000
**116.** 23,390,000  **117.** 61,092,000  **118.** 73,032,000

### Exercise Set D, p. 757

**1.** 3  **2.** 8  **3.** 4  **4.** 1  **5.** 32  **6.** 9  **7.** 7  **8.** 5
**9.** 37  **10.** 5  **11.** 9  **12.** 4  **13.** 6  **14.** 9  **15.** 5
**16.** 8  **17.** 9  **18.** 6  **19.** 3  **20.** 2  **21.** 9  **22.** 2
**23.** 3  **24.** 7  **25.** 8  **26.** 4  **27.** 7  **28.** 2  **29.** 3
**30.** 6  **31.** 1  **32.** 4  **33.** 6  **34.** 3  **35.** 7  **36.** 9
**37.** 0  **38.** Undefined  **39.** Undefined  **40.** 7  **41.** 8
**42.** 7  **43.** 1  **44.** 5  **45.** 0  **46.** 0  **47.** 3  **48.** 1
**49.** 4  **50.** 7  **51.** 1  **52.** 6  **53.** 1  **54.** 5  **55.** 2
**56.** 8  **57.** 0  **58.** 0  **59.** 8  **60.** 3  **61.** 4

**62.** Undefined   **63.** 4   **64.** 1   **65.** 69 R 1   **66.** 199 R 1
**67.** 92 R 1   **68.** 138 R 3   **69.** 21   **70.** 118   **71.** 91 R 1
**72.** 321 R 2   **73.** 1723 R 4   **74.** 2925   **75.** 864 R 1
**76.** 522 R 3   **77.** 964 R 3   **78.** 679 R 5   **79.** 2897
**80.** 1328 R 2   **81.** 43 R 15   **82.** 32 R 27   **83.** 137 R 43

**84.** 248 R 26   **85.** 47 R 12   **86.** 42 R 9   **87.** 91 R 22
**88.** 52 R 13   **89.** 23   **90.** 321 R 18   **91.** 224 R 10
**92.** 27 R 60   **93.** 29   **94.** 55   **95.** 228 R 22   **96.** 20 R 379
**97.** 22 R 85   **98.** 49 R 46   **99.** 383 R 91   **100.** 74 R 440

# Glossary

## A

**Absolute value**   The distance that a number is from 0 on the number line

**Acute angle**   An angle whose measure is greater than 0° and less than 90°

**Acute triangle**   A triangle in which all three angles are acute

**Addends**   In addition, the numbers being added

**Additive identity**   The number 0

**Additive inverse**   A number's opposite; two numbers are additive inverses of each other if their sum is zero.

**Additive inverse of a polynomial**   Two polynomials are additive inverses, or opposites, of each other if their sum is zero.

**Algebraic expression**   A number or variable, or a collection of numbers and variables, on which operations are performed

**Angle**   A set of points consisting of two rays (half-lines) with a common endpoint (vertex)

**Area**   The number of square units that fill a plane region

**Associative law of addition**   The statement that when three numbers are added, regrouping the addends gives the same sum

**Associative law of multiplication**   The statement that when three numbers are multiplied, regrouping the factors gives the same product

**Average**   A center point of a set of numbers found by adding the numbers and dividing by the number of items of data; also called the *mean*

**Axes**   Two perpendicular number lines used to identify points in a plane

## B

**Bar graph**   A graphic display of data using bars proportional in length to the numbers represented

**Base**   In exponential notation, the number being raised to a power

## C

**Celsius**   A temperature scale in which water freezes at 0° and boils at 100°

**Circumference**   The distance around a circle

**Coefficient**   The numeric multiplier of a variable

**Commission**   A percent of total sales paid to a salesperson

**Commutative law of addition**   The statement that when two numbers are added, changing the order in which the numbers are added does not affect the sum

**Commutative law of multiplication**   The statement that when two numbers are multiplied, changing the order in which the numbers are multiplied does not affect the product

**Complementary angles**   Two angles for which the sum of their measures is 90°

**Complex fraction**   A fraction in which the numerator and/or denominator contains one or more fractions

**Composite number**   A natural number, other than 1, that is not prime

**Compound interest**   Interest computed on the sum of an original principal and the interest previously accrued by that principal

**Congruent angles**   Two angles that have the same measure

**Constant**   A number or letter that stands for just one number

**Cross products**   Given an equation with a single fraction on each side, the products formed by multiplying the left numerator and the right denominator, and the left denominator and the right numerator

## D

**Decimal notation**   A representation of a number containing a decimal point

**Denominator**   The number below the fraction bar in a fraction

**Descending order**   When a polynomial is written with the powers of the variable decreasing as read from left to right, it is said to be in descending order.

**Diameter**   A segment that passes through the center of a circle and has its endpoints on the circle

**Difference**   The result of subtracting one number from another

**Digit**   A number 0, 1, 2, 3, 4, 5, 6, 7, 8, or 9 that names a place-value location

**Discount**   The amount subtracted from the original price of an item to find the sale price

**Distributive law**   The statement that multiplying a factor by the sum of two numbers gives the same result as multiplying the factor by each of the two numbers and then adding

**Dividend**   In division, the number being divided

**Divisible**   The number $b$ is said to be divisible by another number $a$ if $b$ is a multiple of $a$.

**Divisor**   In division, the number dividing another number

## E

**Equation**   A number sentence that says that the expressions on either side of the equals sign, $=$, represent the same number

**Equilateral triangle**   A triangle in which all sides are the same length

**Equivalent equations**   Equations with the same solutions

**Equivalent expressions**   Expressions that have the same value for all allowable replacements

**Equivalent fractions**   Fractions that represent the same number

**Evaluate**   To substitute a value for each occurrence of a variable in an expression and calculate the result

**Exponent**   In expressions of the form $a^n$, the number $n$ is an exponent.

**Exponential notation**   A representation of a number using a base raised to a power

**Extrapolation**   The process of estimating a value that goes beyond the given data

## F

**Factor**   *Verb*: to write an equivalent expression that is a product. *Noun*: a multiplier

**Factoring**   Writing an expression as a product

**Factorization**   A number expressed as a product of two or more numbers

**Fahrenheit**   A temperature scale in which water freezes at 32° and boils at 212°

**Fraction notation**   A number written using a numerator and a denominator

## G

**Grade point average (GPA)**   The average of the grade point values for each credit hour taken

## H

**Hypotenuse**   In a right triangle, the side opposite the right angle

## I

**Inequality**   A mathematical sentence using $<$, $>$, $\leq$, $\geq$, or $\neq$

**Integers**   The whole numbers and their opposites

**Interest**   A percentage of an amount invested or borrowed

**Interest rate**   The percent at which interest is calculated on a principal

**Interpolation**   The process of estimating a value between given values

**Isosceles triangle**   A triangle in which two or more sides are the same length

## L

**Least common denominator (LCD)**   The least common multiple of the denominators of two or more fractions

**Least common multiple (LCM)**   The smallest number that is a multiple of two or more numbers

**Legs**   In a right triangle, the two sides that form the right angle

**Like terms**   Terms that have exactly the same variable factors

**Line graph**   A graph in which quantities are represented as points connected by straight-line segments

**Linear equation**   Any equation that can be written in the form $Ax + By = C$, where $x$ and $y$ are variables

## M

**Marked price**   The original price of an item

**Mean**   A center point of a set of numbers found by adding the numbers and dividing the sum of the numbers by the number of items in the set; also called the *average*

**Median**   In a set of data listed in order from smallest to largest, the middle number if there is an odd number of data items, or the average of the two middle numbers if there is an even number of data items

**Minuend**   The number from which another number is being subtracted

**Mixed numeral**   A number represented by a whole number and a fraction less than 1

**Mode**   The number or numbers that occur most often in a set of data

**Monomial**   A constant, a variable, or a product of a constant and one or more variables

**Multiple of a number**   The product of the number and an integer

**Multiplicative identity**   The number 1

## N

**Natural numbers**   The counting numbers: 1, 2, 3, 4, 5, . . .

**Negative integers**   Integers to the left of zero on the number line

**Numerator**   The number above the fraction bar in a fraction

## O

**Obtuse angle**   An angle whose measure is greater than 90° and less than 180°

**Obtuse triangle**   A triangle in which one angle is an obtuse angle

**Opposite**   The opposite, or additive inverse, of a number $x$ is written $-x$. Opposites are the same distance from 0 on the number line but on different sides of 0.

**Opposite of a polynomial**   Two polynomials are opposites, or additive inverses, of each other if their sum is zero.

**Ordered pair**   A pair of numbers of the form $(a, b)$ for which the order in which the numbers are listed is important

**Origin**   The point $(0, 0)$ on a graph where the two axes intersect

**Original price**   The price of an item before a discount is deducted

## P

**Palindrome prime**   A prime number that remains a prime number when its digits are reversed

**Parallelogram**   A four-sided polygon with two pairs of parallel sides

**Percent notation**   A representation of a number as parts per 100; $n\%$

**Perimeter**   The distance around an object or the sum of the lengths of its sides

**Periods**   Groups of three digits, separated by commas

**Pi ($\pi$)**   The number that results when the circumference of a circle is divided by its diameter; $\pi \approx 3.14$, or $\frac{22}{7}$

**Pictograph**   A graphic means of displaying information using symbols to represent the amounts

**Polygon**   A closed geometric figure with three or more lines segments as sides

**Polynomial**   A monomial or a sum of monomials

**Positive integers**   Integers to the right of zero on the number line

**Prime factorization**   A factorization of a composite number as a product of prime numbers

**Prime number**   A natural number that has exactly two different factors: itself and 1

**Principal**   An amount of money that is invested or borrowed

**Product**   The result when one number is multiplied by another

**Proportion**   An equation stating that two ratios are equal

**Protractor**   A device used to measure and draw angles

**Purchase price**   The price of an item before sales tax is added

**Pythagorean equation**   The equation $a^2 + b^2 = c^2$, where $a$ and $b$ are lengths of the legs of a right triangle and $c$ is the length of the hypotenuse

## Q

**Quadrants**   The four regions into which the axes divide a plane

**Quotient**   The result when one number is divided by another

## R

**Radical sign**   The symbol $\sqrt{\phantom{x}}$

**Radius**   A segment with one endpoint on the center of a circle and the other endpoint on the circle

**Rate**   A ratio used to compare two different kinds of measure

**Ratio**   The quotient of two quantities; the ratio of $a$ to $b$ is $\frac{a}{b}$, also written $a:b$

**Rational number**   Any number that can be written as the ratio of two integers $\frac{a}{b}$, where $b \neq 0$

**Ray**   A part of a line consisting of one endpoint and all the points of the line on one side of the endpoint

**Reciprocal**   A multiplicative inverse; two numbers are reciprocals if their product is 1

**Rectangle**   A four-sided polygon with four 90° angles

**Repeating decimal**   A decimal in which a block of digits repeats indefinitely

**Right angle**   An angle whose measure is 90°

**Right triangle**   A triangle that includes a right angle

**Rounding**   Approximating the value of a number; used when estimating

## S

**Sale price**   The price of an item after a discount has been deducted

**Sales tax**   A tax added to the purchase price of an item

**Scalene triangle**   A triangle in which all sides are of different lengths

**Scientific notation**   A representation of a number written in the form $M \times 10^n$, where $n$ is an integer, $1 \leq M < 10$, and $M$ is expressed in decimal notation

**Similar triangles**   Triangles in which corresponding sides are proportional and corresponding angles are congruent

**Simple interest**   A percentage of an amount $P$ invested or borrowed for $t$ years, computed by calculating principal $\times$ interest rate $\times$ time

**Simplify**   To rewrite an expression in an equivalent, abbreviated form

**Solution of an equation**   A replacement for the variable that makes the equation true

**Solve**   To find all solutions of an equation, inequality, or problem

**Sphere**   The set of all points in space that are a given distance radius from a given point center

**Square**   A four-sided polygon with four right angles and all sides of equal length

**Square root**   The number $c$ is a square root of $a$ if $c^2 = a$.

**Standard form of a linear equation**   An equation written in the form $Ax + By = C$

**Statistic**   A number that describes a set of data

**Straight angle**   An angle whose measure is 180°

**Substitute**   To replace a variable with a number

**Subtrahend**   In subtraction, the number being subtracted

**Sum**   The result in addition

**Supplementary angles**   Two angles for which the sum of their measures is 180°

## T

**Table**   A representation of data in rows and columns

**Term**   A number, a variable, or a product or a quotient of numbers and/or variables

**Terminating decimal**   A decimal that can be written using a finite number of decimal places

**Total price**   The sum of the purchase price of an item and the sales tax on the item

**Trapezoid**   A four-sided polygon with exactly two parallel sides

**Triangle**   A three-sided polygon

## U

**Unit price**   The ratio of price to the number of units

**Unit rate**   The ratio of quantity to the number of units

## V

**Value**   The numerical result after a number has been substituted into an expression and the calculations have been carried out

**Variable**   A letter that represents an unknown number

**Vertex**   The common endpoint of the two rays that form an angle

**Vertical angles**   Two angles formed by intersecting lines that have no side in common

**Volume**   The number of unit cubes needed to fill an object

## W

**Whole numbers**   The natural numbers and 0: 0, 1, 2, 3, 4, 5, . . .

# Index

Solving equations *(continued)*
    containing fractions, 243, 250–252, 296
    division principle for, 137–138, 146
    multiplication principle for, 206–208, 216
    selecting approach for, 138–139
    by trial, 48
Solving percent problems, 526–528, 533–534, 572
Solving proportions, 475–477, 503
Spheres, volume of, 616, 664, 667
Square roots, 640
    approximating, 640–641
Squares, perimeter of, 130
Standard form of linear equations, 418–419
Standard notation, 2–11, 80
    converting to expanded notation from, 3–4
    converting between word names and, 4–5
Straight angles, 630
Subtraction, 735–740
    by adding the opposite, 100–101
    basic, 735–736
    with borrowing, 15, 737–740
    without borrowing, 737
    of decimals, 317–319, 382
    of fractions, 241–242, 244, 296
    of integers, 99–101, 145
    with mixed numerals, 265–266, 297
    of polynomials, 701, 715
    repeated, division as, 26
    of whole numbers, 14–16, 81
    zeros in, 739–740
Subtrahends, 14
Sums, 9, 728. *See also* Addition
    estimation of, 354
Supplementary angles, 631–632
Supplements of angles, 631
Symbols
    grouping, order of operations and, 72–75, 114, 288
    inequality, 42

**T**
Tables, 392–394, 446
Temperature-humidity index, 396

Temperature scales, 652–653
    converting between, 653, 664, 668
Ten
    divisibility by, 156
    multiplication by, 745
    multiplication by multiples of, 748
Tens place, borrowing from, 737–738
Ten thousands, 80
Terminating decimals, 345
Terms, 128, 699
    like, combining, 128–129, 146, 319
Third quadrant, 412
Thousands, 2, 80
    carrying, 732
Three, divisibility by, 156
Translating percent problems to proportions, 531–533
Translating word problems
    to algebraic expressions, 366–368
    to equations, 63, 131, 209, 245, 281, 374, 435, 488, 525, 549, 644, 693
Trapezoids, 602–603
    area of, 602–603, 664, 666
Trial, solving equations by, 48
Triangles, 633–634
    acute, 633
    area of, 194–195
    equilateral, 633
    isosceles, 633
    obtuse, 633
    right, 633, 641–642
    scalene, 633
    similar, 494–496, 504
    sum of angle measures of, 634, 667
Trillions, 2
Two, divisibility by, 155

**U**
"Undefined," 28, 112–113
Unit prices (unit rates), 468–469, 503
Unit segments, 582

**V**
Value of algebraic expressions, 120
Variables, 120
Vertex (vertices) of angles, 627
Vertical angles, 632–633
Vertical lines, graphing, 421–422

Volume, 614–616
    of a circular cylinder, 615–616, 664, 667
    of a rectangular solid, 614–615, 664, 666
    of a sphere, 616, 664, 667

**W**
Weight
    mass versus, 649
    units of, 649, 664, 668
Whole numbers, 3
    addition of, 9–10, 80
    division of, 26–31, 81
    as divisors for decimals, 333–337
    multiplication of, 19–23, 81
    order of, 42
    rounding, 37, 81
    subtraction of, 14–16, 81
    word names for, 5
Widmann, Johann, 728
Word names
    converting between standard notation and, 4–5
    decimal notation and, 306–308
Word problems
    translating to algebraic expressions, 366–368
    translating to equations, 63, 131, 209, 245, 281, 374, 435, 488, 525, 549, 644, 693

**Z**
Zero
    addition of, 96, 728
    as denominator, 170–171
    divided by any nonzero real number, 113, 754
    division by, 28, 112–113, 753–754
    as exponent, 678, 686
    fraction notation for, 170–171
    lack of reciprocal for, 201
    multiplication by, 107, 743
    in quotients, 31
    in subtraction, 739–740